A COURSE IN MATHEMATICAL LOGIC

A COURSE IN MATHEMATICAL LOGIC

by

J. L. BELL

London School of Economics and Political Science

and

M. MACHOVER

Chelsea College of Science and Technology

Amsterdam – Lausanne – New York – Oxford – Shannon – Singapore – Tokyo

ELSEVIER SCIENCE B.V.
Sara Burgerhartstraat 25
P.O. Box 211, 1000 AE Amsterdam, The Netherlands

First printing: 1977
Second impression: 1986
Third impression: 1993
Fourth impression: 1997

Library of Congress Cataloging in Publication Data

Bell. John Lane.
A course in mathematical logic.

Bibliography: p. 576
includes index.
1. Logic, Symbolic and mathematical. I. Machover. Moshé, joint author. II. Title.
QA9.B3953 511'. 3 75-33890

ISBN: 0 7204 2844 0

This book is printed on acid-free paper.

Transferred to digital printing 2006

On the contrary, I find nothing in logistic for the discoverer but shackles. It does not help us at all in the direction of conciseness, far from it; and if it requires 27 equations to establish that 1 is a number, how many will it require to demonstrate a real theorem?

H. POINCARÉ

Although it is a distinctly minor issue, we must mention Fibonacci's famous recurring series.... There is an extensive literature, some of it bordering on the eccentric, concerning these numbers.... Some professional and dilettant esthetes have applied Fibonacci's num bers to the mathematical dissection of masterpieces in painting and sculpture with results not always agreeable, although sometimes ludicrous, to creative artists. Others have discovered these protean numbers in religion, phyllotaxis, and the convolutions of sea shells

E. T. BELL

Infinitesimals as explaining continuity must be regarded as unnecessary, erroneous, and self-contradictory.

B. RUSSELL

Dedicated to the memory of
Andrzej Mostowski
and
Abraham Robinson

ACKNOWLEDGEMENTS

In the course of producing this book we have become indebted to many people. To begin with, we would like to put on record our intellectual debt to those logicians and mathematicians whose work we have expounded: in a work of this kind it would not be feasible to attribute each result to its creator, but we hope that the historical references at the end of each chapter will furnish a general (if sketchy) guide to who proved what.

Our students have been very helpful in furnishing comments and criticism; in this connection we would particularly like to thank Michael Bate, Narciso Garcia and Ali al-Nowaihi.

We are also grateful to our colleagues Daniel Leivant, Dirk de Jongh and Brian Rotman for reading sections of the manuscript and making useful suggestions.

The job of typing the manuscript was, to say the least, an arduous and tricky one and for skilfully carrying out this operation we are indebted to Buffy Fennelly, Diane Roberts, Barbara Silver and Marie-Louise Varichon.

We would also like to thank the printers, and the staff of North-Holland Publishing Company, in particular Einar Fredriksson and Thomas van den Heuvel, for the competent and friendly way they have handled all stages in the production of this book.

Finally, we offer our warmest thanks to our wives Mimi and Ilana for, among many other things, their patience and encouragement throughout the long writing period and for their help in preparing the bibliography.

John Bell
Moshé Machover

CONTENTS

INTERDEPENDENCE SCHEME FOR THE CHAPTERS

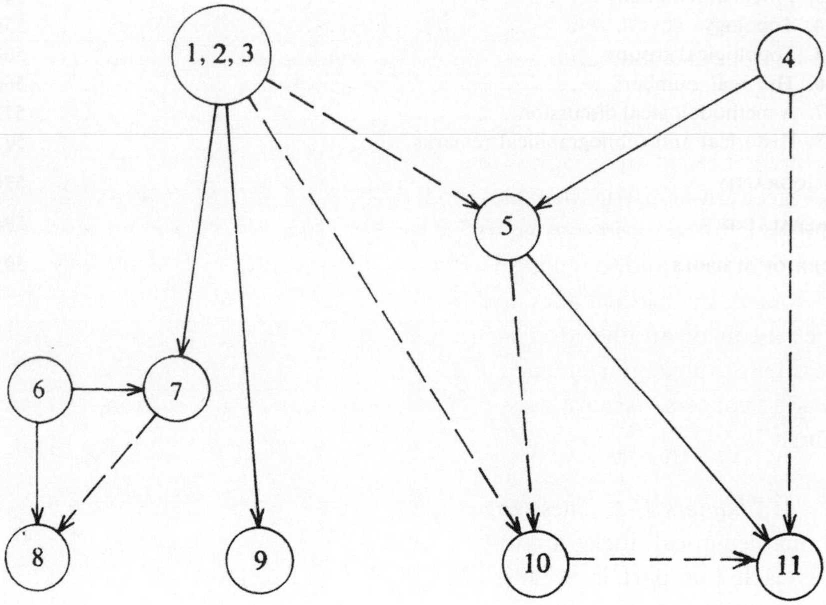

INTRODUCTION

For the past seven years, the authors have conducted a one-year M.Sc. programme in mathematical logic and foundations of mathematics at London University. The present book developed from our lecture notes for this programme, and the student should therefore be able to work through the text in (roughly) one academic year. The main problem that we faced in constructing the programme was the following. First, we wanted it to be an integrated and balanced account of the most important aspects of logic and foundations. But secondly, since parts of our programme are taken by mathematics and philosophy of science students who for one reason or another do not want to cover all the topics we discuss, we were led to arrange it in such a way that parts could be taken as separate smaller courses. Accordingly, the book itself falls naturally into several units:

1. *Chapters 1–3*. These together constitute an elementary introduction to mathematical logic up to the Gödel–Henkin completeness theorem. We teach this part in a fairly leisurely way (four hours per week for ten weeks, including problem classes), and accordingly the pace of the text here is rather gentle. There is one feature which deserves special mention and that is the use of, Smullyan's tableau method. This method serves a dual purpose. First, it is a proof-theoretic instrument that allows us to obtain constructive proofs of various results. In this respect it is equivalent to Gentzen's calculus and to various systems of natural deduction. Secondly, our teaching experience shows that Smullyan's method has the great advantage of being a *practical* tool — after a little practice, it furnishes a quick, efficient and almost computational method of actually detecting the truth or falsehood of formulas. (This efficiency stems in part from the fact that, unlike Gentzen's calculus, it does not require the same formula to be copied again and again.) However, the material

on tableaux has in fact been isolated in separate (starred) sections so that the reader who does not want to use this material can simply ignore it; what remains is a self-contained standard account of first-order logic. A middle course is also possible: a reader wishing to enjoy the *practical* advantages of tableaux but who lacks the time or patience for the somewhat complex constructive proofs of elimination theorems (Ch. 1, §8, and Ch. 2, §§5, 6) can skip the latter because the same results are also obtained in an easier but *non-constructive* way elsewhere (Ch. 1, §9 and Ch. 2, §8). We should like to point out that the somewhat rebarbative complexities of Ch. 2, §5 could have been avoided by using different symbols for free and bound variables (as is often done in texts devoted mainly to proof theory). This, however, would detract slightly from the practical utility of the method and in any case would be contrary to accepted usage in most other branches of logic.

2. *Chapters 4 and 5*. The contents of Chapter 4 are taught for 1 hour a week over 10 weeks, concurrently with the material in Chapters 1–3 (of which it is totally independent). It in fact constitutes a separate short course on Boolean algebras. The material in Chapter 5 — model theory — is taught over the following 10 weeks for 2 hours a week. It depends heavily on Chapter 4 but only slightly on Chapters 1–3 inasmuch as it can be read by anyone having modest acquaintance with the notation and main results of first-order logic.

3. *Chapters 6 and 8*. These two chapters constitute a self-contained course on recursion theory. The material in Chapter 6 is taught for 2 hours a week over 10 weeks, concurrently with Chs. 1–4, of which it is totally independent. There are two points here which call for comment. First, we employ register machines instead of Turing machines, because the former are much closer in spirit to actual digital computers, and are also smoother theoretically. Secondly, this chapter includes a full proof of the Matiyasevich–Robinson–Davis–Putnam (MRDP) theorem that every recursively enumerable relation is diophantine. We believe that — despite the length and tedium of the proof — this result is of such importance that no modern account of recursion theory can afford to omit it. In teaching this part of the chapter, we have found that some of the material in §§13, 14, and the first half of §15 can be omitted in class and given to the student to study at home. As for Chapter 8, it is taught for the following eight weeks at a rate of 2 hours per week. The material here of course depends entirely on Chapter 6, but in this book it appears *after* Chapter 7,

because it is motivated by and illuminates the results contained there. However, no *detailed* knowledge of Chapter 7 is required to understand Chapter 8. In Chapters 6 and 8 we have adopted a somewhat formal approach: in proving that such-and-such a function is recursive, we employ the precise apparatus furnished by the recursion theorem, rather than the intuitive "proof by Church's thesis". We have chosen this course because we believe that the beginning student has not yet developed sufficient experience in the subject to be totally convinced by intuitive proofs which employ Church's thesis.

4. *Chapter 7.* This chapter contains an account of the limitative results about formal mathematical systems. Reliance on the MRDP theorems has enabled us to simplify some of the proofs and obtain somewhat sharper results than usual. The chapter presupposes a good knowledge of first-order logic and some knowledge of recursion theory. However, it can be and is taken by students who have *no* detailed acquaintance with the latter. We have found it feasible to develop all the requisite results from recursion theory — *except* the MRDP theorem — using Church's thesis, the MRDP theorem itself being stated without proof. This approach enables us to teach the material in this chapter intelligibly to students who do not want take a full-fledged course in recursion theory. (The material here is in fact taught concurrently with Chapter 5 for 10 weeks at 2 hours per week.)

5. *Chapter 9.* Here we have an elementary introduction to first-order intuitionistic logic. While neither of the authors claim to be an expert on intuitionism, we nevertheless believe that the constructivist approach to mathematics is of such great importance that some discussion of it is essential. (The material in this chapter is taught concurrently with Chapters 5, 7 and 8 for 10 weeks at 1 hour per week.)

6. *Chapter 10.* This is devoted to an axiomatic investigation of Zermelo–Fraenkel set theory, up to the relative consistency of the axiom of choice and the generalized continuum hypothesis. It requires modest familiarity with first-order logic and with the Löwenheim–Skolem theorem in Chapter 5. (This material is taught over roughly 10 weeks at 2 hours per week at the end of the second term and the beginning of the final term.)

7. *Chapter 11.* This chapter contains an introduction to nonstandard analysis, which is an important method of applying model theory to mathematics. The material here is taught over 10 weeks at 2 hours per week,

during the latter part of the year. Although this chapter presupposes a few results of model theory, these results can be stated concisely without proof for the benefit of those students who wish to study the subject without doing a special course on model theory. In fact it is possible to teach nonstandard analysis to students who have only a slender acquaintance with logic.

As can be seen from the foregoing synopsis, the material in the book can be regarded as forming several relatively independent units. However, the book has been conceived as an organic whole, and provides what is in our view a "balanced diet". We have striven to reveal the interplay between "structural" (i.e. set-theoretical) ideas and "constructive" methods. The latter play a particularly prominent role in mathematical logic, and we have therefore stressed the constructive approach where appropriate but without, we hope, undue fanaticism.

The *problems* constitute an essential part of the book. They are not mere brainteasers, nor should they be too difficult for the student to solve, given the hints that are provided. Many of them contain results which are later employed in proofs of theorems. Accordingly, no unstarred problem should be skipped!

Certain sections and problems are *starred*. This does not necessarily indicate that they are more difficult, but rather that they may be omitted at a first reading. Some problems have been starred because they require more knowledge or skill than is needed for understanding the text in the same section.

Each chapter is divided into sections. When we want to refer to a theorem, problem, definition, etc., within the same chapter, we give the number of the section in which it occurs, followed by its number in that section. Thus, e.g., Def. 2.10.1 is the first numbered statement in §10 of Ch. 2 and within Ch. 2 it is referred to as "Def. 10.1." (or simply "10.1").

We use the convenient abbreviation "iff" for "if and only if". The mark ∎ is used *either* to signify the end of a proof *or*, when it appears immediately after the statement of a result, to indicate that the proof is immediate and is accordingly omitted.

References to the bibliography are given thus: KELLEY [1955]. The overwhelming majority of references to the bibliography are given in a separate section at the end of each chapter.

RECOMMENDED READING

In addition to the special works referred to in the text, we would like to recommend the following books as general reading in logic and the foundations of mathematics.

BARWISE [1977] (A compendious anthology of expository articles on all aspects of mathematical logic.)

BENACERRAF and PUTNAM [1964]. (An excellent anthology of seminal works in the philosophy of mathematics.)

DAVIS [1965]. (An anthology of important papers on limitative theorems and recursion theory.)

FRAENKEL, BAR-HILLEL and LEVY [1973]. (A comprehensive survey of the foundations of mathematics, and set theory in particular.)

VAN HEIJENOORT [1967]. (Another excellent anthology containing many of the classic papers in logic and foundations.)

KNEEBONE [1963]. (An elementary but scholarly general introduction to and survey of the foundations and philosophy of mathematics.)

KREISEL and KRIVINE [1967]. (The appendices to this work contain a penetrating discussion of the philosophical foundations of mathematics.)

LAKATOS [1967]. (A collection of papers on the foundations and philosophy of mathematics delivered at the 1965 London colloquium.)

MOSTOWSKI [1966]. (A wide-ranging survey of the development of mathematical logic from 1930 to the 1960's.)

CHAPTER 0

PREREQUISITES

In this book we assume that the reader is familiar with the basic facts of naive set theory (including the fundamentals of cardinal and ordinal arithmetic) as presented, e.g., in FRAENKEL [1961], HALMOS [1960] or KURATOWSKI–MOSTOWSKI [1968]. Facts about cardinals and ordinals are used at the end of Ch. 3, occasionally in Chs. 4 and 9, and throughout Chs. 5 and 10. In some places (especially in Chs. 4 and 11), we assume a slender acquaintance with the basic notions of general topology as presented, e.g., in Ch. 1 of BOURBAKI [1961] or the first few chapters of KELLEY [1955].

We distinguish between *classes* and *sets*. Except in Chs. 10 and 11 (where the terms "class" and "set" are assigned a more precise technical meaning), a *class* is understood to be an arbitrary collection of objects, while a *set* is a class which can be a member of another class. (Another distinguishing feature of sets is that only they have cardinalities.)

Given an object x and a class X, we write as usual $x \in X$ for "x is a *member (element, point)* of X" and say "X *contains* x" is *in* X". If X contains every member of a class Y, we say "X *includes* Y" and write $Y \subseteq X$. Two classes are regarded as identical if they have the same members.

The set of natural numbers (which contains 0) is denoted by N or ω. Except in Ch. 10, the empty set is denoted by $\emptyset$. If A is a set, the *power set* PA of A is the set of all subsets of A.

Given $n \geqslant 1$ objects $x_1, \ldots, x_n$, we write $\langle x_1, \ldots, x_n \rangle$ for the *ordered n-tuple* of $x_1, \ldots, x_n$. Thus $\langle x, y \rangle$ is the *ordered pair* of x and y. *By convention*, we put $\langle x \rangle = x$ (the ordered singleton of x).

The *Cartesian product* of a finite sequence of classes $A_1, \ldots, A_n$ (with $n \geqslant 1$), denoted by $A_1 \times \ldots \times A_n$, is the collection of all n-tuples $\langle a_1, \ldots, a_n \rangle$ with $a_1 \in A_1, \ldots, a_n \in A_n$. If each A_i is identical with a fixed class A, we write A^n for $A_1 \times \ldots \times A_n$. *By convention*, we set $A^0 = \{\emptyset\}$; thus A^0 has exactly one member, namely $\emptyset$.

For $n \geq 1$, an *n-ary relation* on a class A is a collection of n-tuples of members of A, i.e. a subclass of A^n. A unary relation on A is called a *property*; it is just a subclass of A. The *identity* (or *diagonal*) relation on A is the binary relation

$$\{\langle x,x\rangle: x \in A\}.$$

The *membership* relation on A is the binary relation

$$\{\langle x,y\rangle: x \in A \quad \text{and} \quad y \in A \quad \text{and} \quad x \in y\}.$$

If R is an n-ary relation on A and $B \subseteq A$, the *restriction* of R to B is defined to be the n-ary relation $R \cap B^n$ on B. If R is a binary relation, we often write xRy for $\langle x,y\rangle \in R$.

A *function (map, mapping)* is a class f of ordered pairs such that, whenever $\langle x,y\rangle \in f$ and $\langle x,z\rangle \in f$, we have $y=z$. The *domain* $\mathrm{dom}(f)$ of f is the class

$$\{x: \text{for some } y, \langle x,y\rangle \in f\}$$

and the *range* $\mathrm{ran}(f)$ of f is the class

$$\{y: \text{for some } x, \langle x,y\rangle \in f\}.$$

If f is a function, and $x \in \mathrm{dom}(f)$, then the unique y for which $\langle x,y\rangle \in f$ is denoted (*except in Ch.* 10) by $f(x)$, or sometimes fx, etc., and is called the *value* of f at x. Sometimes we specify a function f in terms of its values; under these conditions we write $x \mapsto f(x)$. (Thus, for example, $x \mapsto x+1$ describes the successor function on N.) If f is a function such that $\mathrm{dom}(f)=A$ and $\mathrm{ran}(f) \subseteq B$, we say that f is a *function from A to (into) B* and write $f: A \to B$. If $f: A \to B$ and $X \subseteq A$, we define the *restriction* $f|X: X \to B$ by

$$(f|X)(x)=f(x) \quad \text{for} \quad x \in X.$$

If $X \subseteq A$ and $Y \subseteq B$, we put

$$f[X]=\{f(x): x \in X\}, \qquad f^{-1}[Y]=\{x: f(x) \in Y\},$$

and, for $y \in Y$, we put

$$f^{-1}(y)=f^{-1}[\{y\}].$$

A function $f: A \to B$ is *one-one* (an *injection*) if $f(x)=f(y)$ implies $x=y$ for all $x,y \in A$; *onto* (a *surjection*) if $f[A]=B$; and a *one-one correspondence* (a *bijection*) between A and B if both of these conditions hold. A and B are said to be *equipollent* if there is a bijection between A and B. If $f: A \to B$ and $g: B \to C$, the *composition* $g \circ f: A \to C$ is defined by $(g \circ f)(x)=g(f(x))$

for $x \in A$. We sometimes omit the $\circ$ and write simply gf instead of $g \circ f$. Observe that, for each class A, the identity relation on A is a bijection between A and A; for this reason it is also called the *identity map* on A. If $A \subseteq B$, the *natural injection* of A into B is the map $i : A \to B$ defined by $i(x) = x$ for $x \in A$.

If A is a class and I is a set, we write A^I for the collection of all functions from I into A. (Notice that this definition implies $A^{\emptyset} = \{\emptyset\} = A^0$.) If $\{A_i\colon i \in I\}$ i an indexed family of sets, we write $\prod_{i \in I} A_i$ for the collection of al functions f with domain I such that $f(i) \in A_i$ for all $i \in I$. The *axiom* ol *choice* asserts that, if each $A_i \neq \emptyset$, then $\prod_{i \in I} A_i \neq \emptyset$.

For any $n \in N$, an *n-ary operation* on a class A is a function from A^n to A. In particular, a 0-ary operation on A is a function from $\{\emptyset\}$ to A, and therefore has a unique value which *we identify with the given* 0*-ary operation*. Thus a 0-ary operation on A is just a member of A. If f is an n-ary operation on A, we write $f(a_1,\ldots,a_n)$ for $f(\langle a_1,\ldots,a_n\rangle)$. A subclass B of A is said to be *closed* or *stable* under f if $f(b_1,\ldots,b_n) \in B$ whenever $b_1,\ldots,b_n \in B$. If B is closed under f, we define the *restriction* $f|B$ of f to B by

$$(f|B)(b_1,\ldots,b_n) = f(b_1,\ldots,b_n) \quad \text{for} \quad b_1,\ldots,b_n \in B.$$

A binary relation R on a class A is called an *equivalence* relation if it satisfies:

(a) xRx for all $x \in A$,

(b) xRy implies yRx for all $x,y \in A$,

(c) xRy and yRz implies xRz for all $x,y,z \in A$.

If R is an equivalence relation on A, for each $x \in A$ the set $x_R = \{y \in A\colon xRy\}$ is called the *R-class* of x. Calling a family $\mathscr{B}$ of subsets of A a *partition* of A if $\bigcup \mathscr{B} = A$ and $X \cap Y = \emptyset$ for any distinct members X,Y of $\mathscr{B}$, we see immediately that, if R is an equivalence relation on A, the family of all R-classes of members of A constitutes a partition of A.

A *partially ordered set* is an ordered pair $\langle A, \leqslant\rangle$ in which A is a set and $\leqslant$ is a binary relation on A satisfying:

(a) $x \leqslant x$ for all $x \in A$,

(b) $x \leqslant y$ and $y \leqslant x$ implies $x = y$ for all $x,y \in A$,

(c) $x \leqslant y$ and $y \leqslant z$ implies $x \leqslant z$ for all $x,y,z \in A$.

If $x \leqslant y$, we say that x is *less than or equal to* y or y is *greater than or equal to* x. We also write "$x < y$" for "$x \leqslant y$ and $x \neq y$". If $\langle A, \leqslant\rangle$ is a partially ordered set, $\leqslant$ is called a *partial ordering* on A. A partially ordered set is said to be *totally* (or *linearly*) *ordered* if in addition

(d) $x \leqslant y$ or $y \leqslant x$ for all $x,y \in A$.

If A is any family of sets, the relation $\subseteq$ of set inclusion is a partial ordering on A. We frequently identify a partially ordered set $\langle A, \leqslant \rangle$ with its underlying set A.

If $\langle A, \leqslant \rangle$ is a partially ordered set and $X \subseteq A$, a member $a \in A$ is an *upper (lower) bound* for X, if $x \leqslant a$ $(a \leqslant x)$ for every $x \in X$. An upper (lower) bound a for X in A is called the *supremum (infimum)* of X if a is less than (greater than) every other upper (lower) bound for X in A. If X has a supremum (infimum) in A, we denote it by $\sup(X)$ $(\inf(X))$. Notice that if $\emptyset$ has a supremum (infimum) in A, then $\sup(\emptyset)$ $(\inf(\emptyset))$ is an elemen $a \in A$ such that $a \leqslant x (x \leqslant a)$ for every $x \in A$. That is, if $\sup(\emptyset)$ $(\inf(\emptyset))$ exists in A, then it must be the least (greatest) element of A.

A *chain* in a partially ordered set $\langle A, \leqslant \rangle$ is a subset X of A such that $\leqslant$ totally orders X, i.e. such that $x \leqslant y$ or $y \leqslant x$ for all $x, y \in X$. $\langle A, \leqslant \rangle$ is said to be *inductive* if each chain in A has an upper bound in A. An element $a \in A$ is *maximal* if whenever $x \in A$ and $a \leqslant x$ we have $x = a$. *Zorn's lemma* (which is equivalent to the axiom of choice) asserts that for each element x of an inductive set $\langle A, \leqslant \rangle$ there is a maximal element $a \in A$ such that $x \leqslant a$.

A partially ordered set $\langle A, \leqslant \rangle$ is *well-ordered* if each non-empty subset X of A contains an element x such that $x \leqslant y$ for every $y \in X$. Assuming the axiom of choice, a totally ordered set $\langle A, \leqslant \rangle$ is well-ordered iff A contains no sequence $a_0, a_1, a_2, \ldots$ such that $a_{n+1} < a_n$ for all n.

We conceive of *ordinals* in such a way that each ordinal is the set of all smaller ordinals, and the *finite* ordinals as being identical with the natural numbers. Each well-ordered set is order-isomorphic to a unique ordinal. A *cardinal* is an ordinal which is not equipollent with a smaller ordinal. The *cardinality* of a set X, denoted by $|X|$, is the unique cardinal equipollent with X. (This needs the axiom of choice.) Notice that $|X| = |Y|$ iff X and Y are equipollent. If α and β are cardinals, then α^β denotes the result of *cardinal* exponentation, i.e. the product of α with itself β times. Thus e.g. the cardinality of $\mathsf{P}A$ is $2^{|A|}$, for any set A.

A set A is said to be *finite* if it is equipollent with n for some $n \in N$, *denumerable* if it is equipollent with N, and *countable* if it is finite or denumerable.

CHAPTER 1

BEGINNING MATHEMATICAL LOGIC

In the first three sections of this chapter we prepare and motivate the systematic development, which begins in §4. In §§4 and 5 we deal with the basic syntax of first-order languages. The rest of the chapter (§§6–15) is devoted to propositional logic.

§ 1. General considerations

We shall not attempt to define mathematical logic. Rather, we shall make a few general observations concerning this discipline. In particular, we shall explain why mathematical logic involves the study of formal languages; we shall then point out some distinctions that must be made in such a study.

Let us start with an example. Consider the inference

(1) *Every tove is slithy*

(2) *Alice is not slithy*

---

(3) *Alice is not a tove*

Here statement (3) is correctly inferred from statements (1) and (2). To recognize the fact that (3) is indeed a consequence of (1) and (2) we do not need to know whether or not any one of these three statements is true. Nor do we have to know what the words "tove" and "slithy" mean, and who (or what) is the person (or thing) named "Alice".[1]

What makes (3) a logical consequence of (1) and (2) is the fact that *if* (1) and (2) are true *then* (3) must be true as well. *Any* interpretation of "tove", "slithy" and "Alice" under which (1) and (2) come out true

[1] However, in order to be sure that (1), (2) and (3) are phrased according to the rules of English syntax, we do (presumably) have to know that "tove" is a common noun, "slithy" an adjective and "Alice" a proper noun.

will make (3) true.[1] Also, if "tove", "slithy" and "Alice" are replaced by other words[2], this will not affect the validity of our inference.

We may say that (3) is a consequence of (1) and (2) by virtue of the *form* — as distinct from the *matter* — of these statements. In this connection the words "every" and "not" must be regarded as part of the form: if they are re-interpreted or replaced by other words, the inference may well become invalid. (Such words are said to be *logical*, while words used as common nouns, adjectives, etc. and whose meanings may be changed without affecting the validity of inferences are *extralogical*.)

Mathematical logic studies inferences of the kind used primarily in mathematics. In the sense suggested above, it is a *formal* discipline. Ordinarily, mathematical statements are formulated in a semi-formal language consisting of some natural language (like English, Japanese, etc.) supplemented by special mathematical symbols. But it is an extremely difficult task to apply precise logical analysis to arguments formulated in a natural language. This is so for two main reasons.

First, all natural languages abound in irregularities, ambiguities and structural inconsistencies which tend to obscure and confuse the logical form of statements made in them.

Secondly, as the discussion of the example above suggests, logical analysis requires a separation between form and matter; we may need, e.g., to assign to the same term different (and rather arbitrary) meanings. But in a natural language words have already got particular (if sometimes rather vague) meanings attached to them.

For these reasons we need to construct artificial formal languages whose structure will be perfectly regular. In such a language some symbols (viz. those that correspond to logical words, like "every" and "not") will have fixed meanings. But other symbols (e.g. those corresponding to proper nouns) will not be given a fixed meaning, and will receive different meanings according to need.

In dealing with a formal language $\mathscr{L}$ we must make a distinction between *syntax* and *semantics*. Syntactic questions are purely formal; they are concerned with expressions of $\mathscr{L}$ as strings of symbols, irrespective of any interpretation. Semantics, on the other hand, is concerned with the

[1] We refer, of course, only to interpretations that respect the parts of speech to which these words belong; so that "tove" must be interpreted as a common noun, etc.

[2] These new words must, of course, belong to the right parts of speech; "tove" can only be replaced by a common noun, etc.

meaning that expressions of $\mathscr{L}$ receive when the symbols occurring in them have been interpreted in some way.

Language serves as a medium of discourse and communication. Discussion of any topic takes place in some language. Now, when the object under discussion is *itself* a language, there are *two* languages involved, on two different levels. First, there is the language that is *being discussed*; this is called the *object language*. Then there is the language *in which* the discussion takes place; this is called the *metalanguage*. The distinction between the two is extremely important and must be constantly borne in mind.[1]

Mathematical logic is *mathematical* in two different senses, corresponding to the two levels just mentioned. The object languages which we shall discuss will be certain formal languages in which portions of mathematics can be expressed. When the symbols of such a language are suitably interpreted, expressions of that language may express propositions of, say, arithmetic or plane geometry. Syntactic and semantic study of such expressions throws light on the logic used in mathematics and on the logical relations between propositions of a given branch of mathematics. This is one sense in which mathematical logic is mathematical.

The second sense is that the study of formal languages, their syntax and semantics, is itself conducted in a mathematical fashion. Facts about formal languages and expressions in them are proved in much the same way as facts about groups are proved in algebra and facts about natural numbers in number theory.

In this book, the metalanguage used for discussing formal languages is English supplemented by additional symbols that will be introduced according to need. These metalinguistic symbols must not be confused with symbols of the object languages. This distinction is especially important in connection with two particular kinds of metalinguistic symbol, whose use will now be explained.

A mathematician writing about, say, real numbers needs symbols which serve as proper names of some particular numbers (or of particular sets of numbers, etc.). Such symbols are called *constants*. For example "1" is the constant used to denote (i.e., as a proper name for) the smallest positive integer, and "*e*" is normally used to denote a certain transcendental number. Here there can be no confusion between constants and their denotations (i.e., what they denote): the number *e* is an abstract entity

[1] Even when the two languages happen to coincide (e.g., when English is studied in English) the distinction between the two levels is important.

which is not a symbol, while the constant "e" is a symbol denoting that number. Numbers and their names are different kinds of things — the former are not part of a language, while the latter are part of the language used to discuss the former.

When the objects to be studied are *themselves* symbols and combinations of symbols (of some object language) the situation is a bit more tricky. A standard device for naming a symbol or a combination of symbols is to enclose it in quotation marks. (We have used this device in the preceding paragraph.) We could follow this procedure in this book, so that when a symbol of an object language is enclosed in quotation marks, it would become a name *in the metalanguage* for that symbol. In practice, however, this tends to become rather cumbersome. Another practice we could follow is to drop the quotation marks and use the same symbol (or combination of symbols) both in the object language and — as a name of itself — in the metalanguage, and to rely on the context to tell the reader in what capacity a symbol is being used. This, however, can be a bit confusing.

Fortunately, there is another way out. In an English book about the grammar of some foreign language it may be undesirable or even impossible to actually display the symbols of that language. (E.g., the script may be technically difficult to print; or the language may not have a written form at all, its symbols being not graphic signs but certain sounds and gestures.) To get round this difficulty, the author of such a book may use *transcription*, i.e., he will include in his metalanguage symbols *denoting* symbols of the object language.

We shall employ a similar device in this book. For our purposes the actual graphic shape (if any) of the symbols of a formal language is immaterial. We shall therefore not exhibit such symbols. On the other hand we shall use certain *metalinguistic* symbols — called *syntactic constants* — to denote particular symbols or combinations of symbols of the object languages. Thus, when we say, e.g. "$\mathscr{L}$ has the implication symbol $\rightarrow$", the arrow-shaped sign used here is to be regarded as a syntactic constant denoting a certain symbol (also called "implication symbol") of the object language $\mathscr{L}$. What the latter symbol actually looks like is of no importance; and the reader may give free rein to his imagination.

Besides constants, a mathematician writing about numbers also needs *variables*. Variables are like constants, except that whereas a constant denotes *one* object, a variable is allowed to *range* over a given *collection* of objects; each object belonging to that collection may serve as a *value* of the variable. Variables are used, e.g., in stating conditions which numbers may or may not satisfy (such as "$x>2$") and in making general statements about numbers (such as "for all x and y, $x+y=y+x$").

Similarly, we shall use metalinguistic symbols called *syntactic variables,* each of which will range over a prescribed collection of symbols or combinations of symbols of some formal language.[1]

§ 2. Structures and formal languages

In this section we discuss the various kinds of symbols and expressions that a formal language may be expected to have, if it is to be used as a medium for expressing mathematical statements.

A great many — some would say all — mathematical statements are about *structures.* A structure consists of the following ingredients:

(1) A non-empty class, called the *universe* or *domain* of the structure. The members of this universe are called the *individuals* of the structure.

(2) Various operations on the universe. These are called the *basic operations* of the structure.

(3) Various relations on the universe. These are called the *basic relations* of the structure.

Here (2) is optional: we admit structures having no basic operations. On the other hand, a structure must have at least one basic relation. The identity relation on the universe is normally taken as one of the basic relations.

Among the basic operations of a structure there may be 0-ary operations. According to our convention (see Ch. 0) such an operation is simply an individual. The 0-ary basic operations of a structure are called its *designated individuals.*

Let us give a few examples of structures associated with various branches of mathematics.

Elementary arithmetic may be defined as the study of one particular structure — *the elementary structure of natural numbers.* It has the set of natural numbers as universe, two designated individuals (viz. 0 and 1) and two basic operations (viz. addition and multiplication). Here the only basic relation is the identity relation.

Set theory is concerned with a structure whose individuals are all sets

[1] In this book syntactic constants and syntactic variables are printed in bold type, unless otherwise stated.

(i.e., its universe is the class of all sets). In addition to the identity relation, this structure has the basic relation of membership: the relation that holds for all pairs $\langle x,y \rangle$ such that x and y are sets and x is a member of y.

Elementary Euclidean plane geometry may be regarded as the study of a structure — *the elementary Euclidean plane* — whose individuals are points and straight lines. Among the basic relations of this structure are the property[1] of being a point, the property of being a straight line, and the ternary relation of "betweenness" which holds for all triples $\langle x,y,z \rangle$ such that x,y and z are collinear points and y lies between x and z.

A *topological space* may be thought of as a structure whose individuals are all points of the space, as well as all sets of points, all sets of sets of points, etc. There is one designated individual — the set of all *open* sets of points (this designated individual is the *topology* of the space); and the basic relations are the property of being a point, the identity relation and the membership relation between individuals.

Similarly, a structure can be associated with each group, ring and with other entities studied in algebra.

Suppose we are given a structure $\mathfrak{A}$ and we want to set up a formal language $\mathscr{L}$ in which statements about $\mathfrak{A}$ are to be expressed. What symbols should $\mathscr{L}$ have?

First, we would like $\mathscr{L}$ to have symbols that may be used as *variables* ranging over the universe of $\mathfrak{A}$. The need for variables is obvious to anyone acquainted with mathematics. Variables ranging over the universe of $\mathfrak{A}$ are used, e.g., in expressing conditions which individuals of $\mathfrak{A}$ may or may not satisfy, and in making general statements about $\mathfrak{A}$.

Next, we expect $\mathscr{L}$ to have symbols that may be used to denote the various basic operations of $\mathfrak{A}$. Such symbols are called *function symbols*. More specifically, a symbol designed to denote an n-ary operation is called an *n-ary function symbol*. In particular, if $\mathfrak{A}$ has designated individuals then $\mathscr{L}$ should have symbols for denoting them. Such symbols are called *individual constants* or, more briefly, just *constants*. Since designated individuals are regarded as 0-ary operations, constants are to be regarded as 0-ary function symbols.

Using variables and function symbols, we can construct expressions called *terms*. Roughly speaking, terms are the nounlike expressions of $\mathscr{L}$. For example, in a formal language suitable for elementary arithmetic we should have variables, say $\mathbf{x},\mathbf{y}$, etc., intended to range over the set

[1] Recall that we have agreed (Ch. 0) that a property is a unary relation.

N of natural numbers; and function symbols, say $\mathbf{0},\mathbf{1},+,\times$, intended to denote the numbers zero and one and the operations of addition and multiplication, respectively. Then $\mathbf{x}$, $\mathbf{1}$, $\mathbf{1}+\mathbf{x}$, $\mathbf{0}\times\mathbf{y}$, $((\mathbf{1}+\mathbf{x})\times(\mathbf{y}+\mathbf{1}))+\mathbf{0}$ are some of the terms we can form.

Of course, different interpretations can be applied to one and the same language. For example, the language described in the preceding paragraph can be re-interpreted by letting its variables range over some arbitrary non-empty class and letting $\mathbf{0},\mathbf{1},+,\times$ denote two arbitrarily chosen members of that class and two arbitrarily chosen binary operations on it.

But suppose we have fixed one particular interpretation for a formal language $\mathscr{L}$, by means of a structure $\mathfrak{U}$. Then those terms of $\mathscr{L}$ that do not contain variables will denote individuals of $\mathfrak{U}$. A term containing variables will not denote any particular individual, but will assume various individuals as values, depending on which individuals are assigned as values to the variables.

For example, in the particular language described above (taken with its originally intended interpretation) the term $\mathbf{1}+\mathbf{1}$ denotes the natural number two, while the term $(\mathbf{1}+\mathbf{1})\times\mathbf{x}+\mathbf{y}$ has as value the number obtained by adding whatever number is assigned as value to $\mathbf{y}$, to twice the number that happens to be assigned as value to $\mathbf{x}$.

Variables and function symbols alone do not suffice for formulating in $\mathscr{L}$ statements about a structure $\mathfrak{U}$. For this, $\mathscr{L}$ must have symbols that can be used to denote the basic relations of $\mathfrak{U}$. A symbol designed to denote an n-ary relation is called an *n-ary predicate symbol.*

If — as is usually the case — the identity relation is one of the basic relations of $\mathfrak{U}$, then $\mathscr{L}$ needs to have a predicate symbol to serve as a name for it. It is convenient to earmark one particular symbol, $=$, for this role.[1]

To illustrate how predicate symbols are used, let us return to our example of the language with function symbols $\mathbf{0},\mathbf{1},+,\times$. Assuming this language to have the predicate symbol $=$, we can write *formulas* like

(1) $\mathbf{1}=\mathbf{1}\times\mathbf{1}$,

(2) $\mathbf{1}+\mathbf{0}=\mathbf{1}\times\mathbf{0}$,

(3) $\mathbf{x}+\mathbf{x}=\mathbf{x}\times\mathbf{x}$,

(4) $\mathbf{x}+\mathbf{x}=\mathbf{y}$.

[1] This amounts to treating $=$ as a logical symbol: it is always interpreted as denoting the identity relation on the universe of discourse.

Formulas (1) and (2) are *sentences*: they express *propositions*. Under the originally intended interpretation, (1) expresses the *true* proposition that the number one is the same as its own square; and (2) expresses the *false* proposition that the sum of one and zero is the same as their product. Thus, under the interpretation which we are assuming, (1) is a *true* sentence and (2) is a *false* sentence. We also say that (1) has the *truth value 'truth'* and (2) has the *truth value 'falsehood'*.

Formulas (3) and (4) are not sentences; they do not express propositions, but *conditions* regarding the values which may be assigned to the variables involved. Thus, (3) expresses the condition that when the value of **x** is added to itself the result is the same as when it is multiplied by itself. It makes no sense to say that (3) is true (or false) outright. Rather, (3) will assume a truth value — either *truth* or *falsehood* — depending on which natural number is assigned to **x** as value. If that value is zero or two, (3) will have the truth value *truth* or, in other words, (3) will be *satisfied*. For all other values of **x**, (3) has the truth value *falsehood*. Similarly, (4) is satisfied (has the truth-value *truth*) iff the value assigned to **y** is double that assigned to **x**.

This example illustrates a general fact. Any *n*-ary predicate symbol of a formal language $\mathscr{L}$ can be combined with *n* terms of $\mathscr{L}$ to form a formula. Once an interpretation of $\mathscr{L}$ is fixed, some formulas of $\mathscr{L}$ (viz. those that are sentences) become true or false outright; all other formulas acquire truth values when the variables have been assigned values (belonging to the universe of the structure used for interpreting $\mathscr{L}$).

Using variables and function symbols we can construct terms; using these and predicate symbols we can construct formulas — but only rather simple ones. In order to combine simpler formulas into more complex ones, we shall require $\mathscr{L}$ to have two new kinds of symbol, called *connectives* and *quantifiers*.

We would like *negation* to be expressible in $\mathscr{L}$. Thus, for any formula $\boldsymbol{\alpha}$ of $\mathscr{L}$ we want $\mathscr{L}$ to have a formula $\neg\boldsymbol{\alpha}$ (read: "*not* $\boldsymbol{\alpha}$" or "*it is not the case that* $\boldsymbol{\alpha}$") which will be true whenever $\boldsymbol{\alpha}$ is false, and false whenever $\boldsymbol{\alpha}$ is true. (Thus $\boldsymbol{\alpha}$ and $\neg\boldsymbol{\alpha}$ always have opposite truth values.)

Next, we want the conjunction *and* to be expressible in $\mathscr{L}$. Thus, for any two formulas $\boldsymbol{\alpha}$, $\boldsymbol{\beta}$ we need a formula $\boldsymbol{\alpha}\wedge\boldsymbol{\beta}$ (read: "$\boldsymbol{\alpha}$ and $\boldsymbol{\beta}$") which will be true iff both $\boldsymbol{\alpha}$ and $\boldsymbol{\beta}$ are true.

Similarly, we want $\mathscr{L}$ to have, for any formulas α and β, a formula $\alpha \vee \beta$ (read: "α *or* β") which is false iff both α and β are false.[1]

Further, we want $\mathscr{L}$ to be capable of expressing conditional statements. Therefore, for any formulas α, β of $\mathscr{L}$ there should be in $\mathscr{L}$ a formula $\alpha \to \beta$ (read: "α *implies* β" or "*if* α, *then* β"). This formula will be false iff α is true but β is false.[2]

Finally, $\mathscr{L}$ should have, for any formulas α and β a formula $\alpha \leftrightarrow \beta$ (read: "α *iff* β") which is true whenever α and β have the same truth values and false whenever their truth values are different.

We could satisfy all these demands by requiring $\mathscr{L}$ to have five logical symbols, called *connectives*, viz. $\neg$, $\wedge$, $\vee$, $\to$, $\leftrightarrow$. But as a matter of fact we can make do with less. To start with, $(\alpha \to \beta) \wedge (\beta \to \alpha)$ behaves in just the way we want $\alpha \leftrightarrow \beta$ to behave; so we do not require $\mathscr{L}$ to have the symbol $\leftrightarrow$. Next, $\neg(\alpha \to \neg\beta)$ is easily seen to behave exactly as $\alpha \wedge \beta$ should; so $\wedge$ too can be eliminated. Finally, $(\neg\alpha) \to \beta$ behaves just like $\alpha \vee \beta$; so we can eliminate $\vee$ as well. Thus $\mathscr{L}$ need only have two connectives $\neg$ and $\to$. (For reasons that will become apparent later on, it is convenient to economize in the stock of logical symbols. No sacrifice is involved, since we can *define* $\alpha \wedge \beta$ to be $\neg(\alpha \to \neg\beta)$ etc.)

The last demand we shall make on $\mathscr{L}$ is that for any formula α and any variable $\mathbf{x}$, $\mathscr{L}$ should have a formula $\forall \mathbf{x} \alpha$ (read: "*for every* [*value of*] $\mathbf{x}$, α") and a formula $\exists \mathbf{x} \alpha$ (read: "*for some* [*value of*] $\mathbf{x}$, α").

To explain the meaning of such formulas, suppose first that α expresses some condition regarding the value of the variable $\mathbf{x}$. Then $\forall \mathbf{x} \alpha$ expresses the proposition that *every* possible value of $\mathbf{x}$ (i.e., all individuals of the structure) satisfy that condition. And $\exists \mathbf{x} \alpha$ says that the condition expressed by α is satisfied by *some* (i.e., at least one) individual.

Thus, e.g., in the particular language discussed above, taken with its intended interpretation, the formula $\mathbf{x} = \mathbf{x} \times \mathbf{x}$ expresses the condition that the value of $\mathbf{x}$ be equal to its own square; $\forall \mathbf{x}(\mathbf{x} = \mathbf{x} \times \mathbf{x})$ then expresses the false proposition that *every* natural number is equal to its own square, and $\exists \mathbf{x}(\mathbf{x} = \mathbf{x} \times \mathbf{x})$ expresses the true proposition that *some* natural number is equal to its own square.[3]

[1] In mathematical usage, "or" has the meaning which jurists express by "and/or". We are following this usage here.

[2] Here too we are following normal mathematical usage, which dictates that whenever α is false $\alpha \to \beta$ should be regarded as (vacuously) true, irrespective of the truth value of β.

[3] In fact, there are of course two such numbers.

More generally, suppose that α expresses a condition regarding the values of the variables $\mathbf{x},\mathbf{y}_1,\ldots,\mathbf{y}_n$. Then $\forall\mathbf{x}\alpha$ and $\exists\mathbf{x}\alpha$ express conditions regarding the values of $\mathbf{y}_1,\ldots,\mathbf{y}_n$. The values $b_1,\ldots,b_n$ (assigned to $\mathbf{y}_1,\ldots,\mathbf{y}_n$ respectively) satisfy [the condition expressed by] $\forall\mathbf{x}\alpha$ iff for *every* possible value a of $\mathbf{x}$ the values $a,b_1,\ldots,b_n$ satisfy α. And the values $b_1,\ldots,b_n$ satisfy $\exists\mathbf{x}\alpha$ iff there is *at least one* value a of $\mathbf{x}$ such that the values $a,b_1,\ldots,b_n$ satisfy α.

Thus, to return to the particular language we have been using as an example, the formula $\mathbf{x}\times\mathbf{y}=\mathbf{z}$ expresses the condition that the product of the values of $\mathbf{x}$ and $\mathbf{y}$ equals that of $\mathbf{z}$. Then $\forall\mathbf{x}(\mathbf{x}\times\mathbf{y}=\mathbf{z})$ expresses the condition that if *any* number be multiplied by the value of $\mathbf{y}$, the result is always equal to the value of $\mathbf{z}$. (This is satisfied iff both $\mathbf{y}$ and $\mathbf{z}$ are assigned the value zero.) And $\exists\mathbf{x}(\mathbf{x}\times\mathbf{y}=\mathbf{z})$ expresses the condition that the value of $\mathbf{z}$ is a multiple of that of $\mathbf{y}$.

We could require $\mathscr{L}$ to have two logical symbols $\forall$ and $\exists$, called *universal quantifier* and *existential quantifier* respectively. But again we can make do with less. As a matter of fact, it is enough if $\mathscr{L}$ only has the symbol $\forall$ (universal quantifier); for the formula $\neg\forall\mathbf{x}\neg\alpha$ behaves just as we want $\exists\mathbf{x}\alpha$ to behave, so that we can *define* the latter to be the former.

Formal languages possessing just the equipment sketched above (with the possible omission of function symbols) are called *first-order* languages.

*§ 3. Higher-order languages

Logicians sometimes deal with (apparently) richer formal languages. For example, a *second-order* language has, in addition to the equipment of a first-order language, another type of variable, ranging not over the individuals of a structure $\mathfrak{U}$ but over all *sets* of individuals (i.e., all subsets of the universe). Among the symbols of such a language there is a special one used to denote the relation of membership between an individual and a set of individuals. Quantifiers are allowed to apply not only to the *individual* variables (those ranging over individuals) but also to the new set variables. Thus, if α is a formula and $\mathbf{X}$ is a set variable, $\forall\mathbf{X}\alpha$ is also a formula.

Similarly, in a third-order language there are variables ranging over sets of sets of individuals, and so on for languages of still higher orders.

However, most logicians agree that such languages are — at least in principle — dispensable. Indeed, let $\mathfrak{U}$ be any structure and let $\mathfrak{B}$ be a structure obtained from $\mathfrak{U}$ in the following way. The universe of $\mathfrak{B}$ consists of all individuals of $\mathfrak{U}$ plus all sets of individuals of $\mathfrak{U}$. The basic operations of $\mathfrak{B}$ are defined in such a way that when they are restricted to the universe of $\mathfrak{U}$ (i.e., when they are applied to individuals of $\mathfrak{U}$) they behave exactly as the corresponding basic operations of $\mathfrak{U}$. Finally, the basic relations of $\mathfrak{B}$ are all the basic relations of $\mathfrak{U}$ plus two additional relations: the property of being an individual of $\mathfrak{U}$, and the relation of membership between an individual of $\mathfrak{U}$ and a set of individuals of $\mathfrak{U}$. Then any statement about $\mathfrak{U}$ expressed in a second-order language with set variables, can easily be "translated" into a statement about $\mathfrak{B}$ in a first-order language. In this sense, second-order languages are dispensable. A similar argument applies to other higher-order languages.

We therefore do not lose much by confining our attention to first-order languages only.

§ 4. Basic syntax

We shall now begin to put into practice some of the ideas discussed in §1 and §2. We proceed to describe an arbitrary *first-order language* $\mathscr{L}$. Throughout this book, unless the contrary is stated, $\mathscr{L}$ will be kept fixed. We shall introduce various notions relating to $\mathscr{L}$. These will be labelled by the prefix "$\mathscr{L}$-" or by phrases "of $\mathscr{L}$", "in $\mathscr{L}$", etc. However, once such a notion has been introduced, we shall omit these labels except where they are needed for emphasis or clarity.

The *symbols* of $\mathscr{L}$ are:

(a) An infinite sequence of *(individual) variables*, namely

$$\mathbf{v}_1, \mathbf{v}_2, \mathbf{v}_3, \ldots .$$

(b) For each natural number n, a set of *n-ary function symbols.*

(c) For each positive natural number n, a set of *n-ary predicate symbols.* For at least one n this set must be non-empty.

(d) The *connectives* $\neg$ *(negation)* and $\rightarrow$ *(implication).*

(e) The *universal quantifier* $\forall$.

The 0-ary function symbols (if any) are called *(individual) constants.*

If $\mathscr{L}$ has the special binary predicate $\boldsymbol{=}$, we say that $\mathscr{L}$ is a language *with equality. We stipulate that if $\mathscr{L}$ has at least one function symbol that is not an individual constant, then $\mathscr{L}$ must be a language with equality.*[1]

[1] Notice the difference between "$\boldsymbol{=}$" and "$=$". We use the former as a syntactic constant denoting the equality symbol of $\mathscr{L}$, while the latter is used in this book (as in most other mathematical texts) as short for "is the same as".

The variables, the connectives, the universal quantifier and $=$ are called *logical* symbols. They are assumed to be the same in all first-order languages (or, in the case of $=$, in all first-order languages with equality). The function symbols and the predicate symbols other than $=$ are called *extralogical* symbols and may differ from one language to another.

Notice that we have fixed a particular ordering of the variables: $\mathbf{v}_n$ is the nth *variable*[1] of $\mathscr{L}$. This ordering is called the *alphabetic* ordering of the variables.

A finite (possibly empty) sequence of $\mathscr{L}$-symbols is called an $\mathscr{L}$-*string*. (A given symbol may, of course, occur several times in the same string.)

The *length* of a string is the total number of (occurrences of) symbols in it. In particular, the empty string has length 0; and any single symbol is a string of length 1.

If S and T are strings, we define ST as the string obtained by concatenating S and T, in this order. Similarly for three or more strings.

If R=ST, where R,S,T are strings, then S is an *initial segment* of R. If T is non-empty, then S is a *proper* initial segment of R. Similarly, T is a *terminal segment* of R, and it is a *proper* one if S is non-empty. Obviously, a string of length n has $n+1$ different initial segments (including the empty string and the entire string itself).

We shall only be interested in two kinds of strings, called *terms* and *formulas*.

$\mathscr{L}$-*terms* are $\mathscr{L}$-strings formed according to the following two rules:

(1) Any $\mathscr{L}$-string consisting of (a single occurrence of) a variable is an $\mathscr{L}$-*term*.

(2) If $\mathbf{f}$ is an n-ary function symbol of $\mathscr{L}$ and $\mathbf{t}_1,\ldots,\mathbf{t}_n$ are $\mathscr{L}$-*terms*, then $\mathbf{f}\mathbf{t}_1\ldots\mathbf{t}_n$ is an $\mathscr{L}$-*term*.

Notice that, for $n=0$, (2) says that any constant is a term. In a term $\mathbf{f}\mathbf{t}_1\ldots\mathbf{t}_n$ formed according to (2), $\mathbf{t}_1,\ldots,\mathbf{t}_n$ are called *arguments* of $\mathbf{f}$.

By the *degree of complexity* of a term $\mathbf{t}$ (briefly deg $\mathbf{t}$) we mean the total number of occurrences of function symbols in $\mathbf{t}$.

The stipulation that in a term formed according to (2) the arguments always follow the function symbol, makes it unnecessary for $\mathscr{L}$ to have punctuation marks to indicate the grouping of arguments. To show this let us define the *weight* of a string S as the sum obtained by adding up -1

[1] In accordance with a convention introduced in §1, the bold arrow "$\boldsymbol{\rightarrow}$", e.g., is a syntactic constant belonging to our metalanguage and denoting a certain connective of $\mathscr{L}$. Similarly, "$\mathbf{v}_1$", e.g., is a syntactic constant denoting the first variable of $\mathscr{L}$.

for each occurrence of a variable in s and $n-1$ for each occurrence of an n-ary function symbol in s. We then have:

4.1. LEMMA. *Each term* **t** *has weight* -1; *but the weight of any proper initial segment of* **t** *is non-negative.*

PROOF. We use induction on deg **t**. If **t** is just a single occurrence of a variable then our claim is obviously true (the only proper initial segment of **t** is the empty string, which has weight 0).

If **t** is $\mathbf{f}\mathbf{t}_1 \ldots \mathbf{t}_n$ where **f** is an n-ary function symbol and $\mathbf{t}_1, \ldots, \mathbf{t}_n$ are terms, then $\deg \mathbf{t}_1, \ldots, \deg \mathbf{t}_n < \deg \mathbf{t}$ and we see by the induction hypothesis that the weight of **t** must be $(n-1)+n\cdot(-1)=-1$. Also, it is clear that the weight of any proper initial segment of **t** is non-negative. ∎

Using this result we see that in a string of the form $\mathbf{t}_1 \ldots \mathbf{t}_n$ (where $\mathbf{t}_1, \ldots, \mathbf{t}_n$ are terms) $\mathbf{t}_1$ is uniquely determined as the shortest initial segment with weight -1. Similarly, $\mathbf{t}_2, \ldots, \mathbf{t}_n$ are uniquely determined. Thus in a term $\mathbf{f}\mathbf{t}_1 \ldots \mathbf{t}_n$ all the arguments are uniquely determined.

$\mathscr{L}$-formulas are $\mathscr{L}$-strings formed according to the following four rules:

(1) If **P** is an n-ary predicate symbol of $\mathscr{L}$ and $\mathbf{t}_1, \ldots, \mathbf{t}_n$ are $\mathscr{L}$-terms, then $\mathbf{P}\mathbf{t}_1 \ldots \mathbf{t}_n$ is an *$\mathscr{L}$-formula.*

(2) If $\boldsymbol{\alpha}$ is an *$\mathscr{L}$-formula,* then so is $\neg\boldsymbol{\alpha}$.

(3) If $\boldsymbol{\alpha}$ and $\boldsymbol{\beta}$ are *$\mathscr{L}$-formulas,* then so is $\rightarrow\boldsymbol{\alpha}\boldsymbol{\beta}$.

(4) If $\boldsymbol{\alpha}$ is an *$\mathscr{L}$-formula* and **x** is a variable, then $\forall\mathbf{x}\boldsymbol{\alpha}$ is an *$\mathscr{L}$-formula.*

A formula $\mathbf{P}\mathbf{t}_1 \ldots \mathbf{t}_n$ formed according to (1) is called an *atomic* formula; the terms $\mathbf{t}_1, \ldots, \mathbf{t}_n$ here are the *arguments* of **P**. An atomic formula whose predicate symbol is $=$ is called an *equation* and its first and second arguments are called its *left-hand side* and *right-hand side,* respectively.

A formula $\neg\boldsymbol{\alpha}$ formed according to (2) is called a *negation* formula; $\neg\boldsymbol{\alpha}$ is the negation *of* $\boldsymbol{\alpha}$.

A formula $\rightarrow\boldsymbol{\alpha}\boldsymbol{\beta}$ formed according to (3) is called an *implication* formula. Here $\boldsymbol{\alpha}$ is the *antecedent* (or *implicans*) and $\boldsymbol{\beta}$ is the *consequent* (or *implicate).*

A formula $\forall\mathbf{x}\boldsymbol{\alpha}$ formed according to (4) is called a *universal* formula. Here **x** is the *variable of quantification,* and the string $\mathbf{x}\boldsymbol{\alpha}$ is the *scope* of the initial symbol $\forall$.

The *degree of complexity* of a formula $\boldsymbol{\alpha}$ (briefly deg $\boldsymbol{\alpha}$) is the sum obtained by adding up 2 for each occurrence of $\rightarrow$ and 1 for each occurrence of $\neg$ and $\forall$ in $\boldsymbol{\alpha}$.

In the case of formulas also, punctuation marks are unnecessary. To see this let us give weight $n-1$ to each occurrence of an n-ary predicate symbol and weights 0, 1, 1 to each occurrence of $\neg$, $\rightarrow$, $\forall$ respectively.

(The variables and function symbols retain their old weights.) Then we have:

4.2. LEMMA. *Each formula* α *has weight* -1; *but the weight of any proper initial segment of* α *is non-negative.*

The proof of this is similar to that of Lemma 4.1 and is left to the reader. ▮

It follows that in an implication formula the antecedent and consequent are uniquely determined.

By *$\mathscr{L}$-expression* we mean $\mathscr{L}$-term or $\mathscr{L}$-formula.

4.3. PROBLEM. Show that if R,S,T are strings such that RS and ST are expressions then S and T cannot both be non-empty. Thus, if two expressions overlap, one of them must be a terminal segment of the other.

If S is a term (or formula) and R,T are strings such that RST is again a term (or formula, respectively), then S is said to be a *subterm* (or *subformula*, respectively) of RST. If moreover R is non-empty[1] then S is a *proper* subterm (or subformula, respectively) of RST.

4.4. PROBLEM. Show that the proper subterms of a term $\mathbf{f}\mathbf{t}_1 \ldots \mathbf{t}_n$ are all the subterms of the arguments $\mathbf{t}_1, \ldots, \mathbf{t}_n$. Also, show that the proper subformulas of

(a) $\neg\alpha$ and $\forall\mathbf{x}\alpha$ are all the subformulas of α,

(b) $\rightarrow\alpha\beta$ are all the subformulas of α and of β.

§ 5. Notational conventions

In this book, unless otherwise stated:

(a) Boldface lower-case Roman letters from the end of the alphabet (especially "**x**", "**y**", "**z**") are used as syntactic variables ranging over the variables of a first-order language.

(b) As syntactic variables ranging over function symbols we use "**f**", "**g**" and "**h**".

(c) As syntactic variables ranging over constants we use "**a**", "**b**" and "**c**".

(d) As syntactic variables ranging over predicate symbols we use "**P**", "**Q**" and "**R**".

(e) As syntactic variables ranging over terms we use "**r**", "**s**" and "**t**".

[1] Note that, by Lemmas 4.1 and 4.2, if R is empty then so is T.

(f) Boldface lower-case and upper-case Greek letters are used as syntactic variables ranging over formulas and sets of formulas, respectively.

In §4 we saw that no punctuation marks are needed in $\mathscr{L}$. In various theoretical contexts this is an advantage, because it simplifies the syntax. However, this advantage has been achieved at the cost of some artificiality, by insisting that function symbols and predicate symbols (including $=$) always precede their arguments and that $\rightarrow$ precede the antecedent.

In practice it will now be convenient to make certain concessions to common usage. The reader must note that these concessions do not in any way constitute a modification of the formation rules of $\mathscr{L}$ laid down in §4; they are just conventions in our *metalanguage*, which are used when we talk *about* $\mathscr{L}$.

Also, for reasons of economy we have allowed $\mathscr{L}$ to have only two connectives and only the universal quantifier. We shall now introduce other connectives and the existential quantifier *metalinguistically* by definition.

5.1. DEFINITION.

(a) $(\mathbf{r}=\mathbf{s}) =_{\text{df}} =\mathbf{rs}$.
(b) $(\mathbf{r}\neq\mathbf{s}) =_{\text{df}} \neg(\mathbf{r}=\mathbf{s})$.
(c) $(\boldsymbol{\alpha} \rightarrow \boldsymbol{\beta}) =_{\text{df}} \rightarrow\boldsymbol{\alpha\beta}$.
(d) $(\boldsymbol{\alpha} \wedge \boldsymbol{\beta}) =_{\text{df}} \neg(\boldsymbol{\alpha} \rightarrow \neg\boldsymbol{\beta})$.
(e) $(\boldsymbol{\alpha} \vee \boldsymbol{\beta}) =_{\text{df}} (\neg\boldsymbol{\alpha} \rightarrow \boldsymbol{\beta})$.
(f) $(\boldsymbol{\alpha} \leftrightarrow \boldsymbol{\beta}) =_{\text{df}} ((\boldsymbol{\alpha} \rightarrow \boldsymbol{\beta}) \wedge (\boldsymbol{\beta} \rightarrow \boldsymbol{\alpha}))$.
(g) $\exists \mathbf{x}\boldsymbol{\alpha} =_{\text{df}} \neg\forall\mathbf{x}\neg\boldsymbol{\alpha}$.

Thus, e.g., "$(\boldsymbol{\alpha} \wedge \boldsymbol{\beta})$" is, by definition, a synonym for "$\neg(\boldsymbol{\alpha} \rightarrow \neg\boldsymbol{\beta})$", which is in turn another name for the formula $\neg\rightarrow\boldsymbol{\alpha}\neg\boldsymbol{\beta}$.

The formula $(\boldsymbol{\alpha}\wedge\boldsymbol{\beta})$ is called a *conjunction formula,* $\boldsymbol{\alpha}$ and $\boldsymbol{\beta}$ being the *first* and *second conjuncts,* respectively. Similarly, $(\boldsymbol{\alpha}\vee\boldsymbol{\beta})$ is a *disjunction formula* with $\boldsymbol{\alpha}$ and $\boldsymbol{\beta}$ as *first* and *second disjuncts.* Also, $(\boldsymbol{\alpha}\leftrightarrow\boldsymbol{\beta})$ is a *bi-implication formula* with $\boldsymbol{\alpha}$ and $\boldsymbol{\beta}$ as *left-hand side* and *right-hand side.* Finally, $\exists\mathbf{x}\boldsymbol{\alpha}$ is an *existential* formula.

In our new metalinguistic notation parentheses *are* needed to prevent ambiguity. To facilitate reading we shall replace some parentheses by brackets of various shapes. Also, we often omit parentheses, subject to the convention that "$\leftrightarrow$", "$\rightarrow$", "$\vee$", "$\wedge$" should be taken in this order of priority (just as according to the rules of English punctuation a full stop has priority over a semi-colon, and the latter has priority over a comma). The ranges of "$\neg$", "$\forall$" and "$\exists$" are to be as short as possible.

Thus, e.g., $\alpha\wedge\beta\rightarrow\gamma\leftrightarrow\neg\alpha\rightarrow\beta\vee\gamma$ is $\{[(\alpha\wedge\beta)\rightarrow\gamma]\leftrightarrow[\neg\alpha\rightarrow(\beta\vee\gamma)]\}$ and $\exists x\alpha\wedge\beta\vee\gamma$ is $[(\exists x\alpha\wedge\beta)\vee\gamma]$. But in "$\alpha\wedge(\beta\vee\gamma)$" and in "$\exists x(\alpha\wedge\beta)$" the parentheses cannot be omitted.

We also agree that where there are several occurrences of "$\rightarrow$" (or "$\wedge$", etc.) the one farthest to the left has highest priority. (This is the convention of *association to the right.*) Thus $\alpha\rightarrow\beta\wedge\gamma\rightarrow\beta\rightarrow\gamma$ is $\alpha\rightarrow\{(\beta\wedge\gamma)\rightarrow(\beta\rightarrow\gamma)\}$ but in "$(\alpha\wedge\beta)\wedge\gamma$" the parentheses must not be omitted.

We stress once more that all the conventions introduced in the present section are merely metalinguistic devices used in referring to $\mathscr{L}$-formulas. In the language $\mathscr{L}$ itself nothing has been changed. It should also be noticed that strictly speaking "$\wedge$", "$\vee$", "$\leftrightarrow$" and "$\exists$" do not, *by themselves*, denote anything at all. For example, it is only the *whole* combination "$(\alpha\wedge\beta)$" that has been given meaning (as denoting the formula $\neg\rightarrow\alpha\neg\beta$) while "$\wedge$" in isolation has not been given any meaning. Nevertheless, we shall occasionally express ourselves a bit loosely, as if $\mathscr{L}$ had connectives $\wedge$ *(conjunction)*, $\vee$ *(disjunction)*, $\leftrightarrow$ *(bi-implication)* as well as an *existential quantifier* $\exists$.

§ 6. Propositional semantics

The rest of this chapter is devoted to *propositional logic* which is (roughly speaking) that part of logic concerned with the meaning of the connectives and with rules for manipulating them.

In the present section we deal with the *semantic* aspects of propositional logic. In the next chapter we shall specify in detail the way in which truth values may be assigned to $\mathscr{L}$-formulas. For the present, however, we take such valuations as given arbitrarily, subject only to the condition that the intended meaning of $\neg$ and $\rightarrow$ is respected.

Let us denote by "$\top$" and "$\bot$" respectively the truth values *truth* and *falsehood.* Then the above explanation motivates the following:

6.1. DEFINITION. A *truth valuation on* $\mathscr{L}$ is a mapping σ assigning to each $\mathscr{L}$-formula α a truth value (i.e, a member of the set $\{\top, \bot\}$) α^σ such that for all $\mathscr{L}$-formulas β and γ

(1) $(\neg\beta)^\sigma=\top$ iff $\beta^\sigma=\bot$,

(2) $(\beta\rightarrow\gamma)^\sigma=\top$ iff $\beta^\sigma=\bot$ or[1] $\gamma^\sigma=\top$.

[1] Here, as usual in mathematical texts, "or" means what jurists mean by "and/or".

("α^σ" should be read as "the value of α under σ" or, briefly, "α under σ".)

We call α a *prime* formula if it is atomic or universal. Note thet Def. 6.1 does not impose any condition on α^σ for prime α. In fact, it is not difficult to see that *any* mapping of the set of all prime formulas into $\{\top, \bot\}$ can be extended in a unique way to a truth valuation. (If α^σ is fixed arbitrarily for all prime formulas α, then conditions (1) and (2) of 6.1 define α^σ for *all* formulas α by recursion on deg α.)

Conditions (1) and (2) of 6.1 can be encapsulated in *truth tables:*

β	$\neg\beta$
$\top$	$\bot$
$\bot$	$\top$

β	γ	$\beta\to\gamma$
$\top$	$\top$	$\top$
$\top$	$\bot$	$\bot$
$\bot$	$\top$	$\top$
$\bot$	$\bot$	$\top$

More generally, given a finite number of formulas $\beta_1,\ldots,\beta_k$ and a formula α compounded from them with $\neg$ and $\to$, we shall show how to construct a truth table for α in terms of $\beta_1,\ldots,\beta_k$.

To be precise, let us say that a formula is a *(propositional) combination* of $\beta_1,\ldots,\beta_k$ if it can be obtained by the following rules:

(a) Each β_i $(i = 1,\ldots,k)$ is a *combination* of $\beta_1,\ldots,\beta_k$.

(b) If γ is a *combination* of $\beta_1,\ldots,\beta_k$ then so is $\neg\gamma$.

(c) If γ and δ are *combinations* of $\beta_1,\ldots,\beta_k$ then so is $\gamma\to\delta$.

To construct a truth table for a formula α in terms of $\beta_1,\ldots,\beta_k$ we start by setting up a rectangular table with k columns — headed "β_1",...,"β_k" respectively — and 2^k rows. In each of the $k\cdot 2^k$ spaces we enter "$\top$" or "$\bot$" so that no two rows are filled out in the same way. (Thus, each of the 2^k different sequences of length k made up of "$\top$"s and "$\bot$"s will appear in exactly one row.) Now, if α is a combination of $\beta_1,\ldots,\beta_k$, we add a new column — headed "α" — and fill it out with "$\top$"s and "$\bot$"s according to the following rules, proceeding by induction on deg α:

(a) If α is β_i $(1 \leq i \leq k)$, we copy the entries of the ith column (headed "β_i") into the new "α" column.

(b) If α is $\neg\gamma$, where γ is a combination of $\beta_1,\ldots,\beta_k$, then by the induction hypothesis (since deg $\gamma <$ deg α) we already know how to construct a "γ" column. Then the "α" column will have "$\top$" in places where the "γ" column has "$\bot$", and "$\bot$" in all other places.

(c) If α is $\gamma\to\delta$, where γ and δ are combinations of $\beta_1,\ldots,\beta_k$, then by the induction hypothesis we know how to construct "γ" and "δ" columns.

Then the "α" column will have "$\perp$" in every row where the "γ" column has "$\top$" but the "δ" column has "$\perp$"; and in all other places the "α" column will have "$\top$".

It should be noted that these rules do not necessarily yield a unique result: we do not exclude the possibility that some of the β_i are themselves combinations of the remaining β_i. In some cases, therefore, we are allowed to choose between applying rule (a) and one of the other two rules. But *any* table with columns headed "β_1",...,"β_k" and "α" and filled out according to the above rules is a truth table for α in terms of $\beta_1,\ldots,\beta_k$.

6.2. PROBLEM. Using Def. 5.1 (clauses (d)–(f)) set up truth tables for $\alpha\wedge\beta$, $\alpha\vee\beta$ and $\alpha\leftrightarrow\beta$ in terms of α and β.

6.3. PROBLEM. In a truth table in terms of two formulas α and β the third column can be filled out in 16 ($=2^4$) different ways. Find 16 combinations of α and β which yield all these different third columns.

6.4. PROBLEM. We have chosen to regard $\neg$ and $\rightarrow$ as primitive (i.e., as symbols of $\mathscr{L}$) and to reduce $\wedge$, $\vee$ and $\leftrightarrow$ to them (Def. 5.1). Truth tables for negation and implication were incorporated into Def. 6.1, while the truth tables for conjunction, disjunction and bi-implication were obtained as a result (Prob. 6.2). Show that, similarly, $\rightarrow$ could be reduced to

(a) $\neg$ and $\wedge$,

(b) $\neg$ and $\vee$.

*6.5. PROBLEM. Show that $\rightarrow$ cannot be reduced to $\neg$ and $\leftrightarrow$.

6.6. PROBLEM. Let $\alpha|\beta$ be defined as $\neg(\alpha\wedge\beta)$. Show that we could have taken | *(Sheffer's stroke)* as a sole primitive connective and reduce both $\neg$ and $\rightarrow$ to it.

6.7. DEFINITION. A truth valuation σ on $\mathscr{L}$ *satisfies* a set Φ of $\mathscr{L}$-formulas (briefly, $\sigma\models\Phi$) if $\varphi^\sigma=\top$ for every φ in Φ.

If Φ consists of just one formula φ we write "$\sigma\models\varphi$" (read: "σ satisfies φ") instead of "$\sigma\models\{\varphi\}$".

If $\sigma\models\alpha$ for every truth valuation σ, then α is called a *tautology*.

6.8. THEOREM. *Consider a truth table for a formula* α *in terms of* $\beta_1,\ldots,\beta_k$. *If the* "α" *column contains only* "$\top$"*s, then* α *is a tautology. The converse is also true, provided* $\beta_1,\ldots,\beta_k$ *are prime and distinct.*

PROOF. Take any row in the given truth table. Let v_i be $\top$ or $\perp$ according as the "β_i" entry in our row (i.e., that coming under the heading "β_i") is "$\top$" or "$\perp$". Similarly, let v be $\top$ or $\perp$ according as the "α" entry in our row is "$\top$" or "$\perp$". It follows easily from Def. 6.1 and the rules for constructing truth tables that if σ is a truth valuation such that

(1) $\beta_i^\sigma = v_i$ for $i=1,\dots,k$,

then also $\alpha^\sigma = v$.

In particular, if the "α" column contains only "T"s, then α must be a tautology, because any truth valuation σ will fulfil (1) for *some* choice of v_i, corresponding to some row in the truth table.

If the β_i are all prime and distinct, then for each of the 2^k different choices of v_i there is a truth valuation σ for which (1) holds (see remark following Def. 6.1). Thus if α is a tautology, the "α" column must contain only "T"s. ∎

For each α we can find a finite set of prime formulas of which α is a combination. The smallest such set is the set of *prime components* of α, defined by induction on deg α as follows: if α is prime, then it is its own sole *prime component*; if $\alpha = \neg\beta$, then α has the same *prime components* as β; if $\alpha = \beta \to \gamma$, then the *prime components* of α are the *prime components* of β plus those of γ. By setting up a truth table for α in terms of its prime components, we can find out in a finite number of steps whether or not α is a tautology.

6.9. REMARK. From the above it is clear that the property of being a tautology is *invariant with respect to language:* if $\mathscr{L}$ and $\mathscr{L}'$ are two first-order languages such that α is both an $\mathscr{L}$-formula and an $\mathscr{L}'$-formula, then α is a tautology in $\mathscr{L}'$ (i.e., satisfied by every truth valuation on $\mathscr{L}'$) iff it is a tautology in $\mathscr{L}$. This follows from the fact that the procedure for checking whether or not α is a tautology is exactly the same, and must yield the same result, in both $\mathscr{L}$ and $\mathscr{L}'$. More directly, it is enough to observe that α^σ depends only on the values assigned by σ to the prime components of α.

6.10. PROBLEM. Using the first part of Thm. 6.8 verify that for any α, β and γ the following are tautologies:

(a) $\alpha \to \beta \to \alpha$,
(b) $(\alpha \to \beta \to \gamma) \to (\alpha \to \beta) \to \alpha \to \gamma$,
(c) $(\neg\alpha \to \beta) \to (\neg\alpha \to \neg\beta) \to \alpha$,
(d) $(\alpha \to \beta) \wedge (\beta \to \gamma) \to \alpha \to \gamma$,
(e) $(\alpha \to \beta) \wedge (\alpha \to \gamma) \to \alpha \to \beta \wedge \gamma$,
(f) $(\alpha \to \gamma) \wedge (\beta \to \gamma) \to \alpha \vee \beta \to \gamma$,
(g) $[\alpha \to (\beta \vee \gamma)] \to (\alpha \to \beta) \vee (\alpha \to \gamma)$.

(h) $(\alpha \wedge \beta \rightarrow \gamma) \rightarrow (\alpha \rightarrow \gamma) \vee (\beta \rightarrow \gamma)$,
(i) $(\alpha \rightarrow \neg \alpha) \rightarrow \neg \alpha$.

6.11. Definition. A formula α is a *tautological consequence* of a set Φ of formulas if $\sigma \models \alpha$ for every truth valuation σ such that $\sigma \models \Phi$. In this case we write "$\Phi \models_0 \alpha$"; and if Φ is empty we write simply "$\models_0 \alpha$".

(It is clear that α is a tautology iff it is a tautological consequence of the empty set, i.e., $\models_0 \alpha$.)

If Φ consists of just one formula φ, we say "tautological consequence of φ" instead of "tautological consequence of $\{\varphi\}$".

Formulas α and β are *tautologically equivalent* if they are tautological consequences of each other, i.e., if $\alpha^\sigma = \beta^\sigma$ for every truth valuation σ.

6.12. Problem. Show that α is a tautological consequence of $\{\varphi_1, \ldots, \varphi_k\}$ iff the formula

$$\varphi_1 \rightarrow \varphi_2 \rightarrow \ldots \rightarrow \varphi_k \rightarrow \alpha$$

is a tautology.

6.13. Problem. Show that α and β are tautologically equivalent iff $\alpha \leftrightarrow \beta$ is a tautology. Verify that for all $\alpha, \beta, \gamma, \varphi_1, \ldots, \varphi_k$ the following pairs are tautologically equivalent:

(a) $\alpha \rightarrow \beta,\ \neg \beta \rightarrow \neg \alpha$;
(b) $\neg(\alpha \rightarrow \beta),\ \alpha \wedge \neg \beta$;
(c) $\neg(\varphi_1 \wedge \varphi_2 \wedge \ldots \wedge \varphi_k),\ \neg \varphi_1 \vee \neg \varphi_2 \vee \ldots \vee \neg \varphi_k$;
(d) $\neg(\varphi_1 \vee \varphi_2 \vee \ldots \vee \varphi_k),\ \neg \varphi_1 \wedge \neg \varphi_2 \wedge \ldots \wedge \neg \varphi_k$;
(e) $\alpha \wedge \beta \wedge \gamma,\ (\alpha \wedge \beta) \wedge \gamma$;
(f) $\alpha \vee \beta \vee \gamma,\ (\alpha \vee \beta) \vee \gamma$;
(g) $\varphi_1 \wedge \varphi_2 \wedge \ldots \wedge \varphi_k \rightarrow \alpha,\ \varphi_1 \rightarrow \varphi_2 \rightarrow \ldots \rightarrow \varphi_k \rightarrow \alpha$.

6.14. Problem. Let $\varphi_1, \ldots, \varphi_k$ be all the different prime components of α Assuming that α is satisfied by at least one truth valuation, show how to. construct a *disjunctive normal form* for α, i.e., a formula β which is tautologically equivalent to α and which has the form $\beta_1 \vee \ldots \vee \beta_n$, where $n \geqq 1$ and each β_j is of the form $\varphi'_1 \wedge \ldots \wedge \varphi'_k$, where, for each i, φ'_i is φ_i or $\neg \varphi_i$.

6.15. Problem. Show that the relations of tautological consequence and tautological equivalence are invariant with respect to language (cf. Remark 6.9).

In the following three sections (§§7–9) we study the *method of propositional tableaux*, which can be used to find out whether there exists a truth valuation satisfying a given finite set of formulas, whether a given formula is a tautological consequence of a given finite set of formulas, and whether a given formula is a tautology. This method is far more efficient and elegant than that of truth tables, and is also of considerable theoretical interest. However, a reader who wishes to take a short cut may skip these three sections as well as §11, which depends on them.

From now on we shall often write, e.g. "$\mathbf{\Phi}, \alpha$" for "$\mathbf{\Phi} \cup \{\alpha\}$".

*§ 7. Propositional tableaux

A *tableau* is a set of elements, called *nodes,* partially ordered and classified into *levels* as explained below. With each node is associated a finite set of formulas. We shall usually identify a given node with its associated set of formulas; this is somewhat imprecise (since in fact the same set of formulas can be associated with different nodes) but will not cause confusion.

Each node belongs to a unique *level,* which is labelled by some natural number. There is just one node of level 0, called the *initial* node of the tableau. Each node of level $n+1$ is a *successor* of a unique node, which must be of level n. (In representing a tableau diagramatically, we put the successors of a given node below it, and join them to it by edges.) A node is *terminal* if it has no successors.

If a tableau has at least one node of level d but no nodes of greater level, we say that d is the *depth* of the tableau.

A *branch* of a tableau is a finite sequence of nodes $\mathbf{\Phi}_0,\ldots,\mathbf{\Phi}_k$ such that $\mathbf{\Phi}_0$ is the initial node of the tableau, and, for $i=1,\ldots,k$, $\mathbf{\Phi}_i$ is a successor of $\mathbf{\Phi}_{i-1}$, and $\mathbf{\Phi}_k$ is terminal. This branch is said to *terminate at* $\mathbf{\Phi}_k$.

Clearly, for each terminal node there is a unique branch terminating at it.

A formula belonging to a node of a branch is said to be a formula *of* that branch. The formulas belonging to the initial node are the *initial formulas* of the tableau.

The nodes of a tableau are partially ordered by the relation of *following:* each node is *followed* by its own successors, by their successors, etc.

A tableau with initial node $\mathbf{\Phi}_0$ is said to be a tableau *for* $\mathbf{\Phi}_0$.

We shall now prescribe how to construct the kind of tableau studied in the present chapter, namely *propositional tableaux.*

Firstly, if $\mathbf{\Phi}_0$ is any finite set of formulas, the tableau consisting of $\mathbf{\Phi}_0$ as sole node is a *propositional tableau* for $\mathbf{\Phi}_0$. Here $\mathbf{\Phi}_0$ is both initial and terminal, and there is just one branch.

Next, having obtained a *propositional tableau* $\boldsymbol{T}$ for $\mathbf{\Phi}_0$, we are allowed to extend it into a new one T' by any one of the following three rules. In each case T' will have all the nodes of T, plus one or two new nodes. *Succession* among the old nodes is the same in T' as in T. The new node(s) will be assigned as *successor(s)* to a node that was terminal in T.

Rule $\neg\neg$: If among the formulas of a branch of T terminating at node $\mathbf{\Phi}$ there is a formula $\neg\neg\alpha$, add a new node $\{\alpha\}$ as successor to $\mathbf{\Phi}$.

Rule $\rightarrow$: If among the formulas of a branch of T terminating at node $\mathbf{\Phi}$ there is a formula $\alpha\rightarrow\beta$, add two new nodes $\{\neg\alpha\}$ and $\{\beta\}$ as successors to $\mathbf{\Phi}$.

Rule $\neg\rightarrow$: If among the formulas of a branch of T terminating at node $\mathbf{\Phi}$ there is a formula $\neg(\alpha\rightarrow\beta)$, add a new node $\{\alpha, \neg\beta\}$ as successor to $\mathbf{\Phi}$.

In applying any one of these rules the respective formula $\neg\neg\alpha$, $\alpha\rightarrow\beta$ or $\neg(\alpha\rightarrow\beta)$ is said to be *used for* that application.

Schematically the rules are represented as follows:

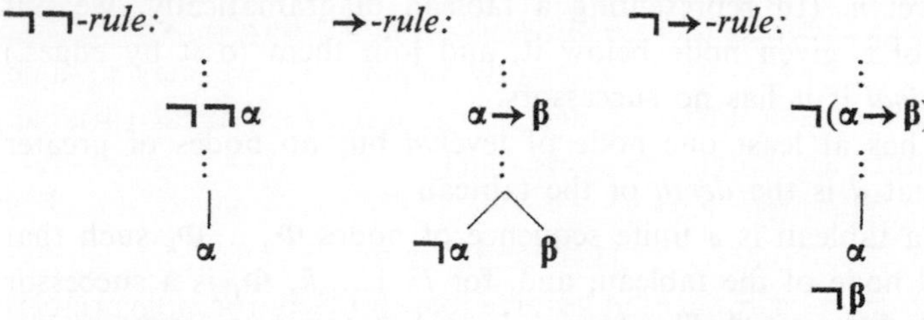

In going over from T to T', one particular branch of T is extended into a longer branch, or (in the case of Rule$\rightarrow$) extended and *split* into two longer branches. All the other branches of T remain unchanged.

In this chapter we shall refer to propositional tableaux briefly as just *tableaux*, since no other kind will be discussed here.

A branch of a tableau is *closed* if there is a prime formula α such that both α and $\neg\alpha$ are formulas of that branch. The formulas α and $\neg\alpha$ in question are said to be *used* for closing the branch. A (propositional) tableau for $\mathbf{\Phi}$ is called a *(propositional) confutation of* $\mathbf{\Phi}$ if all its branches are closed. To *confute* $\mathbf{\Phi}$ is to construct a confutation of $\mathbf{\Phi}$.

The idea behind the method of tableaux is made explicit in the following:

7.1. PROBLEM. Show that the three rules are *(semantically) sound* in the following sense: if a truth valuation σ satisfies [all the formulas of] a given branch in a tableau, and if that branch is extended into a new branch (or extended and split into two new branches) by one of the rules, then σ also satisfies the new branch (or at least one of the two new branches, respectively). Hence show that the tableau method is *(semantically) sound,* in the sense that if we can confute a finite set $\mathbf{\Phi}$ of formulas, there can be no truth valuation satisfying $\mathbf{\Phi}$; in particular, if $\mathbf{\Phi}=\{\neg\varphi\}$ then φ is a tautology. Also show that if $\mathbf{\Phi}, \neg\alpha$ can be confuted then $\mathbf{\Phi}\models_0\alpha$.

In constructing a tableau for $\mathbf{\Phi}$ the reader may think intuitively of $\mathbf{\Phi}$ as a story which he is trying to criticize. At each stage of the construction, the various branches represent alternative (but more specific and detailed) versions of the same story. Moreover, if $\mathbf{\Phi}$ is to be believed then at least one of these versions must be believed as well. A closed branch represents an *unbelievable* version.

Suppose that we have a branch (in some tableau) such that for the set of formulas of that branch we already possess a confutation. Then for all practical purposes that branch is *as good as closed,* because we know that by successively extending it we can eventually get branches all of which are actually closed. In practice, a branch which is closed (or as good as closed) need not be extended any further even if the rules allow us to do so.

7.2. PROBLEM. Show that for every α a confutation of $\{\alpha, \neg\alpha\}$ can be constructed. (Hint: use induction on $\deg\alpha$.) It follows that if a branch has α as well as $\neg\alpha$ then that branch is as good as closed, even if α is *not* prime.

Our rules do not prevent us from using the same formula over and over again indefinitely. But to do this is obviously pointless, since we get nothing essentially new. This motivates the following:

7.3. DEFINITION. A formula φ of a branch is *used up in* that branch, unless one of the following three conditions holds:

(a) φ is $\neg\neg\alpha$ and α is not a formula of the branch.

(b) φ is $\alpha\rightarrow\beta$ and neither $\neg\alpha$ nor β are formulas of the branch.

(c) φ is $\neg(\alpha\rightarrow\beta)$ and $\alpha, \neg\beta$ are not both formulas of the branch.

In practice, a formula used up in a branch need not be used to extend it. Moreover, when a formula is being used, it is good practice to use it at once to extend *all* branches to which it belongs (except those which are as good as closed) and then to tick it off as a sign that it should not be used again.

When dealing with formulas rendered in our metalinguistic notation using "$\wedge$", "$\vee$" and "$\leftrightarrow$" it is in practice unnecessary each time to unpack these *via* Def. 5.1. Instead, we can use the following six rules, which can be seen to be sound directly from Prob. 6.2, or derived from rules $\neg\neg$, $\rightarrow$, $\neg\rightarrow$ *via* Def. 5.1.

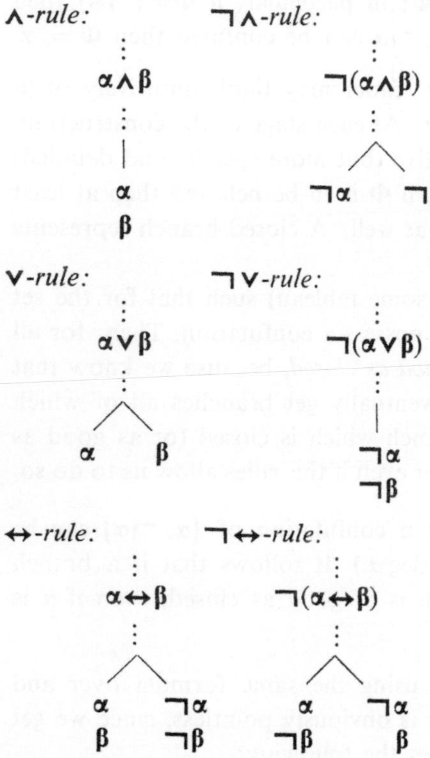

Of course, in all theoretical discussions we work in terms of $\neg$ and $\rightarrow$ only, and recognize only the three old rules.

One last practical remark. In general when two or more rules are applicable at the same time, it is advantageous to apply those that do not lead to a split before those that do.

EXAMPLE. Take the formula of Prob. 6.10(h). To show that it is a tautology we confute its negation.[1] For greater clarity we shall re-copy the tableaux obtained at each step. (In normal practice this is not necessary: each step may be grafted onto the tableau obtained at the previous step.) We put

[1] See Prob. 7.1.

"X" at the bottom of a branch if it is as good as closed. Here are the successive steps:

(1)

$$\neg[(\alpha\wedge\beta\to\gamma)\to(\alpha\to\gamma)\vee(\beta\to\gamma)] \quad \surd$$
$$|$$
$$\alpha\wedge\beta\to\gamma$$
$$\neg[(\alpha\to\gamma)\vee(\beta\to\gamma)]$$

(2)

$$\neg[(\alpha\wedge\beta\to\gamma)\to(\alpha\to\gamma)\vee(\beta\to\gamma)] \quad \surd$$
$$|$$
$$\alpha\wedge\beta\to\gamma$$
$$\neg[(\alpha\to\gamma)\vee(\beta\to\gamma)] \quad \surd$$
$$|$$
$$\neg(\alpha\to\gamma)$$
$$\neg(\beta\to\gamma)$$

(3)

$$\neg[(\alpha\wedge\beta\to\gamma)\to(\alpha\to\gamma)\vee(\beta\to\gamma)] \quad \surd$$
$$|$$
$$\alpha\wedge\beta\to\gamma$$
$$\neg[(\alpha\to\gamma)\vee(\beta\to\gamma)] \quad \surd$$
$$|$$
$$\neg(\alpha\to\gamma) \quad \surd$$
$$\neg(\beta\to\gamma)$$
$$|$$
$$\alpha$$
$$\neg\gamma$$

(4)

$$\neg[(\alpha\wedge\beta\to\gamma)\to(\alpha\to\gamma)\vee(\beta\to\gamma)] \quad \surd$$
$$|$$
$$\alpha\wedge\beta\to\gamma$$
$$\neg[(\alpha\to\gamma)\vee(\beta\to\gamma)] \quad \surd$$
$$|$$
$$\neg(\alpha\to\gamma) \quad \surd$$
$$\neg(\beta\to\gamma) \quad \surd$$
$$|$$
$$\alpha$$
$$\neg\gamma$$
$$|$$
$$\beta$$
$$\neg\gamma$$

(5) $\neg[(\alpha\wedge\beta\to\gamma)\to(\alpha\to\gamma)\vee(\beta\to\gamma)]\ \surd$

$$
\begin{array}{c}
\neg[(\alpha\wedge\beta\to\gamma)\to(\alpha\to\gamma)\vee(\beta\to\gamma)]\ \surd\\
|\\
\alpha\wedge\beta\to\gamma\ \surd\\
\neg[(\alpha\to\gamma)\vee(\beta\to\gamma)]\ \surd\\
|\\
\neg(\alpha\to\gamma)\ \surd\\
\neg(\beta\to\gamma)\ \surd\\
|\\
\alpha\\
\neg\gamma\\
|\\
\beta\\
\neg\gamma\\
\diagup\quad\diagdown\\
\neg(\alpha\wedge\beta)\qquad\gamma\\
\qquad\qquad\times
\end{array}
$$

(6) $\neg[(\alpha\wedge\beta\to\gamma)\to(\alpha\to\gamma)\vee(\beta\to\gamma)]\ \surd$

$$
\begin{array}{c}
\neg[(\alpha\wedge\beta\to\gamma)\to(\alpha\to\gamma)\vee(\beta\to\gamma)]\ \surd\\
|\\
\alpha\wedge\beta\to\gamma\ \surd\\
\neg[(\alpha\to\gamma)\vee(\beta\to\gamma)]\ \surd\\
|\\
\neg(\alpha\to\gamma)\ \surd\\
\neg(\beta\to\gamma)\ \surd\\
|\\
\alpha\\
\neg\gamma\\
|\\
\beta\\
\neg\gamma\\
\diagup\quad\diagdown\\
\neg(\alpha\wedge\beta)\ \surd\qquad\gamma\\
\diagup\quad\diagdown\qquad\qquad\times\\
\neg\alpha\qquad\neg\beta\\
\times\qquad\times
\end{array}
$$

7.4. **Problem.** Show by tableaux that the other formulas of 6.10 are tautologies.

*§ 8. The Elimination Theorem for propositional tableaux

We shall need the following five simple lemmas.

8.1. Lemma. *Given a confutation of* Φ, *we can confute* $\Phi \cup \Psi$, *where* Ψ *is any finite set of formulas.* ∎

8.2. Lemma. *Given a confutation of* $\Phi, \neg\neg\alpha$, *we can confute* Φ, α.
Proof. We imitate the given confutation, except that nodes $\{\alpha\}$ which were obtained by using $\neg\neg\alpha$ must now be cut out; but since we have now got α in the initial node, we can still use α in every branch. ∎

8.3. Lemma. *Given a confutation of* $\Phi, \alpha\rightarrow\beta$, *we can confute* $\Phi, \neg\alpha$ *and* Φ, β.
Proof. To confute $\Phi, \neg\alpha$, we imitate the given confutation except that where $\alpha\rightarrow\beta$ was used to get a configuration of the form

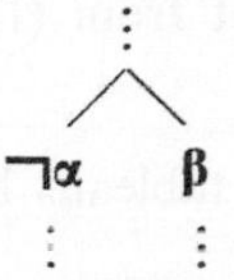

we now cut out the node $\{\neg\alpha\}$, as well as the node $\{\beta\}$ and all the nodes following it. In a similar way we get a confutation of Φ, β. ∎

8.4. Lemma. *Given a confutation of* $\Phi, \neg(\alpha\rightarrow\beta)$, *we can confute* $\Phi, \alpha, \neg\beta$.
Proof. Similar to that of 8.2. ∎

8.5. Lemma. *Given confutations of* Φ, δ *and* $\Phi, \neg\delta$, *where* δ *is prime, we can confute* Φ.
Proof. We imitate the given confutation of Φ, δ, except where δ was used. The only use that could have been made of δ was to close branches in which $\neg\delta$ turned up. But since we are given a confutation of $\Phi, \neg\delta$ and we have Φ as initial node, such a branch is as good as closed even without the help of δ. ∎

We can now generalize Lemma 8.5.

8.6. Elimination Lemma. *Given confutations of* Φ, δ *and* $\Phi, \neg\delta$, *where* δ *is any formula, we can confute* Φ.
Proof. By induction on deg δ. If δ is prime, we appeal to Lemma 8.5. If δ is $\neg\alpha$ then we are given confutations of $\Phi, \neg\alpha$ and $\Phi, \neg\neg\alpha$. From the latter we construct by Lemma 8.2 a confutation of Φ, α. By the induction hypothesis we can now confute Φ, since $\deg\alpha = \deg\delta - 1$.

Finally, if $\boldsymbol{\delta}$ is $\boldsymbol{\alpha} \rightarrow \boldsymbol{\beta}$ then we have got confutations of $\boldsymbol{\Phi}, \boldsymbol{\alpha} \rightarrow \boldsymbol{\beta}$ and $\boldsymbol{\Phi}, \neg(\boldsymbol{\alpha} \rightarrow \boldsymbol{\beta})$. From these we construct (using Lemmas 8.3 and 8.4)

(1) a confutation of $\boldsymbol{\Phi}, \neg\boldsymbol{\alpha}$,

(2) a confutation of $\boldsymbol{\Phi}, \boldsymbol{\beta}$,

(3) a confutation of $\boldsymbol{\Phi}, \boldsymbol{\alpha}, \neg\boldsymbol{\beta}$.

By Lemma 8.1 we can get from (2)

(4) a confutation of $\boldsymbol{\Phi}, \boldsymbol{\alpha}, \boldsymbol{\beta}$.

From (3) and (4) we get by the induction hypothesis (since $\deg \boldsymbol{\beta} < \deg \boldsymbol{\delta}$)

(5) a confutation of $\boldsymbol{\Phi}, \boldsymbol{\alpha}$,

and using the induction hypothesis once more we get from (1) and (5) a confutation of $\boldsymbol{\Phi}$. ▮

Let us try to strengthen the method of propositional tableaux by adding to the three old rules a fourth one:

Rule EM: If $\boldsymbol{\Phi}$ is any terminal node of a tableau T, and $\boldsymbol{\delta}$ is any formula, add two nodes $\{\boldsymbol{\delta}\}$, and $\{\neg\boldsymbol{\delta}\}$ as successors to $\boldsymbol{\Phi}$.

Schematically, this is represented thus:

EM-*rule:*

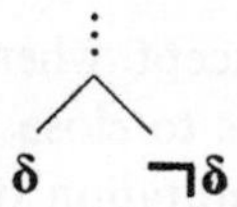

We say that the formula $\boldsymbol{\delta}$ is *introduced via* this application of the rule. "EM" stands for *excluded middle.* The EM-rule can be justified semantically (as the old rules were justified in Prob. 7.1) by observing that any truth valuation must satisfy $\boldsymbol{\delta}$ or $\neg\boldsymbol{\delta}$.

Though the new rule may help us to get confutations more quickly, it is (in principle) redundant.

8.7. Elimination Theorem. *Given a confutation of* $\boldsymbol{\Phi}_0$ *in which the* EM-*rule is used, we can construct an* EM-*free confutation of* $\boldsymbol{\Phi}_0$ *(i.e., a confutation of* $\boldsymbol{\Phi}_0$ not *using the* EM-*rule).*

PROOF. Let k be the deepest (i.e., maximal) level in the given confutation at which there is an application of the EM-rule. Consider an application at this level:

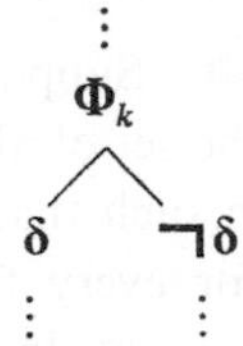

Before this application was made, $\mathbf{\Phi}_k$ was the terminal node of a branch $\mathbf{\Phi}_0,\ldots,\mathbf{\Phi}_k$. Let $\mathbf{\Phi}$ be the set of all formulas of this branch, i.e., $\mathbf{\Phi}=\mathbf{\Phi}_0\cup\ldots\cup\mathbf{\Phi}_k$. By the choice of k, all nodes which follow $\{\delta\}$ in the confutation must be obtained by the three old rules. Thus, by imitating that part of the given confutation we can get an EM-free confutation of $\mathbf{\Phi}, \delta$. Similarly we get an EM-free confutation of $\mathbf{\Phi}, \neg\delta$. By the Elimination Lemma we can now construct an EM-free confutation of $\mathbf{\Phi}$. Thus, before our application of the EM-rule was made, the branch $\mathbf{\Phi}_0,\ldots,\mathbf{\Phi}_k$ was *already* as good as closed, without any need for the EM-rule. We have therefore eliminated one application of the EM-rule from the given confutation. This process can be repeated until we finally get an EM-free confutation of $\mathbf{\Phi}_0$. ∎

We shall *not* regard the EM-rule as one of the rules of the tableau method. However, by virtue of the Elimination Theorem we can still use that rule in practice.

*§ 9. Completeness of propositional tableaux

In this section we shall see (among other things) that the tableau method is *complete* in the sense that if $\mathbf{\Phi}$ is a finite set of formulas which is not satisfied by any truth valuation, then $\mathbf{\Phi}$ can be confuted. (The converse of this, included in Prob. 7.1, is that the method is *sound*.)

If $\mathbf{\Phi}$ is a finite set of formulas, we define $\deg \mathbf{\Phi}$ to be the sum of $\deg \varphi$ for all φ in $\mathbf{\Phi}$.

By the *degree* of a branch of a tableau we mean $\deg \mathbf{\Phi}$, where $\mathbf{\Phi}$ is the set of all formulas of that branch which are not used up in it. A tableau is *exhausted* if all its branches have degree 0.

9.1. PROBLEM. Let $\mathbf{\Phi}_0$ be a finite set of formulas. Let T be a tableau for $\mathbf{\Phi}_0$ constructed in such a way that at each stage of the construction no formula is used to extend a branch if that formula is already used up in the branch. Let $\Phi_0,\ldots,\Phi_k$ be any branch of T. Show that its degree is at most $\deg \Phi_0 - k$. Hence show that an exhausted tableau for Φ_0 can be constructed.

9.2. PROBLEM. Let T be an exhausted tableau for Φ. Suppose that in T there is a branch which is not closed, and let Ψ be the set of all formulas of that branch. Let σ be the (unique) truth valuation such that for every prime formula $\alpha, \alpha^\sigma = \top$ iff α is in Ψ. Show that for every formula φ, $\varphi^\sigma = \top$ if φ is in Ψ and $\varphi^\sigma = \bot$ if $\neg\varphi$ is in Ψ. Hence $\sigma \models \Phi$. In particular, if $\Phi = \{\neg\varphi\}$, then φ is not a tautology.

9.3. THEOREM. *Let Φ be a finite set of formulas. In a finite number of steps we can construct for Φ an exhausted tableau T. T is a confutation iff there is no truth valuation satisfying Φ. In particular, if $\Phi = \{\neg\varphi\}$ then T is a confutation iff φ is a tautology.*

PROOF. Immediate from Probs. 7.1, 9.1 and 9.2. ∎

The method of tableaux can thus be used to find out whether any given formula is a tautology or not. In practice, this method is considerably more efficient than that of truth tables.

9.4. PROBLEM. Using Prob. 7.1 and Thm. 9.3, find an alternative proof for Lemma 8.6.

9.5. PROBLEM. Let Φ be a finite set of formulas. Show that if $\Phi \models_0 \alpha$ then $\Phi, \neg\alpha$ can be confuted. (This is the converse of the last part of Prob. 7.1.)

§ 10. The propositional calculus

The propositional calculus is a formal method for generating tautological consequences (indeed, *all* tautological consequences) of any given set Φ of formulas.

We start with the following simple fact:

10.1. LEMMA. *For any formulas α and β,*

$$\{\alpha, \alpha \to \beta\} \models_0 \beta. \qquad ∎$$

The operation of passing from two formulas α and $\alpha \to \beta$ to the formula β is called *modus ponens*. In this connection, α and $\alpha \to \beta$ are called the *minor premiss* and *major premiss* respectively, and β is called the *conclusion*. Note that the major premiss is an implication formula, whose antecedent and consequent are the minor premiss and conclusion respectively.

Lemma 10.1 shows that *modus ponens* is *(semantically) sound* as a *rule of inference*, in the sense that it "preserves truth": if $\alpha^\sigma = (\alpha \to \beta)^\sigma = \top$, then also $\beta^\sigma = \top$.

We shall now designate certain formulas as *propositional axioms*. We shall then show that if Φ is any set of formulas, then every formula obtained from Φ and these axioms by repeated application of *modus ponens* is a tautological consequence of Φ. (Later on we shall show that *every* tautological consequence of Φ can be obtained in this way.) This machinery for obtaining tautological consequences is the *propositional calculus.*[1]

As *propositional axioms* of $\mathscr{L}$ we take all $\mathscr{L}$-formulas of the following forms:

(Ax. I) $\alpha \to \beta \to \alpha$,

(Ax. II) $(\alpha \to \beta \to \gamma) \to (\alpha \to \beta) \to \alpha \to \gamma$,

(Ax. III) $(\neg\alpha \to \beta) \to (\neg\alpha \to \neg\beta) \to \alpha$,

where α, β, γ are any $\mathscr{L}$-formulas.

Notice that we have got here not three *single* axioms but three *axiom schemes,* each representing infinitely many axioms obtained by all possible choices of formulas α, β, γ.

Let Φ be a set of $\mathscr{L}$-formulas. By a *propositional deduction from* Φ *in* $\mathscr{L}$ we mean a finite non-empty sequence of $\mathscr{L}$-formulas $\varphi_1,\ldots,\varphi_n$ such that, for each k $(1 \leq k \leq n)$, φ_k is a propositional axiom of $\mathscr{L}$, or $\varphi_k \in \Phi$ or φ_k is obtained by *modus ponens* from earlier formulas in the same sequence (i.e., there are $i,j<k$ such that $\varphi_j = \varphi_i \to \varphi_k$).

In this connection Φ is called a set of *hypotheses.*

A propositional deduction in $\mathscr{L}$ from the empty set of hypotheses is called a *propositional proof in* $\mathscr{L}$.

In this chapter we shall often say simply *deduction* and *proof*, omitting the adjective *propositional,* since no other deductions and proofs will be dealt with here. Also, as before, we shall usually omit qualifications like "of $\mathscr{L}$" and "in $\mathscr{L}$".

A deduction (or proof) whose last formula is α is said to be a deduction (or proof, respectively) *of* α.

We write "$\Phi \vdash_0 \alpha$" to assert that α is *deducible* from Φ (i.e., that there is a deduction of α from Φ). If Φ is empty, so that α is *provable* (i.e., there

[1] To be quite precise, this is just *one version* of the propositional calculus. Other versions are based on different choices of axioms and rule(s) of inference.

is a proof of α), we write simply "$\vdash_0 \alpha$". Also, we write, e.g., "$\varphi \vdash_0 \alpha$" instead of "$\{\varphi\} \vdash_0 \alpha$".

10.2. THEOREM. *If* $\Phi \vdash_0 \alpha$, *then* $\Phi \models_0 \alpha$. *In particular, if* $\vdash_0 \alpha$ *then* $\models_0 \alpha$.
PROOF. Let $\varphi_1, \ldots, \varphi_n$ be a deduction of α from Φ. Thus $\varphi_n = \alpha$. By induction on $k = 1, \ldots, n$ we show that $\Phi \models_0 \varphi_k$. (Thus, for $k = n$, we have $\Phi \models_0 \alpha$.)

If φ_k is a propositional axiom, we easily verify that φ_k is a tautology (cf. Prob. 6.10). Thus φ_k is satisfied by *every* truth valuation and is therefore a tautological consequence of *any* set of formulas.

If $\varphi_k \in \Phi$ then clearly $\Phi \models_0 \varphi_k$.

Finally, if for some $i, j < k$ we have $\varphi_j = \varphi_i \rightarrow \varphi_k$, then $\{\varphi_i, \varphi_j\} \models_0 \varphi_k$ by Lemma 10.1. But by the induction hypothesis $\Phi \models_0 \varphi_i$ and $\Phi \models_0 \varphi_j$. Hence clearly $\Phi \models_0 \varphi_k$. ∎

Thm. 10.2 means that the propositional calculus is *(semantically) sound.*

10.3 LEMMA. *For any* α, $\vdash_0 \alpha \rightarrow \alpha$.
PROOF. Here is a propositional proof of $\alpha \rightarrow \alpha$:[1]

$$(\alpha \rightarrow (\alpha \rightarrow \alpha) \rightarrow \alpha) \rightarrow (\alpha \rightarrow \alpha \rightarrow \alpha) \rightarrow \alpha \rightarrow \alpha, \qquad \text{(Ax. II)}$$
$$\alpha \rightarrow (\alpha \rightarrow \alpha) \rightarrow \alpha, \qquad \text{(Ax. I)}$$
$$(\alpha \rightarrow \alpha \rightarrow \alpha) \rightarrow \alpha \rightarrow \alpha, \qquad \text{(m.p.)}$$
$$\alpha \rightarrow \alpha \rightarrow \alpha, \qquad \text{(Ax. I)}$$
$$\alpha \rightarrow \alpha. \qquad \text{(m.p.)}$$
∎

In the sequel we shall make implicit use of the following simple facts about deductions:

If $\Phi \subseteq \Psi$, then any deduction from Φ is also a deduction from Ψ.

If $\varphi_1, \ldots, \varphi_n$ is a deduction from Φ and $1 \leqslant k \leqslant n$, then $\varphi_1, \ldots, \varphi_k$ is also a deduction from Φ.

If $\varphi_1, \ldots, \varphi_m$ is a deduction from Φ and $\psi_1, \ldots, \psi_n$ is a deduction from Ψ, then the concatenation of the two sequences (i.e., $\varphi_1, \ldots, \varphi_m, \psi_1, \ldots, \psi_n$) is a deduction from $\Phi \cup \Psi$.

10.4. DEDUCTION THEOREM. *Given a deduction of* β *from* Φ, α, *we can construct a deduction of* $\alpha \rightarrow \beta$ *from* Φ. *(Thus, if* $\Phi, \alpha \vdash_0 \beta$, *then* $\Phi \vdash_0 \alpha \rightarrow \beta$.*)*
PROOF. Let $\varphi_1, \ldots, \varphi_n (= \beta)$ be the given deduction of β from Φ, α. We show

[1] For convenience we have added justifications in the margin. Thus the first formula is an instance of Ax. II (where we take $\alpha \rightarrow \alpha$ and α for β and γ respectively). In principle such justifications are redundant, since the reader could always check for himself whether a given formula is an instance of one of the axiom schemes, or whether it is obtainable by *modus ponens* from two earlier formulas.

by induction on $k = 1,\dots,n$ that a deduction of $\alpha \to \varphi_k$ from Φ can be constructed. The following cases are possible: φ_k is an axiom, or $\varphi_k \in \Phi$, or $\varphi_k = \alpha$, or φ_k is obtained by *modus ponens* from two earlier formulas φ_i and φ_j.

If φ_k is an axiom then the following is a proof of $\alpha \to \varphi_k$ and *a fortiori* a deduction of $\alpha \to \varphi_k$ from Φ:

$$\begin{array}{ll} \varphi_k, & \text{(Ax.)} \\ \varphi_k \to \alpha \to \varphi_k, & \text{(Ax. I)} \\ \alpha \to \varphi_k. & \text{(m.p.)} \end{array}$$

If $\varphi_k \in \Phi$, the same sequence of three formulas is a deduction of $\alpha \to \varphi_k$ from Φ (except that in this case the justification in the first line should read *hyp.* — short for *hypothesis*).

If $\varphi_k = \alpha$, then in the proof of Lemma 10.3 we had a propositional proof[1] of $\alpha \to \alpha$ ($= \alpha \to \varphi_k$). This is *a fortiori* a deduction of $\alpha \to \varphi_k$ from Φ.

Finally, suppose that for some $i,j < k$ we have $\varphi_j = \varphi_i \to \varphi_k$. Then by the induction hypothesis we have got deductions of $\alpha \to \varphi_i$ and $\alpha \to \varphi_i \to \varphi_k$ from Φ. We concatenate these two deductions and adjoin three new formulas:

$$\begin{array}{ll} \vdots & \\ \alpha \to \varphi_i, & \\ \vdots & \\ \alpha \to \varphi_i \to \varphi_k, & \\ (\alpha \to \varphi_i \to \varphi_k) \to (\alpha \to \varphi_i) \to \alpha \to \varphi_k, & \text{(Ax. II)} \\ (\alpha \to \varphi_i) \to \alpha \to \varphi_k, & \text{(m.p.)} \\ \alpha \to \varphi_k. & \text{(m.p.)} \end{array}$$

We thus have a deduction of $\alpha \to \varphi_k$ from Φ. ∎

Let us note that the converse of the Deduction Theorem is clearly true. For, if we are given a deduction of $\alpha \to \beta$ from Φ, and α is added as an extra hypothesis, we can then employ α with $\alpha \to \beta$ to get β by *modus ponens*. In this way we get a deduction of β from Φ, α.

The Deduction Theorem is of great practical use in constructing deductions. Suppose we want to construct a deduction of $\alpha \to \beta$ from Φ. Then by the Deduction Theorem it is enough to construct a deduction of β from Φ, α. In this way we have both simplified the formula to be deduced

[1] Note the difference in meaning between the two uses of the word *proof* here. The first refers to a (hopefully convincing) argument used to establish a *metamathematical* result (i.e., result *about* $\mathscr{L}$). The second refers to a sequence of formulas *in* $\mathscr{L}$.

(β instead of $\alpha \rightarrow \beta$) *and* gained an extra hypothesis, α. This is a double simplification of the problem: in general it is easier to deal with shorter formulas; and it is also clear that the more hypotheses we have, the easier it is to make deductions from them.

So far we have not made use of Ax. III, which encapsulates the deductive properties of negation. We shall do so now.

10.5. THEOREM. *For any* α *and* β,

(a) $\neg\neg\alpha \vdash_0 \alpha$,

(b) $\alpha \vdash_0 \neg\neg\alpha$,

(c) $\beta, \neg\beta \vdash_0 \alpha$.

PROOF. (a). We take a proof of $\neg\alpha \rightarrow \neg\alpha$ (see Lemma 10.3) and adjoin to it six further formulas, obtaining the following deduction of α from $\neg\neg\alpha$:

$$\begin{array}{ll}
\vdots & \\
\neg\alpha \rightarrow \neg\alpha, & \\
(\neg\alpha \rightarrow \neg\alpha) \rightarrow (\neg\alpha \rightarrow \neg\neg\alpha) \rightarrow \alpha, & \text{(Ax. III)} \\
(\neg\alpha \rightarrow \neg\neg\alpha) \rightarrow \alpha, & \text{(m.p.)} \\
\neg\neg\alpha, & \text{(hyp.)} \\
\neg\neg\alpha \rightarrow \neg\alpha \rightarrow \neg\neg\alpha, & \text{(Ax. I)} \\
\neg\alpha \rightarrow \neg\neg\alpha, & \text{(m.p.)} \\
\alpha. & \text{(m.p.)}
\end{array}$$

(b). By (a) we have a deduction of $\neg\alpha$ from $\neg\neg\neg\alpha$. Using the Deduction Theorem (with Φ empty) we get a propositional proof of $\neg\neg\neg\alpha \rightarrow \neg\alpha$. We take such a proof and by adjoining six further formulas, we obtain a deduction of $\neg\neg\alpha$ from α, as follows:

$$\begin{array}{ll}
\vdots & \\
\neg\neg\neg\alpha \rightarrow \neg\alpha, & \\
(\neg\neg\neg\alpha \rightarrow \alpha) \rightarrow (\neg\neg\neg\alpha \rightarrow \neg\alpha) \rightarrow \neg\neg\alpha, & \text{(Ax. III)} \\
\alpha, & \text{(hyp.)} \\
\alpha \rightarrow \neg\neg\neg\alpha \rightarrow \alpha, & \text{(Ax. I)} \\
\neg\neg\neg\alpha \rightarrow \alpha, & \text{(m.p.)} \\
(\neg\neg\neg\alpha \rightarrow \neg\alpha) \rightarrow \neg\neg\alpha, & \text{(m.p.)} \\
\neg\neg\alpha. & \text{(m.p.)}
\end{array}$$

(c). Here is a deduction of α from $\beta, \neg\beta$:

$$\begin{array}{ll}
(\neg\alpha \rightarrow \beta) \rightarrow (\neg\alpha \rightarrow \neg\beta) \rightarrow \alpha, & \text{(Ax. III)} \\
\beta, & \text{(hyp.)} \\
\beta \rightarrow \neg\alpha \rightarrow \beta, & \text{(Ax. I)} \\
\neg\alpha \rightarrow \beta, & \text{(m.p.)} \\
(\neg\alpha \rightarrow \neg\beta) \rightarrow \alpha, & \text{(m.p.)}
\end{array}$$

$\neg\beta$, (hyp.)
$\neg\beta\to\neg\alpha\to\neg\beta$, (Ax. I)
$\neg\alpha\to\neg\beta$, (m.p.)
α. (m.p.) ∎

In the sequel we shall make implicit use of the following fact. Suppose that $\Phi_i\vdash_0\beta_i$ for $i = 1,\dots,n$ and $\{\beta_1,\dots,\beta_n\}\vdash_0\alpha$. Then $\Phi_1\cup\dots\cup\Phi_n\vdash_0\alpha$. This is so because we can take a sequence of formulas obtained by concatenating deductions of β_i from Φ_i (for $i = 1,\dots,n$) and — using any deduction of α from $\{\beta_1,\dots,\beta_n\}$ — extend this sequence to a deduction of α from $\Phi_1\cup\dots\cup\Phi_n$.

A set Φ of formulas is *propositionally inconsistent* if for some β both $\Phi\vdash_0\beta$ and $\Phi\vdash_0\neg\beta$. Otherwise Φ is *propositionally consistent.* In this chapter we shall often omit the adverb "propositionally".

By virtue of the soundness of the propositional calculus we have:

10.6. THEOREM. *No truth valuation satisfies an inconsistent set of formulas.*
PROOF. Let $\Phi\vdash_0\beta$ and $\Phi\vdash_0\neg\beta$. If $\sigma\models\Phi$ then by Thm. 10.2 we would have both $\sigma\models\beta$ and $\sigma\models\neg\beta$, which is impossible. ∎

It follows that the empty set of formulas is consistent, since this set is satisfied (vacuously) by *every* truth valuation. This fact — the propositional consistency of $\emptyset$ — is usually expressed by saying that *the propositional calculus is consistent.*

In §11 the consistency of the propositional calculus will be proved by another method, using tableaux rather than semantics.

10.7. THEOREM. *A set Φ of formulas is propositionally inconsistent iff $\Phi\vdash_0\alpha$ for every formula α.*
PROOF. If $\Phi\vdash_0\alpha$ for every α, then for some (in fact, for all) β we have $\Phi\vdash_0\beta$ as well as $\Phi\vdash_0\neg\beta$.

Conversely, if for some β both $\Phi\vdash_0\beta$ and $\Phi\vdash_0\neg\beta$, then for any formula α we get $\Phi\vdash_0\alpha$ by Thm. 10.5 (c). ∎

10.8. THEOREM. *For any Φ and α:*

(a) *$\Phi, \neg\alpha$ is propositionally inconsistent iff $\Phi\vdash_0\alpha$,*

(b) *Φ, α is propositionally inconsistent iff $\Phi\vdash_0\neg\alpha$.*

PROOF. (a) If $\Phi, \neg\alpha$ is inconsistent, then for some β we have $\Phi, \neg\alpha\vdash_0\beta$ and $\Phi, \neg\alpha\vdash_0\neg\beta$. By the Deduction Theorem, $\Phi\vdash_0\neg\alpha\to\beta$ and $\Phi\vdash_0\neg\alpha\to\neg\beta$. On the other hand, using

$$(\neg\alpha\to\beta)\to(\neg\alpha\to\neg\beta)\to\alpha \qquad \text{(Ax. III)}$$

we get (with two applications of *modus ponens*) a deduction of α from $\{\neg\alpha\to\beta, \neg\alpha\to\neg\beta\}$. Thus $\Phi \vdash_0 \alpha$.

Conversely, if $\Phi \vdash_0 \alpha$ then $\Phi, \neg\alpha$ is inconsistent because then $\Phi, \neg\alpha \vdash_0 \alpha$ and $\Phi, \neg\alpha \vdash_0 \neg\alpha$.

(b) Using Thm. 10.5(a) it is easy to see that any formula deducible from Φ, α is also deducible from $\Phi, \neg\neg\alpha$. In particular, if Φ, α is inconsistent then $\Phi, \neg\neg\alpha$ is inconsistent as well; so by the first part of the present theorem we have $\Phi \vdash_0 \neg\alpha$.

Conversely, if $\Phi \vdash_0 \neg\alpha$ then Φ, α is inconsistent because both $\Phi, \alpha \vdash_0 \neg\alpha$ and $\Phi, \alpha \vdash_0 \alpha$. ∎

10.9. PROBLEM. Show that if $(\alpha\to\beta)\in\Phi$, and both $\Phi, \neg\alpha$ and Φ, β are inconsistent then Φ is inconsistent. Also show that if $\neg(\alpha\to\beta)\in\Phi$, and $\Phi, \alpha, \neg\beta$ is inconsistent then Φ is inconsistent.

10.10. PROBLEM. Show that a set Φ of formulas is consistent iff every finite subset of Φ is consistent.

*§ 11. The propositional calculus and tableaux

Let Φ be a finite set of formulas and let α be a formula. Consider the following three conditions:

(i) $\Phi \models_0 \alpha$,

(ii) $\Phi \vdash_0 \alpha$,

(iii) $\Phi, \neg\alpha$ can be (propositionally) confuted.

We know (see Prob. 9.5 and Thm. 10.2) that (i)⇒(iii) and (ii)⇒(i) (in fact, the latter holds for infinite Φ as well). It follows that (ii)⇒(iii). In the present section we shall give an alternative proof that (ii)⇒(iii). This proof will be direct; it will not make an excursion through semantics (but will use the results of §8).

Also, we shall give a direct proof that (iii)⇒(ii). It will then follow that (i), (ii) and (iii) are equivalent.

The fact that (iii)⇒(i) has already been proved directly (Prob. 7.1). But (i)⇒(ii) is new. (In §12 we shall prove (i)⇒(ii) directly, and in §13 we shall extend this to infinite as well as finite Φ.)

11.1. THEOREM. *Le Φ be a finite set of formulas. Given a deduction of α from Φ, we can confute $\Phi, \neg\alpha$.*

PROOF. Let $\varphi_1,\ldots,\varphi_n(=\alpha)$ be the given deduction of α from Φ. We show by induction on $k=1,\ldots,n$ how to confute $\Phi, \neg\varphi_k$. (For $k=n$ we get the required confutation of $\Phi, \neg\alpha$.)

If φ_k is a propositional axiom it can easily be verified that $\neg\varphi_k$ can be confuted (see Prob. 7.4). Hence by Lemma 8.1 we can confute $\Phi, \neg\varphi_k$.

If $\varphi_k \in \Phi$ then by Prob. 7.2 and Lemma 8.1 we can confute $\Phi, \neg\varphi_k$.

Finally, if for some $i,j<k$ we have $\varphi_j = \varphi_i \to \varphi_k$, then by the induction hypothesis we can construct

(1) a confutation of $\Phi, \neg\varphi_i$,

(2) a confutation of $\Phi, \neg(\varphi_i \to \varphi_k)$.

We confute $\Phi, \neg\varphi_k$ as follows. First, we apply the EM-rule, introducing the formula $\varphi_i \to \varphi_k$. Next, we apply the $\to$-rule to $\varphi_i \to \varphi_k$ and get

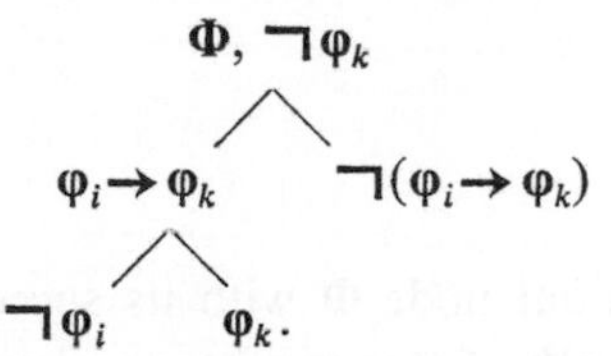

Here the leftmost and rightmost branches are as good as closed, using (1) and (2) respectively. The middle branch is as good as closed because it has both $\neg\varphi_k$ and φ_k (see Prob. 7.2).

By the Elimination Theorem 8.7, we can now eliminate our use of the EM-rule. ∎

11.2. PROBLEM. Using Thm. 11.1 and Prob. 7.1, find an alternative proof for Thm. 10.2. (Warning: note that in Thm. 10.2 Φ may be infinite.)

11.3. THEOREM. *Every finite inconsistent set of formulas can be confuted.*
PROOF. If we are given deductions of formulas β and $\neg\beta$ from a finite set Φ, then by Thm. 11.1 we can confute $\Phi, \neg\beta$ as well as $\Phi, \neg\neg\beta$. Using the Elimination Lemma 8.6, we can confute Φ. ∎

Using Thm. 11.3 we get a new proof for the consistency of the propositional calculus. For, if the empty set were inconsistent, then it could be confuted; but this is impossible since none of the tableau rules can be applied to the empty set.

We now prove the converse of Thm. 11.3.

11.4. THEOREM. *Every finite set of formulas that has a confutation is inconsistent.*

Proof. Given a confutation of $\mathbf{\Phi}$, we show by induction on its depth d that $\mathbf{\Phi}$ is inconsistent.

If $d=0$, this means that the given confutation has just one node, $\mathbf{\Phi}$. Thus there must be some (prime) formula $\boldsymbol{\alpha}$ such that $\boldsymbol{\alpha} \in \mathbf{\Phi}$ and $\neg\boldsymbol{\alpha} \in \mathbf{\Phi}$. Then $\mathbf{\Phi}$ is clearly inconsistent.

Now suppose that $d>0$ and examine level 1 of the given confutation. We consider three cases, corresponding to the three different ways in which the node(s) of this level could have arisen.

Case $\neg\neg$: There is just one node of level 1, and it arises by applying the $\neg\neg$-rule to a formula $\neg\neg\boldsymbol{\beta} \in \mathbf{\Phi}$. Then the given confutation starts thus:

$$\begin{array}{c} \mathbf{\Phi} \\ | \\ \boldsymbol{\beta} \\ \vdots \end{array}$$

Now, if in this tableau we fuse the initial node $\mathbf{\Phi}$ with its successor $\{\boldsymbol{\beta}\}$, we get a confutation of $\mathbf{\Phi}, \boldsymbol{\beta}$ and the depth of this confutation is now $d-1$. By the induction hypothesis $\mathbf{\Phi}, \boldsymbol{\beta}$ must therefore be inconsistent. By Thm. 10.8(b) we have $\mathbf{\Phi} \vdash_0 \neg\boldsymbol{\beta}$, but since $\neg\neg\boldsymbol{\beta} \in \mathbf{\Phi}$ it follows that $\mathbf{\Phi}$ is inconsistent.

Case $\rightarrow$: There are two nodes of level 1, arising by an application of the $\rightarrow$-rule to a formula $\boldsymbol{\alpha} \rightarrow \boldsymbol{\beta} \in \mathbf{\Phi}$. Then the given confutation starts thus:

$$\begin{array}{ccc} & \mathbf{\Phi} & \\ / & & \backslash \\ \neg\boldsymbol{\alpha} & & \boldsymbol{\beta} \\ \vdots & & \vdots \end{array} .$$

Now, if in this confutation we cut out the node $\{\boldsymbol{\beta}\}$ as well as all nodes following it, and we then fuse the node $\mathbf{\Phi}$ with the node $\{\neg\boldsymbol{\alpha}\}$, we get a confutation of $\mathbf{\Phi}, \neg\boldsymbol{\alpha}$ and this has depth $<d$. By the induction hypothesis it follows that $\mathbf{\Phi}, \neg\boldsymbol{\alpha}$ is inconsistent. Similarly, we show that $\mathbf{\Phi}, \boldsymbol{\beta}$ is also inconsistent. Since $\boldsymbol{\alpha} \rightarrow \boldsymbol{\beta} \in \mathbf{\Phi}$ it follows easily that $\mathbf{\Phi}$ must be inconsistent (see Prob. 10.9).

Case $\neg\rightarrow$: There is just one node of level 1, arising by application of the $\neg\rightarrow$-rule to a formula $\neg(\boldsymbol{\alpha} \rightarrow \boldsymbol{\beta}) \in \mathbf{\Phi}$. Then the given confutation starts thus:

$$\begin{array}{c}\Phi\\ |\\ \alpha\\ \neg\beta\\ \vdots\end{array}$$

and as in Case $\neg\neg$ we see that $\Phi, \alpha, \neg\beta$ is inconsistent. Since $\neg(\alpha\to\beta)\in\Phi$, it is easy to show (see Prob. 10.9) that Φ is inconsistent. ∎

11.5. PROBLEM. Show that if Φ is a finite set of formulas such that no truth valuation satisfies Φ, then Φ is propositionally inconsistent. (This is a partial converse to Thm. 10.6.)

11.6. PROBLEM. Show that if Φ is a finite set of formulas such that $\Phi, \neg\alpha$ has a confutation, then $\Phi\vdash_0\alpha$. (This is the converse of Thm. 11.1.)

§ 12. Weak completeness of the propositional calculus

The main result of this section is the following partial converse to Thm. 10.2:

12.1. WEAK COMPLETENESS THEOREM. *If Φ is finite and $\Phi\models_0\alpha$, then $\Phi\vdash_0\alpha$. In particular, if $\models_0\alpha$ then $\vdash_0\alpha$.*

PROOF. The result follows at once from Probs. 9.5 and 11.6. An alternative proof, which is more direct since it does not make an excursion *via* tableaux, is sketched in the following three problems (12.2–12.4). ∎

12.2. PROBLEM. Let α be a propositional combination of $\beta_1,\ldots,\beta_k$. Consider a row in a truth table for α in terms of $\beta_1,\ldots,\beta_k$. For each $i = 1,\ldots,k$ let β_i' be β_i or $\neg\beta_i$ according as the "β_i" entry in this row is "$\top$" or "$\perp$". Similarly, let α' be α or $\neg\alpha$ according as the "α" entry in this row is "$\top$" or "$\perp$". Show that $\{\beta_1',\ldots,\beta_k'\}\vdash_0\alpha'$. (Follow the prescription for constructing truth tables given in §6 and use induction on $\deg\alpha$.)

12.3. PROBLEM. Let α be a tautology and let $\beta_1,\ldots,\beta_k$ be all the different prime components of α. By induction on $p = 0,1,\ldots,k$ show that $\{\beta_1',\ldots,\beta_{k-p}'\}\vdash_0\alpha$, where, for each i, β_i' may be chosen as β_i or $\neg\beta_i$ (and the choice is made independently for different i). Hence (for $p = k$) $\vdash_0\alpha$.

12.4. PROBLEM. Let Φ be finite and $\Phi\models_0\alpha$. Using Probs. 6.12 and 12.3, show that a deduction of α from Φ can be constructed.

12.5. PROBLEM. Using Thm. 12.1, find an alternative solution to Prob. 11.5.

12.6. REMARK. It follows from Thms. 10.2 and 12.1 that the relation $\Phi\vdash_0\alpha$ is *invariant with respect to language.* Let $\mathscr{L}$ and $\mathscr{L}'$ be two different first-order languages such that α as well as all formulas of Φ are both $\mathscr{L}$-formulas and $\mathscr{L}'$-formulas. ($\mathscr{L}$ and $\mathscr{L}'$ must both have all the symbols occurring in Φ and α, but each may have different additional symbols.) Suppose we have a deduction of α from Φ in $\mathscr{L}$. This deduction is not necessarily an $\mathscr{L}'$-deduction, because it may contain $\mathscr{L}$-formulas that are not $\mathscr{L}'$-formulas. Nevertheless, α is deducible from Φ in $\mathscr{L}'$ (possibly *via* a different deduction). To see this we notice that the given deduction of α from Φ in $\mathscr{L}$ can only use a finite number of hypotheses. Thus α is deducible in $\mathscr{L}$ from a *finite* subset Φ_0 of Φ. By Thm. 10.2 it follows that $\Phi_0 \vDash_0 \alpha$. But the relation of tautological consequence *is* invariant with respect to language (see Prob. 6.15). Hence $\Phi_0 \vDash_0 \alpha$ holds also in $\mathscr{L}'$ and by the Weak Completeness Theorem 12.1 (applied to $\mathscr{L}'$) we see that α is deducible from Φ_0 in $\mathscr{L}'$. The same result can also be obtained by tableaux.

Similarly, the property of being a propositionally consistent set of formulas is invariant with respect to language.

12.7. REMARK. All proofs given so far in this chapter have been *constructive:* whenever we have asserted the existence of a deduction (or a tableau, etc.) we actually prescribed how to construct it in a finite number of steps. Also, whenever we have proved a set Φ to be inconsistent, the proof actually tells us how to find a formula β and deductions of β and $\neg\beta$ from Φ.

In particular, both proofs of Thm. 12.1 were constructive.

*§ 13. Strong completeness of the propositional calculus

In this section we give a non-constructive proof of the full converse of Thm. 10.2.

A set Φ of $\mathscr{L}$-formulas is *maximal propositionally consistent in* $\mathscr{L}$ if it is propositionally consistent but is not a proper subset of any propositionally consistent set of $\mathscr{L}$-formulas.

As usual, we allow ourselves to omit the qualifications "propositionally" and "in $\mathscr{L}$". But it must be stressed that the property of being maximal consistent is *not* invariant with respect to language.

13.1. THEOREM. *The set* Φ *is maximal consistent iff the following two conditions both hold:*

(a) Φ *is consistent.*

(b) *For every formula* α, $\alpha \in \Phi$ *or* $\neg\alpha \in \Phi$.

PROOF. Suppose $\mathbf{\Phi}$ is maximal consistent. Then (a) holds by definition. Also, if $\alpha \notin \mathbf{\Phi}$ as well as $\neg\alpha \notin \mathbf{\Phi}$, then by definition $\mathbf{\Phi}, \alpha$ and $\mathbf{\Phi}, \neg\alpha$ must both be inconsistent. Thus by Thm. 10.8 both $\mathbf{\Phi} \vdash_0 \neg\alpha$ and $\mathbf{\Phi} \vdash_0 \alpha$ — contrary to (a).

Conversely, suppose that (a) and (b) hold. We show that if $\mathbf{\Phi}$ is a proper subset of $\mathbf{\Psi}$ then $\mathbf{\Psi}$ is inconsistent. Indeed, let $\alpha \in \mathbf{\Psi}$ but $\alpha \notin \mathbf{\Phi}$. Then, by (b), $\neg\alpha \in \mathbf{\Phi}$; so $\alpha \in \mathbf{\Psi}$ as well as $\neg\alpha \in \mathbf{\Psi}$, making $\mathbf{\Psi}$ inconsistent. ∎

13.2. THEOREM. *Let* $\mathbf{\Phi}$ *be maximal consistent. Then* $\alpha \to \beta \in \mathbf{\Phi}$ *iff* $\neg\alpha \in \mathbf{\Phi}$ *or* $\beta \in \mathbf{\Phi}$.

PROOF. Suppose $\alpha \to \beta \in \mathbf{\Phi}$. If $\neg\alpha \notin \mathbf{\Phi}$ and $\beta \notin \mathbf{\Phi}$ then $\mathbf{\Phi}, \neg\alpha$ and $\mathbf{\Phi}, \beta$ are inconsistent. It is then easy to verify (see Prob. 10.9) that $\mathbf{\Phi}$ itself is inconsistent, contrary to our assumption.

Conversely, let $\neg\alpha \in \mathbf{\Phi}$ or $\beta \in \mathbf{\Phi}$. Then $\mathbf{\Phi}, \alpha$ or $\mathbf{\Phi}, \neg\beta$ must be inconsistent. In either case, $\mathbf{\Phi}, \alpha, \neg\beta$ is inconsistent. If $\neg(\alpha \to \beta) \in \mathbf{\Phi}$, then it is easy to verify (see Prob. 10.9) that $\mathbf{\Phi}$ would be inconsistent. Therefore $\neg(\alpha \to \beta) \notin \mathbf{\Phi}$; so by Thm. 13.1 we must have $\alpha \to \beta \in \mathbf{\Phi}$. ∎

In the proof of the following theorem we apply Zorn's Lemma, which makes the proof non-constructive.

13.3. THEOREM. *Let* $\mathbf{\Phi}$ *be consistent. Then there is a maximal consistent* $\mathbf{\Psi}$ *such that* $\mathbf{\Phi} \subseteq \mathbf{\Psi}$.

PROOF. Consider the family of *all* consistent sets of $\mathscr{L}$-formulas, partially ordered by inclusion $\subseteq$.

If $\{\mathbf{\Phi}_i : i \in I\}$ is an arbitrary *totally* ordered subfamily of that family, then the union $\bigcup\{\mathbf{\Phi}_i : i \in I\}$ is also consistent. To show this we make use of Prob. 10.10. Every *finite* subset of that union is already included in some $\mathbf{\Phi}_i$, and is therefore consistent. So the whole union is consistent.

We apply Zorn's Lemma to obtain the required result. ∎

We can now prove the (full) converse of Thm. 10.6:

13.4. THEOREM. *If* $\mathbf{\Phi}$ *is a propositionally consistent set of formulas, there is a truth valuation* σ *satisfying* $\mathbf{\Phi}$.

PROOF. By Thm. 13.3, we can assume that $\mathbf{\Phi} \subseteq \mathbf{\Psi}$, where $\mathbf{\Psi}$ is maximal consistent.

We determine a truth valuation σ by requiring that

(1) $\varphi^\sigma = \top$ iff $\varphi \in \mathbf{\Psi}$,

for every *prime* formula φ. We shall now show by induction on deg φ that (1) in fact holds for *every* formula φ.

If φ is not prime, then it is either a negation formula or an implication formula.

Suppose $\varphi = \neg\alpha$. Then

$$\begin{aligned} \varphi^\sigma = \top \ &\text{iff } \alpha^\sigma = \bot && \text{(by Def. 6.1)} \\ &\text{iff } \alpha \notin \Psi && \text{(by ind. hyp.)} \\ &\text{iff } \neg\alpha \in \Psi, && \text{(by Thm. 13.1)} \\ &\quad \text{i.e., } \varphi \in \Psi. \end{aligned}$$

Now suppose $\varphi = \alpha \to \beta$. Then

$$\begin{aligned} \varphi^\sigma = \top \ &\text{iff } \alpha^\sigma = \bot \text{ or } \beta^\sigma = \top && \text{(by Def. 6.1)} \\ &\text{iff } \alpha \notin \Psi \text{ or } \beta \in \Psi && \text{(by ind. hyp.)} \\ &\text{iff } \neg\alpha \in \Psi \text{ or } \beta \in \Psi && \text{(by Thm. 13.1)} \\ &\text{iff } \alpha \to \beta \in \Psi, && \text{(by Thm. 13.2)} \\ &\quad \text{i.e., } \varphi \in \Psi. \end{aligned}$$

We have thus verified (1) for *all* formulas. Since $\Phi \subseteq \Psi$ it follows that $\sigma \models \Phi$. ∎

We can now prove the (full) converse of Thm. 10.2:

13.5. STRONG COMPLETENESS THEOREM. *If* $\Phi \models_0 \alpha$ *then* $\Phi \vdash_0 \alpha$.
PROOF. If $\Phi \models_0 \alpha$ then no truth valuation can satisfy $\Phi, \neg\alpha$ (for, if $\sigma \models \Phi$ then $\alpha^\sigma = \top$, so that $(\neg\alpha)^\sigma = \bot$). Hence by Thm. 13.4 $\Phi, \neg\alpha$ is inconsistent. Therefore, by Thm. 10.8(a), $\Phi \vdash_0 \alpha$. ∎

*§ 14. Propositional logic based on $\neg$ and $\wedge$

In some of the later chapters of this book (viz., those devoted to model theory and axiomatic set theory) it will be convenient to assume that the primitive connectives of $\mathscr{L}$ are $\neg$ and $\wedge$ rather than $\neg$ and $\to$. In this section we sketch how the treatment of propositional logic given above can be modified and adapted to that setting.

First, $\alpha \to \beta$ would have to be *defined*, e.g. as $\neg(\alpha \wedge \neg\beta)$ (cf. Prob. 6.4). Then, clause (2) of Def. 6.1 should be modified to read:
$(\beta \wedge \gamma)^\sigma = \top$ iff both $\beta^\sigma = \top$ and $\gamma^\sigma = \top$.
The rules for constructing truth tables should also be modified accordingly.

The treatment of propositional tableaux should be modified by adopting the $\wedge$-rule and $\neg\wedge$-rule (which in the treatment above were recognized only in practice, not in theory) instead of the $\rightarrow$-rule and $\neg\rightarrow$-rule.

All the material presented in §§6–9 carries over, *mutatis mutandis*[1], to the new setting.

In the propositional calculus, we retain the old axiom schemes and *modus ponens,* but add also some new axiom schemes. A possible choice might be the following three axiom schemes:

(Ax. IV) $\alpha\wedge\beta\rightarrow\alpha$,

(Ax. V) $\alpha\wedge\beta\rightarrow\beta$,

(Ax. VI) $\alpha\rightarrow\beta\rightarrow\alpha\wedge\beta$.

These axioms may used be used to solve the modified version of Prob. 10.9. ("Show that if $\alpha\wedge\beta\in\Phi$ and Φ, α, β is inconsistent, then Φ is inconsistent. Also show that if $\neg(\alpha\wedge\beta)\in\Phi$ and both $\Phi, \neg\alpha$ and $\Phi, \neg\beta$ are inconsistent, then Φ is inconsistent.")

All the material of §§10–13 then goes through without any difficulty.

*§ 15. Propositional logic based on $\neg$, $\rightarrow$, $\wedge$ and $\vee$

For some purposes — e.g., for comparing classical logic with intuitionistic logic (see Ch. 9) — it is desirable to regard all four connectives $\neg$, $\rightarrow$, $\wedge$ and $\vee$ as primitive symbols of the language rather than define $\wedge$ and $\vee$ in terms of $\neg$ and $\rightarrow$. In this section we sketch the modifications that have to be made to adapt our treatment to that setting. The details are left to the reader.

First, the prescription for constructing formulas (given just after Lemma 4.1) has to include two additional rules: if α and β are formulas, then so are $\wedge\alpha\beta$ and $\vee\alpha\beta$.

Clauses (d) and (e) in Def. 5.1 have to be amended to define $(\alpha\wedge\beta)$ and $(\alpha\vee\beta)$ as $\wedge\alpha\beta$ and $\vee\alpha\beta$ respectively.

In Def. 6.1 we need two additional clauses dealing with formulas of the forms $\beta\wedge\gamma$ and $\beta\vee\gamma$ in the obvious way. Also, two corresponding rules have to be added to the prescription for constructing truth tables.

[1] For example, Lemma 8.3 should be modified to assert that if we are given a confutation of $\Phi, \alpha\wedge\beta$, then we can confute Φ, α, β.

Propositional tableaux require four additional rules ($\wedge$, $\neg\wedge$, $\vee$ and $\neg\vee$).

The propositional calculus may be based on nine axiom schemes: the three schemes of §10, plus the three schemes of §14, plus the following three schemes:

(Ax. VII) $\alpha \to \alpha \vee \beta$,

(Ax. VIII) $\beta \to \alpha \vee \beta$,

(Ax. IX) $(\alpha \to \gamma) \to (\beta \to \gamma) \to \alpha \vee \beta \to \gamma$.

With these modifications, all the material of §§6–13, *mutatis mutandis*, goes through without difficulty.

§ 16. Historical and bibliographical remarks

For a detailed treatment of the topics touched upon in §§1 and 2, the reader is referred to the Introduction of CHURCH [1956]. Also, §29 of Ch. II in Church's book contains a survey of the earlier literature on propositional logic. (In our present Remarks we only deal with topics not covered by Church's survey.)

The tableau method (for propositional *and* first-order logic) was devised independently by BETH [1955], HINTIKKA [1955] and SCHÜTTE [1956]. It was developed and streamlined by SMULLYAN [1968], to whom our treatment is heavily indebted. Essentially, the tableau method is dual to the method of GENTZEN [1934]. Gentzen's method is essentially a systematic search for *proofs* in *tree* form, while the tableau method is a systematic search for *refutations* in *upside-down tree* form. A node in a Gentzen proof tree roughly corresponds to a branch in a tableau. The Elimination Theorem for tableaux corresponds (and is in fact equivalent) to Gentzen's celebrated *Hauptsatz*.

The method used in §13 for proving strong completeness is due to HENKIN [1949]. Thm. 13.3 is due to Adolf Lindenbaum (see TARSKI [1930]).

CHAPTER 2

FIRST-ORDER LOGIC

Whereas in propositional logic the only symbols of the object language to play an essential role were the connectives, we shall now bring into play the full apparatus of our first-order language $\mathscr{L}$. Much of the present chapter will be devoted to extending the ideas developed in §6–9 of Ch. 1 to this more elaborate setting. In addition, we shall develop a new order of ideas, the need for which arises from the specific role played by the variables; this will be done in §§2–3. The final sections of this chapter (§§9–13) deal with some special topics. These will not be required immediately and the reader who is in a hurry to get on to Ch. 3 may put off reading them.

§ 1. First-order semantics

In order to interpret $\mathscr{L}$-terms and $\mathscr{L}$-formulas it is necessary to fix an *$\mathscr{L}$-structure* (or *$\mathscr{L}$-interpretation*) $\mathfrak{U}$ consisting of the following ingredients:

(1) A non-empty class U, called the *universe of discourse* (or, briefly, *universe*, or *domain*) of $\mathfrak{U}$. The members of U are called *individuals* (of $\mathfrak{U}$).

(2) A mapping which assigns to each function symbol $\mathbf{f}$ of $\mathscr{L}$ an operation $\mathbf{f}^{\mathfrak{U}}$ on U such that if $\mathbf{f}$ is an n-ary function symbol, $\mathbf{f}^{\mathfrak{U}}$ is an n-ary operation on U. In particular, if $\mathbf{a}$ is a constant, then $\mathbf{a}^{\mathfrak{U}}$ is an individual. An individual of this kind — i.e., $\mathbf{a}^{\mathfrak{U}}$ for some constant $\mathbf{a}$ of $\mathscr{L}$ — is said to be *designated.*

(3) A mapping which assigns to each extralogical predicate symbol $\mathbf{P}$ of $\mathscr{L}$ a relation $\mathbf{P}^{\mathfrak{U}}$ on U such that if $\mathbf{P}$ is an n-ary predcate symbol, $\mathbf{P}^{\mathfrak{U}}$ is an n-ary relation on U. In particular, if $\mathbf{P}$ is unary, then $\mathbf{P}^{\mathfrak{U}}$ is a subclass of U.

We make the following notational conventions. If an upper-case German letter denotes a given structure, then the corresponding italic letter denotes the universe of that structure. Individuals will be denoted by lower-case italic letters (especially "a", "b", "c", "u", "v"). In contexts where only one particular structure $\mathfrak{U}$ is considered, the operation (or relation) corresponding to a given function symbol (or predicate symbol, respectively) will sometimes be denoted by the corresponding italic letter. Thus we may write "f" and "P" instead of "$\mathbf{f}^{\mathfrak{U}}$" and "$\mathbf{P}^{\mathfrak{U}}$".

An $\mathscr{L}$-interpretation $\mathfrak{U}$ only suffices to assign a value (which is an individual of $\mathfrak{U}$) to an $\mathscr{L}$-term in which variables do not occur. Also, it only suffices to assign truth values to *some* $\mathscr{L}$-formulas (viz. those that are to be called *$\mathscr{L}$-sentences*). But, as explained in §2 of Ch. 1, the value of a term and the truth value of a formula will in general depend not only on the interpretation but also on values assigned to some $\mathscr{L}$-variables. Thus, to evaluate a given $\mathscr{L}$-expression we need an interpretation $\mathfrak{U}$ plus a particular assignment of values in U to some $\mathscr{L}$-variables. Here there is a slight technical problem: different $\mathscr{L}$-expressions may involve different variables, so that we would have to consider, say, an assignment of values to one set of variables in connection with one formula $\boldsymbol{\alpha}$, an assignment to a different set of variables in connection with another formula $\boldsymbol{\beta}$, and yet another assignment in connection with $\boldsymbol{\alpha} \to \boldsymbol{\beta}$. This is feasible, but technically rather awkward. Instead, we prefer to work with assignments that assign a value in U to *all* variables at once. And we shall arrange matters so that in evaluating any given expression the values assigned to variables which the expression does not involve will not in fact make any difference.

We therefore define an *$\mathscr{L}$-valuation* σ to be an $\mathscr{L}$-structure $\mathfrak{U}$ together with an assignment of a value $\mathbf{x}^\sigma \in U$ to each variable $\mathbf{x}$.

The reader must note that an $\mathscr{L}$-valuation is not the same thing as a truth valuation on $\mathscr{L}$, although (as will transpire shortly) each valuation gives rise in a natural way to a truth valuation.

By definition, each valuation σ involves a particular structure $\mathfrak{U}$. We refer to $\mathfrak{U}$ as the structure *underlying* σ. If $\mathfrak{U}$ is the structure underlying σ, we define $\mathbf{f}^\sigma$ and $\mathbf{P}^\sigma$ to be the operation $\mathbf{f}^{\mathfrak{U}}$ and the relation $\mathbf{P}^{\mathfrak{U}}$, respectively, where $\mathbf{f}$ is any function symbol and $\mathbf{P}$ is any extralogical predicate symbol. The universe U of $\mathfrak{U}$ will also be called the *universe of* σ.

We shall say that two valuations σ and τ *agree* on a given variable $\mathbf{x}$ (or function symbol $\mathbf{f}$, or extralogical predicate symbol $\mathbf{P}$) if σ and τ have the same universe and $\mathbf{x}^\sigma = \mathbf{x}^\tau$ (or $\mathbf{f}^\sigma = \mathbf{f}^\tau$, or $\mathbf{P}^\sigma = \mathbf{P}^\tau$, respectively).

Let σ be a valuation with universe U and let $u \in U$. We define $\sigma(\mathbf{x}/u)$ to be the valuation which agrees with σ on every variable other than

$\mathbf{x}$ as well as on every extralogical symbol, while $\mathbf{x}^{\sigma(\mathbf{x}/u)} = u$. Thus, in particular, the structure underlying $\sigma(\mathbf{x}/u)$ is the same as that underlying σ.

Given an $\mathscr{L}$-valuation σ with universe U, we now define, for each $\mathscr{L}$-term $\mathbf{t}$, *the value of* $\mathbf{t}$ *under* σ (briefly, $\mathbf{t}^\sigma$) in such a way that $\mathbf{t}^\sigma \in U$. Also for each $\mathscr{L}$-formula $\boldsymbol{\alpha}$ we define *the value of* $\boldsymbol{\alpha}$ *under* σ (briefly, $\boldsymbol{\alpha}^\sigma$) so that $\boldsymbol{\alpha}^\sigma$ is either $\top$ or $\bot$. This is done by recursion on $\deg \mathbf{t}$ and $\deg \boldsymbol{\alpha}$ respectively, as follows:

1.1. Basic Semantic Definition.

(T1). If $\mathbf{x}$ is a variable, then $\mathbf{x}^\sigma$ is already defined.

(T2). If $\mathbf{f}$ is an n-ary function symbol of $\mathscr{L}$ and $\mathbf{t}_1,\ldots,\mathbf{t}_n$ are $\mathscr{L}$-terms, then

$$(\mathbf{f}\mathbf{t}_1\ldots\mathbf{t}_n)^\sigma = \mathbf{f}^\sigma(\mathbf{t}_1^\sigma,\ldots,\mathbf{t}_n^\sigma).$$

(F1). If $\mathbf{P}$ is an n-ary extralogical predicate symbol of $\mathscr{L}$ and $\mathbf{t}_1,\ldots,\mathbf{t}_n$ are $\mathscr{L}$-terms, then

$$(\mathbf{P}\mathbf{t}_1\ldots\mathbf{t}_n)^\sigma = \begin{cases} \top & \text{if } \langle \mathbf{t}_1^\sigma,\ldots,\mathbf{t}_n^\sigma\rangle \in \mathbf{P}^\sigma, \\ \bot & \text{otherwise.} \end{cases}$$

(F1$^=$). If $\mathbf{s}$ and $\mathbf{t}$ are $\mathscr{L}$-terms and $\mathscr{L}$ is a language with equality, then

$$(\mathbf{s}=\mathbf{t})^\sigma = \begin{cases} \top & \text{if } \mathbf{s}^\sigma = \mathbf{t}^\sigma, \\ \bot & \text{otherwise.} \end{cases}$$

(F2). For every $\mathscr{L}$-formula $\boldsymbol{\beta}$,

$$(\neg\boldsymbol{\beta})^\sigma = \begin{cases} \top & \text{if } \boldsymbol{\beta}^\sigma = \bot, \\ \bot & \text{otherwise.} \end{cases}$$

(F3). For every $\mathscr{L}$-formula $\boldsymbol{\beta}$ and $\mathscr{L}$-formula $\boldsymbol{\gamma}$,

$$(\boldsymbol{\beta}\rightarrow\boldsymbol{\gamma})^\sigma = \begin{cases} \top & \text{if } \boldsymbol{\beta}^\sigma = \bot \text{ or } \boldsymbol{\gamma}^\sigma = \top, \\ \bot & \text{otherwise.} \end{cases}$$

(F4). For every $\mathscr{L}$-formula $\boldsymbol{\beta}$ and variable $\mathbf{x}$,

$$(\forall\mathbf{x}\boldsymbol{\beta})^\sigma = \begin{cases} \top & \text{if } \boldsymbol{\beta}^{\sigma(\mathbf{x}/u)} = \top \text{ for every } u \in U, \\ \bot & \text{otherwise,} \end{cases}$$

where U is the universe of σ.

The above definition will be referred to as "BSD". It must be stressed that what the BSD defines is not the valuation σ itself — which must be given *in advance,* by specifying a structure $\mathfrak{U}$ and an assignment of value $\mathbf{x}^\sigma \in U$ to each variable $\mathbf{x}$ — but two mappings *induced* by σ.

The first of these, defined in (T1) and (T2), maps the set of all terms into U. It is well defined because in (T2) its value for a term $\mathbf{ft}_1 \ldots \mathbf{t}_n$ is reduced to its values for the terms $\mathbf{t}_1, \ldots, \mathbf{t}_n$ whose degrees of complexity are smaller.

The second mapping induced by σ is defined in (F1)–(F4). It maps the set of all formulas into $\{\top, \bot\}$. In (F1) and ($F1^{=}$) its values for atomic formulas are given explicitly in terms of the first induced mapping. In (F2)–(F4) its values for non-atomic formulas are reduced to the values assigned by it (or — in the case of universal formulas, covered in (F4) — by similar mappings induced by other valuations) to formulas with smaller degrees of complexity. Because of clauses (F2) and (F3), this second induced mapping is clearly a truth valuation (see Def. 1.6.1).

There is no need to be overawed by the BSD. It merely spells out with precision what is pretty obvious once the intended meaning of the logical symbols has been understood, and the extralogical symbols have been provided with references by a valuation (which is merely a kind of dictionary). The BSD just puts on record the way in which $\mathscr{L}$-expressions are supposed to be understood.

In particular, clause (F1) is hardly more startling or profound than, e.g., the fact that the Latin sentence "*Catullus uxorem Celeris amavit*" is true iff Catullus did really love Celer's wife. This fact can easily be verified by consulting any standard Latin dictionary and grammar. It obviously has little epistemological value. The question whether that Latin sentence *is* actually true is of course much more exciting — but belongs to historical gossip-mongering, not to logic or philosophy.

The first explicit formulation of (a version of) the BSD was given in 1933 by Tarski.[1] For this reason it is often referred to as *Tarski's truth definition* or *Tarski's definition of satisfaction*.

We conclude this discussion of the BSD with:

1.2. Remark. Because of clause (F4), the BSD is strongly non-constructive: if U is infinite, (F4) does not provide us with a method for computing the value $(\forall \mathbf{x} \boldsymbol{\beta})^{\sigma}$ in a finite number of steps, for it presupposes the values $\boldsymbol{\beta}^{\sigma(\mathbf{x}/u)}$ for infinitely many u. This non-constructive character is inherited by all the semantic definitions given below, which are based on the BSD. Indeed, one of our main tasks will be to obtain a more constructive characterization of the concepts thus defined.

[1] Cf. §14 below.

1.3. Problem. Using Def. 1.5.1(g) show that $(\exists x\alpha)^\sigma = \top$ iff $\alpha^{\sigma(x/u)} = \top$ for some $u \in U$, where U is the universe of the valuation σ.

1.4. Problem. Show that $\exists$ could have been taken as primitive (i.e., as a symbol of $\mathscr{L}$) instead of $\forall$. (Replace BSD (F4) by the statement of Prob. 1.3 and replace Def. 1.5.1(g) by a definition of $\forall$ from which the original (F4) can be derived.)

We have noted above that, for every valuation σ, the induced mapping defined in clauses (F1)–(F4) of the BSD is a truth valuation. We shall say that σ *satisfies* a formula φ (or a set Φ of formulas) — briefly, $\sigma \models \varphi$ (or $\sigma \models \Phi$, respectively) — if the truth valuation induced by σ satisfies φ (or Φ, respectively). Thus $\sigma \models \varphi$ iff $\varphi^\sigma = \top$; and $\sigma \models \Phi$ iff $\varphi^\sigma = \top$ for every $\varphi \in \Phi$.

It must be noted, however, that not every truth valuation is induced by some valuation. For example, if α is any formula then the prime formula $\forall x(\alpha \rightarrow \alpha)$ is satisfied by every valuation but not by every truth valuation. Thus a truth valuation not satisfying this prime formula cannot be induced by any valuation.

1.5. Definition. If every valuation satisfying a set Φ of formulas also satisfies a formula α, we say that α is a *logical consequence of* Φ (or α *follows logically from* Φ, or Φ *logically entails* α). We write this briefly as "$\Phi \models \alpha$". As usual, we shall write "$\varphi \models \alpha$" instead of "$\{\varphi\} \models \alpha$" and say that α is a logical consequence of φ. If α is satisfied by *every* valuation (i.e., if α follows logically from the empty set of formulas), then we say that α is *logically true* (or *logically valid*) and we write "$\models \alpha$". If $\alpha \models \beta$ as well as $\beta \models \alpha$ (i.e., $\alpha^\sigma = \beta^\sigma$ for every valuation σ), we say that α and β are *logically equivalent*. We say that a formula φ (or a set Φ of formulas) is *satisfiable* if $\sigma \models \varphi$ (or $\sigma \models \Phi$, respectively) for some valuation σ.

1.6. Theorem. *If $\Phi \models_0 \alpha$ then $\Phi \models \alpha$. In particular, if $\models_0 \alpha$ then $\models \alpha$; and if α and β are tautologically equivalent then they are logically equivalent.* ∎

The converse of Thm. 1.6 is false. For example, $\forall x(\alpha \rightarrow \alpha)$ is logically true but not tautologically true; and it is logically (but not tautologically) equivalent to $\forall x(\alpha \leftrightarrow \alpha)$.

1.7. Problem. Show that $\Phi, \alpha \models \beta$ iff $\Phi \models \alpha \rightarrow \beta$. Hence show that $\{\varphi_1, \ldots, \varphi_n\} \models \beta$ iff $\models \varphi_1 \rightarrow \ldots \rightarrow \varphi_n \rightarrow \beta$. Also show that α and β are logically equivalent iff $\models \alpha \leftrightarrow \beta$.

1.8. Problem. Let $\boldsymbol{\alpha}$ be a subformula of $\boldsymbol{\beta}$, and let $\boldsymbol{\alpha}$ and $\boldsymbol{\alpha}'$ be logically equivalent. Let $\boldsymbol{\beta}'$ be obtained from $\boldsymbol{\beta}$ by replacing zero or more occurrences of $\boldsymbol{\alpha}$ by occurrences of $\boldsymbol{\alpha}'$ (i.e., $\boldsymbol{\beta}=\tau_0\boldsymbol{\alpha}\tau_1\boldsymbol{\alpha}\tau_2\ldots\boldsymbol{\alpha}\tau_k$, where $\tau_0,\ldots,\tau_k$ are strings, and $\boldsymbol{\beta}'=\tau_0\boldsymbol{\alpha}_1\tau_1\boldsymbol{\alpha}_2\tau_2\ldots\boldsymbol{\alpha}_k\tau_k$, where, for $i=1,\ldots,k$, $\boldsymbol{\alpha}_i$ is $\boldsymbol{\alpha}$ or $\boldsymbol{\alpha}'$). Show by induction on deg $\boldsymbol{\beta}$ that $\boldsymbol{\beta}$ and $\boldsymbol{\beta}'$ are logically equivalent.

§ 2. Freedom and bondage

The value of a given expression (i.e., term or formula) under a valuation σ does not in fact depend on the whole of σ. For terms the situation is very simple, as we now go on to show.

2.1. Theorem. *Let* $\mathbf{t}$ *be a term, and let* σ *and* τ *be valuations which agree on all variables and function symbols occurring in* $\mathbf{t}$. *Then* $\mathbf{t}^\sigma=\mathbf{t}^\tau$.

Proof. By straightforward induction on deg $\mathbf{t}$. If $\mathbf{t} = \mathbf{x}$ then, since $\mathbf{x}$ is a variable occurring in $\mathbf{t}$, we must have $\mathbf{x}^\sigma=\mathbf{x}^\tau$, i.e., $\mathbf{t}^\sigma=\mathbf{t}^\tau$.

If $\mathbf{t}=\mathbf{f}\mathbf{t}_1\ldots\mathbf{t}_n$, where $\mathbf{f}$ is an n-ary function symbol and $\mathbf{t}_1,\ldots,\mathbf{t}_n$ are terms, then by assumption $\mathbf{f}^\sigma$ and $\mathbf{f}^\tau$ are the same. Also, since every variable or function symbol occurring in one of the arguments $\mathbf{t}_1,\ldots,\mathbf{t}_n$ occurs also in $\mathbf{t}$, we have by the induction hypothesis

$$\mathbf{t}_1^\sigma=\mathbf{t}_1^\tau,\ldots,\mathbf{t}_n^\sigma=\mathbf{t}_n^\tau.$$

Thus

$$\begin{aligned}\mathbf{t}^\sigma &=(\mathbf{f}\mathbf{t}_1\ldots\mathbf{t}_n)^\sigma\\ &=\mathbf{f}^\sigma(\mathbf{t}_1^\sigma,\ldots,\mathbf{t}_n^\sigma) && \text{(by BSD (T2))}\\ &=\mathbf{f}^\tau(\mathbf{t}_1^\tau,\ldots,\mathbf{t}_n^\tau) && \text{(by ind. hyp.)}\\ &=(\mathbf{f}\mathbf{t}_1\ldots\mathbf{t}_n)^\tau && \text{(by BSD (T2))}\\ &=\mathbf{t}^\tau. \end{aligned}$$

∎

We say that a term $\mathbf{t}$ is *closed* if it does not contain any variable. By Thm. 2.1 it follows that in this case $\mathbf{t}^\sigma$ depends only on the structure $\mathfrak{U}$ underlying σ (and not on the values assigned by σ to the variables). If $\mathbf{t}$ is a closed term we define $\mathbf{t}^\mathfrak{U}$ to be the value of $\mathbf{t}$ under some (and hence every) valuation σ whose underlying structure is $\mathfrak{U}$.

For formulas we can prove a result analogous to Thm. 2.1, but in fact we shall need a stronger result. For example, though the variable $\mathbf{x}$ occurs in $\forall\mathbf{x}\boldsymbol{\beta}$, we should not expect $(\forall\mathbf{x}\boldsymbol{\beta})^\sigma$ to depend on $\mathbf{x}^\sigma$. This may be understood from the informal explanation given in §2 of Ch. 1. More precisely, it can be seen from clause (F4) of the BSD. For this clause defines $(\forall\mathbf{x}\boldsymbol{\beta})^\sigma$ in terms of the values $\boldsymbol{\beta}^{\sigma(\mathbf{x}/u)}$ for all u in the universe U of σ; so that here $\mathbf{x}^\sigma$ makes no difference.

This leads us to a distinction between two kinds of occurrence of a variable $\mathbf{x}$ in a formula $\boldsymbol{\alpha}$: a given occurrence of $\mathbf{x}$ in $\boldsymbol{\alpha}$ is *bound* if this occurrence is within a subformula of $\boldsymbol{\alpha}$ having the form $\forall \mathbf{x}\boldsymbol{\beta}$ (i.e., a universal subformula of $\boldsymbol{\alpha}$ which has $\mathbf{x}$ as variable of quantification); all other occurrences of $\mathbf{x}$ in $\boldsymbol{\alpha}$ are *free*. To be quite precise, we define these concepts by recursion on $\deg \boldsymbol{\alpha}$:

2.2. DEFINITION. A given occurrence of a variable $\mathbf{x}$ in a formula $\boldsymbol{\alpha}$ is *free* in $\boldsymbol{\alpha}$ iff it is not *bound* in $\boldsymbol{\alpha}$. Moreover:

(1) If $\boldsymbol{\alpha}$ is atomic, then every occurrence of $\mathbf{x}$ in $\boldsymbol{\alpha}$ is *free* in $\boldsymbol{\alpha}$.

(2) If $\boldsymbol{\alpha} = \neg\boldsymbol{\beta}$, then a given occurrence of $\mathbf{x}$ in $\boldsymbol{\alpha}$ is *free* in $\boldsymbol{\alpha}$ iff the same occurrence is *free* in $\boldsymbol{\beta}$.

(3) If $\boldsymbol{\alpha} = \boldsymbol{\beta} \rightarrow \boldsymbol{\gamma}$, then a given occurrence of $\mathbf{x}$ in $\boldsymbol{\alpha}$ is *free* in $\boldsymbol{\alpha}$ iff that occurrence is a *free* occurrence of $\mathbf{x}$ in $\boldsymbol{\beta}$ or in $\boldsymbol{\gamma}$.

(4) If $\boldsymbol{\alpha} = \forall \mathbf{x}\boldsymbol{\beta}$, then every occurrence of $\mathbf{x}$ in $\boldsymbol{\alpha}$ is *bound* in $\boldsymbol{\alpha}$, but if $\boldsymbol{\alpha} = \forall \mathbf{y}\boldsymbol{\beta}$, where $\mathbf{y}$ is a variable other than $\mathbf{x}$, then a given occurrence of $\mathbf{x}$ in $\boldsymbol{\alpha}$ is *free* in $\boldsymbol{\alpha}$ iff that occurrence is *free* in $\boldsymbol{\beta}$.

Note that the same variable may have both free and bound occurrences in the same formula. For example, in the formula

$$\forall \mathbf{x}\, [\mathbf{x} = \mathbf{y} \wedge \exists \mathbf{y}(\mathbf{y} = \mathbf{x})] \rightarrow \exists \mathbf{z}(\mathbf{x} \neq \mathbf{z}),$$

the first three occurrences of $\mathbf{x}$ are bound and the last is free; the first occurrence of $\mathbf{y}$ is free and the other two are bound; and both occurrences of $\mathbf{z}$ are bound. (Here we are assuming that $\mathbf{x}$,$\mathbf{y}$ and $\mathbf{z}$ are distinct variables.)

We say that $\mathbf{x}$ is *free in* $\boldsymbol{\alpha}$ if $\mathbf{x}$ has at least one free occurrence in $\boldsymbol{\alpha}$. Note that $\mathbf{x}$ may *also* have bound occurrences in $\boldsymbol{\alpha}$; thus both $\mathbf{x}$ and $\mathbf{y}$ (but not $\mathbf{z}$) are free in the formula of the above example.

The *free variables of* $\boldsymbol{\alpha}$ are the variables which are free in $\boldsymbol{\alpha}$.

2.3. THEOREM. *Let σ and τ be valuations which have the same universe U and which agree on every free variable of $\boldsymbol{\alpha}$ as well as on every extra-logical symbol occurring in $\boldsymbol{\alpha}$. Then $\boldsymbol{\alpha}^\sigma = \boldsymbol{\alpha}^\tau$.*

PROOF. By induction on $\deg \boldsymbol{\alpha}$. We deal here only with the case where $\boldsymbol{\alpha}$ is universal, leaving the other (and easier) cases to the reader.

Let $\boldsymbol{\alpha} = \forall \mathbf{x}\boldsymbol{\beta}$. Then, by the BSD, $\boldsymbol{\alpha}^\sigma = \top$ iff $\boldsymbol{\beta}^{\sigma(\mathbf{x}/u)} = \top$ for every u in the universe U of σ. Now, the extralogical symbols of $\boldsymbol{\beta}$ are exactly those of $\boldsymbol{\alpha}$. Also, the free variables of $\boldsymbol{\beta}$ are either exactly those of $\boldsymbol{\alpha}$, or they are those plus $\mathbf{x}$. But (for every $u \in U$) $\sigma(\mathbf{x}/u)$ and $\tau(\mathbf{x}/u)$ clearly agree not only on the free variables and extralogical symbols of $\boldsymbol{\alpha}$, but also on $\mathbf{x}$.

Since $\deg \boldsymbol{\beta} < \deg \boldsymbol{\alpha}$, it follows from the induction hypothesis that $\boldsymbol{\beta}^{\sigma(\mathbf{x}/u)} = \boldsymbol{\beta}^{\tau(\mathbf{x}/u)}$. Thus $\boldsymbol{\alpha}^{\sigma} = \top$ iff $\boldsymbol{\beta}^{\tau(\mathbf{x}/u)} = \top$ for all $u \in U$, i.e., iff $\boldsymbol{\alpha}^{\tau} = \top$. ∎

2.4. **Problem.** Show that if $\mathbf{x}$ is not free in $\boldsymbol{\alpha}$, then $\boldsymbol{\alpha}$, $\forall \mathbf{x} \boldsymbol{\alpha}$ and $\exists \mathbf{x} \boldsymbol{\alpha}$ are logically equivalent.

A formula which has no free variables (so that all occurrences of variables in it, if any, are bound) is called a *sentence*. It follows from Thm. 2.3 that if $\boldsymbol{\alpha}$ is a sentence then the value $\boldsymbol{\alpha}^{\sigma}$ depends only on the structure $\mathfrak{U}$ underlying σ. In this case we define $\boldsymbol{\alpha}^{\mathfrak{U}}$ to be that value (i.e., $\boldsymbol{\alpha}^{\mathfrak{U}} = \boldsymbol{\alpha}^{\sigma}$ for any valuation σ which $\mathfrak{U}$ underlies).

If $\boldsymbol{\alpha}^{\mathfrak{U}} = \top$, we say that the structure $\mathfrak{U}$ *satisfies* the sentence $\boldsymbol{\alpha}$ (or $\boldsymbol{\alpha}$ *holds* in $\mathfrak{U}$, or $\mathfrak{U}$ is a *model* for $\boldsymbol{\alpha}$), briefly, $\mathfrak{U} \models \boldsymbol{\alpha}$. If $\mathfrak{U} \models \boldsymbol{\varphi}$ for every $\boldsymbol{\varphi}$ in a set $\boldsymbol{\Phi}$ of sentences, we say that $\mathfrak{U}$ is a *model* for $\boldsymbol{\Phi}$.

More generally, let $\boldsymbol{\alpha}$ be a formula such that all the free variables of $\boldsymbol{\alpha}$ are among the first k variables of $\mathscr{L}$, namely $\mathbf{v}_1, \ldots, \mathbf{v}_k$. Then, by Thm 2.3, $\boldsymbol{\alpha}^{\sigma}$ depends only on the structure $\mathfrak{U}$ underlying σ and on $\mathbf{v}_i^{\sigma}$ for $i = 1, \ldots, k$. We write

$$\mathfrak{U} \models \boldsymbol{\alpha}\, [u_1, \ldots, u_k]$$

when we wish to assert that $\sigma \models \boldsymbol{\alpha}$ for some (hence for every) valuation σ such that $\mathfrak{U}$ underlies σ and such that $\mathbf{v}_i^{\sigma} = u_i$ for $i = 1, \ldots, k$.

2.5. **Problem.** Construct a sentence $\boldsymbol{\alpha}$ containing only logical symbols (i.e., no function symbol and no predicate symbol other than $=$) such that $\boldsymbol{\alpha}$ holds in a structure $\mathfrak{U}$ iff U has

(a) at least three members,

(b) at most three members,

(c) exactly three members.

2.6. **Problem.** Using just one binary predicate symbol (but no other predicate symbols and no function symbols) construct a sentence $\boldsymbol{\alpha}$ such that $\boldsymbol{\alpha}$ has no finite model (i.e., no model with finite universe); but if U is any infinite set then $\boldsymbol{\alpha}$ has a model whose universe is U.

2.7. **Remark.** From Thm. 2.3 it follows that the various semantic concepts defined in Def. 1.5 are invariant with respect to language. For, if $\mathscr{L}$ and $\mathscr{L}'$ are two first-order languages and σ is an $\mathscr{L}$-valuation then there is an $\mathscr{L}'$-valuation σ' which agrees with σ on the symbols which $\mathscr{L}$ and $\mathscr{L}'$ have in common. Any formula $\boldsymbol{\alpha}$ belonging to both $\mathscr{L}$ and $\mathscr{L}'$ will then get the same value under σ and σ'. Thus, e.g., if $\boldsymbol{\alpha}$ is satisfiable as an $\mathscr{L}$-formula (i.e., satisfied by some $\mathscr{L}$-valuation) it is also satisfiable as an $\mathscr{L}'$-formula.

§ 3. Substitution

Let $\mathbf{s}$ and $\mathbf{t}$ be terms. We define $\mathbf{s}(\mathbf{x}/\mathbf{t})$ as the term obtained from $\mathbf{s}$ when an occurrence of $\mathbf{t}$ is substituted for each occurrence of $\mathbf{x}$ in $\mathbf{s}$. In detail, $\mathbf{s}(\mathbf{x}/\mathbf{t})$ is defined by recursion on deg $\mathbf{s}$ as follows:

3.1. DEFINITION. If $\mathbf{s}=\mathbf{x}$ then $\mathbf{s}(\mathbf{x}/\mathbf{t})=\mathbf{t}$; but if $\mathbf{s}=\mathbf{y}$, where $\mathbf{y}$ is a variable other than $\mathbf{x}$, then $\mathbf{s}(\mathbf{x}/\mathbf{t})=\mathbf{y}$. If $\mathbf{s}=\mathbf{f}\mathbf{s}_1...\mathbf{s}_n$, where $\mathbf{f}$ is an n-ary function symbol and $\mathbf{s}_1,...,\mathbf{s}_n$ are terms, then $\mathbf{s}(\mathbf{x}/\mathbf{t})=\mathbf{f}\mathbf{s}_1(\mathbf{x}/\mathbf{t})...\mathbf{s}_n(\mathbf{x}/\mathbf{t})$.

We would now like to investigate the semantic behaviour of $\mathbf{s}(\mathbf{x}/\mathbf{t})$. To this end, let us fix a valuation σ. Then the value of $\mathbf{s}$ is $\mathbf{s}^\sigma$. Now let us assign to $\mathbf{x}$ various values u in the universe U of σ, while the rest of σ is held fixed. We get various valuations $\sigma(\mathbf{x}/u)$ and the corresponding values of $\mathbf{s}$ will be $\mathbf{s}^{\sigma(\mathbf{x}/u)}$. We have thus got a function f (mapping U into itself) defined by

(1) $$f(u)=\mathbf{s}^{\sigma(\mathbf{x}/u)}, \quad u\in U.$$

In particular, since $\sigma(\mathbf{x}/\mathbf{x}^\sigma)$ is σ itself, we have

$$\mathbf{s}^\sigma=f(\mathbf{x}^\sigma).$$

Now, in $\mathbf{s}(\mathbf{x}/\mathbf{t})$, $\mathbf{t}$ has taken the place that $\mathbf{x}$ had in $\mathbf{s}$. We would therefore *expect* that $\mathbf{s}(\mathbf{x}/\mathbf{t})^\sigma$ will depend on $\mathbf{t}^\sigma$ in the same way as $\mathbf{s}^\sigma$ depends on $\mathbf{x}^\sigma$; that is, we conjecture

(2) $$\mathbf{s}(\mathbf{x}/\mathbf{t})^\sigma=f(\mathbf{t}^\sigma).$$

Combining (2) with the identity (1) which defines f, we can put our conjecture in the form

$$\mathbf{s}(\mathbf{x}/\mathbf{t})^\sigma=\mathbf{s}^{\sigma(\mathbf{x}/t)},$$

where $t=\mathbf{t}^\sigma$. In fact, this conjecture is correct.

3.2. THEOREM. *If* $\mathbf{s}$ *and* $\mathbf{t}$ *are terms,* $\mathbf{x}$ *a variable and* σ *a valuation, then*

$$\mathbf{s}(\mathbf{x}/\mathbf{t})^\sigma=\mathbf{s}^{\sigma(\mathbf{x}/t)},$$

where $t=\mathbf{t}^\sigma$.

PROOF. By induction on deg $\mathbf{s}$. Each of the three cases in Def. 3.1 has to be treated separately. The details are routine and are left to the reader. ∎

We now want to do the same kind of thing with formulas. Thus, for given formula $\boldsymbol{\alpha}$ and valuation σ, we define a mapping f of U into $\{\top,\bot\}$ by

(1′) $$f(u)=\boldsymbol{\alpha}^{\sigma(\mathbf{x}/u)}, \quad u\in U.$$

And we should like to define $\alpha(\mathbf{x}/\mathbf{t})$ in such a way that

(2′) $\alpha(\mathbf{x}/\mathbf{t})^\sigma = f(\mathbf{t}^\sigma)$.

But here there are two snags. First, it follows from Thm. 2.3 (and, indeed, directly from clause (F4) of the BSD) that $f(u)$ defined in (1′) depends on u only through the *free* occurrences of $\mathbf{x}$ in α. Thus, in order to have (2′) *we must define* $\alpha(\mathbf{x}/\mathbf{t})$ *in such a way that* $\mathbf{t}$ *is only substituted for the free occurrences of* $\mathbf{x}$ in α, not for the bound occurrences.[1]

But even if we take this precaution we come across another snag. For example, let α be $\forall\mathbf{y}(\mathbf{x}=\mathbf{y})$, where $\mathbf{x}$ and $\mathbf{y}$ are different variables. Using clauses (F4) and ($F1^=$) of the BSD) it is easy to verify that in this case

$$\alpha^{\sigma(\mathbf{x}/u)} = \top \quad \text{iff} \quad u = v \text{ for all } v \in U,$$

where U is the universe of σ. (More intuitively, α says: "the value of $\mathbf{x}$ is equal to every individual".) Thus $\alpha^{\sigma(\mathbf{x}/u)} = \top$ iff U has exactly one member, and (1′) becomes

$$f(u) = \begin{cases} \top & \text{if } U \text{ has exactly one member,} \\ \bot & \text{otherwise.} \end{cases}$$

On the other hand, if we take $\mathbf{t}$ to be $\mathbf{y}$ and substitute it for the occurrence of $\mathbf{x}$ in α (which is free!) we get $\forall\mathbf{y}(\mathbf{y}=\mathbf{y})$. This sentence is logically true: we have

$$[\forall\mathbf{y}(\mathbf{y}=\mathbf{y})]^\sigma = \top$$

for every σ. If U has more than one member, we therefore find

$$[\forall\mathbf{y}(\mathbf{y}=\mathbf{y})]^\sigma \neq f(\mathbf{y}^\sigma).$$

So we cannot take $\forall\mathbf{y}(\mathbf{y}=\mathbf{y})$ to be $\alpha(\mathbf{x}/\mathbf{y})$, if we want (2′) to hold.

More intuitively, we want $\alpha(\mathbf{x}/\mathbf{t})$ to say about the value of $\mathbf{t}$ what α says about the value of $\mathbf{x}$. But in our example $\forall\mathbf{y}(\mathbf{x}=\mathbf{y})$ says that the value of $\mathbf{x}$ is equal to every individual, whereas $\forall\mathbf{y}(\mathbf{y}=\mathbf{y})$ says nothing about the value of $\mathbf{y}$ but merely states that every individual is equal to itself.

What went wrong? Clearly the snag was that when we substituted an occurrence of $\mathbf{y}$ for the free occurrence of $\mathbf{x}$ in $\forall\mathbf{y}(\mathbf{x}=\mathbf{y})$, this new occurrence of $\mathbf{y}$ was *captured* — it fell within the scope of a $\mathbf{y}$-quantifier.

[1] Besides, if $\mathbf{t}$ is substituted for *all* occurrences of $\mathbf{x}$ in α, then the result is not always a formula. Thus, e.g., from $\forall\mathbf{x}(\mathbf{x}=\mathbf{y})$ we would get $\forall\mathbf{t}(\mathbf{t}=\mathbf{y})$ — which is not a formula, unless $\mathbf{t}$ happens to be a variable.

What are we to do? Let us take a clue from manipulations used in the integral calculus. The value of, say, the integral

$$\int_0^1 xy^2\,dy$$

depends on the value of x but *not* on the value of y. (A variable of integration behaves analogously to a variable of quantification!) If we want to substitute for x in this integral an expression containing y, then *we first have to change the variable of integration* getting, say,

$$\int_0^1 xz^2\,dz$$

where z does not occur in the expression we want to substitute. This integral has exactly the same value as the one we had before, but now we can perform our substitution quite safely.

Similarly, instead of substituting **y** for **x** in $\forall\mathbf{y}(\mathbf{x}=\mathbf{y})$ we must, it seems first change $\forall\mathbf{y}(\mathbf{x}=\mathbf{y})$ into $\forall\mathbf{z}(\mathbf{x}=\mathbf{z})$ — it is easy to see that these two formulas are logically equivalent — and only *then* substitute **y** for **x**. We now get $\forall\mathbf{z}(\mathbf{y}=\mathbf{z})$, a formula which says that the value of **y** is equal to every individual — just what we wanted.

We shall first define $\boldsymbol{\alpha}(\mathbf{x}/\mathbf{t})$ only in those cases where the substitution of **t** for **x** in $\boldsymbol{\alpha}$ does not lead to "capture" and thus does not require any change of the variable of quantification. Later we shall also define $\boldsymbol{\alpha}(\mathbf{x}/\mathbf{t})$ in the remaining cases, by prescribing the changes that must be made in $\boldsymbol{\alpha}$ before the substitution may take place.

We shall say that **t** is *free to be substituted for* **x** *in* $\boldsymbol{\alpha}$ (briefly, *free for* **x** *in* $\boldsymbol{\alpha}$) if no free occurrence of **x** in $\boldsymbol{\alpha}$ is within a subformula of $\boldsymbol{\alpha}$ having the form $\forall\mathbf{y}\boldsymbol{\beta}$, where **y** occurs in **t**.

If **t** is free for **x** in $\boldsymbol{\alpha}$, we shall define $\boldsymbol{\alpha}(\mathbf{x}/\mathbf{t})$ as the result of substituting an occurrence of **t** for each free occurrence of **x** in $\boldsymbol{\alpha}$. (Note that because **t** is assumed to be free for **x** in $\boldsymbol{\alpha}$, all occurrences of variables that have been introduced *via* the substitution are free in $\boldsymbol{\alpha}(\mathbf{x}/\mathbf{t})$.)

We now spell out precisely, by recursion on $\deg\boldsymbol{\alpha}$, the conditions under which **t** is free for **x** in $\boldsymbol{\alpha}$; and simultaneously we define $\boldsymbol{\alpha}(\mathbf{x}/\mathbf{t})$ under these conditions.

3.3. Definition. If $\boldsymbol{\alpha}$ is an atomic formula $\mathbf{P}\mathbf{s}_1\ldots\mathbf{s}_n$, then **t** is *free for* **x** *in* $\boldsymbol{\alpha}$. And $\boldsymbol{\alpha}(\mathbf{x}/\mathbf{t})$ is defined as $\mathbf{P}\mathbf{s}_1(\mathbf{x}/\mathbf{t})\ldots\mathbf{s}_n(\mathbf{x}/\mathbf{t})$. (Here, for $n=2$, **P** may also be the logical predicate symbol $=$.)

If $\alpha = \neg\beta$, then $\mathbf{t}$ is *free for* $\mathbf{x}$ *in* α iff $\mathbf{t}$ is *free for* $\mathbf{x}$ *in* β; if this is the case, $\alpha(\mathbf{x}/\mathbf{t})$ is defined to be $\neg[\beta(\mathbf{x}/\mathbf{t})]$.

If $\alpha = \beta \rightarrow \gamma$, then $\mathbf{t}$ is *free for* $\mathbf{x}$ *in* α iff $\mathbf{t}$ is *free for* $\mathbf{x}$ *in* both β and γ; if this is the case we define $\alpha(\mathbf{x}/\mathbf{t})$ as $\beta(\mathbf{x}/\mathbf{t}) \rightarrow \gamma(\mathbf{x}/\mathbf{t})$.

If $\alpha = \forall \mathbf{y}\beta$, then $\mathbf{t}$ is *free for* $\mathbf{x}$ *in* α iff one of the following conditions holds:

(a) $\mathbf{x}$ is not free in α,

(b) $\mathbf{x}$ is free in α (hence, in particular, $\mathbf{x} \neq \mathbf{y}$), and $\mathbf{t}$ is *free for* $\mathbf{x}$ *in* β, and $\mathbf{y}$ does not occur in $\mathbf{t}$.

In case (a) we define $\alpha(\mathbf{x}/\mathbf{t})$ to be α. In case (b) we define $\alpha(\mathbf{x}/\mathbf{t})$ to be $\forall \mathbf{y}\,[\beta(\mathbf{x}/\mathbf{t})]$.

It is easy to verify that if no variable occurring in $\mathbf{t}$ has a bound occurrence in α, then $\mathbf{t}$ is free for $\mathbf{x}$ in α. Also, $\mathbf{x}$ is always free for itself in α, and $\alpha(\mathbf{x}/\mathbf{x}) = \alpha$.

3.4. THEOREM. *If* $\mathbf{t}$ *is free for* $\mathbf{x}$ *in* α *then, for every valuation* σ,

$$\alpha(\mathbf{x}/\mathbf{t})^{\sigma} = \alpha^{\sigma(\mathbf{x}/t)},$$

where $t = \mathbf{t}^{\sigma}$.

PROOF. By induction on $\deg \alpha$. We distinguish various cases, corresponding to the cases in Def. 3.2. Here we only deal with the case $\alpha = \forall \mathbf{y}\beta$, leaving the other (easier) cases to the reader.

First suppose that $\mathbf{x}$ is not free in α. Then $\alpha(\mathbf{x}/\mathbf{t}) = \alpha$. Also, by Thm. 2.3, $\alpha^{\sigma} = \alpha^{\sigma(\mathbf{x}/t)}$. Thus

$$\alpha(\mathbf{x}/\mathbf{t})^{\sigma} = \alpha^{\sigma} = \alpha^{\sigma(\mathbf{x}/t)}.$$

Now suppose that $\mathbf{x}$ is free in α and $\mathbf{t}$ is free for $\mathbf{x}$ in β and $\mathbf{y}$ does not occur in $\mathbf{t}$. Then we have

(1) $$\alpha(\mathbf{x}/\mathbf{t})^{\sigma} = (\forall \mathbf{y}\,[\beta(\mathbf{x}/\mathbf{t})])^{\sigma}.$$

By the BSD,

(2) $$(\forall \mathbf{y}\,[\beta(\mathbf{x}/\mathbf{t})])^{\sigma} = \top \quad \text{iff} \quad \beta(\mathbf{x}/\mathbf{t})^{\sigma(\mathbf{y}/u)} = \top \quad \text{for all} \quad u \in U,$$

where U is the universe of σ. Since $\deg \beta < \deg \alpha$, the induction hypothesis yields

(3) $$\beta(\mathbf{x}/\mathbf{t})^{\sigma(\mathbf{y}/u)} = \beta^{\sigma(\mathbf{y}/u)(\mathbf{x}/t')},$$

where $t' = \mathbf{t}^{\sigma(\mathbf{y}/u)}$. But $\mathbf{y}$ does not occur in $\mathbf{t}$. Hence by Thm. 2.1

$$t' = \mathbf{t}^{\sigma(\mathbf{y}/u)} = \mathbf{t}^{\sigma} = t.$$

Also, **x** and **y** are different (otherwise **x** could not be free in **α**); hence

$$\sigma(\mathbf{y}/u)(\mathbf{x}/t)=\sigma(\mathbf{x}/t)(\mathbf{y}/u).$$

For, it makes no difference whether we *first* change the value of **x** from $\mathbf{x}^{\sigma}$ to t and *then* change the value of **y** to u, or *vice versa*. (It *would* make a difference if **x** were the same as **y**!) Hence we can rewrite (3) as

(4) $$\boldsymbol{\beta}(\mathbf{x}/\mathbf{t})^{\sigma(\mathbf{y}/u)}=\boldsymbol{\beta}^{\sigma(\mathbf{x}/t)(\mathbf{y}/u)}.$$

Now, by the BSD,

$$\boldsymbol{\beta}^{\sigma(\mathbf{x}/t)(\mathbf{y}/u)}=\top \text{ for all } u\in U \quad\text{iff}\quad [\forall\mathbf{y}\boldsymbol{\beta}]^{\sigma(\mathbf{x}/t)}=\top.$$

Combining this with (1), (2) and (4) we get the required result. ∎

3.5. Definition. If **z** is a variable which is not free in **β** but is free for **x** in **β**, we say that $\forall\mathbf{z}\,[\boldsymbol{\beta}(\mathbf{x}/\mathbf{z})]$ arises from $\forall\mathbf{x}\boldsymbol{\beta}$ by *(correct) alphabetic change*. (Note that if **z** does not occur at all in **β**, then **z** certainly satisfies both of the above conditions.)

An alphabetic change is analogous to a change in the variable of integration. We note that this operation is *reversible*. For, if **z** is not free in **β** but is free for **x** in **β**, then the free occurrences of **z** in **β(x/z)** are *precisely* those that arose from the free occurrences of **x** in **β** in the course of the substitution. Thus no free occurrence of **z** in **β(x/z)** can fall within the scope of an **x**-quantifier; i.e., **x** is free for **z** in **β(x/z)**. Also, **x** cannot occur free in **β(x/z)**, because if it did this could only mean that **x** and **z** are the same variable, hence **β(x/z)** = **β** — but **z** was assumed *not* to be free in **β**. It follows that $\forall\mathbf{x}\boldsymbol{\beta}$ can be retrieved from $\forall\mathbf{z}\,[\boldsymbol{\beta}(\mathbf{x}/\mathbf{z})]$ by alphabetic change.

3.6. Theorem. *If* $\forall\mathbf{z}\,[\boldsymbol{\beta}(\mathbf{x}/\mathbf{z})]$ *arises from* $\forall\mathbf{x}\boldsymbol{\beta}$ *by alphabetic change, then these two formulas are logically equivalent.*

Proof. Take any valuation σ and let U be its universe. Then by the BSD

(1) $$(\forall\mathbf{z}\,[\boldsymbol{\beta}(\mathbf{x}/\mathbf{z})])^{\sigma}=\top \quad\text{iff}\quad \boldsymbol{\beta}(\mathbf{x}/\mathbf{z})^{\sigma(\mathbf{z}/u)}=\top \quad \text{for all} \quad u\in U.$$

By Thm. 3.4 we have

(2) $$\boldsymbol{\beta}(\mathbf{x}/\mathbf{z})^{\sigma(\mathbf{z}/u)}=\boldsymbol{\beta}^{\sigma(\mathbf{z}/u)(\mathbf{x}/z')},$$

where $z'=\mathbf{z}^{\sigma(\mathbf{z}/u)}$. But $\mathbf{z}^{\sigma(\mathbf{z}/u)}=u$ by definition of $\sigma(\mathbf{z}/u)$. Using this, as well as the fact that **z** is not free in **β**, we obtain

(3) $$\boldsymbol{\beta}^{\sigma(\mathbf{z}/u)(\mathbf{x}/z')}=\boldsymbol{\beta}^{\sigma(\mathbf{x}/u)}.$$

By the BSD,

$$\beta^{\sigma(x/u)} = \top \quad \text{for all} \quad u \in U \quad \text{iff} \quad (\forall x \beta)^{\sigma} = \top.$$

Combining this with (1), (2) and (3) we get the required result. ∎

Consider a given formula α. Suppose α has a universal subformula, say $\forall y \beta$. Let us replace one occurrence of $\forall y \beta$ in α by an occurrence of a formula $\forall z\,[\beta(y/z)]$ arising from $\forall y \beta$ by alphabetic change (i.e., z is not free in β, but is free for y in β). We shall say that α' is a *variant* of α (briefly, $\alpha \sim \alpha'$) if α can be transformed into α' by a finite number of applications of steps like the one just described. (We include the case where the number of such steps is 0, so that $\alpha \sim \alpha$.)

More precisely, we can define *being a variant of* α by recursion on deg α, as follows:

3.7. DEFINITION. If α is atomic, then α is its own sole *variant*.

If $\alpha = \neg \beta$, then the *variants* of α are all formulas of the form $\neg(\beta')$, where β' is a *variant* of β.

If $\alpha = \beta \to \gamma$, then the *variants* of α are all formulas of the form $\beta' \to \gamma'$, where β' and γ' are *variants* of β and γ respectively.

If $\alpha = \forall y \beta$, then the *variants* of α are all formulas $\forall y \beta'$, where β' is a *variant* of β, as well as all formulas $\forall z\,[\beta'(y/z)]$ obtained from such $\forall y \beta'$ by alphabetic change.

It is easy to check that $\sim$ is an equivalence relation (e.g., the symmetry of $\sim$ follows at once from the fact that alphabetic changes are reversible).

It can also be readily verified that if $\alpha \sim \alpha'$ then the only difference between α and α' is that *bound occurrences* of some variables in α may be replaced by *bound occurrences* of other variables in α'. But if the ith symbol[1] of α is a free occurrence of a variable, then the ith symbol of α' is a free occurrence of the same variable. Similarly, if a symbol other than a variable occurs in α at the ith place, then, the same symbol also occurs in α' at the same place.

Moreover, we have:

3.8. THEOREM. *If $\alpha \sim \alpha'$ then α and α' are logically equivalent.*

PROOF. Immediate from Thm. 3.6 and Prob. 1.8. ∎

[1] Recall that α is a string, i.e., a finite sequence of symbols. The ith symbol of α is simply the ith member of this sequence.

For any formula α and a finite number of variables $\mathbf{y}_1,\ldots,\mathbf{y}_n$, we can always find a variant α' of α such that $\mathbf{y}_1,\ldots,\mathbf{y}_n$ do not occur bound in α'. Suppose, e.g., that α has a subformula $\forall \mathbf{y}_1 \beta$. Then this can be replaced by $\forall \mathbf{z}\, [\beta(\mathbf{y}_1/\mathbf{z})]$, where $\mathbf{z}$ does not occur *at all* in β (and hence is free for $\mathbf{y}_1$ in β) and is different from $\mathbf{y}_1,\ldots,\mathbf{y}_n$. After a finite number of such replacements, α is transformed into an α' with the desired property. In particular, if all the variables occurring in a given term $\mathbf{t}$ are among the $\mathbf{y}_i$, then $\mathbf{t}$ is free for $\mathbf{x}$ in α'.

We are now in a position to define $\alpha(\mathbf{x}/\mathbf{t})$ even when $\mathbf{t}$ is not free for $\mathbf{x}$ in α. We simply choose some variant α' of α such that $\mathbf{t}$ *is* free for $\mathbf{x}$ in α' and define $\alpha(\mathbf{x}/\mathbf{t})$ to be $\alpha'(\mathbf{x}/\mathbf{t})$. To make $\alpha(\mathbf{x}/\mathbf{t})$ uniquely defined, we have to choose α' by some definite rule. We do this now, by recursion on $\deg \alpha$.

3.9. DEFINITION. For given $\mathbf{x}$ and $\mathbf{t}$, define α' for each α as follows:

If α is atomic then $\alpha' = \alpha$.

If $\alpha = \neg\beta$, then $\alpha' = \neg(\beta')$.

If $\alpha = \beta \rightarrow \gamma$, then $\alpha' = \beta' \rightarrow \gamma'$.

If $\alpha = \forall \mathbf{y} \beta$, we distinguish three cases:

(a) If $\mathbf{x}$ is not free in α, then $\alpha' = \alpha$.

(b) If $\mathbf{x}$ is free in α and $\mathbf{y}$ does not occur in $\mathbf{t}$, then $\alpha' = \forall \mathbf{y} \beta'$.

(c) If $\mathbf{x}$ is free in α but $\mathbf{y}$ *does* occur in $\mathbf{t}$, then $\alpha' = \forall \mathbf{z}\, [\beta'(\mathbf{y}/\mathbf{z})]$, where $\mathbf{z}$ is the first variable (i.e., the $\mathbf{v}_i$ with least index) such that $\mathbf{z}$ does not occur in $\mathbf{t}$ and such that α' arises from $\forall \mathbf{y} \beta'$ by alphabetic change.

We then define $\alpha(\mathbf{x}/\mathbf{t})$ to be $\alpha'(\mathbf{x}/\mathbf{t})$, where the latter is already defined in Def. 3.3.

Comparing this definition of α' with Defs. 3.3 and 3.7, it is easy to check that α' is indeed a variant of α, and $\mathbf{t}$ is free for $\mathbf{x}$ in α'. Also, if $\mathbf{t}$ is already free for $\mathbf{x}$ in α itself, then $\alpha' = \alpha$. So Def. 3.9 is consistent with Def. 3.3, and $\alpha(\mathbf{x}/\mathbf{t})$ is now well defined for all α, $\mathbf{x}$ and $\mathbf{t}$.

The following theorem is a stonger version of Thm. 3.4 and shows that $\alpha(\mathbf{x}/\mathbf{t})$ has the required semantic property.

3.10. THEOREM. *For all* $\alpha,\mathbf{x},\mathbf{t}$ *and* σ,

$$\alpha(\mathbf{x}/\mathbf{t})^{\sigma} = \alpha^{\sigma(x/t)},$$

where $t = \mathbf{t}^{\sigma}$.

PROOF. By Def. 3.9, $\alpha(\mathbf{x}/\mathbf{t}) = \beta(\mathbf{x}/\mathbf{t})$, where $\beta(=\alpha')$ is some particular variant of $\boldsymbol{\alpha}$ such that $\mathbf{t}$ is free for $\mathbf{x}$ in $\boldsymbol{\beta}$. Hence $\boldsymbol{\alpha}(\mathbf{x}/\mathbf{t})^\sigma = \boldsymbol{\beta}(\mathbf{x}/\mathbf{t})^\sigma$. By Thm. 3.4, $\boldsymbol{\beta}(\mathbf{x}/\mathbf{t})^\sigma = \boldsymbol{\beta}^{\sigma(\mathbf{x}/t)}$, but since $\boldsymbol{\alpha} \sim \boldsymbol{\beta}$ we have $\boldsymbol{\beta}^{\sigma(\mathbf{x}/t)} = \boldsymbol{\alpha}^{\sigma(\mathbf{x}/t)}$ by Thm. 3.8. ∎

As an application, we prove:

3.11. THEOREM. *Let $\boldsymbol{\Phi}$ be a set of formulas, and let $\neg\forall\mathbf{x}\boldsymbol{\alpha} \in \boldsymbol{\Phi}$. Let $\mathbf{y}$ be a variable that is not free in $\boldsymbol{\Phi}$ (i.e., not free in any $\boldsymbol{\varphi} \in \boldsymbol{\Phi}$). If $\boldsymbol{\Phi}$ is satisfiable, then so is*[1] $\boldsymbol{\Phi}, \neg\boldsymbol{\alpha}(\mathbf{x}/\mathbf{y})$.

PROOF. Let $\sigma \models \boldsymbol{\Phi}$, and let U be the universe of σ. Since $\neg\forall\mathbf{x}\boldsymbol{\alpha} \in \boldsymbol{\Phi}$ we have $(\neg\forall\mathbf{x}\boldsymbol{\alpha})^\sigma = \top$. Therefore, by the BSD, $\boldsymbol{\alpha}^{\sigma(\mathbf{x}/u)} = \bot$ for some $u \in U$.

Consider $\sigma(\mathbf{y}/u)$. Since $\mathbf{y}$ is not free in $\boldsymbol{\Phi}$, it follows from Thm. 2.3 that $\sigma(\mathbf{y}/u) \models \boldsymbol{\Phi}$. Also, by Thm. 3.10,

$$\boldsymbol{\alpha}(\mathbf{x}/\mathbf{y})^{\sigma(\mathbf{y}/u)} = \boldsymbol{\alpha}^{\sigma(\mathbf{y}/u)(\mathbf{x}/y')},$$

where $y' = \mathbf{y}^{\sigma(\mathbf{y}/u)} = u$. Therefore

(1) $$\boldsymbol{\alpha}(\mathbf{x}/\mathbf{y})^{\sigma(\mathbf{y}/u)} = \boldsymbol{\alpha}^{\sigma(\mathbf{y}/u)(\mathbf{x}/u)}.$$

If $\mathbf{y}$ is the same variable as $\mathbf{x}$, then $\sigma(\mathbf{y}/u)(\mathbf{x}/u) = \sigma(\mathbf{x}/u)$, so that

(2) $$\boldsymbol{\alpha}^{\sigma(\mathbf{y}/u)(\mathbf{x}/u)} = \boldsymbol{\alpha}^{\sigma(\mathbf{x}/u)} = \bot.$$

On the other hand, if $\mathbf{y} \neq \mathbf{x}$ then $\mathbf{y}$ cannot be free in $\boldsymbol{\alpha}$ (because $\mathbf{y}$ is not free in $\neg\forall\mathbf{x}\boldsymbol{\alpha}$) and then (2) holds by Thm. 2.3. Thus in any case we can conclude from (1) and (2) that $\boldsymbol{\alpha}(\mathbf{x}/\mathbf{y})^{\sigma(\mathbf{y}/u)} = \bot$, i.e., $\sigma(\mathbf{y}/u) \models \neg\boldsymbol{\alpha}(\mathbf{x}/\mathbf{y})$. ∎

3.12. PROBLEM. Let $\boldsymbol{\Phi}$ be a set of formulas and let $\neg\forall\mathbf{x}\boldsymbol{\alpha} \in \boldsymbol{\Phi}$. Let $\mathbf{a}$ be a constant that does not occur in $\boldsymbol{\Phi}$. Show that if $\boldsymbol{\Phi}$ is satisfiable, then so is $\boldsymbol{\Phi}, \neg\boldsymbol{\alpha}(\mathbf{x}/\mathbf{a})$.

In the sequel we shall sometimes need to substitute a term $\mathbf{t}$ for a constant $\mathbf{c}$ (rather than for a variable).

3.13. DEFINITION. $\boldsymbol{\alpha}(\mathbf{c}/\mathbf{t})$ is the formula obtained by substituting an occurrence of $\mathbf{t}$ for each occurence of $\mathbf{c}$ in $\boldsymbol{\alpha}$.

In practice we shall use $\boldsymbol{\alpha}(\mathbf{c}/\mathbf{t})$ only when no occurrence of $\mathbf{c}$ in $\boldsymbol{\alpha}$ is within the scope of a quantifier whose variable of quantification occurs in $\mathbf{t}$ (e.g., when $\mathbf{t}$ is a closed term).

We conclude this section with a treatment of *simultaneous* substitution.

[1] Notice that, by Defs. 3.3 and 3.9, $[\neg\boldsymbol{\alpha}](\mathbf{x}/\mathbf{t}) = \neg[\boldsymbol{\alpha}(\mathbf{x}/\mathbf{t})]$. Hence we can write simply "$\neg\boldsymbol{\alpha}(\mathbf{x}/\mathbf{t})$".

Suppose that we wish to substitute terms $\mathbf{t}_1,\ldots,\mathbf{t}_n$ *simultaneously* for variables[1] $\mathbf{x}_1,\ldots,\mathbf{x}_n$ in α. Of course, we wish do to so *correctly*, so that an appropriate generalization of Thm. 3.10 should hold. In general, it will not be good enough to substitute the $\mathbf{t}_i$ one by one. For example, $\mathbf{t}_1$ may contain $\mathbf{x}_2$; so, if we first substitute $\mathbf{t}_1$ for $\mathbf{x}_1$, the formula $\alpha(\mathbf{x}_1/\mathbf{t}_1)$ will have new free occurrences of $\mathbf{x}_2$ which have been introduced by this substitution. And if we now substitute $\mathbf{t}_2$ for $\mathbf{x}_2$, then $\mathbf{t}_2$ will replace these additional free occurrences of $\mathbf{x}_2$, which is contrary to our intention. We therefore have to be a bit more cautious.

We shall now define $\alpha(\mathbf{x}_1/\mathbf{t}_1,\ldots,\mathbf{x}_n/\mathbf{t}_n)$ — the result of substituting $\mathbf{t}_1,\ldots,\mathbf{t}_n$ in α simultaneously for $\mathbf{x}_1,\ldots,\mathbf{x}_n$ respectively — by recursion on n. The case $n=1$ has already been treated above (Def. 3.9). For $n>1$, we assume as our inductive hypothesis that $\alpha(\mathbf{x}_1/\mathbf{s}_1,\ldots,\mathbf{x}_{n-1}/\mathbf{s}_{n-1})$ has already been defined for all terms $\mathbf{s}_1,\ldots,\mathbf{s}_{n-1}$. We now define:

3.14. DEFINITION. If $\mathbf{x}_n$ is not free in α, then

$$\alpha(\mathbf{x}_1/\mathbf{t}_1,\ldots,\mathbf{x}_n/\mathbf{t}_n)=\alpha(\mathbf{x}_1/\mathbf{t}_1,\ldots,\mathbf{x}_{n-1}/\mathbf{t}_{n-1}).$$

If $\mathbf{x}_n$ *is* free in α, then

$$\alpha(\mathbf{x}_1/\mathbf{t}_1,\mathbf{x}_2/\mathbf{t}_2,\ldots,\mathbf{x}_n/\mathbf{t}_n)=\alpha(\mathbf{x}_1/\mathbf{s}_1,\ldots,\mathbf{x}_{n-1}/\mathbf{s}_{n-1})(\mathbf{x}_n/\mathbf{t}_n)(\mathbf{z}/\mathbf{x}_n),$$

where $\mathbf{z}$ is the first variable which occurs neither in α nor in $\mathbf{t}_1,\ldots,\mathbf{t}_n$ and $\mathbf{s}_i=\mathbf{t}_i(\mathbf{x}_n/\mathbf{z})$ for $i=1,\ldots,n-1$.

Thus we first substitute $\mathbf{z}$ for $\mathbf{x}_n$ in $\mathbf{t}_1,\ldots,\mathbf{t}_{n-1}$. Then we substitute the resulting terms $\mathbf{s}_1,\ldots,\mathbf{s}_{n-1}$ simultaneously for $\mathbf{x}_1,\ldots,\mathbf{x}_{n-1}$ in α (the induction hypothesis being that we know how to do this). Next, we substitute $\mathbf{t}_n$ for $\mathbf{x}_n$. Finally, we replace $\mathbf{z}$ again by $\mathbf{x}_n$. The role of $\mathbf{z}$ is to replace $\mathbf{x}_n$ at points at which we do not want $\mathbf{t}_n$ to enter. After $\mathbf{t}_n$ is safely substituted in the right places for $\mathbf{x}_n$, we put $\mathbf{x}_n$ back in place of $\mathbf{z}$.

We now show rigorously that this definition has the required semantic property.

3.15. THEOREM. $\alpha(\mathbf{x}_1/\mathbf{t}_1,\ldots,\mathbf{x}_n/\mathbf{t}_n)^\sigma=\alpha^{\sigma(\mathbf{x}_1/t_1)\ldots(\mathbf{x}_n/t_n)}$, *where* $t_1=\mathbf{t}_1^\sigma,\ldots,t_n=\mathbf{t}_n^\sigma$.

PROOF. By induction on n. For $n=1$ the result has already been proved (Thm. 3.10). Now let $n>1$. The case in which $\mathbf{x}_n$ is not free in α is left to the reader.

[1] Here, and in the rest of this section, $\mathbf{x}_1,\ldots,\mathbf{x}_n$ are always assumed to be *n distinct* variables.

If $\mathbf{x}_n$ is free in $\boldsymbol{\alpha}$, we have, by Def. 3.14 and Thm. 3.10,

$$\boldsymbol{\alpha}(\mathbf{x}_1/\mathbf{t}_1,\ldots,\mathbf{x}_n/\mathbf{t}_n)^\sigma = \boldsymbol{\alpha}(\mathbf{x}_1/\mathbf{s}_1,\ldots,\mathbf{x}_{n-1}/\mathbf{s}_{n-1})(\mathbf{x}_n/\mathbf{t}_n)^{\sigma(\mathbf{z}/x_n)},$$

where $x_n = \mathbf{x}_n{}^\sigma$. Applying Thm. 3.10 again, we have

$$\boldsymbol{\alpha}(\mathbf{x}_1/\mathbf{t}_1,\ldots,\mathbf{x}_n/\mathbf{t}_n)^\sigma = \boldsymbol{\alpha}(\mathbf{x}_1/\mathbf{s}_1,\ldots,\mathbf{x}_{n-1}/\mathbf{s}_{n-1})^{\sigma(\mathbf{z}/x_n)(\mathbf{x}_n/t')},$$

where $t' = \mathbf{t}_n{}^{\sigma(\mathbf{z}/x_n)}$. But since $\mathbf{z}$ was chosen so as not to occur in $\mathbf{t}_n$, we have in fact $t' = \mathbf{t}_n{}^\sigma = t_n$. We now apply our induction hypothesis and get

$$(1) \qquad \boldsymbol{\alpha}(\mathbf{x}_1/\mathbf{t}_1,\ldots,\mathbf{x}_n/\mathbf{t}_n)^\sigma = \boldsymbol{\alpha}^{\sigma(\mathbf{z}/x_n)(\mathbf{x}_n/t_n)(\mathbf{x}_1/s_1)\ldots(\mathbf{x}_{n-1}/s_{n-1})},$$

where

$$s_i = \mathbf{s}_i{}^{\sigma(\mathbf{z}/x_n)(\mathbf{x}_n/t_n)} \quad \text{for} \quad i=1,\ldots,n-1.$$

But $\mathbf{s}_i = \mathbf{t}_i(\mathbf{x}_n/\mathbf{z})$. Hence by Thm. 3.2 we get

$$(2) \qquad s_i = \mathbf{t}_i{}^{\sigma(\mathbf{z}/x_n)(\mathbf{x}_n/t_n)(\mathbf{x}_n/z')}, \quad i=1,\ldots,n-1,$$

where

$$(3) \qquad z' = \mathbf{z}^{\sigma(\mathbf{z}/x_n)(\mathbf{x}_n/t_n)}.$$

Clearly, (2) can be simplified:

$$(4) \qquad s_i = \mathbf{t}_i{}^{\sigma(\mathbf{z}/x_n)(\mathbf{x}_n/z')}, \quad i=1,\ldots,n-1.$$

Since $\mathbf{z}$ is different from $\mathbf{x}_n$, from (3) we clearly have $z' = x_n$, and hence (4) yields

$$s_i = \mathbf{t}_i{}^{\sigma(\mathbf{z}/x_n)(\mathbf{x}_n/x_n)}.$$

But since $\mathbf{z}$ does not occur in $\mathbf{t}_i$ we have

$$s_i = \mathbf{t}_i{}^{\sigma(\mathbf{x}_n/x_n)} = \mathbf{t}_i{}^\sigma = t_i.$$

We can now rewrite (1) as

$$\boldsymbol{\alpha}(\mathbf{x}_1/\mathbf{t}_1,\ldots,\mathbf{x}_n/\mathbf{t}_n)^\sigma = \boldsymbol{\alpha}^{\sigma(\mathbf{z}/x_n)(\mathbf{x}_1/t_1)(\mathbf{x}_2/t_2)\ldots(\mathbf{x}_n/t_n)},$$

and since $\mathbf{z}$ is not free in $\boldsymbol{\alpha}$ we have, in fact,

$$\boldsymbol{\alpha}(\mathbf{x}_1/\mathbf{t}_1,\ldots,\mathbf{x}_n/\mathbf{t}_n)^\sigma = \boldsymbol{\alpha}^{\sigma(\mathbf{x}_1/t_1)\ldots(\mathbf{x}_n/t_n)}$$

as claimed. ∎

In some chapters of this book we shall adopt the convention of writing "$\boldsymbol{\alpha}(\mathbf{t}_1,\ldots,\mathbf{t}_n)$" instead of "$\boldsymbol{\alpha}(\mathbf{x}_1/\mathbf{t}_1,\ldots,\mathbf{x}_n/\mathbf{t}_n)$", provided the choice of $\mathbf{x}_1,\ldots,\mathbf{x}_n$ is determined unambiguously by the context or by an explicit agreement (e.g., we may agree that $\mathbf{x}_1,\ldots,\mathbf{x}_n$ are the first n variables of $\mathscr{L}$, i.e., $\mathbf{v}_1,\ldots,\mathbf{v}_n$ respectively).

The following three sections (§§4–6) as well as §8 are devoted to the study of *first-order tableaux*. The material covered in them bears to first-order logic much the same relation as the material in §§7–9 of Ch. 1 bore to propositional logic. (More specifically, §4 parallels §7 of Ch. 1; §5 contains technical lemmas needed in §6; and §6 parallels §8 of Ch. 1. The material of §7 does not deal with tableaux; it is inserted here because it will be used in §8 *in connection* with tableaux, but it will also be used for another purpose in Ch. 3. Finally, §8 parallels §9 of Ch. 1).

A reader who does not wish to study first-order tableaux at all should proceed as follows: First, do Probs. 4.3–4.6, which can be solved (albeit somewhat less efficiently) without tableaux. Then read §7 and from there proceed to §9 or (if one wants to skip §§9–13 as well) directly to Ch. 3. In Ch. 3, §2 must also be omitted.

A reader who would like to benefit from the practical advantages of the tableau method, but who does not have the time or the patience for the painstaking work of §§5–6, may skip these two sections only. In §8 (see Prob. 8.6) the tools are provided for an alternative — easier, but non-constructive and indirect — proof of the main result of §6.

*§ 4. First-order tableaux

Propositional tableaux were designed as a strategy for showing that a given finite set of formulas is not satisfied by any *truth valuation*. Now we shall introduce the method of *first-order tableaux* as a strategy for showing that a given finite set of formulas is not satisfiable (i.e., not satisfied by any *valuation*).

The construction of first-order tableaux is similar to that of propositional tableaux, except that in addition to the three old *propositional rules*

($\neg\neg$, $\rightarrow$ and $\neg\rightarrow$) we now admit two *quantifier rules* represented schematically as follows:

$$\forall\text{-}rule: \quad \begin{array}{c} \vdots \\ \forall \mathbf{x}\alpha \\ \vdots \\ | \\ \alpha(\mathbf{x}/\mathbf{t}) \end{array} \qquad \neg\forall\text{-}rule: \quad \begin{array}{c} \vdots \\ \neg\forall \mathbf{x}\alpha \\ \vdots \\ | \\ \neg\alpha(\mathbf{x}/\mathbf{y}) \end{array} \quad (\mathbf{y} \text{ restricted})$$

In the $\forall$-rule, $\mathbf{t}$ may be any $\mathscr{L}$-term. In the $\neg\forall$-rule, $\mathbf{y}$ may be any variable *which is not free in any formula of the branch which is being extended.* The particular $\mathbf{y}$ used in a given application of the $\neg\forall$-rule is said to be the *critical* variable of that application.

If $\mathscr{L}$ is a language with equality, we also admit three *equality rules,* represented schematically as follows:

$$\text{SI-}rule: \quad \begin{array}{c} \vdots \\ | \\ \mathbf{t}=\mathbf{t} \end{array}$$

where $\mathbf{t}$ is any $\mathscr{L}$-term;

$$\text{SF-}rule: \quad \begin{array}{c} \vdots \\ | \\ \mathbf{t}_1=\mathbf{t}_{n+1}\rightarrow\mathbf{t}_2=\mathbf{t}_{n+2}\rightarrow\ldots\rightarrow\mathbf{t}_n=\mathbf{t}_{2n}\rightarrow\mathbf{f}\mathbf{t}_1\ldots\mathbf{t}_n=\mathbf{f}\mathbf{t}_{n+1}\ldots\mathbf{t}_{2n} \end{array}$$

where $\mathbf{f}$ is any n-ary function symbol of $\mathscr{L}$ and $\mathbf{t}_1,\ldots,\mathbf{t}_{2n}$ are any $\mathscr{L}$-terms;

$$\text{SP-}rule: \quad \begin{array}{c} \vdots \\ | \\ \mathbf{t}_1=\mathbf{t}_{n+1}\rightarrow\mathbf{t}_2=\mathbf{t}_{n+2}\rightarrow\ldots\rightarrow\mathbf{t}_n=\mathbf{t}_{2n}\rightarrow\mathbf{P}\mathbf{t}_1\ldots\mathbf{t}_n\rightarrow\mathbf{P}\mathbf{t}_{n+1}\ldots\mathbf{t}_{2n} \end{array}$$

where $\mathbf{P}$ is any n-ary predicate symbol of $\mathscr{L}$ and $\mathbf{t}_1,\ldots,\mathbf{t}_{2n}$ are any $\mathscr{L}$-terms. (For $n=2$, $\mathbf{P}$ may be $=$.)

The names of these rules are abbreviations of "*self-identity*", "*substitutivity of equals in functions*" and "*substitutivity of equals in predicates*".

Note that the equality rules can always be applied to extend any branch, irrespective of what formulas belong to that branch.

The last difference between first-order tableaux and propositional tableaux is that a branch of a first-order tableau is *closed* if there is an atomic[1] formula α such that both α and $\neg\alpha$ belong to that branch. A first-order tableau for Φ whose branches are all closed is called *a first-order confutation of* Φ.

In this chapter we shall often say briefly "tableau" instead of "first-order tableau", etc.

4.1. THEOREM. *The tableau method is semantically sound: if a finite set* Φ *of formulas can be confuted, then* Φ *is not satisfiable. In particular, if* $\Phi=\{\neg\varphi\}$ *then* φ *is logically true.*

PROOF. We show that if a branch of a tableau is satisfiable (i.e., if the set of formulas belonging to the branch is satisfiable) and if that branch is extended (or extended and split) by any one of the eight rules, then the new branch (or, in the case of the $\rightarrow$-rule, at least one of the two new branches) is also satisfiable.

In the case of the three propositional rules, the proof is easy and is left to the reader (cf. Prob. 1.7.1).

In the case of the equality rules, it is enough to observe that the formulas introduced by these rules are all logically true, as the reader can easily check.

Now consider the $\forall$-rule. Let Ψ be the set of formulas of the branch which is about to be extended. Let $\forall x\alpha \in \Psi$ and let $\mathbf{t}$ be any term. Suppose $\sigma \models \Psi$. Then $(\forall x\alpha)^\sigma = \top$, and by the BSD $\alpha^{\sigma(x/u)} = \top$ for every u in the universe U of σ. On the other hand, by Thm. 3.10, $\alpha(x/\mathbf{t})^\sigma = \alpha^{\sigma(x/t)}$, where $t = \mathbf{t}^\sigma$. Since $\mathbf{t}^\sigma \in U$, it follows that $\alpha(x/\mathbf{t})^\sigma = \top$. Thus $\sigma \models \Psi, \alpha(x/\mathbf{t})$.

The case of the $\neg\forall$-rule is treated by Thm. 3.11.

Since any tableau for Φ is constructed by successive extensions from the tableau which has Φ as its sole node, it follows that if Φ is satisfiable, any tableau for Φ must have at least one satisfiable branch. But a closed branch is clearly not satisfiable. Thus if Φ can be confuted, it is not satisfiable. In particular, if $\Phi=\{\neg\varphi\}$ then $\neg\varphi$ is not satisfiable, i.e., φ is logically true. ∎

4.2. THEOREM. *If* Φ *is a finite set of formulas and if for some formula* α *both* $\alpha \in \Phi$ *and* $\neg\alpha \in \Phi$, *then* Φ *can be confuted. Hence, in any tableau, a branch that has both* α *and* $\neg\alpha$ *is as good as closed even if* α *is not atomic.*

PROOF. By induction on $\deg\alpha$. If α is atomic, then the tableau having Φ as its sole node is already a confutation.

[1] N.B.: Not just prime.

The cases $\alpha = \neg\beta$ and $\alpha = \beta \to \gamma$ are left to the reader.

Finally, let $\alpha = \forall x \beta$. We choose a variable y which is not free in Φ, and apply the $\neg\forall$-rule to $\neg\alpha$ with y as critical variable. Then we apply the $\forall$-rule to α with y as the term t. We get

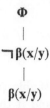

$$\begin{array}{c} \Phi \\ | \\ \neg\beta(x/y) \\ | \\ \beta(x/y) \end{array}$$

By the induction hypothesis this is as good as closed. ∎

Here are some practical hints for constructing conputations efficiently.

As in the propositional case, a branch which is seen to be as good as closed should not be extended any further.

Also, when a formula is being used, it is usually best to use it at once to extend all branches to which it belongs (except those that are as good as closed).

In addition to the eight rules, we may in practice use the six "unofficial" propositional rules (see §7 of Ch. 1) as well as the following "unofficial" quantifier rules:

$$\exists\textit{-rule:}\quad \begin{array}{c} \vdots \\ \exists x\alpha \\ \vdots \\ | \\ \alpha(x/y)\ \ (y \text{ restricted}) \end{array} \qquad\qquad \neg\exists\textit{-rule:}\quad \begin{array}{c} \vdots \\ \neg\exists x\alpha \\ \vdots \\ | \\ \neg\alpha(x/t) \end{array}$$

Here t may be any term, while y is subject to the same restriction as in the $\neg\forall$-rule. These two rules can easily be justified *via* Def. 1.5.1.

In general, it is best to apply the restricted quantifier rules ($\neg\forall$ and $\exists$) as soon as possible, and to delay applying the unrestricted quantifier rules ($\forall$ and $\neg\exists$) as long as possible. (For example, if in the proof of Thm. 4.2 we had applied the $\forall$-rule *first* to obtain $\beta(x/y)$, we could then no longer use y as critical variable for the $\neg\forall$-rule.)

EXAMPLE. We show that if x is not free in β, then $(\forall x\alpha \to \beta) \to \exists x(\alpha \to \beta)$ is logically true. By Thm. 4.1 it is enough to confute the negation of that formula:

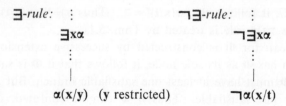

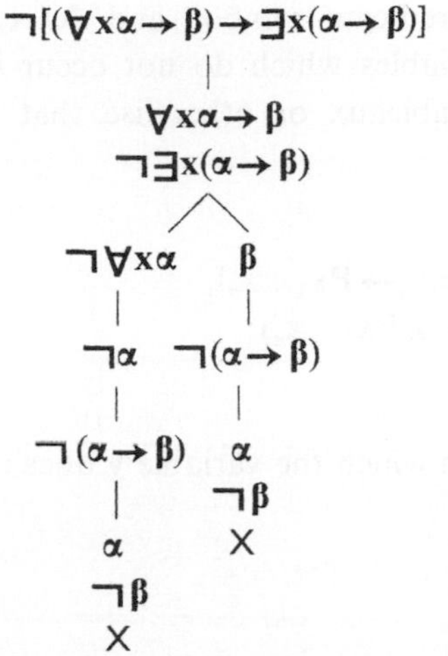

Here the $\neg\forall$-rule has been applied to $\neg\forall x\alpha$, to yield $\neg\alpha(x/x)=\neg\alpha$. (This is legitimate, since x is not free in β.) The $\neg\exists$-rule has been applied to $\neg\exists x(\alpha\rightarrow\beta)$ to yield $\neg(\alpha\rightarrow\beta)(x/x)=\neg(\alpha\rightarrow\beta)$. The rest is self-explanatory.

4.3. Problem. Show by tableaux or otherwise that the following formulas are logically true:

(a) $\forall x(\alpha\rightarrow\beta)\rightarrow\forall x\alpha\rightarrow\forall x\beta$;
(b) $\alpha\rightarrow\forall x\alpha$, where x is not free in α;
(c) $\forall x\alpha\rightarrow\alpha(x/t)$,
(d) $\alpha(x/t)\rightarrow\exists x\alpha$,

where, in (c) and (d), t is any term;

(e) $\forall x(\alpha\wedge\beta)\leftrightarrow\forall x\alpha\wedge\beta$,
(f) $\exists x(\alpha\wedge\beta)\leftrightarrow\exists x\alpha\wedge\beta$,
(g) $\forall x(\alpha\vee\beta)\leftrightarrow\forall x\alpha\vee\beta$,
(h) $\exists x(\alpha\vee\beta)\leftrightarrow\exists x\alpha\vee\beta$,
(i) $\forall x(\alpha\rightarrow\beta)\leftrightarrow\exists x\alpha\rightarrow\beta$,
(j) $\exists x(\alpha\rightarrow\beta)\leftrightarrow\forall x\alpha\rightarrow\beta$,
(k) $\forall x(\beta\rightarrow\alpha)\leftrightarrow\beta\rightarrow\forall x\alpha$,
(l) $\exists x(\beta\rightarrow\alpha)\leftrightarrow\beta\rightarrow\exists x\alpha$,

where in (e)–(l) x is not free in β.

4.4. PROBLEM. Let $\mathbf{P}$ be an n-ary predicate symbol, and let $\mathbf{t}_1,\ldots,\mathbf{t}_n$ be terms. Let $\mathbf{x}_1,\ldots,\mathbf{x}_n$ be n distinct variables which do not occur in any of the $\mathbf{t}_i$. Using Prob. 1.7, show by tableaux or otherwise that the three formulas

$$\mathbf{Pt}_1\ldots\mathbf{t}_n,$$
$$\forall\mathbf{x}_1\ldots\forall\mathbf{x}_n(\mathbf{t}_1=\mathbf{x}_1\rightarrow\ldots\rightarrow\mathbf{t}_n=\mathbf{x}_n\rightarrow\mathbf{Px}_1\ldots\mathbf{x}_n),$$
$$\exists\mathbf{x}_1\ldots\exists\mathbf{x}_n(\mathbf{t}_1=\mathbf{x}_1\wedge\ldots\wedge\mathbf{t}_n=\mathbf{x}_n\wedge\mathbf{Px}_1\ldots\mathbf{x}_n)$$

are logically equivalent.

4.5. PROBLEM. Let $\mathbf{s}$ and $\mathbf{t}$ be terms in which the variable $\mathbf{y}$ does not occur. Show that the three formulas

$$\mathbf{s}=\mathbf{t},$$
$$\forall\mathbf{y}(\mathbf{s}=\mathbf{y}\rightarrow\mathbf{t}=\mathbf{y}),$$
$$\exists\mathbf{y}(\mathbf{s}=\mathbf{y}\wedge\mathbf{t}=\mathbf{y})$$

are logically equivalent.

4.6. PROBLEM. Let $\mathbf{s},\mathbf{t}_1,\ldots,\mathbf{t}_n$ be terms, and let $\mathbf{f}$ be an n-ary function symbol. Let $\mathbf{x}_1,\ldots,\mathbf{x}_n$ be distinct variables which do not occur in any of the terms $\mathbf{s},\mathbf{t}_1,\ldots,\mathbf{t}_n$. Show that the three formulas

$$\mathbf{ft}_1\ldots\mathbf{t}_n=\mathbf{s},$$
$$\forall\mathbf{x}_1\ldots\forall\mathbf{x}_n(\mathbf{t}_1=\mathbf{x}_1\rightarrow\ldots\rightarrow\mathbf{t}_n=\mathbf{x}_n\rightarrow\mathbf{fx}_1\ldots\mathbf{x}_n=\mathbf{s}),$$
$$\exists\mathbf{x}_1\ldots\exists\mathbf{x}_n(\mathbf{t}_1=\mathbf{x}_1\wedge\ldots\wedge\mathbf{t}_n=\mathbf{x}_n\wedge\mathbf{fx}_1\ldots\mathbf{x}_n=\mathbf{s})$$

are logically equivalent.

4.7. PROBLEM. Prove by tableaux that the formulas $\forall\mathbf{x}\forall\mathbf{y}(\mathbf{x}=\mathbf{y})$ and $\exists\mathbf{x}\forall\mathbf{y}(\mathbf{x}=\mathbf{y})$ are logically equivalent.

*§ 5. Some "book-keeping" lemmas

In practice, the method of first-order tableaux is very easy to use. However, the theoretical study of this method is technically a bit tricky because in applying the $\forall$-rule to a formula $\forall\mathbf{x}\boldsymbol{\alpha}$, the term $\mathbf{t}$ that we need to substitute may not be free for $\mathbf{x}$ in $\boldsymbol{\alpha}$; thus, the substitution may involve alphabetic changes inside $\boldsymbol{\alpha}$. (A similar situation arises in connection with the $\neg\forall$-rule.) If several such substitutions are made in the same tableau, it is rather difficult to keep track of all these alphabetic changes.

Let T be a tableau for $\boldsymbol{\Phi}$. We shall say that T is *pure* if the following two conditions hold:

(1) The terms **t** used in the applications of the $\forall$-rule in T do not contain any variable that occurs bound in Φ.

(2) The critical variables of the applications of the $\neg\forall$-rule in T do not occur bound in Φ.

It is easy to see that in a pure tableau no alphabetic changes are ever made.

Instead of trying directly to keep track of the alphabetic changes that are made in an impure tableau, we propose to reduce the general case to that of pure tableaux.

Let T and T' be tableaux. We shall say that T' is a *variant* of T (briefly, $T \sim T'$) if T can be transformed into T' by replacing each formula α in T by a variant α', in such a way that each application of the $\forall$-rule in T is transformed into an application of the $\forall$-rule with *the same term* **t**, and each application of the $\neg\forall$-rule in T is transformed into an application of the $\neg\forall$-rule with *the same critical variable*.[1]

Similarly, we say that $\Phi \sim \Phi'$ if Φ can be transformed into Φ' by replacing each $\varphi \in \Phi$ by a variant φ'.

Our first aim is to show that, given a tableau T for Φ, we can construct a variant T' of T for any variant Φ' of Φ. For this we shall need

5.1. LEMMA. *If* $\alpha \sim \alpha'$, *then* $\alpha(\mathbf{x}/\mathbf{t}) \sim \alpha'(\mathbf{x}/\mathbf{t})$.

PROOF. By induction on $\deg \alpha$. We leave the easy cases to the reader and deal here only with the case where α is universal.

Let $\alpha = \forall \mathbf{y} \beta$. Then (by Def. 3.7) $\alpha' = \forall \mathbf{y} \beta'$ or $\alpha' = \forall \mathbf{z} [\beta'(\mathbf{y}/\mathbf{z})]$, where $\beta \sim \beta'$ and **z** is not free in β' but is free for **y** in β'. We may assume that **x** is free in α (hence also in α') because otherwise we have

$$\alpha(\mathbf{x}/\mathbf{t}) = \alpha, \qquad \alpha'(\mathbf{x}/\mathbf{t}) = \alpha',$$

and the assertion of our lemma is trivial. Also, *for the time being,* we shall assume that **t** is free for **x** in both α and α'.

By Def. 3.3, **y** cannot occur in **t**, and

$$\alpha(\mathbf{x}/\mathbf{t}) = \forall \mathbf{y} \, [\beta(\mathbf{x}/\mathbf{t})].$$

If $\alpha' = \forall \mathbf{y} \beta'$, then Def. 3.3 similarly implies that

$$\alpha'(\mathbf{x}/\mathbf{t}) = \forall \mathbf{y} \, [\beta'(\mathbf{x}/\mathbf{t})],$$

and our assertion follows at once from the induction hypothesis.

[1] As a matter of fact the last two conditions (beginning with "in such a way...") are redundant, but it is convenient to include them nevertheless.

Now suppose $\boldsymbol{\alpha}' = \forall \mathbf{z}\,[\boldsymbol{\beta}'(\mathbf{y}/\mathbf{z})]$. Then Def. 3.3 implies that $\mathbf{z}$ cannot occur in $\mathbf{t}$, $\mathbf{t}$ is free for $\mathbf{x}$ in $\boldsymbol{\beta}'(\mathbf{y}/\mathbf{z})$, and

$$\boldsymbol{\alpha}'(\mathbf{x}/\mathbf{t}) = \forall \mathbf{z}\,[\boldsymbol{\beta}'(\mathbf{y}/\mathbf{z})(\mathbf{x}/\mathbf{t})].$$

Since $\mathbf{z}$ is free for $\mathbf{y}$ in $\boldsymbol{\beta}'$ and $\mathbf{t}$ is free for $\mathbf{x}$ in $\boldsymbol{\beta}'(\mathbf{y}/\mathbf{z})$ it follows that in going from $\boldsymbol{\beta}'$ to $\boldsymbol{\beta}'(\mathbf{y}/\mathbf{z})(\mathbf{x}/\mathbf{t})$ no alphabetic changes are made. Thus $\boldsymbol{\beta}'$ and $\boldsymbol{\beta}'(\mathbf{y}/\mathbf{z})(\mathbf{x}/\mathbf{t})$ have exactly the same bound occurrences of variables. Next, we observe that

$$\boldsymbol{\beta}'(\mathbf{y}/\mathbf{z})(\mathbf{x}/\mathbf{t}) = \boldsymbol{\beta}'(\mathbf{x}/\mathbf{t})(\mathbf{y}/\mathbf{z}).$$

For, $\mathbf{x}$ is different from both $\mathbf{y}$ and $\mathbf{z}$ ($\mathbf{x}$ is supposed to be free in $\boldsymbol{\alpha}$ and $\boldsymbol{\alpha}'$) and $\mathbf{y}$ does not occur in $\mathbf{t}$; so it makes no difference whether we *first* substitute $\mathbf{z}$ for $\mathbf{y}$ and *then* $\mathbf{t}$ for $\mathbf{x}$, or *vice versa*. Thus

$$\boldsymbol{\alpha}'(\mathbf{x}/\mathbf{t}) = \forall \mathbf{z}\,[\boldsymbol{\beta}'(\mathbf{x}/\mathbf{t})(\mathbf{y}/\mathbf{z})].$$

Now, $\mathbf{z}$ is not free in $\boldsymbol{\beta}'(\mathbf{x}/\mathbf{t})$, because $\mathbf{z}$ was not free in $\boldsymbol{\beta}'$ and $\mathbf{t}$ does not contain $\mathbf{z}$. Also, $\mathbf{z}$ is free for $\mathbf{y}$ in $\boldsymbol{\beta}'(\mathbf{x}/\mathbf{t})$, because $\mathbf{z}$ was free for $\mathbf{y}$ in $\boldsymbol{\beta}'$ and the substitution of $\mathbf{t}$ for $\mathbf{x}$ in $\boldsymbol{\beta}'$ cannot change matters in this respect. (Recall that $\mathbf{t}$ does not contain $\mathbf{y}$ and the substitution does not involve any alphabetic change within $\boldsymbol{\beta}'$!) Therefore our $\boldsymbol{\alpha}'(\mathbf{x}/\mathbf{t})$ arises by alphabetic change from $\forall \mathbf{y}\,[\boldsymbol{\beta}'(\mathbf{x}/\mathbf{t})]$ and we have

$$\begin{aligned}\boldsymbol{\alpha}'(\mathbf{x}/\mathbf{t}) &\sim \forall \mathbf{y}\,[\boldsymbol{\beta}'(\mathbf{x}/\mathbf{t})]\\ &\sim \forall \mathbf{y}\,[\boldsymbol{\beta}(\mathbf{x}/\mathbf{t})],\end{aligned}$$

by the induction hypothesis.

We have therefore proved the assertion of our lemma under the assumption that $\mathbf{t}$ is free for $\mathbf{x}$ in $\boldsymbol{\alpha}$ and $\boldsymbol{\alpha}'$. However, even if this assumption does *not* hold, then by Def. 3.9 we have at any rate

$$\boldsymbol{\alpha}(\mathbf{x}/\mathbf{t}) = \gamma(\mathbf{x}/\mathbf{t}), \qquad \boldsymbol{\alpha}'(\mathbf{x}/\mathbf{t}) = \gamma'(\mathbf{x}/\mathbf{t}),$$

where γ and γ' are certain variants of $\boldsymbol{\alpha}$ and $\boldsymbol{\alpha}'$ respectively, and $\mathbf{t}$ *is* free for $\mathbf{x}$ in γ and γ'. Thus, by what we have already proved,

$$\boldsymbol{\alpha}(\mathbf{x}/\mathbf{t}) = \gamma(\mathbf{x}/\mathbf{t}) \sim \gamma'(\mathbf{x}/\mathbf{t}) = \boldsymbol{\alpha}'(\mathbf{x}/\mathbf{t}). \qquad \blacksquare$$

5.2. LEMMA. *Let* $\boldsymbol{\Phi} \sim \boldsymbol{\Phi}'$. *Given a tableau* T *for* $\boldsymbol{\Phi}$, *we can construct a tableau* T' *for* $\boldsymbol{\Phi}'$ *such that* $T \sim T'$.

PROOF. By induction on the number of nodes in T. If T has $\boldsymbol{\Phi}$ as its only node, we take T' to be the tableau having $\boldsymbol{\Phi}'$ as its only node.

If T has more than one node, then T is obtained by extending a tableau T_0 by means of one of the eight rules. By the induction hypothesis, we have got a tableau T_0' for $\mathbf{\Phi}'$ such that $T_0 \sim T_0'$. We shall show that the required T' can be obtained by extending T_0' using the same rule which yielded T from T_0. Eight cases have to be treated, corresponding to the eight rules.

The cases of the propositional and equality rules are straightforward and are left to the reader. We shall deal here with the quantifier rules.

Rule $\forall$. Suppose some branch of T_0 has a formula $\forall \mathbf{x}\alpha$ and T was obtained by adjoining $\{\alpha(\mathbf{x}/\mathbf{t})\}$ as successor to the terminal node of that branch.

The corresponding branch of T_0' must have some variant $(\forall \mathbf{x}\alpha)'$ of $\forall \mathbf{x}\alpha$. By Def. 3.7, we have

$$(\forall \mathbf{x}\alpha)' = \forall \mathbf{x}\alpha' \quad \text{or} \quad (\forall \mathbf{x}\alpha)' = \forall \mathbf{z}\,[\alpha'(\mathbf{x}/\mathbf{z})],$$

where $\alpha \sim \alpha'$ and $\mathbf{z}$ is not free in α' but is free for $\mathbf{x}$ in α'.

If $(\forall \mathbf{x}\alpha)' = \forall \mathbf{x}\alpha'$, then we can apply the $\forall$-rule to $\forall \mathbf{x}\alpha'$ and extend T_0' to T' by adjoining $\{\alpha'(\mathbf{x}/\mathbf{t})\}$ as successor to the terminal node of the branch in question. By Lemma 5.1, $\alpha(\mathbf{x}/\mathbf{t}) \sim \alpha'(\mathbf{x}/\mathbf{t})$, so that $T \sim T'$ as required.

Now suppose $(\forall \mathbf{x}\alpha)' = \forall \mathbf{z}\,[\alpha'(\mathbf{x}/\mathbf{z})]$. Again, we apply to this formula the $\forall$-rule with the same term $\mathbf{t}$. It remains to show that

(1) $\qquad \alpha(\mathbf{x}/\mathbf{t}) \sim \alpha'(\mathbf{x}/\mathbf{z})(\mathbf{z}/\mathbf{t}).$

As a matter of fact, it can be proved that

(2) $\qquad \alpha'(\mathbf{x}/\mathbf{z})(\mathbf{z}/\mathbf{t}) = \alpha'(\mathbf{x}/\mathbf{t}),$

hence, again by Lemma 5.1, we have (1). However, to prove (2) rigorously is a bit tricky. (The substitution of $\mathbf{z}$ for $\mathbf{x}$ in α' involves no alphabetic changes, since $\mathbf{z}$ is free for $\mathbf{x}$ in α'; but one has to verify that the substitution of $\mathbf{t}$ for $\mathbf{z}$ in $\alpha'(\mathbf{x}/\mathbf{z})$ involves exactly the same alphabetic changes as the substitution of $\mathbf{t}$ for $\mathbf{x}$ in α'.)

Instead of appealing to (2), we can proceed as follows. We take any formula α'' such that $\alpha' \sim \alpha''$ (hence $\alpha \sim \alpha''$) and such that $\mathbf{z}$ and the variables occurring in $\mathbf{t}$ do not occur bound in α''. Then the substitution of $\mathbf{z}$ in α'' and of $\mathbf{t}$ in both α'' and $\alpha''(\mathbf{x}/\mathbf{z})$ do not involve any alphabetic changes. Thus it is easy to see that

(3) $\qquad \alpha''(\mathbf{x}/\mathbf{z})(\mathbf{z}/\mathbf{t}) = \alpha''(\mathbf{x}/\mathbf{t}).$

Moreover, by Lemma 5.1,

(4) $\quad \boldsymbol{\alpha}'(\mathbf{x}/\mathbf{z})(\mathbf{z}/\mathbf{t}) \sim \boldsymbol{\alpha}''(\mathbf{x}/\mathbf{z})(\mathbf{z}/\mathbf{t})$,

(5) $\quad \boldsymbol{\alpha}(\mathbf{x}/\mathbf{t}) \sim \boldsymbol{\alpha}''(\mathbf{x}/\mathbf{t})$.

Using (3), (4) and (5) we obtain (1).

Rule $\neg\forall$. This is treated in the same way as the $\forall$-rule. Here one also needs to observe that if $\mathbf{y}$ is not free in any formula of a given branch of T_0, then $\mathbf{y}$ cannot be free in any formula of the corresponding branch of T_0'. ∎

It is clear that if T is a confutation and $T \sim T'$ then T' is a confutation as well. (Recall that an atomic formula has no variants other than itself.)

Also, if T is a tableau for $\boldsymbol{\Phi}$, we can always choose $\boldsymbol{\Phi}'$ such that $\boldsymbol{\Phi} \sim \boldsymbol{\Phi}'$ and such that the tableau T' of Lemma 5.2 is pure. (We obtain $\boldsymbol{\Phi}'$ from $\boldsymbol{\Phi}$ by suitable alphabetic changes inside each $\boldsymbol{\varphi} \in \boldsymbol{\Phi}$ in such a way that the terms $\mathbf{t}$ used in the applications of the $\forall$-rule in T do not contain any variable bound in $\boldsymbol{\Phi}'$, and the critical variables of T also do not occur bound in $\boldsymbol{\Phi}'$.)

We can now prove:

5.3. Lemma. *Let* $\mathbf{y}_1, \ldots, \mathbf{y}_n$ *be any n variables. Given a confutation of* $\boldsymbol{\Phi}$, *we can construct for* $\boldsymbol{\Phi}$ *a confutation in which none of the variables* $\mathbf{y}_1, \ldots, \mathbf{y}_n$ *is used as a critical variable.*

Proof. We may assume that $\mathbf{y}_1, \ldots, \mathbf{y}_n$ are not free in $\boldsymbol{\Phi}$, because a variable which *is* free in $\boldsymbol{\Phi}$ cannot in any case be used as a critical variable in a tableau for $\boldsymbol{\Phi}$.

Let T be the given confutation. We choose $\boldsymbol{\Phi}'$ such that $\boldsymbol{\Phi} \sim \boldsymbol{\Phi}'$ and such that the tableau T' constructed in Lemma 5.2 is pure. Moreover, we can choose $\boldsymbol{\Phi}'$ so that $\mathbf{y}_1, \ldots, \mathbf{y}_n$ do not occur bound — hence do not occur *at all* — in $\boldsymbol{\Phi}'$. Because T' is pure, $\mathbf{y}_1, \ldots, \mathbf{y}_n$ do not occur bound in any formula in T'. Since T is a confutation, T' is a confutation as well.

We choose n variables $\mathbf{z}_1, \ldots, \mathbf{z}_n$ which are different from all the $\mathbf{y}_i$ and do not occur at all in T'. Clearly, $\mathbf{z}_i$ could have been used just as well as $\mathbf{y}_i$. More precisely, let T'' be obtained from T' by putting $\mathbf{z}_1, \ldots, \mathbf{z}_n$ everywhere for $\mathbf{y}_1, \ldots, \mathbf{y}_n$ respectively. (In particular, wherever $\mathbf{y}_i$ is used in T' as a critical variable, it will now be replaced by $\mathbf{z}_i$.) Clearly, T'' is also a confutation of $\boldsymbol{\Phi}'$ and $\mathbf{y}_1, \ldots, \mathbf{y}_n$ are not used in T'' as critical variables (in fact, they do not occur at all in T'').

Finally, using Lemma 5.2 again, we can obtain for $\boldsymbol{\Phi}$ a confutation T''' such that $T'' \sim T'''$ Clearly, T''' has the required properties. ∎

From now on we put

$$\Phi(x/t) = \{\varphi(x/t) : \quad \varphi \in \Phi\}.$$

We now come to the final result of this section.

5.4. LEMMA. *Given a confutation of* Φ, *we can confute* $\Phi(z/s)$.

PROOF. Let T be the given confutation. By Lemma 5.3 we may assume that no variable occurring in the term s serves in T as a critical variable. Also, *for the time being* we shall assume that

(1) T is pure,

(2) no variable occurring in s occurs bound in Φ.

From (1) and (2) it is easy to see that s is free for z (indeed, for *any* variable) in every formula of T. In fact, no variable occurring in s can be bound in any formula of T.

Let T^* be obtained from T by replacing each formula α in T by $\alpha(z/s)$. We shall show that T^* is a confutation of $\Phi(z/s)$.

When we go from T to T^*, the initial node Φ of T becomes $\Phi(z/s)$. We shall verify that each application of a rule in T is transformed into an application of the same rule in T^*. For the propositional and equality rules this is quite obvious (e.g., $\alpha \to \beta$ is transformed into $\alpha(z/s) \to \beta(z/s)$ and $t = t$ is transformed into $t(z/s) = t(z/s)$). We shall deal in detail with the quantifier rules.

Rule $\forall$. Suppose $\forall x \alpha$ is used in T to yield $\alpha(x/t)$. When T is transformed into T^*, $\forall x \alpha$ becomes $[\forall x \alpha](z/s)$ and $\alpha(x/t)$ becomes $\alpha(x/t)(z/s)$.

If z happens to be x, then clearly

$$[\forall x \alpha](z/s) = \forall x \alpha,$$
$$\alpha(x/t)(z/s) = \alpha(x/t)(x/s) = \alpha(x/t(x/s)).$$

Thus $\alpha(x/t)(z/s)$ can be obtained from $[\forall x \alpha](z/s)$ by the $\forall$-rule.

If $z \neq x$, then we have

$$[\forall x \alpha](z/s) = \forall x\, [\alpha(z/s)],$$

and since by (1) and (2) x does not occur in s we also have

$$\alpha(x/t)(z/s) = \alpha(z/s)(x/t(z/s)).$$

Here again $\alpha(x/t)(z/s)$ can be obtained by the $\forall$-rule from $[\forall x \alpha](z/s)$.

Rule $\neg\forall$. Suppose $\neg\forall x \alpha$ is used in T to yield $\neg\alpha(x/y)$. Upon going from T to T^*, $\neg\forall x \alpha$ becomes $[\neg\forall x \alpha](z/s)$ and $\neg\alpha(x/y)$ becomes $\neg\alpha(x/y)(z/s)$.

If **z** happens to be **x**, then

$$[\neg\forall \mathbf{x}\alpha](\mathbf{z}/\mathbf{s}) = \neg\forall \mathbf{x}\alpha.$$

Also, since *T* is assumed to be pure, the critical variable **y** is not bound in **Φ** and hence cannot be bound in any formula of *T* (for, in a pure tableau no alphabetic changes are ever made). Hence **y** must be different from **x** and we have

$$\neg\alpha(\mathbf{x}/\mathbf{y})(\mathbf{z}/\mathbf{s}) = \neg\alpha(\mathbf{x}/\mathbf{y})(\mathbf{x}/\mathbf{s}) = \neg\alpha(\mathbf{x}/\mathbf{y}).$$

Thus, if $\mathbf{z}=\mathbf{x}$, the transformation from *T* to T^* does not affect $\neg\forall\mathbf{x}\alpha$ and $\neg\alpha(\mathbf{x}/\mathbf{y})$. We still have to check that **y** is not free in any formula preceding $\neg\alpha(\mathbf{x}/\mathbf{y})$ in T^*. But this is actually the case. For **y** was not free in any formula preceding $\neg\alpha(\mathbf{x}/\mathbf{y})$ in *T*; it is true that these formulas have been transformed by substituting **s** for **z**, but we have assumed in the beginning of this proof that **y** does not occur in **s** either.

Finally, suppose that $\mathbf{z}\neq\mathbf{x}$. Then

$$[\neg\forall \mathbf{x}\alpha](\mathbf{z}/\mathbf{s}) = \neg\forall \mathbf{x}\,[\alpha(\mathbf{z}/\mathbf{s})].$$

Also, we may assume that **z** is free in **Φ**, otherwise the assertion of our lemma is trivial. Hence **y**, being a critical variable in *T*, must be different from **z**, so that $\mathbf{y}(\mathbf{z}/\mathbf{s})=\mathbf{y}$. And since, by (1) and (2), **x** does not occur in **s**, we have

$$\neg\alpha(\mathbf{x}/\mathbf{y})(\mathbf{z}/\mathbf{s}) = \neg\alpha(\mathbf{z}/\mathbf{s})(\mathbf{x}/\mathbf{y}(\mathbf{z}/\mathbf{s})) = \neg\alpha(\mathbf{z}/\mathbf{s})(\mathbf{x}/\mathbf{y}).$$

As before, we know that **y** cannot be free in any formula of T^* preceding $\neg\alpha(\mathbf{x}/\mathbf{y})(\mathbf{z}/\mathbf{s})$. Thus in this case too $\neg\alpha(\mathbf{x}/\mathbf{y})(\mathbf{z}/\mathbf{s})$ can be obtained from $[\neg\forall\mathbf{x}\alpha](\mathbf{z}/\mathbf{s})$ using the $\neg\forall$-rule.

It follows that T^* is a tableau for $\mathbf{\Phi}(\mathbf{z}/\mathbf{s})$; and it is clear that every branch of T^* is closed, because an atomic formula in *T* is transformed into an atomic formula in T^*.

So far we have assumed that (1) and (2) hold. If this is not the case, we can find a variant **Φ′** of **Φ** such that no variable of **s** occurs bound in **Φ′** and such that the tableau *T′* constructed from *T* in Lemma 5.2 is pure. By what we have proved so far, we can construct a confutation of $\mathbf{\Phi}'(\mathbf{z}/\mathbf{s})$. But by Lemma 5.1 we have $\mathbf{\Phi}(\mathbf{z}/\mathbf{s})\sim\mathbf{\Phi}'(\mathbf{z}/\mathbf{s})$; so, using Lemma 5.2 once more, we get a confutation of $\mathbf{\Phi}(\mathbf{z}/\mathbf{s})$. ∎

*§ 6. The Elimination Theorem for first-order tableaux

We start by proving four simple lemmas analogous to Lemmas 1.8.1–1.8.4. (The wording is actually *identical*, except that now "confutation" means first-order confutation, not propositional confutation.)

6.1. LEMMA. *Given a confutation of* Φ, *we can confute* $\Phi \cup \Psi$, *where* Ψ *is any finite set of formulas.*

PROOF. This is not quite as trivial as the analogous result for propositional tableaux (Lemma 1.8.1). In general, we cannot simply adjoin Ψ to the initial node of the given tableau, because in that tableau variables that are free in Ψ might have been used as critical variables. However, by Lemma 5.3 we can construct for Φ a confutation in which no free variable of Ψ is used as a critical variable. In *this* tableau we can adjoin Ψ to the initial node, getting a confutation of $\Phi \cup \Psi$. ∎

It follows from Lemma 6.1 that if we possess a confutation for a subset of the set of formulas belonging to a given branch, then this branch is as good as closed. Below we shall make implicit use of this fact.

6.2. LEMMA. *Given a confutation of* $\Phi, \neg\neg\alpha$, *we can confute* Φ, α. ∎

6.3. LEMMA. *Given a confutation of* $\Phi, \alpha \to \beta$, *we can confute* $\Phi, \neg\alpha$ *and* Φ, β. ∎

6.4. LEMMA. *Given a confutation of* $\Phi, \neg(\alpha \to \beta)$, *we can confute* $\Phi, \alpha, \neg\beta$. ∎

We shall also need the following:

6.5. LEMMA. *Given a confutation of* $\Phi, \neg\forall x\alpha$, *we can confute* $\Phi, \neg\alpha(x/t)$, *where* t *is any term.*

PROOF. Let T be the given confutation. We proceed by induction on the depth d of T. By Lemma 5.3 we may assume that no variable occurring in t is used as a critical variable in T.

If $d=0$, then for some atomic formula β both $\beta \in \Phi$ and $\neg\beta \in \Phi$. Then clearly $\Phi, \neg\alpha(x/t)$ is also a confutation (with a single node).

If $d>0$, we examine how the node(s) of level 1 in T could have been obtained.

First, suppose that T begins thus:

$$\begin{array}{l} \Phi, \neg\forall x\alpha \\ | \\ \beta \\ \vdots \end{array}$$

where **β** is obtained by an equality rule, or by applying a quantifier or propositional rule — except the →-rule — to a formula $\varphi \in \Phi$. If in *T* we fuse the initial node with the node {**β**} of level 1, we get a confutation of depth $d-1$ for **Φ**, **β**, **¬∀xα**. By the induction hypothesis, we can then confute **Φ**, **β**, **¬α(x/t)**. We now start a tableau for **Φ**, **¬α(x/t)** thus:

$$\begin{array}{c} \Phi,\ \neg\alpha(x/t) \\ | \\ \beta \end{array}$$

where **β** is obtained in the same way as in *T*. But, since we already possess a confutation of **Φ**, **β**, **¬α(x/t)**, the single branch of our tableau (1) is as good as closed.

Next, suppose that *T* begins thus:

$$\begin{array}{c} \Phi,\ \neg\forall x\alpha \\ \swarrow \quad \searrow \\ \beta \qquad \gamma \\ \vdots \qquad \vdots \end{array}$$

where **β** and γ are obtained by applying the →-rule to some $\varphi \in \Phi$. If we fuse the initial node with the node {**β**}, we get a confutation of depth $<d$ for **Φ**, **β**, **¬∀xα**. Hence by the induction hypothesis we can confute **Φ**, **β**, **¬α(x/t)**. Similarly, we confute **Φ**, γ, **¬α(x/t)**. We can now start a tableau for **Φ**, **¬α(x/t)** thus:

$$\begin{array}{c} \Phi,\ \neg\alpha(x/t) \\ \swarrow \quad \searrow \\ \beta \qquad \gamma \end{array}$$

and both branches of this tableau are as good as closed.

Finally, we have to consider the case where level 1 of *T* is obtained by applying the ¬∀-rule to **¬∀xα**. Then *T* begins thus:

$$\begin{array}{c} \Phi,\ \neg\forall x\alpha \\ | \\ \neg\alpha(x/y) \\ \vdots \end{array}$$

and by Lemma 5.3 we may assume that **y** does not occur at all in **¬∀xα**. By fusing the initial node with that of level 1, we get a confutation of depth $d-1$ for **Φ**, **¬α(x/y)**, **¬∀xα**. By the induction hypothesis (with **y** used instead of **t**) we can confute **Φ**, **¬α(x/y)**, **¬α(x/y)**, i.e., **Φ**, **¬α(x/y)**.

By Lemma 5.4, we can now confute $\mathbf{\Phi(y/t)}, \neg\boldsymbol{\alpha}\mathbf{(x/y)(y/t)}$. But **y**, being a critical variable in *T*, cannot be free in $\mathbf{\Phi}$. Hence $\mathbf{\Phi(y/t)=\Phi}$. Also, it can be shown that

(2) $$\neg\boldsymbol{\alpha}\mathbf{(x/y)(y/t)} = \neg\boldsymbol{\alpha}\mathbf{(x/t)}$$

because **y** does not occur at all in $\boldsymbol{\alpha}$. However, instead of appealing to (2), which is a bit tricky to prove rigorously, we may argue as follows.[1] We choose $\boldsymbol{\alpha}'$ such that $\boldsymbol{\alpha} \sim \boldsymbol{\alpha}'$ and such that **y** and the variables of **t** do not occur bound in $\boldsymbol{\alpha}'$ (hence **y** cannot occur at all in $\boldsymbol{\alpha}'$). Then it is easy to see that

(3) $$\neg\boldsymbol{\alpha}'\mathbf{(x/y)(y/t)} = \neg\boldsymbol{\alpha}'\mathbf{(x/t)},$$

since these substitutions involve no alphabetic changes. On the other hand, by Lemma 5.1 we have

(4) $$\neg\boldsymbol{\alpha}\mathbf{(x/y)(y/t)} \sim \neg\boldsymbol{\alpha}'\mathbf{(x/y)(y/t)}.$$

Thus, by (3) and (4), and another appeal to Lemma 5.1,

$$\neg\boldsymbol{\alpha}\mathbf{(x/y)(y/t)} \sim \neg\boldsymbol{\alpha}'\mathbf{(x/t)} \sim \neg\boldsymbol{\alpha}\mathbf{(x/t)}.$$

Since we already possess a confutation of $\mathbf{\Phi}, \neg\boldsymbol{\alpha}\mathbf{(x/y)(y/t)}$, we can use Lemma 5.2 to confute $\mathbf{\Phi}, \neg\boldsymbol{\alpha}\mathbf{(x/t)}$. ∎

We remark that Lemma 6.5 is rather powerful. It implies that if in a branch of a tableau we have both $\neg\forall\mathbf{x}\boldsymbol{\alpha}$ and $\neg\boldsymbol{\alpha}\mathbf{(x/t)}$, where **t** is any term, then $\neg\forall\mathbf{x}\boldsymbol{\alpha}$ can be considered *used up* in that branch and need not be used to extend it. Intuitively speaking, the reason for this is that $\neg\boldsymbol{\alpha}\mathbf{(x/t)}$ conveys at least as much information as $\neg\forall\mathbf{x}\boldsymbol{\alpha}$. (On the other hand, for any finite number of terms $\mathbf{t}_1,\ldots,\mathbf{t}_n$, the formulas $\boldsymbol{\alpha}(\mathbf{x}/\mathbf{t}_1),\ldots,\boldsymbol{\alpha}(\mathbf{x}/\mathbf{t}_n)$ jointly do not in general convey as much information as $\forall\mathbf{x}\boldsymbol{\alpha}$. Therefore $\forall\mathbf{x}\boldsymbol{\alpha}$ may never be used up in a branch.)

As a last preparatory lemma we get in the same way as 1.8.5:

6.6. LEMMA. *Given confutations of* $\mathbf{\Phi}, \boldsymbol{\delta}$ *and* $\mathbf{\Phi}, \neg\boldsymbol{\delta}$, *where* $\boldsymbol{\delta}$ *is atomic, we can confute* $\mathbf{\Phi}$. ∎

We can now prove:

6.7. ELIMINATION LEMMA. *Given confutations of* $\mathbf{\Phi}, \boldsymbol{\delta}$ *and* $\mathbf{\Phi}, \neg\boldsymbol{\delta}$, *where* $\boldsymbol{\delta}$ *is any formula, we can confute* $\mathbf{\Phi}$.

[1] Cf. the proof of Lemma 5.2.

PROOF. By induction on deg δ. If δ is atomic, we appeal to Lemma 6.6. The cases where $\delta = \neg\alpha$ and $\delta = \alpha \rightarrow \beta$ are treated exactly as in the proof of the Elimination Lemma for propositional tableaux (Lemma 1.8.6), using Lemmas 6.1–6.4.

The remaining case is $\delta = \forall x\alpha$. Let T be the given confutation of Φ, $\forall x\alpha$. We shall now transform T by eliminating from it all nodes obtained by applying the $\forall$-rule to $\forall x\alpha$.

Suppose T has a node $\{\alpha(x/t)\}$ obtained by applying the $\forall$-rule to $\forall x\alpha$. We choose a node of this form at the deepest possible level in T, i.e., such that no node of a deeper level is obtained in a similar way. Let this level be $k+1$, and let $\Phi_0,\ldots,\Phi_k$ (in this order) be the nodes of T preceding our $\{\alpha(x/t)\}$. In particular, $\Phi_0 = \Phi, \forall x\alpha$. Let

$$\Psi = \Phi \cup \Phi_1 \cup \ldots \cup \Phi_k.$$

(Note that $\forall x\alpha$ has been *excluded* from Ψ.)

Since $\forall x\alpha$ is not used in T below level $k+1$, it is clear that by adjoining Ψ to our node $\{\alpha(x/t)\}$ and by considering the nodes of T that follow this node, we get a confutation of $\Psi, \alpha(x/t)$.

On the other hand, we were given a confutation of $\Phi, \neg\forall x\alpha$; so by Lemma 6.5 we can confute $\Phi, \neg\alpha(x/t)$. Since $\Phi \subseteq \Psi$, we can use Lemma 6.1 to confute $\Psi, \neg\alpha(x/t)$.

Now we have got confutations of both $\Psi, \alpha(x/t)$ and $\Psi, \neg\alpha(x/t)$. Since $\deg \alpha(x/t) = \deg \alpha = \deg \delta - 1$, we can (by the induction hypothesis) confute Ψ.

We transform T as follows. First, we cut out our node $\{\alpha(x/t)\}$ as well as all nodes following it. We are now left with the "truncated" branch $\Phi_0, \Phi_1, \ldots, \Phi_k$. But Ψ is a subset of the set of formulas of this branch, and we possess a confutation of Ψ. Therefore this branch is as good as closed, without making any further use of $\forall x\alpha$.

Thus we can transform T into a confutation of Φ, $\forall x\alpha$ in which $\forall x\alpha$ is used fewer times than in T. Repeating this process a finite number of times, we obtain a confutation of Φ, $\forall x\alpha$ in which $\forall x\alpha$ is never used. We can now delete $\forall x\alpha$ from the initial node, getting the required confutation of Φ. ∎

As in §8 of Ch. 1, we now (temporarily) admit the EM-rule in addition to our eight rules. However, we have again:

6.8. ELIMINATION THEOREM. *Given a confutation of* $\mathbf{\Phi}$ *in which the* EM-*rule is used, we can construct for* $\mathbf{\Phi}$ *an* EM-*free confutation.* ∎

Therefore we shall not regard the EM-rule as one of the tableau rules, but we can still use it in practice.

§ 7. Hintikka sets

In proving the completeness of the method of first-order tableaux (see next section), as well as in proving the completeness of the predicate calculus (in Ch. 3), we shall make use of *Hintikka sets.*

7.1. DEFINITION. A set $\mathbf{\Psi}$ of $\mathscr{L}$-formulas is a *Hintikka set in* $\mathscr{L}$ if the following conditions hold:

(1) If $\boldsymbol{\varphi}$ is atomic, then $\boldsymbol{\varphi}$ and $\neg\boldsymbol{\varphi}$ do not both belong to $\mathbf{\Psi}$ (i.e., if $\boldsymbol{\varphi}\in\mathbf{\Psi}$, then $\neg\boldsymbol{\varphi}\notin\mathbf{\Psi}$).

(2) If $\neg\neg\boldsymbol{\alpha}\in\mathbf{\Psi}$, then $\boldsymbol{\alpha}\in\mathbf{\Psi}$.

(3) If $\boldsymbol{\alpha}\rightarrow\boldsymbol{\beta}\in\mathbf{\Psi}$, then $\neg\boldsymbol{\alpha}\in\mathbf{\Psi}$ or $\boldsymbol{\beta}\in\mathbf{\Psi}$.

(4) If $\neg(\boldsymbol{\alpha}\rightarrow\boldsymbol{\beta})\in\mathbf{\Psi}$, then both $\boldsymbol{\alpha}\in\mathbf{\Psi}$ and $\neg\boldsymbol{\beta}\in\mathbf{\Psi}$.

(5) If $\forall\mathbf{x}\boldsymbol{\alpha}\in\mathbf{\Psi}$, then $\boldsymbol{\alpha}(\mathbf{x}/\mathbf{t})\in\mathbf{\Psi}$ for every $\mathscr{L}$-term $\mathbf{t}$.

(6) If $\neg\forall\mathbf{x}\boldsymbol{\alpha}\in\mathbf{\Psi}$, then $\neg\boldsymbol{\alpha}(\mathbf{x}/\mathbf{t})\in\mathbf{\Psi}$ for some $\mathscr{L}$-term $\mathbf{t}$.

If $\mathscr{L}$ is a language with equality, we also require:

(7) For every $\mathscr{L}$-term $\mathbf{t}$, $(\mathbf{t}=\mathbf{t})\in\mathbf{\Psi}$.

(8) If $\mathbf{f}$ is an n-ary function symbol of $\mathscr{L}$ and $\mathbf{t}_1,\dots,\mathbf{t}_{2n}$ are $\mathscr{L}$-terms, then the formula

$$\mathbf{t}_1=\mathbf{t}_{n+1}\rightarrow\mathbf{t}_2=\mathbf{t}_{n+2}\rightarrow\dots\rightarrow\mathbf{t}_n=\mathbf{t}_{2n}\rightarrow\mathbf{f}\mathbf{t}_1\dots\mathbf{t}_n=\mathbf{f}\mathbf{t}_{n+1}\dots\mathbf{t}_{2n}$$

belongs to $\mathbf{\Psi}$.

(9) If $\mathbf{P}$ is an n-ary predicate symbol of $\mathscr{L}$ and $\mathbf{t}_1,\dots,\mathbf{t}_{2n}$ are terms then the formula

$$\mathbf{t}_1=\mathbf{t}_{n+1}\rightarrow\mathbf{t}_2=\mathbf{t}_{n+2}\rightarrow\dots\rightarrow\mathbf{t}_n=\mathbf{t}_{2n}\rightarrow\mathbf{P}\mathbf{t}_1\dots\mathbf{t}_n\rightarrow\mathbf{P}\mathbf{t}_{n+1}\dots\mathbf{t}_{2n}$$

belongs to $\mathbf{\Psi}$. (For $n=2$, $\mathbf{P}$ can be $=$.)

Throughout the present section we let $\mathbf{\Psi}$ be a fixed (but arbitrary) Hintikka set. Also, throughout this section we shall refer to the nine conditions of Def. 7.1 simply as (1), (2) etc. instead of 7.1.(1), 7.1.(2) etc.

Our aim is to show that $\mathbf{\Psi}$ is satisfiable; but this will require some preliminary work.

We begin by defining a binary relation E between $\mathscr{L}$-terms.

If $\mathscr{L}$ is a language without equality, we simply take E to be the identity relation. In other words, $\mathbf{s}E\mathbf{t}$ iff $\mathbf{s}$ and $\mathbf{t}$ are the same term.[1]

If $\mathscr{L}$ has equality, we define $\mathbf{s}E\mathbf{t}$ to mean that the equation $\mathbf{s}{=}\mathbf{t}$ belongs to $\boldsymbol{\Psi}$.

7.2. LEMMA. *E is an equivalence relation (i.e., it is reflexive, symmetric and transitive).*

PROOF. If $\mathscr{L}$ does not have equality, the assertion of our lemma is trivial. Therefore we assume that $\mathscr{L}$ does have equality.

The reflexivity of E follows at once from (7).

To show that E is symmetric, let us assume $\mathbf{s}E\mathbf{t}$; we shall show that $\mathbf{t}E\mathbf{s}$. By definition, $\mathbf{s}E\mathbf{t}$ means that $(\mathbf{s}{=}\mathbf{t}) \in \boldsymbol{\Psi}$. Now we make use of (9), where we take $n{=}2$ and choose $\mathbf{P}, \mathbf{t}_1, \mathbf{t}_2, \mathbf{t}_3, \mathbf{t}_4$ to be $=, \mathbf{s}, \mathbf{s}, \mathbf{t}, \mathbf{s}$ respectively. Then (9) says the formula

$$\mathbf{s}{=}\mathbf{t} \to \mathbf{s}{=}\mathbf{s} \to \mathbf{s}{=}\mathbf{s} \to \mathbf{t}{=}\mathbf{s}$$

belongs to $\boldsymbol{\Psi}$. It now follows from (3) that (at least) one of the formulas

$$\mathbf{s}{\neq}\mathbf{t}, \qquad \mathbf{s}{=}\mathbf{s} \to \mathbf{s}{=}\mathbf{s} \to \mathbf{t}{=}\mathbf{s}$$

must belong to $\boldsymbol{\Psi}$. But, since we already have $(\mathbf{s}{=}\mathbf{t}) \in \boldsymbol{\Psi}$, it follows from (1) that $(\mathbf{s}{\neq}\mathbf{t}) \notin \boldsymbol{\Psi}$. Thus the formula

$$\mathbf{s}{=}\mathbf{s} \to \mathbf{s}{=}\mathbf{s} \to \mathbf{t}{=}\mathbf{s}$$

belongs to $\boldsymbol{\Psi}$. Again, (3) implies that one of the formulas

$$\mathbf{s}{\neq}\mathbf{s}, \qquad \mathbf{s}{=}\mathbf{s} \to \mathbf{t}{=}\mathbf{s}$$

must belong to $\boldsymbol{\Psi}$. But, by (7), $(\mathbf{s}{=}\mathbf{s}) \in \boldsymbol{\Psi}$, hence, by (1), $(\mathbf{s}{\neq}\mathbf{s}) \notin \boldsymbol{\Psi}$. Thus the formula

$$\mathbf{s}{=}\mathbf{s} \to \mathbf{t}{=}\mathbf{s}$$

belongs to $\boldsymbol{\Psi}$. Using (3), (7) and (1) once more, we finally see that $(\mathbf{t}{=}\mathbf{s}) \in \boldsymbol{\Psi}$, i.e., $\mathbf{t}E\mathbf{s}$.

To show that E is transitive, we assume that $\mathbf{r}E\mathbf{s}$ and $\mathbf{s}E\mathbf{t}$; we shall show that $\mathbf{r}E\mathbf{t}$. Since E is symmetric, $\mathbf{r}E\mathbf{s}$ implies $\mathbf{s}E\mathbf{r}$, i.e., $(\mathbf{s}{=}\mathbf{r}) \in \boldsymbol{\Psi}$. Also,

[1] By the way, recall that in §4 of Ch. 1 it was stipulated that a first-order language without equality cannot have function symbols other than constants. Thus the only terms of such a language are the variables and the constants.

$s E t$ means that $(\mathbf{s}=\mathbf{t}) \in \mathbf{\Psi}$. We use (9) with $n=2$, choosing $\mathbf{P}, \mathbf{t}_1, \mathbf{t}_2, \mathbf{t}_3, \mathbf{t}_4$ to be $=, \mathbf{s}, \mathbf{s}, \mathbf{r}, \mathbf{t}$ respectively. Then (9) says that the formula

$$\mathbf{s}=\mathbf{r} \to \mathbf{s}=\mathbf{t} \to \mathbf{s}=\mathbf{s} \to \mathbf{r}=\mathbf{t}$$

belongs to $\mathbf{\Psi}$. From (3), (1) and the fact that $(\mathbf{s}=\mathbf{r}) \in \mathbf{\Psi}$ we infer that the formula

$$\mathbf{s}=\mathbf{t} \to \mathbf{s}=\mathbf{s} \to \mathbf{r}=\mathbf{t}$$

belongs to $\mathbf{\Psi}$. It follows from (3), (1) and the fact that $(\mathbf{s}=\mathbf{t}) \in \mathbf{\Psi}$ that the formula

$$\mathbf{s}=\mathbf{s} \to \mathbf{r}=\mathbf{t}$$

belongs to $\mathbf{\Psi}$. Using (3), (7) and (1) we finally get $(\mathbf{r}=\mathbf{t}) \in \mathbf{\Psi}$, i.e., $\mathbf{r}E\mathbf{t}$. ∎

By Lemma 7.2, the set of all $\mathscr{L}$-terms is partitioned into mutually exclusive E-classes. Terms $\mathbf{s}$ and $\mathbf{t}$ belong to the same E-class iff $\mathbf{s}E\mathbf{t}$. For each term $\mathbf{t}$ we put

$$[\mathbf{t}] = \{\mathbf{s} : \mathbf{s}E\mathbf{t}\},$$

i.e., $[\mathbf{t}]$ is the E-class to which $\mathbf{t}$ belongs. Clearly,

$$[\mathbf{s}] = [\mathbf{t}] \quad \text{iff} \quad \mathbf{s}E\mathbf{t}.$$

7.3. LEMMA. *Let* $\mathbf{t}_1, \ldots, \mathbf{t}_{2n}$ *be* $\mathscr{L}$*-terms such that* $[\mathbf{t}_i] = [\mathbf{t}_{n+i}]$ *for* $i=1,\ldots,n$. *Then:*

(a) *For any n-ary function symbol* $\mathbf{f}$ *of* $\mathscr{L}$,

$$[\mathbf{f}\mathbf{t}_1 \ldots \mathbf{t}_n] = [\mathbf{f}\mathbf{t}_{n+1} \ldots \mathbf{t}_{2n}].$$

(b) *For every n-ary predicate symbol* $\mathbf{P}$ *of* $\mathscr{L}$, *if* $\mathbf{P}\mathbf{t}_1 \ldots \mathbf{t}_n \in \mathbf{\Psi}$ *then* $\mathbf{P}\mathbf{t}_{n+1} \ldots \mathbf{t}_{2n} \in \mathbf{\Psi}$.

PROOF. If $\mathscr{L}$ is without equality then the assertion of our lemma is trivial. (In this case the assumption of the lemma is that $\mathbf{t}_i = \mathbf{t}_{n+i}$ for $i=1,\ldots,n$. Incidentally, here (a) is vacuous for $n>0$, because the only function symbols in such a language are the constants.)

We therefore assume that $\mathscr{L}$ is a language with equality. Then our assumption is that

$$(\mathbf{t}_i = \mathbf{t}_{n+i}) \in \mathbf{\Psi}, \quad i=1,\ldots,n.$$

We use (8) and apply (3) and (1) n times; we then get

$$(\mathbf{f}\mathbf{t}_1 \ldots \mathbf{t}_n = \mathbf{f}\mathbf{t}_{n+1} \ldots \mathbf{t}_{2n}) \in \mathbf{\Psi},$$

which proves (a). Similarly, (b) follows from (9) by applying (3) and (1) $n+1$ times. ∎

We define an $\mathscr{L}$-valuation σ as follows.

As universe we take the set U of all E-classes of $\mathscr{L}$-terms, i.e.,

$$U=\{[\mathbf{t}]:\ \ \mathbf{t} \text{ is an } \mathscr{L}\text{-term}\}.$$

If $\mathbf{f}$ is an n-ary function symbol of $\mathscr{L}$, we define the n-ary operation $\mathbf{f}^\sigma$ on U by putting

$$\mathbf{f}^\sigma([\mathbf{t}_1],\ldots,[\mathbf{t}_n])=[\mathbf{f}\mathbf{t}_1\ldots\mathbf{t}_n].$$

Note that this definition is legitimate by Lemma 7.3.(a), which guarantees that $[\mathbf{f}\mathbf{t}_1\ldots\mathbf{t}_n]$ is independent of the particular choice of "representatives" $\mathbf{t}_1,\ldots,\mathbf{t}_n$ for the respective E-classes $[\mathbf{t}_1],\ldots,[\mathbf{t}_n]$.

If $\mathbf{P}$ is an n-ary extralogical predicate symbol of $\mathscr{L}$, we define the n-ary relation $\mathbf{P}^\sigma$ on U by putting

$$\langle[\mathbf{t}_1],\ldots,[\mathbf{t}_n]\rangle\in\mathbf{P}^\sigma \quad\text{iff}\quad \mathbf{P}\mathbf{t}_1\ldots\mathbf{t}_n\in\mathbf{\Psi}.$$

Again, this definition is legitimate by Lemma 7.3.(b).

Finally, for each variable $\mathbf{x}$ we put

$$\mathbf{x}^\sigma=[\mathbf{x}].$$

This completes the definition of σ. We note that the cardinality of the universe U of σ is not greater than the cardinality of the set of all $\mathscr{L}$-terms.

7.4. LEMMA. *For each $\mathscr{L}$-term* $\mathbf{t}$,

$$\mathbf{t}^\sigma=[\mathbf{t}].$$

PROOF. By induction on $\deg\mathbf{t}$. If $\mathbf{t}$ is a variable $\mathbf{x}$, then $\mathbf{x}^\sigma=[\mathbf{x}]$ by the definition of σ. If $\mathbf{t}=\mathbf{f}\mathbf{t}_1\ldots\mathbf{t}_n$, then $\deg\mathbf{t}_i<\deg\mathbf{t}$ for $i=1,\ldots,n$ and we have

$$\begin{aligned}
\mathbf{t}^\sigma&=(\mathbf{f}\mathbf{t}_1\ldots\mathbf{t}_n)^\sigma\\
&=\mathbf{f}^\sigma(\mathbf{t}_1{}^\sigma,\ldots,\mathbf{t}_n{}^\sigma) && \text{(by the BSD)}\\
&=\mathbf{f}^\sigma([\mathbf{t}_1],\ldots,[\mathbf{t}_n]) && \text{(by ind. hyp.)}\\
&=[\mathbf{f}\mathbf{t}_1\ldots\mathbf{t}_n] && \text{(by def. of } \mathbf{f}^\sigma)\\
&=[\mathbf{t}].
\end{aligned}$$
∎

We can now prove the main result of this section:

7.5. THEOREM. $\sigma\models\mathbf{\Psi}$.

PROOF. We show by induction on deg φ that

(a) if $\varphi \in \Psi$ then $\varphi^\sigma = \top$,

(b) if $\neg\varphi \in \Psi$ then $\varphi^\sigma = \bot$.

Case 1: $\varphi = \mathbf{P}\mathbf{t}_1 \ldots \mathbf{t}_n$, where $\mathbf{P}$ is an extralogical predicate symbol.

If $\mathbf{P}\mathbf{t}_1 \ldots \mathbf{t}_n \in \Psi$,
then $\langle [\mathbf{t}_1], \ldots, [\mathbf{t}_n] \rangle \in \mathbf{P}^\sigma$, (by def. of $\mathbf{P}^\sigma$)
then $\langle \mathbf{t}_1^\sigma, \ldots, \mathbf{t}_n^\sigma \rangle \in \mathbf{P}^\sigma$, (by Lemma 7.4)
then $(\mathbf{P}\mathbf{t}_1 \ldots \mathbf{t}_n)^\sigma = \top$. (by BSD)

On the other hand,

if $\neg\mathbf{P}\mathbf{t}_1 \ldots \mathbf{t}_n \in \Psi$,
then $\mathbf{P}\mathbf{t}_1 \ldots \mathbf{t}_n \notin \Psi$, (by (1))
then $\langle [\mathbf{t}_1], \ldots, [\mathbf{t}_n] \rangle \notin \mathbf{P}^\sigma$, (by def. of $\mathbf{P}^\sigma$)
then $\langle \mathbf{t}_1^\sigma, \ldots, \mathbf{t}_n^\sigma \rangle \notin \mathbf{P}^\sigma$, (by Lemma 7.4)
then $(\mathbf{P}\mathbf{t}_1 \ldots \mathbf{t}_n)^\sigma = \bot$. (by BSD)

Case $1^=$: $\varphi = (\mathbf{s} = \mathbf{t})$.

If $(\mathbf{s} = \mathbf{t}) \in \Psi$,
then $[\mathbf{s}] = [\mathbf{t}]$,
then $\mathbf{s}^\sigma = \mathbf{t}^\sigma$, (by Lemma 7.4)
then $(\mathbf{s} = \mathbf{t})^\sigma = \top$. (by BSD)

On the other hand,

if $(\mathbf{s} \neq \mathbf{t}) \in \Psi$,
then $(\mathbf{s} = \mathbf{t}) \notin \Psi$, (by (1))
then $[\mathbf{s}] \neq [\mathbf{t}]$,
then $\mathbf{s}^\sigma \neq \mathbf{t}^\sigma$, (by Lemma 7.4)
then $(\mathbf{s} = \mathbf{t})^\sigma = \bot$. (by BSD)

Case 2: $\varphi = \neg\alpha$. Here deg $\alpha <$ deg φ.

If $\neg\alpha \in \Psi$,
then $\alpha^\sigma = \bot$, (by ind. hyp.)
then $(\neg\alpha)^\sigma = \top$. (by BSD)

On the other hand,

if $\neg\neg\alpha \in \Psi$,
then $\alpha \in \Psi$, (by (2))
then $\alpha^\sigma = \top$, (by ind. hyp.)
then $(\neg\alpha)^\sigma = \bot$. (by BSD)

Case 3: $\varphi = \alpha \to \beta$. Here $\deg \alpha$ and $\deg \beta$ are $< \deg \varphi$.

If $\alpha \to \beta \in \Psi$,
then $\neg\alpha \in \Psi$ or $\beta \in \Psi$, (by (3))
then $\alpha^\sigma = \bot$ or $\beta^\sigma = \top$, (by ind. hyp.)
then $(\alpha \to \beta)^\sigma = \top$. (by BSD)

On the other hand,

if $\neg(\alpha \to \beta) \in \Psi$,
then $\alpha \in \Psi$ and $\neg\beta \in \Psi$, (by (4))
then $\alpha^\sigma = \top$ and $\beta^\sigma = \bot$, (by ind. hyp.)
then $(\alpha \to \beta)^\sigma = \bot$. (by BSD)

Case 4: $\varphi = \forall x \alpha$. Here $\deg \alpha < \deg \varphi$.

If $\forall x \alpha \in \Psi$,
then $\alpha(x/t) \in \Psi$ for every term t, (by (5))
then $\alpha(x/t)^\sigma = \top$ for every t, (by ind. hyp.)
then $\alpha^{\sigma(x/t^\sigma)} = \top$ for every t, (by 3.10)
then $\alpha^{\sigma(x/[t])} = \top$ for every t, (by 7.4)
then $\alpha^{\sigma(x/u)} = \top$ for every $u \in U$, (by def. of U)
then $(\forall x \alpha)^\sigma = \top$. (by BSD)

On the other hand,

if $\neg\forall x \alpha \in \Psi$,
then $\neg\alpha(x/t) \in \Psi$ for some t, (by (6))
then $\alpha(x/t)^\sigma = \bot$ for some t, (by ind. hyp.)
then $\alpha^{\sigma(x/t^\sigma)} = \bot$ for some t, (by 3.10)
then $\alpha^{\sigma(x/[t])} = \bot$ for some t, (by 7.4)
then $\alpha^{\sigma(x/u)} = \bot$ for some $u \in U$, (by def. of U)
then $(\forall x \alpha)^\sigma = \bot$. (by BSD) ∎

Thus every Hintikka set in $\mathscr{L}$ is satisfied by a valuation whose universe has cardinality not greater than the cardinality of the set of $\mathscr{L}$-terms.

*§ 8. Completeness of first-order tableaux

In this section we shall describe a procedure by which one can systematically search for a confutation of any given finite set Φ of formulas. Moreover, we shall show that if this systematic search does not yield a confutation of Φ, then Φ is satisfiable, so that by Thm. 4.1 Φ cannot be confuted.

For any set Φ of formulas we define $\mathscr{L}_\Phi$ as the "poorest" first-order language such that Φ is still in $\mathscr{L}_\Phi$. More precisely, the extralogical symbols of $\mathscr{L}_\Phi$ are exactly those extralogical symbols that actually occur in Φ. If $=$ occurs in Φ, or if some function symbol (other than a constant) occurs in Φ, then we let $\mathscr{L}_\Phi$ be a language with equality. Clearly, $\mathscr{L}_\Phi$ is a *sublanguage* of our original language $\mathscr{L}$, in the (obvious) sense that every $\mathscr{L}_\Phi$-expression is also an $\mathscr{L}$-expression.

For the rest of this section we let Φ be a fixed (but arbitrary) *finite* set of formulas.

The set of symbols of $\mathscr{L}_\Phi$ is denumerable: it consists of finitely many function symbols, finitely many predicate symbols (possibly including $=$), two connectives, the universal quantifier, and the variables $\mathbf{v}_1, \mathbf{v}_2, \ldots$.

Since every $\mathscr{L}_\Phi$-formula is a string (i.e., a finite sequence) of $\mathscr{L}_\Phi$-symbols, we can enumerate these formulas by some effective rule. (For example, suppose that the symbols of $\mathscr{L}_\Phi$, excluding the variables, are k in number. We assign to them "lexical numbers" $1,\ldots,k$ respectively. And to each variable $\mathbf{v}_i$ we assign the "lexical number" $k+i$. Let Δ_n be the set of all formulas whose length is at most n and in which there occur only symbols whose lexical number is at most n. Then Δ_n is clearly finite and we can arrange it in lexicographic order. Finally, we can arrange all the Δ_n among themselves in increasing order of n. The resulting enumeration of $\mathscr{L}_\Phi$-formulas is *effective*, in the sense that for any m we can find in a finite number of steps the m-th formula in that enumeration.)

For the rest of this section, we fix some particular effective enumeration

$$\{\varphi_n \colon n=0,1,\ldots\}$$

of all $\mathscr{L}_\Phi$-formulas. Without loss of generality *we shall assume that each $\mathscr{L}_\Phi$-formula α is repeated infinitely often, i.e., $\alpha=\varphi_n$ for infinitely many n.* (For the particular enumeration suggested above, this is actually the case. But even if this were not so, we could always replace the enumeration $\varphi_0, \varphi_1, \varphi_2$ etc. by $\varphi_0, \varphi_1, \varphi_0, \varphi_1, \varphi_2, \varphi_0, \varphi_1, \varphi_2, \varphi_3$ etc.)

Similarly, we fix some effective enumeration

$$\{\mathbf{t}_n \colon n=0,1,\ldots\}$$

of all $\mathscr{L}_\Phi$-terms. (Here it does not matter how many times each term is repeated.)

Finally, if $\mathscr{L}_\Phi$ has $=$, we fix some effective enumeration

$$\{\psi_n \colon n=0,1,\ldots\}$$

of all $\mathscr{L}_\Phi$-formulas which can be introduced into a tableau by the equality rules (i.e., all formulas of the forms mentioned in clauses (7), (8), (9) of Def. 7.1, except that now we confine ourselves to $\mathscr{L}_\Phi$).

For every natural number n, we now prescribe how to construct a tableau T_n for $\mathbf{\Phi}$. We proceed by recursion on n.

T_0 is the tableau having $\mathbf{\Phi}$ as its sole node.

Suppose T_n has been constructed. We first construct a tableau $T_n{}^*$ as follows:

(a) If φ_n occurs in T_n and if φ_n can be used for any one of the four rules $\neg\neg$, $\rightarrow$, $\neg\rightarrow$ and $\neg\forall$, we extend every branch of T_n having φ_n by applying to it the appropriate rule. (In the case of the $\neg\forall$-rule we use as critical variable the first variable which can be so used.) The resulting tableau is $T_n{}^*$.

(b) If φ_n occurs in T_n and has the form $\forall \mathbf{x}\alpha$, we extend $n+1$ times each branch of T_n having φ_n by adjoining successively the nodes $\{\alpha(\mathbf{x}/\mathbf{t}_0)\}$,… …,$\{\alpha(\mathbf{x}/\mathbf{t}_n)\}$. The resulting tableau is $T_n{}^*$.

(c) In the remaining case — where φ_n is not in T_n, or where φ_n is an atomic formula or a negation of an atomic formula — we let $T_n{}^*$ be T_n.

Having got $T_n{}^*$, we now construct T_{n+1}. If $\mathscr{L}_\Phi$ does not have $=$, we let T_{n+1} be $T_n{}^*$. If $\mathscr{L}_\Phi$ has $=$, we obtain T_{n+1} by adjoining the node $\{\psi_n\}$ to each branch of $T_n{}^*$.

If for some n all branches of T_n are closed, then, by Thm. 4.1, $\mathbf{\Phi}$ is unsatisfiable.[1]

On the other hand, suppose that each T_n has at least one open branch. By construction, if $n \geqslant m$ then every (open) branch of T_n is an extension of — i.e., obtained by zero or more successive extensions from — one and only one (open) branch of T_m. We now prove:

8.1. LEMMA. *If each* T_n *has an open branch, then there exists a sequence* $\{B_n: n=0,1,\ldots\}$ *such that for all* n B_n *is an open branch of* T_n *and* B_{n+1} *is an extension of* B_n.

PROOF. Let B be a branch of T_n. We say that B is *good* if for infinitely many $m>n$ T_m has an open branch which is an extension of B. (It follows of course that B must be open.)

If B is a branch of T_n, there are finitely many branches of T_{n+1} extending B, say $B^1,\ldots,B^k$. If B is good, it follows that at least one of these B^i

[1] The fact that here we are working with $\mathscr{L}_\Phi$ rather than $\mathscr{L}$ makes no difference; see 2.7.

must be good as well, because each open branch of T_m (where $m>n$) which extends B must also be an extension of some B^i.

Now, T_0 has just one branch; call it B_0. Every branch of every T_n is an extension of B_0. Since each T_n has an open branch, B_0 is good. Therefore at least one branch of T_1 which extends B_0 must be good as well. Choose such a branch and call it B_1. Similarly, T_2 has a good branch B_2 extending B_1, etc. ∎

Lemma 8.1 is (a version of) the result commonly known as *Koenig's Tree Lemma.* The reader should note that the proof was *not* constructive: having chosen a good B_n we can only show that one of the extensions of B_n in T_{n+1} must be good. But we have no way of telling *which* of these extensions is good — because for this we would have to examine *all* T_m with $m>n$. However, the proof of Lemma 8.1 does *not* require the axiom of choice: the branches of each T_n can be ordered "from left to right", and for each n we can take B_{n+1} as the *leftmost* branch of T_{n+1} which is a good extension of B_n.

The main result of this section is contained in:

8.2. THEOREM. *If each T_n has an open branch, then Φ is satisfied by a valuation with countable (i.e., finite or denumerable) universe.*

PROOF. Let $\{B_n: n=0,1,\ldots\}$ be a sequence as described in Lemma 8.1. Let Φ_n be the set of all formulas of B_n. In particular, $\Phi_0=\Phi$. Now let Ψ be the union of all the Φ_n. From the construction of the T_n and the properties of the B_n it can readily be verified that Ψ is a Hintikka set in $\mathscr{L}_\Phi$. Thus, by Thm. 7.5, Ψ is satisfied by a valuation with countable universe. (Recall that the set of $\mathscr{L}_\Phi$-terms is denumerable.) But $\Phi=\Phi_0\subseteq\Psi$. ∎

8.3. COROLLARY. *A finite set of formulas can be confuted iff that set is unsatisfiable. Every satisfiable finite set of formulas is satisfied by some valuation whose universe is a countable set.*

PROOF. Thms. 4.1 and 8.2. ∎

8.4. REMARK. The above results do *not* provide us with an effective method for deciding whether or not a given finite set Φ is satisfiable. The definition of T_n *is* constructive, and we may proceed to construct T_0, T_1, T_2 etc., knowing that if Φ is unsatisfiable then *sooner or later* we shall find a confutation T_n. But there is in general no way of telling in advance how large is the first n for which T_n will be a confutation. Thus, if we have constructed T_m for some huge m and find that it has an open branch, there are still two possibilities: either $\mathbf{\Phi}$ is in fact satisfiable, or $\mathbf{\Phi}$ is unsatisfiable but the first n for which T_n is a confutation is (perhaps) much larger than our m.

Contrast this with the results of §9 of Ch. 1. There we could tell in advance how far we have to go to obtain an exhausted propositional tableau for $\mathbf{\Phi}$ and thus to decide whether or not $\mathbf{\Phi}$ is satisfied by some truth valuation.

8.5. PROBLEM. Show that if $\mathbf{\Phi}$ is finite then $\mathbf{\Phi} \models \boldsymbol{\alpha}$ iff $\mathbf{\Phi}, \neg\boldsymbol{\alpha}$ can be confuted.

8.6. PROBLEM Using Cor. 8.3 find an alternative proof[1] for the Elimination Lemma 6.7.

8.7. PROBLEM. Let $\mathbf{\Phi}$ be a finite set of formulas in which $\neg$ does not occur. Show that in any tableau for $\mathbf{\Phi}$ there must be a branch in which $\neg$ does not occur. Hence show that $\mathbf{\Phi}$ is satisfiable.

An $\forall$-*formula* is a formula of the form

$$\forall \mathbf{x}_1 \ldots \forall \mathbf{x}_k \boldsymbol{\beta},$$

where $k \geqslant 0$ and $\boldsymbol{\beta}$ is quantifier-free (i.e., $\forall$ does not occur in $\boldsymbol{\beta}$).

8.8. PROBLEM. Let $\mathscr{L}$ be a language without equality, and let $\mathbf{\Phi}$ be a finite set of $\forall$-formulas in $\mathscr{L}$. Let $\{\mathbf{t}_1,\ldots,\mathbf{t}_m\}$ be a finite non-empty set of $\mathscr{L}$-terms (i.e., variables and constants) which contains every variable free in $\mathbf{\Phi}$ and every constant occurring in $\mathbf{\Phi}$. Let $\mathbf{\Phi}^*$ be the smallest set of formulas such that $\mathbf{\Phi} \subseteq \mathbf{\Phi}^*$ and such that whenever $\forall \mathbf{x} \boldsymbol{\alpha} \in \mathbf{\Phi}^*$ then $\boldsymbol{\alpha}(\mathbf{x}/\mathbf{t}_i) \in \mathbf{\Phi}^*$ for $i=1,\ldots,m$. ($\mathbf{\Phi}^*$ can clearly be constructed from $\mathbf{\Phi}$ in a finite number of steps.) Let $\mathbf{\Phi}_0$ be the set of all quantifier-free formulas belonging to $\mathbf{\Phi}^*$. Show that $\mathbf{\Phi}$ is unsatisfiable iff a closed *propositional* tableau can be constructed for $\mathbf{\Phi}_0$. (Let T be a propositional tableau for $\mathbf{\Phi}_0$ which is exhausted in the sense of §9 of Ch. 1. If T is a propositional confutation, then $\mathbf{\Phi}$ is easily seen to be unsatisfiable. Otherwise, choose an open branch of T and let $\mathbf{\Psi}$ be the set of all formulas of that branch. Define a valuation σ as follows. Let the universe U of σ be any set of m objects, $U=\{u_1,\ldots,u_m\}$. Put $\mathbf{t}_i^\sigma = u_i$ for $i=1,\ldots,m$. If $\mathbf{P}$ is an n-ary predicate symbol of $\mathscr{L}$, let

$$\langle u_{i_1},\ldots,u_{i_n}\rangle \in \mathbf{P}^\sigma \quad \text{iff} \quad \mathbf{P}\mathbf{t}_{i_1}\ldots\mathbf{t}_{i_n} \in \mathbf{\Psi}.$$

Show that $\sigma \models \mathbf{\Phi}$.)

[1] However, note that the direct constructive proof in §6 is more informative: given confutations of $\mathbf{\Phi}, \boldsymbol{\delta}$ and $\mathbf{\Phi}, \neg\boldsymbol{\delta}$ that proof gave an *explicit* construction of a confutation of $\mathbf{\Phi}$. Using Cor. 8.3 (or Thm. 8.2) we can only say that, for *some n*, T_n will be a confutation, without being able to estimate how large such an n might be.

An $\exists$-*formula* is a formula of the form

$$\exists \mathbf{x}_1 \ldots \exists \mathbf{x}_k \boldsymbol{\beta},$$

where $k \geqq 0$ and $\boldsymbol{\beta}$ is quantifier-free.

8.9 PROBLEM. Let $\boldsymbol{\Phi}$ be a finite set of $\exists$-formulas in a language without equality. Show how to find out, in a finite number or steps, whether or not $\boldsymbol{\Phi}$ is satisfiable. (Consider a variant $\boldsymbol{\Phi}'$ of $\boldsymbol{\Phi}$ such that no bound variable of any formula occurs in any other formula of $\boldsymbol{\Phi}'$. Let $\boldsymbol{\Phi}^*$ be obtained from $\boldsymbol{\Phi}'$ by removing all existential quantifiers, leaving only the quantifier-free parts. Show that $\boldsymbol{\Phi}$ is unsatisfiable iff a *propositional* confutation can be constructed for $\boldsymbol{\Phi}^*$.)

*§ 9. Prenex and Skolem forms

A formula is said to be *prenex* if it has the form

$$\mathsf{Q}_1\mathbf{x}_1\mathsf{Q}_2\mathbf{x}_2 \ldots \mathsf{Q}_k\mathbf{x}_k\boldsymbol{\beta},$$

where $k \geqslant 0$, and for each i, Q_i is $\forall$ or $\exists$, and $\boldsymbol{\beta}$ is quantifier-free. In this connection, the string $\mathsf{Q}_1\mathbf{x}_1 \ldots \mathsf{Q}_k\mathbf{x}_k$ is called the *prefix* and $\boldsymbol{\beta}$ is called the *matrix*. A *normal* prenex formula is a prenex formula such that the variables $\mathbf{x}_1, \ldots, \mathbf{x}_k$ in its prefix are distinct, and all of them occur in the matrix $\boldsymbol{\beta}$.

A *prenex form for* $\boldsymbol{\alpha}$ is a prenex formula logically equivalent to $\boldsymbol{\alpha}$.

The following problem yields a procedure whereby we can find a prenex form for any given formula.

9.1. PROBLEM. Let $\boldsymbol{\varphi}$ be a formula which contains $n+1$ quantifiers. Show how to find a formula of the form $\mathsf{Q}\mathbf{x}\boldsymbol{\psi}$ — where Q is $\forall$ or $\exists$, and $\boldsymbol{\psi}$ contains only n quantifiers — which is logically equivalent to $\boldsymbol{\varphi}$. (Proceed by induction on deg $\boldsymbol{\varphi}$. Clearly, $\boldsymbol{\varphi}$ cannot be atomic. The cases where $\boldsymbol{\varphi}$ is a negation formula or a universal formula are easy. If $\boldsymbol{\varphi}$ is $\boldsymbol{\alpha} \rightarrow \boldsymbol{\beta}$, then by the induction hypothesis we may assume that $\boldsymbol{\alpha}$ has the form $\mathsf{Q}\mathbf{x}\boldsymbol{\gamma}$ or $\boldsymbol{\beta}$ has the form $\mathsf{Q}\mathbf{y}\boldsymbol{\delta}$, and by alphabetic change we can arrange that $\mathbf{x}$ is not free in $\boldsymbol{\beta}$ and $\mathbf{y}$ is not free in $\boldsymbol{\alpha}$. Then use Prob. 4.3.)

Using recursion on n, we can now find a prenex form for $\boldsymbol{\varphi}$.

Note that by Prob. 2.4. we can always delete quantifiers in a prenex form for $\boldsymbol{\varphi}$, until a *normal* prenex form for $\boldsymbol{\varphi}$ is obtained.

Let $\mathscr{L}'$ be a first-order language which is an extension of $\mathscr{L}$ (i.e., every $\mathscr{L}$-symbol is also an $\mathscr{L}'$-symbol). Then for each $\mathscr{L}'$-valuation σ' there is a unique $\mathscr{L}$-valuation σ which agrees with σ' on all $\mathscr{L}$-symbols.[1] We say that σ is *the* $\mathscr{L}$*-reduction* of σ'. Also, σ' is said to be an $\mathscr{L}'$*-expansion* of σ. (Note that in general an $\mathscr{L}$-valuation has more than one $\mathscr{L}'$-expansion.)

Let $\boldsymbol{\alpha}$ be an $\mathscr{L}$-formula. By an S*-form*[2] *for* $\boldsymbol{\alpha}$ we mean a formula $\boldsymbol{\alpha}'$ in some extension $\mathscr{L}'$ of $\mathscr{L}$, such that any $\mathscr{L}$-valuation σ satisfies $\boldsymbol{\alpha}$ iff σ has an $\mathscr{L}'$-expansion σ' satisfying $\boldsymbol{\alpha}'$.

It is clear that if $\boldsymbol{\alpha}'$ is an S-form for $\boldsymbol{\alpha}$, then $\boldsymbol{\alpha}'$ is satisfiable iff $\boldsymbol{\alpha}$ is. Also, if $\boldsymbol{\alpha}''$ is an S-form for $\boldsymbol{\alpha}'$ (in some extension $\mathscr{L}''$ of $\mathscr{L}'$) then $\boldsymbol{\alpha}''$ is an S-form for $\boldsymbol{\alpha}$.

By a V*-form*[3] *for* $\boldsymbol{\alpha}$ we mean a formula $\boldsymbol{\alpha}'$ in some extension $\mathscr{L}'$ of $\mathscr{L}$, such that any $\mathscr{L}$-valuation σ satisfies $\boldsymbol{\alpha}$ iff every $\mathscr{L}'$-expansion σ' of σ satisfies $\boldsymbol{\alpha}'$. (It follows, in particular, that $\boldsymbol{\alpha}'$ is logically valid iff $\boldsymbol{\alpha}$ is.)

9.2. PROBLEM. Show that $\boldsymbol{\alpha}'$ is a V-form for $\boldsymbol{\alpha}$ iff $\neg\boldsymbol{\alpha}'$ is an S-form for $\neg\boldsymbol{\alpha}$. Similarly, $\boldsymbol{\alpha}'$ is an S-form for $\boldsymbol{\alpha}$ iff $\neg\boldsymbol{\alpha}'$ is a V-form for $\neg\boldsymbol{\alpha}$.

The rest of this section will be concerned with finding various kinds of S-forms and V-forms for a given formula.

9.3. LEMMA. *Let* $\boldsymbol{\alpha}$ *be an* $\mathscr{L}$*-formula and let* $\mathbf{x}_1,\ldots,\mathbf{x}_k,\mathbf{y}$ *be* $k+1$ *different variables, where* $k\geq 0$. *Let* $\mathbf{P}$ *be a* $(k+1)$*-ary extralogical predicate symbol which is not in* $\mathscr{L}$. *Then*

(1) $$\forall\mathbf{x}_1\ldots\forall\mathbf{x}_k\,[\exists\mathbf{y}\mathbf{P}\mathbf{x}_1\ldots\mathbf{x}_k\mathbf{y}\wedge\forall\mathbf{y}(\mathbf{P}\mathbf{x}_1\ldots\mathbf{x}_k\mathbf{y}\rightarrow\boldsymbol{\alpha})]$$

is an S*-form for*

(2) $$\forall\mathbf{x}_1\ldots\forall\mathbf{x}_k\exists\mathbf{y}\boldsymbol{\alpha}.$$

PROOF. We take $\mathscr{L}'$ to be any extension of $\mathscr{L}$ such that $\mathbf{P}$ is an $\mathscr{L}'$-symbol.

Let σ be an $\mathscr{L}$-valuation with universe U. First, suppose that σ satisfies formula (2). Then it follows from the BSD that for all $u_1,\ldots,u_k$ in U there exists some v in U such that

$$\sigma(\mathbf{x}_1/u_1)\ldots(\mathbf{x}_k/u_k)(\mathbf{y}/v)\models\boldsymbol{\alpha}.$$

[1] Thus σ and σ' have the same universe; and $\mathbf{x}^\sigma=\mathbf{x}^{\sigma'}$ for every variable $\mathbf{x}$, $\mathbf{f}^\sigma=\mathbf{f}^{\sigma'}$ for every function symbol $\mathbf{f}$ of $\mathscr{L}$, and $\mathbf{P}^\sigma=\mathbf{P}^{\sigma'}$ for every extralogical predicate symbol $\mathbf{P}$ of $\mathscr{L}$.

[2] This is short for *satisfiability form.*

[3] Short for *validity form.*

Let σ' be an expansion of σ such that

$$\langle u_1,\ldots,u_k,v\rangle \in \mathbf{P}^{\sigma'} \quad \text{iff} \quad \sigma(\mathbf{x}_1/u_1)\ldots(\mathbf{x}_k/u_k)(\mathbf{y}/v) \models \boldsymbol{\alpha}.$$

Then for all $u_1,\ldots,u_k$ in U there is some v in U such that

$$\sigma'(\mathbf{x}_1/u_1)\ldots(\mathbf{x}_k/u_k)(\mathbf{y}/v) \models \mathbf{P}\mathbf{x}_1\ldots\mathbf{x}_k\mathbf{y} \tag{3}$$

and for each such v we also have

$$\sigma'(\mathbf{x}_1/u_1)\ldots(\mathbf{x}_k/u_k)(\mathbf{y}/v) \models \boldsymbol{\alpha}. \tag{4}$$

It follows from the BSD that σ' satisfies formula (1).

Conversely, if σ' is *any* expansion of σ and σ' satisfies formula (1), then for all $u_1,\ldots,u_k$ in U there is some v in U such that (3) holds, and, for every such v, (4) holds as well. It follows that σ' satisfies formula (2). But since σ agrees with σ' on all $\mathscr{L}$-symbols, σ too satisfies formula (2). ∎

An $\forall\exists$-*formula* is a prenex formula in whose prefix all universal quantifiers precede all existential quantifiers.

By a PS-*Skolem form*[1] *for* an $\mathscr{L}$-formula $\boldsymbol{\varphi}$ we mean a formula $\boldsymbol{\psi}$ such that

(a) $\boldsymbol{\psi}$ is in a language $\mathscr{L}'$ obtained from $\mathscr{L}$ by adding new extralogical predicate symbols only,

(b) $\boldsymbol{\psi}$ is an $\forall\exists$-formula,

(c) $\boldsymbol{\psi}$ is an S-form for $\boldsymbol{\varphi}$.

$\exists\forall$-*formula* is defined like $\forall\exists$-formula, but with the roles of $\forall$ and $\exists$ reversed. Also, PV-*Skolem form* is defined like PS-Skolem form, but with "$\forall\exists$" replaced by "$\exists\forall$" and "S-form" replaced by "V-form".

9.4. THEOREM. *Given any* $\mathscr{L}$-*formula* $\boldsymbol{\varphi}$, *we can find a* PS-*Skolem form for* $\boldsymbol{\varphi}$.

PROOF. By Prob. 9.1, we can assume without loss of generality that $\boldsymbol{\varphi}$ is a normal prenex formula. If $\boldsymbol{\varphi}$ is an $\forall\exists$-formula, then $\boldsymbol{\varphi}$ is a PS-Skolem form for itself, and we are through.

Otherwise $\boldsymbol{\varphi}$ has the form

$$\forall\mathbf{x}_1\ldots\forall\mathbf{x}_k\exists\mathbf{y}\mathbf{Q}_1\mathbf{z}_1\ldots\mathbf{Q}_m\mathbf{z}_m\boldsymbol{\beta},$$

where $k\geqslant 0$, $m>0$, each of the $\mathbf{Q}_j$ is $\forall$ or $\exists$ and at least one $\mathbf{Q}_j$ is $\forall$, $\boldsymbol{\beta}$ is quantifier-free and $\mathbf{x}_1,\ldots,\mathbf{x}_k,\mathbf{y},\mathbf{z}_1,\ldots,\mathbf{z}_m$ are all different.

[1] Here "P" is short for "*predicate*" and "S" (as before) is short for "*satisfiability*".

For brevity we put

$$\boldsymbol{\alpha} = \mathbf{Q}_1\mathbf{z}_1 \ldots \mathbf{Q}_m\mathbf{z}_m\boldsymbol{\beta}.$$

Using Lemma 9.3 we obtain for $\boldsymbol{\varphi}$ the S-form

(1) $$\forall \mathbf{x}_1 \ldots \forall \mathbf{x}_k [\exists \mathbf{y}\mathbf{P}\mathbf{x}_1 \ldots \mathbf{x}_k\mathbf{y} \wedge \forall \mathbf{y}(\mathbf{P}\mathbf{x}_1 \ldots \mathbf{x}_k\mathbf{y} \rightarrow \boldsymbol{\alpha})],$$

where **P** is an extralogical predicate symbol not belonging to $\mathscr{L}$. We now choose a variable **z** which does not occur in $\boldsymbol{\varphi}$ and replace $\exists \mathbf{y}\mathbf{P}\mathbf{x}_1 \ldots \mathbf{x}_k\mathbf{y}$ in (1) by the logically equivalent formula $\exists \mathbf{z}\mathbf{P}\mathbf{x}_1 \ldots \mathbf{x}_k\mathbf{z}$. Next, to the formula $\mathbf{P}\mathbf{x}_1 \ldots \mathbf{x}_k\mathbf{y} \rightarrow \boldsymbol{\alpha}$ in (1) we apply the last two clauses of Prob. 4.3. We repeat this m times, until $\forall \mathbf{y}(\mathbf{P}\mathbf{x}_1 \ldots \mathbf{x}_k\mathbf{y} \rightarrow \boldsymbol{\alpha})$ is replaced by the logically equivalent

$$\forall \mathbf{y}\mathbf{Q}_1\mathbf{z}_1 \ldots \mathbf{Q}_k\mathbf{z}_k(\mathbf{P}\mathbf{x}_1 \ldots \mathbf{x}_k\mathbf{y} \rightarrow \boldsymbol{\beta}).$$

We now apply clauses (e) and (f) of Prob. 4.3 with the quantifiers

$$\forall \mathbf{y}, \mathbf{Q}_1\mathbf{z}_1, \ldots, \mathbf{Q}_m\mathbf{z}_m, \exists \mathbf{z},$$

in this order. At the end of all this, (1) is transformed into the logically equivalent formula

(2) $$\forall \mathbf{x}_1 \ldots \forall \mathbf{x}_k \forall \mathbf{y}\mathbf{Q}_1\mathbf{z}_1 \ldots \mathbf{Q}_m\mathbf{z}_m \exists \mathbf{z} [\mathbf{P}\mathbf{x}_1 \ldots \mathbf{x}_k\mathbf{z} \wedge (\mathbf{P}\mathbf{x}_1 \ldots \mathbf{x}_k\mathbf{y} \rightarrow \boldsymbol{\beta})].$$

Since (2) is logically equivalent to (1), (2) is an S-form for $\boldsymbol{\varphi}$. If (2) is an $\forall\exists$-formula, we are through. If not, we can apply to (2) the same process again. Each time, a new predicate symbol is introduced, an existential quantifier in the middle of the prefix is replaced by a universal quantifier, and a fresh existential quantifier crops up at the *end* of the prefix.

After a finite number of iterations, we clearly get a PS-Skolem form for $\boldsymbol{\varphi}$. ∎

9.5. Problem. Show that we can find a PV-Skolem form for any $\mathscr{L}$-formula $\boldsymbol{\varphi}$. (Use Prob. 9.2.)

9.6. Problem. Let $\boldsymbol{\alpha}$ be an $\mathscr{L}$-formula. For $k \geqslant 0$, let $\mathbf{x}_1, \ldots, \mathbf{x}_k, \mathbf{y}$ be distinct variables, and let **f** be a k-ary function symbol not belonging to $\mathscr{L}$. Show (using the axiom of choice) that

$$\forall \mathbf{x}_1 \ldots \forall \mathbf{x}_k [\boldsymbol{\alpha}(\mathbf{y}/\mathbf{f}\mathbf{x}_1 \ldots \mathbf{x}_k)]$$

is an S-form for $\forall \mathbf{x}_1 \ldots \forall \mathbf{x}_k \exists \mathbf{y}\boldsymbol{\alpha}$.

We say that ψ is an FS-*Skolem form*[1] *for* an $\mathscr{L}$-formula φ if

(a) ψ is in a language $\mathscr{L}'$ obtained from $\mathscr{L}$ by adding new function symbols only,

(b) ψ is an $\forall$-formula,

(c) ψ is an S-form for φ.

FV-*Skolem form* is defined similarly, but with "$\exists$-formula" and "V-form" instead of "$\forall$-formula" and "S-form".

9.7. PROBLEM. Show that, given any $\mathscr{L}$-formula φ, we can find an FS-Skolem form and an FV-Skolem form for φ.

*§ 10. Elimination of function symbols

In this section we shall show that function symbols are not indispensable in a first-order language with equality. Roughly speaking, whatever can be expressed using an n-ary function symbol can also be expressed using an $(n+1)$-ary predicate symbol.

It will be convenient to use the following abbreviation.

10.1. DEFINITION. If $\boldsymbol{\alpha}$ is a formula and $\mathbf{x}$ is a variable, then

$$\exists!\mathbf{x}\boldsymbol{\alpha}$$

is (an abbreviated notation for) the formula[2]

$$\exists\mathbf{y}\forall\mathbf{x}(\boldsymbol{\alpha}\leftrightarrow\mathbf{x}=\mathbf{y}),$$

where $\mathbf{y}$ is the first variable which is different from $\mathbf{x}$ and is not free in $\boldsymbol{\alpha}$.

10.2. PROBLEM. Verify that $\sigma\models\exists!\mathbf{x}\boldsymbol{\alpha}$ iff in the universe U of σ there is exactly one individual u such that $\sigma(\mathbf{x}/u)\models\boldsymbol{\alpha}$.

Consider two first-order languages $\mathscr{L}$ and $\mathscr{L}'$. Let σ be an $\mathscr{L}$-valuation, and let σ' be an $\mathscr{L}'$-valuation. We shall say that σ and σ' are *compatible* if they agree on all symbols which are common to $\mathscr{L}$ and $\mathscr{L}'$. (In particular, σ and σ' have the same universe, and agree on all the variables.) Note that if $\mathscr{L}'$ is an extension of $\mathscr{L}$ then σ' is compatible with σ iff σ' is an expansion of σ.

Let $\boldsymbol{\alpha}$ and $\boldsymbol{\alpha}'$ be formulas of $\mathscr{L}$ and $\mathscr{L}'$ respectively. We shall say that $\boldsymbol{\alpha}$ and $\boldsymbol{\alpha}'$ are *co-satisfiable* if for each $\mathscr{L}$-valuation σ satisfying $\boldsymbol{\alpha}$ there

[1] Here "F" is short for "function".

[2] Here we tacitly assume $\mathscr{L}$ to be a language with equality. Def. 10.1 will only be used for such languages.

exists an $\mathscr{L}'$-valuation σ' compatible with σ that satisfies $\boldsymbol{\alpha}'$, and also conversely: for each $\mathscr{L}'$-valuation σ' satisfying $\boldsymbol{\alpha}'$ there exists a compatible $\mathscr{L}$-valuation σ satisfying $\boldsymbol{\alpha}$. Note that if $\mathscr{L}'$ is an extension of $\mathscr{L}$ then $\boldsymbol{\alpha}$ and $\boldsymbol{\alpha}'$ are co-satisfiable iff $\boldsymbol{\alpha}'$ is an S-form for $\boldsymbol{\alpha}$.

We shall say that $\boldsymbol{\alpha}$ and $\boldsymbol{\alpha}'$ are *co-valid* if whenever σ and σ' are a compatible $\mathscr{L}$-valuation and $\mathscr{L}'$-valuation respectively, then $\sigma \models \boldsymbol{\alpha}$ iff $\sigma' \models \boldsymbol{\alpha}'$.

10.3. PROBLEM. Verify that $\boldsymbol{\alpha}$ and $\boldsymbol{\alpha}'$ are co-valid (co-satisfiable) iff $\neg\boldsymbol{\alpha}$. and $\neg\boldsymbol{\alpha}'$ are co-satisfiable (co-valid).

For the rest of this section we assume that $\mathscr{L}$ is a language with equality.

10.4. LEMMA. *For each $\mathscr{L}$-formula $\boldsymbol{\alpha}$ we can find an $\mathscr{L}$-formula $\boldsymbol{\alpha}^*$ logically equivalent to $\boldsymbol{\alpha}$, such that each atomic subformula of $\boldsymbol{\alpha}^*$ which contains a function symbol has the form*

$$\mathbf{f}\mathbf{x}_1 \ldots \mathbf{x}_n = \mathbf{y},$$

where $\mathbf{x}_1,\ldots,\mathbf{x}_n,\mathbf{y}$ are distinct variables.

PROOF. Suppose $\boldsymbol{\alpha}$ has an atomic subformula of the form

(1) $$\mathbf{P}\mathbf{t}_1 \ldots \mathbf{t}_n,$$

where $\mathbf{P}$ is extralogical. If not all the arguments $\mathbf{t}_1,\ldots,\mathbf{t}_n$ are variables, we choose n different variables $\mathbf{x}_1,\ldots,\mathbf{x}_n$ which do not occur in any of these arguments, and we recall (see Prob. 4.4) that (1) is logically equivalent to

(2) $$\forall \mathbf{x}_1 \ldots \forall \mathbf{x}_n(\mathbf{t}_1 = \mathbf{x}_1 \rightarrow \ldots \rightarrow \mathbf{t}_n = \mathbf{x}_n \rightarrow \mathbf{P}\mathbf{x}_1 \ldots \mathbf{x}_n)$$

as well as to

(3) $$\exists \mathbf{x}_1 \ldots \exists \mathbf{x}_n(\mathbf{t}_1 = \mathbf{x}_1 \wedge \ldots \wedge \mathbf{t}_n = \mathbf{x}_n \wedge \mathbf{P}\mathbf{x}_1 \ldots \mathbf{x}_n).$$

Thus (1) can be replaced by (2) or (3). In this way we transform $\boldsymbol{\alpha}$ into a logically equivalent formula $\boldsymbol{\alpha}_1$ such that each atomic subformula of $\boldsymbol{\alpha}_1$ is an equation or has the form $\mathbf{P}\mathbf{x}_1 \ldots \mathbf{x}_n$.

Next, suppose $\boldsymbol{\alpha}_1$ has an atomic subformula which is an equation $\mathbf{s}=\mathbf{t}$ in which the right-hand side $\mathbf{t}$ contains a function symbol. We choose a variable $\mathbf{y}$ which does not occur in $\mathbf{s}=\mathbf{t}$ and recall (see Prob. 4.5) that $\mathbf{s}=\mathbf{t}$ is logically equivalent to

(4) $$\forall \mathbf{y}(\mathbf{s}=\mathbf{y} \rightarrow \mathbf{t}=\mathbf{y})$$

as well as to

(5) $\exists \mathbf{y}(\mathbf{s}=\mathbf{y} \wedge \mathbf{t}=\mathbf{y})$.

Thus $\mathbf{s}=\mathbf{t}$ can be replaced by (4) or (5), and we can transform $\boldsymbol{\alpha}_1$ into a logically equivalent formula $\boldsymbol{\alpha}_2$ in which function symbols only occur on the left-hand side of equations.

Finally, suppose $\boldsymbol{\alpha}_2$ has an atomic subformula of the form $\mathbf{ft}_1\ldots\mathbf{t}_n=\mathbf{y}$, where the arguments $\mathbf{t}_1,\ldots,\mathbf{t}_n$ are not distinct variables different from $\mathbf{y}$. Then we choose n distinct variables $\mathbf{x}_1,\ldots,\mathbf{x}_n$ which are different from $\mathbf{y}$ and do not occur in any of the arguments $\mathbf{t}_1,\ldots,\mathbf{t}_n$. By Prob. 4.6, $\mathbf{ft}_1\ldots\mathbf{t}_n=\mathbf{y}$ is logically equivalent to (and hence can be replaced by) each of the two formulas

$$\forall \mathbf{x}_1\ldots\forall \mathbf{x}_n(\mathbf{t}_1=\mathbf{x}_1\rightarrow\ldots\rightarrow\mathbf{t}_n=\mathbf{x}_n\rightarrow\mathbf{fx}_1\ldots\mathbf{x}_n=\mathbf{y}),$$
$$\exists \mathbf{x}_1\ldots\exists \mathbf{x}_n(\mathbf{t}_1=\mathbf{x}_1\wedge\ldots\wedge\mathbf{t}_n=\mathbf{x}_n\wedge\mathbf{fx}_1\ldots\mathbf{x}_n=\mathbf{y}).$$

Repeating this process, we can transform $\boldsymbol{\alpha}_2$ to a formula $\boldsymbol{\alpha}^*$ with the desired property. ∎

As before, we assume $\mathscr{L}$ to be a language with equality. Select an n-ary function symbol $\mathbf{f}$ of $\mathscr{L}$, and let $\mathscr{L}'$ be obtained from $\mathscr{L}$ by excluding $\mathbf{f}$ and introducing a new $(n+1)$-ary predicate symbol $\mathbf{P}$. We prove:

10.5. THEOREM. *For any $\mathscr{L}$-formula $\boldsymbol{\alpha}$ we can find an $\mathscr{L}'$-formula which is co-satisfiable with $\boldsymbol{\alpha}$ and an $\mathscr{L}'$-formula which is co-valid with $\boldsymbol{\alpha}$.*

PROOF. By Lemma 10.4, we may assume without loss of generality that $\mathbf{f}$ only occurs in $\boldsymbol{\alpha}$ in equations of the form $\mathbf{ft}_1\ldots\mathbf{t}_n=\mathbf{s}$, where $\mathbf{t}_1,\ldots,\mathbf{t}_n,\mathbf{s}$ do not contain $\mathbf{f}$. Let $\boldsymbol{\alpha}'$ be obtained from $\boldsymbol{\alpha}$ by replacing each such equation by $\mathbf{Pt}_1\ldots\mathbf{t}_n\mathbf{s}$. We show that the formula

(1) $\forall \mathbf{v}_1\ldots\forall \mathbf{v}_n\exists!\mathbf{v}_{n+1}\mathbf{Pv}_1\ldots\mathbf{v}_n\mathbf{v}_{n+1}\wedge\boldsymbol{\alpha}'$

is co-satisfiable with $\boldsymbol{\alpha}$. Indeed, if σ is an $\mathscr{L}$-valuation satisfying $\boldsymbol{\alpha}$, we let σ' be the $\mathscr{L}'$-valuation which is compatible with σ and such that for all $u_1,\ldots,u_n,v$ in the universe U of σ,

(2) $\langle u_1,\ldots,u_n,v\rangle\in\mathbf{P}^{\sigma'}$ iff $\mathbf{f}^{\sigma}(u_1,\ldots,u_n)=v$.

Then it is easy to see that σ' satisfies (1).

Conversely, suppose that σ' is an $\mathscr{L}'$-valuation satisfying (1). In particular, σ' must satisfy the first conjunct of (1). It follows that for all $u_1,\ldots,u_n$ in U there is a unique v in U such that $\langle u_1,\ldots,u_n,v\rangle\in\mathbf{P}^{\sigma'}$. We let σ be the $\mathscr{L}$-valuation which is compatible with σ' and such that (2) holds. It is easy to see that $\sigma\models\boldsymbol{\alpha}$. Thus $\boldsymbol{\alpha}$ and (1) are co-satisfiable.

Next, consider the formula

(3) $\forall v_1 \ldots \forall v_n \exists ! v_{n+1} P v_1 \ldots v_n v_{n+1} \rightarrow \alpha'$.

The negation of (3) is tautologically equivalent to

$$\forall v_1 \ldots \forall v_n \exists ! v_{n+1} P v_1 \ldots v_n v_{n+1} \wedge \neg \alpha',$$

which — by what we already proved — is co-satisfiable with $\neg\alpha$. Thus, by Prob. 10.3, α and (3) are co-valid. ∎

By Thm. 10.5, a first-order language with equality loses virtually nothing in expressiveness if we eliminate from it a function symbol and introduce a suitable extralogical predicate instead. Indeed, using the same method we can eliminate *all* function symbols in favour of predicate symbols.

Therefore in certain theoretical contexts we lose little by confining our attention to languages without function symbols, or without function symbols other than constants.[1] Such languages possess the advantage of not having complex terms (the only terms being the variables, or the variables and constants).

However, it can readily be seen that the formula (1) obtained in Thm. 10.5 is in general much longer than the given formula α. (Note that α must first be transformed by the method of Lemma 10.4!) Also, (1) will usually have many more quantifiers than α. Therefore languages *with* function symbols express statements much more concisely.

Besides, in many branches of mathematics it is more natural to consider structures *with* basic operations.[2] For example, in arithmetic it is more natural to think of addition and multiplication as binary operations rather than ternary relations. Similarly, in group theory the multiplication in a group is normally thought of as a binary operation.

For these reasons it is in general desirable to deal with languages having function symbols.

[1] In this book we shall do so, e.g., in Ch. 5.

[2] It may be argued that part of the reason for this is precisely the fact that mathematical statements can often be expressed much more concisely in terms of operations rather than relations.

*§ 11. Elimination of equality

In this section we shall investigate the expressive power of a first-order language without equality, compared to that of a language with equality.

11.1. PROBLEM. Let $\mathscr{L}$ be a language without equality. Let σ be an $\mathscr{L}$-valuation with universe U. Let V be an arbitrary class such that $U \subseteq V$. Show that there is an $\mathscr{L}$-valuation τ with universe V such that, for every $\mathscr{L}$-formula $\boldsymbol{\alpha}$, $\sigma \models \boldsymbol{\alpha}$ iff $\tau \models \boldsymbol{\alpha}$. (Take any mapping f of V onto U such that $f(u) = u$ for every $u \in U$. For any n-ary predicate symbol $\mathbf{P}$, put

$$\mathbf{P}^\tau = \{\langle v_1, \ldots, v_n \rangle : \langle f(v_1), \ldots, f(v_n) \rangle \in \mathbf{P}^\sigma\}.$$

For every variable $\mathbf{x}$, let $\mathbf{x}^\tau = \mathbf{x}^\sigma$, and for every constant $\mathbf{a}$, let $\mathbf{a}^\tau = \mathbf{a}^\sigma$.)

Thus, a first-order formula without equality (and even an arbitrary set of such formulas) cannot impose an upper bound on the size of the universe. Contrast this with Prob. 2.5.

For the rest of this section, let $\mathscr{L}$ be a first-order language with equality but without function symbols other than constants. Let $\mathscr{L}'$ be obtained from $\mathscr{L}$ by excluding $=$ and introducing a new *extralogical* binary predicate symbol $\mathbf{E}$.

Let $\boldsymbol{\alpha}$ be a given $\mathscr{L}$-formula. We define $\boldsymbol{\alpha}^*$ as the $\mathscr{L}'$-formula obtained from $\boldsymbol{\alpha}$ when each occurrence of $=$ is replaced by an occurrence of $\mathbf{E}$. Let $\mathbf{P}_1, \ldots, \mathbf{P}_k$ be all the predicate symbols occurring in $\boldsymbol{\alpha}^*$. If $\mathbf{P}_i$ is n_i-ary, let $\boldsymbol{\delta}_i$ be the sentence

$$\forall \mathbf{v}_i \ldots \forall \mathbf{v}_{2n_i} (\mathbf{E}\mathbf{v}_1\mathbf{v}_{n_i+1} \to \mathbf{E}\mathbf{v}_2\mathbf{v}_{n_i+2} \to \ldots \to \mathbf{E}\mathbf{v}_{n_i}\mathbf{v}_{2n_i} \to \\ \to \mathbf{P}_i\mathbf{v}_1 \ldots \mathbf{v}_{n_i} \to \mathbf{P}_i\mathbf{v}_{n_i+1} \ldots \mathbf{v}_{2n_i}).$$

Let $\boldsymbol{\alpha}'$ be the formula

$$\forall \mathbf{v}_1 \mathbf{E}\mathbf{v}_1\mathbf{v}_1 \wedge \boldsymbol{\delta}_1 \wedge \ldots \wedge \boldsymbol{\delta}_k \wedge \boldsymbol{\alpha}^*.$$

Then we have:

11.2. THEOREM. *If σ is an $\mathscr{L}$-valuation satisfying $\boldsymbol{\alpha}$, then there is an $\mathscr{L}'$-valuation σ' which is compatible with σ and which satisfies $\boldsymbol{\alpha}'$. If $\boldsymbol{\alpha}'$ is satisfiable, then so is $\boldsymbol{\alpha}$.*

PROOF. The first part is trivial: if σ' is the $\mathscr{L}'$-valuation which is compatible with σ and for which $\mathbf{E}^{\sigma'}$ is the identity relation on the universe, then clearly $\sigma' \models \boldsymbol{\alpha}'$.

To prove the second part, we may assume that $=$ occurs in $\boldsymbol{\alpha}$; otherwise $\boldsymbol{\alpha}^*$ is $\boldsymbol{\alpha}$ and the assertion is trivial. Thus $\mathbf{E}$ is one of the $\mathbf{P}_i$.

Let σ' be any $\mathscr{L}'$-valuation satisfying $\boldsymbol{\alpha}'$. From the fact that $\forall \mathbf{v}_1 \mathbf{E} \mathbf{v}_1 \mathbf{v}_1, \boldsymbol{\delta}_1, \ldots, \boldsymbol{\delta}_k$ are satisfied by σ' we can easily draw two conclusions. First, if V is the universe of σ' then $\mathbf{E}^{\sigma'}$ is an equivalence relation on V. Second, if we denote by $[v]$ the $\mathbf{E}^{\sigma'}$-class of $v \in V$, then we have:

If $[v_j]=[v_{n_i+j}]$ for $j=1,\ldots,n_i$ and if $\langle v_1,\ldots,v_{n_i}\rangle \in \mathbf{P}_i^{\sigma'}$ then $\langle v_{n_i+1},\ldots,v_{2n_i}\rangle \in \mathbf{P}_i^{\sigma'}$.

We define an $\mathscr{L}$-valuation σ as follows. The universe of σ will be[1]

$$U=\{[v] : v \in V\}.$$

If $\mathbf{P}_i$ is an extralogical predicate symbol occurring in $\boldsymbol{\alpha}$ we put

$$\mathbf{P}_i^{\sigma}=\{\langle [v_1],\ldots,[v_{n_i}]\rangle : \langle v_1,\ldots,v_{n_i}\rangle \in \mathbf{P}_i^{\sigma'}\}$$

(and for any extralogical predicate $\mathbf{P}$ that does not occur in $\boldsymbol{\alpha}$, $\mathbf{P}^{\sigma}$ may be defined arbitrarily, e.g. as the empty set).

Finally, for any variable $\mathbf{x}$ and any constant $\mathbf{a}$ we put $\mathbf{x}^{\sigma}=[\mathbf{x}^{\sigma'}]$ and $\mathbf{a}^{\sigma}=[\mathbf{a}^{\sigma'}]$.

Then it is easy to see that $\sigma \models \boldsymbol{\alpha}$. ∎

*§ 12. Relativization

For the purpose of this section we select for a special role some formula $\boldsymbol{\varphi}$ and some variable $\mathbf{z}$. Any variable which is free in $\boldsymbol{\varphi}$ but is different from $\mathbf{z}$ will be called a *parameter*.

We define, for any formula $\boldsymbol{\alpha}$, a formula $\boldsymbol{\alpha}^*$ called the *relativization of* $\boldsymbol{\alpha}$ *to* $\boldsymbol{\varphi}$ *(with* $\mathbf{z}$ *as chosen variable).* We proceed by recursion on deg $\boldsymbol{\alpha}$.

12.1. Definition.

(a) For atomic $\boldsymbol{\alpha}$, $\boldsymbol{\alpha}^*=\boldsymbol{\alpha}$.

(b) $(\neg\boldsymbol{\beta})^*=\neg(\boldsymbol{\beta}^*)$.

(c) $(\boldsymbol{\beta}\to\boldsymbol{\gamma})^*=\boldsymbol{\beta}^*\to\boldsymbol{\gamma}^*$.

(d) If $\mathbf{x}$ is not a parameter, then

$$(\forall \mathbf{x}\boldsymbol{\beta})^*=\forall \mathbf{x}[\boldsymbol{\varphi}(\mathbf{z}/\mathbf{x})\to\boldsymbol{\beta}^*].$$

(e) If $\mathbf{x}$ is a parameter, then

$$(\forall \mathbf{x}\boldsymbol{\beta})^*=\{\forall \mathbf{y}\,[\boldsymbol{\beta}(\mathbf{x}/\mathbf{y})]\}^*,$$

where $\mathbf{y}$ is the first variable which is not a parameter, and is not free in $\boldsymbol{\beta}$ but is free for $\mathbf{x}$ in $\boldsymbol{\beta}$.

[1] This definition of U is legitimate only if all the equivalence classes $[v]$ are sets. This may not be the case if V is a proper class (i.e., a class that is not a set). But by Cor. 8.3 we can always assume V to be a countable *set*.

To investigate the semantic behaviour of $\boldsymbol{\alpha}^*$, we shall need a few auxiliary definitions.

If $\mathbf{f}$ is an n-ary function symbol, we define $\boldsymbol{\varphi}\{\mathbf{f}\}$ to be the formula

$$\forall \mathbf{x}_1 \ldots \forall \mathbf{x}_n \, [\boldsymbol{\varphi}(\mathbf{z}/\mathbf{x}_1) \rightarrow \ldots \rightarrow \boldsymbol{\varphi}(\mathbf{z}/\mathbf{x}_n) \rightarrow \boldsymbol{\varphi}(\mathbf{z}/\mathbf{f}\mathbf{x}_1 \ldots \mathbf{x}_n)],$$

where $\mathbf{x}_1,\ldots,\mathbf{x}_n$ are the first n variables which are not parameters. (In particular, if $\mathbf{a}$ is a constant then $\boldsymbol{\varphi}\{\mathbf{a}\}$ is $\boldsymbol{\varphi}(\mathbf{z}/\mathbf{a})$.)

Let $\mathbf{f}_1,\ldots,\mathbf{f}_m$ be any function symbols, and let $\mathbf{u}_1,\ldots,\mathbf{u}_k$ be any variables. We shall say that a valuation σ *has the closure property for* $\{\mathbf{f}_1,\ldots,\mathbf{f}_m;\, \mathbf{u}_1,\ldots,\mathbf{u}_k\}$ if the following three conditions hold:

(1) $\sigma \models \exists \mathbf{z} \boldsymbol{\varphi}$;
(2) $\sigma \models \boldsymbol{\varphi}\{\mathbf{f}_i\}$ for $i=1,\ldots,m$;
(3) $\sigma \models \boldsymbol{\varphi}(\mathbf{z}/\mathbf{u}_j)$ for $j=1,\ldots,k$.

12.2. PROBLEM. Let U be the universe of σ, and let

$$U^* = \{u : u \in U \quad \text{and} \quad \sigma(\mathbf{z}/u) \models \boldsymbol{\varphi}\}.$$

Verify that conditions (1), (2) and (3) are respectively equivalent to
(1′) $U^* \neq \emptyset$;
(2′) U^* is closed under the operation $\mathbf{f}_i^\sigma$ for $i=1,\ldots,m$;
(3′) $\mathbf{u}_j^\sigma \in U^*$ for $j=1,\ldots,k$.

Suppose that σ has the closure property for $\{\mathbf{f}_1,\ldots,\mathbf{f}_m;\, \mathbf{u}_1,\ldots,\mathbf{u}_k\}$, and let U^* be the class defined in Prob. 12.2. We shall say that a valuation σ^* is a *restriction of* σ *for* $\{\mathbf{f}_1,\ldots,\mathbf{f}_m;\, \mathbf{u}_1,\ldots,\mathbf{u}_k\}$ if the following four conditions hold:
(a) The universe of σ^* is U^*.
(b) For every extralogical predicate symbol $\mathbf{P}$, $\mathbf{P}^{\sigma^*}$ is the restriction of $\mathbf{P}^\sigma$ to U^*.
(c) For $i=1,\ldots,m$, $\mathbf{f}_i^{\sigma^*}$ is the restriction of $\mathbf{f}_i^\sigma$ to U^*.
(d) For $j=1,\ldots,k$, $\mathbf{u}_j^{\sigma^*} = \mathbf{u}_j^\sigma$.

We can now prove the main result about relativization:

12.3. THEOREM. *Let* $\boldsymbol{\alpha}$ *be a formula such that all the function symbols occurring in* $\boldsymbol{\alpha}$ *are among* $\mathbf{f}_1,\ldots,\mathbf{f}_m$ *and all the free variables of* $\boldsymbol{\alpha}$ *are among* $\mathbf{u}_1,\ldots,\mathbf{u}_k$. *Let* σ *have the closure property for* $\{\mathbf{f}_1,\ldots,\mathbf{f}_m;\, \mathbf{u}_1,\ldots,\mathbf{u}_k\}$ *and let* σ^* *be a restriction of* σ *for* $\{\mathbf{f}_1,\ldots,\mathbf{f}_m;\, \mathbf{u}_1,\ldots,\mathbf{u}_k\}$. *Then* $\sigma \models \boldsymbol{\alpha}^*$ *iff* $\sigma^* \models \boldsymbol{\alpha}$.

PROOF. By induction on deg $\boldsymbol{\alpha}$. We distinguish five cases (a)–(e), corresponding to the five clauses of Def. 12.1. We outline the proof, leaving the details to the reader.

Cases (a), (b) and (c) are routine. Case (e) is easily reduced to (d) *via* Thm. 3.6.

Case (d): $\boldsymbol{\alpha} = \forall \mathbf{x}\boldsymbol{\beta}$, where $\mathbf{x}$ is not a parameter. Then all the function symbols occurring in $\boldsymbol{\beta}$ are among $\mathbf{f}_1,\ldots,\mathbf{f}_m$ and all the free variables of $\boldsymbol{\beta}$ are among $\mathbf{u}_1,\ldots,\mathbf{u}_k,\mathbf{x}$. Let u^* be any member of U^*. It can readily be verified that $\sigma(\mathbf{x}/u^*)$ has the closure property for $\{\mathbf{f}_1,\ldots,\mathbf{f}_m;\ \mathbf{u}_1,\ldots,\mathbf{u}_k,\mathbf{x}\}$ and that $\sigma^*(\mathbf{x}/u^*)$ is a restriction of $\sigma(\mathbf{x}/u^*)$ for $\{\mathbf{f}_1,\ldots,\mathbf{f}_m;\ \mathbf{u}_1,\ldots,\mathbf{u}_k,\mathbf{x}\}$. By the induction hypothesis it follows that $\sigma(\mathbf{x}/u^*) \models \boldsymbol{\beta}^*$ iff $\sigma^*(\mathbf{x}/u^*) \models \boldsymbol{\beta}$. Using this and the definition of $\boldsymbol{\alpha}^*$, it is easy to check that $\sigma \models \boldsymbol{\alpha}^*$ iff $\sigma^* \models \boldsymbol{\alpha}$. ∎

12.4. Corollary. *Let* $\boldsymbol{\alpha}$ *be as in Thm.* 12.3. *If the formula*

$$\exists \mathbf{z}\boldsymbol{\varphi} \wedge \boldsymbol{\varphi}\{\mathbf{f}_1\} \wedge \ldots \wedge \boldsymbol{\varphi}\{\mathbf{f}_m\} \wedge \boldsymbol{\varphi}(\mathbf{z}/\mathbf{u}_1) \wedge \ldots \wedge \boldsymbol{\varphi}(\mathbf{z}/\mathbf{u}_k) \wedge \boldsymbol{\alpha}^*$$

is satisfiable, then $\boldsymbol{\alpha}$ *is satisfiable as well. Also, if* $\boldsymbol{\alpha}$ *is logically valid, then so is the formula*

$$\exists \mathbf{z}\boldsymbol{\varphi} \rightarrow \boldsymbol{\varphi}\{\mathbf{f}_1\} \rightarrow \ldots \rightarrow \boldsymbol{\varphi}\{\mathbf{f}_m\} \rightarrow \boldsymbol{\varphi}(\mathbf{z}/\mathbf{u}_1) \rightarrow \ldots \rightarrow \boldsymbol{\varphi}(\mathbf{z}/\mathbf{u}_k) \rightarrow \boldsymbol{\alpha}^*. \quad ∎$$

*§ 13. Virtual terms

We have noted in §10 that languages without function symbols (other than constants) possess the advantage of having simple terms only. On the other hand, in practice function symbols are needed for making statements in a more concise way. In this section we describe a way of getting the best of both these worlds.

Throughout this section we assume that $\mathscr{L}$ is a language with equality. We choose a particular $\mathscr{L}$-formula $\boldsymbol{\alpha}$ having exactly one free variable $\mathbf{z}$. We shall keep $\boldsymbol{\alpha}$ fixed throughout this section.

We shall be interested only in those $\mathscr{L}$-valuations σ for which $\sigma \models \exists!\mathbf{z}\boldsymbol{\alpha}$. Accordingly, we say that such valuations are *admissible*. (As a matter of fact, since $\exists!\mathbf{z}\boldsymbol{\alpha}$ is a sentence, the admissibility of an $\mathscr{L}$-valuation depends only on its underlying $\mathscr{L}$-structure.)

If σ is an admissible valuation, then we define a_σ to be the unique individual in the universe U of σ for which $\sigma(\mathbf{z}/a_\sigma) \models \boldsymbol{\alpha}$. (See Prob. 10.2.)

Now let $\boldsymbol{\varphi}$ be any $\mathscr{L}$-formula and let $\mathbf{y}$ be a variable. For the time being, we hold $\boldsymbol{\varphi}$ and $\mathbf{y}$ fixed as well. Let $\mathbf{x}_1,\ldots,\mathbf{x}_n$ be all the different free variables of $\boldsymbol{\varphi}$ other than $\mathbf{y}$ (i.e., all the free variables of $\forall \mathbf{y}\boldsymbol{\varphi}$). To be quite definite, we suppose that the $\mathbf{x}_i$ are taken in alphabetic order: $\mathbf{x}_i = \mathbf{v}_{j_i}$, where $j_1 < \ldots < j_n$.

Next, let $\mathscr{L}'$ be the extension of $\mathscr{L}$ obtained by adding a new n-ary function symbol $\mathbf{f}$. We define $\varphi\{\alpha,\mathbf{f},\mathbf{y}\}$ to be the $\mathscr{L}'$-sentence

$$\forall \mathbf{x}_1 \ldots \forall \mathbf{x}_n [\exists ! \mathbf{y}\varphi \wedge \varphi(\mathbf{y}/\mathbf{f}\mathbf{x}_1 \ldots \mathbf{x}_n) \vee \neg \exists ! \mathbf{y}\varphi \wedge \alpha(\mathbf{z}/\mathbf{f}\mathbf{x}_1 \ldots \mathbf{x}_n)].$$

If σ is an admissible $\mathscr{L}$-valuation with universe U, we define an n-ary operation f_σ on U as follows. For any $u_1,\ldots,u_n \in U$, if there is a unique $v \in U$ such that $\sigma(\mathbf{x}_1/u_1)\ldots(\mathbf{x}_n/u_n)(\mathbf{y}/v) \models \varphi$ then we put $f_\sigma(u_1,\ldots,u_n)=v$; otherwise we put $f_\sigma(u_1,\ldots,u_n)=a_\sigma$.

It is easy to see that for any admissible $\mathscr{L}$-valuation σ there is a unique $\mathscr{L}'$-expansion σ' of σ such that $\sigma' \models \varphi\{\alpha,\mathbf{f},\mathbf{y}\}$. Namely, it is the one for which $\mathbf{f}^{\sigma'}$ is our f_σ.

We now associate with each $\mathscr{L}'$-formula ψ an $\mathscr{L}$-formula ψ_0 as follows. If $\mathbf{f}$ does not occur in ψ (i.e., if ψ is itself an $\mathscr{L}$-formula), we put $\psi_0=\psi$. Otherwise, we first use the method of Lemma 10.4 (more precisely, one of the alternative methods provided in the proof of 10.4) to transform ψ into an $\mathscr{L}'$-formula ψ^* which is logically equivalent to ψ and in which $\mathbf{f}$ occurs only in equations of the form

(1) $\mathbf{f}\mathbf{u}_1 \ldots \mathbf{u}_n = \mathbf{v}$,

where $\mathbf{u}_1,\ldots,\mathbf{u}_n,\mathbf{v}$ are distinct variables. We let ψ_0 be the $\mathscr{L}$-formula resulting from ψ^* when each equation of the form (1) is replaced by

(2) $\exists ! \mathbf{v}\varphi(\mathbf{x}_1/\mathbf{u}_1,\ldots,\mathbf{x}_n/\mathbf{u}_n,\mathbf{y}/\mathbf{v}) \wedge \varphi(\mathbf{x}_1/\mathbf{u}_1,\ldots,\mathbf{x}_n/\mathbf{u}_n,\mathbf{y}/\mathbf{v}) \vee$
$\vee \neg \exists ! \mathbf{v}\varphi(\mathbf{x}_1/\mathbf{u}_1,\ldots,\mathbf{x}_n/\mathbf{u}_n,\mathbf{y}/\mathbf{v}) \wedge \alpha(\mathbf{z}/\mathbf{v})$.

Note that in going from ψ to ψ_0 the only changes that occur are that some of the atomic parts of ψ are replaced by other formulas. In fact we have

$$(\neg\psi)_0 = \neg(\psi)_0, \qquad (\psi \rightarrow \chi)_0 = (\psi)_0 \rightarrow (\chi)_0, \qquad (\forall \mathbf{x}\psi)_0 = \forall \mathbf{x}(\psi)_0.$$

We now prove:

13.1. Lemma. *For each $\mathscr{L}'$-formula ψ,*

$$\exists ! \mathbf{z}\alpha,\ \varphi\{\alpha,\mathbf{f},\mathbf{y}\} \models \psi \leftrightarrow \psi_0.$$

Proof. Let σ' be a $\mathscr{L}'$-valuation such that $\sigma' \models \exists ! \mathbf{z}\alpha,\ \varphi\{\alpha,\mathbf{f},\mathbf{y}\}$. Then σ' must be an expansion of an admissible $\mathscr{L}'$-valuation σ, and by what we have seen above $\mathbf{f}^{\sigma'}$ must be f_σ. But from this it follows without difficulty that the formulas (1) and (2) above have the same truth value under σ'. ∎

13.2. Theorem. *Let Φ be a set of $\mathscr{L}$-formulas such that $\Phi \models \exists ! \mathbf{z}\alpha$. Then for any $\mathscr{L}'$-formula ψ we have $\Phi \models \psi_0$ iff $\Phi,\ \varphi\{\alpha,\mathbf{f},\mathbf{y}\} \models \psi$.*

Proof. Since $\Phi \models \exists ! \mathbf{z}\alpha$, it follows from Lemma 13.1. that

(*) $\Phi,\ \varphi\{\alpha,\mathbf{f},\mathbf{y}\} \models \psi \leftrightarrow \psi_0$.

Thus if $\Phi \models \psi_0$ we must have $\Phi,\ \varphi\{\alpha,\mathbf{f},\mathbf{y}\} \models \psi$.

Conversely, suppose $\Phi, \varphi\{\alpha,f,y\} \models \psi$. Then by (*) we get

$$(**) \qquad \Phi, \varphi\{\alpha,f,y\} \models \psi_0.$$

Now, if σ is an $\mathscr{L}$-valuation such that $\sigma \models \Phi$, then σ is admissible because $\Phi \models \exists ! z\alpha$. Hence there is a unique $\mathscr{L}'$-expansion σ' of σ such that $\sigma' \models \varphi\{\alpha,f,y\}$ and by (**) we have also $\sigma' \models \psi_0$. But since ψ_0 is an $\mathscr{L}$-formula, it follows that $\sigma \models \psi_0$. Thus $\Phi \models \psi_0$. ∎

Note, in particular, that if in Thm. 13.2 we take ψ to be an $\mathscr{L}$-formula, then $\psi_0 = \psi$, and we have $\Phi \models \psi$ iff $\Phi, \varphi\{\alpha,f,y\} \models \psi$. We say that the addition of $\varphi\{\alpha,f,y\}$ to Φ is *conservative,* since the logical consequences of $\Phi, \varphi\{\alpha,f,y\}$ in $\mathscr{L}$ are exactly the same as those of Φ.

Thm. 13.2 is applied in the following way. Suppose that Φ is as in the theorem and that we want to derive logical consequences of Φ in $\mathscr{L}$. In practice it may be more convenient to work in $\mathscr{L}'$ (since we can then use an additional function symbol f) and add $\varphi\{\alpha,f,y\}$ to Φ. If we discover an $\mathscr{L}'$-formula ψ such that $\Phi, \varphi\{\alpha,f,y\} \models \psi$, then by our theorem it follows that $\Phi \models \psi_0$. (In particular, if ψ happens to be in $\mathscr{L}$ then $\Phi \models \psi$.)

If we like, we may regard each $\mathscr{L}'$-formula ψ merely as an abbreviation of the corresponding $\mathscr{L}$-formula ψ_0. This has the advantage that while in practice we work in $\mathscr{L}'$, which has a greater facility of expression, we can pretend that all our formulas belong to the simpler language $\mathscr{L}$. If we adopt this point of view, we say that terms containing the new function symbol f are *virtual* terms: they can be used in practice but ignored in theory, since any formula that contains them is regarded as an abbreviation for a formula that does not. In particular, the virtual term $fx_1 \ldots x_n$ is denoted by "$\iota_\alpha y\varphi$". Since we have fixed α once and for all, we shall omit the subscript "α" and write, more briefly, "$\iota y\varphi$". (Thus, if w is a variable other than y, or χ is an $\mathscr{L}$-formula other than φ, then the virtual term $\iota w\chi$ is not in $\mathscr{L}'$ but in a language obtained in a similar way, with α playing the *same* role and with w and χ playing the roles of y and φ. Of course, every formula of this other language is also regarded as an abbreviation for a suitable $\mathscr{L}$-formula.)

With the new notation, the sentence $\varphi\{\alpha,f,y\}$ can be written in the form

$$\forall x_1 \ldots \forall x_n\, [\exists ! y\varphi \wedge \varphi(y/\iota y\varphi) \vee \neg \exists ! y\varphi \wedge \alpha(z/\iota y\varphi)].$$

Note also that since $\varphi\{\alpha,f,y\}$ itself is now regarded as an abbreviation for — and can therefore be identified with — the corresponding $\mathscr{L}$-sentence $(\varphi\{\alpha,f,y\})_0$, it follows from Thm. 13.2 that

$$\Phi \models \varphi\{\alpha,f,y\}.$$

This procedure of adding virtual terms can be iterated several times. If φ' is any $\mathscr{L}'$-formula and $\mathbf{y}'$ is any variable, we may extend $\mathscr{L}'$ to $\mathscr{L}''$ by introducing a suitable new function symbol $\mathbf{f}'$. Any $\mathscr{L}''$-formula may be regarded as an abbreviation for an appropriate $\mathscr{L}'$-formula, which is in turn itself regarded as an abbreviation for an $\mathscr{L}$-formula. We then have $\Phi \models \varphi'\{\alpha,\mathbf{f}',\mathbf{y}'\}$. And so on.

13.3. PROBLEM. Let Φ, φ and $\mathbf{y}$ be as in Thm. 13.2. Regarding $\mathscr{L}'$-formulas as abbreviations for the corresponding $\mathscr{L}$-formulas, verify that

$$\Phi \models \forall \mathbf{x}_1 \ldots \forall \mathbf{x}_n \, [\exists ! \mathbf{y}\varphi \rightarrow \varphi(\mathbf{y}/\iota\mathbf{y}\varphi)],$$

$$\Phi \models \forall \mathbf{x}_1 \ldots \forall \mathbf{x}_n \, [\neg \exists ! \mathbf{y}\varphi \rightarrow \alpha(\mathbf{z}/\iota\mathbf{y}\varphi)].$$

Also, taking the special case where φ and $\mathbf{y}$ happen to be the same as α and $\mathbf{z}$ respectively, show that

$$\Phi \models \alpha(\mathbf{z}/\iota\mathbf{z}\alpha).$$

Hence for *any* φ and $\mathbf{y}$ show that

$$\Phi \models \forall \mathbf{x}_1 \ldots \forall \mathbf{x}_n \, [\neg \exists ! \mathbf{y}\varphi \rightarrow \iota\mathbf{y}\varphi = \iota\mathbf{z}\alpha].$$

13.4. PROBLEM. Let Φ, φ and $\mathbf{y}$ be as before. The *free variables* of $\iota\mathbf{y}\varphi$ are defined to be the same as the free variables of $\forall\mathbf{y}\varphi$, i.e., $\mathbf{x}_1,\ldots,\mathbf{x}_n$.

(i) If $\mathbf{w}$ is not a free variable of $\iota\mathbf{y}\varphi$, show that

$$\Phi \models \iota\mathbf{y}\varphi = \iota\mathbf{w}\,[\varphi(\mathbf{y}/\mathbf{w})].$$

(ii) Let $\mathbf{t}$ be any term (virtual or otherwise) that does not contain the variable $\mathbf{y}$, and let $\mathbf{x}$ be any variable other than $\mathbf{y}$. Show that

$$\Phi \models \iota\mathbf{y}\,[\varphi(\mathbf{x}/\mathbf{t})] = [\iota\mathbf{y}\varphi](\mathbf{x}/\mathbf{t}).$$

§ 14. Historical and bibliographical remarks

Historical notes to some of the topics covered in this chapter can be found in §49 of Ch. IV in CHURCH [1956].

The first explicit formulation of the BSD was contained in a paper published by Tarski in Polish in 1933, of which TARSKI [1935] is a German translation. (English translation in TARSKI [1956].)

For remarks on the method of tableaux see §16 of Ch. 1.

CHAPTER 3

FIRST-ORDER LOGIC (CONTINUED)

This chapter is devoted to the first-order predicate calculus. We shall do for first-order logic what we did in §§10–13 of Chapter 1 for propositional logic.

§ 1. The first-order predicate calculus

By a *generalization* of $\boldsymbol{\alpha}$ we mean any formula of the form $\forall \mathbf{x}_1 \ldots \forall \mathbf{x}_k \boldsymbol{\alpha}$, where $k \geqslant 1$ and $\mathbf{x}_1, \ldots, \mathbf{x}_k$ are any variables, not necessarily distinct.

As *first-order axioms* of $\mathscr{L}$ we take all $\mathscr{L}$-formulas of the following eight groups:

(Ax.1) All propositional axioms of $\mathscr{L}$.

(Ax.2) $\forall \mathbf{x}(\boldsymbol{\alpha} \rightarrow \boldsymbol{\beta}) \rightarrow \forall \mathbf{x} \boldsymbol{\alpha} \rightarrow \forall \mathbf{x} \boldsymbol{\beta}$,
where $\boldsymbol{\alpha}$, $\boldsymbol{\beta}$ are any $\mathscr{L}$-formulas and $\mathbf{x}$ is any variable.

(Ax.3) $\boldsymbol{\alpha} \rightarrow \forall \mathbf{x} \boldsymbol{\alpha}$,
where $\boldsymbol{\alpha}$ is any $\mathscr{L}$-formula and the variable $\mathbf{x}$ is not free in $\boldsymbol{\alpha}$.

(Ax.4) $\forall \mathbf{x} \boldsymbol{\alpha} \rightarrow \boldsymbol{\alpha}(\mathbf{x}/\mathbf{t})$,
where $\boldsymbol{\alpha}$ is any $\mathscr{L}$-formula and $\mathbf{t}$ is any $\mathscr{L}$-term free for $\mathbf{x}$ in $\boldsymbol{\alpha}$.

(Ax.5) $\mathbf{t}=\mathbf{t}$,
where $\mathbf{t}$ is any $\mathscr{L}$-term.

(Ax.6) $\mathbf{t}_1=\mathbf{t}_{n+1} \rightarrow \ldots \rightarrow \mathbf{t}_n=\mathbf{t}_{2n} \rightarrow \mathbf{f}\mathbf{t}_1 \ldots \mathbf{t}_n=\mathbf{f}\mathbf{t}_{n+1} \ldots \mathbf{t}_{2n}$,
where $\mathbf{f}$ is any n-ary function symbol of $\mathscr{L}$ and $\mathbf{t}_1, \ldots, \mathbf{t}_{2n}$ are any $\mathscr{L}$-terms.

(Ax.7) $\mathbf{t}_1=\mathbf{t}_{n+1} \rightarrow \ldots \rightarrow \mathbf{t}_n=\mathbf{t}_{2n} \rightarrow \mathbf{P}\mathbf{t}_1 \ldots \mathbf{t}_n \rightarrow \mathbf{P}\mathbf{t}_{n+1} \ldots \mathbf{t}_{2n}$,
where $\mathbf{P}$ is any n-ary predicate symbol of $\mathscr{L}$ and $\mathbf{t}_1, \ldots, \mathbf{t}_{2n}$ are any $\mathscr{L}$-terms.

(Ax.8) All generalizations of axioms of the preceding groups.

If $\mathscr{L}$ is without equality then (Ax.5), (Ax.6) and (Ax.7) are omitted. As rule of inference we again take *modus ponens.*

First-order deduction and *first-order proof* are defined like the corresponding propositional notions (§10 of Ch. 1) except that now the axioms are the first-order axioms instead of just the propositional axioms. We use "$\vdash$" for first-order deducibility as "$\vdash_0$" was used for propositional deducibility.

From now on we shall often drop the label "first-order" and say briefly, e.g., "deduction" instead of "first-order deduction".

1.1. THEOREM. *If* $\mathbf{\Phi} \vdash_0 \boldsymbol{\alpha}$ *then* $\mathbf{\Phi} \vdash \boldsymbol{\alpha}$. *In particular, if* $\vdash_0 \boldsymbol{\alpha}$ *then* $\vdash \boldsymbol{\alpha}$. ∎

The following theorem states that the predicate calculus is *semantically sound:*

1.2. THEOREM. *If* $\mathbf{\Phi} \vdash \boldsymbol{\alpha}$ *then* $\mathbf{\Phi} \models \boldsymbol{\alpha}$. *In particular, if* $\vdash \boldsymbol{\alpha}$ *then* $\models \boldsymbol{\alpha}$.

PROOF. Similar to Thm. 1.10.2. We first verify that our first-order axioms are logically true. For (Ax.1) this follows from the fact that the propositional axioms are tautologies. For (Ax.2)–(Ax.7) we can use tableaux (see Thm. 2.4.2 and Prob. 2.4.3) or verify directly that they are logically true. For (Ax.8) we merely need to observe that if $\boldsymbol{\alpha}$ is logically true then so is $\forall \mathbf{x} \boldsymbol{\alpha}$, and hence so is any generalization of $\boldsymbol{\alpha}$.

The rest is exactly as in the proof of Thm. 1.10.2. ∎

1.3. DEDUCTION THEOREM. *Given a deduction of* $\boldsymbol{\beta}$ *from* $\mathbf{\Phi}, \boldsymbol{\alpha}$, *we can construct a deduction of* $\boldsymbol{\alpha} \to \boldsymbol{\beta}$ *from* $\mathbf{\Phi}$. *(Hence, if* $\mathbf{\Phi}, \boldsymbol{\alpha} \vdash \boldsymbol{\beta}$ *then* $\mathbf{\Phi} \vdash \boldsymbol{\alpha} \to \boldsymbol{\beta}$.*)*

PROOF. Exactly the same as for the propositional calculus (Thm. 1.10.4). ∎

Here, as in the propositional calculus, the converse of the Deduction Theorem is easily seen to hold.

As in Ch. 2, we say that a variable $\mathbf{x}$ is free in a set $\mathbf{\Phi}$ of formulas, if $\mathbf{x}$ is free in some formula of $\mathbf{\Phi}$. Similarly, we say that $\mathbf{x}$ is free in a deduction $\boldsymbol{D}$ if $\mathbf{x}$ is free in some formula of $\boldsymbol{D}$.

1.4. THEOREM. *Let* $\mathbf{x}$ *be a variable which is not free in* $\mathbf{\Phi}$. *Given a deduction* $\boldsymbol{D}$ *of* $\boldsymbol{\alpha}$ *from* $\mathbf{\Phi}$, *we can construct a deduction* $\boldsymbol{D}'$ *of* $\forall \mathbf{x} \boldsymbol{\alpha}$ *from* $\mathbf{\Phi}$ *such that*

(a) $\mathbf{x}$ *is not free in* $\boldsymbol{D}'$,

(b) *every variable free in* $\boldsymbol{D}'$ *is free in* $\boldsymbol{D}$ *as well.*

PROOF. Let $\boldsymbol{\varphi}_1, \ldots, \boldsymbol{\varphi}_n$ be the given deduction $\boldsymbol{D}$. So $\boldsymbol{\varphi}_n = \boldsymbol{\alpha}$. By recursion on k ($k = 1, \ldots, n$) we construct a deduction $\boldsymbol{D}_k$ of $\forall \mathbf{x} \boldsymbol{\varphi}_k$ from $\mathbf{\Phi}$, such that

$\mathbf{x}$ is not free in D_k and such that each variable free in D_k is free also in D. Then D_n is the required D'.

If φ_k is an axiom, then $\forall \mathbf{x}\varphi_k$ is an axiom as well — see (Ax.8) — so $\forall \mathbf{x}\varphi_k$ by itself is the required D_k.

If φ_k is in Φ, then $\mathbf{x}$ is not free in φ_k. We take D_k to be

$$\varphi_k, \qquad \text{(hyp.)}$$

$$\varphi_k \to \forall \mathbf{x}\varphi_k, \qquad \text{(Ax.3)}$$

$$\forall \mathbf{x}\varphi_k. \qquad \text{(m.p.)}$$

Finally, if for some $i,j<k$ we have $\varphi_j = \varphi_i \to \varphi_k$, then by the induction hypothesis we already possess D_i and D_j. We let D_k be the deduction obtained by concatenating D_i and D_j and adding three more formulas, as follows:

$$D_i\begin{cases}\vdots\\ \forall \mathbf{x}\varphi_i,\end{cases}$$

$$D_j\begin{cases}\vdots\\ \forall \mathbf{x}(\varphi_i \to \varphi_k),\end{cases}$$

$$\forall \mathbf{x}(\varphi_i \to \varphi_k) \to \forall \mathbf{x}\varphi_i \to \forall \mathbf{x}\varphi_k, \qquad \text{(Ax.2)}$$

$$\forall \mathbf{x}\varphi_i \to \forall \mathbf{x}\varphi_k, \qquad \text{(m.p.)}$$

$$\forall \mathbf{x}\varphi_k. \qquad \text{(m.p.)}$$

This completes the proof. ∎

1.5. Remark. By Thm. 1.4 we have the following *law of generalization on variables*: if $\Phi \vdash \alpha$ and $\mathbf{x}$ is not free in Φ, then $\Phi \vdash \forall \mathbf{x}\alpha$. Two modified forms of this law are stated in the following problem.

1.6. Problem. Assuming that $\mathbf{x}$ is not free in Φ, β, show that

(a) if $\Phi \vdash \beta \to \alpha$, then $\Phi \vdash \beta \to \forall \mathbf{x}\alpha$,

(b) if $\Phi \vdash \alpha \to \beta$, then $\Phi \vdash \exists \mathbf{x}\alpha \to \beta$.

(To prove (b), begin by observing that[1] $\alpha \to \beta \vdash_0 \neg\beta \to \neg\alpha$.)

1.7. Lemma. *Given a deduction D of α from Φ, we can construct a deduction D^* of α from Φ such that every variable free in D^* is free in Φ or in α.*

[1] Here and in the sequel, whenever we make an assertion of the form $\Phi \vdash_0 \alpha$, where Φ is finite, that assertion may be verified e.g. by constructing a propositional confutation of Φ, $\neg\alpha$, or by truth tables (using Prob. 1.6.12 and Thm. 1.12.1).

PROOF. Let $\mathbf{x}_1,\ldots,\mathbf{x}_n$ be all the variables which are free in $\boldsymbol{D}$ but are free in neither $\boldsymbol{\Phi}$ nor $\boldsymbol{\alpha}$. By Thm. 1.4, we can construct a deduction $\boldsymbol{D}'$ of $\forall\mathbf{x}_1\boldsymbol{\alpha}$ from $\boldsymbol{\Phi}$ such that $\mathbf{x}_1$ is not free in $\boldsymbol{D}'$ but any variable free in $\boldsymbol{D}'$ is also free in $\boldsymbol{D}$. We extend $\boldsymbol{D}'$ by adding the two formulas $\forall\mathbf{x}_1\boldsymbol{\alpha}\rightarrow\boldsymbol{\alpha}$ (this is (Ax.4), with $\mathbf{x}_1$ as the term $\mathbf{t}$) and $\boldsymbol{\alpha}$ (by *modus ponens*). Thus we have obtained for $\boldsymbol{\alpha}$ a deduction $\boldsymbol{D}_1$ from $\boldsymbol{\Phi}$ such that $\mathbf{x}_1$ is not free in $\boldsymbol{D}_1$ but each variable free in $\boldsymbol{D}_1$ is also free in $\boldsymbol{D}$.

Continuing in this way, we successively eliminate $\mathbf{x}_2,\ldots,\mathbf{x}_n$ and get $\boldsymbol{D}^*$ as required. ∎

1.8. THEOREM. *If $\boldsymbol{\Phi}\vdash\boldsymbol{\alpha}$ and $\boldsymbol{\alpha}$ as well as all members of $\boldsymbol{\Phi}$ are sentences, then there is a deduction $\boldsymbol{D}$ of $\boldsymbol{\alpha}$ from $\boldsymbol{\Phi}$ such that $\boldsymbol{D}$ is made up of sentences only.* ∎

We say that formulas $\boldsymbol{\alpha}$ and $\boldsymbol{\beta}$ are *provably equivalent* if both $\boldsymbol{\alpha}\vdash\boldsymbol{\beta}$ and $\boldsymbol{\beta}\vdash\boldsymbol{\alpha}$. By the Deduction Theorem, this is tantamount to saying that both $\vdash\boldsymbol{\alpha}\rightarrow\boldsymbol{\beta}$ and $\vdash\boldsymbol{\beta}\rightarrow\boldsymbol{\alpha}$. But it can easily be verified that $\{\boldsymbol{\alpha}\rightarrow\boldsymbol{\beta},\ \boldsymbol{\beta}\rightarrow\boldsymbol{\alpha}\}\vdash_0\boldsymbol{\alpha}\leftrightarrow\boldsymbol{\beta}$, $\boldsymbol{\alpha}\leftrightarrow\boldsymbol{\beta}\vdash_0\boldsymbol{\alpha}\rightarrow\boldsymbol{\beta}$ and $\boldsymbol{\alpha}\leftrightarrow\boldsymbol{\beta}\vdash_0\boldsymbol{\beta}\rightarrow\boldsymbol{\alpha}$. Therefore $\boldsymbol{\alpha}$ and $\boldsymbol{\beta}$ are provably equivalent iff $\vdash\boldsymbol{\alpha}\leftrightarrow\boldsymbol{\beta}$.

It is easy to verify that provable equivalence is indeed an equivalence relation between formulas.

1.9. LEMMA. *Let $\boldsymbol{\alpha}$ be a subformula of $\boldsymbol{\beta}$, and let $\boldsymbol{\alpha}'$ be provably equivalent to $\boldsymbol{\alpha}$. Let $\boldsymbol{\beta}'$ be obtained from $\boldsymbol{\beta}$ by replacing (zero or more) occurrences of $\boldsymbol{\alpha}$ by occurrences of $\boldsymbol{\alpha}'$. Then $\boldsymbol{\beta}'$ is provably equivalent to $\boldsymbol{\beta}$.*

PROOF. By induction on $\deg\boldsymbol{\beta}$. If $\boldsymbol{\alpha}=\boldsymbol{\beta}$, then $\boldsymbol{\beta}'$ is $\boldsymbol{\alpha}$ or $\boldsymbol{\alpha}'$ and there is nothing to prove. From now on we shall assume that $\boldsymbol{\alpha}$ is a *proper* subformula of $\boldsymbol{\beta}$, and in particular that $\boldsymbol{\beta}$ is not atomic. We shall make use of Prob. 1.4.4.

If $\boldsymbol{\beta}=\neg\boldsymbol{\gamma}$, then $\boldsymbol{\beta}'=\neg(\boldsymbol{\gamma}')$, where $\boldsymbol{\gamma}'$ is obtained from $\boldsymbol{\gamma}$ by replacing occurrences of $\boldsymbol{\alpha}$ by occurrences of $\boldsymbol{\alpha}'$. By the induction hypothesis we have $\vdash\boldsymbol{\gamma}\leftrightarrow\boldsymbol{\gamma}'$. But it is easy to verify that

$$\boldsymbol{\gamma}\leftrightarrow\boldsymbol{\gamma}'\vdash_0\neg\boldsymbol{\gamma}\leftrightarrow\neg(\boldsymbol{\gamma}');$$

hence

$$\vdash\neg\boldsymbol{\gamma}\leftrightarrow\neg(\boldsymbol{\gamma}').$$

If $\boldsymbol{\beta}=\boldsymbol{\gamma}\rightarrow\boldsymbol{\delta}$, then $\boldsymbol{\beta}'=\boldsymbol{\gamma}'\rightarrow\boldsymbol{\delta}'$, where $\boldsymbol{\gamma}'$ is obtained from $\boldsymbol{\gamma}$ as above, and $\boldsymbol{\delta}'$ is obtained from $\boldsymbol{\delta}$ in a similar way. By the induction hypothesis we have $\vdash\boldsymbol{\gamma}\leftrightarrow\boldsymbol{\gamma}'$ and $\vdash\boldsymbol{\delta}\leftrightarrow\boldsymbol{\delta}'$. But it is easy to verify that

$$\{\boldsymbol{\gamma}\leftrightarrow\boldsymbol{\gamma}',\ \boldsymbol{\delta}\leftrightarrow\boldsymbol{\delta}'\}\vdash_0\boldsymbol{\gamma}\rightarrow\boldsymbol{\delta}\leftrightarrow\boldsymbol{\gamma}'\rightarrow\boldsymbol{\delta}'.$$

Hence

$$\vdash \gamma \to \delta \leftrightarrow \gamma' \to \delta'.$$

If $\beta = \forall x \gamma$, then $\beta' = \forall x \gamma'$, where γ' is obtained from γ as before. By the induction hypothesis we have $\vdash \gamma \to \gamma'$ and $\vdash \gamma' \to \gamma$. Generalizing on x (see 1.5) we get $\vdash \forall x(\gamma \to \gamma')$ and using

$$\forall x(\gamma \to \gamma') \to \forall x \gamma \to \forall x \gamma' \qquad \text{(Ax.2)}$$

we obtain

$$\vdash \forall x \gamma \to \forall x \gamma'.$$

Similarly,

$$\vdash \forall x \gamma' \to \forall x \gamma. \quad ∎$$

1.10. THEOREM. *If $\beta \sim \beta'$ then β and β' are provably equivalent.*

PROOF. By Lemma 1.9 it is enough to show that if $\forall z\,[\alpha(x/z)]$ is obtained from $\forall x \alpha$ by alphabetic change then these two formulas are provably equivalent.

Since z is free for x in α, the formula $\forall x \alpha \to \alpha(x/z)$ is an instance of (Ax.4). Thus $\forall x \alpha \vdash \alpha(x/z)$. Since z is not free in α, it cannot be free in $\forall x \alpha$ either; hence we may generalize on z (see 1.5) and obtain

$$\forall x \alpha \vdash \forall z\,[\alpha(x/z)].$$

Since an alphabetic change is reversible, $\forall x \alpha$ is obtainable from $\forall z\,[\alpha(x/z)]$ by alphabetic change, and we can prove similarly that $\forall z\,[\alpha(x/z)] \vdash \forall x \alpha$. ∎

Note that Lemma 1.9 and Thm. 1.10 provide new proofs for Thms. 2.3.6. and 2.3.8. (By Thm. 1.2, formulas that are provably equivalent are also logically equivalent.)

1.11. THEOREM. *For every formula α, variable x and term t,*

$$\vdash \forall x \alpha \to \alpha(x/t), \qquad \vdash \alpha(x/t) \to \exists x \alpha.$$

PROOF. By Def. 2.3.9, $\alpha(x/t) = \alpha'(x/t)$, where α' a certain variant of α such that t is free for x in α'. By (Ax.4) we have $\forall x \alpha' \vdash \alpha(x/t)$ and by Thm. 1.10 we also have $\forall x \alpha \vdash \forall x \alpha'$. Hence $\forall x \alpha \vdash \alpha(x/t)$ and by the Deduction Theorem

$$\vdash \forall x \alpha \to \alpha(x/t).$$

By what we have just proved, we have $\vdash \forall x \neg\alpha \rightarrow \neg\alpha(x/t)$. But (as can easily be verified)

$$\forall x \neg\alpha \rightarrow \neg\alpha(x/t) \vdash_0 \alpha(x/t) \rightarrow \neg\forall x \neg\alpha,$$

hence $\vdash \alpha(x/t) \rightarrow \neg\forall x \neg\alpha$, i.e.,

$$\vdash \alpha(x/t) \rightarrow \exists x\alpha. \qquad \blacksquare$$

1.12. THEOREM. *Let* **c** *be a constant which occurs neither in* Φ *nor in* α. *Given a deduction* D *of* $\alpha(x/c)$ *from* Φ *we can construct a deduction* D' *of* $\forall x\alpha$ from Φ.

PROOF. We choose a variable **y** which does not occur in D. Let D_1 be obtained from D by replacing every formula β by $\beta(c/y)$.

It is not difficult to verify that if β is an axiom then so is $\beta(c/y)$. The hypotheses (formulas of Φ) used in D are left unchanged because **c** does not occur in Φ. Also, every application of *modus ponens* is transformed into an application of *modus ponens*. Since **c** does not occur in α,

$$\alpha(x/c)(c/y) = \alpha(x/y).$$

Thus D_1 is a deduction of $\alpha(x/y)$ from Φ. Moreover, if Φ_0 is the subset of Φ consisting of those hypotheses that are actually used in D_1, then **y** does not occur in Φ_0 and D_1 is a deduction of $\alpha(x/y)$ from Φ_0. By Thm. 1.4 we get from D_1 a deduction D_2 of $\forall y[\alpha(x/y)]$ from Φ_0 and hence from Φ.

Clearly, $\forall y[\alpha(x/y)]$ is $\forall x\alpha$ or is obtained from $\forall x\alpha$ by alphabetic change. Therefore (either trivially, or as in Thm. 1.10) we obtain a deduction D_3 of $\forall x\alpha$ from $\forall y[\alpha(x/y)]$. From D_2 and D_3 we get the required D'. $\blacksquare$

1.13. REMARK. By Thm. 1.12 we have the following *law of generalization on constants:*

If $\Phi \vdash \alpha(x/c)$ and **c** does not occur in Φ nor in α, then $\Phi \vdash \forall x\alpha$.

1.14. PROBLEM. Using Thm. 1.11 show that if **x** does not occur in the term **t** then $\vdash \exists x(t=x)$.

1.15. PROBLEM. Show that if **x** does not occur in **t**, then $\vdash \exists! x(t=x)$. (For the definition of $\exists!$ see 2.10.1.)

A set Φ of formulas is *first-order inconsistent* if for some β both $\Phi \vdash \beta$ and $\Phi \vdash \neg\beta$. Otherwise Φ is *first-order consistent.*

We shall usually omit the qualification "first-order" and say just "inconsistent" and "consistent".

If Φ is an inconsistent set, then some finite subset of Φ must be inconsistent, because in deducing β and $\neg\beta$ from Φ we can only make use of a finite number of hypotheses (i.e., formulas of Φ). Conversely, if some subset of Φ is inconsistent, then Φ is clearly inconsistent. Thus a set Φ is consistent iff every finite subset of Φ is consistent. (Cf. Prob. 1.10.10.)

We can now show that the empty set is consistent. This fact is expressed by saying that the *first-order predicate calculus is consistent*.

1.16. PROBLEM. Let α^* be defined by recursion on deg α as follows:

$$(\mathbf{P}\mathbf{t}_1\mathbf{t}_2\ldots\mathbf{t}_n)^* = \mathbf{P}\mathbf{v}_1\mathbf{v}_1\ldots\mathbf{v}_1,$$

$$(\neg\beta)^* = \neg(\beta^*),$$

$$(\beta\to\gamma)^* = \beta^*\to\gamma^*,$$

$$(\forall \mathbf{x}\beta)^* = \beta^*.$$

Show that if α is a first-order axiom then α^* is either a tautology or the formula $\mathbf{v}_1 = \mathbf{v}_1$. Hence prove the consistency of the first-order predicate calculus by showing that if both $\vdash\beta$ and $\vdash\neg\beta$ then the empty set or $\{\mathbf{v}_1 = \mathbf{v}_1\}$ would be *propositionally* inconsistent.

1.17. THEOREM. *An inconsistent set of formulas is unsatisfiable.*
PROOF. Immediate from Thm. 1.2. ∎

From Thm. 1.17 we get another proof for the consistency of the first-order predicate calculus. For, the empty set of formulas is satisfied by every valuation. However, note that this proof, unlike that of Prob. 1.16, makes use of the highly non-constructive notion of satisfiability (see Remark 2.1.2).

1.18. THEOREM. *A set Φ of formulas is inconsistent iff $\Phi\vdash\alpha$ for every formula α.*
PROOF. Similar to that of Thm. 1.10.7. If Φ is inconsistent, then for some β both $\Phi\vdash\beta$ and $\Phi\vdash\neg\beta$. But for every α we have $\{\beta, \neg\beta\}\vdash_0\alpha$; hence $\Phi\vdash\alpha$. The converse is obvious. ∎

1.19. THEOREM. *For any Φ and α,*

(a) *$\Phi, \neg\alpha$ is inconsistent iff $\Phi\vdash\alpha$,*

(b) *Φ, α is inconsistent iff $\Phi\vdash\neg\alpha$.*

PROOF. (a) If $\Phi, \neg\alpha$ is inconsistent, then, by Thm. 1.18, $\Phi, \neg\alpha\vdash\alpha$.

Therefore, by the Deduction Theorem, $\Phi \vdash \neg\alpha \rightarrow \alpha$. But one can easily verify (e.g., by a propositional tableau) that $\neg\alpha \rightarrow \alpha \vdash_0 \alpha$. Hence $\Phi \vdash \alpha$. The converse is obvious.

(b) Similarly, if Φ, α is inconsistent, one shows that $\Phi \vdash \alpha \rightarrow \neg\alpha$. Also, it is readily verified that $\alpha \rightarrow \neg\alpha \vdash_0 \neg\alpha$. Hence $\Phi \vdash \neg\alpha$. The converse is again obvious. ∎

* § 2. The first-order predicate calculus and tableaux

In this section we shall do for first-order logic what we did for propositional logic in §11 of Ch. 1. As in Ch. 2, we say *tableau* when we mean *first-order tableau.*

2.1. THEOREM. *Let Φ be a finite set of formulas. Given a deduction of α from Φ, we can confute Φ, $\neg\alpha$.*

PROOF. The existence of such a confutation follows at once from Prob. 2.8.5 and Thm. 1.2. However, as in the proof of Thm. 1.11.1, we can get a direct construction of a confutation of Φ, $\neg\alpha$ from the given deduction of α from Φ. We proceed as in the proof of Thm. 1.11.1, but now using the Elimination Theorem 2.6.8.

We only have to verify that for each axiom α we can confute $\{\neg\alpha\}$. This is easily done. (For (Ax.2)–(Ax.4) see Prob. 2.4.3. For (Ax.8) observe that if we can confute $\{\neg\alpha\}$ then we can also confute $\{\neg\forall x\alpha\}$, because the $\neg\forall$-rule may be applied to $\neg\forall x\alpha$, yielding $\neg\alpha$.)

Observe that from Thm. 2.1 and Prob. 2.8.5 we get a new (and indirect) proof for Thm. 1.2. (Cf. Prob. 1.11.2.)

2.2. THEOREM. *Every finite inconsistent set of formulas can be confuted.*

PROOF. Similar to Thm. 1.11.3. ∎

Using Thm. 2.2, we obtain yet another constructive proof for the consistency of the first-order predicate calculus. We only need to verify that the empty set of formulas cannot be confuted. If $\mathscr{L}$ is without equality, this is obvious since no tableau rule can be applied to the empty set. If $\mathscr{L}$ does have equality, then the equality rules can always be applied. But it is easy to see that if Φ is a finite set of formulas in which $\neg$ does not occur (in particular, if Φ is empty) then in every tableau for Φ there must be at least one branch in which $\neg$ does not occur. Hence Φ cannot be confuted. (See Prob. 2.8.7.)

In order to prove the converse of Thm. 2.2 we shall need:

2.3. LEMMA. *If Φ is any set of formulas, each of the following six conditions is sufficient for Φ to be inconsistent:*

(a) *For some α, $\neg\neg\alpha \in \Phi$ and Φ, α is inconsistent.*

(b) *For some α and β, $\alpha \to \beta \in \Phi$ and both $\Phi, \neg\alpha$ and Φ, β are inconsistent.*

(c) *For some α and β, $\neg(\alpha \to \beta) \in \Phi$ and $\Phi, \alpha, \neg\beta$ is inconsistent.*

(d) *For some α, $\mathbf{x}$ and $\mathbf{t}$, $\forall \mathbf{x}\alpha \in \Phi$ and $\Phi, \alpha(\mathbf{x}/\mathbf{t})$ is inconsistent.*

(e) *For some α, $\mathbf{x}$ and $\mathbf{y}$, $\neg\forall \mathbf{x}\alpha \in \Phi$, and $\mathbf{y}$ is not free in Φ, and $\Phi, \neg\alpha(\mathbf{x}/\mathbf{y})$ is inconsistent.*

(f) *For some axiom α, Φ, α is inconsistent.*

PROOF. (a) Since Φ, α is inconsistent, we have $\Phi \vdash \neg\alpha$ by Thm. 1.19. But $\neg\neg\alpha \in \Phi$, hence $\Phi \vdash \neg\neg\alpha$, and Φ is inconsistent.

(b) Since $\Phi, \neg\alpha$ is inconsistent, we have $\Phi \vdash \alpha$ by Thm. 1.19. But $\alpha \to \beta \in \Phi$, hence $\Phi \vdash \beta$ (using *modus ponens*). On the other hand, $\Phi \vdash \neg\beta$ because Φ, β is inconsistent. Thus Φ is inconsistent.

(c) Since $\Phi, \alpha, \neg\beta$ is inconsistent, it follows that $\Phi, \alpha \vdash \beta$. Hence, by the Deduction Thm., $\Phi \vdash \alpha \to \beta$. But $\neg(\alpha \to \beta) \in \Phi$.

(d) Since $\Phi, \alpha(\mathbf{x}/\mathbf{t})$ is inconsistent, $\Phi \vdash \neg\alpha(\mathbf{x}/\mathbf{t})$. But $\forall \mathbf{x}\alpha \in \Phi$, hence, by Thm. 1.11., $\Phi \vdash \alpha(\mathbf{x}/\mathbf{t})$.

(e) Since $\Phi, \neg\alpha(\mathbf{x}/\mathbf{y})$ is inconsistent, we have $\Phi \vdash \alpha(\mathbf{x}/\mathbf{y})$. Because $\mathbf{y}$ is not free in Φ, we may generalize on $\mathbf{y}$ (see Remark 1.5) and obtain $\Phi \vdash \forall \mathbf{y}\,[\alpha(\mathbf{x}/\mathbf{y})]$. If $\mathbf{y}$ is $\mathbf{x}$, then this means $\Phi \vdash \forall \mathbf{x}\alpha$, and since $\neg\forall \mathbf{x}\alpha \in \Phi$, clearly Φ is inconsistent. Now suppose $\mathbf{y}$ is not $\mathbf{x}$. Then by Def. 2.3.9, $\alpha(\mathbf{x}/\mathbf{y}) = \alpha'(\mathbf{x}/\mathbf{y})$, where α' is a certain variant of α such that $\mathbf{y}$ is free for $\mathbf{x}$ in α'. Moreover, since $\mathbf{y}$ is not free in Φ, and $\neg\forall \mathbf{x}\alpha \in \Phi$, it follows that $\mathbf{y}$ is not free in α (because $\mathbf{y} \neq \mathbf{x}$) and hence $\mathbf{y}$ cannot be free in α'. Thus $\forall \mathbf{y}\,[\alpha'(\mathbf{x}/\mathbf{y})]$ is obtained from $\forall \mathbf{x}\alpha'$ by alphabetic change and we have

$$\forall \mathbf{x}\alpha \sim \forall \mathbf{x}\alpha' \sim \forall \mathbf{y}\,[\alpha'(\mathbf{x}/\mathbf{y})] = \forall \mathbf{y}\,[\alpha(\mathbf{x}/\mathbf{y})],$$

so, by Thm. 1.10,

$$\forall \mathbf{y}\,[\alpha(\mathbf{x}/\mathbf{y})] \vdash \forall \mathbf{x}\alpha.$$

Thus again $\Phi \vdash \forall \mathbf{x}\alpha$, and Φ is inconsistent.

(f) This is obvious, because if α is an axiom then whatever can be deduced from Φ, α can also be deduced from Φ. ∎

We can now prove the converse of Thm. 2.2:

2.4. THEOREM. *If Φ is a finite set of formulas having a confutation, then Φ is inconsistent.*

Proof. Similar to the proof of Thm. 1.11.4. We proceed by induction on the depth d of a given confutation of Φ.

If $d=0$, then Φ must contain both α and $\neg\alpha$ for some (atomic) α and hence Φ is inconsistent.

If $d>0$, we distinguish cases according to the rule that gave rise to level 1 of the given confutation. These are dealt with as in the proof of Thm. 1.11.4. In the cases corresponding to the three propositional rules we use parts (a), (b) and (c) of Lemma 2.3. For the quantifier rules we use parts (d) and (e) of that lemma. Finally for the equality rules we use part (f) of the lemma. ∎

2.5. **Problem.** Prove the converse of Thm. 2.1.

§ 3. Completeness of the first-order predicate calculus

We begin this section by proving the weak completeness of the first-order predicate calculus. Then we go on to prove the strong completeness of that calculus. The proof of the former result (but not that of the latter) will depend on §2. The reader who has skipped §2 should now proceed directly to Def. 3.4. Remark 3.2 should in this case be read after Thm. 3.13.

3.1. **Theorem.** *Every finite consistent set of formulas is satisfied by some valuation with a countable universe.*

Proof. If Φ is finite and consistent, then by Thm. 2.4 Φ cannot be confuted. Hence by Cor. 2.8.3 Φ is satisfied by a valuation with a countable universe. ∎

3.2. **Remark.** We can now see that consistency is invariant with respect to language. For, a set Φ is consistent iff every finite subset of Φ is consistent; and by Thms. 1.17 and 3.1 this is the same as every finite subset of Φ being satisfiable. But we know that satisfiablity is invariant with respect to language (see Remark 2.2.7). It follows that deducibility too is invariant with respect to language, for $\Phi \vdash \alpha$ iff $\Phi, \neg\alpha$ is inconsistent (see Thm. 1.19).

3.3. **Weak Completeness Theorem.** *If Φ is finite and $\Phi \models \alpha$, then $\Phi \vdash \alpha$. In particular, if $\models \alpha$ then $\vdash \alpha$.*

Proof. If $\Phi \models \alpha$, then $\Phi, \neg\alpha$ is unsatisfiable. Hence, by Thm. 3.1, $\Phi, \neg\alpha$ is inconsistent, so, by Thm. 1.19, $\Phi \vdash \alpha$. ∎

Thm. 3.3 is due to Gödel (see §6). Both this theorem and Thm. 3.1 are known as *Gödel's Completeness Theorem.*

The rest of this section is mainly devoted to strengthening Thms. 3.1 and 3.3. Whereas our proof of Thm. 3.1 makes an excursion *via* tableaux, the stronger version will be proved directly.

3.4. DEFINITION. A set $\mathbf{\Phi}$ of $\mathscr{L}$-formulas is *maximal first-order consistent in* $\mathscr{L}$ if $\mathbf{\Phi}$ is first-order consistent but is not a proper subset of any first-order consistent set of $\mathscr{L}$-formulas.

As usual, we shall omit the qualification "first-order". When only the language $\mathscr{L}$ is being considered, we may also omit the qualification "in $\mathscr{L}$", but it must be stressed that maximal consistency is *not* invariant with respect to language.

3.5. THEOREM. *A set* $\mathbf{\Phi}$ *is maximal consistent iff both of the following conditious hold:*

(a) $\mathbf{\Phi}$ *is consistent.*

(b) *For every formula* $\boldsymbol{\alpha}$, $\boldsymbol{\alpha} \in \mathbf{\Phi}$ *or* $\neg\boldsymbol{\alpha} \in \mathbf{\Phi}$.

PROOF. Similar to Thm. 1.13.1. ∎

3.6. THEOREM. *If* $\mathbf{\Phi}$ *is maximal consistent and* $\mathbf{\Phi} \vdash \boldsymbol{\alpha}$, *then* $\boldsymbol{\alpha} \in \mathbf{\Phi}$.

PROOF. Otherwise we would have $\neg\boldsymbol{\alpha} \in \mathbf{\Phi}$, by Thm. 3.5. Hence $\mathbf{\Phi} \vdash \neg\boldsymbol{\alpha}$, making $\mathbf{\Phi}$ inconsistent. ∎

3.7. THEOREM. *Let* $\mathbf{\Psi}$ *be maximal consistent in* $\mathscr{L}$. *Then conditions* (1)–(5) *of Def.* 2.7.1. *hold. If* $\mathscr{L}$ *is a language with equality, then conditions* (7)–(9) *of Def.* 2.7.1 *hold as well.*

PROOF. (1) We cannot have both $\boldsymbol{\varphi} \in \mathbf{\Psi}$ and $\neg\boldsymbol{\varphi} \in \mathbf{\Psi}$ for atomic — or indeed for any — formula $\boldsymbol{\varphi}$, because $\mathbf{\Psi}$ is consistent.

(2) If $\neg\neg\boldsymbol{\alpha} \in \mathbf{\Psi}$, then $\boldsymbol{\alpha} \in \mathbf{\Psi}$ by Thm. 3.6, because $\neg\neg\boldsymbol{\alpha} \vdash_0 \boldsymbol{\alpha}$.

(3) Suppose $\boldsymbol{\alpha} \rightarrow \boldsymbol{\beta} \in \mathbf{\Psi}$. Then if $\neg\boldsymbol{\alpha} \notin \mathbf{\Psi}$ we have $\boldsymbol{\alpha} \in \mathbf{\Psi}$ by Thm. 3.5. Therefore (using *modus ponens*) $\mathbf{\Psi} \vdash \boldsymbol{\beta}$ and hence $\boldsymbol{\beta} \in \mathbf{\Psi}$ by Thm. 3.6. Thus $\neg\boldsymbol{\alpha} \in \mathbf{\Psi}$ or $\boldsymbol{\beta} \in \mathbf{\Psi}$.

(4) If $\neg(\boldsymbol{\alpha} \rightarrow \boldsymbol{\beta}) \in \mathbf{\Psi}$, then $\mathbf{\Psi} \vdash \boldsymbol{\alpha}$ and $\mathbf{\Psi} \vdash \neg\boldsymbol{\beta}$ because $\neg(\boldsymbol{\alpha} \rightarrow \boldsymbol{\beta}) \vdash_0 \boldsymbol{\alpha}$ and $\neg(\boldsymbol{\alpha} \rightarrow \boldsymbol{\beta}) \vdash_0 \neg\boldsymbol{\beta}$. Therefore $\boldsymbol{\alpha} \in \mathbf{\Psi}$ and $\neg\boldsymbol{\beta} \in \mathbf{\Psi}$ by Thm. 3.6.

(5) If $\forall \mathbf{x}\boldsymbol{\alpha} \in \mathbf{\Psi}$, then for every $\mathscr{L}$-term $\mathbf{t}$ we have $\mathbf{\Psi} \vdash \boldsymbol{\alpha}(\mathbf{x}/\mathbf{t})$ by Thm. 1.11. Hence $\boldsymbol{\alpha}(\mathbf{x}/\mathbf{t}) \in \mathbf{\Psi}$ by Thm. 3.6.

(7)–(9) Every axiom of $\mathscr{L}$ — and in particular every instance of (Ax.5), (Ax.6) and (Ax.7) — is provable, and hence deducible from $\mathbf{\Psi}$. Thus by Thm. 3.6 every axiom belongs to $\mathbf{\Psi}$. ∎

The reader will have noticed that condition (6) of Def. 2.7.1 has been omitted in Thm. 3.7. Indeed, this condition does not necessarily hold for a maximal consistent set.

EXAMPLE. Let $\mathscr{L}$ be the language with equality but without any extra-logical symbols. Let U be a set having two members u and v. Let σ be the valuation with U as universe, such that $\mathbf{x}^\sigma = u$ for every variable $\mathbf{x}$. Define Ψ as the set of all $\mathscr{L}$-formulas α such that $\sigma \models \alpha$. Then Ψ is consistent by Thm. 1.17 because $\sigma \models \Psi$. Also, for each α, either $\alpha \in \Psi$ or $\neg\alpha \in \Psi$, so, by Thm. 3.5, Ψ is maximal consistent.

Now consider the formula $\neg\forall\mathbf{x}(\mathbf{x}=\mathbf{y})$, where $\mathbf{x}$ and $\mathbf{y}$ are distinct variables. It is easy to see that this formula is satisfied by σ. (Indeed, $\sigma(\mathbf{x}/v)$ does not satisfy $\mathbf{x}=\mathbf{y}$, hence σ does not satisfy $\forall\mathbf{x}(\mathbf{x}=\mathbf{y})$.)

On the other hand, the only terms in $\mathscr{L}$ are the variables and clearly σ does not satisfy $\mathbf{z}\neq\mathbf{y}$ for any variable $\mathbf{z}$.

3.8. DEFINITION. Ψ is a *Henkin set in* $\mathscr{L}$ if Ψ is maximal consistent in $\mathscr{L}$ and whenever $\neg\forall\mathbf{x}\alpha \in \Psi$ then for some $\mathscr{L}$-term $\mathbf{t}$ also $\neg\alpha(\mathbf{x}/\mathbf{t}) \in \Psi$.

Thus a Henkin set is a maximal consistent set for which condition (6) of Def. 2.7.1 *does* hold. It follows at once from Thm. 3.7 that every Henkin set in $\mathscr{L}$ is a Hintikka set in $\mathscr{L}$.

3.9. LEMMA. *If Ψ is a Henkin set in $\mathscr{L}$, then Ψ is satisfied by some valuation whose universe has cardinality not greater than the cardinality of the set of all $\mathscr{L}$-terms.*

PROOF. Immediate from Thm. 2.7.5. ∎

3.10. DEFINITION. By the *cardinality of* $\mathscr{L}$, (briefly, $\|\mathscr{L}\|$) we mean the cardinality of the set of all $\mathscr{L}$-symbols.

For the rest of this section we put $\|\mathscr{L}\| = \lambda$.

3.11. THEOREM. *The set of all $\mathscr{L}$-terms has cardinality $\leqslant\lambda$ and the set of all $\mathscr{L}$-formulas has cardinality λ.*

PROOF. Each term or formula is an $\mathscr{L}$-string, i.e., a finite sequence of $\mathscr{L}$-symbols. Since λ is clearly infinite (there are denumerably many variables!) it follows that the set of $\mathscr{L}$-strings has cardinality λ. Thus both the set of $\mathscr{L}$-terms and the set of $\mathscr{L}$-formulas have cardinality $\leqslant\lambda$.

It remains to show that the set of $\mathscr{L}$-formulas has cardinality $\geqslant\lambda$. Since λ is infinite, at least one of the following sets must have cardinality λ:

(a) The set of all variables and constants.

(b) The set of all function symbols other than constants.

(c) The set of all predicate symbols.

In each of these cases it is easy to see that the set of all atomic formulas of the form $\mathbf{Ptt}\ldots\mathbf{t}$ has cardinality λ. ∎

We now extend $\mathscr{L}$ to $\mathscr{L}'$ by adding new individual constants:

$$\{\mathbf{c}_\xi: \xi<\lambda\}.$$

(Below we shall continue to refer to these as *the new constants*.) It is clear that $\|\mathscr{L}'\|=\|\mathscr{L}\|=\lambda$. Also, the set of $\mathscr{L}'$-terms has cardinality λ.

3.12. THEOREM. *Let* $\mathbf{\Phi}$ *be a consistent set of* $\mathscr{L}$*-formulas. Then there exists a set* $\mathbf{\Psi}$ *such that* $\mathbf{\Phi}\subseteq\mathbf{\Psi}$ *and* $\mathbf{\Psi}$ *is a Henkin set in* $\mathscr{L}'$.

PROOF. By Thm. 3.11, the set of all $\mathscr{L}'$-formulas has cardinality λ. We fix a well-ordering

$$\{\boldsymbol{\varphi}_\xi: \xi<\lambda\}$$

of this set. By transfinite recursion, we define for every $\xi\leqslant\lambda$ a set $\mathbf{\Phi}_\xi$ of $\mathscr{L}'$-formulas such that

(1) $\mathbf{\Phi}_\eta\subseteq\mathbf{\Phi}_\xi$ for all $\eta<\xi$,

(2) $\mathbf{\Phi}_\xi$ is consistent,

(3) Only finitely many, or at most $|\xi|$ new constants occur in $\mathbf{\Phi}_\xi$.

We put $\mathbf{\Phi}_0=\mathbf{\Phi}$. Then for $\xi=0$ condition (1) holds vacuously, (3) holds because $\mathbf{\Phi}$ is in $\mathscr{L}$ and has *no* new constants, and (2) holds by assumption.[1]

Now suppose that $0<\zeta\leqslant\lambda$ and that for all $\xi<\zeta$ the $\mathbf{\Phi}_\xi$ have been defined in accordance with (1), (2) and (3). If ζ is a limit ordinal (in particular if $\zeta=\lambda$), we put $\mathbf{\Phi}_\zeta=\bigcup\{\mathbf{\Phi}_\xi: \xi<\zeta\}$. Then by definition $\mathbf{\Phi}_\xi\subseteq\mathbf{\Phi}_\zeta$ for all $\xi<\zeta$. Also, $\mathbf{\Phi}_\zeta$ is consistent because every finite subset of $\mathbf{\Phi}_\zeta$ is included in some $\mathbf{\Phi}_\xi$, where $\xi<\zeta$, and $\mathbf{\Phi}_\xi$ is consistent by (2). Finally, since (3) holds for all $\xi<\zeta$, it is easy to see that (3) holds for ζ as well.

If ζ is a successor ordinal, say $\zeta=\xi+1$, we distinguish three cases.

Case 1. If $\mathbf{\Phi}_\xi\cup\{\boldsymbol{\varphi}_\xi\}$ is inconsistent, we put $\mathbf{\Phi}_{\xi+1}=\mathbf{\Phi}_\xi$. Then (1), (2) and (3) clearly hold for $\xi+1$.

Case 2. If $\mathbf{\Phi}_\xi\cup\{\boldsymbol{\varphi}_\xi\}$ is consistent and $\boldsymbol{\varphi}_\xi$ is not of the form $\neg\forall\mathbf{x}\boldsymbol{\alpha}$, we put $\mathbf{\Phi}_{\xi+1}=\mathbf{\Phi}_\xi\cup\{\boldsymbol{\varphi}_\xi\}$. Again, (1), (2) and (3) clearly hold for $\xi+1$.

Case 3. If $\mathbf{\Phi}_\xi\cup\{\boldsymbol{\varphi}_\xi\}$ is consistent and $\boldsymbol{\varphi}_\xi$ has the form $\neg\forall\mathbf{x}\boldsymbol{\alpha}$, then by (3) there exists a new constant $\mathbf{c}$ which does not occur in $\mathbf{\Phi}_\xi$ nor in $\boldsymbol{\varphi}_\xi$.

[1] $\mathbf{\Phi}$ was assumed consistent in $\mathscr{L}$. To see that it is also consistent in $\mathscr{L}'$ we can use Remark 3.2 or — in order to make the present proof independent of the method of tableaux — we can argue directly as follows. If $\mathbf{\Phi}$ were inconsistent in $\mathscr{L}'$ then for every $\mathscr{L}$-formula $\boldsymbol{\gamma}$ we would have a deduction of $\boldsymbol{\gamma}$ from $\mathbf{\Phi}$ in $\mathscr{L}'$. This deduction may contain new constants, but these may be replaced (as in the proof of Thm. 1.12) by variables which do not already occur in the deduction. Thus we would obtain a deduction of $\boldsymbol{\gamma}$ from $\mathbf{\Phi}$ in $\mathscr{L}$, so $\mathbf{\Phi}$ would be inconsistent in $\mathscr{L}$.

We take such **c** and put $\mathbf{\Phi}_{\xi+1}=\mathbf{\Phi}_\xi \cup \{\varphi_\xi, \neg\alpha(\mathbf{x}/\mathbf{c})\}$. Then (1) and (3) clearly hold for $\xi+1$. To see that (2) holds as well, suppose $\mathbf{\Phi}_{\xi+1}$ were inconsistent. Then by Thm. 1.19 we have $\mathbf{\Phi}_\xi, \varphi_\xi \vdash \alpha(\mathbf{x}/\mathbf{c})$ and hence $\mathbf{\Phi}_\xi, \varphi_\xi \vdash \forall \mathbf{x}\alpha$ by Rem. 1.13. Since $\varphi_\xi = \neg\forall\mathbf{x}\alpha$, this contradicts our assumption that $\mathbf{\Phi}_\xi \cup \{\varphi_\xi\}$ is consistent.

We put $\mathbf{\Psi}=\mathbf{\Phi}_\lambda$. Then by (1) we have $\mathbf{\Phi}=\mathbf{\Phi}_0 \subseteq \mathbf{\Psi}$, and, by (2), $\mathbf{\Psi}$ is consistent.

To see that $\mathbf{\Psi}$ is *maximal* consistent in $\mathscr{L}'$, let α be any $\mathscr{L}'$-formula not belonging to $\mathbf{\Psi}$. Then $\alpha=\varphi_\xi$ for some $\xi<\lambda$. Since $\alpha \notin \mathbf{\Psi}$, it follows that $\mathbf{\Phi}_{\xi+1}$ must have been defined by Case 1, for in the other two cases we have $\alpha=\varphi_\xi \in \mathbf{\Phi}_{\xi+1} \subseteq \mathbf{\Psi}$. Thus $\mathbf{\Phi}_\xi \cup \{\alpha\}$ is inconsistent, and $\mathbf{\Psi} \cup \{\alpha\}$ must also be inconsistent.

Finally, if $\neg\forall\mathbf{x}\alpha \in \mathbf{\Psi}$ then $\neg\forall\mathbf{x}\alpha=\varphi_\xi$ for some $\xi<\lambda$. $\mathbf{\Phi}_\xi \cup \{\neg\forall\mathbf{x}\alpha\}$ is a subset of $\mathbf{\Psi}$, and is therefore consistent. It follows that $\mathbf{\Phi}_{\xi+1}$ must have been defined by Case 3 and hence for some **c** we have $\neg\alpha(\mathbf{x}/\mathbf{c}) \in \mathbf{\Phi}_{\xi+1}$.

Thus $\mathbf{\Psi}$ has all the required properties. ∎

3.13. THEOREM. *If $\mathbf{\Phi}$ is a consistent set of $\mathscr{L}$-formulas, then $\mathbf{\Phi}$ is satisfied by some valuation whose universe has cardinality* $\leqslant \|\mathscr{L}\|$.
PROOF. Immediate from Lemma 3.9, Thm. 3.11, and Thm. 3.12. ∎

3.14. STRONG COMPLETENESS THEOREM. *If $\mathbf{\Phi} \models \alpha$, then $\mathbf{\Phi} \vdash \alpha$.*
PROOF. If $\mathbf{\Phi} \models \alpha$ then $\mathbf{\Phi}, \neg\alpha$ is unsatisfiable. Hence, by Thm. 3.13, $\mathbf{\Phi}, \neg\alpha$ is inconsistent. Then, by Thm. 1.19, $\mathbf{\Phi} \vdash \alpha$. ∎

Both Thm. 3.14 and Thm. 3.13 are known as *Henkin's Completeness Theorem*.

We end this section with two important semantic consequences of Thm. 3.13.

3.15. THEOREM. *If $\mathbf{\Phi}$ is a satisfiable set of $\mathscr{L}$-formulas, then $\mathbf{\Phi}$ is satisfied by some valuation whose universe has cardinality* $\leqslant \|\mathscr{L}\|$.
PROOF. Immediate from Thm. 1.17 and Thm. 3.13. ∎

3.16. COMPACTNESS THEOREM. *If every finite subset of $\mathbf{\Phi}$ is satisfiable, then $\mathbf{\Phi}$ is satisfiable.*
PROOF. By Thm. 1.17, every finite subset of $\mathbf{\Phi}$ is consistent. Hence $\mathbf{\Phi}$ is consistent. It follows from Thm. 3.13 that $\mathbf{\Phi}$ is satisfiable. ∎

Note that both Theorem 3.15 and the Compactness Theorem are purely semantic and do not directly involve any notion about the predicate calculus (such as consistency, deducibility, etc.). It is therefore natural to ask whether they can be proved without making such a long excursion through the predicate calculus. This is indeed the case, and in Ch. 5 we shall prove these theorems by purely semantic means.

*§ 4. First-order logic based on $\exists$

In some parts of this book it will be convenient to regard $\exists$ as primitive, and define $\forall x$ as an abbreviation for $\neg\exists x\neg$ (cf. Prob. 2.1.4). All the material covered in Ch. 2 can easily be adapted to this setting. Only slight (and obvious) modifications are needed. For example, in the tableau method we have to take the $\neg\exists$ and $\exists$ rules instead of the $\forall$ and $\neg\forall$ rules.

In Def. 2.7.1, conditions (5) and (6) should be replaced by:

(5′) If $\neg\exists x\alpha \in \Psi$, then $\neg\alpha(x/t) \in \Psi$ for every $\mathscr{L}$-term t.

(6′) If $\exists x\alpha \in \Psi$, then $\alpha(x/t) \in \Psi$ for some $\mathscr{L}$-term t.

In the predicate calculus we may add a new axiom scheme, e.g.,

$$\forall x(\alpha \to \beta) \to \exists x\alpha \to \exists x\beta,$$

and modify (Ax.8) to cover generalizations of these new axioms as well. The material of the present chapter up to Lemma 1.9 requires no further change. In the last part of the proof of Lemma 1.9 we can use the new axiom scheme to show that $\exists x\gamma$ and $\exists x\gamma'$ are provably equivalent. The proof of Thm. 1.10 can be modified in a similar way. The rest of the material of the present chapter requires only slight and obvious changes.

§ 5. What have we achieved?

One of the main tasks of logic is to clarify and characterize the notions of logical consequence and logical validity. Let us now look back and see to what extent we have accomplished this task. Of course, since we have confined ourselves to first-order languages, we can consider logical consequence and logical validity only for such languages.

In §1 of Ch. 2 we explicated these two notions by reducing them to the notion of satisfaction. This, in turn, was characterized in the BSD. Now, the BSD is direct and natural because it merely spells out the intended meaning of the logical symbols. But it is highly non-constructive.

In the course of Ch. 2 and the present chapter, we have been able to discover more constructive (and more informative) characterizations of satisfiability, logical consequence and logical validity.

The method of tableaux can be used for detecting unsatisfiability. Given a finite set Φ of formulas, we can systematically search for a confutation of Φ (e.g., as described in §8 of Ch. 2). While this does not constitute an effective procedure for deciding whether or not Φ is satisfiable, we know at least that if Φ is unsatisfiable this fact will eventually be discovered.

Since α is logically valid iff $\neg\alpha$ is unsatisfiable, we also have a method for detecting logical validity.

Another method of detecting logical validity is provided by the first-order predicate calculus. By the Completeness Theorem, $\models\alpha$ iff $\vdash\alpha$. So, if we want to know whether α is logically valid, we can search systematically for a first-order proof of α. *If* such a proof exists, we shall sooner or later discover it. (On the other hand, if no such proof exists, this fact may never be revealed by our procedure.)

More generally, suppose Φ is a countable set of hypotheses and we are given an enumeration of Φ,

$$\Phi=\{\varphi_i:\ i=0,1,2,\ldots\}.$$

Let α be any formula. Then $\Phi\models\alpha$ iff there is a deduction of α from Φ in the language $\mathscr{L}_{\Phi,\alpha}$ (which is the "poorest" first-order language in which α and all members of Φ are formulas).

The language $\mathscr{L}=\mathscr{L}_{\Phi,\alpha}$ is denumerable (i.e., $\|\mathscr{L}\|=\aleph_0$), so we can effectively enumerate one by one all deductions from Φ in $\mathscr{L}$. (For example, we can first fix some effective enumeration of all axioms in $\mathscr{L}$. Now, let $\mathscr{D}_n$ be the set of all deductions whose length is at most n and in which not more than the first n axioms and the first n hypotheses are used. Then $\mathscr{D}_n$ is finite and we can fix some ordering in $\mathscr{D}_n$. The different $\mathscr{D}_n$ can be arranged in order of increasing n). If $\Phi\models\alpha$, then sooner or later we shall discover a deduction of α from Φ. (On the other hand, if α is not a logical consequence of Φ, we may never discover this.)

The procedures described above may not be efficient. However, the important fact is that they show *in principle* that, e.g., every logical truth can eventually be discovered.

Let us point out another consequence of the Completeness Theorem (or, more precisely, of Thm. 3.15). Initially, the notion of satisfiability was defined with reference to valuations whose universes may be arbitrary *classes*. From Thm. 3.15 it follows, however, that it is enough to consider universes which are *sets*. Moreover, only sets whose cardinality is $\leqslant\|\mathscr{L}\|$

need be considered. In particular, if $\mathscr{L}$ is denumerable, then a set $\mathbf{\Phi}$ of $\mathscr{L}$-formulas is satisfiable iff it is satisfied by some valuation with countable universe.

The results mentioned in this section constitute considerable achievements of the formal method (and the formal approach to mathematics). However, in Ch. 7 we shall see that — just because of these positive results — the formal method has some very crucial inherent limitations.

§ 6. Historical and bibliographical remarks

The predicate calculus is essentially due to FREGE [1879]. (The version used in this book differs from Frege's in notation as well as in the choice of axioms and rules of inference.) Frege allowed quantification not only over individual variables but also over predicate symbols, in a way which made his system inconsistent. The flaw is corrected in WHITEHEAD and RUSSELL [1910]. GÖDEL [1930] proved the Weak Completeness Theorem for the first-order predicate calculus (in the version of Whitehead and Russell). In the same paper he also proved the Strong Completeness Theorem and the Compactness Theorem for the case in which the set $\mathbf{\Phi}$ is denumerable. The full Strong Completeness Theorem is due to HENKIN [1949] and the full Compactness Theorem is due to MAL'CEV [1936]. Theorem 3.15 was first proved by LÖWENHEIM [1915] for single formulas, but his proof had several gaps. SKOLEM [1920] gave the first complete proof and generalized the theorem to denumerable sets of formulas.

CHAPTER 4

BOOLEAN ALGEBRAS

For any first order language $\mathscr{L}$, we may regard the logical symbols $\wedge$, $\vee$, $\neg$ as *operations* on the set of all formulas of $\mathscr{L}$. If we identify provably equivalent formulas, then these operations define an algebraic structure B whose properties are determined by $\mathscr{L}$. Accordingly, a knowledge of the properties of B will yield information about $\mathscr{L}$. B is an example of a special kind of algebraic structure called a *Boolean algebra*: the general theory of such structures forms the theme of this chapter. Later on (Ch. 5, §§3, 5, 6) we shall employ some of the results we obtain in proving important theorems in model theory. Some of the topics treated in this chapter (those to be found in sections marked with a *) will not be used in the sequel; they have been included to make the chapter a reasonably broad introduction to the theory.

The content of this chapter is entirely independent of that of earlier chapters.

§ 1. Lattices

A *lattice* is a (non-empty) partially ordered set[1] $\langle L, \leqslant \rangle$ in which each pair of elements x,y has a supremum — denoted by $x \vee y$ — and an infimum — denoted by $x \wedge y$. $x \vee y$ and $x \wedge y$ are often called, respectively, the *join* and *meet* of x and y. Clearly, we have, for any elements x,y of a lattice,

$$x \leqslant y \Leftrightarrow x \wedge y = x \Leftrightarrow x \vee y = y.$$

1.1. PROBLEM. Show that the following identities hold in any lattice:

$$x \wedge y = y \wedge x, \qquad x \vee y = y \vee x;$$
$$x \vee (y \vee z) = (x \vee y) \vee z, \quad x \wedge (y \wedge z) = (x \wedge y) \wedge z;$$
$$(x \wedge y) \vee y = y, \qquad (x \vee y) \wedge y = y.$$

[1] When discussing a partially ordered set $\langle X, \leqslant \rangle$ we shall frequently commit the pardonable sin of identifying it with its underlying set X.

1.2. EXAMPLES. (i) Any totally ordered set $\langle X, \leqslant \rangle$ is a lattice; clearly in this case we have

$$x \wedge y = \min(x,y), \qquad x \vee y = \max(x,y).$$

(ii) For any set Y, the power set $\mathsf{P}Y$ of Y is a lattice under the partial ordering of set inclusion. In this case we have, for any $A,B \in \mathsf{P}Y$,

$$A \wedge B = A \cap B, \qquad A \vee B = A \cup B.$$

(iii) The families of open sets and closed sets, respectively, of a topological space each form a lattice under the partial ordering of set inclusion. In these lattices joins and meets are the same as in example (ii).

(iv) Define a partial ordering $|$ on the set ω of natural numbers by stipulating that $m|n$ iff m is a divisor of n. Then $\langle \omega, | \rangle$ is a lattice in which

$$m \wedge n = \text{g.c.d.}(m,n), \qquad m \vee n = \text{l.c.m.}(m,n).$$

(v) Let **C** be the set of complex numbers; define a partial ordering $\preccurlyeq$ on **C** by setting $a+ib \preccurlyeq c+id$ (for real a, b, c, d) iff $a \leqslant c$ and $b \leqslant d$. Then $\langle \mathbf{C}, \preccurlyeq \rangle$ is a lattice. What are the join and meet of a pair of elements of **C**?

(vi) Let V be a vector space over a field F. Then the set $\mathscr{S}$ of all vector subspaces of V is a lattice under the partial ordering of inclusion. In this lattice the join of any two elements A and B is the subspace of V generated by $A \cup B$, while the meet of A and B is just $A \cap B$.

A mapping h from a lattice L to a lattice L' is called a (lattice) *homomorphism* if for all $x,y \in L$ we have

$$h(x \wedge y) = h(x) \wedge h(y), \quad h(x \vee y) = h(x) \vee h(y).$$

If h is one–one and onto, then h is called an *isomorphism* of L onto L', and L' is said to be *isomorphic* to L. A *sublattice* of a lattice L is a subset L' of L which is closed under the meet and join operations of L, i.e., if $x,y \in L'$, then $x \wedge y \in L'$ and $x \vee y \in L'$.

1.3. PROBLEM. Show that, if h is a homomorphism of a lattice L into a lattice L', then $h[L]$ is a sublattice of L' (called the *image* of L under h).

A lattice L is said to be *distributive* if the following conditions are satisfied by all $x,y,z \in L$:

$$x \wedge (y \vee z) = (x \wedge y) \vee (x \wedge z),$$

$$x \vee (y \wedge z) = (x \vee y) \wedge (x \vee z).$$

It is a curious fact that each of the above conditions implies the other. For example, suppose that the first condition holds. Then we have, using Prob. 1.1.,

$$\begin{aligned}(x \vee y) \wedge (x \vee z) &= [x \wedge (x \vee z)] \vee [y \wedge (x \vee z)] \\ &= x \vee [(y \wedge x) \vee (y \wedge z)] \\ &= [x \vee (y \wedge x)] \vee (y \wedge z) \\ &= x \vee (y \wedge z).\end{aligned}$$

The converse is proved similarly.

It is easy to show by induction that any (non-empty) *finite* subset $X = \{x_1, \ldots, x_n\}$ of a lattice has both a supremum and an infimum. We denote the supremum (or *join*) of X by $x_1 \vee \ldots \vee x_n$ or $\bigvee_{i=1}^n x_i$ and its infimum (or *meet*) by $x_1 \wedge \ldots \wedge x_n$ or $\bigwedge_{i=1}^n x_i$.

Notice that an *infinite* subset of a lattice need not have an infimum or a supremum. For instance, in the (totally ordered) lattice **Z** of integers, the set of *even* integers has neither. If, however, every subset[1] of a lattice L *does* have an infimum and a supremum, then L is said to be *complete*. Another curious fact about lattices is that, for a lattice to be complete, it suffices for each subset to have a supremum (or for each subset to have an infimum). We accord this fact the status of a theorem.

1.4. THEOREM. *Let L be a partially ordered set in which each subset has a supremum (or in which each subset has an infimum). Then L is a complete lattice.*

PROOF. Suppose that each subset of L has a supremum. Then to show that L is a complete lattice we have to prove that each subset of L has an infimum. So let X be a subset of L, and let Y be the set of *lower bounds* of X in L, i.e.

$$Y = \{z \in L : \forall x \in X\, [z \leqslant x]\}.$$

Then Y has a supremum y, and it is not hard to see that y is the infimum of X. A similar argument goes through in the case in which each subset of L has an infimum. ∎

If L is a lattice (not necessarily complete) and $X \subseteq L$, we shall denote the supremum or *join* of X (if it is has one) by $\bigvee X$ and the infimum or

[1] This includes the empty subset $\emptyset$ of L. Recall (Ch. 0) that the supremum of $\emptyset$ (if it has one) must be the *least* element of L, and that the infimum of $\emptyset$ (if it has one) must be the *greatest* element of L.

meet of X (if it has one) by $\bigwedge X$. If we suppose that X is indexed as $\{x_i\colon i\in I\}$, then we sometimes write the join and meet of X as $\bigvee_{i\in I}x_i$, $\bigwedge_{i\in I}x_i$ respectively.

1.5. EXAMPLES. (i) The power set lattice $\mathbf{P}Y$ of any set Y is a complete lattice, in which joins and meets coincide with set-theoretic unions and intersections respectively.

(ii) The lattices $\mathcal{O}$ and $\mathcal{C}$ of open sets and closed sets of a topological space X are both complete. If $\{A_i\colon i\in I\}\subseteq\mathcal{O}$, then in the lattice $\mathcal{O}$,

$$\bigwedge_{i\in I}A_i=(\bigcap_{i\in I}A_i)^\circ, \qquad \bigvee_{i\in I}A_i=\bigcup_{i\in I}A_i.$$

If $\{B_i\colon i\in I\}\subseteq\mathcal{C}$, then, in the lattice $\mathcal{C}$,

$$\bigwedge_{i\in I}B_i=\bigcap_{i\in I}B_i, \qquad \bigvee_{i\in I}B_i=(\bigcup_{i\in I}B_i)^-.$$

(Here A° and A^- denote the interior and closure, respectively, of a subset A of a topological space.)

Let L be a lattice. An element x of L is said to be the *least* element of L if $x\leqslant y$ for all $y\in L$, and the *greatest* element of L if $y\leqslant x$ for all $y\in L$. Clearly the least and greatest elements of a lattice — if they exist — must be unique. Accordingly we agree always to write 0 for the least and 1 for the greatest element of a given lattice, assuming they exist.

A lattice L is said to be *complemented* if it has a least and a greatest element and for each $x\in L$ there is $y\in L$ such that

$$x\vee y=1, \qquad x\wedge y=0.$$

Such an element y is called a *complement* of x. In general (cf. Prob. 1.7), an element of a lattice may have more than one complement — or none whatsoever. But the situation in a *distributive* lattice is much more pleasant as we now show.

1.6. THEOREM. *An element of a distributive lattice has at most one complement.*

PROOF. Let L be a distributive lattice, let $x\in L$ and suppose that y, y' are both complements of x. Then

$$x\vee y=1, \qquad x\vee y'=1,$$
$$x\wedge y=0, \qquad x\wedge y'=0,$$

and we have

$$y = y \vee 0 = y \vee (x \wedge y') = (y \vee x) \wedge (y \vee y')$$
$$= 1 \wedge (y \vee y')$$
$$= y \vee y'.$$

Similarly, $y' = y \vee y'$ so that $y = y'$. ∎

It follows from this result that in a complemented distributive lattice L we can define a mapping $x \mapsto x^*$ which sends each $x \in L$ onto its *unique* complement x^* in L. Clearly we have $x^{**} = x$ for each $x \in L$.

1.7. PROBLEM. (i) Consider the lattice $\mathscr{S}$ of vector subspaces of Euclidean 2-space E_2. Consider the following subspaces of E_2:

$$V_1 = \{\langle x, y\rangle \in E_2 \colon x = 0\},$$
$$V_2 = \{\langle x, y\rangle \in E_2 \colon y = 0\},$$
$$V_3 = \{\langle x, y\rangle \in E_2 \colon x = y\}.$$

Show that V_2 and V_3 are both complements of V_1 in $\mathscr{S}$, and deduce that $\mathscr{S}$ is not distributive.

(ii) Let L be a totally ordered set with a 0 and a 1. Show that, if $x \in L$ and $0 \neq x \neq 1$, then x has no complement in L.

§ 2. Boolean algebras

The stage is now set for us to introduce the central notion of the chapter.

A *Boolean algebra* is a *complemented distributive lattice*.

By the results and definitions of §1, the following identities hold in any Boolean algebra:

(i)	$x \vee y = y \vee x$,	$x \wedge y = y \wedge x$;
(ii)	$x \vee (y \vee z) = (x \vee y) \vee z$,	$x \wedge (y \wedge z) = (x \wedge y) \wedge z$;
(iii)	$(x \vee y) \wedge y = y$,	$(x \wedge y) \vee y = y$;
(iv)	$(x \vee y) \wedge z = (x \wedge z) \vee (y \wedge z)$,	$(x \wedge y) \vee z = (x \vee z) \wedge (y \vee z)$;
(v)	$x \vee x^* = 1$,	$x \wedge x^* = 0$.

The join, meet and complementation mappings in a Boolean algebra are called *Boolean operations*.

2.1. PROBLEM. Let $\langle B, \wedge, \vee, {}^*, 0, 1\rangle$ be a structure consisting of a set B, two binary operations $\wedge$, $\vee$ and one unary operation * on B, and two designated elements 0, 1 of B. Suppose that the identities (i)–(v) hold in this structure. Define a relation $\leqslant$ on B by

$$x \leqslant y \Leftrightarrow x \wedge y = x.$$

Show that $\langle B, \leqslant \rangle$ is a Boolean algebra in which $\wedge$, $\vee$ and * are the meet, join and complementation operations, and 0, 1 are the least and greatest elements.

Prob. 2.1 shows that a Boolean algebra may alternatively be defined as a structure $\langle B, \wedge, \vee, {}^*, 0, 1 \rangle$ satisfying identities (i)–(v). We shall use this characterization occasionally in the sequel.

Let **P** be a statement about Boolean algebras which involves just the Boolean operations $\wedge$, $\vee$, * and the two elements 0 and 1. The *dual* of **P** is obtained from **P** by interchanging $\wedge$ with $\vee$, and 0 with 1. The *principle of duality* for Boolean algebras is the observation that, if **P** holds in all Boolean algebras, then so does its dual. This can be proved as follows. Suppose that **P** holds in all Boolean algebras, and let $\langle B, \leqslant \rangle$ be any Boolean algebra. Define a new relation $\leqslant'$ on B by putting

$$x \leqslant' y \Leftrightarrow y \leqslant x.$$

It is then easy to verify that $\langle B, \leqslant' \rangle$ is a Boolean algebra. Moreover, the meet (join) operation in $\langle B, \leqslant' \rangle$ coincides with the join (meet) operation in $\langle B, \leqslant \rangle$, the least (greatest) element of $\langle B, \leqslant' \rangle$ with the greatest (least) element of $\langle B, \leqslant \rangle$, and complementation in $\langle B, \leqslant' \rangle$ with complementation in $\langle B, \leqslant \rangle$. Now, since **P** was assumed to hold in all Boolean algebras, it holds in $\langle B, \leqslant' \rangle$, and by the preceding remark, this means that the dual of **P** holds in $\langle B, \leqslant \rangle$. This establishes the principle of duality; we shall frequently employ it without further comment.

For example, it is easy to show that in any Boolean algebra we have $(x \wedge y)^* = x^* \vee y^*$. The principle of duality allows us to infer that $(x \vee y)^* = x^* \wedge y^*$.

2.2. Examples of Boolean algebras. (i) The power set lattice $\mathsf{P}X$ of a set X is a Boolean algebra called the *power set algebra* of X. In this algebra we have

$$A \wedge B = A \cap B, \quad A \vee B = A \cup B,$$

$$A^* = X - A,$$

$$0 = \emptyset, \quad 1 = X.$$

(ii) Let $\mathscr{L}$ be a first-order language, and let $\boldsymbol{\Phi}$ be the set of all formulas of $\mathscr{L}$. For $\boldsymbol{\alpha},\boldsymbol{\beta}\in\boldsymbol{\Phi}$ write $\boldsymbol{\alpha}\approx\boldsymbol{\beta}$ for $\vdash_0\boldsymbol{\alpha}\leftrightarrow\boldsymbol{\beta}$ (Ch. 1, §10). Then $\approx$ is an equivalence relation on $\boldsymbol{\Phi}$. For each $\boldsymbol{\alpha}\in\boldsymbol{\Phi}$ let

$$|\boldsymbol{\alpha}|=\{\boldsymbol{\beta}\in\boldsymbol{\Phi} : \boldsymbol{\alpha}\approx\boldsymbol{\beta}\}$$

be the $\approx$-class of $\boldsymbol{\alpha}$. Let B be the set of all $\approx$-classes. Define a relation $\leqslant$ on B by putting

$$|\boldsymbol{\alpha}|\leqslant|\boldsymbol{\beta}| \quad \text{iff} \quad \vdash_0\boldsymbol{\alpha}\rightarrow\boldsymbol{\beta}.$$

Then $\langle B,\leqslant\rangle$ is a Boolean algebra called the *propositional Lindenbaum algebra* of $\mathscr{L}$. The Boolean operations on B are defined by

$$|\boldsymbol{\alpha}|\wedge|\boldsymbol{\beta}|=|\boldsymbol{\alpha}\wedge\boldsymbol{\beta}|, \qquad |\boldsymbol{\alpha}|\vee|\boldsymbol{\beta}|=|\boldsymbol{\alpha}\vee\boldsymbol{\beta}|,$$

$$|\boldsymbol{\alpha}|^*=|\neg\boldsymbol{\alpha}|.$$

The greatest and least elements of B are given by

$$1=|\boldsymbol{\alpha}| \quad \text{for any } \boldsymbol{\alpha} \text{ such that } \vdash_0\boldsymbol{\alpha},$$

$$0=|\boldsymbol{\beta}| \quad \text{for any } \boldsymbol{\beta} \text{ such that } \vdash_0\neg\boldsymbol{\beta}.$$

(These matters are taken up in further detail in Ch. 5.)

(iii) Let X be a topological space, and let $\mathsf{C}X$ be the family of all simultaneously closed and open *(clopen)* subsets of X. Then, with the partial ordering of inclusion, $\mathsf{C}X$ is a Boolean algebra called the *clopen algebra* of X.

(iv) Let $\mathsf{F}X$ consist of all finite subsets and complements of finite subsets of a set X. Then with the partial ordering of inclusion, $\mathsf{F}X$ is a Boolean algebra called the *finite-cofinite* algebra of X.

(v) Define a total ordering $\leqslant$ on the two element set $2=\{0, 1\}$ by putting $0\leqslant 0$, $0\leqslant 1$, $1\leqslant 1$. Then $\langle 2,\leqslant\rangle$ is a Boolean algebra called the *minimal algebra.*

A *subalgebra* of a Boolean algebra B is a non-empty subset B' which is closed under the Boolean operations in B. It is clear that each subalgebra of B contains 0 and 1.

At this point we decree that a Boolean algebra *must have at least two elements,* which is equivalent to asserting that $0\neq 1$ in any Boolean algebra we consider. With this proviso, it is clear that each Boolean algebra includes as a subalgebra a copy of the minimal algebra, namely, the two element subset $\{0, 1\}$.

2.3. PROBLEM. Show that for any elements x,y of a Boolean algebra,

$$x \leq y \Leftrightarrow y^* \leq x^* \Leftrightarrow x \wedge y^* = 0 \Leftrightarrow x^* \vee y = 1;$$

$$(x \wedge y)^* = x^* \vee y^*, \qquad (x \vee y)^* = x^* \wedge y^*.$$

2.4. PROBLEM. (i) Show that the intersection of a family of subalgebras of a Boolean algebra is a subalgebra, and deduce that each subset X of a Boolean algebra B is included in a smallest subalgebra A (under inclusion). A is called the subalgebra *generated* by X. If $A = B$, then X is said to *generate* B.

(ii) Show that the subalgebra generated by a subset X of a Boolean algebra B consists of all elements of B of the form $\bigvee_{j=1}^{n} \bigwedge_{k=1}^{m_j} x_{jk}$ (or of all elements of the form $\bigwedge_{j=1}^{n} \bigvee_{k=1}^{m_j} x_{jk}$), where for all j,k, either $x_{jk} \in X$ or $x_{jk}^* \in X$. Deduce that if B' is a subalgebra of B, and $x \in B$, then the subalgebra of B generated by $B' \cup \{x\}$ consists of all elements of B of the form $(a \wedge x) \vee (b \wedge x^*)$ with $a,b \in B'$.

(iii) Let X be a subset of a Boolean algebra B, and let A be the subalgebra generated by X. Show that, if X is finite, then $|A| \leq 2^{2^{|X|}}$, and that, if X is infinite, then $|A| = |X|$. (Use (ii).)

2.5. PROBLEM. A *Boolean ring* is a ring R with identity such that $x^2 = x$ for all $x \in R$.

(i) Show that for any elements x,y of a Boolean ring R we have $xy = yx$ and $x + x = 0$.

(ii) Let R be a Boolean ring. Show that the relation $\leq$ on R defined by putting $x \leq y \Leftrightarrow xy = x$ is a partial ordering on R and that $\langle R, \leq \rangle$ is a Boolean algebra. (In this Boolean algebra, $x \vee y = x + y + xy$, $x \wedge y = xy$.)

(iii) Conversely, let B be a Boolean algebra, and define

$$x + y = (x \wedge y^*) \vee (x^* \wedge y), \qquad x \cdot y = x \wedge y \qquad \text{for all } x,y \in B.$$

Show that $\langle B, +, \cdot \rangle$ is a Boolean ring.

2.6. PROBLEM. Let B a Boolean algebra and let $\{x_i\colon i \in I\}$ be a subset of B for which $\bigvee_{i \in I} x_i$ exists. Show that $\bigwedge_{i \in I} x_i^*$ exists and is equal to $(\bigvee_{i \in I} x_i)^*$. Dualize. Show also that, for each $y \in B$, $\bigvee_{i \in I}(y \wedge x_i)$ exists and is equal to $y \wedge \bigvee_{i \in I} x_i$. Dualize.

§ 3. Filters and homomorphisms

A *filter* in a Boolean algebra B is a *non-empty* subset F of B satisfying the following conditions:

(i) $x, y \in F \Rightarrow x \wedge y \in F$,

(ii) $x \in F \,\&\, x \leq y \Rightarrow y \in F$,

(iii) $0 \notin F$.

For reasons which will become apparent later on, filters will play an important role in our investigation of the structure of Boolean algebras. We now give a necessary and sufficient condition for a subset of a Boolean algebra to be included in a filter.

A subset X of a Boolean algebra B is said to have the *finite meet property*[1] (*f.m.p.* for short) if whenever $x_1, \ldots, x_n \in X$ we have $x_1 \wedge \ldots \wedge x_n \neq 0$.

3.1. THEOREM. *A subset X of a Boolean algebra B is included in some filter in B iff X has the finite meet property.*

PROOF. Suppose that X is included in a filter F; then

$$x_1, \ldots, x_n \in X \Rightarrow x_1 \wedge \ldots \wedge x_n \in F \Rightarrow x_1 \wedge \ldots \wedge x_n \neq 0$$

since $0 \notin F$.

Conversely, suppose that X has the f.m.p. If $X \neq \emptyset$, define

$$X^+ = \{y \in B\colon \exists x_1 \ldots \exists x_n \in X\, [x_1 \wedge \ldots \wedge x_n \leq y]\}.$$

Clearly $X \subseteq X^+$; we claim that X^+ is a filter. Obviously X^+ satisfies clause (ii) of the filter condition. To verify clause (i), suppose that $y, y' \in X^+$. Then there exist $x_1, \ldots, x_n, x_1', \ldots, x_m' \in X$ such that

$$x_1 \wedge \ldots \wedge x_n \leq y, \qquad x_1' \wedge \ldots \wedge x_m' \leq y'.$$

Hence

$$x_1 \wedge \ldots \wedge x_n \wedge x_1' \wedge \ldots \wedge x_m' \leq y \wedge y'.$$

Therefore $y \wedge y' \in X^+$ and clause (i) is satisfied. Finally, since X has the f.m.p., $0 \notin X^+$. Accordingly X^+ is a filter including X.

If, on the other hand, $X = \emptyset$, then we put $X^+ = \{1\}$. ∎

The filter X^+ constructed in the proof of the preceding theorem is easily seen to be the *smallest* filter including X. It is called the filter *generated* by X. In this connection, a filter is said to be *principal* if it is generated by (the singleton of) a single (non-zero) element. Evidently

[1] If B is a subalgebra of a power set algebra, then meets in B are set intersections and accordingly we use the term *finite intersection property*.

the principal filter generated by an element x is $\{y: x \leqslant y\}$; x is called the *generating* element of this filter.

We turn now to the important notion of a homomorphism of Boolean algebras. A *homomorphism* of a Boolean algebra B into a Boolean algebra B' is a map $h: B \to B'$ such that for all $x, y \in B$ we have

$$h(x \wedge y) = h(x) \wedge h(y),$$
$$h(x \vee y) = h(x) \vee h(y),$$
$$h(x^*) = h(x)^*.$$

If h is one–one and onto, it is said to be an *isomorphism* of B onto B'; in this situation B and B' are said to be *isomorphic*, and we write $B \cong B'$.

3.2. PROBLEM. Show that, if h is a homomorphism of a Boolean algebra B into a Boolean algebra B', the image $h[B]$ of B is a subalgebra of B'. Show also that h is one–one iff $h^{-1}(0) = \{0\}$ or $h^{-1}(1) = \{1\}$.

3.3. PROBLEM. Let h be a map of a Boolean algebra B into a Boolean algebra B'.

(i) Show that the following conditions are equivalent:

(a) h is a homomorphism;

(b) for all $x, y \in B$,

$$h(x \wedge y) = h(x) \wedge h(y), \qquad h(x^*) = h(x)^*;$$

(c) for all $x, y \in B$,

$$h(x \vee y) = h(x) \vee h(y), \qquad h(x^*) = h(x)^*.$$

(ii) Suppose that h is one–one and onto. Show that h is an isomorphism iff for all $x, y \in B$,

$$x \leqslant y \Leftrightarrow h(x) \leqslant h(y).$$

Filters and homomorphisms are closely related. First, observe that if $h: B \to B'$ is a homomorphism of Boolean algebras, then

$$h^{-1}(1) = \{x \in B: h(x) = 1\}$$

is a filter in B'. (PROBLEM: Prove this.) This filter is called the *hull* of h. Thus each homomorphism of a Boolean algebra into another can be associated with a filter. We now show how to establish the converse.

For each pair of elements x, y of a Boolean algebra B we put

$$x \to y = x^* \vee y, \qquad x \leftrightarrow y = (x \to y) \wedge (y \to x).$$

Given a filter F in B, we define the binary relation $\sim_F$ on B by putting

$$x \sim_F y \Leftrightarrow x \leftrightarrow y \in F.$$

Then $\sim_F$ is an equivalence relation on B. (PROBLEM: Prove this.) Moreover, $\sim_F$ is a *congruence relation* with respect to the Boolean operations in B. This means that if $x \sim_F y$ and $x' \sim_F y'$, then

$$x \vee x' \sim_F y \vee y', \qquad x \wedge x' \sim_F y \wedge y',$$

$$x^* \sim_F y^*.$$

(PROBLEM: Prove this.) It follows that the set B/F of $\sim_F$-classes of members of B can be turned into a Boolean algebra by means of the prescription

$$\tilde{x} \vee \tilde{y} = (x \vee y)^\sim, \qquad \tilde{x} \wedge \tilde{y} = (x \wedge y)^\sim,$$

$$\tilde{x}^* = (x^*)^\sim,$$

where, for each $x \in B$,

$$\tilde{x} = \{y \colon x \sim_F y\}$$

is the $\sim_F$-class containing x. The resulting Boolean algebra B/F is called the *quotient* of B by the filter F. Evidently, if we define $h\colon B \to B/F$ by $h(x) = \tilde{x}$, then h is a homomorphism of B onto B/F (called the *canonical homomorphism*). Moreover, F is the hull of h. (PROBLEM: Prove this.) Thus with each filter F we can associate a Boolean algebra $B' = B/F$ and a homomorphism h of B onto B' such that F is the hull of h.

3.4. PROBLEM. Let $h\colon B \to B'$ be a homomorphism of Boolean algebras and let F be the hull of h. Show that the image $h[B]$ of B is isomorphic to B/F.

We have seen that filters arise as the hulls of homomorphisms. Especially important are those which are associated with homomorphisms onto the *minimal algebra* 2. Such homomorphisms are called 2-*valued*. It is possible (and useful) to given a particularly simple description of these filters; this we proceed to do.

A filter F in a Boolean algebra is called an *ultrafilter* if it is not properly included in any filter in B. That is, F is an ultrafilter if whenever F' is a filter including F, then $F = F'$. Later on we shall show that ultrafilters exist; for the present we prove:

3.5. THEOREM. *Let F be a filter in a Boolean algebra B. Then the following conditions are equivalent:*

(i) $B/F \cong 2$;

(ii) *F is the hull of a 2-valued homomorphism h on B;*

(iii) *F is an ultrafilter;*

(iv) *for all $x,y \in B$, if $x \vee y \in F$, then $x \in F$ or $y \in F$;*

(v) *for each $x \in B$, either $x \in F$ or $x^* \in F$.*

PROOF. (i)⇒(ii). If (i) holds, then the canonical homomorphism of B onto $B/F \cong 2$ meets the requirements imposed in (ii).

(ii)⇒(iii). Suppose (ii) holds, and let F' be a filter including F. Then, if $x \in F'$, we must have $x^* \notin F$ since F' is a filter and $F \subseteq F'$. Therefore $h(x^*) \neq 1$, whence $h(x^*)=0$. But then $h(x)=1$ so that, since F is the hull of h, we have $x \in F$. Hence $F'=F$ and F is an ultrafilter.

(iii)⇒(iv). Suppose (iii) holds and that $x \vee y \in F$. If $x \notin F$, it is easy to see that $G=\{z\colon x \vee z \in F\}$ is a filter which includes F, and so, since F is an ultrafilter, $F=G$. But since $x \vee y \in F$, it follows that $y \in G$ and hence that $y \in F$.

(iv)⇒(v). This is immediate since for any $x \in B$ we have $x \vee x^* = 1 \in F$.

(v)⇒(i). Assume (v), and for each $x \in B$ let $\tilde{x}$ be the image of x under the canonical homomorphism of B onto B/F. Suppose that $\tilde{x} \neq 1$ in B/F. Then $x \notin F$, so that $x^* \in F$. But then $\tilde{x}^* = (x^*)^\sim = 1$. It follows that $\tilde{x}=0$. Therefore B/F contains just two elements; in other words, it is isomorphic to 2. ∎

Thus ultrafilters are precisely the hulls of homomorphisms onto the minimal algebra. The question now arises as to whether ultrafilters actually exist. Assuming Zorn's lemma, the answer is in the affirmative. In fact we can establish the existence of an extremely rich class of ultrafilters in any Boolean algebra.

3.6. ULTRAFILTER THEOREM. *Each filter in a Boolean algebra is included in an ultrafilter.*

PROOF. Let $\mathscr{F}$ be the set of all filters in a Boolean algebra B; $\mathscr{F}$ can be partially ordered by inclusion. We will show that, with respect to this ordering, chains in $\mathscr{F}$ have upper bounds in $\mathscr{F}$.

Let $\mathscr{C}$ be a chain in $\mathscr{F}$, and let $C=\bigcup\mathscr{C}$. If $x,y \in C$, then for some $D,E \in \mathscr{C}$, $x \in D$ and $y \in E$. Since $\mathscr{C}$ is a chain, either $D \subseteq E$ or $E \subseteq D$; suppose the latter case obtains. Then $x,y \in D$ and because D is a filter we have $x \wedge y \in D \subseteq C$. If $z \in B$ and $x \leqslant z$, then *ipso facto* $z \in D \subseteq C$. Since $0 \notin D$

for all $D \in \mathscr{C}$, it follows that $0 \notin C$. Therefore C is a filter and is the required upper bound for $\mathscr{C}$ in F.

We may accordingly invoke Zorn's Lemma to conclude that, for every filter F in B, $\mathscr{F}$ contains a maximal member, i.e. an ultrafilter, which includes F. ∎

3.7. COROLLARY. *A subset of a Boolean algebra is included in an ultrafilter iff it has the f.m.p.*

PROOF. This follows immediately from Thms. 3.1 and 3.6. ∎

This gives instantly:

3.8. COROLLARY. *Each non-zero element of a Boolean algebra is contained in an ultrafilter.* ∎

3.9. COROLLARY. *For any pair of distinct elements of a Boolean algebra there is an ultrafilter containing one but not the other.*

PROOF. Let x and y be distinct elements of a Boolean algebra B. Then either $x \not\leqslant y$ or $y \not\leqslant x$; suppose the former case obtains. Then, by Prob. 2.3, $x \wedge y^* \neq 0$. Hence, by Cor. 3.8, there is an ultrafilter containing $x \wedge y^*$; this ultrafilter clearly contains x but not y. ∎

The correspondence between ultrafilters and 2-valued homomorphisms set up in Thm. 3.5 enables one to reformulate Thm. 3.6 and its corollaries in terms of such homomorphisms. Thus, e.g. Cor. 3.9 becomes: if x, y are distinct elements of a Boolean algebra B, then there is a 2-valued homomorphism h on B such that $h(x) \neq h(y)$. In other words, there are enough 2-valued homomorphisms to distinguish points of a Boolean algebra.

Let F be a filter in a Boolean algebra B. A subset X of F is called a *base* for F if for each $y \in F$ there is $x \in X$ such that $x \leqslant y$. If for each subset X of B we put $X^\wedge$ for the set of all finite non-empty meets of members of X, it is clear that, if X is non-empty and generates a filter F, then $X^\wedge$ is a base for F. Moreover, it is obvious that if X is a base for a filter F, then F is the (unique) filter generated by X.

A base X for a filter F is said to be *strong* if X is closed under finite meets, i.e. if $X^\wedge = X$. Evidently each filter F has at least one strong base, namely F itself.

3.10. PROBLEM. Let B be a Boolean algebra.

(i) Show that a non-empty subset X of a filter F in B is a base for F iff X generates F and for all $x, y \in X$ there exists $z \in X$ such that $z \leqslant x \wedge y$. Deduce that a non-empty subset X of B is a base for a (unique) filter in B iff $0 \notin X$ and for all $x, y \in X$ there is $z \in X$ such that $z \leq x \wedge y$.

(ii) Show that a subset X of B is a base for a (unique) *ultrafilter* in B iff X has the f.m.p. and for all $y \in B$ there is $x \in X$ such that $x \leq y$ or $x \leq y^*$.

(iii) Show that, if X is a strong base for a filter F, then, for any $x \in X$, the set $\{y \in X: y \leq x\}$ is also a strong base for F.

(iv) Show that a filter in B is principal iff it has a finite base, or equivalently, iff it has a finite strong base.

In general a filter may have many strong bases. For example, let B be a Boolean algebra with more than two elements, and let a be an element of B with $0 \neq a \neq 1$. Then both $\{a\}$ and $\{a, 1\}$ are strong bases for the principal filter generated by a. On the other hand, it is easy to see that all strong bases for a given principal filter F must contain the generating element of F, so that a principal filter *cannot have disjoint strong bases.*

However, the situation in this regard is quite different for *non-principal* filters. For example, consider the family $\mathscr{C}$ of all cofinite sets (i.e., complements of finite sets) of natural numbers. This is obviously a non-principal filter in the power set algebra $\mathsf{P}\omega$ of the set ω of natural numbers. Moreover, if for each $n \in \omega$ we put $S_n = \{m \in \omega: n \leq m\}$, then it is easy to see that $\{S_{2n}: n \in \omega\}$ and $\{S_{2n+1}: n \in \omega\}$ are both strong bases for $\mathscr{C}$, and that they are disjoint. Our next result, which will be useful in our discussion of non-standard analysis, shows that this conclusion is satisfied by *all* non-principal filters.

*3.11. LEMMA. *Any non-principal filter F in a Boolean algebra B has at least two disjoint strong bases.*

PROOF. Among the strong bases for F choose one, say X, which has minimal cardinality, say $\varkappa$. Then $\varkappa$ must be infinite by Prob. 3.10(iv).

Let X be well-ordered in the form $X = \{x_\alpha: \alpha < \varkappa\}$. By transfinite recursion we define y_α and z_α for each $\alpha < \varkappa$ in such a way that

(i) $y_\alpha, z_\alpha \in X$, $y_\alpha, z_\alpha \leq x_\alpha$;

(ii) $\{y_\beta: \beta \leq \alpha\}^\wedge \cap \{z_\beta: \beta \leq \alpha\}^\wedge = \emptyset$.

Suppose that y_α and z_α have been defined in conformity with (i) and (ii) for all $\alpha < \gamma$, where $\gamma < \varkappa$. We show how y_γ and z_γ may be defined.

Clearly, the set $\{y_\alpha: \alpha < \gamma\}^\wedge$ is not a strong base for F, since its cardinality is less than $\varkappa$. Thus there must be some $x \in X$ such that $y \notin \{y_\alpha: \alpha < \gamma\}^\wedge$ for all $y \leq x$. Choose the first such x in the wellordering of X, and let $z_\gamma = x \wedge x_\gamma$. Then $z_\gamma \in X$, $z_\gamma \leq x_\gamma$ and $y \notin \{y_\alpha: \alpha < \gamma\}^\wedge$ for all $y \leq z_\gamma$.

Similarly, we can define y_γ in such a way that $y_\gamma \leqslant x_\gamma$, $y_\gamma \in X$ and $y \notin \{z_\alpha : \alpha \leqslant \gamma\}^\wedge$ for all $y \leqslant y_\gamma$.

It is now not difficult to verify that

$$\{y_\alpha : \alpha \leqslant \gamma\}^\wedge \cap \{z_\alpha : \alpha \leqslant \gamma\}^\wedge = \emptyset.$$

The sets $\{y_\alpha : \alpha < \varkappa\}^\wedge$ and $\{z_\alpha : \alpha < \varkappa\}^\wedge$ are disjoint strong bases for F. ∎

3.12. PROBLEM. (Throughout this problem we adopt the notation and terminology of Ex. 2.2.(ii)). Let $\mathscr{L}$ be a first-order language, and let B be the propositional Lindenbaum algebra of $\mathscr{L}$.

(i) Let $\boldsymbol{\Sigma}$ be a set of formulas of $\mathscr{L}$. Show that $\boldsymbol{\Sigma}$ is propositionally consistent iff $\{|\boldsymbol{\alpha}| : \boldsymbol{\alpha} \in \boldsymbol{\Sigma}\}$ has the f.m.p. in B. Suppose that, whenever $\boldsymbol{\alpha} \in \boldsymbol{\Sigma}$, we have $|\boldsymbol{\alpha}| \subseteq \boldsymbol{\Sigma}$. Show that $\boldsymbol{\Sigma}$ is maximal propositionally consistent (see §13 of Ch. 1) iff $\{|\boldsymbol{\alpha}| : \boldsymbol{\alpha} \in \boldsymbol{\Sigma}\}$ is an ultrafilter in B.

(ii) Let $\boldsymbol{\Sigma}$ be a propositionally consistent set of formulas of $\mathscr{L}$. Let U be an ultrafilter in B including $\{|\boldsymbol{\alpha}| : \boldsymbol{\alpha} \in \boldsymbol{\Sigma}\}$, and let h be the canonical homomorphism of B onto $B/U \cong 2$. Define the mapping $\sigma : \boldsymbol{\Phi} \to \{\top, \bot\}$ by

$$\sigma(\boldsymbol{\alpha}) = \top \;\Leftrightarrow\; h(|\boldsymbol{\alpha}|) = 1, \qquad \sigma(\boldsymbol{\alpha}) = \bot \;\Leftrightarrow\; h(|\boldsymbol{\alpha}|) = 0.$$

Show that σ is a truth valuation satisfying $\boldsymbol{\Sigma}$. (This gives an alternative proof of Thm. 1.13.4. For an extension of this method to the full predicate calculus, see Ch. 5.)

3.13. PROBLEM. An *ideal* in a Boolean algebra B is a subset I of B such that

(a) $x, y \in I \Rightarrow x \vee y \in I$,

(b) $x \in I$ and $y \leqslant x \Rightarrow y \in I$,

(c) $1 \notin I$.

(i) For each subset $X \subseteq B$ let $X^* = \{x^* : x \in X\}$. Show that X is a filter (ideal) iff X^* is an ideal (filter).

(ii) Reformulate the results of this section in terms of ideals instead of filters.

3.14. PROBLEM. Show that each filter in a Boolean algebra is the intersection of the collection of all ultrafilters which include it.

*3.15. PROBLEM (Sikorski). (i) Let B be a Boolean algebra, let A be a subalgebra of B, and let C be a complete Boolean algebra. Show that any homomorphism of A into C can be extended to a homomorphism of B into C. (Let h: $A \to C$ be the given homomorphism. By Zorn's lemma, there is a maximal pair $\langle f, D \rangle$, where D is a subalgebra of B including A and f is a homomorphism of D into C extending h. Then

$D = B$, for if $a \in B - D$ then f can be extended to the subalgebra

$$\{(x \wedge a) \vee (y \wedge a^*):\ x, y \in D\}$$

generated by $D \cup \{a\}$ by setting

$$f((x \wedge a) \vee (y \wedge a^*)) = (f(x) \wedge b) \vee (f(y) \wedge b^*),$$

where b is any element of C satisfying

$$\bigvee \{f(x):\ x \leqslant a \,\&\, x \in D\} \leqslant b \leqslant \bigwedge \{f(y):\ a \leqslant y \,\&\, y \in D\}.)$$

(ii) A subalgebra A of a Boolean algebra B is said to be *dense* in B if for every $x \in B$, $x \neq 0$, there is $y \in A$, such that $y \neq 0$ and $y \leqslant x$. Show that, if A is a dense subalgebra of B, then every one–one homomorphism of A into a complete Boolean algebra C can be extended to a one-one homomorphism of B into C. (Use (i).)

3.16. Problem. If X is a set, a filter in the power set algebra $\mathsf{P}X$ is called a filter *over* X.

(i) Show that the following conditions on an ultrafilter $\mathscr{U}$ over X are equivalent:

(a) $\mathscr{U}$ is principal.

(b) $\mathscr{U}$ is generated by a singleton.

(c) $\mathscr{U}$ contains a finite subset of X.

(ii) Show that, if X is infinite, there is a non-principal ultrafilter over X. (Show that the family of cofinite subsets of X has the f.m.p., and use Cor. 3.7.)

(iii) Show that, if $\mathscr{U}$ is an ultrafilter over X and $\{Y_1, \ldots, Y_k\}$ is a partition of a member of $\mathscr{U}$, then exactly one of the Y_i is a member of $\mathscr{U}$. Extend this result to ultrafilters in arbitrary Boolean algebras.

(iv) An ultrafilter $\mathscr{U}$ over X is said to be *uniform* if $|Y| = |X|$ for all $Y \in \mathscr{U}$. Show that, if X is infinite, there is a uniform ultrafilter over X. Show also that an ultrafilter over a denumerable set is uniform iff it is non-principal.

(v) An ultrafilter $\mathscr{U}$ over X is said to be *countably incomplete* if there is a denumerable subset $\{X_n:\ n \in \omega\}$ of $\mathscr{U}$ such that $\bigcap_{n \in \omega} X_n \notin \mathscr{U}$. Show that, if $\mathscr{U}$ is a countably incomplete ultrafilter over X, there is a descending chain $Y_0 \supseteq Y_1 \supseteq Y_2 \supseteq \ldots$ such that each $Y_n \in \mathscr{U}$ and $\bigcap_{n \in \omega} Y_n = \emptyset$. Show also that every non-principal ultrafilter over ω is countably incomplete.

3.17. Problem. Let X be a topological space and for each $x \in X$ let $\mathscr{F}_x$ be the family of neighbourhoods of x.

(i) Show that $\mathscr{F}_x$ is a filter over X.

(ii) A filter $\mathscr{F}$ over X is said to *converge* to a point $x \in X$ if $\mathscr{F}_x \subseteq \mathscr{F}$. Show that X is Hausdorff iff each ultrafilter over X converges to at most one point, and that X is compact iff each ultrafilter over X converges to at least one point.

3.18. PROBLEM. A subset X of a Boolean algebra B is said to be an *antichain* if whenever $x, y \in X$ and $x \neq y$, we have $x \wedge y = 0$. B is said to satisfy the *countable chain condition* if any antichain in B is countable. Show that the following conditions are equivalent:

(i) B satisfies the countable chain condition;

(ii) for each subset X of B, there is a countable subset Y of X such that X and Y have the same set of upper bounds.

(For (i) ⇒ (ii), consider a maximal antichain in the ideal I generated by X, i.e. $I = \{y \in B\colon y \leqslant x_1 \vee \ldots \vee x_n \text{ for some } x_1, \ldots, x_n \in X\}$.)

3.19. PROBLEM. A *positive measure* on a Boolean algebra B is a function μ on B to the non-negative real numbers such that

(i) $\mu(x \vee y) = \mu(x) + \mu(y)$ for all $x, y \in B$ such that $x \wedge y = 0$,

(ii) $\mu(1) = 1$,

(iii) $\mu(x) = 0$ iff $x = 0$.

Show that, if B admits a positive measure, then B satisfies the countable chain condition.

§ 4. The Stone Representation Theorem

The very first example of a Boolean algebra that occurred to us was the power set algebra of a set. Now it is easy to construct Boolean algebras which are not isomorphic to any power set algebra. In fact, the finite–cofinite algebra Fω of the set ω of natural numbers is such an algebra, since Fω has cardinality $\aleph_0$, while evidently no power set algebra can have this cardinality. Nonetheless, we are going to show that each Boolean algebra is isomorphic to a *subalgebra* of a power set algebra, or, in other words, each Boolean algebra can be *represented* as a subalgebra of a power set algebra.

Let us define a *field of sets* to be a subalgebra of a power set algebra. In particular, a *field of subsets* of a set X is a subalgebra of PX.

If B is a Boolean algebra, we denote by SB the set of all ultrafilters in B. Then we have:

4.1. STONE REPRESENTATION THEOREM. *Each Boolean algebra B is isomorphic to a field of subsets of* SB.

Proof. Let B be a Boolean algebra. Define a mapping $u: B \to \mathsf{P}\mathsf{S}B$ by putting

$$u(x) = \{F \in \mathsf{S}B : x \in F\}$$

for each $x \in B$. Thus $u(x)$ is the set of all ultrafilters containing x.

We claim that u is a homomorphism of B into $\mathsf{PS}B$. For suppose $x, y \in B$; then, if $F \in \mathsf{S}B$, we have

$$F \in u(x \wedge y) \Leftrightarrow x \wedge y \in F \Leftrightarrow x \in F \quad \& \quad y \in F \Leftrightarrow F \in u(x) \cap u(y).$$

Hence $u(x \wedge y) = u(x) \cap u(y)$. Also, we have

$$F \in u(x^*) \Leftrightarrow x^* \in F \Leftrightarrow x \notin F \text{ (by Thm. 3.5(iv))} \Leftrightarrow F \in \mathsf{S}B - u(x).$$

Accordingly $u(x^*) = \mathsf{S}B - u(x)$, so that, by Prob. 3.3, u is a homomorphism.

We also note that u is one–one, for if $x \neq y$ then by Cor. 3.9 there is an ultrafilter F containing x, say, but not y. Then $F \in u(x)$ and $F \notin u(y)$, so that $u(x) \neq u(y)$.

We have therefore shown that u is an *isomorphism* of B onto the subalgebra $u[B]$ of $\mathsf{PS}B$, which proves the theorem. ∎

The Stone Representation Theorem can be phrased much more suggestively within the framework of general topology.

Recall that a simultaneously closed and open subset of a topological space is called a *clopen* set. Let us define a *Boolean* space to be a compact Hausdorff space with a base of clopen sets. (Spaces satisfying this latter condition are sometimes called *totally disconnected* for reasons which are revealed in Prob. 4.14.)

If X is a topological space, then the clopen subsets of X form a field of sets (cf. Ex. 2.2(iii)). This field is called the *clopen algebra of* X and is written $\mathsf{C}X$.

We shall show that we can assign a natural topology to the set $\mathsf{S}B$ of all ultrafilters in B in such a way that it becomes a Boolean space and its clopen algebra is isomorphic to B.

4.2. Lemma. *Let X be a compact space, and let $\mathscr{A}$ be a field of subsets of X which is also a base for the topology on X. Then $\mathscr{A} = \mathsf{C}X$.*

Proof. The elements of $\mathscr{A}$ are open subsets of X since they form a base for the topology. Since $\mathscr{A}$ is a field of sets it is stable under the formation of complements and so its members are also closed. Thus $\mathscr{A} \subseteq \mathsf{C}X$.

Now suppose that $Y \in \mathsf{C}X$. Y is open and $\mathscr{A}$ is a base, so there is some family $\{A_i: i \in I\}$ of members of $\mathscr{A}$ whose union is Y. But Y is also a closed subset of the compact space X, so Y is itself compact, and therefore the open cover $\{A_i: i \in I\}$ of Y has a finite subcover $\{A_{i_1},\ldots,A_{i_n}\}$, i.e. $Y = A_{i_1} \cup \ldots \cup A_{i_n}$. However, $\mathscr{A}$, being a field of sets, is stable under finite unions. Hence $Y \in \mathscr{A}$, and our proof is complete. ▮

Consider the family $u[B] = \{u(x): x \in B\}$, where u is defined as in 4.1 by

$$u(x) = \{F \in SB: x \in F\} \quad \text{for } x \in B;$$

$u[B]$ is a field of subsets of SB and is therefore stable under finite intersections. Accordingly $u[B]$ forms a base for a unique topology on SB. The resulting topological space is called the *Stone space* of B.

Our next result is of great importance.

4.3. THEOREM. *The Stone space* SB *of a Boolean algebra* B *is a Boolean space and* B *is isomorphic to the clopen algebra* CSB *of* SB.

PROOF. SB is a Hausdorff space. For suppose $F, G \in SB$ and $F \neq G$. Then for some $x \in B$, we have $x \in F$ but $x \notin G$; so that $x^* \in G$. Then $F \in u(x)$, $G \in u(x^*)$, and $u(x)$, $u(x^*)$ are disjoint open subsets of SB.

SB is a compact space. To see this it is sufficient to show that each cover of SB by basic open sets has a finite subcover. Suppose that $\{u(x_i): i \in I\}$ $(x_i \in B)$ is such a cover which has no finite subcover. Then, for each finite subset I_0 of I, we have $\bigcup_{i \in I_0} u(x_i) \neq SB$, so that (recalling that u is a homomorphism!)

$$u\Big(\bigwedge_{i \in I_0} x_i^*\Big) = \bigcap_{i \in I_0} u(x_i^*) = \bigcap_{i \in I_0} [SB - u(x_i)] = SB - \bigcup_{i \in I_0} u(x_i) \neq$$
$$\neq \emptyset = u(0).$$

Therefore, since u is one-one, we have $\bigwedge_{i \in I_0} x_i^* \neq 0$. Thus $\{x_i^*: i \in I\}$ has the f.m.p. and can therefore (Cor. 3.7) be extended to an ultrafilter F. Since $x_i^* \in F$ for all $i \in I$, it follows that $F \notin \bigcup_{i \in I} u(x_i)$, which contradicts the assumption that $\{u(x_i): i \in I\}$ covers SB. Hence SB is compact.

Since $u(x)$ is the complement of $u(x^*)$, each member of the base $\{u(x): x \in B\}$ is closed, and so clopen. By 4.1, u is an isomorphism of B onto $u[B]$, while by 4.2, $u[B] = CSB$. Thus u is an isomorphism of B onto CSB. ▮

The upshot of Thm. 4.3 is that *each Boolean algebra* B *may be identified with the clopen algebra of its Stone space* SB. Accordingly, the structure of B is completely reflected in the structure of SB; in other words, each algebraic property of B corresponds to a topological property of SB, and conversely. What, for example, is the topological property of SB which corresponds to *completeness* of B? The answer is somewhat weird, as we shall see.

Let us call a topological space *extremally disconnected* if the closure of every open set is open (and hence clopen!). Then we have:

4.4. THEOREM. *A Boolean algebra is complete iff its Stone space is extremally disconnected.*

PROOF. Identify the given Boolean algebra B with the clopen algebra of SB. Suppose that B is complete, and let U be an open set in SB. Let $\mathscr{A}$ be the family of all members of B included in U; then, since B is a base for SB, we have $U = \bigcup\mathscr{A}$. Since B is complete, $\mathscr{A}$ has a supremum V in B which must by definition be a clopen subset of SB. We claim that $\overline{U} = V$. Since V is an upper bound for $\mathscr{A}$, certainly $U = \bigcup\mathscr{A} \subseteq V$, so that $\overline{U} \subseteq V$ since V is closed. If $V - \overline{U} \neq \emptyset$, then $V - \overline{U}$ is a non-empty open set which must, since SB is a Boolean space, include a non-empty clopen set W. But then $V - W$ is a clopen set which includes $\bigcup\mathscr{A}$ and is properly included in V. This contradicts the choice of V as the supremum of $\mathscr{A}$. Therefore $\overline{U} = V$ as claimed, so *a fortiori* $\overline{U}$ is open.

Conversely, suppose that SB is extremally disconnected, and let $\mathscr{A}$ be a subfamily of B. Then $U = \bigcup\mathscr{A}$ is an open subset of SB (recall that we are identifying B with CSB!) and so, since SB is extremally disconnected, $\overline{U}$ is clopen and hence in B. We claim that $\overline{U} = \bigvee\mathscr{A}$ in B. Certainly $\overline{U}$ is an upper bound for $\mathscr{A}$ in B; on the other hand, if V is a member of B which includes each member of $\mathscr{A}$, then $U = \bigcup\mathscr{A} \subseteq V$ so that $\overline{U} \subseteq V$ since V is clopen. Therefore $\overline{U} = \bigvee\mathscr{A}$ as claimed, and B is complete. ∎

As well as providing curious and amusing characterizations of algebraic properties, such as in Thm. 4.4, appeal to the theory of Boolean spaces can often yield simple proofs of algebraic results whose algebraic proofs would turn out to be somewhat hairy. Witness, for example:

4.5. THEOREM. (i) *Let B be a finite Boolean algebra. Then B is isomorphic to PSB, and hence $|B| = 2^{|SB|}$.*

(ii) *Any two finite Boolean algebras of the same cardinality are isomorphic.*

PROOF. (i) Let B be a finite Boolean algebra of cardinality m. Then SB is a finite Hausdorff space, hence a discrete space, i.e. every subset of SB is clopen. Therefore, $CSB = PSB$, and since $B \cong CSB$, we have (i).

(ii) Let B and B' be two finite Boolean algebras of the same cardinality m. Let SB have p and SB' q elements. Then, by (i), we have $2^p = m = 2^q$, so that $p=q$. Therefore B and B' are both isomorphic to the power set algebra of a set of p elements, and so B and B' are themselves isomorphic. ∎

*4.6. EXAMPLES. (i) Let X be an infinite discrete space. It is not hard to show that the *one-point compactification* $X \cup \{\infty\}$ obtained by adding a new "point at infinity" ∞ to X is a Boolean space which turns out to be the Stone space of the finite–cofinite algebra $\mathsf{F}X$ of X. For each $x \in X$ let $\mathscr{U}_x$ be the ultrafilter in $\mathsf{F}X$ generated by $\{x\}$, and let $\mathscr{U}$ be the ultrafilter in $\mathsf{F}X$ consisting of all cofinite subsets of X. (Notice that $\mathscr{U}$ is *not* an ultrafilter in $\mathsf{P}X$ but it *is* one in $\mathsf{F}X$!) Then $\{\mathscr{U}_x\colon x \in X\} \cup \{\mathscr{U}\}$ consists of *all* ultrafilters in $\mathsf{F}X$, that is,

$$\mathsf{SF}X = \{\mathscr{U}_x\colon x \in X\} \cup \{\mathscr{U}\}.$$

It is now quite easy to see that $\mathsf{SF}X$ is (homeomorphic to) the one-point compactification of X. The subset $\{\mathscr{U}_x\colon x \in X\}$ is a homeomorphic copy of X, and $\mathscr{U}$ functions as the "point at infinity".

(ii) Again let X be an infinite discrete space; this time consider the Stone space of the *power set algebra* of X. This turns out to be (homeomorphic to) the *Stone–Čech compactification* of X.

A natural question that arises is the following: given a Boolean algebra, what is the cardinality of its Stone space? It follows immediately from Thm. 4.5 that, if B is *finite*, then $|\mathsf{S}B| < |B|$. But the situation is entirely different when B is *infinite*, as we shall see. Let X be a set and let $\mathscr{A}$ be a field of subsets of X. $\mathscr{A}$ is said to be *separating* if for each pair of distinct points x, y of X there is $A \in \mathscr{A}$ such that $x \in A$ and $y \notin A$.

4.7. LEMMA. *If X is a compact Hausdorff space and $\mathscr{A}$ is a separating field of clopen subsets of X, then $\mathscr{A}$ is the field of* all *clopen subsets of X.*

PROOF. By Lemma 4.2 we need only show that $\mathscr{A}$ is a base for X. Let Y be a closed subset of X and let $a \notin Y$. For each $y \in Y$ we can find $A_y \in \mathscr{A}$ such that $y \in A_y$ and $a \notin A_y$. Now the family $\{A_y\colon y \in Y\}$ constitutes an open cover of Y; and Y, as a closed subset of a compact space, is itself compact. Accordingly $\{A_y\colon y \in Y\}$ has a finite subcover, the union of which is a member of $\mathscr{A}$ including Y but not containing a. Since a was an arbitrary member of $X - Y$, it follows that Y is the intersection of all members of $\mathscr{A}$ which include it. Since Y was an arbitrary closed set, we infer that any closed set is the intersection of a subfamily of $\mathscr{A}$. Dually, each open set is the union of a subfamily of $\mathscr{A}$, i.e. $\mathscr{A}$ is a base for X. ∎

We can now prove

4.8. THEOREM. *For each infinite Boolean algebra B, we have* $|B| \leq |\mathsf{S}B|$.
PROOF. Let B be an infinite Boolean algebra, and let X be the Stone space of B. Then by Thm. 4.3 we may identify B with the clopen algebra of X. We have to show that $|B| \leq |X|$. Suppose, on the contrary, that $|X| < |B|$. Let

$$Y = \{\langle x, y\rangle \in X \times X :\ x \neq y\}.$$

Then, since B is infinite and $|X| < |B|$, it follows immediately that $|Y| < |B|$. Since X is a Boolean space, for each $z = \langle x, y\rangle \in Y$ there is $V_z \in B$ such that $x \in V_z$ and $y \notin V_z$. Let A be the subalgebra of B generated by $\{V_z :\ z \in Y\}$. It follows from Prob. 2.4(iii) that $|A| < |B|$. But, by construction, A is a separating field of clopen subsets of X, so that, by Lemma 4.7, $A = B$. This contradiction shows that the assumption $|X| < |B|$ was wrong, so that $|B| \leq |X|$. ∎

Thus, if B is an infinite Boolean algebra, $|\mathsf{S}B|$ is bounded below by $|B|$. Also, it is clear that, for any Boolean algebra B, $|\mathsf{S}B|$ is bounded above by $2^{|B|}$. These bounds may, in fact, be attained. For example, it is clear from Ex. 4.6(i) that if B is the finite–cofinite algebra of an infinite set, then $|B| = |\mathsf{S}B|$, while a well-known result of TARSKI [1939] asserts that if B is an infinite power set algebra, then $|\mathsf{S}B| = 2^{|B|}$.

4.9. PROBLEM. Show that a Boolean algebra is finite iff its Stone space is discrete, and hence iff its Stone space is finite.

4.10. PROBLEM. Let B be a Boolean algebra, and let A be a subalgebra of B. Show that A is a proper subalgebra of B iff there are distinct ultrafilters U, U' in B such that $U \cap A = U' \cap A$. (Let A be a proper subalgebra of B; let u be the isomorphism of B onto $\mathsf{CS}B$. Then $u[A]$ is a proper subalgebra of $\mathsf{CS}B$; by Lemma 4.7, $u[A]$ is not separating; the conclusion now follows easily.)

4.11. PROBLEM. (i) Show that, for each infinite Boolean space X, there is a (countably) infinite collection of mutually disjoint clopen subsets of X.

(ii) Deduce from (i) that each infinite Boolean algebra has a (countably) infinite antichain. (Prob. 3.18.)

4.12. PROBLEM. Prove the so-called *dual* form of the Stone representation theorem: each Boolean space is homeomorphic to the Stone space of its clopen algebra. (Let X be a Boolean space. Show that the map $v : X \to \mathsf{SC}X$ defined by $v(x) = \{A \in \mathsf{C}X :\ x \in A\}$ for each $x \in X$ is a homeomorphism.)

4.13. PROBLEM. A Boolean algebra is said to be *σ-complete* if each countable subset has a join and a meet. A subset of a Boolean space is called *σ-open* if it is the union of countably many clopen sets. A Boolean space is said to be *σ-disconnected* if the closure of each σ-open set is open. Show that a Boolean algebra is σ-complete iff its Stone space is σ-disconnected. (Argue as in the proof of Thm. 4.4.)

4.14. PROBLEM. A topological space is said to be *totally disconnected* if no connected subset has more than one point. Show that any Boolean space is totally disconnected.[1]

*4.15. PROBLEM. A subset U of a topological space X is said to be *regular open* if $U=(U^-)^\circ$, i.e. if U coincides with the interior of its closure. Show that the family $\mathrm{R}X$ of all regular open subsets of X forms a complete Boolean algebra under the partial ordering of inclusion. (If $\{U_i\colon i\in I\}\subseteq \mathrm{R}X$, show that, in $\mathrm{R}X$, $\bigvee_{i\in I}U_i=((\bigcup_{i\in I}U_i)^-)^\circ$ and $\bigwedge_{i\in I}U_i=(\bigcap_{i\in I}U_i)^\circ$, while $U^*=X-\bar{U}$ for $U\in \mathrm{R}X$.) $\mathrm{R}X$ is called the *regular open algebra of* X.

(ii) Show that, if X is a Boolean space, then $\mathrm{C}X$ is a dense subalgebra (Prob. 3.15(ii)) of $\mathrm{R}X$. Deduce that, if $\{U_i\colon i\in I\}$ is a subfamily of $\mathrm{C}X$ which has a supremum (infimum) in $\mathrm{C}X$, then this supremum (infimum) is the same as the supremum (infimum) of $\{U_i\colon i\in I\}$ in $\mathrm{R}X$.

(iii) Let X be a topological space in which every family of disjoint open sets is countable, and let $\mathscr{B}$ be a base for X. Show that $|\mathrm{R}X|\leqq|\mathscr{B}|^{\aleph_0}$. (Using Zorn's lemma, for each $U\in \mathrm{R}X$ let $\mathscr{B}_U$ be a maximal family of disjoint members of $\mathrm{P}U\cap\mathscr{B}$. Now show that $U=((\bigcup\mathscr{B}_U)^-)^\circ$, and observe that there are at most $|\mathscr{B}|^{\aleph_0}$ families of the form $\mathscr{B}_U$.)

(iv) If B is a Boolean algebra satisfying the countable chain condition (Prob. 3.18) show that $|\mathrm{RS}B|\leqq|B|^{\aleph_0}$. (Use (v).)

*4.16. PROBLEM. Show that a complete Boolean algebra is isomorphic to the regular open algebra of an extremally disconnected space. (Use Thm. 4.4 and Prob. 4.15.)

*4.17. PROBLEM. Let B be a Boolean algebra, and let X be a subset of B. Then B is a said to be *freely generated by* X, and X is said to be a *free set of generators for* B, if X generates B and any mapping f of X into any Boolean algebra B' can be extended to a homomorphism f' of B into B'. A Boolean algebra is said to be *free* if it is freely generated by some subset.

(i) Show that, if B is freely generated by X and f is a mapping from X

[1] The converse holds for any compact Hausdorff space; see GILLMAN and JERISON [1960], Thm. 16.17.

into a Boolean algebra B', then the homomorphism f' extending f is *unique*. (Let f'' be another extension; consider $\{x\in B\colon f'(x)=f''(x)\}$.)

(ii) Let B and B' be Boolean algebras freely generated by X and X' respectively. Show that B is isomorphic to B' iff $|X|=|X'|$. (Use (i), and Prob. 2.4 (iii).)

*4.18. PROBLEM *(Cantor spaces)*. Let X be a set. The *Cantor space* over X is the product space 2^X with the product topology (where $2=\{0, 1\}$ is assigned the discrete topology).

(i) Show that any Cantor space is a Boolean space. (Use Tychonoff's theorem that the product of compact spaces is compact.)

(ii) Show that the *Cantor ternary set* (i.e. the set of real numbers in the closed unit interval, which have a triadic expansion without the digit 1, with the topology inherited from the closed unit interval) is homeomorphic to the Cantor space 2^ω, where ω is the set of natural numbers. (Map each $f\in 2^\omega$ onto the real number $\sum_{i=1}^{\infty} 2f(i)3^{-i}$.)

(iii) Show that, for each X, the clopen algebra of the Cantor space 2^X is freely generated by a set of the same cardinality as X. In particular, the minimal algebra 2 is freely generated by $\emptyset$. (Let B be the clopen algebra of 2^X, and for each $x\in X$ define

$$j(x)=\{f\in 2^X\colon f(x)=1\}.$$

Let $Y=\{j(x)\colon x\in X\}$; then $|Y|=|X|$, and Y generates B. To show that Y freely generates B, suppose that h is a mapping of Y into a Boolean algebra B'. Identify B' with the clopen algebra of its Stone space SB'. Define $f\colon SB'\to 2^X$ by

$$f(y)(x)=\begin{cases}1 & \text{if } y\in h(j(x)),\\ 0 & \text{if } y\notin h(j(x)).\end{cases}$$

Show that f is continuous, and that, if one defines $h'\colon B\to B'$ by $h'(F)= =f^{-1}[F]$, then h' is a homomorphism extending h to B.)

(iv) Show that each Boolean algebra is the image of a free algebra under a homomorphism. (Use (iii) to show that B is a homomorphic image of the clopen algebra of 2^B.)

(v) Show that, if n is finite, the free Boolean algebra generated by a set of n elements has 2^{2^n} elements. (Use (ii).) Deduce that a finite Boolean algebra is free iff it has 2^{2^n} elements for some n. (For sufficiency, use Thm. 4.5(ii).)

(vi) Let X be a subset of a Boolean algebra B which generates B. Show that the following conditions are equivalent:

(a) B is freely generated by X;

(b) each mapping of X into 2 can be extended to a homomorphism of B into 2;

(c) for any distinct elements $x_1,\dots,x_n$ of X, we have $x_1'\wedge\dots\wedge x_n'\neq 0$, where x_i' is either x_i or x_i^*.

(For (c)⇒(a), show that, using Prob. 2.4(ii), B is isomorphic to the free Boolean algebra generated by a set of the same cardinality as X whose existence is proved in (iii).)

(vii) Let $\mathscr{L}$ be a first-order language. Show that the propositional Lindenbaum algebra of $\mathscr{L}$ is freely generated by the set

$$\{|\alpha| : \alpha \text{ is a prime formula of } \mathscr{L}\}.$$

(Use (vi).)

*4.19. Problem. Let B be a Boolean algebra freely generated by a subset X.

(i) Let $Y\subseteq X$, $x\in X-Y$, let B' be the subalgebra of B generated by Y, and let B'' be the subalgebra of B generated by $B'\cup\{x\}$. Show that each element y of B'' can be uniquely expressed in the form

$$y=(a\wedge x)\vee(b\wedge x^*)$$

with $a,b\in B'$. (Use Probs. 2.4(ii) and 4.18(vi) (c).)

(ii) With the assumptions and notation of (i), show that each positive measure μ on B' can be extended to a positive measure μ' on B''. (Let r be any real number such that $0<r<1$, and for $y\in B'$ put

$$\mu'(y)=r\cdot\mu(a)+(1-r)\cdot\mu(b),$$

where $y=(a\wedge x)\vee(b\wedge x^*)$ with $a,b\in B'$.)

(iii) Show that B admits a positive measure. (Apply Zorn's lemma and (ii) to the set $\{\mu : \mu$ is a positive measure on the subalgebra of B generated by a subset of $X\}$, partially ordered by inclusion.)

(iv) Show that any free Boolean algebra — in particular, the clopen algebra of any Cantor space — satisfies the countable chain condition. (Use (iii) and Prob. 3.19.)

(v) Let I be an infinite set, and let X be the Cantor space 2^I. Show that $|\mathrm{R}X|\leq|I|^{\aleph_0}$. (For the definition of $\mathrm{R}X$, see Prob. 4.15. Use (iv) and Prob 4.15 (iv).)

§ 5. Atoms

An *atom* in a Boolean algebra B is a minimal non-zero element. In other words, an element x of B is an atom iff $x \neq 0$ and for all $y \in B$, if $y \leq x$, then $y = x$ or $y = 0$. It is easy to see that x is an atom in B iff for each $y \in B$ *exactly one* of $x \leq y$ or $x \leq y^*$ holds.

5.1. LEMMA. *A non-zero element x of a Boolean algebra B is an atom iff the filter $\{y \in B: x \leq y\}$ generated by x is an ultrafilter.* ∎

5.2. THEOREM. *Let U be an ultrafilter in a Boolean algebra B. Then the following conditions are equivalent:*

(i) *U is generated by an atom;*

(ii) *U is an isolated point in SB (i.e. $\{U\}$ is open in SB).*

PROOF. (i)⇒(ii). Suppose $U = \{x \in B: a \leq x\}$ for some atom a. Let u be the isomorphism of B onto the clopen algebra of SB defined in Thm. 4.1. Then, using Thm. 5.1, we see immediately that $u(a) = \{U\}$, so that $\{U\}$ is open and U is isolated.

(ii)⇒(i). Suppose that U is an isolated point of SB. Then $\{U\}$ is a clopen subset of SB so that $\{U\} = u(a)$ for some $a \in B$. Let $F = \{x \in B: a \leq x\}$ be the filter generated by a. We claim that $F = U$. Clearly $F \subseteq U$. On the other hand, if $F \neq U$, then there is $x \in U$ such that $a \not\leq x$, so that $a \wedge x^* \neq 0$. Accordingly there is an ultrafilter U' in B containing $a \wedge x^*$, hence both a and x^*. But then $U' \in u(a)$ and $U' \neq U$, contradicting the fact that $u(a) = \{U\}$. Hence $F = U$. By Thm. 5.1, a is then an atom and it generates U. ∎

5.3. COROLLARY. *A Boolean algebra B is finite iff every ultrafilter in B is principal.*

PROOF. B is finite ⇔ (by Prob. 4.9) SB is discrete ⇔ every point of SB is isolated ⇔ (by Thm. 5.2) every ultrafilter in B is principal. ∎

A Boolean algebra B is said to be *atomic* if for each $x \in B$, $x \neq 0$, there is an atom $a \in B$ such that $a \leq x$. At the other extreme, B is said to be *atomless* if it contains no atoms at all.

Notice that any power set algebra is atomic. (What are the atoms in such an algebra?) Also, it is clear that every *finite* Boolean algebra is atomic. On the other hand, the field of subsets of the real line generated by intervals of the form $[x, +\infty)$ is easily seen to be an *atomless* Boolean algebra.

We pointed out at the beginning of §4 that not every Boolean algebra B is isomorphic to a power set algebra. But notice that if B is isomorphic to such an algebra, it must be complete and atomic, since power set algebras have both these properties. We now prove the converse.

5.4. THEOREM. *Any complete atomic Boolean algebra is isomorphic to a power set algebra.*

PROOF. Let B be a complete atomic algebra, and let A be its set of atoms. We show that $B \cong \mathsf{P}A$. To this end, define h: $B \to \mathsf{P}A$ by

$$h(x) = \{a \in A : a \leqslant x\}$$

for each $x \in B$. We claim that h is an isomorphism of B onto $\mathsf{P}A$.

First, h is a homomorphism. For if $x, y \in B$, then for all $a \in A$ we have

$$a \in h(x \wedge y) \Leftrightarrow a \leqslant x \wedge y \Leftrightarrow a \leqslant x \ \&\ a \leqslant y$$

$$\Leftrightarrow a \in h(x) \cap h(y).$$

Hence $h(x \wedge y) = h(x) \cap h(y)$. Also,

$$a \in h(x^*) \Leftrightarrow a \leqslant x^* \Leftrightarrow \text{(since } a \text{ is an atom)}\ a \nleqslant x \Leftrightarrow a \in A - h(x).$$

Hence $h(x^*) = A - h(x)$. It follows that h is a homomorphism.

Next, h is one-one. For if $x \neq y$ in B, then either $x \nleqslant y$ or $y \nleqslant x$; assume the latter. Then $x^* \wedge y \neq 0$, so, since B is atomic, there is $a \in A$ such that $a \leqslant x^* \wedge y$. Hence $a \leqslant x^*$ and $a \leqslant y$, so $a \nleqslant x$ and $a \leqslant y$. It follows that $a \in h(y) - h(x)$, so that $h(x) \neq h(y)$.

Finally, h is onto. For if $X \in \mathsf{P}A$, let $x = \bigvee X$. (Recall that we have assumed that B is complete!) We claim that $X = h(x)$. For if $a \in X$, then by definition $a \leqslant x$ so that $a \in h(x)$, and accordingly $X \subseteq h(x)$. Conversely, if $a \in A - X$, then, by the definition of an atom we have $a \wedge a' = 0$ for all $a' \in X$, so that $a' \leqslant a^*$ for all $a' \in X$. Therefore $x = \bigvee X \leqslant a^*$, so, since $a \neq 0$, we have $a \nleqslant x$, whence $a \notin h(x)$. Therefore $h(x) \subseteq X$ and our claim is established, completing the proof. ∎

5.5. PROBLEM. Let B be a Boolean algebra. Show that a non-zero element a of B is an atom iff whenever $a = x \vee y$ with $x \wedge y = 0$,then $x = 0$ or $y = 0$. (The term "atom" derives from this property.)

5.6. PROBLEM. Let B be a Boolean algebra, and let A be the set of all atoms in B. Show that B is atomic iff 1 is the only upper bound for A in B.

*5.7. PROBLEM. Show that the regular open algebra (Prob. 4.15) of the closed unit interval [0, 1] is atomless. (This furnishes an example of an atomless complete Boolean algebra.)

*5.8. PROBLEM. A complete Boolean algebra B is said to be *completely distributive* if for every doubly indexed subset $\{x_{ij}: \langle i,j\rangle \in I\times J\}$ of B we have

$$\bigwedge_{i\in I}\bigvee_{j\in J} x_{ij} = \bigvee_{f\in J^I}\bigwedge_{i\in I} x_{if(i)},$$

where J^I is the set of all mappings of all mappings of I into J. Show that the following conditions on a Boolean algebra B are equivalent:

(i) B is complete and completely distributive;

(ii) B is complete and atomic;

(iii) B is isomorphic to a power set algebra.

(To establish (i)⇒(ii), let $B=\{x_i: i\in I\}$ and for each $i\in I$ define $x_{i0}=x_i$ and $x_{i1}=x_i^*$. Observe that $1=\bigwedge_{i\in I}(x_{i0}\vee x_{i1})=\bigvee_{f\in 2^I}\bigwedge_{i\in I}x_{if(i)}$ and that $\bigwedge_{i\in I}x_{if(i)}$ is either 0 or an atom for each $f\in 2^I$, and apply the result of Prob. 5.6.)

5.9. PROBLEM. Show that a Boolean algebra is atomic iff its Stone space has a dense subset of isolated points.

5.10. PROBLEM. Let $\{B_i: i\in I\}$ be a family of Boolean algebras. The *product* of the family is the set $\prod_{i\in I}B_i$ with the Boolean operations defined pointwise, i.e. for all $f,g\in\prod_{i\in I}B_i$, $f\wedge g, f\vee g, f^*$ are defined by

$$(f\wedge g)(i)=f(i)\wedge g(i),$$

$$(f\vee g)(i)=f(i)\vee g(i),$$

$$f^*(i)=f(i)^*,$$

for all $i\in I$.

(i) Show that, if each B_i is complete, or σ-complete, then so is the product $\prod_{i\in I}B_i$.

(ii) Show that the product of atomic Boolean algebras is atomic. Is the converse true? That is, if a product is atomic, must each factor be atomic?

(iii) Show that each Boolean algebra is isomorphic to a subalgebra of a product of 2-element algebras. (This is the Stone representation theorem in disguise. Show that any Boolean algebra B can be embedded in 2^{SB}.)

5.11. PROBLEM. Give a *topological* proof that each complete atomic Boolean algebra B is isomorphic to a power set algebra. (Use the fact that the Stone space X of B is extremally disconnected (4.4) and has a dense subset I of isolated points to show that the map $U\mapsto U\cap I$, for $U\in \mathsf{C}X$, is an isomorphism of $\mathsf{C}X$ with $\mathsf{P}I$.)

5.12. PROBLEM. A one-one homomorphism between Boolean algebras is called a *monomorphism*. If $\varkappa$ is a cardinal, a Boolean algebra B is said to be *$\varkappa$-universal* if for each Boolean algebra A of cardinality $\leqslant\varkappa$ there is a monomorphism of A into B.

(i) Let B be a complete Boolean algebra, and let $\varkappa$ be an infinite cardinal. Show that the following conditions are equivalent:

(a) B is $\varkappa$-universal;

(b) for each set X of cardinality $\leqslant\varkappa$, there is a monomorphism of the finite-cofinite algebra (2.2(iv)) of X into B;

(c) B has an antichain (Prob. 3.18) of cardinality $\varkappa$.

(To prove (c)$\Rightarrow$(a), first show that, assuming (c), B has a antichain $\{b_\xi:\ \xi<\varkappa\}$ such that $b_\xi\neq 0$ for all $\xi<\varkappa$ and $\bigvee_{\xi<\varkappa} b_\xi=1$. Let A be a Boolean algebra of cardinality $\leqslant\varkappa$; let $\{a_\xi:\ \xi<\varkappa\}$ be an enumeration (possibly with repetitions) of the non-zero elements of A, and for each $\xi<\varkappa$ let U_ξ be an ultrafilter in A containing a_ξ. Show the map h: $A\rightarrow B$ defined by $h(x)=\bigvee\{b_\xi:\ x\in U_\xi\}$ for $x\in A$ is a monomorphism.)

(ii) Deduce that every infinite complete Boolean algebra is $\aleph_0$-universal. (Use Prob. 4.11 and (i).)

5.13. PROBLEM. Show that, if the Boolean algebra B is not atomic, then $|SB|\geqslant 2^{\aleph_0}$. Thus, if SB is countable, B is atomic. (If B is not atomic, then there is an element $b\neq 0$ of B such that no element $x\leqslant b$ of B is an atom. Using this fact and Prob. 5.5, construct for each $f\in 2^{\omega}$ a subset $X_f\subseteq\{x: x\leqslant b\}$ such that X_f has the f.m.p. and for $f\neq f'$ there exists $x\in X_f$ and $y\in X_{f'}$ such that $x\wedge y=0$.)

*§ 6. Duality for homomorphisms and continuous mappings

In this section we show that there is a natural correspondence between homomorphisms of Boolean algebras and continuous mappings of Boolean spaces.

Consider two Boolean algebras B, B', and a homomorphism h of B into B'. For each ultrafilter U in B', it is easy to see that $h^{-1}[U]$ is an ultrafilter in B, so we can define a mapping $h_\#: SB'\rightarrow SB$ by putting

$$h_\#(U)=h^{-1}[U]$$

for each $U\in SB'$. The mapping $h_\#$ is called the *dual* of the homomorphism h.

Let u and u' be the isomorphisms of B,B' onto the clopen algebras of SB, SB', respectively.

6.1. THEOREM. *If h is a homomorphism of B into B', $h_\#$ is a continuous mapping of SB' into SB.*

PROOF. It suffices to show that the inverse image under $h_{\#}$ of a clopen set in $\mathsf{S}B$ is clopen in $\mathsf{S}B'$. Now each clopen set in $\mathsf{S}B$ is of the form $u(x)$ for $x\in B$, and we have

$$\begin{aligned} h_{\#}^{-1}[u(x)] &= \{U\in \mathsf{S}B' : h_{\#}(U)\in u(x)\} \\ &= \{U\in \mathsf{S}B' : h^{-1}[U]\in u(x)\} \\ &= \{U\in \mathsf{S}B' : x\in h^{-1}[U]\} \\ &= \{U\in \mathsf{S}B' : h(x)\in U\} \\ &= u'(h(x)), \end{aligned}$$

which is, by definition, clopen in $\mathsf{S}B'$. ∎

Now consider two Boolean spaces X, X', and a continuous mapping φ of X into X'. If V is a clopen subset of X', then $\varphi^{-1}[V]$ is a clopen subset of X, so we can define a mapping $\varphi_{\natural} : \mathsf{C}X' \to \mathsf{C}X$ by putting

$$\varphi_{\natural}(V) = \varphi^{-1}[V]$$

for each $V\in \mathsf{C}X'$. The mapping $\varphi_{\natural}$ is called the *dual* of φ; it is easy to see that $\varphi_{\natural}$ is a *homomorphism of* $\mathsf{C}X'$ *into* $\mathsf{C}X$.

If h is a homomorphism of a Boolean algebra B into a Boolean algebra B', then, by Thm. 6.1, its dual $h_{\#}$ is a continuous mapping of $\mathsf{S}B'$ into $\mathsf{S}B$, and so the dual $h_{\#\natural}$ of $h_{\#}$ is a homomorphism of $\mathsf{CS}B$ into $\mathsf{CS}B'$. We thus obtain a diagram

$$\text{(6.2)} \qquad \begin{array}{ccc} B & \xrightarrow{\;h\;} & B' \\ {\scriptstyle u}\downarrow & & \downarrow{\scriptstyle u'} \\ \mathsf{CS}B & \xrightarrow{\;h_{\#\natural}\;} & \mathsf{CS}B' \end{array}$$

By the proof of Thm. 6.1 and the definition of $h_{\#\natural}$ we have, for each $x\in B$,

$$h_{\#\natural}(u(x)) = h_{\#}^{-1}[u(x)] = u'(h(x)),$$

so that (6.2) *commutes*. Therefore, if we identify B with $\mathsf{CS}B$ (via u) and B' with $\mathsf{CS}B'$ (via u'), *h is to be identified with its "second dual"* $h_{\#\natural}$.

The situation for Boolean spaces is similar. Thus let X and X' be Boolean spaces, let φ be a continuous mapping of X into X' and let v and v' be the homeomorphisms of X and X' onto $\mathsf{SC}X$ and $\mathsf{SC}X'$ given in Prob. 4.11. (These homeomorphisms are defined by

$$v(x) = \{V\in \mathsf{C}X : x\in V\},$$
$$v'(x') = \{V'\in \mathsf{C}X' : x'\in V'\}$$

for $x \in X$ and $x' \in X'$.) The dual $\varphi_\flat$ of φ is a homomorphism of $\mathrm{C}X'$ into $\mathrm{C}X$ and so, by Thm. 6.1, the dual $\varphi_{\flat\#}$ of $\varphi_\flat$ is a continuous mapping of $\mathrm{SC}X$ into $\mathrm{SC}X'$. We therefore obtain a diagram

$$\begin{array}{ccc} X & \xrightarrow{\ \varphi\ } & X' \\ {\scriptstyle v}\big\downarrow & & \big\downarrow{\scriptstyle v'} \\ \mathrm{SC}X & \xrightarrow{\ \varphi_{\flat\#}\ } & \mathrm{SC}X' \end{array} \tag{6.3}$$

By the definitions of $\varphi_{\flat\#}$, v and v', we have, for each $x \in X$,

$$\begin{aligned} \varphi_{\flat\#}(v(x)) &= \varphi_\flat^{-1}[v(x)] \\ &= \varphi_\flat^{-1}[\{V \in \mathrm{C}X\colon\ x \in V\}] \\ &= \{V' \in \mathrm{C}X'\colon\ x \in \varphi_\flat(V')\} \\ &= \{V' \in \mathrm{C}X'\colon\ x \in \varphi^{-1}[V']\} \\ &= \{V' \in \mathrm{C}X'\colon\ \varphi(x) \in V'\} \\ &= v'(\varphi(x)). \end{aligned}$$

Accordingly (6.3) *commutes*. Therefore, if we identify X with $\mathrm{SC}X$ (via v) and X' with $\mathrm{SC}X'$ (via v'), *then φ is to be identified with its "second dual"* $\varphi_{\flat\#}$.

We now prove:

6.4. THEOREM. (i) *Let h be a homomorphism of a Boolean algebra B into a Boolean algebra B'. Then h is one–one iff $h_\#$ is onto, and h is onto iff $h_\#$ is one–one.*

(ii) *Let φ be a continuous mapping of a Boolean space X into a Boolean space X'. Then φ is one–one iff $\varphi_\flat$ is onto, and φ is onto iff $\varphi_\flat$ is one–one.*

PROOF. (i) For simplicity put $\mathrm{S}B = X$, $\mathrm{S}B' = Y$. Then each of the following statements is equivalent to its neighbours:

(a) $h_\#$ is onto;
(b) $X - h_\#[Y] = \emptyset$;
(c) $X - h_\#[Y]$ includes no non-empty clopen set;
(d) if $A \subseteq X$ is clopen and $h_\#^{-1}[A] = \emptyset$, then $A = \emptyset$;
(e) if $A \in \mathrm{C}X$ and $h_{\#\flat}(A) = \emptyset$, then $A = \emptyset$;
(f) $h_{\#\flat}$ is one–one;
(g) h is one–one.

The only equivalences here which require justification are (b) $\Leftrightarrow$ (c) and (e) $\Leftrightarrow$ (f). For the first of these we merely observe that Y is compact, so $h_{\#}[Y]$ is compact, hence closed, so that $X-h_{\#}[Y]$ is open. For the second we appeal to the fact that $h_{\#\flat}$ is a homomorphism and to Prob. 3.2.

This proves the first part of (i). As for the second, again notice that each of the following statements is equivalent to its neighbours:

(a′) $h_{\#}$ is one–one;

(b′) $\{h_{\#}^{-1}[A]: A\in \mathrm{C}X\}$ is a separating field of (clopen) subsets of Y;

(c′) $\mathrm{C}Y=\{h_{\#}^{-1}[A]: A\in \mathrm{C}X\}$;

(d′) $\mathrm{C}Y=\{h_{\#\flat}(A): A\in \mathrm{C}X\}$;

(e′) $h_{\#\flat}$ is onto;

(f′) h is onto.

To see that (a′) ⇔ (b′), first assume (a′). Then if $x\neq y$ in Y, we have $h_{\#}(x)\neq h_{\#}(y)$, so that, since X is a Boolean space, there is $A\in \mathrm{C}X$ such that $h_{\#}(x)\in A$ and $h_{\#}(y)\notin A$. Hence $x\in h_{\#}^{-1}[A]$, $y\notin h_{\#}^{-1}[A]$ and (b′) holds. Conversely, assume (b′). Then if $x\neq y$ in Y, there is $A\in \mathrm{C}X$ such that $x\in h_{\#}^{-1}[A]$ and $y\notin h_{\#}^{-1}[A]$, so that, *a fortiori*, $h_{\#}(x)\neq h_{\#}(y)$, and (a′) holds.

(b′) ⇔ (c′) is an immediate consequence of Lemma 4.7, and the other equivalences are clear. This proves (i).

(ii) is a simple consequence of (i). For φ is one–one (resp. onto) iff $\varphi_{\flat\#}$ is one-one (resp. onto), and by (i) this latter condition holds iff $\varphi_{\flat}$ is onto (resp. one–one). ∎

We may sum up the results of this section rather concisely using the language of category theory. Let $\mathscr{B}$ be the category whose objects are all Boolean algebras and whose morphismes are all homomorphisms between these algebras, and let $\mathscr{S}$ be the category whose objects are all Boolean spaces and whose morphisms are all continuous mappings between these spaces. Let D (for duality!) be the mapping of $\mathscr{B}$ into $\mathscr{S}$ which assigns $\mathrm{S}B$ to each object B of $\mathscr{B}$ and $h_{\#}$ to each morphism h of $\mathscr{B}$, and let D' be the mapping of $\mathscr{S}$ into $\mathscr{B}$ which assigns $\mathrm{C}X$ to each object X of $\mathscr{S}$ and $\varphi_{\flat}$ to each morphism φ of $\mathscr{S}$. It is easy to verify that, if h, h' are two composable morphisms of $\mathscr{B}$, and φ, φ' are two composable morphisms of $\mathscr{S}$, then

$$(h'\circ h)_{\#}=h_{\#}\circ h'_{\#},$$

$$(\varphi'\circ\varphi)_{\flat}=\varphi_{\flat}\circ\varphi'_{\flat}.$$

Thus D and D' are *contravariant functors* of $\mathscr{B}$ into $\mathscr{S}$ and $\mathscr{S}$ into $\mathscr{B}$ respectively. Thm. 6.4. implies that D and D' transform monomorphisms into epimorphisms, and conversely. Moreover, the commutativity of diagrams (6.2) and (6.3) shows that D and D' are *mutually quasi-inverse* functors, so that $\mathscr{B}$ and $\mathscr{S}$ are *anti-equivalent* categories.

6.5. PROBLEM. Let B and B' be Boolean algebras.

(i) Show that B is isomorphic to a subalgebra of B' iff $\mathrm{S}B$ is a continuous image of $\mathrm{S}B'$.

(ii) Show that B is isomorphic to a quotient of B' iff $\mathrm{S}B$ is homeomorphic to a (necessarily closed) subspace of $\mathrm{S}B'$. (Apply Thm. 6.4.)

6.6. PROBLEM. Let F be a filter in a Boolean algebra B. The *dual* $\tilde{F}$ of F is the closed subset $\bigcap\{u(x)\colon x\in F\}$ of $\mathrm{S}B$, where u is the natural isomorphism between B and $\mathrm{CS}B$.

(i) Show that $F\mapsto\tilde{F}$ is a one–one mapping of the set of all filters in B onto the family of all non-empty closed sets in $\mathrm{S}B$.

(ii) Show that $\tilde{F}$, regarded as a subspace of $\mathrm{S}B$, is homeomorphic to $\mathrm{S}(B/F)$. (Apply Thm. 6.4 to the canonical homomorphism $h\colon B\twoheadrightarrow B/F$.)

§ 7. The Rasiowa–Sikorski Theorem

We conclude our discussion of Boolean algebras with a result which has important applications in model theory (Ch. 5, §§5, 6).

Let B be a Boolean algebra, and let T be a subset of B which has a join $\bigvee T$. An ultrafilter U in B is said to *respect* T (or the join $\bigvee T$) if we have

$$\bigvee T\in U\Rightarrow T\cap U\neq\emptyset.$$

Clearly, for *any* ultrafilter (in fact any filter) U we have

$$T\cap U\neq\emptyset\Rightarrow\bigvee T\in U.$$

Thus U respects T iff

$$\bigvee T\in U\ \Leftrightarrow\ T\cap U\neq\emptyset.$$

If $\mathscr{T}$ is a family of subsets of B, each member of which has a join, we say that U *respects* $\mathscr{T}$ (or the family of joins $\{\bigvee T\colon T\in\mathscr{T}\}$) if U respects each member of $\mathscr{T}$.

7.1. PROBLEM. Show that, if each member of $\mathscr{T}$ is finite, then every ultrafilter respects $\mathscr{T}$.

Let h be the canonical homomorphism of B onto $B/U\cong 2$. It is easy to see that U respects $\mathscr{T}$ iff $h(\bigvee T)=\bigvee h[T]$ for all $T\in\mathscr{T}$. A two-valued homomorphism h satisfying this condition is called a *$\mathscr{T}$-complete* homomorphism.

We now ask: given a Boolean algebra B and a family $\mathscr{T}$ of subsets of B each member of which has a join, is there always an ultrafilter in B which respects $\mathscr{T}$? In general the answer is no, in view of:

7.2. THEOREM. *Let B be a complete Boolean algebra. Then there is a natural one–one correspondence between atoms of B and ultrafilters in B which respect $\mathsf{P}B$, and hence also between atoms of B and $\mathsf{P}B$-complete 2-valued homomorphisms on B.*

PROOF. We give the merest sketch. For each atom $a \in B$, let U_a be the ultrafilter generated by a; then it is easy to verify that U_a respects $\mathsf{P}B$. Conversely, let U be an ultrafilter which respects $\mathsf{P}B$ and let $a = \bigwedge U$. Setting $U^* = \{x^*: x \in U\}$, we see that, by Problem 2.6, $a^* = \bigvee U^*$. Since $U \cap U^* = \emptyset$ and U respects $\mathsf{P}B$, it follows that $a^* \notin U$, whence $a \in U$. Therefore a generates U, so that, by Lemma 5.1, a is an atom. This shows that the mapping $a \mapsto U_a$ is a bijection of the set of atoms of B onto the set of ultrafilters in B which respect $\mathsf{P}B$. ∎

It follows from Thm. 7.2 that if B is an atomless complete Boolean algebra (e.g. the regular open algebra of $[0, 1]$; cf. Probs. 4.11 and 5.7), then there is no ultrafilter in B which respects $\mathsf{P}B$. However, since a complete atomless Boolean algebra B must be infinite (every finite Boolean algebra being atomic), $\mathsf{P}B$ is *uncountable*. Thus, in general, there may be no ultrafilter respecting a given uncountable family $\mathscr{T}$ of subsets of a Boolean algebra. But the situation is quite different when $\mathscr{T}$ is *countable*, as our final result shows.

7.3. THE RASIOWA–SIKORSKI THEOREM. *Let B be a Boolean algebra and let $\mathscr{T}$ be a countable family of subsets of B, each member of which has a join in B. Then there is an ultrafilter in B which respects $\mathscr{T}$.*

PROOF. Enumerate $\mathscr{T}$ as $\{T_n: n \in \omega\}$, and let $t_n = \bigvee T_n$ for each $n \in \omega$. We define by induction a sequence $\{b_n: n \in \omega\}$ of elements of B such that, for each $n \in \omega$, $b_n \in T_n$ and the set $\{t_0^* \vee b_0, \ldots, t_n^* \vee b_n\}$ has a non-zero meet. Suppose that $n \in \omega$ and that for each $m < n$ we have found b_m to satisfy these conditions. If $n = 0$, let

$$y = 1,$$

and if $n > 0$ let

$$y = (t_0^* \vee b_0) \wedge \ldots \wedge (t_{n-1}^* \vee b_{n-1}).$$

Then if $n=0$ we have $y\neq 0$ by assumption, and if $n>0$ we have $y\neq 0$ by induction hypothesis. Suppose now, for contradiction's sake, that $y\wedge(t_n^*\vee b)=0$ for all $b\in T_n$. Then

$$0=(y\wedge t_n^*)\vee(y\wedge b)$$

so that $y\wedge t_n^*=0$ and $y\wedge b=0$ for all $b\in T_n$. It follows that $y\leqslant b^*$ for all $b\in T_n$ so that by Prob. 2.6

$$y\leqslant\bigwedge\{b^*:\ b\in T_n\}=(\bigvee T_n)^*=t_n^*.$$

Thus $y=y\wedge t_n^*=0$, contradicting the induction hypothesis.

Accordingly, we can find b_n to satisfy the required conditions, and therefore such a b_n can be found for each $n\in\omega$. Then the set $\{t_0^*\vee b_0,\ldots,t_n^*\vee b_n,\ldots\}$ has the f.m.p. and is therefore included in an ultrafilter U in B. We show that U respects $\mathscr{T}$. If $t_n\in U$ then, since $t_n^*\vee b_n\in U$ by construction, it follows that

$$b_n=t_n\wedge b_n=t_n\wedge(t_n^*\vee b_n)\in U,$$

so that $T_n\cap U\neq\emptyset$. Thus U respects $\mathscr{T}$ and we are finished. ∎

*7.4. Problem. The *Baire category theorem* asserts that, if $\mathscr{V}$ is a countable family of dense open subsets of a compact Hausdorff space, then $\bigcap\mathscr{V}$ is dense. Derive the following strong form of the Rasiowa–Sikorski Theorem from the Baire category theorem: for any Boolean algebra B, any countable family $\mathscr{T}$ of subsets of B, each member of which has a join in B, and any non-zero x in B, there is an ultrafilter in B which contains x and respects $\mathscr{T}$. (For each $n\in\omega$ let $Q_n=\{U\in SB:\ U \text{ respects } T_n\}$; where $\mathscr{T}=\{T_n:\ n\in\omega\}$; show that Q_n is dense and open in SB and apply the Baire category theorem.)

§ 8. Historical and bibliographical remarks

Boolean algebras are named after the English mathematician George Boole (1815–1864). In 1847 he made the first successful attempt to apply mathematical techniques to logic. The equivalence between Boolean algebras and complemented distributive lattices was first formulated by Huntington in 1904. The Ultrafilter Theorem is due to Tarski [1930]; it is often stated in terms of ideals and is then known as the *Boolean prime ideal theorem.* Lemma 3.11 is due to Hirschfeld and appears in

MACHOVER and HIRSCHFELD [1969]. The Stone Representation Theorem is to be found in STONE [1936] and the first application of topological methods to the theory of Boolean algebras in STONE [1937]. Thm. 4.8 is due to MAKINSON [1969]; the proof we give was suggested by Brian Rotman. The Rasiowa–Sikorski Theorem was first proved in RASIOWA and SIKORSKI [1951], using topological methods (see Prob. 7.4). The simple proof we give is due to Tarski (see FEFERMAN [1952]).

Good introductions to the theory of Boolean algebras are to be found in DWINGER [1961] and HALMOS [1963]. More advanced treatises include BIRKHOFF [1967] and SIKORSKI [1964].

CHAPTER 5

MODEL THEORY

The theme of this chapter — *model theory* — is the relationship between sets of first-order sentences and the structures in which they are satisfied, i.e. their *models*. In particular we shall be concerned with the various methods by which models with prescribed properties can be constructed. In §1 we set up a natural framework for discussing models; and in §2 we prove the basic results on the existence of models of prescribed cardinalities. In §3 we introduce an important method of constructing models — the *ultraproduct construction* — which is a modification of the familiar algebraic procedure of forming direct products. In §4 we discuss the extent to which a set of sentences determines the properties of its models, and apply our results to specific formal mathematical theories. In §5 we show how the theory of Boolean algebras is connected with the theory of models, and in §6 we discuss the role played in the theory by formulas with free variables. Finally, in §7, we introduce models generated by Skolem functions and prove the existence of models with many automorphisms. (§§6 and 7 are of a more specialized character than the other sections and may be omitted at a first reading.) Acquaintance with the contents of Chs. 1–3 is required for an understanding of this chapter. However, only §§3, 5 and 6 assume familiarity with the results and methods of Ch. 4.

§ 1. Basic ideas of model theory

Throughout this chapter — with the exception of §7 — we shall use the symbol $\mathscr{L}$ to denote a first-order language with equality but with no function symbols apart from individual constants.[1] We shall take conjunction ($\wedge$), negation ($\neg$) and the existential quantifier ($\exists$) as the

[1] This restriction is not essential but is made for the sake of simplicity. The results of this chapter can be extended in a straightforward way to languages with function symbols.

primitive logical symbols[1] of $\mathscr{L}$ and regard the other logical symbols ($\vee$, $\rightarrow$, $\leftrightarrow$, $\forall$) as being defined in terms of these. We shall assume that the individual variables of $\mathscr{L}$ are enumerated in a fixed alphabetic sequence[2] $\mathbf{v}_0, \mathbf{v}_1, \ldots, \mathbf{v}_n, \ldots$. Moreover, we assume that the predicate symbols and constant symbols of $\mathscr{L}$ are given in the form of indexed sets $\{\mathbf{R}_i: i \in I\}$, $\{\mathbf{c}_j: j \in J\}$ respectively. For each $i \in I$ we let $\lambda(i)$ be the number of argument places in the predicate $\mathbf{R}_i$. Thus λ is a mapping of the set I into the set of positive integers; it is called the *signature* of $\mathscr{L}$.

The notion of an *$\mathscr{L}$-structure* has already been introduced in §1 of Ch. 2. There we allowed the domain of a structure to be an arbitrary (non-empty) *class*. In model theory, however, we confine our attention to structures whose domains are *sets*. (By the remark of the end of §5 of Ch. 3, there is no loss of generality in making this restriction, at least as far as satisfiability of formulas is concerned.)

It is clear that any $\mathscr{L}$-structure whose domain is a set may be regarded as an ordered triple

$$\mathfrak{A} = \langle A, \mathscr{R}, \mathfrak{c} \rangle,$$

where

(1) A is a non-empty set called the *domain* or *universe* of $\mathfrak{A}$;

(2) $\mathscr{R}$ is a mapping of I into the set of all relations on A such that for each $i \in I$, $\mathscr{R}(i)$ is a $\lambda(i)$-ary relation;

(3) $\mathfrak{c}$ is a mapping of J into A.

For each $i \in I$ and each $j \in J$ we often write R_i for $\mathscr{R}(i)$ and c_j for $\mathfrak{c}(j)$, and we also write

(4) $\mathfrak{A} = \langle A, \langle R_i \rangle_{i \in I}, \langle c_j \rangle_{j \in J} \rangle$.

The R_i and the c_j are called the *relations* and *designated individuals* of $\mathfrak{A}$, respectively. We shall sometimes write $\mathbf{R}_i^{\mathfrak{A}}$ for R_i and $\mathbf{c}_j^{\mathfrak{A}}$ for c_j, in order to emphasize the fact that R_i is the interpretation of $\mathbf{R}_i$, and c_j that of $\mathbf{c}_j$, in $\mathfrak{A}$.

We shall always use upper-case German letters to denote structures. If we are given a structure denoted by a German letter, we agree to use the corresponding upper-case italic letter to denote its domain. Thus A is the domain of $\mathfrak{A}$, B that of $\mathfrak{B}$, etc.

If $\mathfrak{A}$ is an $\mathscr{L}$-structure, we often call $\mathscr{L}$ the *language for* $\mathfrak{A}$.

[1] Thus the degree of complexity deg φ of an $\mathscr{L}$-formula φ will be the number obtained by adding up 1 for each occurrence of $\exists$, $\neg$, and $\wedge$ in φ.

[2] This is a departure from the convention adopted in previous chapters in which we assumed the enumeration of the individual variables to start with $\mathbf{v}_1$.

Given an $\mathscr{L}$-structure of the form (4), we obtain an *$\mathscr{L}$-valuation* (Chapter 2, §1) by further specifying a sequence

$$\mathfrak{a}=\langle a_0,a_1,\ldots\rangle$$

of members of A as an assignment of values to the variables $\mathbf{v}_0,\mathbf{v}_1,\ldots$ of $\mathscr{L}$. We shall call such a sequence an *assignment in* $\mathfrak{A}$.

If $\mathfrak{a}$ is an assignment in an $\mathscr{L}$-structure $\mathfrak{A}$ and $b\in A$, we define $\mathfrak{a}(n|b)$ to be the assignment which assigns the same values to the variables as does $\mathfrak{a}$, *except* that it assigns the value b to the variable $\mathbf{v}_n$. Thus

$$\mathfrak{a}(n|b)=\langle a_0,a_1,\ldots,a_{n-1},b,a_{n+1},\ldots\rangle.$$

Notice that the value assigned to $\mathbf{v}_n$ by $\mathfrak{a}(n|b)$ is *independent* of the value assigned to $\mathbf{v}_n$ by $\mathfrak{a}$.

For convenience we now restate the Basic Semantic Definition (2.1.1) in the form best suited to our present purpose.

Let $\mathscr{L}$ be a language with predicate symbols $\{\mathbf{R}_i\colon\ i\in I\}$, constant symbols $\{\mathbf{c}_j\colon\ j\in J\}$ and signature λ. Let

$$\mathfrak{A}=\langle A,\langle R_i\rangle_{i\in I},\langle c_j\rangle_{j\in J}\rangle$$

be an $\mathscr{L}$-structure, and let $\mathfrak{a}=\langle a_0,a_1,\ldots\rangle$ be an assignment in $\mathfrak{A}$. For all $\mathscr{L}$-formulas φ we define the relation $\mathfrak{a}$ *satisfies* φ *in* $\mathfrak{A}$, which we write $\mathfrak{A}\models_{\mathfrak{a}}\varphi$, by induction on deg φ:

(1) For terms $\mathbf{t}_1$, $\mathbf{t}_2$ of $\mathscr{L}$,

$$\mathfrak{A}\models_{\mathfrak{a}}\mathbf{t}_1=\mathbf{t}_2 \;\Leftrightarrow\; b_1=b_2,$$

where if $\mathbf{t}_n$ ($n=1,2$) is the variable $\mathbf{v}_k$ then b_n is a_k, while if $\mathbf{t}_n$ is the constant $\mathbf{c}_j$ then b_n is c_j.

(2) For $i\in I$ and terms $\mathbf{t}_1,\ldots,\mathbf{t}_{\lambda(i)}$ of $\mathscr{L}$,

$$\mathfrak{A}\models_{\mathfrak{a}}\mathbf{R}_i\mathbf{t}_1\ldots\mathbf{t}_{\lambda(i)} \;\Leftrightarrow\; \langle b_1,\ldots,b_{\lambda(i)}\rangle\in R_i,$$

where if $\mathbf{t}_n$ ($n=1,\ldots,\lambda(i)$) is the variable $\mathbf{v}_k$ then b_n is a_k, while if $\mathbf{t}_n$ is the constant $\mathbf{c}_j$ then b_n is c_j.

(3) $\mathfrak{A}\models_{\mathfrak{a}}\neg\varphi \;\Leftrightarrow\;$ not $\mathfrak{A}\models_{\mathfrak{a}}\varphi$.

(4) $\mathfrak{A}\models_{\mathfrak{a}}\varphi\wedge\psi \;\Leftrightarrow\; \mathfrak{A}\models_{\mathfrak{a}}\varphi$ and $\mathfrak{A}\models_{\mathfrak{a}}\psi$.

(5) $\mathfrak{A}\models_{\mathfrak{a}}\exists v_n\varphi \;\Leftrightarrow\; \mathfrak{A}\models_{\mathfrak{a}(n|b)}\varphi$ for some $b\in A$.

It should be clear that the above definition does not differ essentially from that given in 2.1.1. In fact, it is easy to verify that if $\mathfrak{A}$ is an $\mathscr{L}$-structure and $\mathfrak{a}$ is an assignment in $\mathfrak{A}$, then, if σ is the $\mathscr{L}$-valuation determined by

$\mathfrak{A}$, $\mathfrak{a}$, we have, for each $\mathscr{L}$-formula $\boldsymbol{\varphi}$,

$$\mathfrak{A} \models_{\mathfrak{a}} \boldsymbol{\varphi} \Leftrightarrow \boldsymbol{\varphi}^{\sigma} = \top .$$

The following facts are clear:

(a) $\mathfrak{A} \models_{\mathfrak{a}} \forall \mathbf{v}_n \boldsymbol{\varphi} \Leftrightarrow \mathfrak{A} \models_{\mathfrak{a}(n|b)} \boldsymbol{\varphi}$ for all $b \in A$;

(b) if $\boldsymbol{\varphi}$ is a formula and $\mathfrak{a}$, $\mathfrak{a}'$ are assignments in $\mathfrak{A}$ such that $a_n = a'_n$ whenever $\mathbf{v}_n$ occurs free in $\boldsymbol{\varphi}$, then $\mathfrak{A} \models_{\mathfrak{a}} \boldsymbol{\varphi} \Leftrightarrow \mathfrak{A} \models_{\mathfrak{a}'} \boldsymbol{\varphi}$. (See Thm. 2.2.3.)

In view of fact (b), the truth of $\mathfrak{A} \models_{\mathfrak{a}} \boldsymbol{\varphi}$, insofar as it depends on $\mathfrak{a}$, depends only on the values $\mathfrak{a}$ assigns to the *free* variables of $\boldsymbol{\varphi}$. Accordingly we make the following definition: if $\boldsymbol{\varphi}$ is a formula all of whose free variables are among $\mathbf{v}_0, \ldots, \mathbf{v}_n$ and $a_0, \ldots, a_n \in A$, we say that the finite sequence $a_0, \ldots, a_n$ *satisfies* $\boldsymbol{\varphi}$ in $\mathfrak{A}$ and write

$$\mathfrak{A} \models \boldsymbol{\varphi}\, [a_0, \ldots, a_n]$$

if $\mathfrak{A} \models_{\mathfrak{a}'} \boldsymbol{\varphi}$ for some assignment $\mathfrak{a}'$ in A such that $a'_0 = a_0, \ldots, a'_n = a_n$. It follows immediately from (b) that $\mathfrak{A} \models \boldsymbol{\varphi}\, [a_0, \ldots, a_n]$ iff $\mathfrak{A} \models_{\mathfrak{a}'} \boldsymbol{\varphi}$ for *all* assignments $\mathfrak{a}'$ in $\mathfrak{A}$ such that $a'_0 = a_0, \ldots, a'_n = a_n$.

If $\boldsymbol{\sigma}$ is a *sentence*, i.e. a formula *without* free variables, we say that $\boldsymbol{\sigma}$ is *valid* or *holds* in $\mathfrak{A}$, or that $\mathfrak{A}$ is a *model* of $\boldsymbol{\sigma}$, and write

$$\mathfrak{A} \models \boldsymbol{\sigma},$$

if $\mathfrak{A} \models_{\mathfrak{a}} \boldsymbol{\sigma}$ for some assignment — and hence, in view of (b), all assignments — $\mathfrak{a}$ in $\mathfrak{A}$. If $\boldsymbol{\Sigma}$ is a set of sentences, we say that $\mathfrak{A}$ is a *model* of $\boldsymbol{\Sigma}$ and write

$$\mathfrak{A} \models \boldsymbol{\Sigma}$$

if $\mathfrak{A}$ is a model of each sentence in $\boldsymbol{\Sigma}$.

Let $\mathscr{L}'$ be a language which is an extension of $\mathscr{L}$, so that, in addition to the predicate symbols and constant symbols of $\mathscr{L}$, $\mathscr{L}'$ contains a set $\{\mathbf{R}_i\colon i \in I'\}$ of predicate symbols and a set $\{\mathbf{c}_j\colon j \in J'\}$ of constants. Given an $\mathscr{L}'$-structure

$$\mathfrak{A}' = \langle A, \langle R_i \rangle_{i \in I \cup I'}, \langle c_j \rangle_{j \in J \cup J'} \rangle,$$

the $\mathscr{L}$-structure

$$\mathfrak{A} = \langle A, \langle R_i \rangle_{i \in I}, \langle c_j \rangle_{j \in J} \rangle$$

is called the *$\mathscr{L}$-reduction of* $\mathfrak{A}'$, and $\mathfrak{A}'$ is called an *$\mathscr{L}'$-expansion* of $\mathfrak{A}$ (cf. Ch. 2, §9). (Notice that in general an $\mathscr{L}$-structure has more than one $\mathscr{L}'$-expansion.)

Let

$$\mathfrak{A} = \langle A, \langle R_i \rangle_{i \in I}, \langle c_j \rangle_{j \in J} \rangle,$$

$$\mathfrak{A}' = \langle A', \langle R_i' \rangle_{i \in I}, \langle c_j' \rangle_{j \in J} \rangle$$

be $\mathscr{L}$-structures. We say that $\mathfrak{A}$ is a *substructure* of $\mathfrak{A}'$ and write $\mathfrak{A} \subseteq \mathfrak{A}'$ if $A \subseteq A'$, for each $j \in J$, $c_j = c_j'$, and, for each $i \in I$, R_i is the restriction of R_i' to A, i.e. $R_i = R_i' \cap A^{\lambda(i)}$. If B is a non-empty subset of A which contains all the designated individuals c_j of $\mathfrak{A}$, we define the *restriction* $\mathfrak{A}|B$ of $\mathfrak{A}$ to B by

$$\mathfrak{A}|B = \langle B, \langle R_i \cap B^{\lambda(i)} \rangle_{i \in I}, \langle c_j \rangle_{j \in J} \rangle.$$

It is clear that for any subset B of A which contains all the designated individuals of $\mathfrak{A}$ we have $\mathfrak{A}|B \subseteq \mathfrak{A}$.

An *embedding* of $\mathfrak{A}$ into $\mathfrak{A}'$ is a one–one mapping f of A into A' such that

(i) $f(c_j) = c_j'$ for all $j \in J$;
(ii) $\langle a_1, \ldots, a_{\lambda(i)} \rangle \in R_i \Leftrightarrow \langle f(a_1), \ldots, f(a_{\lambda(i)}) \rangle \in R_i'$

for all $i \in I$ and all $a_1, \ldots, a_{\lambda(i)} \in A$.

If f is an embedding of $\mathfrak{A}$ into $\mathfrak{A}'$, it follows from (i) that $f[A]$ contains all the designated individuals of $\mathfrak{A}'$, so that we can form the restriction $\mathfrak{A}'|f[A]$. This is written $f[\mathfrak{A}]$ and is called the *image* of $\mathfrak{A}$ under f.

An *isomorphism* of $\mathfrak{A}$ onto $\mathfrak{A}'$ is an embedding of $\mathfrak{A}$ *onto* $\mathfrak{A}'$. If there is an isomorphism of $\mathfrak{A}$ onto $\mathfrak{A}'$, we say that $\mathfrak{A}$ and $\mathfrak{A}'$ are *isomorphic* and write $\mathfrak{A} \cong \mathfrak{A}'$. Clearly, if f is an embedding of $\mathfrak{A}$ into $\mathfrak{A}'$, we have $\mathfrak{A} \cong f[\mathfrak{A}]$.

$\mathfrak{A}$ and $\mathfrak{A}'$ are said to be ($\mathscr{L}$-)*elementarily equivalent*, and we write $\mathfrak{A} \equiv \mathfrak{A}'$, if for any $\mathscr{L}$-sentence $\boldsymbol{\sigma}$ we have $\mathfrak{A} \models \boldsymbol{\sigma} \Leftrightarrow \mathfrak{A}' \models \boldsymbol{\sigma}$. Thus two $\mathscr{L}$-structures are elementarily equivalent if they cannot be distinguished by an $\mathscr{L}$-sentence.

1.1. PROBLEM. Let f be an isomorphism of $\mathfrak{A}$ onto $\mathfrak{A}'$. Show by induction on deg $\boldsymbol{\varphi}$ that for any formula $\boldsymbol{\varphi}$ all of whose free variables are among $\mathbf{v}_0, \ldots, \mathbf{v}_n$ and all $a_0, \ldots, a_n \in A$ we have

$$\mathfrak{A} \models \boldsymbol{\varphi}\,[a_0, \ldots, a_n] \Leftrightarrow \mathfrak{A}' \models \boldsymbol{\varphi}\,[f(a_0), \ldots, f(a_n)].$$

Infer that, *if* $\mathfrak{A} \cong \mathfrak{A}'$, *then* $\mathfrak{A} \equiv \mathfrak{A}'$. (We shall see later on that the converse is false.)

$\mathfrak{A}$ is said to be an ($\mathscr{L}$-)*elementary substructure* of $\mathfrak{A}'$, and $\mathfrak{A}'$ an ($\mathscr{L}$-)*elementary extension* of $\mathfrak{A}$ if $\mathfrak{A} \subseteq \mathfrak{A}'$ and for any $\mathscr{L}$-formula $\boldsymbol{\varphi}$ all of whose free variables are among $\mathbf{v}_0,\ldots,\mathbf{v}_n$ we have

$$\mathfrak{A} \models \varphi\,[a_0,\ldots,a_n] \;\Leftrightarrow\; \mathfrak{A}' \models \varphi\,[a_0,\ldots,a_n]$$

for all $a_0,\ldots,a_n \in A$. In this situation we write $\mathfrak{A} \prec \mathfrak{A}'$.

It is clear that $\mathfrak{A} \prec \mathfrak{A}' \Rightarrow \mathfrak{A} \equiv \mathfrak{A}'$. Our next problem shows that the converse is false.

1.2. PROBLEM. Let $\mathfrak{A} = \langle \omega - \{0\}, < \rangle$, $\mathfrak{A}' = \langle \omega, < \rangle$ where $<$ is the usual ordering of the natural numbers. Show that $\mathfrak{A} \cong \mathfrak{A}'$, $\mathfrak{A} \subseteq \mathfrak{A}'$, but not $\mathfrak{A} \prec \mathfrak{A}'$.

1.3. PROBLEM. Show that if $\mathfrak{A} \prec \mathfrak{A}'$, $\mathfrak{A}'' \prec \mathfrak{A}'$ and $\mathfrak{A} \subseteq \mathfrak{A}''$, then $\mathfrak{A} \prec \mathfrak{A}''$.

An embedding f of $\mathfrak{A}$ into $\mathfrak{A}'$ is called an ($\mathscr{L}$-)*elementary embedding* if for any $\mathscr{L}$-formula φ all of whose free variables are among $\mathbf{v}_0,\ldots,\mathbf{v}_n$ we have

$$\mathfrak{A} \models \varphi\,[a_0,\ldots,a_n] \;\Leftrightarrow\; \mathfrak{A}' \models \varphi\,[f(a_0),\ldots,f(a_n)]$$

for all $a_0,\ldots,a_n \in A$.

1.4. PROBLEM. (i) Let f be an embedding of $\mathfrak{A}$ into $\mathfrak{A}'$. Show that f is an elementary embedding iff $f[\mathfrak{A}] \prec \mathfrak{A}'$.

(ii) Let $\mathfrak{A} \subseteq \mathfrak{A}'$. Show that $\mathfrak{A} \prec \mathfrak{A}'$ iff the natural injection of $\mathfrak{A}$ into $\mathfrak{A}'$ is an elementary embedding of $\mathfrak{A}$ into $\mathfrak{A}'$.

(iii) Let f be any mapping of A into A' such that, for all $a_0,\ldots,a_n \in A$ and all formulas φ with free variables among $\mathbf{v}_0,\ldots,\mathbf{v}_n$, $\mathfrak{A} \models \varphi\,[a_0,\ldots,a_n] \Leftrightarrow$ $\Leftrightarrow \mathfrak{A}' \models \varphi\,[f(a_0),\ldots,f(a_n)]$. Show that f is one–one and hence an elementary embedding of $\mathfrak{A}$ into $\mathfrak{A}'$.

$\mathfrak{A}$ is said to be *elementarily embeddable* in $\mathfrak{A}'$ if there is an elementary embedding of $\mathfrak{A}$ into $\mathfrak{A}'$. Clearly $\mathfrak{A}$ is elementarily embeddable in $\mathfrak{A}'$ iff $\mathfrak{A}$ is isomorphic to an elementary substructure of $\mathfrak{A}'$. Evidently, also, if $\mathfrak{A}$ is elementarily embeddable in $\mathfrak{A}'$, then $\mathfrak{A}$ is elementarily equivalent to $\mathfrak{A}'$.

We now prove some lemmas which will be very useful later.

1.5. LEMMA. *Suppose that* $\mathfrak{A} \subseteq \mathfrak{A}'$. *Then the following two assertions are equivalent.*

(i) $\mathfrak{A} \prec \mathfrak{A}'$;

(ii) *for any* n, *any* $\mathscr{L}$-*formula* φ *whose free variables are all among* $\mathbf{v}_0,\ldots,\mathbf{v}_n$, *and any* $a_0,\ldots,a_{n-1} \in A$, *if there is* $a' \in A'$ *for which* $\mathfrak{A}' \models \varphi\,[a_0,\ldots,a_{n-1},a']$, *then there is* $a \in A$ *for which* $\mathfrak{A}' \models \varphi\,[a_0,\ldots,a_{n-1},a]$.

PROOF. (i)$\Rightarrow$(ii). Suppose $\mathfrak{A} \prec \mathfrak{A}'$ and $\mathfrak{A}' \models \varphi\,[a_0,\ldots,a_{n-1},a']$ with $a' \in A'$. Then $\mathfrak{A}' \models \exists \mathbf{v}_n \varphi\,[a_0,\ldots,a_{n-1}]$, so that, since $\mathfrak{A} \prec \mathfrak{A}'$, $\mathfrak{A} \models \exists \mathbf{v}_n \varphi\,[a_0,\ldots,a_{n-1}]$.

Hence $\mathfrak{A} \models \varphi\, [a_0,\ldots,a_{n-1},a]$ for some $a \in A$, so that, since $\mathfrak{A} \prec \mathfrak{A}'$, $\mathfrak{A}' \models \varphi\, [a_0,\ldots,a_{n-1},a]$.

(ii)$\Rightarrow$(i). Assume (ii). We have to show that

(1) $$\mathfrak{A} \models \varphi\, [a_0,\ldots,a_n] \;\Leftrightarrow\; \mathfrak{A}' \models \varphi\, [a_0,\ldots,a_n]$$

for any $\mathscr{L}$-formula φ with free variables among $\mathbf{v}_0,\ldots,\mathbf{v}_n$ and any $a_0,\ldots,a_n \in A$. We prove (1) by induction on deg φ. That (1) holds for atomic φ follows immediately from the assumption that $\mathfrak{A} \subseteq \mathfrak{A}'$. The induction steps for $\wedge$ and $\neg$ are trivial, so it remains to establish the induction step for $\exists$.

Let φ be $\exists \mathbf{v}_k \psi$, and suppose that the free variables of φ are all among $\mathbf{v}_0,\ldots,\mathbf{v}_n$. We want to prove (1); clearly it will be enough to prove

$$\mathfrak{A} \models \varphi\, [a_0,\ldots,a_m] \;\Leftrightarrow\; \mathfrak{A}' \models \varphi\, [a_0,\ldots,a_m]$$

for some $m \geqslant n$. Therefore we may assume without loss of generality that n is greater than the indices of all the variables occurring — free or bound — in φ; in particular, $n \geqslant k$ and the free variables of ψ are all among $\mathbf{v}_0,\ldots,\mathbf{v}_n$. If

$$\mathfrak{A} \models (\exists \mathbf{v}_k \psi)\, [a_0,\ldots,a_n],$$

then we have $\mathfrak{A} \models \psi\, [a_0,\ldots,a_{k-1},a,a_{k+1},\ldots,a_n]$ for some $a \in A$, so that, by inductive hypothesis,

$$\mathfrak{A}' \models \psi\, [a_0,\ldots,a_{k-1},a,a_{k+1},\ldots,a_n],$$

whence $\mathfrak{A}' \models (\exists \mathbf{v}_k \psi)\, [a_0,\ldots,a_n]$.

Conversely, suppose that $\mathfrak{A}' \models (\exists \mathbf{v}_k \psi)\, [a_0,\ldots,a_n]$. Put $\chi = \psi(\mathbf{v}_k/\mathbf{v}_{n+1})$; then $\exists \mathbf{v}_k \psi$ and $\exists \mathbf{v}_{n+1} \chi$ are variants — hence logically equivalent — so that

$$\mathfrak{A}' \models (\exists \mathbf{v}_{n+1} \chi)\, [a_0,\ldots,a_n].$$

Thus there is some $a' \in A'$ for which $\mathfrak{A}' \models \chi\, [a_0,\ldots,a_n,a']$. Hence, by (ii), there is $a \in A$ such that $\mathfrak{A}' \models \chi\, [a_0,\ldots,a_n,a]$. Since deg $\chi <$ deg φ and the free variables of χ are all among $\mathbf{v}_0,\ldots,\mathbf{v}_{n+1}$, it follows from the inductive hypothesis that

$$\mathfrak{A} \models \chi\, [a_0,\ldots,a_n,a]$$

and so $\mathfrak{A} \models (\exists \mathbf{v}_{n+1} \chi)\, [a_0,\ldots,a_n]$. Therefore

$$\mathfrak{A} \models (\exists \mathbf{v}_k \psi)\, [a_0,\ldots,a_n].$$

This completes the induction step and the proof. ∎

Before stating the next lemma we need to set up some more notation. Let K be a set such that $J \cap K = \emptyset$ (recalling that J is the set indexing the constants of $\mathscr{L}$). We define $\mathscr{L}_K$ to be the language obtained from $\mathscr{L}$ by adding a set of entirely new distinct constants $\{\mathfrak{c}_k\colon k \in K\}$.

Now let $\mathfrak{A} = \langle A, \mathscr{R}, \mathfrak{c}\rangle$ be an $\mathscr{L}$-structure, where, as usual, $\mathfrak{c}$ is a mapping of J onto the set of designated individuals of $\mathfrak{A}$. Given a mapping $\mathfrak{a}$ of K into A, we define

$$(\mathfrak{A}, \mathfrak{a}) = \langle A, \mathscr{R}, \mathfrak{c} \cup \mathfrak{a}\rangle.$$

Clearly $(\mathfrak{A}, \mathfrak{a})$ is an $\mathscr{L}_K$-expansion of $\mathfrak{A}$ in which $\mathfrak{c}_k^{(\mathfrak{A},\mathfrak{a})} = a_k$ for each $k \in K$.

If $\mathfrak{a}$ maps K *onto* A, we say that $\mathfrak{a}$ is an *indexing* of A by K.

We can now state:

1.6. LEMMA. *Let $\mathfrak{A}$ and $\mathfrak{A}'$ be $\mathscr{L}$-structures, and let $\mathfrak{a}$ be an indexing of A by a set K. Then $\mathfrak{A}$ is elementarily embeddable in $\mathfrak{A}'$ iff there is a mapping $\mathfrak{a}'$ of K into A' such that $(\mathfrak{A}, \mathfrak{a})$ and $(\mathfrak{A}', \mathfrak{a}')$ are ($\mathscr{L}_K$-) elementarily equivalent.*

PROOF. We merely sketch the proof, leaving the details to be filled in by the reader.

If f is an elementary embedding of $\mathfrak{A}$ into $\mathfrak{A}'$, define $\mathfrak{a}' : K \to A'$ by $\mathfrak{a}'(k) = f(\mathfrak{a}(k))$ for $k \in K$. It is then easy to verify that $(\mathfrak{A}, \mathfrak{a}) \equiv (\mathfrak{A}', \mathfrak{a}')$.

Conversely, if $\mathfrak{a}' : K \to A'$ is such that $(\mathfrak{A}, \mathfrak{a}) \equiv (\mathfrak{A}', \mathfrak{a}')$, define $f : A \to A'$ by $f(\mathfrak{a}(k)) = \mathfrak{a}'(k)$ for $k \in K$. Then f is an elementary embedding of $\mathfrak{A}$ into $\mathfrak{A}'$. ∎

1.7. PROBLEM. Show that, if $\mathfrak{A}$ and $\mathfrak{A}'$ are $\mathscr{L}$-structures, then $\mathfrak{A} \cong \mathfrak{A}'$ iff there is a set K and indexings $\mathfrak{a}$ and $\mathfrak{a}'$ of A and A', respectively, by K such that $(\mathfrak{A}, \mathfrak{a}) \equiv (\mathfrak{A}', \mathfrak{a}')$.

1.8. PROBLEM. (i) Let $\mathfrak{A}$ and $\mathfrak{A}'$ be $\mathscr{L}$-structures such that $\mathfrak{A} \subseteq \mathfrak{A}'$. Suppose that for any finite subset $\{a_1, \ldots, a_k\}$ of A and any $a' \in A'$ there is an isomorphism of $\mathfrak{A}'$ onto itself which leaves each a_i fixed and carries a' into A. Show that $\mathfrak{A} \prec \mathfrak{A}'$. (Use Prob. 1.1 and Lemma 1.5.)

(ii) Let $\mathfrak{Q}$ be the ordered set of rationals and $\mathfrak{R}$ the ordered set of reals. Deduce from (i) that $\mathfrak{Q} \prec \mathfrak{R}$.

§ 2. The Löwenheim–Skolem Theorems

In this section we show that any infinite structure has elementary substructures and extensions in a wide range of cardinalities. In particular, in Thm. 2.2 we derive a strengthened version of Thm. 3.3.15 for sets of sentences. The proof of Thm. 2.2 is entirely semantic, i.e. makes no use of the machinery of the predicate calculus.

If $\mathfrak{A}$ is an $\mathscr{L}$-structure, we put $\|\mathfrak{A}\| = |A|$. $\|\mathfrak{A}\|$ is called the *cardinality* of $\mathfrak{A}$. Notice that $\|\mathfrak{A}\|$ depends only on the domain of $\mathfrak{A}$; it has nothing to do with the relations and designated individuals of $\mathfrak{A}$.

Recall that the cardinality $\|\mathscr{L}\|$ of the language $\mathscr{L}$ is the cardinality of the set of all $\mathscr{L}$-symbols. Clearly $\|\mathscr{L}\|=\max\{|I|, |J|, \aleph_0\}$, where I and J are the sets indexing the predicate symbols and constants, respectively, of $\mathscr{L}$.

Our first result asserts the existence of "small" elementary substructures of a given infinite structure.

2.1. THEOREM. *Let $\mathfrak{A}$ be an infinite $\mathscr{L}$-structure, and let $X\subseteq A$. Then for any cardinal α satisfying* $\max\{|X|, \|\mathscr{L}\|\}\leqslant\alpha\leqslant\|\mathfrak{A}\|$ *there is an elementary substructure $\mathfrak{B}$ of $\mathfrak{A}$ such that $\|\mathfrak{B}\|=\alpha$ and $X\subseteq B$.*

PROOF. Let h be a choice function for the non-empty subsets of A, i.e. such that $h(Y)\in Y$ whenever $Y\subseteq A$ and $Y\neq\emptyset$; the existence of such a function is ensured by the axiom of choice. Define the sequence $B_0, B_1, \ldots$ of subsets of A as follows: B_0 is any subset of A such that $X\subseteq B_0$ and $|B_0|=\alpha$ (such subsets of A exist in view of the conditions on α!) and for each $n\in\omega$,

$$B_{n+1}=\{h(Y)\text{: for some } m, \text{ some } \mathscr{L}\text{-formula } \varphi \text{ all of whose free variables are among } \mathbf{v}_0,\ldots,\mathbf{v}_m \text{ and some } a_0,\ldots,a_{m-1}\in B_n,$$

$$\emptyset\neq Y=\{x\in A:\ \mathfrak{A}\models\varphi\,[a_0,\ldots,a_{m-1},x]\}\}.$$

We claim first that B_1 contains all the designated individuals of $\mathfrak{A}$. For let c_j be a designated individual of $\mathfrak{A}$, and let φ be the formula $\mathbf{v}_0=\mathbf{c}_j$. Then

$$\{c_j\}=\{x\in A:\ \mathfrak{A}\models\varphi\,[x]\},$$

so that $c_j=h(\{c_j\})\in B_1$ by the definition of B_1.

Secondly, $B_n\subseteq B_{n+1}$ for each $n\in\omega$. For if $a_0\in B_n$, then, putting φ for the formula $\mathbf{v}_0=\mathbf{v}_1$, we have

$$\{a_0\}=\{x\in A:\ \mathfrak{A}\models\varphi\,[a_0,x]\}$$

so that $a_0=h(\{a_0\})\in B_{n+1}$ by the definition of B_{n+1}.

Thirdly, $|B_n|=\alpha$ for each $n\in\omega$. This is proved by induction on n. We have $|B_0|=\alpha$ by assumption, so assume that $n>0$ and $|B_{n-1}|=\alpha$. We already know that $B_{n-1}\subseteq B_n$, so certainly $|B_n|\geqslant\alpha$. To prove the reverse inequality we observe that each member of B_n is determined by an $\mathscr{L}$-formula and an m-tuple of members of B_{n-1}, for some $m\in\omega$. It follows that, if we write Φ for the set of $\mathscr{L}$-formulas (so that by Thm. 3.3.11 $|\Phi|=\|\mathscr{L}\|$),

$$|B_n|\leqslant\Big|\bigcup_{m\in\omega}\Phi\times(B_{n-1})^m\Big|\leqslant\sum_{m\in\omega}|\Phi|\cdot|B_{n-1}|^m=\|\mathscr{L}\|\cdot\sum_{m\in\omega}\alpha^m.$$

Since $\alpha\geqslant\|\mathscr{L}\|\geqslant\aleph_0$, it follows that

$$\|\mathscr{L}\|\cdot\sum_{m\in\omega}\alpha^m=\|\mathscr{L}\|\cdot\alpha=\alpha.$$

Therefore $|B_n|\leqslant\alpha$ and so $|B_n|=\alpha$.

Now put $B=\bigcup_{n\in\omega}B_n$. Then

$$|B|\leqslant\sum_{m\in\omega}|B_n|=\aleph_0\cdot\alpha=\alpha.$$

Accordingly $|B|=\alpha$. Moreover, since B_1 contains the designated individuals of $\mathfrak{A}$, so does B, and we may therefore put $\mathfrak{B}=\mathfrak{A}|B$.

We claim that $\mathfrak{B}$ satisfies the required conditions. Certainly we have $X\subseteq\mathfrak{B}$, and $\|\mathfrak{B}\|=\alpha$ has been proved above. It remains to show that $\mathfrak{B}\prec\mathfrak{A}$. To do this, we apply Lemma 1.5. Suppose then that $a_0,\ldots,a_{n-1}\in B$, φ is an $\mathscr{L}$-formula whose free variables are all among $\mathbf{v}_0,\ldots,\mathbf{v}_n$, and there is $a'\in A$ for which $\mathfrak{A}\models\varphi\,[a_0,\ldots,a_{n-1},a']$. Now we have already shown that the B_n form an increasing chain, so there is $m\in\omega$ such that $a_0,\ldots,a_{n-1}\in B_m$. The set

$$Y=\{x\in A:\ \mathfrak{A}\models\varphi\,[a_0,\ldots,a_{n-1},x]\}$$

is non-empty by assumption, so, if we put $a=h(Y)$, we have $a\in B_{m+1}\subseteq B$ and $\mathfrak{A}\models\varphi\,[a_0,\ldots,a_{n-1},a]$. By Lemma 1.5, it follows that $\mathfrak{B}\prec\mathfrak{A}$, and the proof is finished. ∎

From this we infer the following important result (cf. Thm. 3.3.15).

2.2. Downward Löwenheim–Skolem Theorem. *Let Σ be a set of sentences of $\mathscr{L}$ with an infinite model of cardinality $\alpha\geqslant|\Sigma|$. Then Σ has a model of any cardinality β such that* $\max(|\Sigma|,\aleph_0)\leqslant\beta\leqslant\alpha$.

Proof. Let $\gamma=\max(|\Sigma|,\aleph_0)$. Then at most γ extralogical symbols occur in the sentences of Σ so that $\|\mathscr{L}_\Sigma\|=\gamma$, where $\mathscr{L}_\Sigma$ is the "poorest" first order language (with equality) in which Σ is still formulable (Ch. 2, §8). Let $\mathfrak{A}$ be an infinite model of Σ of cardinality $\alpha\geqslant|\Sigma|$. Then the $\mathscr{L}_\Sigma$-reduction $\mathfrak{A}'$ of $\mathfrak{A}$ is also a model of Σ of cardinality α. Since $\gamma\leqslant\beta\leqslant\alpha$, by Thm. 2.1 there is an ($\mathscr{L}_\Sigma$-)elementary substructure $\mathfrak{B}'$ of $\mathfrak{A}'$ of cardinality β. $\mathfrak{B}'$ is then a model of Σ, and if $\mathfrak{B}$ is any $\mathscr{L}$-expansion of $\mathfrak{B}'$, then $\mathfrak{B}$ is a model of Σ which is an $\mathscr{L}$-structure of cardinality β. ∎

As a special case of Thm. 2.2, we have:

2.3. COROLLARY. *Any countable set of sentences with an infinite model has a countable model.* ∎

We now turn to the problem of showing that each infinite structure has elementary *extensions* of prescribed cardinalities. The next lemma, although quite simple, is very useful in this connection.

2.4. LEMMA. *Let $\mathfrak{A}$ be an $\mathscr{L}$-structure. Then, for any cardinal $\alpha \geqslant \|\mathfrak{A}\|$, $\mathfrak{A}$ has an elementary extension of cardinality α iff $\mathfrak{A}$ is elementarily embeddable in a structure of cardinality α.*

PROOF. Necessity follows immediately from Prob. 1.4(ii). Conversely, let $\mathfrak{A}'$ be an $\mathscr{L}$-structure of cardinality $\alpha \geqslant \|\mathfrak{A}\|$, and let f be an elementary embedding of $\mathfrak{A}$ into $\mathfrak{A}'$. If B is a set of cardinality α including A, let g be a bijection of B onto A' which extends f and then use g^{-1} to "transfer the structure" of $\mathfrak{A}'$ onto B. The resulting structure $\mathfrak{B}$ is an elementary extension of $\mathfrak{A}$ of cardinality α. ∎

For convenience we restate the compactness theorem (Thm. 3.3.16) in the form most suitable for our present purpose.[1]

2.5. COMPACTNESS THEOREM. *If each finite subset of a set Σ of sentences of $\mathscr{L}$ has a model, then Σ has a model.* ∎

We now employ the compactness theorem in the proof of:

2.6. THEOREM. *Let $\mathfrak{A}$ be an infinite $\mathscr{L}$-structure. Then $\mathfrak{A}$ has an elementary extension of any cardinality $\alpha \geqslant \max(\|\mathfrak{A}\|, \|\mathscr{L}\|)$.*

PROOF. Let $\mathfrak{a} : K \to A$ be an indexing of A; let K' be a set of cardinality α disjoint from both K and J. Let Σ be the set of all sentences of $\mathscr{L}_K$ which hold in $(\mathfrak{A},\mathfrak{a})$, and let

$$\Sigma' = \Sigma \cup \{(\mathbf{c}_{k'} \neq \mathbf{c}_{k''}) : k', k'' \in K' \text{ and } k' \neq k''\}.$$

Then clearly Σ' is a set of sentences of $\mathscr{L}_{K \cup K'}$ of cardinality α. Moreover, each finite subset Σ_0 of Σ' has a model. For let $\{k'_1,\ldots,k'_n\}$ be the finite set of members k' of K' such that $\mathbf{c}_{k'}$ occurs in a sentence of Σ_0. Since A is infinite, we can choose n distinct members $a'_1,\ldots,a'_n$ of A. Define the mapping $\mathfrak{a}' : K' \to A$ by putting

$$\mathfrak{a}'(k'_i) = a'_i \quad \text{for } 1 \leqslant i \leqslant n,$$

$$\mathfrak{a}'(k') = a'_1 \quad \text{for } k' \in K' - \{k'_1,\ldots,k'_n\}.$$

[1] In §3 we shall give a direct model-theoretic proof of the compactness theorem which does not depend on the results of Chs. 1–3.

Then it is clear that $(\mathfrak{A}, \mathfrak{a} \cup \mathfrak{a}')$ is a model of Σ_0. By the compactness theorem, Σ' has a model which must be of cardinality $\geqslant \alpha$ since Σ' contains the sentences $\mathbf{c}_{k'} \neq \mathbf{c}_{k''}$ for all $k', k'' \in K$ such that $k' \neq k''$. Since $|\Sigma'| = \alpha$, by Thm. 2.2 Σ' has a model $\mathfrak{B}'$ of cardinality α. Now $\mathfrak{B}'$, as an $\mathscr{L}_{K \cup K'}$-structure, must be of the form $(\mathfrak{B}, \mathfrak{b} \cup \mathfrak{b}')$, where $\mathfrak{B}$ is an $\mathscr{L}$-structure, and $\mathfrak{b}, \mathfrak{b}'$ are mappings of K, K', respectively, into B. Since $\mathfrak{B}'$ is a model of Σ', $(\mathfrak{B}, \mathfrak{b})$ is a model of Σ and therefore $(\mathfrak{A}, \mathfrak{a}) \equiv (\mathfrak{B}, \mathfrak{b})$. Since $\mathfrak{a}$ was taken to be an indexing of A, it follows from Lemma 1.6 that $\mathfrak{A}$ is elementarily embeddable in $\mathfrak{B}$. Since $\|\mathfrak{B}'\| = \|\mathfrak{B}\| = \alpha$, Lemma 2.4 implies that $\mathfrak{A}$ has an elementary extension of cardinality α. ∎

We can now prove the following important theorems.

2.7. Upward Löwenheim–Skolem Theorem. *Let Σ be a set of $\mathscr{L}$-sentences with a model of cardinality $\alpha \geqslant \aleph_0$. Then Σ has a model of any cardinality* $\geqslant \max(\alpha, |\Sigma|)$.

Proof. Let $\mathfrak{A}$ be a model of Σ of cardinality α. Then the $\mathscr{L}_\Sigma$-reduction $\mathfrak{A}'$ of $\mathfrak{A}$ is also a model of Σ of cardinality α. Let $\beta \geqslant \max(\alpha, |\Sigma|)$; then since $\|\mathscr{L}_\Sigma\| = \max(\aleph_0, |\Sigma|)$, we have $\beta \geqslant \max(\alpha, \|\mathscr{L}_\Sigma\|)$, so that, by Thm. 2.6, $\mathfrak{A}'$ has an elementary extension $\mathfrak{B}'$ of cardinality β. $\mathfrak{B}'$ is then a model of Σ, and any $\mathscr{L}$-expansion of $\mathfrak{B}'$ is an $\mathscr{L}$-structure of cardinality β which is a model of Σ. ∎

2.8. Löwenheim–Skolem Theorem. *Let Σ be a set of $\mathscr{L}$-sentences with an infinite model. Then Σ has a model of each cardinality* $\geqslant \max(|\Sigma|, \aleph_0)$.

Proof. Suppose Σ has a model of cardinality $\alpha \geqslant \aleph_0$. Let $\beta \geqslant \max(|\Sigma|, \aleph_0)$. If $\beta \leqslant \alpha$ then Σ has a model of cardinality β by Thm. 2.2, while if $\alpha \leqslant \beta$ then $\beta \geqslant \max(\alpha, |\Sigma|)$, so that Σ has a model of cardinality β by Thm 2.7. ∎

2.9. Theorem. *If Σ is a consistent set of $\mathscr{L}$-sentences, then either Σ has a finite model or Σ has a model of any cardinality* $\geqslant \max(|\Sigma|, \aleph_0)$.

Proof. Let Σ be a consistent set of sentences of $\mathscr{L}$. Then, by Thm. 3.3.13, Σ has a model. If this model is infinite, then, by Thm. 2.8, Σ has a model of any cardinality $\geqslant \max(|\Sigma|, \aleph_0)$. ∎

2.10. Problem. Let $\mathscr{A}$ be a family of $\mathscr{L}$-structures, where $\mathscr{L}$ has no constant symbols. The *union* of the family $\mathscr{A}$ is the structure

$$\bigcup \mathscr{A} = \langle \bigcup_{\mathfrak{A} \in \mathscr{A}} A, \langle \bigcup_{\mathfrak{A} \in \mathscr{A}} \mathbf{R}_i^{\mathfrak{A}} \rangle_{i \in I} \rangle.$$

Thus the domain of $\bigcup \mathscr{A}$ is the union of the domains of the members of $\mathscr{A}$, and for each $i \in I$ the interpretation of $\mathbf{R}_i$ in $\bigcup \mathscr{A}$ is $\bigcup_{\mathfrak{A} \in \mathscr{A}} \mathbf{R}_i^{\mathfrak{A}}$.

$\mathscr{A}$ is said to be a *chain* (resp. *elementary chain*) if for all $\mathfrak{A}, \mathfrak{A}' \in \mathscr{A}$ we have $\mathfrak{A} \subseteq \mathfrak{A}'$ or $\mathfrak{A}' \subseteq \mathfrak{A}$ (resp. $\mathfrak{A} \prec \mathfrak{A}'$ or $\mathfrak{A}' \prec \mathfrak{A}$). Show that the union of a chain of structures is an extension of each member of the chain, and that the union of an elementary chain is an elementary extension of each member of the chain.

2.11. PROBLEM. (i) Use the compactness theorem to show that if a set of sentences $\boldsymbol{\Sigma}$ has arbitrarily large finite models, then it has an infinite model. (Show that each finite subset of the set $\boldsymbol{\Sigma} \cup \{\boldsymbol{\sigma}_n \colon n \in \omega\}$ has a model, where $\boldsymbol{\sigma}_n$ is a sentence which asserts that there are at least n distinct individuals.)

(ii) Let $\mathscr{L}$ be a language containing a binary predicate $\mathbf{R}$. Show that there is no set $\boldsymbol{\Sigma}$ of $\mathscr{L}$-sentences with at least one infinite model such that $\mathbf{R}^{\mathfrak{A}}$ is a well-ordering of A for each infinite model $\mathfrak{A}$ of $\boldsymbol{\Sigma}$. (Let $\{\mathbf{c}_n \colon n \in \omega\}$ be a set of new constants, and put $\boldsymbol{\Sigma}' = \boldsymbol{\Sigma} \cup \{\mathbf{R}\mathbf{c}_{n+1}\mathbf{c}_n \wedge \mathbf{c}_{n+1} \neq \mathbf{c}_n \colon n \in \omega\}$. Show that each finite subset of $\boldsymbol{\Sigma}'$ has a model, and apply compactness to obtain an infinite model of $\boldsymbol{\Sigma}$ in which the interpretation of $\mathbf{R}$ is not a wellordering.)

2.12. PROBLEM. Let $\mathfrak{A}$ be an $\mathscr{L}$-structure. A set $\boldsymbol{\Gamma}$ of $\mathscr{L}$-formulas each member of which has precisely one free variable, $\mathbf{v}_0$ say, is said to be *finitely satisfiable* in $\mathfrak{A}$ if for each finite subset $\{\boldsymbol{\varphi}_1, \ldots, \boldsymbol{\varphi}_n\}$ of $\boldsymbol{\Gamma}$ there is an $a \in A$ such that $\mathfrak{A} \models (\boldsymbol{\varphi}_1 \wedge \ldots \wedge \boldsymbol{\varphi}_n)[a]$. $\boldsymbol{\Gamma}$ is said to be *satisfiable* in $\mathfrak{A}$ if there is $a \in A$ such that $\mathfrak{A} \models \boldsymbol{\varphi}[a]$ for *all* $\boldsymbol{\varphi} \in \boldsymbol{\Gamma}$.

(i) Show that for each structure $\mathfrak{A}$ there is an elementary extension $\mathfrak{A}'$ of $\mathfrak{A}$ such that every set of $\mathscr{L}$-formulas finitely satisfiable in $\mathfrak{A}$ is satisfiable in $\mathfrak{A}'$.

(ii) Let $\mathfrak{A}$ be a finite structure, and let $\boldsymbol{\Gamma}$ be a set of $\mathscr{L}$-formulas with one free variable. Show that, if $\boldsymbol{\Gamma}$ is finitely satisfiable in $\mathfrak{A}$, then $\boldsymbol{\Gamma}$ is satisfiable in $\mathfrak{A}$.

(iii) Let $\mathfrak{A}$ be a finite structure, and let $\mathfrak{B}$ be a structure such that $\mathfrak{A} \equiv \mathfrak{B}$. Show that $\mathfrak{A} \cong \mathfrak{B}$. (First show that A and B have the same finite number n of elements. Let $A = \{a_1, \ldots, a_n\}$. Use (ii) to construct by induction a sequence of n distinct elements $\{b_1, \ldots, b_n\}$ of B such that $(\mathfrak{A}, \langle a_1, \ldots, a_i \rangle) \equiv$ $\equiv (\mathfrak{B}, \langle b_1, \ldots, b_i \rangle)$ for each i, $1 \leqslant i \leqslant n$.)

§ 3. Ultraproducts

To facilitate the exposition, in this section we confine our attention to structures $\langle A,R\rangle$ consisting of a non-empty set A and a *single binary relation* R on A. The language appropriate for these structures will be denoted by $\mathscr{L}$; thus $\mathscr{L}$ is a first-order language with equality and another binary predicate symbol **R** but with no constant symbols. It will be clear that everything we do can be extended to structures with arbitrarily many relations (and operations) and hence to first-order languages with arbitrarily many predicate (and function) symbols, merely by complicating the notation. We leave it to the reader to make these extensions.

Let I be a fixed but arbitrary non-empty index set, and for each $i\in I$ let $\mathfrak{A}_i=\langle A_i,R_i\rangle$ be an $\mathscr{L}$-structure. Also let $\prod_{i\in I}A_i=A$ be the Cartesian product[1] of the sets A_i. We shall use f, g, h, f', g', h' to denote elements of A.

The *direct product* $\prod_{i\in I}\mathfrak{A}_i$ of the family $\{\mathfrak{A}_i: i\in I\}$ is the structure $\langle\prod_{i\in I}A_i,S\rangle$, where S is the set of all pairs $\langle f,g\rangle$ such that $\langle f(i),g(i)\rangle\in R_i$ for all $i\in I$. The direct product is a natural construction which finds application in many branches of mathematics. But from the logician's point of view it suffers from the drawback that $\prod_{i\in I}\mathfrak{A}_i$ may not share the first-order properties of the $\mathfrak{A}_i$. In other words, there may be a sentence $\boldsymbol{\sigma}$ such that each $\mathfrak{A}_i$ is a model of $\boldsymbol{\sigma}$ but $\prod\mathfrak{A}_i$ is not. For example, suppose that each $\mathfrak{A}_i$ is a totally ordered set with at least two elements, and let $\boldsymbol{\sigma}$ be the sentence $\forall\mathbf{v}_0\forall\mathbf{v}_1[\mathbf{R}\mathbf{v}_0\mathbf{v}_1\vee\mathbf{R}\mathbf{v}_1\mathbf{v}_0]$. Then $\mathfrak{A}_i\models\boldsymbol{\sigma}$ for each $i\in I$, but, if I has at least two elements, $\prod\mathfrak{A}_i\models\neg\boldsymbol{\sigma}$. (In less technical terms, the product of totally ordered sets is not in general totally ordered.)

We are going to introduce a modification of the direct product construction which does not suffer from the drawback mentioned above and which is accordingly of great usefulness in model theory.

First we define two mappings E and R of $A\times A$ into $\mathsf{P}I$ by putting, for $f,g\in A$,

$$E(f,g)=\{i\in I: f(i)=g(i)\},$$

$$R(f,g)=\{i\in I: \langle f(i),g(i)\rangle\in R_i\}.$$

Now let $\mathscr{L}(A)$ be language obtained from $\mathscr{L}$ by adding a new constant

[1] Where the index set is clear from the context, we shall in future write Π_iA_i or ΠA_i and similarly for other expressions of the same kind.

symbol $\mathbf{f}$ for each $f \in A$. We define the mapping $\boldsymbol{\sigma} \mapsto \|\boldsymbol{\sigma}\|$ of the sentences of $\mathscr{L}(A)$ into $\mathsf{P}I$ by induction on deg $\boldsymbol{\sigma}$ as follows:

$$\|\mathbf{f}=\mathbf{g}\| =_{\mathrm{df}} E(f,g),$$

$$\|\mathbf{Rfg}\| =_{\mathrm{df}} R(f,g);$$

for $\mathscr{L}(A)$-sentences $\boldsymbol{\sigma}, \boldsymbol{\sigma}'$,

$$\|\boldsymbol{\sigma} \wedge \boldsymbol{\sigma}'\| =_{\mathrm{df}} \|\boldsymbol{\sigma}\| \cap \|\boldsymbol{\sigma}'\|$$

$$\|\neg \boldsymbol{\sigma}\| =_{\mathrm{df}} I - \|\boldsymbol{\sigma}\|;$$

and for each variable $\mathbf{x}$ and each $\mathscr{L}(A)$-formula $\boldsymbol{\varphi}$ with at most $\mathbf{x}$ free,

$$\|\exists \mathbf{x} \boldsymbol{\varphi}\| =_{\mathrm{df}} \bigcup_{f \in A} \|\boldsymbol{\varphi}(\mathbf{x}/\mathbf{f})\|.$$

We may think of the mapping $\|\cdot\|$ as assigning "Boolean truth values" in the Boolean algebra $\mathsf{P}I$ to the sentences of $\mathscr{L}(A)$, and the triple $\langle A,E,R \rangle$ as a "Boolean-valued" $\mathscr{L}$-structure, in which the "truth value" of an $\mathscr{L}$-sentence $\boldsymbol{\sigma}$ is now $\|\boldsymbol{\sigma}\|$, and no longer simply one of the two truth values $\top$ or $\bot$ (cf. Prob. 5.14).

In order to simplify the notation employed in the sequel, we make the following conventions. If $\boldsymbol{\varphi}$ is any $\mathscr{L}$-formula all of whose free variables are among $\mathbf{v}_0,\ldots,\mathbf{v}_n$ and $f_0,\ldots,f_n \in A$, we agree to write $\boldsymbol{\varphi}(\mathbf{f}_0,\ldots,\mathbf{f}_n)$ for $\boldsymbol{\varphi}(\mathbf{v}_0/\mathbf{f}_0,\ldots,\mathbf{v}_n/\mathbf{f}_n)$. Also, if $\boldsymbol{\varphi}$ has at most the variable $\mathbf{x}$ free and $f \in A$, we write $\boldsymbol{\varphi}(\mathbf{f})$ for $\boldsymbol{\varphi}(\mathbf{x}/\mathbf{f})$.

3.1. THEOREM. *Let $\boldsymbol{\varphi}$ be any $\mathscr{L}$-formula whose free variables are all among $\mathbf{v}_0,\ldots,\mathbf{v}_n$. Then for any $f_0,\ldots,f_n \in A$ we have*[1]

$$\|\boldsymbol{\varphi}(\mathbf{f}_0,\ldots,\mathbf{f}_n)\| = \{i \in I\colon\ \mathfrak{A}_i \models \boldsymbol{\varphi}\,[f_0(i),\ldots,f_n(i)]\}.$$

PROOF. We argue by induction on deg $\boldsymbol{\varphi}$. For atomic $\boldsymbol{\varphi}$ the result is true by definition.

Suppose now that the result holds for all $\boldsymbol{\psi}$ with deg $\boldsymbol{\psi} <$ deg $\boldsymbol{\varphi}$. We show that it holds for $\boldsymbol{\varphi}$. There are three cases to consider.

[1] Notice that, under these conditions $\boldsymbol{\varphi}(\mathbf{f}_0, \ldots, \mathbf{f}_n)$ is an $\mathscr{L}(A)$-*sentence*, so $\|\boldsymbol{\varphi}(\mathbf{f}_0, \ldots, \mathbf{f}_n)\|$ is defined.

(a) φ is $\psi\wedge\chi$. We have

$$\begin{aligned}\|\varphi(\mathbf{f}_0,\ldots,\mathbf{f}_n)\| &= \|\psi(\mathbf{f}_0,\ldots,\mathbf{f}_n)\|\cap\|\chi(\mathbf{f}_0,\ldots,\mathbf{f}_n)\|\\ &= \{i\in I:\ \mathfrak{A}_i\models\psi\,[f_0(i),\ldots,f_n(i)]\}\\ &\quad\cap\{i\in I:\ \mathfrak{A}_i\models\chi\,[f_0(i),\ldots,f_n(i)]\}\\ &= \{i\in I:\ \mathfrak{A}_i\models(\psi\wedge\chi)\,[f_0(i),\ldots,f_n(i)]\}\\ &= \{i\in I:\ \mathfrak{A}_i\models\varphi\,[f_0(i),\ldots,f_n(i)]\},\end{aligned}$$

and so in this case the result holds for φ.

(b) φ is $\neg\chi$. We have

$$\begin{aligned}\|\varphi(\mathbf{f}_0,\ldots,\mathbf{f}_n)\| &= I-\|\chi(\mathbf{f}_0,\ldots,\mathbf{f}_n)\|\\ &= I-\{i\in I:\ \mathfrak{A}_i\models\chi\,[f_0(i),\ldots,f_n(i)]\}\\ &= \{i\in I:\ \mathfrak{A}_i\models\neg\chi\,[f_0(i),\ldots,f_n(i)]\}\\ &= \{i\in I:\ \mathfrak{A}_i\models\varphi\,[f_0(i),\ldots,f_n(i)]\},\end{aligned}$$

so the result holds for φ in this case as well.

(c) φ is $\exists \mathbf{v}_k\psi$. Without loss of generality we may assume that $k\leqslant n$, and we have

$$\begin{aligned}&\|\varphi(\mathbf{f}_0,\ldots,\mathbf{f}_n)\| =\\ &\quad= \|(\exists\mathbf{v}_k\psi)(\mathbf{f}_0,\ldots,\mathbf{f}_n)\|\\ &\quad= \bigcup_{f\in A}\|\psi(\mathbf{f}_0,\ldots,\mathbf{f}_{k-1},\mathbf{f},\mathbf{f}_{k+1},\ldots,\mathbf{f}_n)\|\\ &\quad= \bigcup_{f\in A}\{i\in I:\ \mathfrak{A}_i\models\psi\,[f_0(i),\ldots,f_{k-1}(i),f(i),f_{k+1}(i),\ldots,f_n(i)]\}\\ &\quad= \{i\in I:\ \text{for some } f\in A,\ \mathfrak{A}_i\models\psi\,[f_0(i),\ldots,f_{k-1}(i),f(i),\\ &\qquad\qquad\qquad\qquad f_{k+1}(i),\ldots,f_n(i)]\}\\ &\quad= X,\ \text{say}.\end{aligned}$$

Clearly

$$X\subseteq\{i\in I:\ \mathfrak{A}_i\models(\exists\mathbf{v}_k\psi)\,[f_0(i),\ldots,f_n(i)]\}.$$

On the other hand, if $i\in I$ is such that

$$\mathfrak{A}_i\models(\exists\mathbf{v}_k\psi)[f_0(i),\ldots,f_n(i)]$$

then there is some $a\in A_i$ such that

$$\mathfrak{A}_i\models\psi\,[f_0(i),\ldots,f_{k-1}(i),a,f_{k+1}(i),\ldots,f_n(i)].$$

By taking $f \in A$ so that $f(i)=a$, we see immediately that $i \in X$. Hence

$$X=\{i \in I: \ \mathfrak{A}_i \models (\exists \mathbf{v}_k \psi)\,[f_0(i),\ldots,f_n(i)]\}$$
$$=\{i \in I: \ \mathfrak{A}_i \models \varphi\,[f_0(i),\ldots,f_n(i)]\},$$

and so the result holds for $\boldsymbol{\varphi}$ in this case too. ∎

3.2. PROBLEM. (i) Show that, if $\boldsymbol{\sigma}$ is any $\mathscr{L}(A)$-sentence such that $\vdash \boldsymbol{\sigma}$, then $\|\boldsymbol{\sigma}\|=I$. (Use Thm. 3.1.)

(ii) Suppose that each $\mathfrak{A}_i$ is identical with a fixed structure $\mathfrak{B}$. Then $\prod A_i=B^I$. For each $b \in B$ let $\hat{b}$ be the function on I with constant value b. Show that, for each formula $\boldsymbol{\varphi}$ with free variables among $\mathbf{v}_0,\ldots\mathbf{v}_n$, and all $b_0,\ldots,b_n \in B$,

$$\mathfrak{B} \models \varphi\,[b_0,\ldots,b_n] \ \Leftrightarrow \ \|\boldsymbol{\varphi}(\hat{\mathbf{b}}_0,\ldots,\hat{\mathbf{b}}_n)\|=I.$$

(Use Thm. 3.1).

3.3. PROBLEM. Let $\boldsymbol{\varphi}$ be any $\mathscr{L}(A)$-formula with at most the variable x free, and let $f,g \in A$. Show that

$$\|\mathbf{f}=\mathbf{g}\| \cap \|\boldsymbol{\varphi}(\mathbf{f})\| \subseteq \|\boldsymbol{\varphi}(\mathbf{g})\|.$$

(Use Thm. 3.1; alternatively, argue by induction on deg $\boldsymbol{\varphi}$.)

Recall that when we defined the mapping $\|\cdot\|$ we insisted that

$$\|\exists \mathbf{x}\boldsymbol{\varphi}\| = \bigcup_{f \in A} \|\boldsymbol{\varphi}(\mathbf{f})\|,$$

i.e. $\|\exists \mathbf{x}\boldsymbol{\varphi}\|$ is the supremum — in $\mathsf{P}I$ — of the $\|\boldsymbol{\varphi}(\mathbf{f})\|$ for $f \in A$. We now show that this supremum is actually *attained* by at least one of the $\|\boldsymbol{\varphi}(\mathbf{f})\|$.

3.4. LEMMA. *Let $\boldsymbol{\varphi}$ be an $\mathscr{L}(A)$-formula with at most the variable $\mathbf{x}$ free. Then there is $f \in A$ such that.*

$$\|\exists \mathbf{x}\boldsymbol{\varphi}\| = \|\boldsymbol{\varphi}(\mathbf{f})\|.$$

PROOF. Well-order A in the form $\{f_\xi : \xi < \alpha\}$ for some ordinal α. For each $\xi < \alpha$, put

$$X_\xi = \|\boldsymbol{\varphi}(\mathbf{f}_\xi)\| - \bigcup_{\eta<\xi} \|\boldsymbol{\varphi}(\mathbf{f}_\eta)\|.$$

Then we have

(1) $$\|\exists \mathbf{x}\boldsymbol{\varphi}\| = \bigcup_{f \in A} \|\boldsymbol{\varphi}(\mathbf{f})\| = \bigcup_{\xi<\alpha} \|\boldsymbol{\varphi}(\mathbf{f}_\xi)\| = \bigcup_{\xi<\alpha} X_\xi.$$

Also, if $\xi \neq \eta$, $X_\xi \cap X_\eta = \emptyset$, so we can choose $f \in A$ to satisfy $f|X_\xi = f_\xi|X_\xi$ for all $\xi < \alpha$. Then $X_\xi \subseteq \|\mathbf{f}=\mathbf{f}_\xi\|$ and hence, by 3.3,

$$\|\varphi(\mathbf{f})\| \supseteq \|\mathbf{f}=\mathbf{f}_\xi\| \cap \|\varphi(\mathbf{f}_\xi)\| \supseteq X_\xi$$

for all $\xi < \alpha$. Hence

$$\bigcup_{\xi<\alpha} X_\xi \subseteq \|\varphi(\mathbf{f})\|$$

and so $\|\exists \mathbf{x}\varphi\| \subseteq \|\varphi(\mathbf{f})\|$ by (1). The reverse inclusion is an immediate consequence of the definition of $\|\exists \mathbf{x}\varphi\|$, and the result follows. ∎

We next show how to reduce the "Boolean-valued structure" $\langle A,E,R\rangle$ to an ordinary $\mathscr{L}$-structure.

Let $\mathscr{F}$ be a subset of PI, and define the relations $\sim_{\mathscr{F}}$, $R_{\mathscr{F}}$ on A by putting

$$f \sim_{\mathscr{F}} g \Leftrightarrow \|\mathbf{f}=\mathbf{g}\| \in \mathscr{F},$$

$$\langle f,g\rangle \in R_{\mathscr{F}} \Leftrightarrow \|\mathbf{Rfg}\| \in \mathscr{F}.$$

We now have:

3.5. Lemma. *Suppose that* $\mathscr{F}$ *is a* filter *over* I (Prob. 4.3.16). *Then:*

(i) $\sim_{\mathscr{F}}$ *is an equivalence relation on* A;

(ii) *if* $f \sim_{\mathscr{F}} f'$ *and* $g \sim_{\mathscr{F}} g'$, *then* $\langle f,g\rangle \in R_{\mathscr{F}}$ *implies* $\langle f',g'\rangle \in R_{\mathscr{F}}$.

Proof. (i) By definition,

$$\|\mathbf{f}=\mathbf{g}\| = \{i \in I: f(i) = g(i)\}.$$

Since $I \in \mathscr{F}$, $\sim_{\mathscr{F}}$ is reflexive. Clearly $\sim_{\mathscr{F}}$ is symmetric, so it remains to establish its transitivity. Thus suppose that $f \sim_{\mathscr{F}} g$ and $g \sim_{\mathscr{F}} h$; then $\|\mathbf{f}=\mathbf{g}\| \in \mathscr{F}$ and $\|\mathbf{g}=\mathbf{h}\| \in \mathscr{F}$. Accordingly, since $\mathscr{F}$ is a filter,

$$\|\mathbf{f}=\mathbf{g}\| \cap \|\mathbf{g}=\mathbf{h}\| \in \mathscr{F}.$$

But clearly

$$\|\mathbf{f}=\mathbf{g}\| \cap \|\mathbf{g}=\mathbf{h}\| \subseteq \|\mathbf{f}=\mathbf{h}\|.$$

Hence $\|\mathbf{f}=\mathbf{g}\| \in \mathscr{F}$, so $f \sim_{\mathscr{F}} h$. Thus $\sim_{\mathscr{F}}$ is transitive, and (i) is proved.

The proof of (ii) is similar to that of (i), and is entrusted to the reader. ∎

From now on we will assume that $\mathscr{F}$ is a fixed but arbitrary *ultrafilter* over I. Intuitively an ultrafilter over I consists of "large" subsets[1] of I

[1] In the sense that, if $\mathscr{F}$ is an ultrafilter over I, and h is the canonical homomorphism (Ch. 4, §3) of PI onto 2, then $h(X) = 1$ if and only if $X \in \mathscr{F}$.

and we may think of the statement $f \sim_{\mathscr{F}} g$ as asserting that f and g agree on a "large" subset of I (or that the Boolean truth value $\|\mathbf{f}=\mathbf{g}\|$ of the assertion that $f=g$ is close to the largest element I of $\mathsf{P}I$).

For each $f \in A$ we let $f/\mathscr{F}$ be the $\sim_{\mathscr{F}}$-class of f and we put

$$A/\mathscr{F} = \prod_{i \in I} A_i/\mathscr{F} = \{f/\mathscr{F}: f \in A\}.$$

The relation $R_{\mathscr{F}}$ on A naturally induces the relation $R/\mathscr{F}$ on $A/\mathscr{F}$ defined by

$$\langle f/\mathscr{F}, g/\mathscr{F} \rangle \in R/\mathscr{F} \Leftrightarrow \langle f,g \rangle \in R_{\mathscr{F}}.$$

By 3.5(ii), this is a sound definition.

Now define $\prod_{i \in I} \mathfrak{A}_i/\mathscr{F}$ to be the structure[1] $\langle A/\mathscr{F}, R/\mathscr{F} \rangle$. Then $\prod_{i \in I} \mathfrak{A}_i/\mathscr{F}$ is an $\mathscr{L}$-structure called the *ultraproduct* of the family $\{\mathfrak{A}_i: i \in I\}$ with respect to the ultrafilter $\mathscr{F}$. If each $\mathfrak{A}_i$ is identical with some fixed structure $\mathfrak{B}$, the ultraproduct is written $\mathfrak{B}^I/\mathscr{F}$ and is called the *ultrapower* of $\mathfrak{B}$ with respect to $\mathscr{F}$.

Our next result is basic.

3.6. THEOREM. *Let* φ *be an* $\mathscr{L}$*-formula all of whose free variables are among* $\mathbf{v}_0, \ldots, \mathbf{v}_n$. *Then for any* $f_0, \ldots, f_n \in A$ *we have*

$$\prod \mathfrak{A}_i/\mathscr{F} \models \varphi\, [f_0/\mathscr{F}, \ldots, f_n/\mathscr{F}] \Leftrightarrow \|\varphi(\mathbf{f}_0, \ldots, \mathbf{f}_n)\| \in \mathscr{F}.$$

PROOF. The proof is by induction on $\deg \varphi$. For atomic φ the result holds by definition.

Suppose now that the result holds for all ψ with $\deg \psi < \deg \varphi$. We show that it holds for φ. As usual, there are three cases to consider:

(a) φ is $\psi \wedge \chi$. We have, using the inductive hypothesis and the fact that $\mathscr{F}$ is a filter,

$$\begin{aligned}
&\|\varphi(\mathbf{f}_0, \ldots, \mathbf{f}_n)\| \in \mathscr{F} \Leftrightarrow \\
&\Leftrightarrow \|(\psi \wedge \chi)(\mathbf{f}_0, \ldots, \mathbf{f}_n)\| \in \mathscr{F} \\
&\Leftrightarrow \|\psi(\mathbf{f}_0, \ldots, \mathbf{f}_n)\| \in \mathscr{F} \;\&\; \|\chi(\mathbf{f}_0, \ldots, \mathbf{f}_n)\| \in \mathscr{F} \\
&\Leftrightarrow \prod \mathfrak{A}_i/\mathscr{F} \models \psi\, [f_0/\mathscr{F}, \ldots, f_n/\mathscr{F}] \;\&\; \prod \mathfrak{A}_i/\mathscr{F} \models \chi\, [f_0/\mathscr{F}, \ldots, f_n/\mathscr{F}] \\
&\Leftrightarrow \prod \mathfrak{A}_i/\mathscr{F} \models (\psi \wedge \chi)\, [f_0/\mathscr{F}, \ldots, f_n/\mathscr{F}] \\
&\Leftrightarrow \prod \mathfrak{A}_i/\mathscr{F} \models \varphi\, [f_0/\mathscr{F}, \ldots, f_n/\mathscr{F}]
\end{aligned}$$

so in this case the result holds for φ.

[1] $\langle A/\mathscr{F}, R/\mathscr{F} \rangle$ is the *quotient* of the Boolean-valued structure $\langle A,E,R \rangle$ by the ultrafilter $\mathscr{F}$.

(b). φ is $\neg\psi$. We have, using the inductive hypothesis and the fact that $\mathscr{F}$ is an ultrafilter,

$$\begin{aligned}\|\varphi(\mathbf{f}_0,\ldots,\mathbf{f}_n)\|\in\mathscr{F} &\Leftrightarrow \|\neg\chi(\mathbf{f}_0,\ldots,\mathbf{f}_n)\|\in\mathscr{F}\\ &\Leftrightarrow I-\|\chi(\mathbf{f}_0,\ldots,\mathbf{f}_n)\|\in\mathscr{F}\\ &\Leftrightarrow \|\chi(\mathbf{f}_0,\ldots,\mathbf{f}_n)\|\notin\mathscr{F}\\ &\Leftrightarrow \textit{not}\ \prod\mathfrak{A}_i/\mathscr{F}\models\chi\,[f_0/\mathscr{F},\ldots,f_n/\mathscr{F}]\\ &\Leftrightarrow \prod\mathfrak{A}_i/\mathscr{F}\models\neg\chi\,[f_0/\mathscr{F},\ldots,f_n/\mathscr{F}]\\ &\Leftrightarrow \prod\mathfrak{A}_i/\mathscr{F}\models\varphi\,[f_0/\mathscr{F},\ldots,f_n/\mathscr{F}],\end{aligned}$$

so the result holds for φ in this case as well.

(c). φ is $\exists\mathbf{v}_k\psi$. As usual we may assume without loss of generality that $k\leqslant n$. Then we have

$$\begin{aligned}&\|\varphi(\mathbf{f}_0,\ldots,\mathbf{f}_n)\|\in\mathscr{F}\\ &\Leftrightarrow \|\exists\mathbf{v}_k\psi(\mathbf{f}_0,\ldots,\mathbf{f}_n)\|\in\mathscr{F}\\ &\Leftrightarrow \text{for some } f\in A,\ \|\psi(\mathbf{f}_0,\ldots,\mathbf{f}_{k-1},\mathbf{f},\mathbf{f}_{k+1},\ldots,\mathbf{f}_n)\|\in\mathscr{F}\ \text{(by Lemma 3.4)}\\ &\Leftrightarrow \text{for some } f\in A,\ \prod\mathfrak{A}_i/\mathscr{F}\models\psi\,[f_0/\mathscr{F},\ldots,f_{k-1}/\mathscr{F},f/\mathscr{F},f_{k+1}/\mathscr{F},\ldots,f_n/\mathscr{F}]\\ &\Leftrightarrow \prod\mathfrak{A}_i/\mathscr{F}\models(\exists\mathbf{v}_k\psi)\,[f_0/\mathscr{F},\ldots,f_n/\mathscr{F}]\\ &\Leftrightarrow \prod\mathfrak{A}_i/\mathscr{F}\models\varphi\,[f_0/\mathscr{F},\ldots,f_n/\mathscr{F}].\end{aligned}$$

Thus in this case, too, the result holds for φ, and the proof is complete. ∎

Putting Thms. 3.1 and 3.6 together we obtain the following fundamental result:

3.7. Łoś' Theorem. *For any $\mathscr{L}$-formula φ whose free variables are all among $\mathbf{v}_0,\ldots,\mathbf{v}_n$ and any $f_0,\ldots,f_n\in\prod A_i$ we have*

$$\prod\mathfrak{A}_i/\mathscr{F}\models\varphi\,[f_0/\mathscr{F},\ldots,f_n/\mathscr{F}] \Leftrightarrow \{i\in I:\ \mathfrak{A}_i\models\varphi\,[f_0(i),\ldots,f_n(i)]\}\in\mathscr{F}.$$

In particular, for any $\mathscr{L}$-sentence σ we have

$$\prod\mathfrak{A}_i/\mathscr{F}\models\sigma \Leftrightarrow \{i\in I:\ \mathfrak{A}_i\models\sigma\}\in\mathscr{F}.$$ ∎

Now let $\mathfrak{A}$ be an $\mathscr{L}$-structure and let $\mathscr{F}$ be an ultrafilter over I. For each $a\in A$ let $\hat{a}\in A^I$ be defined by $\hat{a}(i)=a$ for each $i\in I$; $\hat{a}$ is called the *constant function* (on I) *with value* a. Define the mapping $d\colon A \to A^I/\mathscr{F}$ by putting

$$d(a) = \hat{a}/\mathscr{F} \quad \text{for each } a \in A.$$

It is easy to verify that d is an embedding of $\mathfrak{A}$ into $\mathfrak{A}^I/\mathscr{F}$; it is called the *canonical embedding* of $\mathfrak{A}$ into $\mathfrak{A}^I/\mathscr{F}$.

3.8. THEOREM. *The canonical embedding d is an elementary embedding of $\mathfrak{A}$ into $\mathfrak{A}^I/\mathscr{F}$. Hence $\mathfrak{A}$ and $\mathfrak{A}^I/\mathscr{F}$ are elementarily equivalent.*

PROOF. Let φ be a formula whose free variables are all among $\mathbf{v}_0,\ldots,\mathbf{v}_n$, and let $a_0,\ldots,a_n$ be any elements of A. Then by Łoś' Theorem we have

$$\begin{aligned}\mathfrak{A}^I/\mathscr{F} \models \varphi\,[d(a_0),\ldots,d(a_n)] &\Leftrightarrow \mathfrak{A}^I/\mathscr{F} \models \varphi\,[\hat{a}_0/\mathscr{F},\ldots,\hat{a}_n/\mathscr{F}]\\ &\Leftrightarrow \{i \in I\colon \mathfrak{A} \models \varphi\,[\hat{a}_0(i),\ldots,\hat{a}_n(i)]\} \in \mathscr{F}\\ &\Leftrightarrow \mathfrak{A} \models \varphi\,[a_0,\ldots,a_n].\end{aligned}$$

∎

3.9. PROBLEM. (i) Show that if F is principal and generated by $i_0 \in I$, then $\prod \mathfrak{A}_i/\mathscr{F} \cong \mathfrak{A}_{i_0}$. (Define $h\colon \prod A_i/\mathscr{F} \to A_{i_0}$ by $h(f/\mathscr{F}) = f(i_0)$; h is the required isomorphism.)

(ii) Deduce from (i) that if I is finite, then $\mathfrak{A}^I/\mathscr{F} \cong \mathfrak{A}$.

3.10. PROBLEM. Show that, if A is finite, then $\mathfrak{A}^I/\mathscr{F} \cong \mathfrak{A}$. (In this case d maps A onto $A^I/\mathscr{F}$.)

3.11. PROBLEM. (i) For each $n \in \omega$ let $\mathfrak{A}_n = \langle A_n, <_n\rangle$ be an infinite well-ordered set, and let $\mathscr{F}$ be a non-principal ultrafilter over ω. Show that $\prod \mathfrak{A}_n/\mathscr{F}$ is not well-ordered. (We may assume that $\langle \omega, <\rangle$ is an initial segment of each $\mathfrak{A}_n$. Show that the elements $\langle 0,1,2,3,\ldots\rangle/\mathscr{F}$, $\langle 0,0,1,2,\ldots\rangle/\mathscr{F}$, $\langle 0,0,0,1,\ldots\rangle/\mathscr{F}$, etc., constitute an infinite descending chain in $\prod \mathfrak{A}_n/\mathscr{F}$.)

(ii) Let $\mathfrak{N} = \langle \omega, +, \times, 0, 1\rangle$, and let $\mathscr{F}$ be a non-principal ultrafilter over ω. Show that $\mathfrak{N}^\omega/\mathscr{F}$ is not isomorphic to $\mathfrak{N}$. (Use (i).)

We now apply Łoś' Theorem to obtain a proof of the compactness theorem which is purely semantic and independent of the results of Chapters 1–3.

3.12. COMPACTNESS THEOREM.[1] *Let Σ be a set of $\mathscr{L}$-sentences. If each finite subset of Σ has a model, then Σ has a model.*

PROOF. Suppose that each finite subset of Σ has a model; let I be the family of all finite subsets of Σ. For each $\Delta \in I$ let $\mathfrak{A}_\Delta$ be a model of Δ,

[1] Once we have given a purely semantic proof of the compactness theorem, it will be clear that the proofs of Thms. 2.6, 2.7 and 2.8 also become purely semantic. A different semantic proof of Thm. 2.6 is given in Prob. 3.16.

and define

$$\tilde{\Delta}=\{\Delta'\in I\colon\ \Delta\subseteq\Delta'\}.$$

If $\{\Delta_1,\ldots,\Delta_n\}\subseteq I$, then clearly

$$\Delta_1\cup\ldots\cup\Delta_n\in\tilde{\Delta}_1\cap\ldots\cap\tilde{\Delta}_n,$$

so that $\{\tilde{\Delta}\colon\ \Delta\in I\}$ is a subset of PI with the finite intersection property. Let $\mathscr{F}$ be an ultrafilter over I containing each $\tilde{\Delta}$. We claim that the ultraproduct $\prod_{\Delta\in I}\mathfrak{A}_\Delta/\mathscr{F}$ is a model of $\boldsymbol{\Sigma}$. For suppose that $\boldsymbol{\sigma}\in\boldsymbol{\Sigma}$. Then $\{\boldsymbol{\sigma}\}\in I$, and if $\{\boldsymbol{\sigma}\}\subseteq\Delta\in I$ then $\mathfrak{A}_\Delta\models\boldsymbol{\sigma}$, so that

$$\{\boldsymbol{\sigma}\}^{\sim}=\{\Delta\in I\colon\ \{\boldsymbol{\sigma}\}\subseteq\Delta\}\subseteq\{\Delta\in I\colon\ \mathfrak{A}_\Delta\models\boldsymbol{\sigma}\}\in\mathscr{F},$$

since $\{\boldsymbol{\sigma}\}^{\sim}\in\mathscr{F}$ by construction. It follows from Łoś' Theorem that $\prod\mathfrak{A}_\Delta/\mathscr{F}\models\boldsymbol{\sigma}$. This completes the proof. ∎

REMARK. *How the Compactness Theorem got its name.* The term "compact" is topological, but there is no mention of topology either in the Compactness Theorem or its proof. Why, then, do we call it this?

The answer is really quite simple. Let $\mathscr{S}$ be the the class of all $\mathscr{L}$-structures, and for each $\mathscr{L}$-sentence $\boldsymbol{\sigma}$ and each set $\boldsymbol{\Sigma}$ of $\mathscr{L}$-sentences let

$$\mathrm{Mod}(\boldsymbol{\sigma})=\{\mathfrak{A}\in\mathscr{S}\colon\ \mathfrak{A}\models\boldsymbol{\sigma}\},\qquad \mathrm{Mod}(\boldsymbol{\Sigma})=\{\mathfrak{A}\in\mathscr{S}\colon\ \mathfrak{A}\models\boldsymbol{\Sigma}\}.$$

Classes of structures of the form Mod($\boldsymbol{\sigma}$) or Mod($\boldsymbol{\Sigma}$) are called *elementary* and *generalized elementary* classes, respectively. It is easy to see that the elementary classes form a field of subclasses of $\mathscr{S}$; accordingly they form a base for a topology on $\mathscr{S}$, called the *elementary* topology. Since the generalized elementary classes are precisely the intersections of elementary classes (as is easily verified) and the elementary classes are *clopen* subsets of $\mathscr{S}$, the generalized elementary classes constitute the family of all *closed* sets in the elementary topology. Using this remark, one easily shows that the Compactness Theorem means that the elementary topology is compact!

3.13. PROBLEM. Let $\mathscr{S}$ be the class of all $\mathscr{L}$-structures, and let $\mathscr{X}\subseteq\mathscr{S}$.

(i) Show that $\mathscr{X}$ is an elementary class iff both $\mathscr{X}$ and $\mathscr{S}-\mathscr{X}$ are generalized elementary classes.

(ii) Show that $\mathscr{X}$ is a generalized elementary class iff every ultraproduct of members of $\mathscr{X}$ is in $\mathscr{X}$ and every structure elementarily equivalent to a member of $\mathscr{X}$ is in $\mathscr{X}$. (Let $\boldsymbol{\sigma}$ be a variable ranging over *sentences*. For the "if" part, let

$$\Sigma = \{\sigma:\ \sigma \text{ holds in all } \mathfrak{A} \in \mathscr{X}\}.$$

Then $\mathscr{X} \subseteq \mathrm{Mod}(\Sigma)$. If $\mathfrak{A} \in \mathrm{Mod}(\Sigma)$, let

$$\Delta = \{\sigma:\ \mathfrak{A} \models \sigma\},$$

and for each $\sigma \in \Delta$ show that one can choose $\mathfrak{B}_\sigma \in \mathscr{X}$ so that $\mathfrak{B}_\sigma \models \sigma$. For each $\sigma \in \Delta$ let $\Gamma_\sigma = \{\tau \in \Delta:\ \mathfrak{B}_\tau \models \sigma\}$; show that $\{\Gamma_\sigma:\ \sigma \in \Delta\}$ can be extended to an ultrafilter $\mathscr{F}$ over Δ and that $\prod_{\sigma \in \Delta} \mathfrak{B}_\sigma / \mathscr{F} \equiv \mathfrak{A}$.)

*3.14. Problem. (i) Let $\mathfrak{A}$ and $\mathfrak{B}$ be $\mathscr{L}$-structures. Show that $\mathfrak{A} \equiv \mathfrak{B}$ iff $\mathfrak{A}$ is elementarily embeddable in an ultrapower of $\mathfrak{B}$. (For the "only if" part, suppose that $\mathfrak{A} \equiv \mathfrak{B}$, and let $\mathfrak{a} = \langle a_k:\ k \in K\rangle$ be an indexing of A. Let Σ be the set of all $\mathscr{L}_K$-sentences which hold in $(\mathfrak{A}, \mathfrak{a})$. For each $\sigma \in \Sigma$ show that there is a sequence $\mathfrak{b}_\sigma \in B^K$ such that $(\mathfrak{B}, \mathfrak{b}_\sigma) \models \sigma$; then put

$$\Delta_\sigma = \{\tau \in \Sigma:\ (\mathfrak{B}, \mathfrak{b}_\tau) \models \sigma\}.$$

Show that $\{\Delta_\sigma:\ \sigma \in \Sigma\}$ can be extended to an ultrafilter $\mathscr{F}$ over Σ and that $\mathfrak{A}$ is elementarily embeddable in $\mathfrak{B}^\Sigma / \mathscr{F}$.)

(ii) Deduce from (i) that if A is finite and $\mathfrak{A} \equiv \mathfrak{B}$ then $\mathfrak{A} \cong \mathfrak{B}$.

*3.15. Problem. Let $\{\mathfrak{A}_i:\ i \in I\}$ be a family of fields, and let $\mathscr{F}$ be an ultrafilter over I. For each $f \in \prod A_i$ let

$$Z(f) = \{i \in I:\ f(i) = 0\}$$

and let

$$M = \{f \in \prod A_i:\ Z(f) \in \mathscr{F}\}.$$

Show that $\prod \mathfrak{A}_i$ is a ring in which M is a maximal ideal, and that the ultraproduct $\prod \mathfrak{A}_i / \mathscr{F}$ is isomorphic to the quotient ring $\prod \mathfrak{A}_i / M$.

*3.16. Problem. For each set I let $\mathsf{P}_\omega(I)$ be the family of all finite subsets of I. An ultrafilter $\mathscr{F}$ over I is said to be *regular* if there is a one–one map f of I onto $\mathsf{P}_\omega(I)$ such that for each $i \in I$,

$$\{j \in I:\ i \in f(j)\} \in \mathscr{F}.$$

(i) Show that, if I is infinite, there is a regular ultrafilter over I. (Let f be a one–one map of I onto $\mathsf{P}_\omega(I)$. For each $i \in I$ let

$$E_i = \{j \in I:\ i \in f(j)\}.$$

Show that $\{E_i:\ i \in I\}$ has the finite intersection property and that any ultrafilter over I which contains each E_i is regular.)

(ii) Let A,I be infinite sets, and let $\mathscr{F}$ be a regular ultrafilter over I. Show that $|A^I/\mathscr{F}|=|A|^{|I|}$. (It suffices to construct a one–one map h of A^I into $\mathscr{B}^I/\mathscr{F}$, where $\mathscr{B}$ is the set of all finite sequences of members of A. Let g be a one–one map of I onto $\mathsf{P}_\omega(I)$ such that, for each $j\in I$,

$$\{i\in I:\ j\in g(i)\}\in\mathscr{F}.$$

Let $<$ be a total ordering of I. For each $f\in A^I$ define $f^*\in\mathscr{B}^I$ as follows. For $i\in I$,

$$g(i)=\{i_0,\ldots,i_k\}\in\mathsf{P}_\omega(I)$$

with $i_0<\ldots<i_k$. Put

$$f^*(i)=\langle f(i_0),\ldots,f(i_k)\rangle.$$

Now define $h\colon A^I\to\mathscr{B}^I/\mathscr{F}$ by setting $h(f)=f^*/\mathscr{F}$ for each $f\in A^I$. Show that h is one–one.)

(iii) Deduce Thm. 2.6 from (i) and (ii).

*3.17. Problem. (i) Let $\mathscr{L}$ be a countable language, let $\mathscr{F}$ be a countably incomplete ultrafilter (Prob. 2.3.16) over a set I, let $\{\mathfrak{A}_i:\ i\in I\}$ be a family of $\mathscr{L}$-structures, and let $\mathbf{\Gamma}$ be a set of $\mathscr{L}$-formulas each of which has exactly one free variable $\mathbf{v}_0$. Show that, if $\mathbf{\Gamma}$ is finitely satisfiable (Prob. 2.12) in $\prod\mathfrak{A}_i/\mathscr{F}$, then $\mathbf{\Gamma}$ is satisfiable in $\prod\mathfrak{A}_i/\mathscr{F}$. (Let $\mathbf{\Gamma}=\{\boldsymbol{\varphi}_n:\ n\in\omega\}$, let $I_0\supseteq I_1\supseteq I_2\supseteq\ldots$ be a descending chain of members of $\mathscr{F}$ such that $\bigcap I_n=\emptyset$ (Prob. 2.3.16), for each $n\in\omega$ let

$$Y_n=I_n\cap\{i\in I:\ \mathfrak{A}_i\models\exists\mathbf{v}_0(\boldsymbol{\varphi}_0\wedge\ldots\wedge\boldsymbol{\varphi}_n)\},$$

and for each $i\in I_0$ let $\mu(i)$ be the greatest n such that $i\in Y_n$. Define $f\in\prod A_i$ by setting $f(i)$ to be an arbitrary element of I if $i\notin Y_0$, while if $i\in Y_0$ choose $f(i)\in A_i$ so that

$$\mathfrak{A}_i\models(\boldsymbol{\varphi}_1\wedge\ldots\wedge\boldsymbol{\varphi}_{\mu(i)})\,[f(i)].$$

Show, using Łoś' Theorem, that $f/\mathscr{F}$ satisfies $\mathbf{\Gamma}$ in $\prod\mathfrak{A}_i/\mathscr{F}$.)

(ii) Use (i) to give another solution for Prob. 2.12(i).

§ 4. Completeness and categoricity

Throughout this section we let $\mathscr{L}$ be a *countable* first-order language. By a *theory* in $\mathscr{L}$ we shall mean a set $\mathbf{\Sigma}$ of $\mathscr{L}$-sentences which is closed under deducibility, i.e. such that for each $\mathscr{L}$-sentence $\boldsymbol{\sigma}$, if $\mathbf{\Sigma}\vdash\boldsymbol{\sigma}$, then $\boldsymbol{\sigma}\in\mathbf{\Sigma}$. A subset $\mathbf{\Gamma}$ of a theory $\mathbf{\Sigma}$ is called a *set of postulates* for $\mathbf{\Sigma}$ if $\mathbf{\Gamma}\vdash\boldsymbol{\sigma}$ for every $\boldsymbol{\sigma}\in\mathbf{\Sigma}$. It is clear that each set of sentences $\mathbf{\Gamma}$ is a set of postulates for a unique theory Σ, namely,

$$\Sigma = \{\sigma : \sigma \text{ is an } \mathscr{L}\text{-sentence and } \Gamma \vdash \sigma\}.$$

A property P of $\mathscr{L}$-structures is called a *first-order* property if there is an $\mathscr{L}$-sentence σ such that, for any $\mathscr{L}$-structure $\mathfrak{A}$,

$$\mathfrak{A} \text{ has property } P \Leftrightarrow \mathfrak{A} \models \sigma.$$

In this case we say that P is *induced* by σ. To what extent does a given theory Σ determine the first-order properties of its models? The completeness theorem tells us that the first-order properties shared by *all* models of Σ are precisely those induced by the sentences in Σ. But this still gives us very little information about the extent to which the first-order properties of a *specific* model of Σ are determined by Σ. And in general the models of a given theory can be very different.

Consider, for example, the (first-order) theory of partially ordered sets, **PO**. This theory is formulated in a language $\mathscr{L}$ having one binary predicate symbol **R**. Its postulates are

$$\forall x Rxx,$$

$$\forall x \forall y\, [Rxy \wedge Ryx \rightarrow x = y],$$

$$\forall x \forall y \forall z\, [Rxy \wedge Ryz \rightarrow Rxz].$$

An $\mathscr{L}$-structure $\langle A, R \rangle$ is then a model of **PO** iff it is a partially ordered set. Since a partially ordered set can have many different first-order properties, e.g. it can be a lattice, or a Boolean algebra, or a totally ordered set, etc., it is clear that **PO** does not precisely determine the first-order properties of its models.

Let us call a theory Σ *complete* if it is consistent and the first-order properties of any model of Σ are just those induced by the sentences in Σ. More precisely, if for each $\mathscr{L}$-structure $\mathfrak{A}$ we define $\mathbf{Th}(\mathfrak{A})$, *the theory of* $\mathfrak{A}$, to be the set of all $\mathscr{L}$-sentences σ such that $\mathfrak{A} \models \sigma$, then Σ is complete iff Σ is consistent and $\mathbf{Th}(\mathfrak{A}) = \Sigma$ for each model $\mathfrak{A}$ of Σ.

4.1. LEMMA. *The following conditions on a consistent theory* Σ *are equivalent:*

(i) Σ *is complete.*

(ii) *Any pair of models of* Σ *are elementarily equivalent.*

(iii) *For any* $\mathscr{L}$*-sentence* σ*, either* $\sigma \in \Sigma$ *or* $\neg\sigma \in \Sigma$.

PROOF. (i)$\Rightarrow$(ii). Suppose that Σ is complete and that $\mathfrak{A}$ and $\mathfrak{B}$ are models of Σ. Then $\mathbf{Th}(\mathfrak{A}) = \Sigma = \mathbf{Th}(\mathfrak{B})$, so that $\mathfrak{A} \equiv \mathfrak{B}$.

(ii)⇒(iii). Assume (ii) and let σ be an $\mathscr{L}$-sentence. If $\sigma \notin \Sigma$, then by Thm. 3.1.14. $\Sigma \cup \{\neg\sigma\}$ is consistent and therefore, by Thm. 2.9, has a model $\mathfrak{A}$. Since $\neg\sigma$ holds in $\mathfrak{A}$, it must hold in every model of Σ and so, by the Strong Completeness Theorem 3.3.14, $\Sigma \vdash \neg\sigma$. Since Σ is a theory, $\neg\sigma \in \Sigma$.

(iii)⇒(i). Assume (iii), and let $\mathfrak{A}$ be any model of Σ. Then $\Sigma \subseteq \mathbf{Th}(\mathfrak{A})$. If $\sigma \notin \Sigma$, then $\neg\sigma \in \Sigma$. Therefore $\mathfrak{A} \models \neg\sigma$, whence $\neg\sigma \in \mathbf{Th}(\mathfrak{A})$, so that $\sigma \notin \mathbf{Th}(\mathfrak{A})$. Accordingly $\mathbf{Th}(\mathfrak{A}) \subseteq \Sigma$, so that $\mathbf{Th}(\mathfrak{A}) = \Sigma$ and Σ is complete. ∎

4.2. Problem. Show that $\mathbf{Th}(\mathfrak{A})$ is a complete theory for any $\mathscr{L}$-structure $\mathfrak{A}$, and that a consistent set Σ of $\mathscr{L}$-sentences is a complete theory iff $\Sigma = \mathbf{Th}(\mathfrak{A})$ for some $\mathscr{L}$-structure $\mathfrak{A}$.

A natural strengthening of the condition for completeness given in 4.1(ii) is to insist that each pair of models of Σ be *isomorphic*. Under these conditions Σ is said to be *categorical*. Certainly any categorical theory is complete. On the other hand, no theory Σ with an infinite model can be categorical, for by the Löwenheim–Skolem theorem Σ must then have models of many cardinalities, and no pair of models of different cardinalities can be isomorphic. However, it is still possible that any two models of Σ of the *same* (infinite) cardinality are isomorphic. This suggests the following weakening of the notion of categoricity.

Let α be an infinite cardinal. A theory Σ is said to be *α-categorical* if any pair of models of Σ of cardinality α are isomorphic. We now give some examples.

4.3. Examples. (i) Let $\mathscr{L}$ be the language with no extralogical symbols, and let Σ be the set of all $\mathscr{L}$-sentences which hold in every $\mathscr{L}$-structure. *Then Σ is α-categorical for every α.*

(ii) Let $\mathscr{L}$ be a language with no extralogical symbols except for one unary predicate symbol, $\mathbf{P}$, and let Σ be the set of $\mathscr{L}$-sentences which hold in every $\mathscr{L}$-structure. *Then Σ is not α-categorical for any α.*

(iii) Let $\mathscr{L}$ be as in (ii) and for each natural number m let σ_m be the sentence which asserts that there are at least m individuals having the property P and at least m individuals not having the property P. Let Σ be the theory with $\{\sigma_m : m \in \omega\}$ as postulates. It is then easy to see that Σ is *$\aleph_0$-categorical but not α-categorical for any uncountable α.* (We give a more interesting example later on.)

(iv) Let $\mathscr{L}$ be a language whose only extralogical symbols are countably many constants $\mathbf{c}_0, \mathbf{c}_1, \ldots$, and let Σ be the theory whose set of postulates is $\{c_m \neq c_n : m \neq n\}$. It is then easily verified that Σ *is* α*-categorical for every uncountable* α *but not* $\aleph_0$*-categorical.* (We give a more interesting example later on.)

Notice that we have not given an example of a (countable) theory which is α-categorical for *some* but not *all* uncountable α. The reason for this omission is that no such theory exists, for MORLEY [1965] has shown that if a countable theory is α-categorical for *some* uncountable α, then it is α-categorical for *all* uncountable α. The proof of this deep result is too lengthy to be included here.

The relevance of α-categoricity to completeness is contained in the following theorem.

4.4. THEOREM. *Let* Σ *be a consistent theory with no finite models, and which is* α*-categorical for some infinite* α. *Then* Σ *is complete.*

PROOF. Suppose that Σ is not complete. Then there is a sentence σ such that $\sigma \notin \Sigma$ and $\neg\sigma \notin \Sigma$. It follows that $\Sigma \cup \{\sigma\}$ and $\Sigma \cup \{\neg\sigma\}$ are both consistent and hence have models, which must both be infinite since Σ has no finite models. Therefore, by the Löweinheim-Skolem Theorem, $\Sigma \cup \{\sigma\}$ and $\Sigma \cup \{\neg\sigma\}$ have models of cardinality α. Since σ holds in one of these models but not in the other, Σ is not α-categorical. This contradiction proves the theorem. ∎

We now apply Thm. 4.4. to establish the completeness of various specific theories.

First consider the theory **UDO** of *unbounded dense (total) orderings.* Let $\mathscr{L}$ be a language with one binary predicate symbol **R**. **UDO** is the theory having the following postulates:

$$(\mathrm{UDO}_1)\quad \forall x Rxx \wedge \forall x \forall y\,[Rxy \wedge Ryx \rightarrow x = y] \wedge \forall x \forall y \forall z [Rxy \wedge Ryz \rightarrow Rxz] \wedge \forall x \forall y\,[Rxy \vee Ryx]$$

$$(\mathrm{UDO}_2)\quad \forall x \forall y\,[Rxy \wedge x \neq y \rightarrow \exists z\,[x \neq z \wedge y \neq z \wedge Rxz \wedge Rzy]]$$

$$(\mathrm{UDO}_3)\quad \forall x \exists y \exists z\,[x \neq y \wedge x \neq z \wedge Ryx \wedge Rxz].$$

UDO_1 asserts that R is a total ordering, UDO_2 asserts that R is dense, and UDO_3 that it is unbounded both below and above. Natural examples of models of Σ are $\mathfrak{Q}$ and $\mathfrak{R}$, the sets of rational numbers and real numbers, with their natural orderings.

Our next result is a classical theorem of Cantor. The proof provides a first example of a so-called "back-and-forth" construction.

4.5. THEOREM. **UDO** *is $\aleph_0$-categorical.*

PROOF. Let $\mathfrak{A}=\langle A,\leqslant\rangle$ and $\mathfrak{B}=\langle B,\leqslant\rangle$ be two denumerable models of **UDO**. Thus each is an unbounded densely ordered set. Let $A=\{a_n: n\in\omega\}$ and $B=\{b_n: n\in\omega\}$. We define two sequences $c_0,c_1,\ldots,$ and $d_0,d_1,\ldots,$ by recursion as follows. First, put $c_0=a_0$ and $d_0=b_0$. Now suppose $k>0$; then we consider two cases.

(i) $k=2m$ is even. In this case we put $c_k=a_m$. If, for some $j<k$, $c_k=c_j$, then we put $d_k=d_j$. Otherwise we let d_k be some element of B which bears the same order relations to the elements of $\{d_0,\ldots,d_{k-1}\}$ as does c_k to the elements of $\{c_0,\ldots,c_{k-1}\}$; that is, for each $j<k$, if[1] $c_k<c_j$ then $d_k<d_j$, and if $c_j<c_k$ then $d_j<d_k$. Since B is dense and unbounded it is clear that we can always find such a d_k.

(ii) $k=2m+1$ is odd. In this case we put $d_k=b_m$. If, for some $j<k$, $d_k=d_j$, then we put $c_k=c_j$. Otherwise we let c_k be some element of A which bears the same order relations to the elements of $\{c_0,\ldots,c_{k-1}\}$ as does d_k to the elements of $\{d_0,\ldots,d_{k-1}\}$. Again such a c_k can always be found.

This completes our recursive definition. We now define $h: A\to B$ by putting $h(c_n)=d_n$ for each $n\in\omega$. It is clear from our construction that h is an isomorphism of $\mathfrak{A}$ onto $\mathfrak{B}$. ∎

From 4.4 and 4.5 we imediately obtain

4.6. COROLLARY. **UDO** *is a complete theory.* ∎

In particular, since $\mathfrak{Q}$ and $\mathfrak{R}$ are models of **UDO** it follows from 4.6 that $\mathfrak{Q}\equiv\mathfrak{R}$ (cf. Prob. 1.8(ii)), and that **UDO** is a set of postulates for both **Th**($\mathfrak{Q}$) and **Th**($\mathfrak{R}$).

4.7. PROBLEM. Show that **UDO** is not α-categorical for any *uncountable* α. (Let $\mathfrak{A}$ be the ordered sum of α copies of $\mathfrak{Q}$, and let $\mathfrak{B}$ the ordered sum of $\alpha+1$ copies of $\mathfrak{Q}$. Then $\mathfrak{A}$ and $\mathfrak{B}$ are models of **UDO** but $\mathfrak{A}$ is not isomorphic to $\mathfrak{B}$ because every initial segment of $\mathfrak{A}$ has cardinality $<\alpha$ while this is not true of $\mathfrak{B}$.)

The theory that we are going to consider next is most naturally formulated in a language with function symbols.

Let $\mathscr{L}_F$ be the *language for fields*, that is, $\mathscr{L}_F$ is a first-order language with constants **0**,**1** and binary function symbols **+**, **·**. The theory **FT** of *fields* has the following postulates (where we write **x+y** and **x·y** for

[1] We write "$a<b$" for "$a\leqslant b$ and $a\neq b$" as usual.

$+\mathbf{xy}$ and $\cdot\mathbf{xy}$):

$$\forall x\forall y\forall z\,[(x+y)+z=x+(y+z)],$$

$$\forall x(x+0=x),$$

$$\forall x\forall y(x+y=y+x),$$

$$\forall x\exists y(x+y=0),$$

$$\forall x\forall y\forall z\,[(x\cdot y)\cdot z=x\cdot(y\cdot z)],$$

$$\forall x(x\cdot 1=x),$$

$$\forall x\forall y(x\cdot y=y\cdot x),$$

$$\forall x\,[x\neq 0\rightarrow\exists y(x\cdot y=1)],$$

$$\forall x\forall y\forall z\,[x\cdot(y+z)=(x\cdot y)+(x\cdot z)],$$

$$0\neq 1.$$

If we add the postulate

$$p1=0$$

where p is a prime number and $p1=1+\ldots+1$ with p summands, we get the theory $\mathbf{FT}(p)$ of *fields of characteristic* p. If on the other hand we add the infinite set of postulates

$$\{p1\neq 0:\ p \text{ a prime}\}$$

we get the theory $\mathbf{FT}(0)$ of *fields of characteristic* 0. We now put $\mathbf{x}^n$ for the expression $\mathbf{x}\cdot(\mathbf{x}\cdot(\cdots\mathbf{x})\cdots)$, with n factors. The infinite list of postulates, one for each $n\geq 1$,

$$\forall x_0\ldots\forall x_n\,[x_n\neq 0\rightarrow\exists y(x_n\cdot y^n+x_{n-1}\cdot y^{n-1}+\ldots+x_1\cdot y+x_0=0)]$$

when added to the theory of fields gives us the theory **ACF** of *algebraically closed fields.* If we add the postulate $p1=0$ or the set of postulates $\{p1\neq 0:$ p a prime$\}$ to **ACF** we get the theories $\mathbf{ACF}(p)$ and $\mathbf{ACF}(0)$ of *algebraically closed fields of characteristic* p and *characteristic* 0, respectively.

Notice that $\mathbf{ACF}(0)$ is not $\aleph_0$-categorical. For the field $\mathfrak{F}$ of algebraic numbers and the algebraic closure of the field $\mathfrak{F}[\pi]$ obtained by adjoining the transcendental π to $\mathfrak{F}$ are countable non-isomorphic models of $\mathbf{ACF}(0)$. On the other hand, a classical theorem of Steinitz asserts that any pair of algebraically closed fields of the same characteristic and the same *uncountable* cardinality are isomorphic, so it follows from Thm. 4.4 that **ACF**(0) and **ACF**(p) are complete theories[1]. Since the field $\mathfrak{C}$ of complex numbers is a model of **ACF**(0), it follows that **ACF**(0) is a set of postulates for **Th**($\mathfrak{C}$). We also see that any first-order property possessed by $\mathfrak{C}$ is shared by all algebraically closed fields of characteristic 0. (This is known as *Lefschetz' principle.*)

This method of establishing completeness also applies to the following theories, each of which is α-categorical for some infinite α, and hence complete:

(1) The theory of atomless Boolean algebras: $\aleph_0$-categorical.

(2) The theory of infinite Abelian groups in which all non-zero elements are of the same prime order: α-categorical for all infinite α.

(3) The theory of infinite, divisible, torsion-free Abelian groups: α-categorical for all uncountable α.

4.8. PROBLEM. Show that if $\boldsymbol{\sigma}$ is a sentence in the language for fields which has models of arbitrarily high finite characteristic, then $\boldsymbol{\sigma}$ has a model of characteristic 0. (Use the Compactness Theorem.)

4.9. PROBLEM. A theory is said to be *finitely axiomatizable* if it has a finite set of postulates.

(i) Show that the theory **FT**(0) is not finitely axiomatizable. (Suppose the contrary; then there is a sentence $\boldsymbol{\sigma}$ such that, for all fields $\mathfrak{F}$, $\mathfrak{F} \models \boldsymbol{\sigma}$ iff $\mathfrak{F}$ is of characteristic 0. Now apply Prob. 4.8 to $\neg\boldsymbol{\sigma}$.) Deduce that the property of being a field of finite non-zero characteristic is not a first-order property.

(ii) Assume the following result: for each $n \in \omega$, there is a field $\mathfrak{F}_n$ which is not algebraically closed but in which all polynomials of degree $\leqslant n$ have zeros. Show that **ACF** is not finitely axiomatizable. (Like 4.8.)

*4.10. PROBLEM. An *ordered field* is a structure of the form $\mathfrak{F} = \langle F, +, \cdot, 0, 1, \leqslant \rangle$, where $\langle F, +, \cdot, 0, 1 \rangle$ is a field and $\leqslant$ is a total ordering on F such that for all $x, y, z \in F$ we have

$$x \leqslant y \Rightarrow x + z \leqslant y + z,$$

$$x \leqslant y,\ z \geqslant 0 \Rightarrow x \cdot z \leqslant y \cdot z.$$

An ordered field is said to be *archimedean* if for any $x, y \geqslant 0$ there is a natural number n such that $y \leqslant n \cdot x$ (where $n \cdot x = x + \ldots + x$ with n summands).

[1] It is well-known that these theories have no finite models.

Show that each ordered field (in particular, the real field) is elementarily equivalent to a non-archimedean ordered field. (Use the Compactness Theorem.) Deduce that the property of being an archimedean ordered field is not a first-order property.

4.11. PROBLEM. Show that every group which has elements of arbitrarily high finite order is elementarily equivalent to a group which has an element of infinite order. (Use the Compactness Theorem.)

§ 5. Lindenbaum algebras

In this section we show that each consistent set of first-order sentences Σ gives rise to a Boolean algebra which can be used to investigate the model-theoretic properties of Σ.

Let $\mathscr{L}$ be a fixed language (not necessarily countable), let Φ be the set of all $\mathscr{L}$-formulas, and let Σ be a consistent set of $\mathscr{L}$-sentences. We define the relation $\approx$ on Φ by putting

$$\varphi \approx \psi \Leftrightarrow \Sigma \vdash \varphi \leftrightarrow \psi$$

for all $\varphi, \psi \in \Phi$. The relation $\approx$ is evidently an *equivalence relation** on Φ. For each $\varphi \in \Phi$, let $|\varphi|$ be the $\approx$-class[1] of φ. Thus

$$|\varphi| = \{\psi \in \Phi:\ \varphi \approx \psi\}.$$

Let

$$B = \{|\varphi|:\ \varphi \in \Phi\},$$

and let $\leqslant$ be the relation on B defined by putting

$$|\varphi| \leqslant |\psi| \Leftrightarrow \Sigma \vdash \varphi \rightarrow \psi. \tag{5.1}$$

5.2. PROBLEM. Show that (5.1) is a sound definition, i.e. that if $|\varphi_1| = |\varphi_2|$ and $|\psi_1| = |\psi_2|$ then

$$\Sigma \vdash \varphi_1 \rightarrow \psi_1 \Leftrightarrow \Sigma \vdash \varphi_2 \rightarrow \psi_2.$$

Our next result is crucial.

5.3. THEOREM. *$\langle B, \leqslant \rangle$ is a Boolean algebra. In $\langle B, \leqslant \rangle$ we have*

$$|\varphi| = 1 \Leftrightarrow \Sigma \vdash \varphi,$$

$$|\varphi| = 0 \Leftrightarrow \Sigma \vdash \neg\varphi.$$

[1] It is important to keep in mind that $\approx$ and $|\varphi|$ *depend on* Σ *as well as* $\mathscr{L}$.

PROOF. We have to show that $\langle B, \leqslant \rangle$ is a complemented distributive lattice with at least 2 elements.

Since $\varphi \rightarrow \varphi$ is a propositional tautology, we have $\Sigma \vdash \varphi \rightarrow \varphi$; hence $|\varphi| \leqslant |\varphi|$, and $\leqslant$ is reflexive. By definition, $\leqslant$ is antisymmetric. To show that it is transitive it is enough to notice that, since $\varphi \rightarrow \chi$ is a tautological consequence of the set of formulas $\{\varphi \rightarrow \psi, \psi \rightarrow \chi\}$, we have

$$\Sigma \vdash \varphi \rightarrow \psi \quad \& \quad \Sigma \vdash \psi \rightarrow \chi \Rightarrow \Sigma \vdash \varphi \rightarrow \chi.$$

Thus $\leqslant$ is a partial ordering of B.

We now show that $\langle B, \leqslant \rangle$ is a lattice, in other words, that any two elements $|\varphi|$, $|\psi|$ have an infimum and a supremum in B. In fact we claim that

$$\inf\{|\varphi|, |\psi|\} = |\varphi \wedge \psi|, \qquad \sup\{|\varphi|, |\psi|\} = |\varphi \vee \psi|.$$

To prove the first equality, observe that we have $\Sigma \vdash \varphi \wedge \psi \rightarrow \varphi$ and $\Sigma \vdash \varphi \wedge \psi \rightarrow \psi$ since these formulas are propositional tautologies; hence

$$|\varphi \wedge \psi| \leqslant |\varphi|, \qquad |\varphi \wedge \psi| \leqslant |\psi|.$$

Now suppose that $|\chi|$ is any lower bound for $\{|\varphi|, |\psi|\}$. Then $\Sigma \vdash \chi \rightarrow \varphi$ and $\Sigma \vdash \chi \rightarrow \psi$. Since $\chi \rightarrow (\varphi \wedge \psi)$ is a tautological consequence of the set of formulas $\{\chi \rightarrow \varphi, \chi \rightarrow \psi\}$, it follows that $\Sigma \vdash \chi \rightarrow (\varphi \wedge \psi)$. Thus

$$|\chi| \leqslant |\varphi \wedge \psi|.$$

Hence $|\varphi \wedge \psi| = \inf\{|\varphi|, |\psi|\}$ as claimed. The other equality is proved similarly.

In accordance with the lattice-theoretic notation introduced in Ch. 4, §1, we write $|\varphi| \wedge |\psi|$ for $\inf\{|\varphi|, |\psi|\}$ and $|\varphi| \vee |\psi|$ for $\sup\{|\varphi|, |\psi|\}$. Thus

$$|\varphi| \wedge |\psi| = |\varphi \wedge \psi|, \qquad |\varphi| \vee |\psi| = |\varphi \vee \psi|.$$

We have, for any formulas φ, ψ, χ,

$$\Sigma \vdash ((\varphi \vee \psi) \wedge \chi) \leftrightarrow ((\varphi \wedge \chi) \vee (\psi \wedge \chi)),$$

since the formula on the right is a propositional tautology. Therefore

$$(|\varphi| \vee |\psi|) \wedge |\chi| = (|\varphi| \wedge |\chi|) \vee (|\psi| \wedge |\chi|),$$

so that one of the distributive laws holds in B. Therefore, by a result in Ch. 4, §1, the other distributive law holds as well, and so $\langle B, \leqslant \rangle$ is a distributive lattice.

We now show that $\langle B,\leqslant\rangle$ has a largest element 1 and a smallest element 0. Let χ be any tautology; then $\Sigma\vdash\chi$ and for any formula ψ we have $\Sigma\vdash\psi\to\chi$; hence $|\psi|\leqslant|\chi|$. It follows that $|\chi|$ is the largest element of $\langle B,\leqslant\rangle$. Similarly, $|\neg\chi|$ is the smallest element of $\langle B,\leqslant\rangle$. Since Σ is consistent, it is easy to see that $0\neq 1$.

If $\Sigma\vdash\varphi$, then for any formula ψ, $\Sigma\vdash\psi\to\varphi$ and so $|\psi|\leqslant|\varphi|$. Hence $|\varphi|=1$. Conversely, if $|\varphi|=1$, then, for any formula ψ, $|\psi|\leqslant|\varphi|$, so that $\Sigma\vdash\psi\to\varphi$. Choosing ψ so that $\Sigma\vdash\psi$ (e.g. take ψ to be a tautology), we obtain $\Sigma\vdash\varphi$ by modus ponens. Therefore

$$|\varphi|=1 \text{ iff } \Sigma\vdash\varphi.$$

Similarly

$$|\varphi|=0 \text{ iff } \Sigma\vdash\neg\varphi.$$

For any formula φ we have

$$\Sigma\vdash\varphi\vee\neg\varphi, \qquad \Sigma\vdash\neg(\varphi\wedge\neg\varphi),$$

since $\varphi\vee\neg\varphi$ and $\neg(\varphi\wedge\neg\varphi)$ are propositional tautologies. Hence

$$|\varphi|\vee|\neg\varphi|=|\varphi\vee\neg\varphi|=1, \qquad |\varphi|\wedge|\neg\varphi|=|\varphi\wedge\neg\varphi|=0.$$

Accordingly the complement of $|\varphi|$ in $\langle B,\leqslant\rangle$ is $|\neg\varphi|$. ∎

The Boolean algebra[1] B is often written $B(\Sigma)$ to point up its dependence on Σ. $B(\Sigma)$ is called the *Lindenbaum algebra* of Σ. The algebra[2] $B(\emptyset)$ is called the *Lindenbaum algebra* of the language $\mathscr{L}$, and is usually denoted by $B(\mathscr{L})$.

In the proof of Thm. 5.3 we saw that the mapping $|\cdot|$ behaves like a *truth valuation* on $\mathscr{L}$ except that it assigns "truth values" in the Boolean algebra $B(\Sigma)$, rather than in the 2-element algebra $\{\top,\bot\}$. (For a more general discussion of such "truth valuations", see the problems at the end of this section.) We now investigate the effect that this "truth valuation" has on the *quantifiers*.

5.4. THEOREM. *Let Σ be a consistent set of $\mathscr{L}$-sentences, and let $\mathbf{T}'$ be a set of $\mathscr{L}$-terms which contains infinitely many variables. Then for each $\mathscr{L}$-formula φ and each variable $\mathbf{v}_k$ we have, in $B(\Sigma)$,*

$$|\exists\mathbf{v}_k\varphi|=\bigvee_{\mathbf{t}\in\mathbf{T}'}|\varphi(\mathbf{v}_k/\mathbf{t})|.$$

[1] As usual we write B for $\langle B,\leqslant\rangle$.
[2] We know already (Prob. 3.1.16.) that $\emptyset$ is a consistent set of sentences of $\mathscr{L}$.

PROOF. By Thm. 3.1.11 we have, for any $\mathbf{t}\in\mathbf{T}'$,

$$\Sigma\vdash\varphi(\mathbf{v}_k/\mathbf{t})\rightarrow\exists\mathbf{v}_k\varphi,$$

so that, for any $\mathbf{t}\in\mathbf{T}'$,

$$|\varphi(\mathbf{v}_k/\mathbf{t})|\leqslant|\exists\mathbf{v}_k\varphi|.$$

Accordingly $|\exists\mathbf{v}_k\varphi|$ is an upper bound in $B(\Sigma)$ for $\{|\varphi(\mathbf{v}_k/\mathbf{t})|:\ \mathbf{t}\in\mathbf{T}'\}$.

Now suppose that $|\psi|$ is any upper bound in $B(\Sigma)$ for $\{|\varphi(\mathbf{v}_k/\mathbf{t})|:\ \mathbf{t}\in\mathbf{T}'\}$. Then, in particular, choosing q so that $\mathbf{v}_q\in\mathbf{T}'$ and $\mathbf{v}_q$ occurs in neither φ nor ψ, we have

$$\Sigma\vdash\varphi(\mathbf{v}_k/\mathbf{v}_q)\rightarrow\psi.$$

Hence, by Prob. 3.1.6,

(1) $$\Sigma\vdash\exists\mathbf{v}_q\varphi(\mathbf{v}_k/\mathbf{v}_q)\rightarrow\psi.$$

But $\exists\mathbf{v}_q\varphi(\mathbf{v}_k/\mathbf{v}_q)$ is a variant of $\exists\mathbf{v}_k\varphi$ so that, by Thm. 3.1.10, we have

(2) $$\Sigma\vdash\exists\mathbf{v}_q\varphi(\mathbf{v}_k/\mathbf{v}_q)\leftrightarrow\exists\mathbf{v}_k\varphi.$$

Now (1) and (2) give

$$\Sigma\vdash\exists\mathbf{v}_k\varphi\rightarrow\psi.$$

Hence $|\exists\mathbf{v}_k\varphi|\leqslant|\psi|$ in $B(\Sigma)$. This shows that $|\exists\mathbf{v}_k\varphi|$ is the *supremum* of $\{|\varphi(\mathbf{v}_k/\mathbf{t})|:\ \mathbf{t}\in\mathbf{T}'\}$, as required. ∎

5.5. PROBLEM. With the assumptions and notation of Thm. 5.4, show that, in $B(\Sigma)$,

$$|\forall\mathbf{v}_k\varphi|=\bigwedge_{\mathbf{t}\in\mathbf{T}'}|\varphi(\mathbf{v}_k/\mathbf{t})|.$$

5.6. PROBLEM. Let U be an ultrafilter in $B(\Sigma)$. Show that

$$\{\varphi:\ |\varphi|\in U\}$$

is a maximal consistent set of formulas (cf. Def. 3.3.4).

We now show how certain types of ultrafilter in $B(\Sigma)$ give rise to models of Σ. Our construction is similar to that employed in Ch. 2, §7.

Let Σ be a consistent set of $\mathscr{L}$-sentences and let U be an ultrafilter in $B(\Sigma)$. Define a relation E on the set $\mathbf{T}$ of all $\mathscr{L}$-terms by

$$\mathbf{t}\,E\,\mathbf{t}'\ \Leftrightarrow\ |\mathbf{t}=\mathbf{t}'|\in U.$$

It is left to the reader as an exercise to show that E is an *equivalence relation* on $\mathbf{T}$. (Using Prob. 5.6 and Thm. 3.3.7 proceed as in the proof of Lemma 2.7.2.) For each $\mathbf{t}\in\mathbf{T}$ let

$$|\mathbf{t}|=\{\mathbf{t}'\in\mathbf{T}:\ \mathbf{t}\,E\,\mathbf{t}'\}$$

be the E-class of $\mathbf{t}$, and let

$$A(U)=\{|\mathbf{t}|:\ \mathbf{t}\in\mathbf{T}\}$$

be the set of all E-classes.

Now suppose that $\mathscr{L}$ is of signature λ and that $\{\mathbf{R}_i:\ i\in I\}$ and $\{\mathbf{c}_j:\ j\in J\}$ are lists of the predicate symbols and constant symbols, respectively, of $\mathscr{L}$. For each $i\in I$ we define the $\lambda(i)$-ary relation R_i on $A(U)$ by putting

$$(5.7)\qquad \langle|\mathbf{t}_1|,\ldots,|\mathbf{t}_{\lambda(i)}|\rangle\in R_i \Leftrightarrow |\mathbf{R}_i\mathbf{t}_1\ldots\mathbf{t}_{\lambda(i)}|\in U$$

for $\mathbf{t}_1,\ldots,\mathbf{t}_{\lambda(i)}\in\mathbf{T}$.

As in the proof of Lemma 2.7.3, one verifies that (5.7) is a sound definition, i.e. the left-hand side is independent of the choice of "representatives" $\mathbf{t}_1,\ldots,\mathbf{t}_{\lambda(i)}$ for $|\mathbf{t}_1|,\ldots,|\mathbf{t}_{\lambda(i)}|$.

We now define the $\mathscr{L}$-structure $\mathfrak{A}(U)$ by

$$(5.8)\qquad \mathfrak{A}(U)=\langle A(U),\ \langle R_i\rangle_{i\in I},\ \langle|\mathbf{c}_j|\rangle_{j\in J}\rangle.$$

$\mathfrak{A}(U)$ is called the *canonical $\mathscr{L}$-structure determined by U.*

Let $\mathbf{\Gamma}(U)$ be the set of all $\mathscr{L}$-sentences $\boldsymbol{\sigma}$ such that $|\boldsymbol{\sigma}|\in U$. Then $\mathbf{\Sigma}\subseteq\mathbf{\Gamma}(U)$, for if $\boldsymbol{\sigma}\in\mathbf{\Sigma}$ then $|\boldsymbol{\sigma}|=1\in U$ in $B(\mathbf{\Sigma})$, so that $\boldsymbol{\sigma}\in\mathbf{\Gamma}(U)$.

It is natural to ask whether $\mathfrak{A}(U)$ is a model of $\mathbf{\Gamma}(U)$, or, at least, of $\mathbf{\Sigma}$. In general the answer is no, as the following example — an adaptation of the example given in Ch. 3, §3 — shows.

Let $\mathscr{L}$ be a language with equality but without any extralogical symbols. Let A be a set with two members a_0 and a_1, and let $\mathfrak{A}$ be the unique $\mathscr{L}$-structure with domain A. Let $\mathbf{\Sigma}$ consist of the single sentence $\exists\mathbf{v}_0\exists\mathbf{v}_1(\mathbf{v}_0\neq\mathbf{v}_1)$. Let a be the assignment in $\mathfrak{A}$ with constant value a_0, and put

$$U=\{|\boldsymbol{\psi}|\in B(\mathbf{\Sigma}):\ \mathfrak{A}\models_a\boldsymbol{\psi}\}.$$

Using the obvious fact that $\mathfrak{A}$ is a model of $\mathbf{\Sigma}$, it is easy to verify that U is an ultrafilter in $B(\mathbf{\Sigma})$. Also, it is clear that $\mathfrak{A}\models_a\mathbf{v}_m=\mathbf{v}_n$ fo rall $m,n\in\omega$, so that $|\mathbf{v}_m=\mathbf{v}_n|\in U$, whence $|\mathbf{v}_m|=|\mathbf{v}_n|$, and so $\mathfrak{A}(U)$ has exactly one element. It follows immediately that $\mathfrak{A}(U)$ is not a model of $\mathbf{\Sigma}$.

It is clear that the unhappy state of affairs exemplified above would not have arisen if U had satisfied the following condition: for any $\mathscr{L}$-formula $\boldsymbol{\varphi}$ and any variable $\mathbf{v}_k$,

(5.9) $$|\exists \mathbf{v}_k \boldsymbol{\varphi}| \in U \Leftrightarrow |\boldsymbol{\varphi}(\mathbf{v}_k/\mathbf{t})| \in U \quad \text{for some } \mathbf{t} \in \mathbf{T}.$$

(Recall that $\mathbf{T}$ is the set of all $\mathscr{L}$-terms.) By Thm. 5.4, (5.9) is equivalent to the condition that U respects (Ch. 4, §7) the family of joins

$$\{\bigvee_{\mathbf{t}\in\mathbf{T}} |\boldsymbol{\varphi}(\mathbf{v}_k/\mathbf{t})|: \boldsymbol{\varphi} \in \boldsymbol{\Phi}, k \in \omega\},$$

where $\boldsymbol{\Phi}$ is the set of all $\mathscr{L}$-formulas. Moreover, using Prob. 5.6 and Def. 3.3.8, it is easy to see that, if U satisfies (5.9), then $\{\boldsymbol{\varphi}: |\boldsymbol{\varphi}| \in U\}$ is a *Henkin set* in $\mathscr{L}$. We are now going to show that condition (5.9) is *sufficient* for $\mathfrak{A}(U)$ to be a model of $\boldsymbol{\Gamma}(U)$ and hence, since $\boldsymbol{\Sigma} \subseteq \boldsymbol{\Gamma}(U)$, also of $\boldsymbol{\Sigma}$.

Let us call an ultrafilter U in $B(\boldsymbol{\Sigma})$ *perfect* if it respects the family of joins

$$\{\bigvee_{\mathbf{t}\in\mathbf{T}} |\boldsymbol{\varphi}(\mathbf{v}_k/\mathbf{t})|: \boldsymbol{\varphi} \in \boldsymbol{\Phi}, k \in \omega\}.$$

Then we have:

5.10. THEOREM. *Let U be a perfect ultrafilter in $B(\boldsymbol{\Sigma})$. Then:*

(i) *For each $\mathscr{L}$-formula $\boldsymbol{\varphi}$ whose free variables are all among $\mathbf{v}_0, \ldots, \mathbf{v}_n$, and each $(n+1)$-tuple $\mathbf{t}_0, \ldots, \mathbf{t}_n$ of members of $\mathbf{T}$, we have*

$$\mathfrak{A}(U) \models \boldsymbol{\varphi}\,[|\mathbf{t}_0|, \ldots, |\mathbf{t}_n|] \Leftrightarrow |\boldsymbol{\varphi}(\mathbf{v}_0/\mathbf{t}_0, \ldots, \mathbf{v}_n/\mathbf{t}_n)| \in U;$$

(ii) *$\mathfrak{A}(U)$ is a model of $\boldsymbol{\Gamma}(U)$ (and hence also of $\boldsymbol{\Sigma}$.)*

PROOF. Notice first that (ii) is an immediate consequence of (i). For if $\boldsymbol{\sigma} \in \boldsymbol{\Gamma}(U)$, then $|\boldsymbol{\sigma}| \in U$, so that, by (i), $\mathfrak{A}(U) \models \boldsymbol{\sigma}$.

It thus remains to prove (i). Put

$$\boldsymbol{\Psi} = \{\boldsymbol{\varphi}: |\boldsymbol{\varphi}| \in U\}.$$

Then, as we remarked a little earlier on, $\boldsymbol{\Psi}$ is a *Henkin set* — hence a *Hintikka set* — in $\mathscr{L}$. We now observe that the $\mathscr{L}$-valuation σ defined in Ch. 2, §7 has the following properties, all of which are verifiable by inspection:

(1) its universe is $A(U)$,

(2) $\mathbf{R}_i^\sigma = R_i$ for $i \in I$;

(3) $\mathbf{t}^\sigma = |\mathbf{t}|$ for $\mathbf{t} \in \mathbf{T}$.

Since $\mathbf{\Psi}$ is a Henkin set, it follows immediately from Thm. 2.7.5 (a) and (b) that, for any formula φ,

$$\varphi^\sigma = \top \Leftrightarrow \varphi \in \mathbf{\Psi}.$$

Hence, if $\mathbf{t}_0,\ldots,\mathbf{t}_n \in \mathbf{T}$, we obtain, using Thm. 2.3.15,

$$\varphi^{\sigma(\mathbf{v}_0/\mathbf{t}_0^\sigma)\ldots(\mathbf{v}_n/\mathbf{t}_n^\sigma)} = \top \Leftrightarrow \varphi(\mathbf{v}_0/\mathbf{t}_0,\ldots,\mathbf{v}_n/\mathbf{t}_n) \in \mathbf{\Psi}$$

i.e.

$$(4) \qquad \varphi^{\sigma(\mathbf{v}_0/\mathbf{t}_0^\sigma)\ldots(\mathbf{v}_n/\mathbf{t}_n^\sigma)} = \top \Leftrightarrow |\varphi(\mathbf{v}_0/\mathbf{t}_0,\ldots,\mathbf{v}_n/\mathbf{t}_n)| \in U.$$

Now suppose that the free variables of φ are all among $\mathbf{v}_0,\ldots,\mathbf{v}_n$. Then (1)—(3) show that the assertion

$$\varphi^{\sigma(\mathbf{v}_0/\mathbf{t}_0^\sigma)\ldots(\mathbf{v}_n/\mathbf{t}_n^\sigma)} = \top$$

is merely an alternative way of expressing the assertion

$$\mathfrak{A}(U) \models \varphi\ [|\mathbf{t}_0|,\ldots,|\mathbf{t}_n|].$$

This fact, together with the equivalence (4) yields the required conclusion. ∎

Theorem 5.10 shows that perfect ultrafilters in $B(\Sigma)$ give rise to canonical models of Σ. We are now going to establish a converse.

Let Q be an equivalence relation on $\mathbf{T}$; for each $\mathbf{t} \in \mathbf{T}$ let $\mathbf{t}/Q$ be the Q-class of $\mathbf{t}$, and let $\mathbf{T}/Q$ be the set of all Q-classes of members of $\mathbf{T}$. An $\mathscr{L}$-structure of the form

$$\mathfrak{A} = \langle \mathbf{T}/Q, \langle R_i \rangle_{i \in I}, \langle \mathbf{c}_j/Q \rangle_{j \in J} \rangle$$

is called a *basic* $\mathscr{L}$-structure.

Clearly $\mathfrak{A}(U)$ is a basic $\mathscr{L}$-structure for each ultrafilter U in $B(\Sigma)$.

5.11. THEOREM. *There is a natural one–one correspondence between the class of perfect ultrafilters in* $B(\Sigma)$ *and the class of basic* $\mathscr{L}$*-structures which are models of* Σ.

PROOF. Let $\mathscr{U}$ be the class of perfect ultrafilters in $B(\Sigma)$ and let $\mathscr{M}$ be the class of basic $\mathscr{L}$-structures which are models of Σ. We define two mappings f and g with domains $\mathscr{U}$ and $\mathscr{M}$, respectively, as follows. For each $U \in \mathscr{U}$, $f(U)$ is $\mathfrak{A}(U)$, the canonical $\mathscr{L}$-structure determined by U. For each $\mathfrak{A} \in \mathscr{M}$, there is an equivalence relation Q on $\mathbf{T}$ such that the domain of $\mathfrak{A}$ is $\mathbf{T}/Q$. Let $\mathfrak{v} = \langle \mathbf{v}_0, \mathbf{v}_1, \ldots \rangle$ be the sequence of all variables of $\mathscr{L}$, and let $\mathfrak{v}/Q = \langle \mathbf{v}_0/Q, \mathbf{v}_1/Q, \ldots \rangle$. Then we put

$$g(\mathfrak{A}) = \{|\varphi| : \mathfrak{A} \models_{\mathfrak{v}/Q} \varphi\}.$$

By construction, $f(U)$ is a basic $\mathscr{L}$-structure for each $U \in \mathscr{U}$ and by Thm. 5.10(ii) it is a model of Σ. Hence $f(U) \in \mathscr{M}$, so that f maps $\mathscr{U}$ into $\mathscr{M}$.

For each $\mathfrak{A} \in \mathscr{M}$, it is easy to see that $g(\mathfrak{A})$ is an ultrafilter in $B(\Sigma)$. We claim that $g(\mathfrak{A}) \in \mathscr{U}$, i.e. that $g(\mathfrak{A})$ is perfect. To prove this, suppose that $|\exists \mathbf{v}_k \varphi| \in g(\mathfrak{A})$. Then, by the definition of $g(\mathfrak{A})$, we have $\mathfrak{A} \models_{\mathfrak{v}/Q} \exists \mathbf{v}_k \varphi$, so that, for some $\mathbf{t} \in \mathbf{T}$,

$$\mathfrak{A} \models_{\mathfrak{v}/Q(k|\mathbf{t}/Q)} \varphi.$$

But this means (cf. Thm. 2.3.10) that $\mathfrak{A} \models_{\mathfrak{v}/Q} \varphi(\mathbf{v}_k/\mathbf{t})$ i.e. that $|\varphi(\mathbf{v}_k/\mathbf{t})| \in g(\mathfrak{A})$. Accordingly $g(\mathfrak{A})$ is perfect.

We now claim that $f \circ g$ is the identity on $\mathscr{M}$ and that $g \circ f$ is the identity on $\mathscr{U}$, from which it immediately follows that f is a bijection of $\mathscr{U}$ onto $\mathscr{M}$. The first of these assertions is left as a simple exercise to the reader. To show that $g \circ f$ is the identity on $\mathscr{U}$, observe that, if $U \in \mathscr{U}$, then, by the definitions of f and g we have for any $\varphi \in \Phi$, putting $|\mathfrak{v}| = \langle |\mathbf{v}_0|, |\mathbf{v}_1|, \ldots \rangle$,

$$|\varphi| \in g(f(U)) \Leftrightarrow \mathfrak{A}(U) \models_{|\mathfrak{v}|} \varphi.$$

If the free variables of φ are all among $\mathbf{v}_0, \ldots, \mathbf{v}_n$, then, since U is perfect, we have, by Thm. 5.10(i),

$$\mathfrak{A}(U) \models_{|\mathfrak{v}|} \varphi \Leftrightarrow |\varphi(\mathbf{v}_0/\mathbf{v}_0, \ldots, \mathbf{v}_n/\mathbf{v}_n)| \in U.$$

But clearly $\varphi(\mathbf{v}_0/\mathbf{v}_0, \ldots, \mathbf{v}_n/\mathbf{v}_n) = \varphi$, so that

$$\mathfrak{A}(U) \models_{|\mathfrak{v}|} \varphi \Leftrightarrow |\varphi| \in U.$$

Putting these equivalences together, we obtain

$$|\varphi| \in g(f(U)) \Leftrightarrow |\varphi| \in U,$$

so that $g(f(U)) = U$. Thus $g \circ f$ is the identity on $\mathscr{U}$, and the proof is finished. ∎

5.12. PROBLEM. Recall that we write $\mathsf{S}B(\Sigma)$ for the Stone space of $B(\Sigma)$. Define h: $\mathsf{S}B(\Sigma) \to \mathsf{S}B(\Sigma)$ by putting $h(U) = g\mathfrak{A}(U)$ for each $U \in \mathsf{S}B(\Sigma)$, where g is defined as in Thm. 5.11. Show that U is a perfect ultrafilter in $B(\Sigma)$ iff $h(U) = U$; in other words, *the perfect ultrafilters in* $B(\Sigma)$ *are precisely the fixed points of the mapping* h.

5.13. EXAMPLES. (i) Let $\mathscr{L}$ be a language with only *countably* many constant symbols but with an *uncountable* family of unary predicate symbols $\{\mathbf{P}_i: i\in I\}$. Let Σ be the set of $\mathscr{L}$-sentences

$$\{\exists \mathbf{v}_0 \mathbf{P}_i \mathbf{v}_0: i\in I\} \cup \{\forall \mathbf{v}_0 \neg(\mathbf{P}_i\mathbf{v}_0 \wedge \mathbf{P}_j\mathbf{v}_0): i,j\in I, i\neq j\};$$

then clearly any model of Σ must be *uncountable*. On the other hand, each basic $\mathscr{L}$-structure is *countable* since $\mathscr{L}$ has only countably many terms. Accordingly no basic $\mathscr{L}$-structure is a model of Σ and hence, by 5.11, there are *no* perfect ultrafilters in $B(\Sigma)$.

(ii) Suppose that $\mathscr{L}$ is countable and let Σ be a consistent set of $\mathscr{L}$-sentences. Then the family of joins

$$\{\bigvee_{\mathbf{t}\in \mathbf{T}} |\boldsymbol{\varphi}(\mathbf{v}_k/\mathbf{t})|: \boldsymbol{\varphi}\in\boldsymbol{\Phi}, k\in\omega\}$$

is countable and so, by the Rasiowa–Sikorski theorem 4.7.3, there is a perfect ultrafilter U in $B(\Sigma)$. $\mathfrak{A}(U)$ is then a model of Σ. This gives a new proof of the assertion that every countable consistent set of sentences has a model, and hence of the Gödel Completeness Theorem for sentences. (This proof is due to RASIOWA and SIKORSKI [1951].)

Throughout Probs. 5.14–5.18 we make the following assumptions. $\mathscr{L}$ is a first-order language whose only extralogical symbol is a binary predicate $\mathbf{R}$. (This assumption is not essential, we make it solely for the sake of simplicity.) If $\mathscr{L}'$ is any extension of $\mathscr{L}$ obtained by adding constants, $\boldsymbol{\varphi}$ is an $\mathscr{L}'$-formula all of whose free variables are among $\mathbf{v}_0,\ldots,\mathbf{v}_n$, and $\mathbf{a}_0,\ldots,\mathbf{a}_n$ are constants of $\mathscr{L}'$, we write $\boldsymbol{\varphi}(\mathbf{a}_0,\ldots,\mathbf{a}_n)$ for $\boldsymbol{\varphi}(\mathbf{v}_0/\mathbf{a}_0,\ldots,\mathbf{v}_n/\mathbf{a}_n)$. Also, if $\boldsymbol{\varphi}$ is an $\mathscr{L}'$-formula with at most the variable $\mathbf{x}$ free, and $\mathbf{a}$ is a constant of $\mathscr{L}'$, we write $\boldsymbol{\varphi}(\mathbf{a})$ for $\boldsymbol{\varphi}(\mathbf{x}/\mathbf{a})$.

*5.14. PROBLEM. Let B be a complete Boolean algebra. A *B-valued $\mathscr{L}$-pre-structure* is a triple $\mathfrak{A}=\langle A,E,R\rangle$ where A is a non-empty set and E,R are mappings of $A\times A$ into B. Let $\mathscr{L}(A)$ be the extension of $\mathscr{L}$ obtained by adding a new constant symbol $\mathbf{a}$ for each $a\in A$. Define the mapping $\|\cdot\|_{\mathfrak{A}}$ from the set of all $\mathscr{L}(A)$-sentences into B by induction as follows:

$$\|\mathbf{a}=\mathbf{b}\|_{\mathfrak{A}}=E(a,b), \qquad \|\mathbf{Rab}\|_{\mathfrak{A}}=R(a,b),$$

$$\|\boldsymbol{\sigma}\wedge\boldsymbol{\sigma}'\|_{\mathfrak{A}}=\|\boldsymbol{\sigma}\|_{\mathfrak{A}}\wedge\|\boldsymbol{\sigma}'\|_{\mathfrak{A}}, \qquad \|\neg\boldsymbol{\sigma}\|_{\mathfrak{A}}=\|\boldsymbol{\sigma}\|_{\mathfrak{A}}{}^*,$$

$$\|\exists\mathbf{x}\boldsymbol{\varphi}\|_{\mathfrak{A}}=\bigvee_{a\in A}\|\boldsymbol{\varphi}(\mathbf{x}/\mathbf{a})\|_{\mathfrak{A}}.$$

$\mathfrak{A}$ is called a *B*-valued $\mathscr{L}$-*structure* if for all $a,b,c,d \in A$ we have

$$\|\mathbf{a}=\mathbf{a}\|_{\mathfrak{A}}=1;$$

$$\|\mathbf{a}=\mathbf{b}\to\mathbf{c}=\mathbf{d}\to\mathbf{Rac}\to\mathbf{Rbd}\|_{\mathfrak{A}}=1;$$

$$\|\mathbf{a}=\mathbf{b}\to\mathbf{c}=\mathbf{d}\to\mathbf{a}=\mathbf{c}\to\mathbf{b}=\mathbf{d}\|_{\mathfrak{A}}=1.$$

Let φ be an $\mathscr{L}$-formula whose free variables are all among $\mathbf{v}_0,\ldots,\mathbf{v}_n$. φ is said to be *B-valid* if for any *B*-valued $\mathscr{L}$-structure $\mathfrak{A}$ and any $a_0,\ldots$ $\ldots,a_n \in A$ we have $\|\varphi(\mathbf{a}_0,\ldots,\mathbf{a}_n)\|_{\mathfrak{A}}=1$. Show that the following are equivalent:

(i) $\vdash\varphi$;

(ii) φ is 2-valid;

(iii) φ is *B*-valid for some complete Boolean algebra *B*;

(iv) φ is *B*-valid for every complete Boolean algebra *B*.

(For (i) $\Leftrightarrow$ (ii), observe that a 2-valued $\mathscr{L}$-structure is essentially just an $\mathscr{L}$-structure and apply the Completeness Theorem. For (i) $\Rightarrow$ (iv), argue as in the proof of Thm. 3.3.12. For (iii) $\Rightarrow$ (ii), notice that each 2-valued structure is a *B*-valued structure.)

*5.15. PROBLEM. Let $\mathfrak{A}=\langle A,R\rangle$ be an $\mathscr{L}$-structure, and let *B* be a complete Boolean algebra. Put

$$A^{(B)}=\{f\in B^A\colon \text{for all } a,b\in A,\ a\neq b\Rightarrow f(a)\wedge f(b)=0 \text{ and } \bigvee_{a\in A} f(a)=1\},$$

and define the mappings E, R' from $A^{(B)}\times A^{(B)}$ into *B* by setting, for $f,g\in A^{(B)}$,

$$E(f,g)=\bigvee_{a\in A} f(a)\wedge g(a),$$

$$R'(f,g)=\bigvee\{f(a)\wedge g(b)\colon \langle a,b\rangle\in R\}.$$

Then, as is easily verified, $\mathfrak{A}^{(B)}=\langle A^{(B)},E,R'\rangle$ is a *B*-valued $\mathscr{L}$-structure; it is called the *B-extension of* $\mathfrak{A}$. For each sentence σ of $\mathscr{L}'=\mathscr{L}(\mathfrak{A}^{(B)})$ write $\|\sigma\|$ for $\|\sigma\|_{\mathfrak{A}^{(B)}}$ (Prob. 5.14).

(i) If $f,g\in A^{(B)}$, show that $\|\mathbf{f}=\mathbf{g}\|=1$ iff $f=g$.

(ii) Let φ be an $\mathscr{L}'$-formula with at most the variable $\mathbf{x}$ free. Show that, for any $f,g\in A^{(B)}$,

$$\|\mathbf{f}=\mathbf{g}\|\wedge\|\varphi(\mathbf{f})\|\leqslant\|\varphi(\mathbf{g})\|.$$

(iii) Let φ be an $\mathscr{L}$-formula whose free variables are all among $\mathbf{v}_0,\ldots,\mathbf{v}_n$, and let $f_0,\ldots,f_n \in A^{(B)}$. Show that[1]

$$\|\boldsymbol{\varphi}(\mathbf{f}_0,\ldots,\mathbf{f}_n)\| = \bigvee\Big\{\bigwedge_{i=0}^{n} f_i(a_i):\ \mathfrak{A} \models \boldsymbol{\varphi}\,[a_0,\ldots,a_n],\ a_0,\ldots,a_n \in A\Big\}.$$

(Argue by induction on deg $\boldsymbol{\varphi}$, using Prob. 4.2.6.)

(iv) For each $a \in A$ define $\hat{a} \in A^{(B)}$ by putting

$$\hat{a}(x) = \begin{cases} 1 & \text{if } a = x, \\ 0 & \text{if } a \neq x. \end{cases}$$

Show that[2], for any $\mathscr{L}$-formula whose free variables are all among $\mathbf{v}_0,\ldots,\mathbf{v}_n$, and any $a_0,\ldots,a_n \in A$,

$$\mathfrak{A} \models \boldsymbol{\varphi}\,[a_0,\ldots,a_n] \;\Leftrightarrow\; \|\boldsymbol{\varphi}(\hat{\mathbf{a}}_0,\ldots,\hat{\mathbf{a}}_n)\| = 1.$$

(Use (iii).) Thus, in an obvious sense, the mapping $a \mapsto \hat{a}$ is a *B-valued elementary embedding* of $\mathfrak{A}$ into $\mathfrak{A}^{(B)}$.

*5.16. PROBLEM. We continue to employ the assumptions and notation of Prob. 5.15.

(i) Let $\{b_i:\ i \in I\} \subseteq B$ satisfy $b_i \wedge b_j = 0$ for $i \neq j$, and let $\{f_i:\ i \in I\} \subseteq A^{(B)}$. Show that there exists an $f \in A^{(B)}$ such that $\|\mathbf{f} = \mathbf{f}_i\| \geqslant b_i$ for all $i \in I$. (First show that without loss of generality one can assume $\bigvee_{i \in I} b_i = 1$. Then define $f \in B^A$ by putting

$$f(a) = \bigvee_{i \in I} b_i \wedge f_i(a)$$

for $a \in A$. Show that f meets the requirements.)

(ii) Let $\boldsymbol{\varphi}$ be an $\mathscr{L}'$-formula with at most the variable $\mathbf{x}$ free. Show that there exists an $f \in A^{(B)}$ such that[3]

$$\|\exists \mathbf{x} \boldsymbol{\varphi}\| = \|\boldsymbol{\varphi}(\mathbf{f})\|.$$

(Let $A^{(B)}$ be well-ordered in the form $\{f_\xi:\ \xi < \alpha\}$; put

$$b_\xi = \|\boldsymbol{\varphi}(\mathbf{f}_\xi)\| \wedge [\bigvee_{\eta < \xi} \|\boldsymbol{\varphi}(\mathbf{f}_\eta)\|]^*.$$

Observe that

$$b_\xi \wedge b_\eta = 0 \quad \text{for} \quad \xi \neq \eta,$$

$$\bigvee_{\xi < \alpha} b_\xi = \|\exists \mathbf{x} \boldsymbol{\varphi}\|.$$

[1] Compare Thm. 3.1.

[2] Compare Prob. 3.2.

[3] Compare Lemma 3.4.

By (i), pick $f \in A^{(B)}$ with $\|\mathbf{f}=\mathbf{f}_\xi\| \geqslant b_\xi$ for all $\xi < \alpha$. Using 5.15 (ii), show that $\|\boldsymbol{\varphi}(\mathbf{f})\| \geqslant b_\xi$ for all $\xi < \alpha$.)

*5.17. Problem. We continue to employ the assumptions and notation of Prob. 5.15. Let F be an ultrafilter in B. Define the relation $\sim_F$ on $A^{(B)}$ by

$$f \sim_F g \Leftrightarrow \|\mathbf{f}=\mathbf{g}\| \in F;$$

then $\sim_F$ is an equivalence relation and for each $f \in A^{(B)}$ we let f^F be the $\sim_F$-class of f. Put

$$A^{(B)}/F = \{f^F : f \in A^{(B)}\}.$$

Define the relation R^F on $A^{(B)}/F$ by

$$\langle f^F, g^F \rangle \in R^F \Leftrightarrow \|\mathbf{Rfg}\| \in F.$$

The $\mathscr{L}$-structure $\mathfrak{A}^{(B)}/F = \langle A^{(B)}/F, R^F \rangle$ is called the (B,F)-(Boolean) *ultrapower* of $\mathfrak{A}$.

(i) Let $\boldsymbol{\varphi}$ be an $\mathscr{L}$-formula whose free variables are all among $\mathbf{v}_0, \ldots, \mathbf{v}_n$, and let $f_0, \ldots, f_n \in A^{(B)}$. Show that[1]

$$\mathfrak{A}^{(B)}/F \models \boldsymbol{\varphi}\,[f_0^F, \ldots, f_n^F] \Leftrightarrow \|\boldsymbol{\varphi}(\mathbf{f}_0, \ldots, \mathbf{f}_n)\| \in F.$$

(Argue by induction on deg $\boldsymbol{\varphi}$, using 5.16(ii) to handle the existential case.)

(ii) Show that the mapping $a \to \hat{a}^F$ (defined in (iv)) is an elementary embedding of $\mathfrak{A}$ into $\mathfrak{A}^{(B)}/F$. (Use (i) and 5.15(iv)).

*5.18. Problem. (The assumptions and notation of Probs. 5.15–5.17 are still in force.) Let I be a non-empty set, let B be the power set algebra $\mathsf{P}I$, and let $\mathscr{F}$ be an ultrafilter in B. Define the mapping $j\colon A^I \to (\mathsf{P}I)^A$ by putting, for each $f \in A^I$ and each $a \in A$, $j(f)(a) = f^{-1}(a)$.

(i) Show that j is a bijection of A^I onto $A^{(\mathsf{P}I)}$.

(ii) Let $\boldsymbol{\varphi}$ be an $\mathscr{L}$-formula whose free variables are all among $\mathbf{v}_0, \ldots, \mathbf{v}_n$. Show that,[2] for any $f_0, \ldots, f_n \in A^I$,

$$\{i \in I : \mathfrak{A} \models \boldsymbol{\varphi}\,[f_0(i), \ldots, f_n(i)]\} = \|\boldsymbol{\varphi}(\mathbf{j}(\mathbf{f}_0), \ldots, \mathbf{j}(\mathbf{f}_n))\|.$$

(Use 5.15 (iii).)

(iii) Show that the mapping $f/\mathscr{F} \mapsto (j(f))^{\mathscr{F}}$ is an isomorphism of the ultrapower $\mathfrak{A}^I/\mathscr{F}$ onto the Boolean ultrapower $\mathfrak{A}^{(\mathsf{P}I)}/\mathscr{F}$. (Use (ii).) Thus, *the ultrapower is a special case of the Boolean ultrapower.*

[1] Compare Thm. 3.6.
[2] Compare Thm. 3.1.

§ 6. Element types and $\aleph_0$-categoricity

So far in this chapter our main concern has been with the properties of structures determined by first-order *sentences*. In this section we turn our attention to the role played by formulas with free variables.

Throughout this section we shall assume thet $\mathscr{L}$ is a fixed *countable* first-order language. We write $\mathbf{\Phi}$ for the set of all $\mathscr{L}$-formulas and for each $n \in \omega$ we write $\mathbf{\Phi}_n$ for the set of all $\mathscr{L}$-formulas whose free variables are all among $\mathbf{v}_0,\ldots,\mathbf{v}_{n-1}$. Thus $\mathbf{\Phi}_0$ is the set of all $\mathscr{L}$*-sentences*. If $\mathbf{\Sigma}$ is a consistent set of $\mathscr{L}$-sentences, we put in $B(\mathbf{\Sigma})$

$$B_n(\mathbf{\Sigma}) = \{|\boldsymbol{\varphi}|: \boldsymbol{\varphi} \in \mathbf{\Phi}_n\},$$

where $|\boldsymbol{\varphi}|$ is the $\approx$-class of $\boldsymbol{\varphi}$ (see §5). We also write B_n for $B_n(\emptyset)$. It is clear that $B_n(\mathbf{\Sigma})$ is a *subalgebra* of the Lindenbaum algebra $B(\mathbf{\Sigma})$. B_0 is called the *Lindenbaum algebra of sentences of* $\mathscr{L}$.

6.1. Problem. Let $\mathbf{\Sigma}$ be a consistent theory in $\mathscr{L}$, and put in $B(\emptyset)$

$$U(\mathbf{\Sigma}) = \{|\boldsymbol{\sigma}|: \boldsymbol{\sigma} \in \mathbf{\Sigma}\}.$$

Show that:

(i) $U(\mathbf{\Sigma})$ is a filter in B_0;

(ii) $\mathbf{\Sigma}$ is finitely axiomatizable (Prob. 4.9) iff $U(\mathbf{\Sigma})$ is a principal filter in B_0;

(iii) $\mathbf{\Sigma}$ is complete iff $U(\mathbf{\Sigma})$ is an ultrafilter in B_0.

Let $\mathbf{\Delta} \subseteq \mathbf{\Phi}_n$. An $\mathscr{L}$-structure $\mathfrak{A}$ is said to *realize* $\mathbf{\Delta}$ if there is a finite sequence $\langle a_0,\ldots,a_{n-1}\rangle$ of members of A such that $\mathfrak{A} \models \boldsymbol{\varphi}\,[a_0,\ldots,a_{n-1}]$ for all $\boldsymbol{\varphi} \in \mathbf{\Delta}$. $\mathfrak{A}$ is said to *omit* $\mathbf{\Delta}$ if it does not realize $\mathbf{\Delta}$.

6.2. Examples. (i) Let $\mathfrak{N}$ be the structure $\langle \omega,+,\times,0,1\rangle$, and let $\mathbf{\Delta}$ be the set of formulas

$$\{\mathbf{0} \neq \mathbf{v}_0,\ \mathbf{0}+\mathbf{1} \neq \mathbf{v}_0,\ \mathbf{0}+\mathbf{1}+\mathbf{1} \neq \mathbf{v}_0,\ldots\}$$

in the language for $\mathfrak{N}$. Then up to isomorphism $\mathfrak{N}$ is the unique model of $\mathbf{Th}(\mathfrak{N})$ which omits $\mathbf{\Delta}$.

(ii) Let $\mathbf{\Delta}$ be the set of formulas

$$\{(\mathbf{v}_0 \neq \mathbf{0} \wedge p\mathbf{v}_0 \neq \mathbf{0}): p \text{ a prime}\}$$

in the language for fields (§4). Then the fields which omit $\mathbf{\Delta}$ are precisely those of finite non-zero characteristic. (We know from Prob. 4.9(ii) that this is not a first-order property.)

(iii) Let $\mathfrak{F}$ be an ordered field (Prob. 4.10), and let Δ be the set of formulas

$$\{1<v_0, 1+1<v_0, 1+1+1<v_0, \ldots\}$$

in the language for $\mathfrak{F}$. Then $\mathfrak{F}$ omits Δ iff $\mathfrak{F}$ is archimedean. (We know from Prob. 4.10 that this is not a first-order property.)

Let Σ be a consistent set of $\mathscr{L}$-sentences, and let $\mathfrak{A}$ be a model of Σ. For each n-tuple $\mathfrak{a}=\langle a_0,\ldots,a_{n-1}\rangle$ of elements of A, the *type* of $\mathfrak{a}$ in $\mathfrak{A}$ is the set

$$\Phi(\mathfrak{A},\mathfrak{a})=\{\varphi\in\Phi_n:\ \mathfrak{A}\models\varphi\,[a_0,\ldots,a_{n-1}]\}.$$

It is clear that $\Phi(\mathfrak{A},\mathfrak{a})$ is a maximal consistent subset of Φ_n.

In $B_n(\Sigma)$ we define

$$U(\mathfrak{A},\mathfrak{a})=\{|\varphi|:\ \varphi\in\Phi(\mathfrak{A},\mathfrak{a})\}.$$

It is easy to verify that $U(\mathfrak{A},\mathfrak{a})$ is an *ultrafilter* in $B_n(\Sigma)$. Conversely, we have:

6.3. THEOREM. *For each ultrafilter U in $B_n(\Sigma)$ there is a countable model $\mathfrak{A}$ of Σ and an n-tuple $\mathfrak{a}=\langle a_0,\ldots,a_{n-1}\rangle$ of elements of A such that $U=U(\mathfrak{A},\mathfrak{a})$.*

PROOF. Let $c_0,\ldots,c_{n-1}$ be new constant symbols (i.e. not already in $\mathscr{L}$). We claim that each finite subset of the set of sentences

$$\Sigma'=\Sigma\cup\{\varphi(v_0/c_0,\ldots,v_{n-1}/c_{n-1}):\ \varphi\in\Phi_n,\ |\varphi|\in U\}$$

has a model. For let $\{|\varphi_1|,\ldots,|\varphi_m|\}$ be a finite subset of U, where $\varphi_1,\ldots,\varphi_m\in\Phi_n$. Then since U has the finite meet property, we have $|\varphi_1|\wedge\ldots\wedge|\varphi_m|\neq 0$. Putting

$$\varphi=\varphi_1\wedge\ldots\wedge\varphi_m,$$

it follows that $|\varphi|\neq 0$, so that $\Sigma\nvdash\neg\varphi$. Hence $\Sigma\nvdash\forall v_0\ldots\forall v_{n-1}\neg\varphi$, so that $\Sigma\cup\{\exists v_0\ldots\exists v_{n-1}\varphi\}$ is consistent, and accordingly has a model $\mathfrak{B}$. Let $\langle b_0,\ldots,b_{n-1}\rangle=\mathfrak{b}$ be an n-tuple of members of B such that

$$\mathfrak{B}\models\varphi\,[b_0,\ldots,b_{n-1}].$$

Then clearly $(\mathfrak{B},\mathfrak{b})$ is a model of

$$\Sigma\cup\{\varphi_i(v_0/c_0,\ldots,v_{n-1}/c_{n-1}):\ i=1,\ldots,m\}.$$

This shows that each finite subset of Σ' has a model, and so, by the Compactness Theorem, Σ' has a model $(\mathfrak{A},\mathfrak{a})$, where $\mathfrak{A}$ is an $\mathscr{L}$-structure and $\mathfrak{a}=\langle a_0,\ldots,a_{n-1}\rangle$ is an n-tuple of elements of A. By the Downward

Löwenheim–Skolem Theorem we may take $\mathfrak{A}$ to be countable. By construction we have $U \subseteq U(\mathfrak{A},\mathfrak{a})$, and since U is an ultrafilter we have $U = U(\mathfrak{A},\mathfrak{a})$. ∎

Let us call a subset of $B_n(\Sigma)$ of the form $U(\mathfrak{A},\mathfrak{a})$ a *Σ-reduced n-(element) type* of $\mathfrak{A}$. Then, by Thm. 6.3, the Σ-reduced n-types of (countable) models of Σ are precisely the ultrafilters in $B_n(\Sigma)$, so that (Ch. 4, §4) we have:

6.4. COROLLARY. *The set of Σ-reduced n-types of (countable) models of Σ is the (set of points of the) Stone space of $B_n(\Sigma)$.* ∎

6.5. PROBLEM. Show, that, if $\mathfrak{A}$ and $\mathfrak{B}$ are models of Σ and $\mathfrak{a},\mathfrak{b}$ are n-tuples of elements of A, B, respectively, then

$$U(\mathfrak{A},\mathfrak{a}) = U(\mathfrak{B},\mathfrak{b}) \Leftrightarrow (\mathfrak{A},\mathfrak{a}) \equiv (\mathfrak{B},\mathfrak{b}).$$

6.6. PROBLEM. Let $\mathscr{X}$ be the class of all structures of the form $(\mathfrak{A},\mathfrak{a})$, where $\mathfrak{A}$ is an $\mathscr{L}$-structure which is a model of Σ and $\mathfrak{a}$ is an n-tuple of members of A. For each $(\mathfrak{A},\mathfrak{a}) \in \mathscr{X}$ let

$$(\mathfrak{A},\mathfrak{a})^* = \{(\mathfrak{B},\mathfrak{b}) : (\mathfrak{B},\mathfrak{b}) \in \mathscr{X} \text{ and } (\mathfrak{A},\mathfrak{a}) \equiv (\mathfrak{B},\mathfrak{b})\}.$$

Let

$$\mathscr{X}^* = \{(\mathfrak{A},\mathfrak{a})^* : (\mathfrak{A},\mathfrak{a}) \in \mathscr{X}\}.$$

For each $\varphi \in \mathbf{\Phi}_n$ let

$$\mathsf{M}(\varphi) = \{(\mathfrak{A},\mathfrak{a})^* : \mathfrak{A} \models \varphi\,[a_0,\ldots,a_{n-1}], \text{ where } \mathfrak{a} = \langle a_0,\ldots,a_{n-1}\rangle\}.$$

Show that the $\mathsf{M}(\varphi)$ form a base for a topology on $\mathscr{X}^*$, and that with this topology $\mathscr{X}^*$ is (homeomorphic to) the Stone space of $B_n(\Sigma)$. (Use 6.3, 6.4 and 6.5).

We now establish a criterion for the existence of a countable model of Σ *realizing* a given subset of $\mathbf{\Phi}_n$.

6.7. THEOREM. *Let $\Delta \subseteq \mathbf{\Phi}_n$. Then the following conditions are equivalent:*

(i) *Σ has a countable model realizing Δ;*

(ii) *$\{|\varphi| : \varphi \in \Delta\}$ has the finite meet property.*

PROOF. (i)$\Rightarrow$(ii). Let $\mathfrak{A}$ be a model of Σ realizing Δ. Then there is $\mathfrak{a} = \langle a_0,\ldots,a_{n-1}\rangle \in A^n$ such that $\mathfrak{A} \models \varphi\,[a_0,\ldots,a_{n-1}]$ for every $\varphi \in \Delta$. It follows that

$$\{|\varphi| : \varphi \in \Delta\} \subseteq U(\mathfrak{A},\mathfrak{a}).$$

Since $U(\mathfrak{A},\mathfrak{a})$ has the finite meet property, so does $\{|\varphi| : \varphi \in \Delta\}$.

(ii)$\Rightarrow$(i). If $\{|\varphi| : \varphi \in \Delta\}$ has the finite meet property, then it is included in an ultrafilter U in $B_n(\Sigma)$. By Thm. 6.3, there is a countable model $\mathfrak{A}$ of Σ and an n-tuple $\mathfrak{a}$ of members of A such that $U = U(\mathfrak{A},\mathfrak{a})$. Clearly $\mathfrak{A}$ realizes Δ. ∎

Our next result deals with the *omission* of subsets of Φ. For each set Σ of $\mathscr{L}$-sentences we put

$$\Sigma^c = \{\sigma \in \Phi_0 : \Sigma \vdash \sigma\}.$$

Notice that Σ^c is a theory.

6.8. THEOREM. *Let Σ be a consistent set of $\mathscr{L}$-sentences and let $\Delta \subseteq \Phi_n$. Consider the conditions*

(i) *Σ has a countable model which omits Δ;*

(ii) *$[\Delta] = \{|\varphi| : \varphi \in \Delta\}$ is not included in any principal filter in $B_n(\Sigma)$.*

Then (ii)$\Rightarrow$(i). *Moreover, if Σ^c is complete, then the conditions are equivalent.*

PROOF. We first prove that (i)$\Rightarrow$(ii) when Σ^c is *complete*. Suppose (ii) fails. Then there is a formula ψ in Φ_n such that $|\psi| \neq 0$ and

$$[\Delta] \subseteq \{|\varphi| : |\psi| \leqslant |\varphi|\}.$$

Since $|\psi| \neq 0$, we have $\Sigma \nvdash \neg\psi$, so that $\Sigma \nvdash \forall v_0 \ldots \forall v_{n-1} \neg\psi$, hence

$$\Sigma \nvdash \neg\exists v_0 \ldots \exists v_{n-1} \psi.$$

Since Σ^c is complete, we must have

$$\Sigma \vdash \exists v_0 \ldots \exists v_{n-1} \psi. \tag{1}$$

But we also have $|\psi| \leqslant |\varphi|$ for each $\varphi \in \Delta$, i.e. $\Sigma \vdash \psi \rightarrow \varphi$, whence

$$\Sigma \vdash \forall v_0 \ldots \forall v_{n-1}(\psi \rightarrow \varphi).$$

But this, together with (1), immediately implies that every model of Σ realizes Δ, so that (i) fails.

(ii)$\Rightarrow$(i). Suppose (ii) holds. We claim that, in $B(\Sigma)$, we have, for each sequence of *distinct* variables $x_0, \ldots, x_{n-1}$,

$$\bigwedge_{\varphi \in \Delta} |\varphi(v_0/x_0, \ldots, v_{n-1}/x_{n-1})| = 0. \tag{2}$$

Suppose not; then there is an $\mathscr{L}$-formula ψ such that $|\psi| \neq 0$ and

$$|\psi| \leqq |\varphi(v_0/x_0, \ldots, v_{n-1}/x_{n-1})| \quad \text{for all} \quad \varphi \in \Delta,$$

i.e.

$$\Sigma \vdash \psi \rightarrow \varphi(v_0/x_0, \ldots, v_{n-1}/x_{n-1}), \tag{3}$$

for all $\varphi \in \Delta$.

Let $\varphi \in \Delta$, and let $\mathbf{y}_0,\ldots,\mathbf{y}_k$ be those free variables of ψ which are not among $\mathbf{x}_0,\ldots,\mathbf{x}_{n-1}$. Then, since $\mathbf{y}_0,\ldots,\mathbf{y}_k$ are not free in $\varphi(\mathbf{v}_0/\mathbf{x}_0,\ldots,\mathbf{v}_{n-1}/\mathbf{x}_{n-1})$, it follows from (3), using Prob. 3.1.6, that

$$(4) \qquad \Sigma \vdash \psi' \to \varphi(\mathbf{v}_0/\mathbf{x}_0,\ldots,\mathbf{v}_{n-1}/\mathbf{x}_{n-1}),$$

where ψ' is $\exists \mathbf{y}_0 \ldots \exists \mathbf{y}_k \psi$. We also have $\vdash \psi \to \psi'$, hence $|\psi| \leq |\psi'|$, and since $|\psi| \neq 0$ it follows that $|\psi'| \neq 0$. Observe that the free variables of ψ' are among $\mathbf{x}_0,\ldots,\mathbf{x}_{n-1}$. It follows from (4) and Remark 3.1.5 that

$$(5) \qquad \Sigma \vdash \forall \mathbf{x}_0 \ldots \forall \mathbf{x}_{n-1}(\psi' \to \varphi(\mathbf{v}_0/\mathbf{x}_0,\ldots,\mathbf{v}_{n-1}/\mathbf{x}_{n-1})),$$

so that

$$(6) \qquad \Sigma \vdash \forall \mathbf{v}_0 \ldots \forall \mathbf{v}_{n-1}(\psi'(\mathbf{x}_0/\mathbf{v}_0,\ldots,\mathbf{x}_{n-1}/\mathbf{v}_{n-1}) \to \varphi),$$

since the sentences on the right-hand side of the turnstiles in (5) and (6) are easily seen to be logically equivalent.

Let ψ'' be $\psi'(\mathbf{x}_0/\mathbf{v}_0,\ldots,\mathbf{x}_{n-1}/\mathbf{v}_{n-1})$; then $\psi'' \in \Phi_n$, $|\psi''| \neq 0$ and, by (6), $|\psi''| \leq |\varphi|$ for all $\varphi \in \Delta$. Therefore $[\Delta]$ is contained in the principal filter in $B_n(\Sigma)$ generated by $|\psi''|$, contradicting assumption. This proves (2).

It follows immediately from (2) that

$$(7) \qquad \bigvee_{\varphi \in \Delta} |\neg \varphi(\mathbf{v}_0/\mathbf{x}_0,\ldots,\mathbf{v}_{n-1}/\mathbf{x}_{n-1})| = 1$$

for *distinct* $\mathbf{x}_0,\ldots,\mathbf{x}_{n-1}$.

For each $m \in \omega$ let $I_m = \omega - \{0,\ldots,m\}$. Let $\mathbf{T}$ be the (countable) set of terms — i.e. variables and constants — of $\mathscr{L}$. Then for each $\mathbf{t} \in \mathbf{T}$ and each variable $\mathbf{v}_k$ distinct from $\mathbf{t}$, we have, by Thm. 5.4 and Prob. 3.1.14,

$$(8) \qquad 1 = |\exists \mathbf{v}_k(\mathbf{t} \equiv \mathbf{v}_k)| = \bigvee_{p \in I_m} |\mathbf{t} \equiv \mathbf{v}_p|.$$

Since $\mathscr{L}$ is countable, so is the family of joins

$$\Big\{\bigvee_{\mathbf{t} \in \mathbf{T}} |\varphi(\mathbf{v}_k/\mathbf{t})| \colon \varphi \in \Phi,\ k \in \omega\Big\} \cup \Big\{\bigvee_{p \in I_m} |\mathbf{t} \equiv \mathbf{v}_p| \colon m \in \omega,\ \mathbf{t} \in \mathbf{T}\Big\}$$
$$\cup \Big\{\bigvee_{\varphi \in \Delta} |\neg \varphi(\mathbf{v}_0/\mathbf{x}_0,\ldots,\mathbf{v}_{n-1}/\mathbf{x}_{n-1})| \colon \mathbf{x}_0,\ldots,\mathbf{x}_{n-1} \text{ distinct variables}\Big\}.$$

Hence, by the Rasiowa–Sikorski Theorem 4.7.3, there is an ultrafilter U in $B(\Sigma)$ which respects this family of joins.

Since U respects

$$\Big\{\bigvee_{\mathbf{t} \in \mathbf{T}} |\varphi(\mathbf{v}_k/\mathbf{t})| \colon \varphi \in \Phi,\ k \in \omega\Big\},$$

U is *perfect* (§5) and hence, by Thm. 5.10, the canonical structure $\mathfrak{A}(U)$ is a model of $\mathbf{\Sigma}$. Since $\mathbf{T}$ is countable, so is $\mathfrak{A}(U)$.

We claim that for each term $\mathbf{t}$ there are *infinitely* many variables $\mathbf{v}_p$ such that $|\mathbf{t}|=|\mathbf{v}_p|$. For U respects the family of joins

$$\{\bigvee_{p\in I_m} |\mathbf{t}=\mathbf{v}_p| : m\in\omega, \mathbf{t}\in\mathbf{T}\};$$

taking $m=0$, it follows from this and (8) that there must be $p_0\in I_0$ such that $|\mathbf{t}=\mathbf{v}_{p_0}|\in U$, i.e. $|\mathbf{t}|=|\mathbf{v}_{p_0}|$. Similarly, there is $p_1\in I_{p_0}$, hence $p_1>p_0$, such that $|\mathbf{t}|=|\mathbf{v}_{p_1}|$. Continuing in this way we obtain a sequence of natural numbers $p_0<p_1<p_2<\ldots$ such that $|\mathbf{t}|=|\mathbf{v}_{p_i}|$. This proves the claim.

We can now show that $\mathfrak{A}(U)$ omits $\mathbf{\Delta}$. For let $\langle|\mathbf{t}_0|,\ldots,|\mathbf{t}_{n-1}|\rangle$ be an n-tuple of elements of $A(U)$. Then, by our claim above, we may choose *distinct* variables $\mathbf{x}_0,\ldots,\mathbf{x}_{n-1}$ such that $|\mathbf{t}_i|=|\mathbf{x}_i|$ for all i, $0\leqslant i\leqslant n-1$. Now, since $\mathbf{x}_0,\ldots,\mathbf{x}_{n-1}$ are distinct, U respects the join

$$\bigvee_{\varphi\in\Delta} |\neg\varphi(\mathbf{v}_0/\mathbf{x}_0,\ldots,\mathbf{v}_{n-1}/\mathbf{x}_{n-1})|,$$

so it follows from (7) that there must be a formula $\varphi\in\mathbf{\Delta}$ such that

$$|\neg\varphi(\mathbf{v}_0/\mathbf{x}_0,\ldots,\mathbf{v}_{n-1}/\mathbf{x}_{n-1})|\in U.$$

Hence, by Thm. 5.10,

$$\mathfrak{A}(U)\models\neg\varphi\,[|\mathbf{x}_0|,\ldots,|\mathbf{x}_{n-1}|]$$

and so

$$\mathfrak{A}(U)\models\neg\varphi\,[|\mathbf{t}_0|,\ldots,|\mathbf{t}_{n-1}|].$$

This shows that no n-tuple of elements of $\mathfrak{A}(U)$ satisfies $\mathbf{\Delta}$, so that $\mathfrak{A}(U)$ omits $\mathbf{\Delta}$. ∎

As a corollary to this result we obtain the following important theorem, due to Ryll–Nardzewski, which gives necessary and sufficient conditions for a complete theory to be $\aleph_0$-categorical.

6.9. THEOREM. *Let $\mathbf{\Sigma}$ be a (countable) complete theory having only infinite models. Then the following conditions are equivalent:*

(i) *$\mathbf{\Sigma}$ is $\aleph_0$-categorical;*
(ii) *the Boolean algebra $B_n(\mathbf{\Sigma})$ is finite for each $n\in\omega$;*
(iii) *for each $n\in\omega$, there are only finitely many $\mathbf{\Sigma}$-reduced n-types;*
(iv) *for each $n\in\omega$, every ultrafilter in $B_n(\mathbf{\Sigma})$ is principal.*

PROOF. (ii) ⇔ (iii). We have already remarked (6.4) that the family of $\mathbf{\Sigma}$-reduced n-types is the Stone space of $B_n(\mathbf{\Sigma})$. The equivalence of (ii) and (iii) now follows immediately from Prob. 4.4.7.

(ii) $\Leftrightarrow$ (iv). This is an immediate consequence of Cor. 4.5.3.

(i)$\Rightarrow$(iv). Suppose that (iv) fails. Let U be a non-principal ultrafilter in $B_n(\Sigma)$, and let

$$\Delta = \{\varphi \in \Phi_n\colon\ |\varphi| \in U\}.$$

Then $[\Delta]=U$, so that, in particular, $[\Delta]$ is not included in any principal filter in $B_n(\Sigma)$. Hence, by 6.8, there is a countable model $\mathfrak{A}$ of Σ which omits Δ. Also, since $U=[\Delta]$ is a filter, $[\Delta]$ has the finite meet property, so that, by 6.7, there is a countable model $\mathfrak{B}$ of Σ which realizes Δ. Since $\mathfrak{A}$ omits Δ and $\mathfrak{B}$ realizes it, clearly $\mathfrak{A}$ cannot be isomorphic to $\mathfrak{B}$, so that Σ is not $\aleph_0$-categorical. Thus (i) fails.

(iv)$\Rightarrow$(i). The argument here is a modification of the "back-and-forth" construction used in the proof of Cantor's theorem 4.5. Assume (iv), and let $\mathfrak{A}$ and $\mathfrak{B}$ be models of Σ of cardinality $\aleph_0$. Let A and B be enumerated as $a_0,a_1,\ldots$ and $b_0,b_1,\ldots$ respectively. We show by induction that we can enumerate A and B into new sequences $c_0,c_1,\ldots$ and $d_0,d_1,\ldots$, respectively, in such a way that, for each $k\in\omega$,

$$U(\mathfrak{A}, \mathfrak{c}^k)=U(B, \mathfrak{d}^k)$$

where $\mathfrak{c}^k=\langle c_0,\ldots,c_k\rangle$ and $\mathfrak{d}^k=\langle d_0,\ldots,d_k\rangle$.

First, we put $d_0=b_0$. By assumption $U(\mathfrak{B},\mathfrak{d}^0)$ is principal; let $|\varphi|$ be a generating element with $\varphi\in\Phi_1$. Then $\mathfrak{B}\models\varphi\,[d_0]$, so that $\mathfrak{B}\models\exists v_0\varphi$. Since $\mathfrak{A}$ and $\mathfrak{B}$ are both models of the complete theory Σ, they are elementarily equivalent, so that $\mathfrak{A}\models\exists v_0\varphi$ also. Let c_0 be an element of A for which $\mathfrak{A}\models\varphi\,[c_0]$. Then $U(\mathfrak{A},\mathfrak{c}^0)$ contains the generating element $|\varphi|$ of $U(\mathfrak{B},\mathfrak{d}^0)$, so it follows immediately that $U(\mathfrak{A},\mathfrak{c}^0)=U(\mathfrak{B},\mathfrak{d}^0)$.

Now suppose that $k>0$ and $c_0,\ldots,c_{k-1}$, $d_0,\ldots,d_{k-1}$ have been chosen in such a way as to satisfy the requirements. We now show how to construct c_k and d_k. There are two cases to consider:

(1) $k=2m+1$ is odd. In this case we put

$$c_k=a_m.$$

By assumption, the ultrafilter $U(\mathfrak{A}, \mathfrak{c}^k)$ is principal; let $|\varphi|$ be a generating element with $\varphi\in\Phi_{k+1}$. In particular, $|\varphi|\in U(\mathfrak{A}, \mathfrak{c}^k)$; thus $\mathfrak{A}\models\varphi\,[c_0,\ldots,c_k]$ so that

$$\mathfrak{A}\models\exists v_k\varphi\,[c_0,\ldots,c_{k-1}],$$

and hence

$$|\exists v_k\varphi|\in U(\mathfrak{A}, \mathfrak{c}^{k-1}).$$

By the inductive hypothesis, $U(\mathfrak{A}, \mathfrak{c}^{k-1}) = U(\mathfrak{B}, \mathfrak{d}^{k-1})$, so that

$$|\exists \mathbf{v}_k \varphi| \in U(\mathfrak{B}, \mathfrak{d}^{k-1}).$$

Hence

$$\mathfrak{B} \models \exists \mathbf{v}_k \varphi \, [d_0, \ldots, d_{k-1}]$$

so that we can choose $d_k \in B$ in such a way that $\mathfrak{B} \models \varphi \, [d_0, \ldots, d_k]$. Then $U(\mathfrak{B}, \mathfrak{d}^k)$ contains the generating element $|\varphi|$ of $U(\mathfrak{A}, \mathfrak{c}^k)$, so that $U(\mathfrak{B}, \mathfrak{d}^k) = U(\mathfrak{A}, \mathfrak{c}^k)$. This completes the induction step for the case in which k is odd.

(2) $k = 2m$ is even. In this case we put

$$d_k = b_m,$$

and choose c_k in the same way as we chose d_k in case (1). We obtain $U(\mathfrak{A}, \mathfrak{c}^k) = U(\mathfrak{B}, \mathfrak{d}^k)$ just as before.

This completes our recursive definition.

It is clear from the construction that $A = \{c_0, c_1, \ldots\}$ and $B = \{d_0, d_1, \ldots\}$. Moreover, since $U(\mathfrak{A}, \mathfrak{c}^k) = U(\mathfrak{B}, \mathfrak{d}^k)$ for each $k \in \omega$, it follows from Prob. 6.5 that $(\mathfrak{A}, \mathfrak{c}^k) \equiv (\mathfrak{B}, \mathfrak{d}^k)$ for each $k \in \omega$, and this immediately implies that

$$(\mathfrak{A}, \langle c_0, c_1, \ldots \rangle) \equiv (\mathfrak{B}, \langle d_0, d_1, \ldots \rangle).$$

Thus $\mathfrak{A} \cong \mathfrak{B}$ by Prob. 1.7. Therefore Σ is $\aleph_0$-categorical and the proof is complete. ∎

6.10. PROBLEM. Let Σ be a consistent set of $\mathscr{L}$-sentences, and let $\Delta \subseteq \Phi_n$. Consider the conditions:

(a) Σ has a countable model which omits Δ;

(b) for each formula $\psi \in \Phi_n$ such that $\Sigma \cup \{\psi\}$ is consistent there is a formula $\varphi \in \Delta$ such that

$$\Sigma \nvdash \psi \rightarrow \varphi.$$

Show that (b)⇒(a) and that if Σ^c is complete the two conditions are equivalent. (Show that (b) is equivalent to condition (ii) of Thm. 6.8.)

6.11. PROBLEM. Let $\mathscr{L}$ be a countable first-order language which has a distiguished unary predicate $\mathbf{P}$ and for each $n \in \omega$ a constant symbol $\mathbf{n}$. An $\mathscr{L}$-structure $\mathfrak{A}$ is called an *ω-model* if

$$\mathbf{P}^{\mathfrak{A}} = \{\mathbf{n}^{\mathfrak{A}} : n \in \omega\}.$$

A theory Σ in $\mathscr{L}$ is said to be *ω-complete* if for each $\mathscr{L}$-formula φ and each variable $\mathbf{x}$ we have

$$\Sigma \vdash \varphi(\mathbf{x}/\mathbf{0}),\ \Sigma \vdash \varphi(\mathbf{x}/\mathbf{1}),\ldots \text{ implies } \Sigma \vdash \forall \mathbf{x}(\mathbf{Px} \to \varphi).$$

Let Δ be the set of formulas

$$\{\mathbf{Pv}_0,\ \mathbf{v}_0 \neq \mathbf{0},\ \mathbf{v}_0 \neq \mathbf{1}, \ldots\}.$$

(i) Let Σ be a set of $\mathscr{L}$-sentences. Show that the following conditions on an $\mathscr{L}$-structure $\mathfrak{A}$ are equivalent:

(a) $\mathfrak{A}$ is an ω-model of Σ;

(b) $\mathfrak{A}$ is a model of $\Sigma \cup \{\mathbf{Pn}: n \in \omega\}$ which omits Δ.

(ii) Show that, if $[\Sigma \cup \{\mathbf{Pn}: n \in \omega\}]^c$ is consistent and ω-complete then Σ has an ω-model. (Show that, using Prob. 6.10 (b), $\Sigma \cup \{\mathbf{Pn}: n \in \omega\}$ has a model omitting Δ, and then apply (i).)

6.12. PROBLEM. We adopt the notation of Prob. 6.11. The *ω-rule* is the following infinite rule of proof: from $\varphi(\mathbf{x}/\mathbf{0}), \varphi(\mathbf{x}/\mathbf{1}),\ldots$ infer $\forall \mathbf{x}(\mathbf{Px} \to \varphi)$, where φ is any $\mathscr{L}$-formula. *ω-logic* is obtained by adding the ω-rule to the axioms and rules of inference of the first-order predicate calculus $\mathscr{L}$ and allowing infinitely long deductions.

(i) Let Σ be a set of $\mathscr{L}$-sentences. Prove the *ω-completeness theorem:* Σ has an ω-model iff $\Sigma \cup \{\mathbf{Pn}: n \in \omega\}$ is consistent in ω-logic. (Consider the set of all $\mathscr{L}$-sentences provable from $\Sigma \cup \{\mathbf{Pn}: n \in \omega\}$ in ω-logic, and apply Prob. 6.11(ii).)

(ii) Show that the compactness theorem fails for ω-models of ω-logic.

6.13. PROBLEM. Let $\mathscr{L}$ be a countable language and for each $n \in \omega$ let $\Delta_n \subseteq \Phi_n$. Let Σ be a consistent set of $\mathscr{L}$-sentences, and consider the conditions:

(a) Σ has a countable model which omits each Δ_n;

(b) for each $n \in \omega$, $[\Delta_n] = \{|\varphi|: \varphi \in \Delta\}$ is not included in a principal filter in $B_n(\Sigma)$.

Show that (b)$\Rightarrow$(a) and that, if Σ^c is complete, the two conditions are equivalent. (Argue as in the proof of Thm. 6.8.)

Throughout Probs. 6.14–6.17 we assume that Σ is a (consistent) complete theory without finite models, formulated in a countable first-order language $\mathscr{L}$.

6.14. PROBLEM. A countable model $\mathfrak{A}$ of Σ is said to be *prime* if $\mathfrak{A}$ is elementarily embeddable in every model of Σ.

(i) Show that $\mathfrak{A}$ is a prime model of Σ iff for each n and each $\mathfrak{a} \in A^n$,

$U(\mathfrak{A}, \mathfrak{a})$ is a principal ultrafilter in $B_n(\Sigma)$. (For necessity, show that

$$\{\varphi \in \Phi_n\colon |\varphi| \in U(\mathfrak{A},\mathfrak{a})\}$$

is realized in every model of Σ, and apply Thm. 6.8. For sufficiency, argue as in the proof that (iv)$\Rightarrow$(i) in Thm. 6.9.)

(ii) Show that Σ has a prime model iff $B_n(\Sigma)$ is atomic (Ch. 4, §5) for each $n \in \omega$. (For necessity, show that, if $\mathfrak{A}$ is a prime model of Σ, then, for each $\varphi \in \Phi_n$ such that $|\varphi| \neq 0$ in $B_n(\Sigma)$, there is $\mathfrak{a} \in A^n$ such that $|\varphi| \in U(\mathfrak{A},\mathfrak{a})$; then apply (i). For sufficiency, argue as follows. For each $n \in \omega$ let

$$\Delta_n = \{\varphi \in \Phi_n\colon |\neg\varphi| \text{ is an atom of } B_n(\Sigma)\};$$

show that $[\Delta_n]$ satisfies (b) of Prob. 6.13; conclude that Σ has a countable model $\mathfrak{A}$ which omits each Δ_n and use (i) to show that $\mathfrak{A}$ is prime.)

(iii) Show that any two prime models of Σ are isomorphic. (Use (i), and argue as in the proof that (iv)$\Rightarrow$(i) Thm. 6.9.)

*6.15. PROBLEM. Let $\mathfrak{A}$ be a countable model of Σ, let $\mathfrak{a} \in A^n$, let $\mathscr{L}(\mathfrak{a})$ be the language for $(\mathfrak{A},\mathfrak{a})$, and let $\Phi_m(\mathfrak{a})$ be the set of all formulas of $\mathscr{L}(\mathfrak{a})$ whose free variables are among $\mathrm{v}_0,\ldots,\mathrm{v}_{m-1}$. A subset $\Gamma \subseteq \Phi_m(\mathfrak{a})$ is said to be *compatible* with $(\mathfrak{A},\mathfrak{a})$ if each finite subset of Γ is realized in $(\mathfrak{A},\mathfrak{a})$. $\mathfrak{A}$ is said to be *finitely saturated* if, for each $\mathfrak{a} \in A^n$ and each subset $\Gamma \subseteq \Phi_1(\mathfrak{a})$ which is compatible with $(\mathfrak{A},\mathfrak{a})$, Γ is realized in $(\mathfrak{A},\mathfrak{a})$. In the sequel we shall say "saturated" instead of "finitely saturated".

(i) Show that $\mathfrak{A}$ is saturated iff for each $\mathfrak{a} = \langle a_0,\ldots,a_{n-1}\rangle \in A^n$ and each ultrafilter U in $B_{n+1}(\Sigma)$ such that $U(\mathfrak{A},\mathfrak{a}) \subseteq U$ there is $a_n \in A$ such that $U = U(\mathfrak{A},\mathfrak{a}')$, where $\mathfrak{a}' = \langle a_0,\ldots,a_n\rangle$.

(ii) Let $\mathfrak{A}$ be a saturated model of Σ and let U be an ultrafilter in $B_n(\Sigma)$. Show that there is an $\mathfrak{a} \in A^n$ such that $U = U(\mathfrak{A},\mathfrak{a})$.

(iii) Show that, if Σ has a saturated model, then each $B_n(\Sigma)$ includes only countably many ultrafilters.

(iv) Let $\mathfrak{A}$ be any countable model of Σ, let $\mathfrak{a} = \langle a_0,\ldots,a_{n-1}\rangle \in A^n$, let U be an ultrafilter in $B_{n+1}(\Sigma)$ such that $U(\mathfrak{A},\mathfrak{a}) \subseteq U$, and let $\mathfrak{B}$ be a countable elementary extension of $\mathfrak{A}$. Show that there is a countable elementary extension $\mathfrak{B}(U,\mathfrak{a})$ of $\mathfrak{B}$ and an element a_n of $\mathfrak{B}(U,\mathfrak{a})$ such that $U = U(\mathfrak{B}(U,\mathfrak{a}),\mathfrak{a}')$, where $\mathfrak{a}' = \langle a_0,\ldots,a_n\rangle$. (Use the Compactness Theorem.)

(v) Suppose that, for each n, $B_n(\Sigma)$ includes only countably many ultrafilters. Show that each countable model $\mathfrak{A}$ of Σ has a countable elementary extension $\mathfrak{A}^*$ such that for each $\mathfrak{a} = \langle a_0,\ldots,a_{n-1}\rangle \in A^n$ and each ultrafilter U in $B_{n+1}(\Sigma)$ such that $U(\mathfrak{A},\mathfrak{a}) \subseteq U$ there is $a_n \in A^*$ such that

$U=U(\mathfrak{A}^*,\mathfrak{a}')$, where $\mathfrak{a}'=\langle a_0,\ldots,a_n\rangle$. (Let $\langle U_m,\mathfrak{a}_m\rangle$ be an enumeration of all pairs $\langle U,\mathfrak{a}\rangle$, where $\mathfrak{a}\in A^n$ for some n and U is an ultrafilter in $B_{n+1}(\Sigma)$ such that $U(\mathfrak{A},\mathfrak{a})\subseteq U$. Using (iv), define an elementary chain (Prob. 2.10) $\mathfrak{A}_0, \mathfrak{A}_1,\ldots$ by putting $\mathfrak{A}_0=\mathfrak{A}$ and, for each m, $\mathfrak{A}_{m+1}=\mathfrak{A}_m(U_m,\mathfrak{a}_m)$ (in the notation of (iv)). Show that $\mathfrak{A}^*=\bigcup_{m\in\omega}\mathfrak{A}_m$ meets the requirements.)

(vi) Suppose that for each n, $B_n(\Sigma)$ includes only countably many ultrafilters. Show that each countable model $\mathfrak{A}$ of Σ has a (countable) saturated elementary extension. (Define an elementary chain $\mathfrak{A}_0, \mathfrak{A}_1,\ldots$ by putting $\mathfrak{A}_0=\mathfrak{A}$ and, for each m, $\mathfrak{A}_{m+1}=\mathfrak{A}_m{}^*$ (in the notation of (v)). Show that $\bigcup_{m\in\omega}\mathfrak{A}_m$ meets the requirements.)

(vii) Show that Σ has a saturated model if and only if the Stone space of each $B_n(\Sigma)$ is countable. (Use (ii) and (vi).)

(viii) Show that, if the isomorphism relation $\cong$ partitions the class of countable models of Σ into countably many equivalence classes, then Σ has a saturated model. (Use (vi).)

*6.16. PROBLEM. (i) Show that any two saturated models of Σ are isomorphic. (If $\mathfrak{A}$ and $\mathfrak{B}$ are two saturated models of Σ show, using a "back-and-forth" argument like that in the proof of Thm. 6.9, that A and B can be enumerated as $\{c_0,c_1,\ldots\}$ and $\{d_0,d_1,\ldots\}$, respectively, in such a way that $(\mathfrak{A},\mathfrak{c}^k)\equiv(\mathfrak{B},\mathfrak{d}^k)$ for each $k\in\omega$.)

(ii) A countable model $\mathfrak{A}$ of Σ is said to be (countably) *universal* if each countable model $\mathfrak{B}$ of Σ can be elementarily embedded in $\mathfrak{A}$. Show that each saturated model of Σ is universal. (Like (i).)

(iii) Show that if Σ has a saturated model it has a prime model. (Use Probs. 6.15(vii), 4.5.13 and 6.14(ii).)

(iv) Show that Σ has a saturated model iff it has a universal model. (For one direction, use (ii). For the other, show that, if Σ has a universal model, then $B_n(\Sigma)$ has only countably many ultrafilters for each n, and apply Prob. 6.15(vii).)

*6.17. PROBLEM. Let $\mathscr{M}(\Sigma)$ be the class of all countable models of Σ. Show that the isomorphism relation $\cong$ on $\mathscr{M}(\Sigma)$ does not have exactly 2 equivalence classes.[1] Less precisely but more suggestively, Σ cannot have exactly 2 non-isomorphic countable models. (Argue by contradiction. Assume that Σ has exactly 2 non-isomorphic models. Then, by the results of Probs. 6.15 and 6.16, Σ has a saturated model $\mathfrak{B}$ and a prime model $\mathfrak{A}$. Show that $\mathfrak{A}$ and $\mathfrak{B}$ cannot be isomorphic, so that $\mathfrak{B}$ cannot be prime.

[1] On the other hand, for any finite $n\neq 0, 2$ it is possible to find a complete theory Σ such that $\cong$ divides $\mathscr{M}(\Sigma)$ into exactly n equivalence classes! This result is due to Ehrenfeucht, cf. VAUGHT [1961].

Then, by Prob. 6.14(i), there is $\mathfrak{b}\in B^n$ such that $U(\mathfrak{B},\mathfrak{b})$ is non-principal. Observe that $(\mathfrak{B},\mathfrak{b})$ is saturated, so that the complete theory $\Sigma'=\mathbf{Th}((\mathfrak{B},\mathfrak{b}))$ has a saturated model, and hence a prime model $(\mathfrak{C},\mathfrak{c})$. Now show that $\mathfrak{C}$ is a model of Σ which is not prime, so that $\mathfrak{C}\not\cong\mathfrak{A}$. Next, using the fact that Σ is not $\aleph_0$-categorical and Thm. 6.9, show that no model of Σ' can be both prime and saturated. Conclude that $(\mathfrak{C},\mathfrak{c})$ is not saturated, so nor is $\mathfrak{C}$, and therefore $\mathfrak{C}\not\cong\mathfrak{B}$.)

*§ 7. Indiscernibles and models with automorphisms

So far we have discussed two major methods of constructing models for sets of first-order sentences: Henkin's method in Chapter 3 (which is closely related to the Boolean method discussed in §5) and ultraproducts in §3. (A third method — that of unions of elementary chains — was outlined in Prob. 6.15.) In this section we introduce a new method: the generation of models by *Skolem functions*. This method has a quite different flavour from the others, and many of its applications involve combinatorial set theory in an essential way. We shall discuss one such application: the construction of models with many automorphisms.

Throughout this section we assume that $\mathscr{L}$ is a countable first-order language which may possess a set $\{\mathbf{f}_n\colon n\in\omega\}$ of *function symbols*[1] as well as a set $\{\mathbf{R}_n\colon n\in\omega\}$ of predicate symbols. (Recall that we regard constants as 0-ary function symbols.)

For each $n\in\omega$ let $\lambda(n)$ be the number of argument places in $\mathbf{R}_n$ and $\lambda'(n)$ the number of argument places in $\mathbf{f}_n$. Then an *$\mathscr{L}$-structure* is a system of the form

$$\mathfrak{A}=\langle A, \langle R_n\rangle_{n\in\omega}, \langle f_n\rangle_{n\in\omega}\rangle,$$

where A is a non-empty set and, for each $n\in\omega$, R_n is a $\lambda(n)$-ary relation and f_n a $\lambda'(n)$-ary operation on A. (Note that some of the f_n may be 0-ary functions, i.e. designated elements of A.) The f_n are called the *(basic) operations* of $\mathfrak{A}$.

If $\mathbf{t}$ is an $\mathscr{L}$-term whose variables are all among $\mathbf{v}_0,\dots,\mathbf{v}_n$ and $a_0,\dots,a_n\in A$ we define the *value* $\mathbf{t}^{\mathfrak{A}}[a_0,\dots,a_n]$ *of* $\mathbf{t}$ *at* $\langle a_0,\dots,a_n\rangle$ inductively as follows. If $\mathbf{t}=\mathbf{v}_i$ then

$$\mathbf{t}^{\mathfrak{A}}[a_0,\dots,a_n]=a_i;$$

[1] The results of §§1–6, proved originally for languages with no function symbols other than constants, extend to $\mathscr{L}$ in a straightforward way. We leave it to the reader to make these extensions.

and if $\mathbf{t}=\mathbf{f}_m\mathbf{t}_1\ldots\mathbf{t}_{\lambda'(m)}$ then

$$\mathbf{t}^{\mathfrak{A}}[a_0,\ldots,a_n]=f_m(\mathbf{t}_1^{\mathfrak{A}}[a_0,\ldots,a_n],\ldots,\mathbf{t}_{\lambda'(n)}^{\mathfrak{A}}[a_0,\ldots,a_n]).$$

We assume that the basic semantic definition for $\mathscr{L}$-formulas given in 2.1.1 has been reformulated as a satisfaction definition as was done for a language without function symbols in §1.

The notion of one $\mathscr{L}$-structure being a *substructure* or *elementary substructure* of another is the obvious extension of the corresponding notion introduced in §1. Thus, e.g., if

$$\mathfrak{A}=\langle A, \langle R_n\rangle_{n\in\omega}, \langle f_n\rangle_{n\in\omega}\rangle,$$

$$\mathfrak{A}'=\langle A', \langle R'_n\rangle_{n\in\omega}, \langle f'_n\rangle_{n\in\omega}\rangle,$$

are $\mathscr{L}$-structures, then $\mathfrak{A}\subseteq\mathfrak{A}'$ iff

$$A\subseteq A';$$

for each $n\in\omega$,

$$R_n=R'_n\cap A^{\lambda(n)};$$

and for each $n\in\omega$ and any elements $a_1,\ldots,a_{\lambda'(n)}$ of A,

$$f_n(a_1,\ldots,a_{\lambda'(n)})=f'_n(a_1,\ldots,a_{\lambda'(n)}).$$

In particular, if $\mathfrak{A}'\subseteq\mathfrak{A}$, then A' is closed under the operations of $\mathfrak{A}$.

If B is a non-empty subset of A which is *closed* under the operations of $\mathfrak{A}$, we define the *restriction* of $\mathfrak{A}$ to B by

$$\mathfrak{A}|B=\langle B, \langle R_n\cap B^{\lambda(n)}\rangle_{n\in\omega}, \langle f_n|B\rangle_{n\in\omega}\rangle.$$

Clearly we have $\mathfrak{A}|B\subseteq\mathfrak{A}$.

Let $\mathscr{L}'$ be an extension of $\mathscr{L}$ obtained by adding a countable set $\{\mathbf{f}'_i: i\in I\}$ of new function symbols, where, for each $i\in I$, $\mathbf{f}'_i$ is $\delta(i)$-ary. Let $\mathfrak{A}$ be an $\mathscr{L}$-structure, and for each $i\in I$ let f'_i be a $\delta(i)$-ary operation on A. We write $(\mathfrak{A}, \langle f'_i\rangle_{i\in I})$ for the structure obtained by adjoining the operations $\{f'_i: i\in I\}$ to $\mathfrak{A}$. Obviously $(\mathfrak{A}, \langle f'_i\rangle_{i\in I})=\mathfrak{A}'$ is an $\mathscr{L}'$-structure; it is called an *$\mathscr{L}'$-expansion* of $\mathfrak{A}$, and $\mathfrak{A}$ is the *$\mathscr{L}$-reduction* of $\mathfrak{A}'$.

We denote by $\mathscr{L}^+$ the extension of $\mathscr{L}$ obtained by introducing a new n-ary function symbol $\mathbf{f}_{\varphi,\mathfrak{x}}$ for each formula φ and each sequence $\mathfrak{x}=\langle\mathbf{x}_0,\ldots,\mathbf{x}_n\rangle$ of distinct variables which include all the free variables of φ. We put $\varphi_{\mathfrak{x}}^+$ for the $\mathscr{L}^+$-sentence

$$\forall\mathbf{x}_1\ldots\forall\mathbf{x}_n(\exists\mathbf{x}_0\varphi\rightarrow\varphi(\mathbf{x}_0/\mathbf{f}_{\varphi,\mathfrak{x}}\mathbf{x}_1\ldots\mathbf{x}_n));$$

$\varphi_{\mathfrak{x}}^{+}$ is called the *defining axiom* for $\mathbf{f}_{\varphi,\mathfrak{x}}$. Given a set Σ of $\mathscr{L}$-sentences, we put Σ^{+} for the union of Σ with the defining axioms of all the $\mathbf{f}_{\varphi,\mathfrak{x}}$. Clearly $|\Sigma^{+}|=\aleph_0$.

We iterate the process of obtaining $\mathscr{L}^{+}$ from $\mathscr{L}$ and Σ^{+} from Σ by setting

$$\mathscr{L}_0 = \mathscr{L}, \qquad \Sigma_0 = \Sigma,$$

$$\mathscr{L}_{n+1} = \mathscr{L}_n^{+}, \qquad \Sigma_{n+1} = \Sigma_n^{+}.$$

We put

$$\mathscr{L}^{*} = \bigcup_{n\in\omega} \mathscr{L}_n, \quad \Sigma^{*} = \bigcup_{n\in\omega} \Sigma_n.$$

Clearly $\mathscr{L}^{*}$ is countable and $|\Sigma^{*}|=\aleph_0$.

We now introduce one of the central notions of this section: Σ is called a *Skolem set* in $\mathscr{L}$ if Σ is a consistent set of $\mathscr{L}$-sentences and

(7.1) for each formula φ and each sequence $\mathbf{x}_0,\dots,\mathbf{x}_n$ of distinct variables which include all the free variables of φ there is a term $\mathbf{t}$ whose variables are all among $\mathbf{x}_1,\dots,\mathbf{x}_n$ such that

$$\Sigma \vdash \forall \mathbf{x}_1 \dots \forall \mathbf{x}_n(\exists \mathbf{x}_0 \varphi \rightarrow \varphi(\mathbf{x}_0/\mathbf{t})).$$

($\mathbf{t}$ is called a *Skolem term* for φ.)

7.2. Lemma. *Let Σ be a consistent set of $\mathscr{L}$-sentences. Then Σ^{*} is a Skolem set in $\mathscr{L}^{*}$. In fact, each model of Σ has an $\mathscr{L}^{*}$-expansion which is a model of Σ^{*}.*

Proof. It is clear from the definitions of Σ^{*} and $\mathscr{L}^{*}$ that Σ^{*} satisfies (7.1). Moreover, since Σ is consistent, it has a model. Thus, once we have shown that each model of Σ can be expanded to a model of Σ^{*} it will follow immediately that Σ^{*} is consistent, and hence that it is a Skolem set.

Let $\mathfrak{A}$ be any model of Σ. We show how to expand $\mathfrak{A}$ to an $\mathscr{L}^{*}$-structure $\mathfrak{A}^{*}$ which is a model of Σ^{*}.

Let g be a choice function for $\mathsf{P}A$. For each $\mathscr{L}^{*}$-formula φ and each sequence $\mathfrak{x}=\langle \mathbf{x}_0,\dots,\mathbf{x}_n\rangle$ of distinct variables which includes all the free variables of φ we shall define a function $f_{\varphi,\mathfrak{x}}$ from the appropriate Cartesian power of A into A which will serve as the interpretation of $\mathbf{f}_{\varphi,\mathfrak{x}}$. Let $\Phi_{-1}=\emptyset$ and for each $n\in\omega$ let Φ_n be the set of all $\mathscr{L}_n$-formulas. For each $\mathscr{L}^{*}$-formula φ let $\mathscr{X}(\varphi)$ be the set of all finite sequences of distinct variables which include all the free variables of φ. Now suppose that $f_{\varphi,\mathfrak{x}}$ has been defined for all $\varphi\in\Phi_{n-1}$ and all $\mathfrak{x}\in\mathscr{X}(\varphi)$. Let

$$\mathfrak{A}_n = (\mathfrak{A}, \langle f_{\varphi,\mathfrak{x}}\rangle_{\varphi\in\Phi_{n-1},\, \mathfrak{x}\in\mathscr{X}(\varphi)}).$$

For each formula $\varphi \in \Phi_n - \Phi_{n-1}$ and each $\mathfrak{x} = \langle \mathbf{x}_0, \ldots, \mathbf{x}_m \rangle \in \mathscr{X}(\varphi)$ we put

$$\psi = \varphi(\mathbf{x}_0/\mathbf{v}_0, \ldots, \mathbf{x}_m/\mathbf{v}_m)$$

and define $f_{\varphi,\mathfrak{x}}: A^m \to A$ by setting, for each $\langle a_1, \ldots, a_m \rangle \in A^m$,

$$f_{\varphi,\mathfrak{x}}(a_1, \ldots, a_m) = \begin{cases} g(Y) & \text{if } Y \neq \emptyset, \\ g(A) & \text{if } Y = \emptyset, \end{cases}$$

where

$$Y = \{a \in A : \mathfrak{A}_n \models \psi\, [a, a_1, \ldots, a_m]\}.$$

In this way, by induction, we define $f_{\varphi,\mathfrak{x}}$ for all $\mathscr{L}^*$-formulas φ and all $\mathfrak{x} \in \mathscr{X}(\varphi)$. If we now put

$$\mathfrak{A}^* = (\mathfrak{A}, \langle f_{\varphi,\mathfrak{x}} \rangle_{\varphi \in \Phi^*, \mathfrak{x} \in \mathscr{X}(\varphi)}),$$

it is easy to see that $\mathfrak{A}^*$ meets the required conditions. ∎

Let us call an $\mathscr{L}$-structure $\mathfrak{A}$ a *Skolem ($\mathscr{L}$-)structure* if $\mathbf{Th}(\mathfrak{A})$ (§4) is a Skolem set in $\mathscr{L}$. It follows immediately from the proof of Lemma 7.2 that each $\mathscr{L}$-structure can be expanded to a Skolem $\mathscr{L}^*$-structure.

Let $\mathfrak{A}$ be an $\mathscr{L}$-structure, and let $X \subseteq A$. We put

$$H_{\mathfrak{A}}(X) = \{\mathbf{t}^{\mathfrak{A}}\, [x_0, \ldots, x_n] : \mathbf{t} \text{ is an } \mathscr{L}\text{-term whose variables are among } \mathbf{v}_0, \ldots, \mathbf{v}_n \text{ and } x_0, \ldots, x_n \in X\}.$$

It is easy to verify that $H_{\mathfrak{A}}(X)$ is the smallest subset of A which includes X and is closed under the operations of $\mathfrak{A}$. In particular, we may put

$$\mathfrak{H}_{\mathfrak{A}}(X) = \mathfrak{A} | H_{\mathfrak{A}}(X).$$

$\mathfrak{H}_{\mathfrak{A}}(X)$ is called the *Skolem hull* of X in $\mathfrak{A}$. X is said to *generate* $\mathfrak{A}$ if $\mathfrak{H}_{\mathfrak{A}}(X) = \mathfrak{A}$.

7.3 LEMMA. *Let $\mathfrak{A}$ be a Skolem structure. Then:*

(i) *each substructure of $\mathfrak{A}$ is an elementary substructure of $\mathfrak{A}$;*

(ii) *if $\emptyset \neq X \subseteq A$, then $\mathfrak{H}_{\mathfrak{A}}(X)$ is the smallest elementary substructure of $\mathfrak{A}$ which includes X. Moreover, if X is infinite, then $|H_{\mathfrak{A}}(X)| = |X|$.*

PROOF. (i) follows immediately from (7.1) and Lemma 1.5. As for (ii), since $\mathfrak{H}_{\mathfrak{A}}(X) \subseteq \mathfrak{A}$ by definition, we have $\mathfrak{H}_{\mathfrak{A}}(X) \prec \mathfrak{A}$ by (i). Also, if $\mathfrak{B} \prec \mathfrak{A}$ and $X \subseteq B$, then B must be closed under the operations of $\mathfrak{A}$, so, since $H_{\mathfrak{A}}(X)$ is the least subset of A including X and closed under the operations of $\mathfrak{A}$, it follows that $H_{\mathfrak{A}}(X) \subseteq B$. The final assertion follows easily from the countability of $\mathscr{L}$. ∎

We now introduce the important notion of a set of *indiscernible* elements is a structure.

Let $\mathfrak{A}$ be an $\mathscr{L}$-structure, and let X be a subset of A which carries a (strict) total ordering $<$. ($<$ is *not necessarily* a basic relation of A.) We say that X is a set of *indiscernible elements*[1] in $\mathfrak{A}$ (with respect to $<$) if for each $\mathscr{L}$-formula φ whose free variables are all among $\mathbf{v}_0,\ldots,\mathbf{v}_n$ and each pair of sequences $x_0<\ldots<x_n$ and $y_0<\ldots<y_n$ from X, we have

$$\mathfrak{A}\models\varphi\,[x_0,\ldots,x_n] \;\Leftrightarrow\; \mathfrak{A}\models\varphi\,[y_0,\ldots,y_n].$$

Clearly this condition is equivalent to

$$(\mathfrak{A},\langle x_0,\ldots,x_n\rangle)\equiv(\mathfrak{B},\langle y_0,\ldots,y_n\rangle).$$

Our next task is to establish the existence of models of a given consistent set of sentences with sets of indiscernibles of any prescribed order type. For this we shall need a combinatorial result.

For each set X and each $r\in\omega$ let $[X]^r$ be the family of all subsets of X with exactly r elements. Then we have:

7.4. RAMSEY'S THEOREM. *Let X be an infinite set, let $r\geqslant 1$, and let $\{C_1,\ldots,C_k\}$ be a partition of $[X]^r$ into k pieces. Then there is an infinite subset Y of X and some j, $1\leqslant j\leqslant k$, such that $[Y]^r\subseteq C_j$.*

PROOF. By induction on r. For $r=1$ the result is trivial. Assume then that the theorem holds for r; we show that it holds for $r+1$.

Let $\{C_1,\ldots,C_k\}$ be a partition of $[X]^{r+1}$. For distinct $x_1,\ldots,x_r$ in X, put

$$M_i(\{x_1,\ldots,x_r\})=\{x\in X:\ \{x_1,\ldots,x_r,x\}\in C_i\}$$

($i=1,\ldots,k$). Let $\mathscr{F}$ be a non-principal ultrafilter over X; then by Prob. 4.3.16.(i) each member of $\mathscr{F}$ is infinite. Moreover, it is clear that, for each r-tuple $x_1,\ldots,x_r$ of distinct elements of X, the $M_i(\{x_1,\ldots,x_r\})$ ($i=1,\ldots,k$) form a finite partition of $Z=X-\{x_1,\ldots,x_r\}$. Since $\mathscr{F}$ is non-principal, $Z\in\mathscr{F}$, so that by Prob. 4.3.16(iii) exactly one of the $M_i(\{x_1,\ldots,x_r\})$ is in $\mathscr{F}$. Let $M(\{x_1,\ldots,x_r\})$ be the unique $M_i(\{x_1,\ldots,x_r\})$ which is in $\mathscr{F}$.

We now define a sequence $y_0,y_1,\ldots$ of distinct elements of X as follows. First, choose $y_0,\ldots,y_{r-1}$ to be any r distinct members of X. Then, if $n\geqslant r$ and y_m has been chosen for all $m<n$, pick

$$y_n\in\bigcap\{M(\{y_{m_1},\ldots,y_{m_r}\}):\ m_1<\ldots<m_r<n\}.$$

[1] For some examples of sets of indiscernible elements see Prob. 7.9.

(Notice that this latter set is a finite intersection of members of $\mathscr{F}$, hence is a member of $\mathscr{F}$ and accordingly non-empty.) From the definition it is clear that y_n is different from all of its predecessors.

Now put $Y'=\{y_0,y_1,...\}$; then Y' is infinite and so by inductive hypothesis the theorem holds for Y' and r. Thus, if for each i, $1\leqslant i\leqslant k$, we define

$$C_i'=\{\{y_{m_1},...,y_{m_r}\}: M_i(\{y_{m_1},...,y_{m_r}\})\in\mathscr{F}\},$$

then $\{C_1',...,C_k'\}$ is a partition of $[Y']^r$, so there must be an infinite subset Y of Y' and a j, $1\leqslant j\leqslant k$, for which $[Y]^r\subseteq C_j'$. We claim that $[Y]^{r+1}\subseteq C_j$. For let $y_{m_1},...,y_{m_{r+1}}\in Y$, with $m_1<...<m_{r+1}$. Then $\{y_{m_1},...,y_{m_r}\}\in C_j'$ so that $M_j(\{y_{m_1},...,y_{m_r}\})\in\mathscr{F}$. Hence, by definition,

$$M(\{y_{m_1},...,y_{m_r}\})=M_j(\{y_{m_1},...,y_{m_r}\}).$$

But then, by the definition of $y_{m_{r+1}}$, we have

$$y_{m_{r+1}}\in M(\{y_{m_1},...,y_{m_r}\}),$$

so that $y_{m_{r+1}}\in M_j(\{y_{m_1},...,y_{m_r}\})$, i.e.

$$\{y_{m_1},...,y_{m_r},y_{m_{r+1}}\}\in C_j.$$

This proves the claim, the induction step, and the theorem. ∎

We now use this result in the proof of:

7.5. THEOREM. *Let Σ be a set of sentences of $\mathscr{L}$ with an infinite model, and let $\langle X,<\rangle$ be any totally ordered set. Then there is a model of Σ of cardinality* $\max(\aleph_0,|X|)$ *which includes X as a set of indiscernibles.*

PROOF. Firstly, we observe that without loss of generality we may assume that X is infinite. For otherwise we may extend it to a countably infinite totally ordered set X', and if the theorem holds for X', it holds also for X. Let $\mathscr{L}'$ be the extension of $\mathscr{L}$ obtained by adding the new set of constants $\{\mathbf{c}_x: x\in X\}$, and let Σ' consist of the following $\mathscr{L}'$-sentences:

(i) all members of Σ;

(ii) $\varphi(\mathbf{v}_0/\mathbf{c}_{x_0},...,\mathbf{v}_n/\mathbf{c}_{x_n}) \leftrightarrow \varphi(\mathbf{v}_0/\mathbf{c}_{y_0},...,\mathbf{v}_n/\mathbf{c}_{y_n})$

for each $\mathscr{L}$-formula φ with free variables among $\mathbf{v}_0,...,\mathbf{v}_n$ and each pair of sequences $x_0,...,x_n$, $y_0,...,y_n$ from X such that $x_0<...<x_n$ and $y_0<...<y_n$;

(iii) $\mathbf{c}_{x_1}\neq\mathbf{c}_{x_2}$

for $x_1\neq x_2$ in X.

We show that each finite subset Σ_0 of Σ' has a model.

Let $\mathfrak{A}$ be an infinite model of Σ, and let $<^*$ be a fixed total ordering of A. Let $\varphi_1,\ldots,\varphi_s$ be a list of all formulas playing the role of φ in sentences of type (ii) in Σ_0; suppose that φ_i has n_i free variables. Then φ_1 determines a partition of $[A]^{n_1}$ into two classes: those n_1-element subsets of A that, when put in increasing order under $<^*$, satisfy φ_1 in $\mathfrak{A}$, and those that do not. Since A is infinite, by Ramsey's Theorem there is an infinite subset A_1 of A such that $[A_1]^{n_1}$ is entirely included in one of these classes. Now repeat the procedure, dividing $[A_1]^{n_2}$ into two classes according to satisfaction of φ_2 or $\neg\varphi_2$ in $\mathfrak{A}$; we obtain an infinite subset A_2 of A_1 such that $[A_2]^{n_2}$ is included in one of these classes. Iterate this procedure s times to obtain a sequence $A_1,\ldots,A_s$ such that

(a) $A_s \subseteq A_{s-1} \subseteq \ldots \subseteq A_1 \subseteq A$;

(b) each A_i is infinite;

(c) any two members of $[A_i]^{n_i}$, when their elements are arranged in increasing order under $<^*$, either both satisfy φ_i in $\mathfrak{A}$ or both satisfy $\neg\varphi_i$ in $\mathfrak{A}$.

Putting $Y=A_s$, it follows from (a) and (c) that for all i, $1\leqslant i\leqslant s$, and each pair of increasing sequences $a_1,\ldots,a_{n_i}$, $b_1,\ldots,b_{n_i}$ from Y,

(d) $\mathfrak{A}\models\varphi_i\,[a_1,\ldots,a_{n_i}] \Leftrightarrow \mathfrak{A}\models\varphi_i\,[b_1,\ldots,b_{n_i}]$.

Let Z be the set of elements x of X such that $\mathbf{c}_x$ occurs in a sentence of Σ_0. Since Y is infinite and Z is finite, there is a one–one order-preserving map f of $\langle Z,<\rangle$ into $\langle Y,<^*\rangle$. We claim that

$$\mathfrak{A}'=(\mathfrak{A},\langle f(z)\rangle_{z\in Z})$$

is a model of Σ_0. Since $\mathfrak{A}$ is a model of Σ, all sentences of type (i) hold in $\mathfrak{A}'$, while those of type (iii) in Σ_0 hold since f is one–one. Finally, (d) and the fact that f is order preserving immediately imply that $\mathfrak{A}'$ is a model of those sentences in Σ_0 of type (ii). Thus $\mathfrak{A}'$ is a model of Σ_0.

By the compactness theorem, then, Σ has a model $\mathfrak{B}''$. Since $\mathfrak{B}''$ is a model of all sentences of the form (iii), we may, without loss of generality, identify the interpretations of $\mathbf{c}_x$, for $x\in X$, in $\mathfrak{B}''$, with the elements x themselves. Thus B'' must be infinite. It is now clear that the $\mathscr{L}$-reduction $\mathfrak{B}'$ of $\mathfrak{B}''$ is a model of Σ including X as a set of indiscernibles. By Thm. 2.1 we can find an elementary substructure $\mathfrak{B}$ of $\mathfrak{B}'$ of cardinality $\max(\aleph_0,|X|)$ which includes X. It is easy to see that X is also a set of indiscernibles in $\mathfrak{B}$, so that $\mathfrak{B}$ satisfies the requirements of the theorem. ∎

An *automorphism* of a structure is an isomorphism of the structure onto itself. If $\langle X,<\rangle$ is a totally ordered set, an *order automorphism* of X is an automorphism of the structure $\langle X,<\rangle$. We now show how order automorphisms of sets of indiscernibles give rise to automorphisms of structures.

7.6. THEOREM. *Let $\mathfrak{A}$ be a Skolem structure, and let $\langle X,<\rangle$ be a set of indiscernibles in $\mathfrak{A}$ such that X generates $\mathfrak{A}$. Then each one–one order preserving map $f: X \to X$ can be extended uniquely to an elementary embedding g of $\mathfrak{A}$ into itself. Moreover, if f is an order automorphism of X, then g is an automorphism of $\mathfrak{A}$.*

PROOF. Since X generates $\mathfrak{A}$, each element a of A is of the form $\mathbf{t}^{\mathfrak{A}}[x_0,\ldots,x_n]$, where $\mathbf{t}$ is a term whose variables are all among $\mathbf{v}_0,\ldots,\mathbf{v}_n$ and $x_0,\ldots,x_n \in X$. By making suitable changes of variable in $\mathbf{t}$, if necessary, we may assume that $x_0<\ldots<x_n$. We shall call the equation $a=\mathbf{t}^{\mathfrak{A}}[x_0,\ldots,x_n]$ a *presentation* of a.

Let $a=\mathbf{t}^{\mathfrak{A}}[x_0,\ldots,x_n]$ be a presentation of a. We put

$$g(a)=\mathbf{t}^{\mathfrak{A}}[f(x_0),\ldots,f(x_n)].$$

We first show that $g(a)$ is well-defined. Let $a=\mathbf{s}^{\mathfrak{A}}[y_0,\ldots y_m]$ be any other presentation of a. Then we have

$$\mathbf{t}^{\mathfrak{A}}[x_0,\ldots,x_n] = \mathbf{s}^{\mathfrak{A}}[y_0,\ldots,y_m]. \tag{1}$$

Let $\{z_0,\ldots,z_q\}$ be an enumeration of the set $\{x_0,\ldots,x_n, y_0,\ldots,y_m\}$ in increasing order. Then the equation (1) may be expressed in the form

$$\mathfrak{A} \models \varphi[z_0,\ldots,z_q]$$

for some formula φ. Since X is a set of indiscernibles for $\mathfrak{A}$ and f preserves order, it follows that

$$\mathfrak{A} \models \varphi[f(z_0),\ldots,f(z_q)].$$

This immediately gives

$$\mathbf{t}^{\mathfrak{A}}[f(x_0),\ldots,f(x_n)]=\mathbf{s}^{\mathfrak{A}}[f(y_0),\ldots,f(y_m)].$$

Thus g is well-defined. Also, it is easy to see that it extends f.

We now show that g is an elementary embedding of $\mathfrak{A}$ into itself. Let φ be a formula whose free variables are all among $\mathbf{v}_0,\ldots,\mathbf{v}_k$ and let $a_0,\ldots,a_k \in A$ be such that $\mathfrak{A} \models \varphi[a_0,\ldots,a_k]$. For each i, $0 \leqslant i \leqslant k$, let $a_i=\mathbf{t}_i^{\mathfrak{A}}[x_{i0},\ldots,x_{in_i}]$ be a presentation of a_i. Let $y_0<\ldots<y_q$ be an enumeration of the set $\bigcup\{\{x_{i0},\ldots,x_{in_i}\}: i=0,\ldots,k\}$ in increasing order. By making suitable changes of variable in the $\mathbf{t}_i$ we may assume that our presentations are actually

$$a_i=\mathbf{t}_i^{\mathfrak{A}}[y_0,\ldots,y_q], \quad 0 \leqslant i \leqslant k.$$

Now let

$$\psi = \varphi(\mathbf{v}_0/\mathbf{t}_0,\ldots,\mathbf{v}_k/\mathbf{t}_k).$$

Then we clearly have

$$\mathfrak{A} \models \varphi\,[a_0,\ldots,a_k] \Leftrightarrow \mathfrak{A} \models \psi\,[y_0,\ldots,y_q],$$
$$\mathfrak{A} \models \varphi\,[g(a_0),\ldots,g(a_k)] \Leftrightarrow \mathfrak{A} \models \psi\,[f(y_0),\ldots,f(y_q)].$$

But, since f is order preserving and X is a set of indiscernibles in $\mathfrak{A}$, we have

$$\mathfrak{A} \models \psi\,[y_0,\ldots,y_q] \Leftrightarrow \mathfrak{A} \models \psi\,[f(y_0),\ldots,f(y_q)].$$

Hence

$$\mathfrak{A} \models \varphi\,[a_0,\ldots,a_k] \Leftrightarrow \mathfrak{A} \models \varphi\,[g(a_0),\ldots,g(a_k)],$$

so that, by Prob. 1.4(iii), g is an elementary embedding of $\mathfrak{A}$ into itself.

The proof of the fact that g is unique and that it is an automorphism if f is an order automorphism is left to the reader. ∎

Our final task in this chapter is to establish the existence of models with many automorphisms. To achieve this we shall need:

7.7. LEMMA. *For each infinite cardinal $\varkappa$ there is an ordered set of cardinality $\varkappa$ with exactly $2^\varkappa$ order automorphisms.*

PROOF. We first observe that the ordered set Q of rational numbers has an order automorphism f which differs from the identity, e.g. $f(r)=r+1$. (In fact Q has $2^{\aleph_0}$ automorphisms, but we shall not need this many.) Now let $\varkappa$ be an infinite cardinal, and let $\{Q_\xi\colon \xi<\varkappa\}$ be a disjoint family of copies of Q. The set $X=\bigcup_{\xi<\varkappa} Q_\xi$ may then be ordered in the obvious way: for $x,y\in X$, put $x<y$ if $x\in Q_\xi$ and $y\in Q_\eta$ with $\xi<\eta$, or $x,y\in Q_\xi$ and $x<y$ in Q_ξ. X is then of cardinality $\varkappa$, and so has *at most* $\varkappa^\varkappa=2^\varkappa$ order automorphisms. We now construct a set of $2^\varkappa$ distinct automorphisms of X as follows. Since f is an automorphism of Q, it may be regarded as an automorphism of each Q_ξ. For each $Y\subseteq\varkappa$ define $g_Y : X\to X$ by the condition

$$g_Y|Q_\xi = \begin{cases} f & \text{if } \xi\in Y; \\ \text{identity on } Q_\xi & \text{if } \xi\notin Y. \end{cases}$$

It is now easy to see that $\{g_Y\colon Y\subseteq\varkappa\}$ is a family of $2^\varkappa$ distinct automorphisms ot X. ∎

We can now prove

7.8. THEOREM. *Let Σ be a set of $\mathscr{L}$-sentences with an infinite model. Then for each infinite cardinal $\varkappa$, Σ has a model of cardinality $\varkappa$ with exactly $2^\varkappa$ automorphisms.*

PROOF. Σ is consistent and has an infinite model, so by Lemma 7.2 the Skolem set Σ^* also has an infinite model. By Lemma 7.7 we can find an ordered set X of cardinality $\varkappa$ with $2^\varkappa$ order automorphisms. By Thm. 7.5 there is a model $\mathfrak{A}$ of Σ^* of cardinality $\varkappa$, including X as a set of indiscernibles. Then $\mathfrak{A}$ is evidently a Skolem structure; let $\mathfrak{A}'$ be the Skolem hull of X in $\mathfrak{A}$. By Lemma 7.3, $\mathfrak{A}'$ is an elementary substructure of $\mathfrak{A}$ of cardinality $\varkappa$. Moreover, X generates $\mathfrak{A}'$. Hence, by Thm. 7.6, each order automorphism of X can be extended to an automorphism of $\mathfrak{A}'$, so that $\mathfrak{A}'$ has at least $2^\varkappa$ automorphisms. Let $\mathfrak{B}$ be the $\mathscr{L}$-reduction of $\mathfrak{A}'$; then every automorphism of $\mathfrak{A}'$ is obviously an automorphism of $\mathfrak{B}$ so that $\mathfrak{B}$ has at least $2^\varkappa$, and hence, since $\|\mathfrak{B}\|=\varkappa$, exactly $2^\varkappa$ automorphisms. ∎

Thm. 7.8 has the following interesting consequences. First, let $\mathfrak{R}$ be the field of real numbers. Then there is a field $\mathfrak{A}$ of cardinality $2^{\aleph_0}$ such that $\mathfrak{A}\equiv\mathfrak{R}$ and $\mathfrak{A}$ has $2^{2^{\aleph_0}}$ automorphisms. This should be contrasted with the well-known fact that $\mathfrak{R}$ itself has only *one* automorphism, namely the identity.

Secondly, Thm. 7.8 implies that the complex field $\mathfrak{C}$ has $2^{2^{\aleph_0}}$ automorphisms, for since — as is well-known — the theory of algebraically closed fields (of characteristic 0) is $2^{\aleph_0}$-categorical, the algebraically closed field of cardinality $2^{\aleph_0}$ with $2^{2^{\aleph_0}}$ automorphisms whose existence is implied by Thm. 7.8 must be isomorphic to $\mathfrak{C}$.

7.9. PROBLEM. (i) Let $\mathfrak{A}$ be an $\mathscr{L}$-structures, and let $\langle X,<\rangle$ be a totally ordered subset of A. Suppose that for any pair of increasing n-tuples $x_1<\ldots<x_n$ and $y_1<\ldots<y_n$ from X there is an automorphism f of $\mathfrak{A}$ such that $f(x_1)=y_1,\ldots,f(x_n)=y_n$. Show that X is a set of indiscernibles in $\mathfrak{A}$.

Use this result to show that in each of the following examples X is a set of indiscernibles in the structure $\mathfrak{A}$:

*(i) $\mathfrak{A}$ is the complex field and X is a set of algebraically independent elements in A;

(ii) $\mathfrak{A}$ is the ordered set of reals and $\langle X,<\rangle=\mathfrak{A}$;

(iii) $\mathfrak{A}$ is the free Boolean algebra generated by a set X;

*(iv) $\mathfrak{A}$ is a Boolean algebra and X is the set of all atoms of $\mathfrak{A}$.

7.10. PROBLEM. Let $\mathfrak{C}$ be the complex field. Show that the set of real numbers is not first-order definable in $\mathfrak{C}$; that is, show that there is no formula φ with one free variable in the language for fields such that, for all $a\in C$, $\mathfrak{C}\models\varphi\,[a]$ iff a is real. (Show that, if there were such a formula φ, then any automorphism of $\mathfrak{C}$ must take reals to reals, and hence would have to be either the identity or conjugation. This contradicts the fact — earlier derived — that $\mathfrak{C}$ has more than 2 automorphisms.)

7.11. Problem. Let X be a set of indiscernibles in a Skolem $\mathscr{L}$-structure $\mathfrak{A}$. Show that:

(i) If $Y \subseteq X$, then Y is a set of indiscernibles in $\mathfrak{H}_{\mathfrak{A}}(Y)$ (with respect to the ordering induced by the ordering of X) and $\mathfrak{H}_{\mathfrak{A}}(Y) \prec \mathfrak{H}_{\mathfrak{A}}(X)$.

(ii) If X is infinite and Y is an arbitrary infinite totally ordered set, then there is a structure $\mathfrak{B}$ in which Y is a set of indiscernibles, and the sets of formulas satisfied by increasing sequences from X in $\mathfrak{A}$ and from Y in $\mathfrak{B}$ are the same. (Let Σ be the set of all $\mathscr{L}$-formulas satisfied by increasing sequences from X in $\mathfrak{A}$. Let $\mathscr{L}'$ be obtained from $\mathscr{L}$ by adding constants $\mathbf{c}_y$ for $y \in Y$, and let Σ' be the set of all $\mathscr{L}'$-sentences $\varphi(\mathbf{c}_{y_1},\ldots,\mathbf{c}_{y_n})$, where $\varphi \in \Sigma$ and $y_1 < \ldots < y_n$ in Y. Show that Σ' is consistent and that the $\mathscr{L}$-reduction $\mathfrak{B}$ of any model $\mathfrak{B}'$ of Σ' meets the requirements.)

(iii) If Y is a set of indiscernibles in $\mathfrak{B}$ such that the sets of formulas satisfied by increasing sequences from X in $\mathfrak{A}$ and Y in $\mathfrak{B}$ are the same, then $\mathfrak{B}$ is a Skolem structure and each one–one order preserving map f of X into Y can be extended uniquely to an elementary embedding of $\mathfrak{H}_{\mathfrak{A}}(X)$ into $\mathfrak{H}_{\mathfrak{B}}(Y)$. (Like the proof of Thm. 7.6.)

(iv) If Y, $\mathfrak{B}$ are as in (iii), and X and Y are infinite, then, for any set of formulas Δ whose free variables are included among $\mathbf{v}_0,\ldots,\mathbf{v}_{n-1}$, $\mathfrak{H}_{\mathfrak{A}}(X)$ realizes Δ iff $\mathfrak{H}_{\mathfrak{B}}(Y)$ realizes Δ.

7.12. Problem. (i) Let $\mathfrak{A}$ be a Skolem structure, and let X be an infinite set of indiscernibles in $\mathfrak{A}$ such that X generates $\mathfrak{A}$. Show that, if f is a one–one order-preserving map of X into itself which is not onto, then the unique elementary embedding of $\mathfrak{A}$ into itself given by Thm. 7.6 is not onto. (Let g be the extension of f to an elementary embedding of $\mathfrak{A}$ into itself. Show first that $g[A] = H_{\mathfrak{A}}(f[X])$. Then show that, if $y \in X - f[X]$, then $y \notin H_{\mathfrak{A}}(f[X])$.)

(ii) Let Σ be a set of sentences with an infinite model. Show that, for each infinite cardinal $\varkappa$, Σ has a model of cardinality $\varkappa$ which has an elementary embedding onto a proper substructure of itself. (Take X to be the ordered set ω, and then argue as in the proof of Thm. 7.8, using (i).)

§ 8. Historical and bibliographical remarks

The discovery that a mathematical theory may have more than one model was made in the nineteenth century when Riemann and Klein established the independence of the parallel postulate by constructing a model of the other axioms of geometry in which the parallel postulate fails. Following the formalization of predicate logic by Frege in the late nineteenth century, Löwenheim proved in 1915 that a finitely axiomatizable theory with a model has a countable model. This result was extended to arbitrary countable theories by Skolem in 1920. The general forms of the Löwenheim–Skolem Theorem are due to Tarski, as are most of the basic model-theoretic notions introduced in §§1 and 2. (See TARSKI and VAUGHT [1957].)

The ultraproduct construction is foreshadowed by some early work of Gödel and Skolem in the 1930's. The construction was, in essence, used by HEWITT [1948] in connection with real closed fields. The general reduced product construction was introduced by ŁOŚ [1955], and Thm. 3.7. was proved there. The proof we give was inspired by the theory of Boolean-valued models; cf. MANSFIELD [1971]. The proof of the Compactness Theorem by ultraproducts is due to FRAYNE, MOREL and SCOTT [1962].

Theorem 4.4 is due to ŁOŚ [1954] and VAUGHT [1954]. The completeness of the theory of unbounded dense orderings was first established by Langford in 1927, that of the theory of algebraically closed fields of characteristic 0 by Tarski, and that of algebraically closed fields of arbitrary characteristic by ROBINSON [1951]. Theorem 4.5 is a famous result of Cantor.

The idea of the Lindenbaum algebra of a language was formulated independently by Tarski and the Polish mathematician A. Lindenbaum in 1935, although Lindenbaum's results were never published. Theorem 5.10 and its application to the proof of Gödel's Completeness Theorem are due to RASIOWA and SIKORSKI [1951]. The results in Probs. 5.14–18 are due to MANSFIELD [1971]. Theorem 6.8 is due to Ehrenfeucht (cf. MOSTOWSKI [1958]), and Thm. 6.9 to RYLL–NARDZEWSKI [1959]. The results in Probs. 6.14–6.17 are due to VAUGHT [1961].

The notion of a set of indiscernibles and Thms. 7.5 and 7.8 are due to EHRENFEUCHT and MOSTOWSKI [1956]. Ramsey's Theorem is due to RAMSEY [1930].

The reader wishing to acquaint himself further with the fascinating and extensive subject of model theory may consult BELL and SLOMSON [1969] and CHANG and KEISLER [1973]. For applications of model theory to algebra see ROBINSON [1963]. For an illuminating discussion of the early history of model theory, see VAUGHT [1974].

CHAPTER 6
RECURSION THEORY

The primary task of recursion theory is to characterize and study the class of all *algorithmic* functions of natural numbers. Roughly speaking, a function f is *algorithmic* if there is an *algorithm* (i.e., a *deterministic mechanical procedure*) for calculating the value $f(\mathfrak{a})$ for any $\mathfrak{a}$ belonging to the domain of f.

An important application of this theory is in the study of *decision problems*. A decision problem has the following general form: a set A and a property P are specified and the problem is then to find — or to prove the impossibility of finding — an algorithm by means of which one could tell, for any $\mathfrak{a} \in A$, whether or not $\mathfrak{a}$ has the property P. In this chapter we shall use recursion theory to solve (negatively) an important decision problem in number theory, Hilbert's Tenth Problem. In Ch. 7 we shall make several crucial applications of recursion theory to logic.

This chapter can be read independently of all earlier chapters.

§ 1. Basic notation and terminology

In this chapter, unless the contrary is stated, the word *number* shall mean natural number (i.e., non-negative integer). The set of all numbers is denoted by N.

By *n-ary function* we mean any mapping

$$f : A \to N,$$

where $A \subseteq N^n$, i.e., A is a set of ordered n-tuples of numbers. We call A the *domain* of f and put $A = \text{dom}(f)$. By *function* we mean n-ary function for some number n.

Recall that by convention (see Ch. 0) N^0 is a set having just one member. Thus if f is a 0-ary function, then either $\text{dom}(f)$ has just one member and f has just one number as value, or $\text{dom}(f) = \emptyset$ and f is nowhere defined.

In the former case we identify f with its unique value. The nowhere defined 0-ary function will be denoted by "∞". Thus the set of all 0-ary functions is $N \cup \{\infty\}$.

Unless otherwise stated, lower-case italic letters from the end of the alphabet (especially "x", "y" and "z") will be used as variables ranging over the set N and lower-case italic letters from the beginning of the alphabet will be used as constants denoting members of that set. (We shall refer to these as *numerical* variables and constants).

We use lower-case German letters as abbreviations for n-tuples. Thus $\mathfrak{x} = \langle x_1, x_2, \ldots, x_n \rangle$, $\mathfrak{a} = \langle a_1, a_2, \ldots, a_n \rangle$, etc. Note that by this convention the number of components of an entity denoted by a lower-case German letter is always taken to be n, rather than k or m etc.

The fact that different n-ary functions do not in general have the same domain could lead to some rather awkward formulations. These will be avoided by the following device: we convert each n-ary function into an n-ary operation on $N \cup \{\infty\}$. Allowing $\mathfrak{w}$ to range over $(N \cup \{\infty\})^n$ we put

(1) $\quad f(\mathfrak{w}) = \infty \quad$ whenever $\mathfrak{w} \notin \operatorname{dom}(f)$.

Thus for any n-ary function f we have

(2) $\quad \operatorname{dom}(f) = \{\mathfrak{w} : f(\mathfrak{w}) \neq \infty\}$,

and since $\operatorname{dom}(f) \subseteq N^n$, (2) implies that

(3) $\quad f(\mathfrak{w}) = \infty \quad$ whenever $\mathfrak{w} \notin N^n$.

Conversely, any n-ary operation f on $N \cup \{\infty\}$ that fulfils (3) is now regarded as an n-ary function, and (2) is regarded as the definition of its domain.[1]

Note that to define an n-ary function one needs to specify the values $f(\mathfrak{x})$ for $\mathfrak{x} \in N^n$ only, since the rest is automatically taken care of by (3).

If f is an n-ary function for which we have not only (3) but also its converse

(4) $\quad f(\mathfrak{w}) \neq \infty \quad$ whenever $\mathfrak{w} \in N^n$,

i.e., if $\operatorname{dom}(f) = N^n$, then f is said to be *total*[2].

[1] Our device amounts simply to regarding "$= \infty$" as an abbreviation of "is undefined" and "$\neq \infty$" as an abbreviation of "is defined (and equal to some number)".

[2] In many books dealing with recursion theory the term *function* is reserved for total functions only, while what *we* have called "functions" are termed *partial functions*. But since the class of mappings satisfying (3) — rather than both (3) and (4) — is the more natural one in recursion theory, we prefer to apply the shorter term to it.

Total unary functions can clearly be identified in a natural way with infinite sequences of numbers. Therefore we call such functions *sequences*. We use lower case Greek letters (except ξ, η, ζ, λ and μ) as constants denoting particular sequences.

The letter ξ (and occasionally also η and ζ) will be used as a variable ranging over the set N^N of all sequences.

By an *n-ary functional* we mean a mapping

$$F : A \to N,$$

where $A = \mathrm{dom}(F) \subseteq N^N \times N^n$. Thus F has one sequence argument and n numerical arguments. We shall normally separate the two kinds of argument by a semi-colon (e.g., "$F(\xi; \mathfrak{x})$"). Using the same device as for n-ary functions, we can regard an n-ary functional as a mapping

$$F\colon N^N \times (N \cup \{\infty\})^n \to N \cup \{\infty\}$$

satisfying the condition

(5) $\qquad F(\xi; \mathfrak{w}) = \infty \quad$ whenever $\mathfrak{w} \notin N^n$.

F is *total* if the converse condition

(6) $\qquad F(\xi; \mathfrak{w}) \neq \infty \quad$ whenever $\mathfrak{w} \in N^n$

is satisfied as well. (Here again $\mathfrak{w}$ ranges over $(N \cup \{\infty\})$.)

Unless otherwise stated, *functional* will mean n-ary functional for some n.

When we assert an identity, say "$r = s$", where the values of r and s are in $N \cup \{\infty\}$ and depend on values taken by variables, say $\xi; x_1, \ldots, x_n$, then (unless otherwise stated) we mean that r and s have the same values for all $\xi \in N^N$ and all[1] $\mathfrak{x} \in N^n$. (N.B. Not necessarily for all $\mathfrak{x} \in (N \cup \{\infty\})^n$.)

With each n-ary function f we can associate in a natural way an n-ary functional, simply by adding a fictitious sequence variable. That is, we define a functional F by the identity

$$F(\xi; \mathfrak{x}) = f(\mathfrak{x}).$$

We shall often identify a function with its associated functional, and so regard functions as a special kind of functional.

[1] Here too we depart from the convention of books dealing with recursion theory, which introduce a special symbol (e.g., "$\simeq$") to denote identity in this strong sense. But it seems to us that our use of "$=$" is more in line with common usage in most branches of mathematics. Also, since the strong sense of identity will be required more often than any other, it seems sensible to denote it by the simpler symbol.

In this chapter we shall sometimes use *Church's lambda notation* as follows. Suppose that r is an expression such that the identity

$$f(\mathfrak{x})=r$$

defines an n-ary function f. Then this function f will also be denoted by

$$\lambda x_1 \ldots x_n r$$

or briefly, $\lambda\mathfrak{x}r$. (Note that we have, for any n-ary function f, $\lambda\mathfrak{x}f(\mathfrak{x})=f$.) Thus, e.g., $\lambda x(x+1)$ is the sequence φ satisfying the identity $\varphi(x)=x+1$, and $\lambda xy(x^y)$ is the binary function f satisfying the identity $f(x,y)=x^y$.

More generally, suppose r is an expression such that the identity

$$F(\xi;\mathfrak{x},y_1,\ldots,y_m)=r$$

defines an $(n+m)$-ary functional F, then the meaning of $\lambda\mathfrak{x}r$ is determined by the identity

$$[\lambda\mathfrak{x}r](\mathfrak{z})=F(\xi;\mathfrak{z},y_1,\ldots,y_m).$$

In other words, $\lambda\mathfrak{x}r$ is $F(\xi;\mathfrak{x},y_1,\ldots,y_m)$ "*as a function* of $\mathfrak{x}$", with $\xi,y_1,\ldots,y_m$ as parameters. For example, $\lambda x\xi(x)$ is simply ξ; and if F is a binary functional than $\lambda xF(\xi;x,y)$ is the unary function $f_{\xi;y}$ (depending on ξ and y as parameters) such that $f_{\xi;y}(x)=F(\xi;x,y)$.

To conclude this section we introduce the *least number symbol* "μ". Suppose that r is an expression such that the identity

$$G(\xi;\mathfrak{x},y)=r$$

defines an $(n+1)$-ary functional G. Then the expression μyr has the following meaning. The identity

$$F(\xi;\mathfrak{x})=\mu yr$$

defines an n-ary functional F which behaves as follows. Let φ be any sequence, and let $\mathfrak{a}$ be any n-tuple of numbers. Consider the values $G(\varphi;\mathfrak{a},y)$ for $y=0, 1, 2, \ldots$. There can be at most one number b such that for all $y<b$ the values $G(\varphi;\mathfrak{a}, y)$ are defined and positive (i.e., different from ∞ and 0) and $G(\varphi;\mathfrak{a}, b)=0$. If such b exists, we put $F(\varphi;\mathfrak{a})=b$. But if no such b exists we leave $F(\varphi;\mathfrak{a})$ undefined, i.e., $F(\varphi;\mathfrak{a})=\infty$.

The same notation is also used in connection with functions rather than functionals.

For example, if we define f by

$$f(x,y)=\mu z(y-xz),$$

then for any numbers a and b, $f(a,b)$ is the least number c such that $b=ac$ if such c exists; otherwise $f(a,b)=\infty$.

§ 2. Algorithmic functions and functionals

This section is devoted to an informal discussion of the intuitive notions that underlie and motivate recursion theory. The most basic among these notions is that of *algorithm*.

By *algorithm* mathematicians generally mean a computation procedure whose application leaves nothing to chance and ingenuity, but requires a rigid stepwise mechanical execution of explicitly stated rules. Here are a few examples of algorithms:

(a) The procedure of "long multiplication" for multiplying two numbers which are represented in decimal notation.

(b) The Euclidean algorithm for finding the highest common factor of two positive numbers.

(c) The well-known procedure for differentiating any combination of rational, trigonometric, exponential and logarithmic functions.

The following examples, taken from earlier chapters of this book, are (or can easily be converted into) algorithms:

(d) The method of truth tables (see §6 of Ch. 1) for finding out whether a given formula is a tautology.

(e) The method described in §8 of Ch. 2 for systematically searching for a first-order confutation of a given finite set of formulas.

(f) The method suggested in §5 of Ch. 3 for systematically searching for a first-order proof of a given formula.

An algorithm is presented as a prescription, consisting of a finite number of *instructions*. It can be applied to any one of a set of possible *inputs* — each input being a finite sequence of symbolic expressions. Once any particular input has been specified, the instructions dictate a succession of discrete simple operations, requiring no recourse to ingenuity or chance. The first operation is applied to the input and transforms it into a new finite sequence of symbolic expressions. This outcome is in turn subjected to a second operation (dictated by the instructions of the algorithm) and so on.

It may happen that after a finite number of steps the instructions dictate that the process must be discontinued and an *output* be read off (in some prescribed way) from the outcome of the last step. For some inputs, however, the process may never terminate and there is no last step and hence no output[1].

With any algorithm $\mathfrak{P}$ we may associate a mapping in the following natural way. Let A be the class of all possible inputs of $\mathfrak{P}$. If $\mathfrak{a} \in A$ and the application of $\mathfrak{P}$ to $\mathfrak{a}$ eventually (i.e., after a finite number of steps) yields an output, we let $\mathfrak{P}(\mathfrak{a})$ be that output; but if $\mathfrak{P}$ applied to $\mathfrak{a}$ yields no output we leave $\mathfrak{P}(\mathfrak{a})$ undefined (or, using the convention introduced in §1 we may put $\mathfrak{P}(\mathfrak{a}) = \infty$ in this case).

A particularly important class of algorithms — to which many other cases can easily be reduced — is that in which, for some n, the set of possible inputs of the algorithm $\mathfrak{P}$ includes the set N^n of all n-tuples of numbers (represented in some particular system of notation) and the corresponding outputs of $\mathfrak{P}$, when they exist, are numbers (also represented symbolically). In this case, the mapping associated with $\mathfrak{P}$, when restricted to N^n, yields an n-ary function, say f. We say that this f is *algorithmic* and $\mathfrak{P}$ is an *algorithm for (calculating)* f.

The foregoing ideas can be extended to deal with *functionals* as well. The snag here, however, is that if F is an n-ary functional, then it would seem that an algorithm $\mathfrak{P}$ for calculating F should have inputs representing pairs of the form $\langle \varphi; \mathfrak{a} \rangle$ where φ is a sequence and $\mathfrak{a}$ an n-tuple of numbers. But this cannot be so, since in general we would need an infinite amount of information in order to determine φ completely, and this cannot be encapsulated in a finite symbolic representation.

We shall therefore suppose that in this case the initial input for $\mathfrak{P}$ will represent only the numerical component $\mathfrak{a}$, while representations of values $\varphi(r)$ of φ will be fed in subsequently, "on demand", during the course of the calculation. Thus, at certain steps of the calculation process the instructions of $\mathfrak{P}$ may dictate that a representation of a value $\varphi(r)$ be added (in some prescribed way) to the outcome of the preceding step. Here r itself is to be determined in some prescribed way from the outcome of the preceding step.

It should be stressed that the sequence argument φ is not itself assumed to be algorithmic. The values $\varphi(r)$ are not necessarily outputs of some

[1] This can actually happen in examples (e) and (f) above. Thus, if we start with a formula that is not logically true, the systematic search for a proof will never terminate.

prior calculation process. We simply *require* that they be made available if and when they are called for.

Now, if $\mathfrak{P}$ is an algorithm that can be applied to all pairs $\langle \xi;\mathfrak{x} \rangle$, where the n-tuple $\mathfrak{x}$ of numbers (in some prescribed system of notation) serves as initial input, and values of ξ are fed in "on demand" as outlined above, and if the corresponding outputs of $\mathfrak{P}$, whenever they exist, are numbers (represented in some specified way) then the mapping associated with $\mathfrak{P}$ yields an n-ary functional F. We say that F is *algorithmic* and that $\mathfrak{P}$ is an *algorithm for (calculating) F.*

§ 3. The computer URIM

It may have occurred to the reader that the informal discussion of §2 has quite a lot to do with computers[1]. In fact, any program that can be used to program a given computer may be regarded as an algorithm.

In this section we shall make use of this idea to define a wide class of algorithmic functions and functionals. We describe an imaginary "computer" called *Unlimited Register Ideal Machine* (briefly, URIM) and the programs under which it may be made to "operate". These programs can then be regarded as algorithms[2], with which we may associate algorithmic functions and functionals as outlined in §2.

We assume that our "computer" URIM has an infinite sequence of *registers* R_i $(i=1,2,\ldots)$. We call the positive number i the *address* of the ith register R_i.

The registers are designed to be storage places for the inputs, output and intermediate stages of a computation. We assume that at each moment of time every register stores some natural number. (We may imagine the register to contain a symbolic representation of the number, but for our purposes it will not matter what particular method of representing numbers is used.)

In addition, URIM has a *program counter* K, which also contains, at each moment, some natural number. (This number will not be part of a computation, but merely a "position marker" for book-keeping purposes.)

[1] By *computer* we mean any calculating machine or automaton which operates in a deterministic and discrete step-by-step way. We exclude probabilistic, continuous and analog devices.

[2] This use of a "machine" and its programs to define algorithms is just a heuristic metaphor. As a matter of fact, the algorithms in question could be described directly in purely abstract mathematical terms, without any reference to machines and programs.

We say that a register or the counter K is *empty* when the number stored in it is 0. Thus to *erase* a register is to put 0 in it.

We shall assume that any number, however large, can be stored in each register and in the program counter. From this it would seem that we require URIM to be able to store an infinite amount of information. However, in fact we shall always assume that at any moment almost all (i.e., all but a finite number) of the registers are *empty*. Moreover, each program will only make use of a finite number of registers, whose addresses can be read off directly from the program itself. Thus we really need only an *unlimited* — but in each case *finite* — number of registers; hence the name of our machine.

Even so, the amount of information that can be stored in URIM, albeit finite, is *unbounded*. It is precisely this that makes URIM a purely *ideal* machine, which cannot be realized physically[1]. Every real computer has a finite storage capacity and can therefore only store a bounded amount of information. True, the storage capacity of some real computers can be expanded in case of need by bringing in new auxiliary memory units (e.g., on magnetic tape); but even this has its practical limitations and cannot go on indefinitely.

However, in our theoretical treatment of algorithmic functions and functionals we ignore such limitations of a purely practical, contingent nature. Thus we put no bound on the number of registers available in URIM or on the size of the numbers stored in the program counter and registers.

In order to deal with functionals (rather than just with functions) we introduce hypothetical entities called *oracles*. If φ is any sequence (i.e., a total unary function) then a *φ-oracle* is an agency able to supply the value $\varphi(r)$ for each r. It must be stressed that an oracle is not assumed to be a computer — in fact, it *cannot* be a computer unless φ happens to be algorithmic. We do not consider oracles to be part of URIM but external to it.

Our only assumption concerning oracles is that, for any sequence φ, URIM can be linked up to a φ-oracle and that values $\varphi(r)$ can be transmitted from the oracle and fed into a register of URIM. The circumstances under which any particular value $\varphi(r)$ is called for by URIM, and the determination of the register into which it must be fed, will be explained below.

[1] In all other respects the assumptions we shall make about URIM will be quite realistic.

We assume that URIM can *obey* four kinds of *commands* as follows.

Z *commands*. For each positive i, the command Z_i is obeyed by erasing the ith register (i.e., causing R_i to become empty) and adding 1 to the program counter K (i.e, causing the number k stored in K to be replaced by $k+1$). The other registers remain unchanged.

S *commands*. For each positive i, the command S_i is obeyed by adding 1 to both R_i and K. The other registers remain unchanged.

A *commands*. We suppose that URIM has been linked up to a φ-oracle. Let i be a positive number and suppose that at a given moment the number stored in R_i is r. Then the command A_i is obeyed by putting $\varphi(r)$ in place of r in R_i and adding 1 to K. The other registers remain unchanged. (The number $\varphi(r)$ is of course supposed to be obtained by URIM from the φ-oracle.)

J *commands*. Let i and j be positive and let k be any number. Suppose that at a given moment the numbers stored in R_i and R_j are r_i and r_j. Then if $r_i = r_j$ the command $J_{i,j,k}$ is obeyed by putting k in K instead of the number that was stored there previously. If $r_i \neq r_j$, then $J_{i,j,k}$ is obeyed by adding 1 to K. In either case all registers remain unchanged.

All together we have commands Z_i, S_i, A_i and $J_{i,j,k}$ for all positive i and j and all k.

By a *program* we mean any finite sequence of commands

$$\mathfrak{P} = \langle C_0, C_1, \ldots, C_{h-1} \rangle.$$

Here h is the *length* of $\mathfrak{P}$, i.e., the number of its commands. For any $k < h$, we call C_k the kth command of $\mathfrak{P}$. (Thus, $\mathfrak{P}$ begins with its 0th command, *not* with its 1st.)

The addresses *occurring* in a program $\mathfrak{P}$ are precisely those positive numbers i and j for which Z_i, S_i, A_i or (for some k) $J_{i,j,k}$ are in $\mathfrak{P}$.

We go on to describe how URIM operates under a given program $\mathfrak{P}$. We suppose that URIM has been linked up to some oracle. When "switched on", URIM will go through a (finite or infinite) succession of steps. Each step consists of obeying some command of $\mathfrak{P}$. Suppose that at a given moment the number stored in K is k. If $k <$ length of $\mathfrak{P}$, the next step will be to obey the kth command of $\mathfrak{P}$. However, if $k \geq$ length of $\mathfrak{P}$ (so that $\mathfrak{P}$ has no k^{th} command) then URIM *halts* — it "switches itself off" and does not perform any more steps.

We shall always assume that initially — i.e., before URIM begins to operate — the program counter K is *empty*. Thus, the commands of $\mathfrak{P}$

are obeyed one by one in order; except that when a command $J_{i,j,k}$ is obeyed and the numbers stored in R_i and R_j happen to be equal, then the next command to be obeyed will be the kth command of $\mathfrak{P}$ (or, if $k \geqslant$ length of $\mathfrak{P}$, URIM will halt). In other words, if the numbers in R_i and R_j are equal, a command $J_{i,j,k}$ makes URIM *jump* to the kth command rather than proceed in order.

We shall also assume that initially almost all the registers are *empty*. It is easy to see that if URIM is operating under the program $\mathfrak{P}$, the only registers that may affect the process or be affected by it are those whose addresses occur in $\mathfrak{P}$. Thus almost all the registers remain empty throughout and play no role at all.

3.1. Example. Let i and j be positive. We construct a program $\mathfrak{R}_{i,j}$ whose effect is to copy the number initially stored in R_i into R_j. All registers other than R_j are to be left unchanged. (In particular R_i will retain its initial contents.)

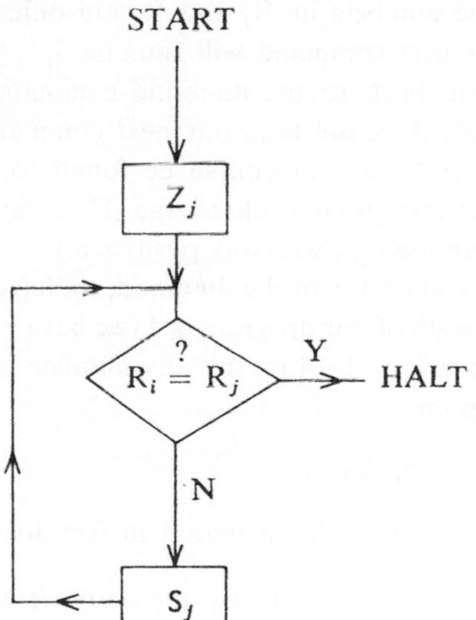

Fig. 1

If $i = j$, then the empty program will do the job. Now let $i \neq j$. We begin by setting up a *flow chart* which shows how we want the required program to work (see Fig.1).

A diamond-shaped box represents a question (in the present case: "are the numbers stored in R_i and R_j equal?"). Two arrows lead away from the diamond, corresponding to the answers "Yes" and "No"; they are labelled accordingly "Y" and "N". The rest is self-explanatory.

By following the chart the reader can see how our program is supposed to work. First, R_j is erased. Then URIM enters a *loop* and goes round and round as long as the numbers in R_i and R_j remain different. Each time round, it adds 1 to R_j. When the number in R_j becomes equal to that in R_i, URIM gets out of the loop and halts.

We now convert our chart into an actual program. We start at the entrance of the chart (marked "START") and work our way along, following the arrows. Our 0th command is Z_j. Next, our 1st command, which corresponds to the diamond, must of course be of the form $J_{i,j,-}$. We leave the third index blank, to be filled in later on; this third index will determine the jump that URIM will have to make when the numbers in R_i and R_j are found to be equal.

In the meantime we follow the arrow labelled "N", which corresponds to the case where the numbers in R_i and R_j are different, and does not involve a jump. Our next command will thus be S_j. After this, we have to make URIM jump back to the diamond command $J_{i,j,-}$ which was our 1st command. We therefore take our next command to be $J_{1,1,1}$: the number in the register R_1 will of course be found to be equal to itself, and URIM will therefore jump back to the 1st command, as required. (Instead of $J_{1,1,1}$ we can use $J_{p,p,1}$ with any positive p.)

Now that we have come back to the diamond, we have all the commands we need. Since the length of our program is 4 (we have got four commands) we can fill the blank in $J_{i,j,-}$ by 4 (or by any number >4). Thus we have constructed the program

$$\mathfrak{R}_{i,j} = \langle Z_j, J_{i,j,4}, S_j, J_{1,1,1} \rangle.$$

The reader should check that this program in fact does what is required.

Let $\mathfrak{P}$ be a program of length h. By the *normalization* of $\mathfrak{P}$ we mean the program $\mathfrak{P}'$ obtained from $\mathfrak{P}$ when every command of the form $J_{i,j,k}$ with $k>h$ is replaced by $J_{i,j,h}$. Clearly, $\mathfrak{P}'$ and $\mathfrak{P}$ work in exactly the same way, except that with $\mathfrak{P}'$ URIM can halt only when the number in K is h, while with $\mathfrak{P}$ that number may be $>h$. If $\mathfrak{P}$ is the same as its normalization, we say that $\mathfrak{P}$ is *normal*.

Let $\mathfrak{P}$ be a program of length h and let $\mathfrak{Q}$ be any program. Then their *concatenation* $\mathfrak{P}\mathfrak{Q}$ is defined to be the program obtained as follows. First, we write the commands of the normalization of $\mathfrak{P}$; then we follow them by the commands of $\mathfrak{Q}$ except that we replace every command $\mathsf{J}_{i,j,k}$ of $\mathfrak{Q}$ by $\mathsf{J}_{i,j,h+k}$. Clearly, under $\mathfrak{P}\mathfrak{Q}$ URIM begins to operate as it would under $\mathfrak{P}$; but when it would come to a halt under $\mathfrak{P}$, it will now go on to operate as under $\mathfrak{Q}$.

Clearly, concatenation is associative: $(\mathfrak{P}\mathfrak{Q})\mathfrak{R}=\mathfrak{P}(\mathfrak{Q}\mathfrak{R})$. So brackets can be omitted.

When drawing up a flow chart, we shall sometimes represent a concatenation, say $\mathfrak{P}\mathfrak{Q}$, by putting $\mathfrak{P}$ and $\mathfrak{Q}$ in two rectangular boxes, with an arrow from one box to the other to show the order of concatenation.

We conclude this section with yet another definition. For any program $\mathfrak{P}$ and number m, we define $\mathfrak{P}^{(m)}$ to be the program obtained from $\mathfrak{P}$ by adding m to every address occurring in it. (Thus, e.g., a command $\mathsf{J}_{i,j,k}$ is to be replaced by $\mathsf{J}_{i+m,j+m,k}$.) It is clear that $\mathfrak{P}^{(m)}$ ignores all registers whose addresses are $\leqq m$, while it treats R_{i+m} (for all positive i) exactly as $\mathfrak{P}$ treats R_i.

§ 4. Computable functionals and functions

Let $\mathfrak{P}$ be a program and let n be any number. We define an n-ary functional $\mathfrak{P}_n$ as follows.

Let φ be any given sequence, and let $\mathfrak{a}$ be any n-tuple of numbers. Suppose that URIM is linked up to a φ-oracle. Let $a_1,\ldots,a_n$ be stored in $\mathrm{R}_1,\ldots,\mathrm{R}_n$ respectively, while all the other registers (as well as K) are empty. Starting from this initial position, let URIM operate under $\mathfrak{P}$.

If after a finite number of steps URIM comes to a halt, and at that moment the number stored in the first register R_1 is b, we define $\mathfrak{P}_n(\varphi;\mathfrak{a})$ to be b.

If, on the other hand, URIM never comes to a halt, then $\mathfrak{P}_n(\varphi;\mathfrak{a})$ is ∞, i.e., it is left undefined.

We say that $\mathfrak{P}$ *computes* the functional $\mathfrak{P}_n$. A functional is *computable* if it is computed by some program.

If $\mathfrak{P}$ is a program having no A commands, then evidently the value of $\mathfrak{P}_n(\xi;\mathfrak{x})$ is independent of the value assigned to the sequence variable ξ. In this case we obtain an n-ary *function* (which we also denote by "$\mathfrak{P}_n$") defined by the identity

$$\mathfrak{P}_n(\mathfrak{x})=\mathfrak{P}_n(\xi;\mathfrak{x}).$$

We say that $\mathfrak{P}$ *computes* the function $\mathfrak{P}_n$. A function is *computable* if it is computed by some program having no A commands.

The reader should note that the term "computable" is used here in a precise technical sense. In the literature the same term is often used in a wider and looser sense, as synonymous with "algorithmic"; but we shall avoid this usage here.

It is clear that each URIM program can in fact be regarded as an algorithm, and every computable function(al) is algorithmic in the sence of § 2. We shall argue later on (see § 7) that it is reasonable to assume that, conversely, every algorithmic function(al) is computable; but this is a matter of reasonable belief, not a theorem.

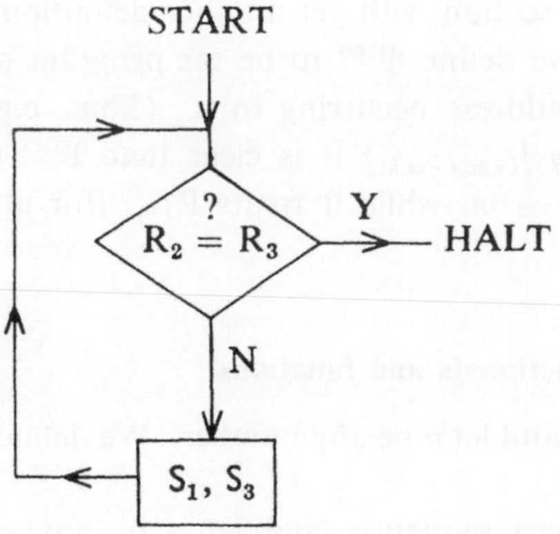

Fig. 2

4.1. Example. We show that addition (i.e., the function f defined by $f(x_1, x_2) = x_1 + x_2$) is computable. We first set up a flow chart (see Fig. 2). We assume that at the start numbers a_1 and a_2 are stored in R_1 and R_2 respectively, and the other registers (in particular R_3) are empty. R_3 is used as a *loop counter*: it shows how many times URIM has been round the loop. Each time round, 1 is added to both R_1 and R_3. This goes on so long as the number in R_3 is less than a_2. When the contents of R_3 reaches a_2, this means that 1 has already been added a_2 times to R_1. At that moment the number in R_1 must be $a_1 + a_2$ and it is time to halt.

Using the same method as in Ex.3.1, we can convert our chart into the program $\langle J_{2,3,4}, S_1, S_3, J_{1,1,0} \rangle$.

Let φ be any given sequence. If $\mathfrak{P}$ is any program, we can get an n-ary function f from the functional $\mathfrak{P}_n$ by giving its sequence variable the fixed value φ. Thus

$$f(\mathfrak{x}) = \mathfrak{P}_n(\varphi; \mathfrak{x}),$$

i.e., f is $\lambda \mathfrak{x} \mathfrak{P}_n(\varphi;\mathfrak{x})$. We say that $\mathfrak{P}$ *computes f relative to* φ (or, briefly, *φ-computes f*). A function is *computable relative to* φ (briefly, *φ-computable*) if it is φ-computed by some program.

From now on we shall often omit the subscript "n" from expressions like "$\mathfrak{P}_n(\varphi;\mathfrak{x})$", "$\mathfrak{P}_n(\mathfrak{x})$" etc., since the n can in any case be determined from the number of numerical arguments shown.

4.2. THEOREM. *A computable function is computable relative to every sequence. A function computable relative to the sequence $\lambda x(x+1)$ is computable.*

PROOF. Let f be an n-ary computable function. Then, by definition, the identity $f(\mathfrak{x})=\mathfrak{P}(\xi;\mathfrak{x})$ holds for some program $\mathfrak{P}$ having no A commands. But then $\mathfrak{P}(\xi;\mathfrak{x})$ is independent of the value assigned to ξ, so for every sequence φ we have $f(\mathfrak{x})=\mathfrak{P}(\varphi;\mathfrak{x})$, and hence f is φ-computable.

Now suppose that f is an n-ary function computable relative to the particular sequence φ for which $\varphi(x)=x+1$. Then for some program $\mathfrak{P}$ the identity $f(\mathfrak{x})=\mathfrak{P}(\varphi;\mathfrak{x})$ must hold. Let $\mathfrak{Q}$ be the program obtained from $\mathfrak{P}$ upon replacing every command A_i by S_i. Then clearly $\mathfrak{Q}$ has no A commands. Also, when URIM is linked up to a φ-oracle, A_i has exactly the same effect as S_i. It follows that $f(\mathfrak{x})=\mathfrak{Q}(\mathfrak{x})$ and f is a computable function. ∎

Every sequence φ is computable relative to itself (it is φ-computed by the program $\langle A_1 \rangle$). On the other hand, since there are only denumerably many programs and non-denumerably many sequences, most sequences are not computable. Thus φ-computability does not in general imply computability.

§ 5. Recursive functionals and functions

In this section we shall define the class of *recursive* functionals. Although this definition will not refer to machines and programs, it will be seen later on that a functional is recursive iff it is computable in the sense of §4.

We begin by introducing an infinite list of formal *symbols:*

$$\mathbf{C}, \mathbf{P}, \mathbf{M}, \mathbf{Z}, \mathbf{S}, \mathbf{A}, \mathbf{I}_{n,i},$$

where n and i are any positive numbers such that $1 \leqslant i \leqslant n$.

Certain expressions made up of these symbols (using also parentheses and commas) will be called *descriptions.* More precisely, for each number n we shall have *n-ary descriptions.*

5.1. DEFINITION. *Descriptions* are expressions formed according to the following seven rules:

(1) The symbol **Z** is a *0-ary description*.

(2) The symbol **S** is a *unary description*.

(3) The symbol **A** is a *unary description*.

(4) For any n and i such that $1 \leq i \leq n$, the symbol $\mathbf{I}_{n,i}$ is an *n-ary description*.

(5) If **G** is a *k-ary description*, with $k \geq 0$, and if $\mathbf{H}_1,\ldots,\mathbf{H}_k$ are *n-ary descriptions*, with $n \geq 0$, then $\mathbf{C}(\mathbf{G}, \mathbf{H}_1,\ldots,\mathbf{H}_k)$ is an *n-ary description*. (Thus if $k=0$ then $\mathbf{C}(\mathbf{G})$ is an *n-ary description* for every n.)

(6) If **G** is an *n-ary description*, with $n \geq 0$, and **H** is an *(n+2)-ary description*, then $\mathbf{P}(\mathbf{G}, \mathbf{H})$ is an *(n+1)-ary description*.

(7) If **G** is an *(n+1)-ary description*, with $n \geq 0$, then $\mathbf{M}(\mathbf{G})$ is an *n-ary description*.

By the *degree of* a description **F** (briefly, deg **F**) we mean the total number of occurrences of the symbols **C**, **P** and **M** in **F**.

To each n-ary description **F** we shall now assign an n-ary functional F; and we shall say that **F** *describes*, or is a description *of* F. (Here and in the sequel we adopt the convention of denoting a description by a bold Roman letter, and the functional described by it is then denoted by the corresponding italic letter.) We proceed by recursion on deg **F**, following the seven cases of Def. 5.1.

5.2. DEFINITION.

(1) The symbol **Z** *describes* the 0-ary functional Z defined by the identity

$$Z(\xi)=0.$$

(2) The symbol **S** *describes* the unary functional S defined by the identity

$$S(\xi;x)=x+1.$$

(3) The symbol **A** *describes* the unary functional A defined by the identity

$$A(\xi;x)=\xi(x).$$

(4) For any n and i, $1 \leq i \leq n$, $\mathbf{I}_{n,i}$ describes the n-ary functional $I_{n,i}$ defined by the identity

$$I_{n,i}(\xi;\mathbf{x})=x_i.$$

(5) If **F** is **C(G, H$_1$,...,H$_k$)**, where **G** *describes* the k-ary functional G and **H$_1$,...,H$_k$** *describe* the n-ary functionals $H_1,\ldots,H_k$ respectively, then **F** *describes* the n-ary functional F defined by the identity

$$F(\xi;\mathfrak{x})=G(\xi;H_1(\xi;\mathfrak{x}),\ldots,H_k(\xi;\mathfrak{x})).$$

We say that F is obtained from G and $H_1,\ldots,H_k$ by *composition*. Note that $F(\varphi;\mathfrak{a})\neq\infty$ iff both $H_i(\varphi;\mathfrak{a})\neq\infty$ for $i=1,\ldots,k$ and

$$G(\varphi;H_1(\varphi;\mathfrak{a}),\ldots,H_k(\varphi;\mathfrak{a}))\neq\infty.$$

(6) If **F** is **P(G,H)**, where **G** *describes* the n-ary functional G and **H** *describes* the $(n+2)$-ary functional H, then **F** *describes* the $(n+1)$-ary functional F defined by the two identities

$$F(\xi;\mathfrak{x},0)=G(\xi;\mathfrak{x}),$$

$$F(\xi;\mathfrak{x},y+1)=H(\xi;\mathfrak{x},y,F(\xi;\mathfrak{x},y)).$$

We say that F is obtained from G and H by *primitive recursion*. (It is easy to verify, by induction on the value assigned to y, that these two identities together determine F uniquely.) Note that if, for given φ, $\mathfrak{a}$ and b, we have $F(\varphi;\mathfrak{a},b)=\infty$, then also $F(\varphi;\mathfrak{a},b')=\infty$ for every $b'>b$.

(7) If **F** is **M(G)**, where **G** *describes* the $(n+1)$-ary functional G, then **F** *describes* the n-ary functional F defined by the identity

$$F(\xi;\mathfrak{x})=\mu y G(\xi;\mathfrak{x},y).$$

We say that F is obtained from G by *minimization*. Note that even if, for given φ and $\mathfrak{a}$, $G(\varphi;\mathfrak{a},y)\neq\infty$ for every value of y, we may still have $F(\varphi;\mathfrak{a})=\infty$.

A functional is *recursive* if it has a description.

Note that while each n-ary description describes a unique n-ary functional, the converse is false. For example, for any description **F**, it is clear that **F** and **C(I$_{1,1}$, F)** describe the same functional. It follows that each recursive functional has infinitely many descriptions.

If **F** is an n-ary description in which the symbol **A** does not occur, then it is easy to verify by induction on deg **F** that the corresponding functional F does not depend on its sequence argument. Thus in this case we can define an n-ary function — which we identify with F and denote by "F" as well — by putting

$$F(\mathfrak{x})=F(\xi;\mathfrak{x}).$$

We then say that **F** *describes* this function F, and that F is a *recursive function.* If **F** is a description in which the symbol **M** does not occur, the functional described by **F** is said to be *primitive recursive* (or, briefly, *p.r.*). If neither **A** nor **M** occur in **F**, then the function described by **F** is *primitive recursive* (or *p.r.*).

It is easy to see that every p.r. function(al) is total.

Let φ be any given sequence. Then, if F is an n-ary recursive functional, the function $f=\lambda\mathfrak{x}F(\varphi;\mathfrak{x})$ is said to be *recursive relative to* φ (briefly, φ*-recursive*). A description of F will also be said to *describe* f *relative to* φ.

5.3. THEOREM. *A recursive function is* φ*-recursive for every sequence* φ. *A function recursive relative to* $\lambda x(x+1)$ *is recursive.*

PROOF. The first part is obvious. To prove the second part, suppose that for $\varphi=\lambda x(x+1)$ we have $f(\mathfrak{x})=F(\varphi;\mathfrak{x})$, where F has a description **F**. Let **G** be obtained from **F** upon replacing the symbol **A** by **S**. It is easy to see (by induction on deg **F**) that **G** describes f. ∎

5.4. PROBLEM. Let G and H be an n-ary and an $(m+1)$-ary recursive functional, and let H be total[1]. Let F be an $(n+m)$-ary functional satisfying the identity

$$F(\xi;\mathfrak{x},z_1,\ldots,z_m)=G(\lambda yH(\xi;y,z_1,\ldots,z_m);\mathfrak{x}).$$

Show that F is recursive. (Use induction on the degree of a given description of G. Verify that the identity

$$A(\lambda yH(\xi;y,z_1,\ldots,z_m);x)=H(\xi;x,z_1,\ldots,z_m)$$

holds.)

We shall now show that every recursive function(al) is computable.

5.5. THEOREM. *Given any description* **F** *of a function(al) we can construct a program which computes that function(al).*

PROOF. We shall deal with functionals rather than functions. However, the reader can easily check that the program we shall construct will contain A instructions only if the symbol **A** occurs in **F**; so the result for functions will follow at once.

Suppose then that **F** describes the n-ary functional F. By induction on deg **F** we show how to write a program $\mathfrak{P}$ that not only computes F,

[1] We require H to be total to make sure that the unary function $\lambda yH(\xi;y,z_1,\ldots,z_m)$ (which depends on $\xi,z_1,\ldots,z_m$ as parameters) is always total and hence eligible to serve as the *sequence* argument of G.

but has the following stronger property:

$$\mathfrak{P}_{n+m}(\xi;\mathfrak{x},y_1,\ldots,y_m)=F(\xi;\mathfrak{x}) \quad \text{for all } m.$$

That is, when URIM operates under $\mathfrak{P}$, the existence and value of the output do not depend on the initial contents of registers other than the first n. Thus, when using $\mathfrak{P}$ to compute F we need not insist that initially all registers other than the first n be empty. Of such $\mathfrak{P}$ we say that it computes F *strongly*.

It will be convenient to use the following notation: for all $m>n$ we let $\mathfrak{C}_{n,m}$ be the program[1]

$$\mathfrak{R}_{1,m}\mathfrak{R}_{2,m+1}\cdots\mathfrak{R}_{n,m+n-1}.$$

It is easy to see that the effect of $\mathfrak{C}_{n,m}$ is to copy the contents of $R_1,\ldots,R_n$ into $R_m,\ldots,R_{m+n-1}$ respectively.

We proceed by cases, corresponding to those in Defs. 5.1 and 5.2.

(1)–(4). The functionals Z,S,A, and $I_{n,i}$ are strongly computed by the programs $\langle Z_1\rangle$, $\langle S_1\rangle$, $\langle A_1\rangle$ and $\mathfrak{R}_{i,1}$ respectively.

(5) Suppose **F** is **C(G, H$_1$,...,H$_k$)** and

$$F(\xi;\mathfrak{x})=G(\xi;H_1(\xi;\mathfrak{x}),\ldots,H_k(\xi;\mathfrak{x})).$$

By the induction hypothesis, we possess programs $\mathfrak{Q},\mathfrak{Q}_1,\ldots,\mathfrak{Q}_k$ that strongly compute $G, H_1,\ldots,H_k$ respectively. We set up the required program $\mathfrak{P}$, which strongly computes F, in the form of a flow chart (see Fig. 3). For the reader's convenience, we show alongside the chart the state of the registers at various key stages in the computation of $F(\varphi;\mathfrak{a})$. Asterisks stand for registers whose contents are irrelevant.

(6) Suppose **F** is **P(G, H)** and

$$F(\xi,\mathfrak{x},0)=G(\xi;\mathfrak{x}),$$

$$F(\xi,\mathfrak{x},y+1)=H(\xi;\mathfrak{x},y,F(\xi,\mathfrak{x},y)).$$

(Here F is $(n+1)$-ary rather than n-ary.) By the induction hypothesis, we possess programs $\mathfrak{Q}$ and $\mathfrak{R}$ which strongly compute G and H respectively. We set up a flow chart (see Fig. 4) which (as in Ex.3.1) can easily be converted into a program $\mathfrak{P}$ that strongly computes F. Again, the state of the registers at various key stages is shown alongside. The reader should note that R_{n+2} is used as a loop counter: at the start it is erased, and from

[1] For the meaning of $\mathfrak{R}_{i,j}$ see Ex. 3.1. Concatenation of programs was also defined in §3.

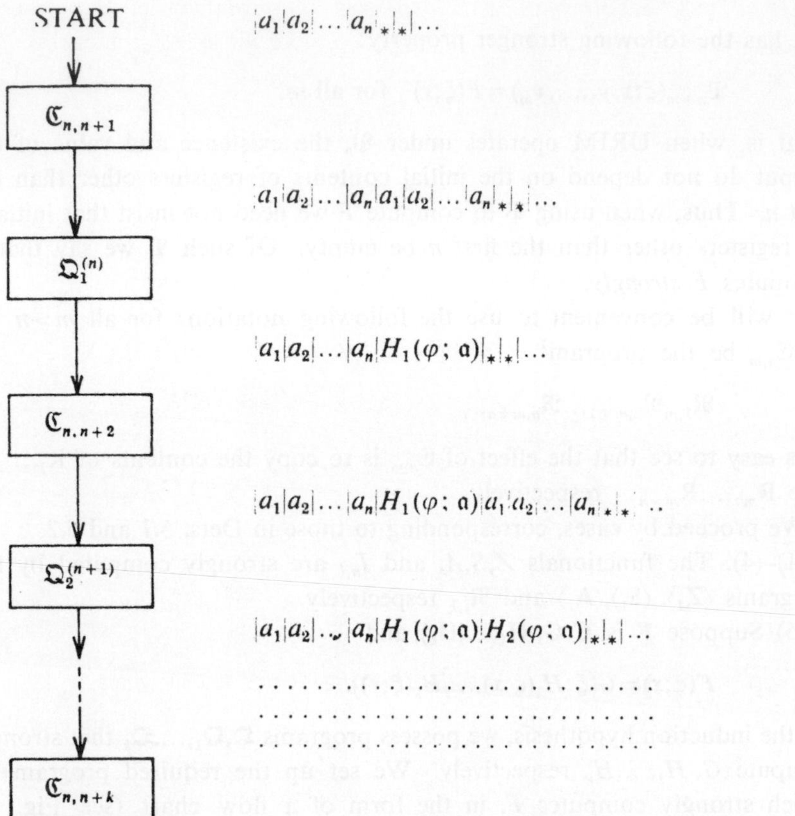
START
$|a_1|a_2|\dots|a_n|_*|_*|\dots$
$\mathfrak{C}_{n,n+1}$
$|a_1|a_2|\dots|a_n|a_1|a_2|\dots|a_n|_*|_*|\dots$
$\mathfrak{Q}_1^{(n)}$
$|a_1|a_2|\dots|a_n|H_1(\varphi;\mathfrak{a})|_*|_*|\dots$
$\mathfrak{C}_{n,n+2}$
$|a_1|a_2|\dots|a_n|H_1(\varphi;\mathfrak{a})|a_1|a_2|\dots|a_n|_*|_*\dots$
$\mathfrak{Q}_2^{(n+1)}$
$|a_1|a_2|\dots|a_n|H_1(\varphi;\mathfrak{a})|H_2(\varphi;\mathfrak{a})|_*|_*|\dots$
$\mathfrak{C}_{n,n+k}$
$\mathfrak{Q}_k^{(n+k-1)}$
$|a_1|a_2|\dots|a_n|H_1(\varphi;\mathfrak{a})|H_2(\varphi;\mathfrak{a})|\dots|H_k(\varphi;\mathfrak{a})|_*|_*|\dots$
$\mathfrak{Q}^{(n)}$
$|a_1|a_2|\dots|a_n|F(\varphi;\mathfrak{a})|_*|_*|\dots$
$\mathfrak{R}_{n+1,1}$
HALT
$F(\varphi;\mathfrak{a})|_*|_*|\dots$

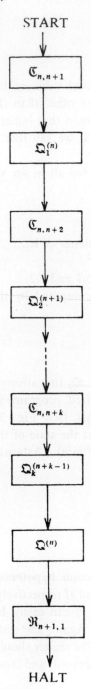

Fig. 3

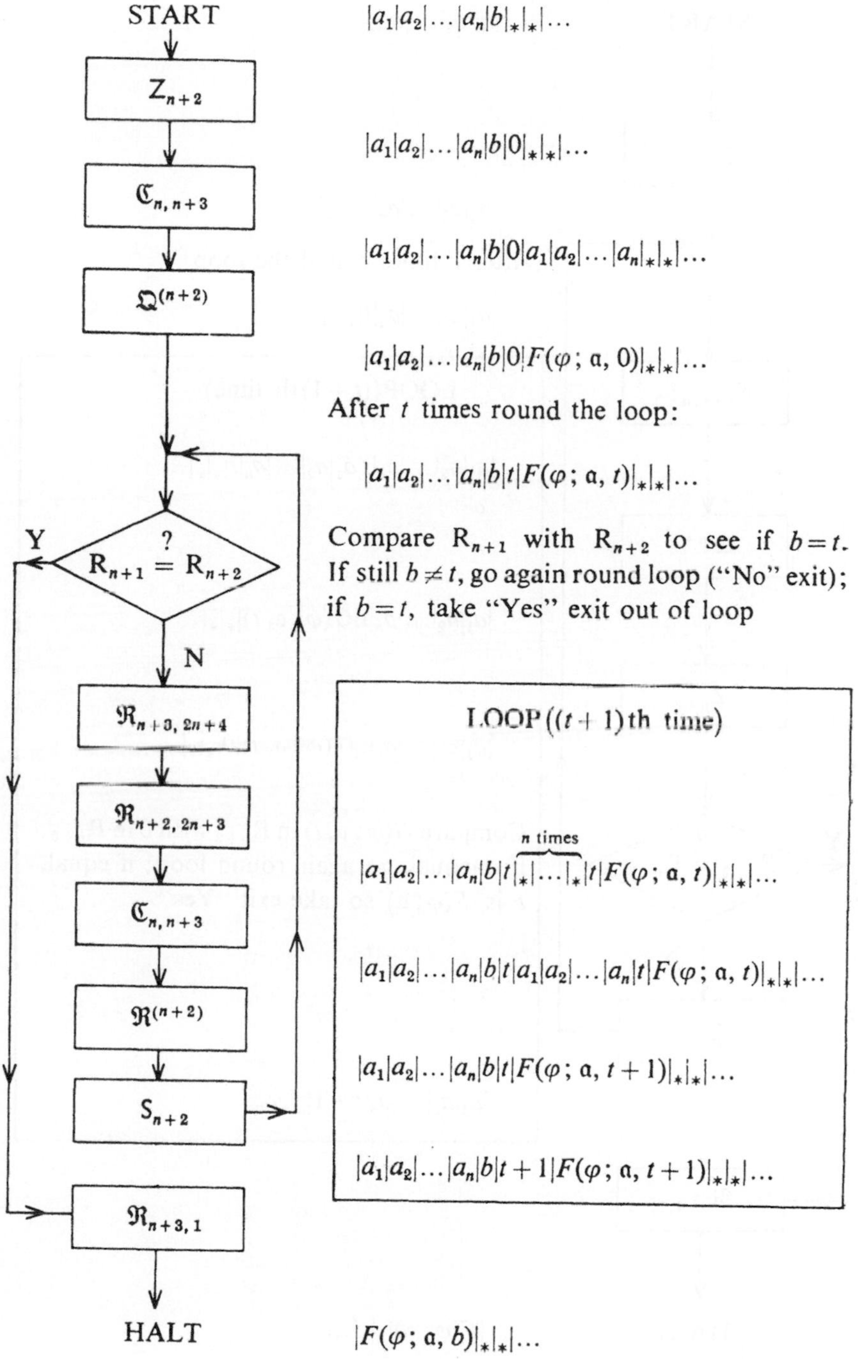
START
$|a_1|a_2|\dots|a_n|b|_*|_*|\dots$
Z_{n+2}
$|a_1|a_2|\dots|a_n|b|0|_*|_*|\dots$
$\mathfrak{C}_{n,n+3}$
$|a_1|a_2|\dots|a_n|b|0|a_1|a_2|\dots|a_n|_*|_*|\dots$
$\mathfrak{Q}^{(n+2)}$
$|a_1|a_2|\dots|a_n|b|0|F(\varphi;\mathfrak{a},0)|_*|_*|\dots$
After t times round the loop:
$|a_1|a_2|\dots|a_n|b|t|F(\varphi;\mathfrak{a},t)|_*|_*|\dots$
Y
$R_{n+1} \overset{?}{=} R_{n+2}$
Compare R_{n+1} with R_{n+2} to see if $b=t$. If still $b\neq t$, go again round loop ("No" exit); if $b=t$, take "Yes" exit out of loop
N
$\mathfrak{R}_{n+3,\,2n+4}$
LOOP ($(t+1)$th time)
$\mathfrak{R}_{n+2,\,2n+3}$
n times
$|a_1|a_2|\dots|a_n|b|t|_*|\dots|_*|t|F(\varphi;\mathfrak{a},t)|_*|_*|\dots$
$\mathfrak{C}_{n,n+3}$
$|a_1|a_2|\dots|a_n|b|t|a_1|a_2|\dots|a_n|t|F(\varphi;\mathfrak{a},t)|_*|_*|\dots$
$\mathfrak{R}^{(n+2)}$
$|a_1|a_2|\dots|a_n|b|t|F(\varphi;\mathfrak{a},t+1)|_*|_*|\dots$
S_{n+2}
$|a_1|a_2|\dots|a_n|b|t+1|F(\varphi;\mathfrak{a},t+1)|_*|_*|\dots$
$\mathfrak{R}_{n+3,\,1}$
HALT
$|F(\varphi;\mathfrak{a},b)|_*|_*|\dots$

Fig. 4

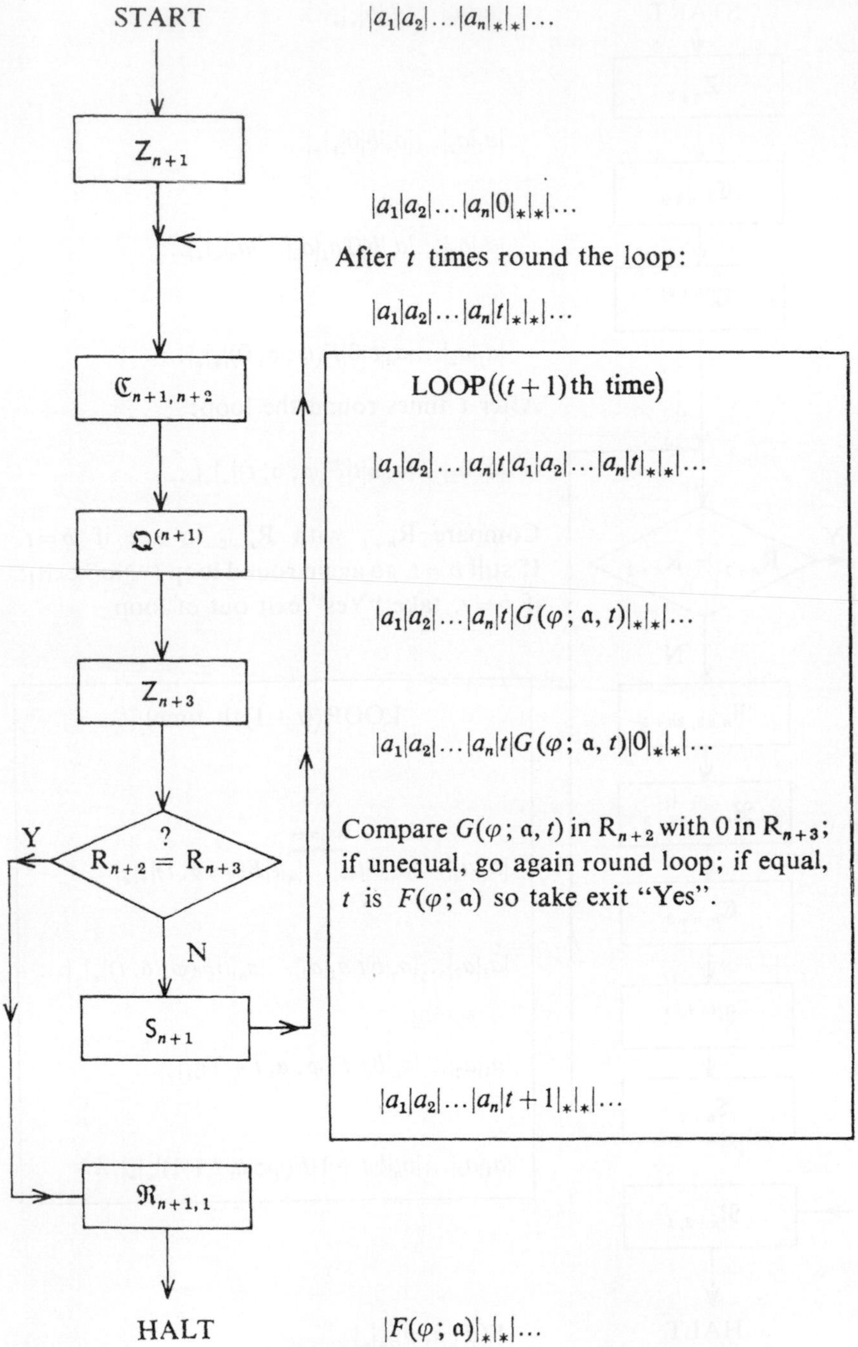
START
$|a_1|a_2|\dots|a_n|_*|_*|\dots$
Z_{n+1}
$|a_1|a_2|\dots|a_n|0|_*|_*|\dots$
After t times round the loop:
$|a_1|a_2|\dots|a_n|t|_*|_*|\dots$
LOOP$((t+1)$th time)
$\mathfrak{C}_{n+1,n+2}$
$|a_1|a_2|\dots|a_n|t|a_1|a_2|\dots|a_n|t|_*|_*|\dots$
$\mathfrak{Q}^{(n+1)}$
$|a_1|a_2|\dots|a_n|t|G(\varphi;\, \mathfrak{a}, t)|_*|_*|\dots$
Z_{n+3}
$|a_1|a_2|\dots|a_n|t|G(\varphi;\, \mathfrak{a}, t)|0|_*|_*|\dots$
Y
$R_{n+2} \overset{?}{=} R_{n+3}$
Compare $G(\varphi;\, \mathfrak{a}, t)$ in R_{n+2} with 0 in R_{n+3}; if unequal, go again round loop; if equal, t is $F(\varphi;\, \mathfrak{a})$ so take exit "Yes".
N
S_{n+1}
$|a_1|a_2|\dots|a_n|t+1|_*|_*|\dots$
$\mathfrak{R}_{n+1,1}$
HALT
$|F(\varphi;\, \mathfrak{a})|_*|_*|\dots$

Fig. 5

then on the number stored in it is the number of times the computation has gone round the loop.

(7) Finally, suppose **F** is **M(G)** and

$$F(\xi;\mathfrak{x})=\mu y G(\xi;\mathfrak{x},y).$$

By the induction hypothesis, we possess a program $\mathfrak{Q}$ that strongly computes G. Again, we set up a flow chart (see Fig. 5) which can easily be converted into a program $\mathfrak{P}$ that strongly computes F. We use R_{n+1} as a loop counter to show how many times the computation has gone through the loop.

This completes our proof. ∎

§ 6. A stockpile of examples

In this section we collect examples of recursive functionals, not merely for illustration but for future use. But first we introduce some new terminology.

Consider an operation that can be applied to given functionals, say $G_1,\ldots,G_k$ (which may either be arbitrary or subject to some special conditions) to yield another functional, say F. We shall say that the operation is *recursive* if for any given descriptions $\mathbf{G}_1,\ldots,\mathbf{G}_k$ of $G_1,\ldots,G_k$, respectively, we can construct a description **F** of the corresponding F such that:

(1) If the descriptions $\mathbf{G}_1,\ldots,\mathbf{G}_k$ do not contain the symbol **A**, then **F** too does not contain **A**.

Clearly, in this case, if $G_1,\ldots,G_k$ are recursive functionals (or functions), then F too is a recursive functional (or function, respectively).

We shall say that the operation in question is *primitive recursive* (briefly, *p.r.*) if in addition to (1) we also have:

(2) If the descriptions $\mathbf{G}_1,\ldots,\mathbf{G}_k$ do not contain the symbol **M**, then **F** too does not contain **M**.

In this case, if $G_1,\ldots,G_k$ are p.r. functionals (or functions) then so is F.

Our treatment can be made to cover *relations* on N as well as functions. In fact, we shall find it convenient to identify an n-ary relation R on N with the n-ary function f defined by

$$f(\mathfrak{x})=\begin{cases}0 & \text{if} \quad \mathfrak{x}\in R,\\ 1 & \text{if} \quad \mathfrak{x}\in N^n-R.\end{cases}$$

This amounts merely to identifying the truth values *truth* and *falsehood* with 0 an 1 respectively.

Thus, for the purposes of recursion theory, we agree that by *first-order n-ary relation* we mean a total n-ary function R such that $R(\mathfrak{x}) \leqslant 1$ for all $\mathfrak{x} \in N^n$.

Similarly, by *second-order n-ary relation* we mean a total n-ary functional R such that $R(\xi;\mathfrak{x}) \leqslant 1$ for all $\xi \in N^N$ and all $\mathfrak{x} \in N^n$.

By *relation* we mean (unless otherwise indicated) a first-order or second-order n-ary relation for some n. Note that, since by our convention functions are a particular kind of functional, first-order relations are a particular kind of second-order relation, and treatment of the former is subsumed under that of the latter.

If R is a relation, we shall often say, e.g., "$R(\xi;\mathfrak{x})$ holds" or "$R(\xi;\mathfrak{x})$ is true" instead of "$R(\xi;\mathfrak{x})=0$".

We are now ready to proceed to our examples.

6.1. EXAMPLE. Let G be any given k-ary functional, and let the n-ary functional F be obtained from G by the identity

$$F(\xi;\mathfrak{x})=G(\xi;x_{i_1},x_{i_2},\dots,x_{i_k}),$$

where $1 \leqslant i_j \leqslant n$ for $j=1, 2,\dots,k$ and the i_j need not be distinct. Then this is a p.r. operation; for, if $\mathbf{G}$ describes G then F is described by

$$\mathbf{C}(\mathbf{G}, \mathbf{I}_{n,i_1}, \mathbf{I}_{n,i_2},\dots,\mathbf{I}_{n,i_k}).$$

There are three important special cases of this operation:

(1) *Permutation of variables*; e.g., $k=n=3$ and F is obtained from G through the identity

$$F(\xi;x,y,z)=G(\xi;y,z,x).$$

(2) *Identification of variables*; e.g., $k=3$, $n=2$ and the identity

$$F(\xi;x,y)=G(\xi;x,y,x).$$

(3) *Introduction of a redundant* (or *fictitious*) *variable*; e.g., $k=1$, $n=2$ and the identity[1]

$$F(\xi;x,y)=G(\xi;x).$$

[1] Note that by the conventions adopted in §1, this identity means that for every sequence φ and *numbers* a and b we must have $F(\varphi;a,b)=G(\varphi;a)$. However, if instead of b we take ∞, then we have $F(\varphi;a,\infty)=\infty$; hence we do *not* have $F(\varphi;a,\infty)=G(\varphi;a)$, unless $G(\varphi;a)$ happens to be ∞.

We shall often apply these operations without special mention. Suppose, e.g., that F is obtained from G, H_1 and H_2 through the identity

$$F(\xi;x,y)=G(\xi;H_1(\xi;x), H_2(\xi;y,x)).$$

Then we conclude at once that this is a p.r. operation and say (stretching the terminology introduced in Def. 5.2) that F is obtained from G, H_1 and H_2 by *composition*. (In fact, we should first obtain two other functionals, say H_1' and H_2', by

$$H_1'(\xi;x,y)=H_1(\xi;x),$$

$$H_2'(\xi;x,y)=H_2(\xi;y,x),$$

i.e., by introduction of a fictitious variable and permutation of variables; and *then* get F from G, H_1' and H_2' by composition in the strict sense of Def. 5.2.)

Similarly, suppose e.g. that F is obtained from G and H through the identities

$$F(\xi;0,y,z) = G(\xi,y),$$

$$F(\xi;x+1,y,z) = H(\xi,F(\xi,x,y,z),z).$$

Then we conclude at once that this is a p.r. operation and (again stretching the terminology of Def. 5.2) say that F is obtained from G and H by *primitive recursion*.

Again, suppose e.g. that F is obtained from G by

$$F(\xi;y) = \mu xG(\xi;x,x,y).$$

Then we conclude at once that this operation is recursive and say that F is obtained from G by *minimization*.

6.2. EXAMPLE. In §1 we agreed to identify any number n with the 0-ary function having n as its (unique) value. (This function is in turn identified with the corresponding 0-ary functional.)

The numbers 0, 1, 2 etc. are then p.r. functions, described by **Z**, **C**(**S**,**Z**), **C**(**S**,**C**(**S**,**Z**)) etc.

As for the 0-ary function ∞, it is recursive, described e.g. by **M**(**S**).

From now on we shall leave it to the reader to construct appropriate descriptions of the functionals we introduce, or at any rate to check that they *could* be constructed.

6.3. EXAMPLE. Let b be any number. If F is obtained from G by putting

$$F(\xi;\mathfrak{x})=G(\xi;\mathfrak{x},b),$$

then by Ex. 6.2 this is a p.r. operation. We shall use this result frequently, without special mention.

(Warning: if φ is a non-recursive sequence[1], then the operation yielding f from G through the identity $f(\mathfrak{x})=G(\varphi;\mathfrak{x})$ is *not* recursive. For example, we have $\varphi(x)=A(\varphi;x)$ although A is recursive and φ is not.)

6.4. EXAMPLE. The *sum* function $\lambda xy(x+y)$ is p.r., since it may be obtained by primitive recursion from $I_{1,1}$ and S, using the identities

$$x+0=x,$$

$$x+(y+1)=(x+y)+1.$$

Product and *power*, i.e., $\lambda xy(xy)$ and $\lambda xy(x^y)$ are now easily seen to be p.r. functions as well.

For typographical reasons we shall often write, e.g., "$\exp(x,y)$" instead of "x^y".

6.5. EXAMPLE. The *cut-off difference* function $\lambda xy(x \dot{-} y)$ is defined by

$$x \dot{-} y=\max(x-y,\ 0).$$

To show that it is p.r., we first consider the unary function f, where $f(x)=x \dot{-} 1$. This f may be obtained by primitive recursion from 0 and $I_{1,1}$:

$$f(0)=0,$$

$$f(x+1)=x.$$

Now cut-off difference, in turn, is obtained by primitive recursion from $I_{1,1}$ and f:

$$x \dot{-} 0=x,$$

$$x \dot{-} (y+1)=f(x \dot{-} y).$$

(The reader should check the correctness of these identities.)

The function max can be defined by

$$\max(x,y)=(x \dot{-} y)+y,$$

[1] Such sequences exist, since there are uncountably many sequences and only countably many of them are recursive.

i.e., it is obtained by composition from $+$ and $\dot{-}$, and is thus a p.r. function.

6.6. EXAMPLE. We introduce the *propositional operations* as p.r. operations.

We define $\neg$ as the unary function $\lambda x(1 \dot{-} x)$, which is clearly a p.r. function. The *operation of negation* is that which when applied to an n-ary functional G yields the functional $\neg G$ obtained by composition of $\neg$ and G, i.e.,

$$\neg G(\xi;\mathfrak{x}) = \neg(G(\xi;\mathfrak{x})).$$

This is evidently a p.r. operation. If G is a relation, then it is easy to verify that $\neg G$ is the relation that takes the value *truth* (i.e., 0) precisely where G takes the value *falsehood* (i.e., 1).

Next, we define $\vee$ to be the product function $\lambda xy(xy)$, which we know from Ex.6.4 to be a p.r. function. We write "$x \vee y$" instead of "$\vee(x,y)$". The *disjunction operation* is that which when applied to n-ary functionals G and H yields the n-ary functional $G \vee H$ such that

$$(G \vee H)(\xi;\mathfrak{x}) = G(\xi;\mathfrak{x}) \vee H(\xi;\mathfrak{x}).$$

This operation is clearly p.r., and if G and H are relations then $G \vee H$ is the relation which is true precisely where G and/or H are true.

Finally, we define $\wedge$ and $\rightarrow$ to be the p.r. functions $\lambda xy(\neg(\neg x \vee \neg y))$ and $\lambda xy(\neg x \vee y)$ respectively. The operations of *conjunction* and *implication* are then defined in the obvious way (and are seen to be p.r.).

6.7. EXAMPLE. The *order relation* $<$ can be defined as the p.r. function $\lambda xy(1 \dot{-} (y \dot{-} x))$. We write "$(x<y)$" instead of "$<(x,y)$". The relation $>$ is p.r. since it is obtained from $<$ by permutation of variables:

$$(x>y) = (y<x).$$

The p.r. relations $\leqslant$ and $\geqslant$ are introduced by

$$(x \leqslant y) = \neg(y<x),$$
$$(x \geqslant y) = (y \leqslant x).$$

The *equality relation* $\doteq$ is defined by

$$(x \doteq y) = (x \leqslant y) \wedge (y \leqslant x),$$

and is thus primitive recursive.

The reader should note that when $\doteq$ (or, for that matter, any other relation) is composed with functionals, the resulting functional may not be a relation. Suppose, e.g., that G and H are n-ary functionals and F

satisfies the identity

$$(1) \qquad F(\xi,\mathfrak{x})=(G(\xi;\mathfrak{x})\doteq H(\xi;\mathfrak{x})).$$

Then clearly F is obtained from G and H by a p.r. operation and is recursive (or p.r.) if G and H are. But, according to our conventions, if $\varphi\in N^N$ and $\mathfrak{a}\in N^n$ are such that $G(\varphi;\mathfrak{a})=\infty$ or $H(\varphi;\mathfrak{a})=\infty$ (or both) then also $F(\varphi;\mathfrak{a})=\infty$; whereas a relation must be total. In fact, F is a relation iff G and H are *total*.

This state of affairs should be contrasted with the following definition. Let r and s be expressions whose values are in the set $N\cup\{\infty\}$ and depend on values assigned to the sequence variable ξ and numerical variables $\mathfrak{x}$. Then the expression

$$(2) \qquad (r=s)$$

will, for any given values of ξ in N^N and $\mathfrak{x}$ in N^n, have the value 0 or 1, according as the corresponding values of r and s are equal (both being the same number or both ∞) or different. It must be stressed that expression (2) must be taken as an indivisible whole, *not* as a result of substituting r and s for x and y respectively in "$(x=y)$".

According to this definition, if G and H are as before and

$$(3) \qquad P(\xi;\mathfrak{x})=(G(\xi;\mathfrak{x})=H(\xi;\mathfrak{x})),$$

then P is certainly a relation. On the other hand, we shall see later that P is not necessarily recursive even if G and H are.

Finally we note that if r and s never take the value ∞ for any values of ξ in N^N and $\mathfrak{x}$ in N^n, then we have

$$(r=s)=(r\doteq s),$$

where "$(r\doteq s)$" *is* the result of substituting r and s for x and y, respectively, in "$(x\doteq y)$". Thus in *this* case we may and shall write "$(r=s)$" instead of "$(r\doteq s)$". In particular, if G and H are total, then F and P of (1) and (3) are the same.

We shall write "$(r\neq s)$" instead of "$\neg(r=s)$".

6.8. Example. The operation of *summation* is applied to an $(n+1)$-ary functional G, yielding another $(n+1)$-ary functional F:

$$F(\xi;\mathfrak{x},y)=\sum_{z<y} G(\xi;\mathfrak{x},z).$$

That this operation is p.r. can be seen from the identities

$$F(\xi;\mathfrak{x},0)=0,$$

$$F(\xi;\mathfrak{x},y+1)=F(\xi;\mathfrak{x},y)+G(\xi;\mathfrak{x},y).$$

Similarly, the *product* operation, where

$$F(\xi;\mathfrak{x},y)=\prod_{z<y} G(\xi;\mathfrak{x},z),$$

is p.r. because

$$F(\xi;\mathfrak{x},0)=1,$$

$$F(\xi;\mathfrak{x},y+1)=F(\xi;\mathfrak{x},y)\cdot G(\xi;\mathfrak{x},y).$$

(We sometimes write, e.g., "$x\leqslant y$" under the summation or product sign, as short for "$x<y+1$".)

6.9. EXAMPLE. *Bounded existential and universal quantifications* are two operations that when applied to an $(n+1)$-ary relation P yield $(n+1)$-ary relations Q and R, where

$$Q(\xi;\mathfrak{x},y)=\exists z<y\, P(\xi;\mathfrak{x},z),$$

$$R(\xi;\mathfrak{x},y)=\forall z<y\, P(\xi;\mathfrak{x},z).$$

Here $Q(\varphi;\mathfrak{a},b)$ holds iff there is *some* $c<b$ for which $P(\varphi;\mathfrak{a},c)$ holds; and $R(\varphi;\mathfrak{a},b)$ holds iff $P(\varphi;\mathfrak{a},c)$ holds for *every* $c<b$. That these operations are p.r. can be seen from the identities

$$Q(\xi;\mathfrak{x},y)=\prod_{z<y} P(\xi;\mathfrak{x},z),$$

$$R(\xi;\mathfrak{x},y)=\neg\exists z<y\,\neg P(\xi;\mathfrak{x},z).$$

We shall write, e.g., "$\exists z\leqslant y$" as short for "$\exists z<y+1$".

6.10. EXAMPLE. The *bounded minimum* operation is applied to an $(n+1)$-ary relation P to yield an $(n+1)$-ary total functional F:

(1) $$F(\xi;\mathfrak{x},y)=\mu z\,[(z<y)\wedge P(\xi;\mathfrak{x},z)\vee(z=y)].$$

The right-hand side of (1) will be written more briefly thus:

$$\min z<y\, P(\xi;\mathfrak{x},z).$$

(The reader should check that if there is some $c<b$ for which $P(\varphi;\mathfrak{a},c)$ holds, then $\min z<b\, P(\varphi;\mathfrak{a},z)$ is the smallest such c; if no such c exists, then $\min z<b\, P(\varphi;\, \mathfrak{a},z) = b$.)

From (1) it follows at once that this operation is recursive. But in fact it is even p.r., in view of the identity

$$\min z<y\, P(\xi;\mathfrak{x},z)=\sum_{z<y}(\exists v\leqslant z\, P(\xi;\mathfrak{x},v)),$$

which we ask the reader to verify.

6.11. EXAMPLE. *Division with remainder* provides us with two p.r. functions q and rm defined by

$$\begin{aligned}\mathrm{q}(x,y) &= \min z<x\,(y(z+1)>x),\\ \mathrm{rm}(x,y) &= x\dot{-}y\cdot\mathrm{q}(x,y).\end{aligned}$$

(If $b\neq 0$, then $\mathrm{q}(a,b)$ and $\mathrm{rm}(a,b)$ are respectively the quotient and remainder obtained when a is divided by b. Also, $\mathrm{q}(a,0)=\mathrm{rm}(a,0)=a$.)

6.12. EXAMPLE. *The sequence of primes* $\lambda x p_x$, where $p_0=2$, $p_1=3$, $p_2=5$ and so on through the prime numbers. Euclid's famous proof that there are infinitely many primes shows that $p_{x+1}\leqslant p_x!+1$. Hence we have[1] $p_{x+1}\leqslant \exp(p_x,p_x)$ and we can obtain our sequence by primitive recursion from functions already known to be p.r.:

$$p_0 = 2,$$

$$p_{x+1} = \min y\leqslant \exp(p_x,p_x)\{(y>p_x)\wedge\forall z<y\,[(z\leqslant 1)\vee(\mathrm{rm}(y,z)>0)]\}$$

Thus this sequence is primitive recursive.

We now define the binary p.r. function $\lambda zx\,[(z)_x]$ by the identity

$$(z)_x = \min y<z\,(\mathrm{rm}(z,p_x^{y+1})>0).$$

Clearly, if $z>0$ then $(z)_x$ is the greatest y for which p_x^y still divides z. (Also, $(0)_x=0$; but we shall not use this.) We call $(z)_x$ the xth *exponent in* z.

We shall write, e.g., "$(z)_{x,y}$" for "$((z)_x)_y$" etc.

Next, we define the p.r. function $\lambda z\mathrm{lh}(z)$ by

$$\mathrm{lh}(z) = \min y<z\,(\prod_{x<y}\exp(p_x,(z)_x)=z).$$

Clearly, for positive z, $\prod_{x<\mathrm{lh}(z)}\exp(p_x,(z)_x)$ is the canonical representation of z as a product of powers of primes. In other words, if e is a positive number, then the equality $e=\prod_{x<h}\exp(p_x,e_x)$ (with $e_{h-1}>0$ if $h>0$) holds iff $h=\mathrm{lh}(e)$ and $e_x=(e)_x$ for each $x<h$. (From the definition it also follows that $\mathrm{lh}(0)=0$, but we shall not use this.) We call $\mathrm{lh}(z)$ the *length of* z.

[1] Recall that $\exp(x,y)=x^y$.

6.13. EXAMPLE. *The pairing function J* is defined as follows. We enumerate all ordered pairs of numbers $\langle x,y\rangle$ in order of their increasing sum, $x+y$, and — among different pairs with the same sum — in order of increasing second component, y. Then $J(x,y)$ is the index of $\langle x,y\rangle$ in this enumeration, beginning with $J(0,0)=0$. Or, in other words, $J(x,y)$ is the number of pairs preceding $\langle x,y\rangle$ in the enumeration.

Now, for each w there are clearly $w+1$ different pairs $\langle u,v\rangle$ with $u+v=w$. Thus for each pair $\langle x,y\rangle$ there are altogether $1+2+\ldots+(x+y)$ different pairs whose sums are $<x+y$. Also, there are y pairs whose sums equal $x+y$ but whose second component is $<y$ (namely, the pairs $\langle x+y-z, z\rangle$ with $z<y$). It follows that

$$J(x,y) = \mathrm{q}((x+y)(x+y+1),2)+y.$$

Hence J is a p.r. function. (From this identity it also follows that $x+y\leqslant J(x,y)$ for all x and y.)

J is a bijection of N^2 onto N. Hence it has a pair of inverses K and L such that

$$K(J(x,y)) = x, \qquad L(J(x,y)) = y, \qquad J(K(z),L(z)) = z.$$

That these inverses are p.r. can be seen from the identities

$$K(z) = \min x\leqslant z\ \exists y\leqslant z\ (J(x,y)=z),$$
$$L(z) = \min y\leqslant z\ (J(K(z),y)=z).$$

6.14. EXAMPLE. *Definition by cases* is an operation that yields an n-ary functional F from n-ary functionals $G_1,\ldots,G_k$ and n-ary relations $R_1,\ldots,R_k$ such that

$$F(\xi;\mathfrak{x})=\begin{cases} G_1(\xi;\mathfrak{x}) & \text{if} \quad R_1(\xi;\mathfrak{x}),\\ G_2(\xi;\mathfrak{x}) & \text{if} \quad R_2(\xi;\mathfrak{x}),\\ \vdots & \\ G_k(\xi;\mathfrak{x}) & \text{if} \quad R_k(\xi;\mathfrak{x}).\end{cases}$$

(Here "$R_i(\xi;\mathfrak{x})$" is short for "$R_i(\xi;\mathfrak{x})$ holds".) The relations R_i are subject to the condition that for each $\varphi\in N^N$ and $\mathfrak{a}\in N^n$ there is exactly one i, $1\leqslant i\leqslant k$, such that $R_i(\varphi;\mathfrak{a})$ holds. Also, for the present we require all the G_i to be *total*. Under these conditions, the operation is p.r. since

$$F(\xi;\mathfrak{x}) = \neg R_1(\xi;\mathfrak{x})\cdot G_1(\xi;\mathfrak{x})+\ldots+\neg R_k(\xi;\mathfrak{x})\cdot G_k(\xi;\mathfrak{x}).$$

We shall see later that if the G_i are not required to be total, definition by cases is a recursive operation.

6.15. COUNTER-EXAMPLE. We know from Ex. 6.3 that if f is a binary recursive function then for each number a the unary functions $\lambda x f(x,a)$ and $\lambda y f(a,y)$ are also recursive. The converse, however, need not be true. For example, let φ be a non-recursive permutation of N, i.e., a sequence that takes each value exactly once. (Such a sequence exists, since there are uncountably many permutations of N.) Let R be the graph of φ, i.e., the binary relation such that

$$R(x,y)=(\varphi(x)=y).$$

For each a we have

$$R(x,a)=(x \doteq \varphi^{-1}(a)),$$
$$R(a,y)=(\varphi(a) \doteq y),$$

where φ^{-1} is the inverse of φ. These two identities show that $\lambda x R(x,a)$ and $\lambda y R(a,y)$ are recursive, and even primitive recursive. But φ can be obtained from R by minimization:

$$\varphi(x)=\mu y R(x,y),$$

so R cannot be recursive.

6.16. PROBLEM. The *Fibonacci sequence* φ is defined by the identities

$$\varphi(0)=0, \quad \varphi(1)=1, \quad \varphi(x+2)=\varphi(x+1)+\varphi(x).$$

Show that it is primitive recursive. (A method which works here and in similar cases is to show first that the sequence $\tilde{\varphi}$, defined by $\tilde{\varphi}(y)= =\prod_{x<y} \exp(p_x,\varphi(x))$ is p.r., and then use the identity $\varphi(y)=(\tilde{\varphi}(y))_y$.)

6.17. PROBLEM. For fixed m, let G_i and H_i be n-ary and $(n+m+2)$-ary functionals respectively, $i=0,\dots,m$. Let F_i be obtained from the G_i and H_i by "*simultaneous primitive recursion*", i.e.,

$$F_i(\xi;\mathfrak{x},0) = G_i(\xi;\mathfrak{x}),$$
$$F_i(\xi;\mathfrak{x},y+1) = H_i(\xi;\mathfrak{x},y,F_0(\xi;\mathfrak{x},y),\dots,F_m(\xi;\mathfrak{x},y)),$$

for $i=0,\dots,m$. Show that each F_i can be obtained by a p.r. operation from $G_0,\dots,G_m, H_0,\dots,H_m$. (Consider functionals G and H defined by

$$G(\xi,\mathfrak{x}) = \prod_{i\leq m} \exp(p_i,G_i(\xi;\mathfrak{x})),$$
$$H(\xi,\mathfrak{x},y,z) = \prod_{i\leq m} \exp(p_i,H_i(\xi;\mathfrak{x},y,(z)_0,\dots,(z)_m)).$$

Obtain a functional F from G and H by primitive recursion; then obtain each F_i from F.)

§ 7. Church's Thesis

Using the methods developed in §6, as well as further methods that can be developed along the same lines, a wide variety of algorithmic functions can be shown to be recursive. As a matter of fact, no one has ever come up with an example of a function *f* such that there is an algorithm (answering the general description of §2) for calculating *f*, and such that *f* could not be shown to be recursive. No one has even outlined a plausible way of producing such a counter-example.

Moreover, during the 1930's as well as since then, many different definitions have been offered, each of which characterizes a class of algorithmic functions. In every case, the class so defined has proved to be a subclass of the class of recursive functions. (In many important cases, the intention was to make the definition as wide as possible. In such cases the class so defined turned out to contain *all* recursive functions, but no others.) These definitions are of three main kinds.

First, there are definitions using imaginary machines. Our own definition of *computable* functions is of this kind, but there are many others, using a wide variety of machines. (Some of these machines, like URIM, have an unlimited storage capacity — or else the class of functions that they can calculate does not contain *all* recursive functions — but in all other respects they are like computers that have been, or could plausibly be, actually built.) In all these cases, the functions calculable by the proposed machine can be shown to be recursive, using the same kind of method we shall employ in §8 to show that all computable functions are recursive. (We shall then point out why that method is likely to work for every machine that might plausibly ever be proposed.)

A second kind of approach is to specify explicitly some basic functions which are obviously algorithmic, and some basic operations which, when applied to algorithmic functions, always yield functions which are in turn algorithmic. The class of functions defined in this way is the smallest class containing the basic functions and closed under the basic operations. (Our definition of the class of recursive functions is of this kind: the basic functions are those described by the symbols $\mathbf{Z}$, $\mathbf{S}$, $\mathbf{I}_{n,i}$, and the basic operations are composition, primitive recursion and minimization.) In all cases where such a definition has been proposed, it is possible to show directly (and for the most part quite easily) that the basic functions are recursive, and that the basic operations are recursive operations in the sense of §6.

A third kind of method proceeds by employing a formal calculus. One takes a suitably chosen formal language (which may or may not be a first-order language of the kind studied in this book) and specifies a procedure for making *formal deductions* in that language. (This is often done by specifying *axioms* and *rules of inference* in much the same way as we have done for the propositional and predicate calculi; other methods resemble the method of tableaux.) Certain deductions in such a calculus may be interpreted (in some natural way) as formal proofs of statements of the form "$f(\mathfrak{a}) = b$". A function f is said to be *representable* in the calculus if the statements of the form "$f(\mathfrak{a}) = b$" (with this particular f) which do have such formal proofs are precisely the correct ones. If the calculus is set up in a suitable way, all functions representable in it will be algorithmic[1]. Many calculi of this kind have been proposed and examined. Some were purpose-built specifically for the task of getting as many algorithmic functions as possible to be representable. Other calculi had arisen in connection with attempts to formalize and axiomatize various portions of mathematics. It turns out that there are rather weak calculi in which all recursive functions are representable. (We shall see examples of this in Ch. 7.) But in every case where the functions representable in a calculus are known to be algorithmic they can be shown to be recursive as well.

To sum up: all the various definitions — using different kinds of approaches — which aimed at characterizing classes of algorithmic functions, have in fact turned out to define subclasses of the class of recursive functions.

This and other evidence has led virtually all logicians to accept the following conclusion, known as *Church's Thesis:*

All algorithmic functions are recursive.

Church's Thesis is a matter of reasonable belief, not a theorem. It cannot be proved mathematically, because the notion of *algorithm* — and hence that of *algorithmic function* — is not precisely *defined*, but merely *described* in intuitive terms. Our grasp of this notion enables us to recognize an algorithm when we see one[2]; but it does not enable us to *prove* the

[1] For this it is sufficient that there should be some effective procedure for enumerating, one by one, all the deductions of the calculus (as we had, e.g., in §5 of Ch. 3). Then, if f is a representable n-ary function and $\mathfrak{a}$ is any n-tuple, we can calculate $f(\mathfrak{a})$ by systematically searching for a formal proof of a statement of the form "$f(\mathfrak{a}) = b$". This yields an algorithm for f.

[2] Thus, we can see that all computable functions are algorithmic, because we easily recognize URIM programs to be algorithms. Since (by Thm. 5.5) all recursive functions rae computable it follows that the converse of Church's thesis is clearly true.

non-existence of an algorithm for calculating a function which we suspect of being non-algorithmic.

When doing recursion theory, we shall not use Church's Thesis in our proofs. These will be based on the definition of recursiveness given in §5. However, Church's Thesis *motivates* recursion theory and gives significance, to its *results.* In particular, if a function is shown to be non-recursive, then by Church's Thesis we conclude that it is non-algorithmic — and this is the only known way to reach that conclusion[1].

We remark that *Church's Thesis for functionals* — "*All algorithmic functionals are recursive*" — is also held to be true, on virtually the same grounds as for functions. However, we do not wish to make much of this, since in this book we shall be concerned with functionals as a means of getting results about functions and not for their own sake.

§ 8. Recursiveness of computable functionals

In this section we shall show that every computable function(al) is recursive. In preparation for this, we assign a *code number* $\#\mathsf{C}$ to each command C as follows:

Command:	Z_i	S_i	A_i	$\mathsf{J}_{i,j,m}$
Code number:	2^i	3^i	5^i	$7^i 11^j 13^m$.

Next, we assign to any program $\mathfrak{P} = \langle \mathsf{C}_0, \ldots, \mathsf{C}_{h-1} \rangle$ the *code number*

$$\#\mathfrak{P} = \prod_{k<h} \exp(p_k, \#\mathsf{C}_k).$$

In particular, $\#$ of the empty program is 1.

We define a property (i.e., unary relation) Prog by the identity

$$\mathrm{Prog}(z) = (z>0) \wedge \forall x<\mathrm{lh}(z)\ \exists u<z\ \exists v<z\ \exists w<z\big[((z)_x=2^{u+1}) \vee ((z)_x=3^{u+1}) \vee ((z)_x=5^{u+1}) \vee ((z)_x=7^{u+1}\cdot 11^{v+1}\cdot 13^w)\big].$$

From this identity it is easy to see that Prog is a p.r. property, and $\mathrm{Prog}(z)$ holds iff $z = \#\mathfrak{P}$ for some program $\mathfrak{P}$.

Next, we define a function $\lambda z(\hat{z})$ as follows:

$$\hat{z} = \begin{cases} z & \text{if } \mathrm{Prog}(z), \\ 1 & \text{if } \neg\mathrm{Prog}(z). \end{cases}$$

[1] On the other hand, to show that a function *is* algorithmic, it is enough to produce an algorithm and prove that it serves for calculating the function.

This function is p.r. (see Ex. 6.14), and $\hat{z}$ is always the # of some program: namely, of the program whose # is z or — if such a program does not exist — of the empty program.

8.1. DEFINITION. $\{z\}$ is the program $\mathfrak{P}$ such that $\# \mathfrak{P} = \hat{z}$.

According to this definition and the conventions of §4, $\{z\}_n$ is the n-ary *functional* computed by $\{z\}$; and if $\{z\}$ has no A commands then $\{z\}_n$ is also the *function* computed by $\{z\}$. As agreed in §4, when "$\{z\}_n$" is written next to the appropriate arguments, the subscipt "n" may be dropped, since n can be determined from the number of numerical arguments shown.

It is easy to see that the length of the program $\{z\}$ (i.e., the number of commands it has) is $\mathrm{lh}(\hat{z})$.

For any $u>0$, we shall say that at a given moment URIM is *in state u* if at that moment the number in the program counter K is $(u)_0$ and, for each positive i, the number in the ith register is $(u)_i$. Note that $(u)_i=0$ for almost all i, in agreement with our decree that at any moment almost all the registers are empty. Note also that at any moment (assuming almost all the registers to be empty) URIM is in some uniquely determined state $u>0$.

From now on until we come to Def. 8.3 below, we assume that URIM is linked up to a ξ-oracle and is operating under the program $\{z\}$.

We say that u is a *halting* state (for the program $\{z\}$) if $u>0$ and $(u)_0 \geqslant \mathrm{lh}(\hat{z})$. (This corresponds to the fact that URIM halts precisely when the number in K is $\geqslant$ the length of the program.)

For each state u there is a unique *next state*, $\mathrm{Nex}(\xi;u,z)$, depending on ξ and z as well as on u. If u is a halting state, then the next state will be u itself (URIM has halted!); otherwise, the next state is that in which URIM will be immediately after obeying the $(u)_0{}^{\text{th}}$ command of $\{z\}$. To make Nex defined also for $u=0$, which is not a state, we put $\mathrm{Nex}(\xi;0,z) = 0$.

We wish to show that Nex is a p.r. functional. To make things a bit easier, we first define auxiliary functions k,i,f and k' as follows:

$$\begin{aligned} k(u) &= (u)_0, \\ i(u,z,x) &= (\hat{z})_{k(u),x}, \\ f(u,z,x) &= \mathrm{q}(u, \exp(p_{i(u,z,x)}, (u)_{i(u,z,x)})), \\ k'(u,z) &= \begin{cases} i(u,z,5) & \text{if } \left((u)_{i(u,z,3)} = (u)_{i(u,z,4)}\right), \\ k(u)+1 & \text{otherwise.} \end{cases} \end{aligned}$$

By the results of §6, these functions are easily seen to be primitive recursive.

To explain what they mean, let us assume that $u>0$ and URIM is in state u. Also, let us assume that u is not a halting state, i.r., $(u)_0<\mathrm{lh}(\hat{z})$.

Then $k(u)$ is the number stored in K. Also, URIM is about to obey the $k(u)^{\text{th}}$ command of $\{z\}$.

Now, $i(u,z,x)$ is the xth exponent in the code number $(\hat{z})_{k(u)}$ of that command. Therefore, recalling that $p_0=2,\ldots,p_5=13$, we observe the following facts.

If $i(u,z,0)>0$, then URIM is about to obey the command $\mathsf{Z}_{i(u,z,0)}$; if $i(u,z,1)>0$, then the command about to be obeyed is $\mathsf{S}_{i(u,z,1)}$; if $i(u,z,2)>0$, then the command about to be obeyed is $\mathsf{A}_{i(u,z,2)}$; and if $i(u,z,3)>0$, then the command about to be obeyed is $\mathsf{J}_{i(u,z,3),\,i(u,z,4),\,i(u,z,5)}$, and in this case the number stored in K *after* that command is obeyed will be $k'(u,z)$.

Finally, $f(u,z,x)$ is a state exactly like u, except that in $f(u,z,x)$ the $i(u,z,x)^{\text{th}}$ register is empty.

On the basis of these explanations, the reader should have no difficulty in verifying the following identity (in which we write, e.g., "$i(0)$" as short for "$i(u,z,0)$" etc.):

$$\mathrm{Nex}(\xi;u,z)=\begin{cases}2\cdot f(u,z,0) & \text{if } (u>0)\wedge(k(u)<\mathrm{lh}(\hat{z}))\wedge(i(0)>0),\\ 2\cdot p_{i(1)}\cdot u & \text{if } (u>0)\wedge(k(u)<\mathrm{lh}(\hat{z}))\wedge(i(1)>0),\\ 2\cdot \exp(p_{i(2)},\ \xi((u)_{i(2)}))\cdot f(u,z,2) & \\ & \text{if } (u>0)\wedge(k(u)<\mathrm{lh}(\hat{z}))\wedge(i(2)>0),\\ 2^{k'(u,z)}\cdot \mathrm{q}(u,2^{k(u)}) & \text{if } (u>0)\wedge(k(u)<\mathrm{lh}(\hat{z}))\wedge(i(3)>0),\\ u & \text{otherwise.}\end{cases}$$

It now follows at once from this definition by cases that Nex is a p.r. functional.

We pause here to make two comments.

First, in showing Nex to be p.r., we have made use of the fact that ξ is not allowed to vary over *all* unary functions but only over sequences. For, Ex. 6.14 (definition by cases) which we have used here was shown to work only when all the functionals on the right-hand side are total. This is just a symptom of the fact that if one allows ξ to vary over all unary functions, the treatment (and even the very notion) of algorithmic functionals becomes a bit problematic. It is precisely for this reason that we have chosen to allow ξ to vary over N^N only.

Second, as we shall soon see, the whole proof that every computable function(al) is recursive hinges on the result that Nex is recursive. Now, it is reasonable to believe that a similar result will hold for any other "computing machine" that may plausibly be proposed for calculating algorithmic functions or functionals. The method of encoding programs, states, etc., by numbers is of general applicability, and so we can be quite sure that for every proposed machine we should be able at least to *define* an appropriate *next-state* function(al). Can this function(al) fail to be recursive? This seems highly unlikely. For, what makes a proposed theoretical device plausible as a "machine" is that the successive steps it is supposed to perform are rather rudimentary and depend in a simple way on the program and on the state of the device at a given moment. The *next-state* function(al) of any plausible machine must certainly be algorithmic, and rather simple at that. However, all the evidence suggests that *if* there is a non-recursive algorithmic function(al) — which is extremely unlikely anyway — it must be very complicated. Our argument is, in other words, that from the virtual certainty that no *simple* algorithmic function can fail to be recursive, it is reasonable to conclude that no non-recursive function can be calculated by *anything* that we might plausibly be ready to call a "computer". Further, since the essential thing about algorithms (as generally understood by mathematicians) is their "mechanical" or "robot-like" character, it is reasonable to believe that every algorithmic function *can* be calculated by some device that one could plausibly regard as a "machine".

Thus, from the virtual certainty that *simple* algorithmic functions are recursive, we arrive by a chain of highly plausible intuitive arguments at Church's thesis that *all* algorithmic functions must be recursive.

Let us now return to our functional Nex. For certain purposes it will be useful to analyse *how* Nex depends on its sequence argument. Let Nex* be the ternary function obtained by a definition by cases exactly like Nex, except that on the right-hand side "$\xi((u)_{i(2)})$" — which is short for "$\xi((u)_{i(u,z,2)})$" — is replaced by "$(v)_{(u)_{i(2)}}$", and on the left hand side "$\text{Nex}(\xi;u,z)$" is replaced by "$\text{Nex}^*(v,u,z)$".

It is then clear that Nex* is a p.r. *function* and

(1) $$\text{Nex}(\xi;u,z) = \text{Nex}^*(v,u,z), \quad \text{provided } \xi((u)_{i(2)}) = (v)_{(u)_{i(2)}}.$$

Next, let us put

(2) $$\bar{\xi}(y) = \prod_{x \leqslant y} p_x^{\xi(x)}.$$

The functional F defined by $F(\xi;y) = \bar{\xi}(y)$ is easily seen to be p.r.: it is obtained by first composing $\lambda x w(p_x^w)$ with the functional A, and then applying the product operation. Also, it follows from (2) that

$$(3) \qquad \xi(x)=(\tilde{\xi}(y))_x, \quad \text{provided } x \leqslant y.$$

We then have:

8.2. LEMMA. Nex* *is a p.r. function, and*

$$\text{Nex}(\xi;(y)_t,z)=\text{Nex}^*(\tilde{\xi}(y),(y)_t,z).$$

PROOF. Let $u=(y)_t$. Then $(u)_{i(2)} \leqslant u \leqslant y$. Therefore, by (3),

$$\xi((u)_{i(2)})=(\tilde{\xi}(y))_{(u)_{i(2)}}.$$

Putting $\tilde{\xi}(y)$ for v in (1) we therefore have $\text{Nex}(\xi;u,z)=\text{Nex}^*(\tilde{\xi}(y),u,z)$ as required. ∎

Let us now take a fixed but arbitrary number n. Let URIM be put in a state u_0 in which numbers $x_1,\ldots,x_n$ are stored in the first n registers $R_1,\ldots,R_n$ respectively, while all the other registers as well as K are empty. Thus,

$$u_0=\exp(p_1,x_1)\ldots\exp(p_n,x_n).$$

(If $n=0$ then this means that $u_0=1$.) Assuming, as before, that URIM has been linked up to a ξ-oracle and is operating under the program $\{z\}$, it will now go through successive states: beginning with the initial state u_0, then $u_1=\text{Nex}(\xi;u_0,z)$, then $u_2=\text{Nex}(\xi;u_1,z)$, etc. We now distinguish two cases regarding this sequence of states:

Case 1. If $\{z\}(\xi;\mathfrak{x})\neq\infty$, then for some t the state u_t is a halting state (i.e., $(u_t)_0 \geqslant \text{lh}(\hat{z})$). We let h be the *first* such t and call the number $\prod_{t\leqslant h}\exp(p_t,u_t)$ the *computation code* for $\langle\xi;z,\mathfrak{x}\rangle$.

Case 2. If $\{z\}(\xi;\mathfrak{x})=\infty$, then there is no t for which u_t is a halting state; in this case *the computation code for* $\langle\xi;z,\mathfrak{x}\rangle$ is undefined (i.e., ∞).

We now come to one of the most important definitions of this chapter. Its meaning will be explained in the theorem which immediately follows it.

8.3. DEFINITION. For each n we define an $(n+3)$-ary relation of the 1st order, T_n^*, and an $(n+2)$-ary relation of the 2nd order, T_n, by the identities

$$T_n^*(v,z,\mathfrak{x},y) = ((y)_0 = \exp(p_1,x_1)\ldots\exp(p_n,x_n))$$
$$\wedge \forall t < \mathrm{lh}(y) \dot{-} 1\, ((y)_{t+1} = \mathrm{Nex}^*(v,(y)_t,z))$$
$$\wedge \forall t < \mathrm{lh}(y) \dot{-} 1\, ((y)_{t,0} < \mathrm{lh}(\hat{z}))$$
$$\wedge ((y)_{\mathrm{lh}(y)\dot{-}1,0} \geq \mathrm{lh}(\hat{z})),$$
$$T_n(\xi;z,\mathfrak{x},y) = T_n^*(\tilde{\xi}(y),z,\mathfrak{x},y).$$

When writing "T_n^*" and "T_n" next to their arguments, the subscript "n" may be omitted, since n is uniquely determined by the context.

8.4. THEOREM. *T_n^* and T_n are primitive recursive. For each sequence ξ and $n+2$ numbers $z,\mathfrak{x}$ and y, $T(\xi;z,\mathfrak{x},y)$ holds iff y is the computation code for $\langle\xi;z,\mathfrak{x}\rangle$.*

PROOF. That T_n^* and T_n are p.r. can be seen at once from their definition.

To see when $T(\xi;z,\mathfrak{x},y)$ holds, substitute $\tilde{\xi}(y)$ for v in the identity defining T_n^*, and use Lemma 8.2. We then see without difficulty that $T(\xi;z,\mathfrak{x},y)$ holds iff the following conditions are satisfied:

y is of the form $\prod_{t\leq h}\exp(p_t, u_t)$, where $u_0 = \exp(p_1,x_1)\ldots\exp(p_n,x_n)$, and for each $t<h$, $u_{t+1} = \mathrm{Nex}(\xi;u_t,z)$ and u_t is not a halting state for the program $\{z\}$; but u_h *is* a halting state for $\{z\}$.

But these are precisely the conditions under which y is the computation code for $\langle\xi;z,\mathfrak{x}\rangle$. ∎

8.5. DEFINITION. U is the unary function defined by $U(y) = (y)_{\mathrm{lh}(y)\dot{-}1,1}$.

Clearly, U is a p.r. function. We now prove:

8.6. THEOREM. $\{z\}(\xi;\mathfrak{x}) = U(\mu y T(\xi;z,\mathfrak{x},y))$.

PROOF. For any given sequence ξ and $n+1$ numbers z and $\mathfrak{x}$, suppose that URIM is linked up to a ξ-oracle and is operating under the program $\{z\}$, from an initial state where $x_1,\ldots,x_n$ are in $R_1,\ldots,R_n$ respectively and the other registers as well as K are empty.

By Thm. 8.4, $\mu y T(\xi;z,\mathfrak{x},y)$ is precisely the computation code for $\langle\xi;z,\mathfrak{x}\rangle$. This is defined (i.e. $\neq\infty$) iff $\{z\}(\xi;\mathfrak{x})\neq\infty$. In case it *is* defined, Def. 8.5. shows that $U(\mu y T(\xi;z,\mathfrak{x},y))$ is the 1st exponent in the last positive exponent in this computation code. But this is precisely the number found in R_1 when URIM comes to a halt, i.e., $\{z\}(\xi;\mathfrak{x})$. ∎

8.7. COROLLARY. *For any given program $\mathfrak{P}$, the functional $\mathfrak{P}_n$ is recursive and we can actually find a description for it. If $\mathfrak{P}$ has no* A *commands, then the function $\mathfrak{P}_n$ is recursive, and we can find a description for it.*

PROOF. We encode $\mathfrak{P}$, obtaining the number $e = \#\mathfrak{P}$. Thus $\mathfrak{P}$ is $\{e\}$ and the functional $\mathfrak{P}_n$ is $\{e\}_n$. By Thm. 8.6, this functional is recursive:

$$\mathfrak{P}(\xi;\mathfrak{x}) = U(\mu y T(\xi;e,\mathfrak{x},y)).$$

Moreover, tracing back the steps leading to the definitions of U and T_n, we can get the required description.

If $\mathfrak{P}$ has no A commands, then in the description obtained for the functional $\mathfrak{P}_n$ we may replace the symbol **A** by **S**, thus getting a description for the *function* $\mathfrak{P}_n$. ∎

From now on we may use the term "*recursive*" in all contexts (including the context "*φ-recursive*") as a synonym for "*computable*". We have the right to do so, for our present results combined with those of §5 show not only that these terms are coextensive, but also provide us with effective methods for converting a description of a function(al) into a program that computes it, and *vice versa.*

From now on, when we say that a recursive function(al) is *given,* we mean that a description of it (or, equivalently, a program that computes it) is specified. Similarly, when we say that such-and-such a recursive function(al) can be *got, or constructed, or found,* we mean that a description of it (or a program computing it) can actually be constructed, given enough time and patience.

Note, by the way, that since U and T_n are p.r., we can find, for any given recursive function(al), a description in which the symbol **M** occurs at most once. (See proof of Cor. 8.7.)

8.8. PROBLEM. Using Thm. 8.6, find a shorter solution to Prob. 5.4. (Note that if $\{e\}$ is a program computing G, then

$$G(\xi;\mathfrak{x}) = U(\mu y T^*(\prod_{t \leqslant y} \exp(p_t, \xi(t)), e, \mathfrak{x}, y)).)$$

*§ 9. Functionals with several sequence arguments

According to the terminology introduced in §1, we have used the term "*functional*" for a mapping having only one sequence argument. In this section we consider a more general case. By an *(m,n)-ary functional* we shall mean here a mapping of a set A into N, where $A \subseteq (N^N)^m \times N^n$. Thus, an n-ary function is a $(0,n)$-ary functional, and an n-ary functional (in the terminology of §1) is a $(1,n)$-ary functional.

We shall indicate how the treatment of this chapter can be extended to cover (m,n)-ary functionals for any m. For simplicity let us consider the case $m=2$; the cases $m=3,4,\dots$ are dealt with by exactly the same methods.

The treatment of computability (§§3 and 4) can be modified by assuming URIM to be linked up to two oracles instead of one. We would then need two kinds of A commands, say A_i^1 and A_i^2, for consulting the first and second oracle respectively.

The treatment of recursiveness (§5) can be modified by introducing, instead of the symbol $\mathbf{A}$, two symbols $\mathbf{A}^1$ and $\mathbf{A}^2$ as descriptions of (2,1)-ary functionals A^1 and A^2, where

$$A^1(\xi,\eta;x) = \xi(x),$$
$$A^2(\xi,\eta;x) = \eta(x).$$

All the rest of our treatment can then be easily adapted to $(2,n)$-ary functionals.

Instead of following these modifications and adaptations step by step, we can cut through them by reducing $(2,n)$-ary recursive functionals to $(1,n)$-ary ones as follows.

Let $[\xi,\eta]$ be the sequence $\lambda x(2^{\xi(x)}\cdot 3^{\eta(x)})$, which depends on the parameters ξ and η. Then one can show quite easily that the $(2,n)$-ary functional F is recursive (or p.r.) iff for some recursive (or p.r., respectively) $(1,n)$-ary functional G we have

$$F(\xi,\eta;\mathfrak{x}) = G([\xi,\eta];\mathfrak{x}).$$

Using this, results about $(1,n)$-ary recursive (or p.r.) functionals can readily be generalized to $(2,n)$-ary functionals.

For this reason, we shall confine ourselves, as before, to dealing with $(1,n)$-ary functionals only.

§ 10. Fundamental theorems

In this section we shall prove three main results, from which virtually the whole of recursion theory can be developed.

We start with a recapitulation of the results of §8.

10.1. Normal Form Theorem. *We have got a p.r. function U and, for each n, an $(n+3)$-ary first-order p.r. relation T_n^* and an $(n+2)$-ary second-order p.r. relation T_n such that*

(1) $$T(\xi;z,\mathfrak{x},y) = T^*(\tilde{\xi}(y),u,\mathfrak{x},y),$$

(2) $$\{z\}(\xi;\mathfrak{x}) = U(\mu y T(\xi;z,\mathfrak{x},y)).$$

Thus for any given n-ary recursive functional F we can find a number e such that

$$F(\xi;\mathfrak{x}) = U(\mu y T(\xi;e,\mathfrak{x},y)).$$

Moreover, for any $\varphi \in N^N$ and numbers e,$\mathfrak{a}$, there is at most one number b for which $T(\varphi;e,\mathfrak{a},b)$ holds.

PROOF. See §5 and §8. ∎

The great power of this result lies mainly in the fact that by (2) (i.e., by Thm. 8.6) the $(n+1)$-ary functional F_n defined by

$$(10.2) \qquad F_n(\xi;\mathfrak{x},z) = \{z\}(\xi;\mathfrak{x})$$

is *recursive*. This is much stronger than the result we had initially set out to prove; namely, that for each number e the computable n-ary functional $\{e\}_n$ is recursive[1]. This recursive functional F_n is *universal for all n-ary recursive functionals,* in the following sense: an n-ary functional F is recursive iff for some number e we have

$$F(\xi;\mathfrak{x}) = F_n(\xi;\mathfrak{x},e).$$

Also, if $\mathfrak{P}$ is a program computing F_n, then $\mathfrak{P}$ is a *universal program for all n-ary computable functionals* in the sense that the program $\{e\}$, applied to the numerical inputs $\mathfrak{a}$ yields exactly the same output (if any) as the program $\mathfrak{P}$ applied to the numerical inputs $\langle\mathfrak{a},e\rangle$. (Intuitively speaking, $\mathfrak{P}$ works as follows: when applied to $n+1$ numerical inputs $\langle\mathfrak{a},e\rangle$, it *decodes* the $(n+1)^{\text{th}}$ input e, and applies the resulting program $\{e\}$ to the first n inputs $\mathfrak{a}$.)

For each n, $\lambda z\mathfrak{x}y\,[T_n(\lambda t(t+1);z,\mathfrak{x},y)]$ is evidently an $(n+2)$-ary *first-order* p.r. relation. We shall denote it by "T_n" (and, when writing it together with its arguments, we shall often omit the subscript "n"). There will be no ambiguity in this, since the reader will always be able to tell from the context which of the two T_n is being referred to — the first-order or the second-order relation.

Also, we shall write, e.g., "$\{z\}(\mathfrak{x})$" instead of "$\{z\}(\lambda t(t+1);\mathfrak{x})$". With this notation we have:

10.3. NORMAL FORM THEOREM (FOR FUNCTIONS). *We have got a p.r. function U and, for each n, an $(n+2)$-ary first-order p.r. relation T_n, such that*

$$\{z\}(\mathfrak{x}) = U(\mu y T(z,\mathfrak{x},y)).$$

[1] The weaker result is all that we needed for Cor. 8.7.

For any given n-ary recursive function f, we can find a number e such that

$$f(\mathfrak{x}) = U(\mu y T(e,\mathfrak{x},y)).$$

Moreover, for any numbers e and $\mathfrak{a}$ *there is at most one number b such that* $T(e,\mathfrak{a},b)$ *holds.*

PROOF. Immediate from Thms 10.1, 4.2 and 5.3. ∎

Here too, the $(n+1)$-ary recursive function f_n defined by

(10.4) $\qquad f_n(\mathfrak{x},z) = \{z\}(\mathfrak{x}),$

is *universal for all n-ary recursive functions,* in the obvious sense.

In future we shall often refer to Thms. 10.1 and 10.3 briefly as "NFT".

We shall say that a number e is an *index* of the n-ary functional F (or of the n-ary function f) if we have the identity $\{e\}(\xi;\mathfrak{x}) = F(\xi;\mathfrak{x})$ (or the identity $\{e\}(\mathfrak{x}) = f(\mathfrak{x})$). The NFT says, among other things, that an index can be found for each recursive function(al). Conversely, a function(al) is recursive only if it has an index.

We shall now present a few applications of the NFT.

10.5. EXAMPLE. In Ex. 6.14 we saw that *definition by cases* is a p.r. operation, provided the given functionals $G_1,\ldots,G_k$ are total. We shall now show that, *without* this proviso, the operation is recursive. Keeping the notation and assumptions of Ex. 6.14 (except that we no longer assume the G_i to be total) we let e_i be an index of G_i for $i=1,\ldots,k$. (If the G_i are given by means of descriptions or programs, we can actually find such e_i.) Then the identity

$$F(\xi;\mathfrak{x})=\{\mu z\,[(z=e_1)\wedge R_1(\xi;\mathfrak{x})\vee\ldots\vee(z=e_k)\wedge R_k(\xi;\mathfrak{x})]\}(\xi;\mathfrak{x})$$

can easily be verified. It shows that F is recursive if the G_i and the R_i are. Moreover, it can be used to get a description of F from those of the G_i and R_i. (Note that we have made essential use of the fact that the universal functional F_n — defined in (10.2) — is recursive.)

10.6. EXAMPLE. The *diagonal function* $\lambda x\,[\{x\}(x)]$ is recursive by the NFT. Consider any unary function g which *extends* the diagonal function, i.e., $g(x)=\{x\}(x)$ for all x such that $\{x\}(x)\neq\infty$. Using our conventions, this can be expressed more succinctly by the identity

$$\{x\}(x)=g(x)+0\cdot\{x\}(x).$$

We show that g cannot be both total and recursive. If g is recursive, then so is $\lambda x\,[g(x)+1]$. Thus by the NFT there is a number e such that

$$g(x)+1=\{e\}(x).$$

Putting e for x in both our identities[1] we obtain

$$g(e)+1=g(e)+0\cdot\{e\}(e).$$

This is possible only if both sides are ∞ (otherwise, $\{e\}(e)$ is some number b, and we would have $b+1=b+0\cdot b=b$). Thus $g(e)=\infty$ and g cannot be total.

With each first-order n-ary relation P we associate a *decision problem:*

> *Find an algorithm whereby, for each $\mathfrak{x}\in N^n$, one could find out whether $P(\mathfrak{x})$ holds or not.*

If such an algorithm exists, the decision problem for P is said to be *solvable* and P itself is said to be *decidable.*

Clearly, Church's thesis implies that P is decidable iff it is recursive. Without committing our technical terminology to Church's thesis, we shall say that the decision problem for P is *recursively solvable*, and P itself is *recursively decidable* if P is a recursive relation; otherwise, we say that the decision problem is *recursively unsolvable* and P is *recursively undecidable*.

10.7. EXAMPLE. The *self-application problem* is the decision problem for the property (i.e., unary relation) $\lambda x(\{x\}(x)=\infty)$. We show that it is recursively unsolvable. Otherwise, this property would be recursive. It would then follow (Ex. 10.5) that the function g defined by

$$g(x)=\begin{cases}\{x\}(x) & \text{if } \{x\}(x)\neq\infty,\\ 0 & \text{if } \{x\}(x)=\infty,\end{cases}$$

is recursive. But then g would be a total recursive extension of the diagonal function $\lambda x\,[\{x\}(x)]$, contrary to Ex. 10.6.

By the way, this justifies our precaution of distinguishing between $=$ and $\doteq$ (see Ex.6.7.). For we see that $\lambda x(\{x\}(x)=\infty)$ is not recursive although the functions $\lambda x\,[\{x\}(x)]$ and ∞ *are* recursive.

It follows at once that the decision problem for $\lambda zx(\{z\}(x)=\infty)$, known as the *halting problem*[2] is recursively unsolvable. For, if $\lambda zx(\{z\}(x)=\infty)$

[1] This is an application of *Cantor's diagonal method.* We shall use this method many times in the sequel.

[2] Because it is concerned with finding, for any given z and x, whether or not URIM will ever halt when it is linked up to a $\lambda t(t+1)$-oracle and is operating under the program $\{z\}$ with x as numerical input.

were recursive, $\lambda x(\{x\}(x)=\infty)$ (which is obtained from it by identification of variables) would also be recursive.

10.8. PROBLEM. The *vanishing problem* is the decision problem for the relation $\lambda zx(\{z\}(x)=0)$. Show that it is recursively unsolvable. (Consider $\lambda x(\{x\}(x)\neq 0)$. If it were recursive, it would have an index e. Get a contradiction by examining $\{e\}(e)$.)

10.9. PROBLEM. Show that there cannot exist a recursive function universal for all recursive *sequences*, i.e., a binary total recursive function f such that for every recursive sequence φ there is a number e for which φ is $\lambda x\,[f(x,e)]$. (If f were such a function, consider the sequence $\lambda x\,[f(x,x)+1]$.)

10.10. PROBLEM. The *totality problem* is the decision problem for the property P such that $P(z)$ holds iff the function $\lambda x\,[\{z\}(x)]$ is total. Prove that it is recursively unsolvable. (Show that if P were recursive this would contradict Prob. 10.9.)

The last two problems illustrate the fact that in recursion theory it is more natural to deal with *all* functions rather than just with total functions. We cannot even have an algorithm whereby we could tell, for any given program without A commands, whether the function computed by it is total or not.

For the second basic result of this section, we shall need the following:

10.11. LEMMA. *For any given program* $\mathfrak{P}$ *we can find a unary p.r. function* $f_{\mathfrak{P}}$ *such that if* $y'>y$ *and* $\mathrm{lh}(y')>\mathrm{lh}(y)$ *then* $f_{\mathfrak{P}}(y')>f_{\mathfrak{P}}(y)$; *and for every normal program*[1] $\mathfrak{Q}$, $f_{\mathfrak{P}}(\#\mathfrak{Q})=\#(\mathfrak{Q}\mathfrak{P})$.

PROOF. Let $e=\#\mathfrak{P}$ and put

$$h(x,y)=\begin{cases}(e)_x\cdot 13^{\mathrm{lh}(y)} & \text{if } (e)_{x,3}>0,\\ (e)_x & \text{otherwise.}\end{cases}$$

Now define $f_{\mathfrak{P}}$ by

$$f_{\mathfrak{P}}(y)=y\cdot\prod_{x<\mathrm{lh}(e)}\exp(p_{x+\mathrm{lh}(y)},h(x,y)).$$

Recalling that $(e)_{x,3}>0$ iff the xth command in $\mathfrak{P}$ is a J command, one can easily verify that $f_{\mathfrak{P}}$ has the required properties. ∎

To continue, it will be convenient to lay down:

10.12. DEFINITION. An n-ary function f will be called *nice* if the following three conditions hold:

[1] For the definitions of *normal* and of the concatenation $\mathfrak{P}\mathfrak{Q}$, see the latter part of §3.

(1) f is primitive recursive;

(2) $f(0,0,\ldots,0)>0$;

(3) f is monotone increasing in each argument, i.e., if $a_i \geqslant b_i$ for $i=1,\ldots,n$ and $a_i > b_i$ for some i, then $f(\mathfrak{a}) > f(\mathfrak{b})$.

We now come to the second main result of this section.

10.13. THEOREM. *Let n and m be any numbers. Then for any given $(n+m)$-ary recursive functional F we can find a nice m-ary function g such that*

$$\{g(y_1,\ldots,y_m)\}(\xi;\mathfrak{x}) = F(\xi;\mathfrak{x},y_1,\ldots,y_m).$$

PROOF. Let $\mathfrak{P}$ be a program that computes F and let $f_{\mathfrak{P}}$ be as in Lemma 10.11. For each m-tuple $\langle y_1,\ldots,y_m\rangle$, let $h(y_1,\ldots,y_m)$ be the code number of the program having $y_1+\ldots+y_m$ commands: y_1 commands S_{n+1}, followed by y_2 commands $\mathsf{S}_{n+2},\ldots$, followed finally by y_m commands S_{n+m}. Now put

$$g(y_1,\ldots,y_m) = f_{\mathfrak{P}}(h(y_1,\ldots,y_m)).$$

It is easy to verify that h is p.r.; and both $h(y_1,\ldots,y_m)$ and $\mathrm{lh}(h(y_1,\ldots,y_m))$ increase when any one of the y_i is increased. Hence g is nice.

If URIM is made to operate under the program $\{g(y_1,\ldots,y_m)\}$ with $x_1,\ldots,x_n$ initially stored in the first n registers and the other registers empty, it will begin by putting $y_1,\ldots,y_m$ in the registers $\mathsf{R}_{n+1},\ldots,\mathsf{R}_{n+m}$ respectively, and then proceed as under $\mathfrak{P}$. ∎

This theorem can be applied to functions in the following way. If f is an $(n+m)$-ary recursive function, we consider the functional F with which f is identified, i.e.,

$$F(\xi;\mathfrak{x},y_1,\ldots,y_m) = f(\mathfrak{x},y_1,\ldots,y_m).$$

We now apply the theorem to F and substitute the sequence $\lambda t(t+1)$ for ξ. Thus we get a nice g such that

$$\{g(y_1,\ldots,y_m)\}(\mathfrak{x}) = f(\mathfrak{x},y_1,\ldots,y_m).$$

Using Thm. 10.13 we now prove:

10.14. COROLLARY. *For any numbers m and n we can find a nice $(m+1)$-ary function S_n^m such that*

$$\{S_n^m(z,y_1,\ldots,y_m)\}(\xi;\mathfrak{x}) = \{z\}(\xi;\mathfrak{x},y_1,\ldots,y_m).$$

PROOF. Put

$$F(\xi;\mathfrak{x},z,y_1,\ldots,y_m) = \{z\}(\xi;\mathfrak{x},y_1,\ldots,y_m).$$

Then F is clearly an $(n+m+1)$-ary recursive functional and by Thm. 10.13 we can find a nice $(m+1)$-ary function — which we now call S_n^m — such that the required identity holds. ∎

We shall refer to Thm. 10.13 as *the S_n^m theorem*[1]. In this connection the variables $y_1,\ldots,y_m$ will be called *parameters*.

We now present a few applications of the S_n^m theorem.

10.15. EXAMPLE. For $m=0$ and for any n we have by Cor. 10.14,

$$\{S_n^0(z)\}(\xi;\mathfrak{x})=\{z\}(\xi;\mathfrak{x}).$$

Thus, for any number e, the n-ary functionals $\{e\}_n$ and $\{S_n^0(e)\}_n$ are the same, i.e., e and $S_n^0(e)$ are indexes of the same n-ary functional. Note that since S_n^0 is nice it follows (by induction on e) that $S_n^0(e)>e$. Applying S_n^0 again and iterating, we get an increasing infinite sequence of indexes of the same n-ary functional $\{e\}_n$.

10.16. EXAMPLE. Let d be any number, and let P be the property $\lambda z(\{z\}(d)=\infty)$. We show that P is recursively undecidable (i.e., not recursive).

Let f be the recursive function defined by the identity $f(y,x)= =\{x\}(x+y\dot{-}d)$. Applying the S_n^m thm. to f, with x as parameter, we get a nice sequence[2] g satisfying the identity $\{g(x)\}(y)=\{x\}(x+y\dot{-}d)$. In particular, giving y the value d, we have the identity

$$\{g(x)\}(d)=\{x\}(x).$$

From this and the definition of P we have

$$P(g(x))=(\{x\}(x)=\infty).$$

Now, if P were recursive then the composition of P and g would also be recursive — contradicting the fact (see Ex. 10.7) that $\lambda x(\{x\}(x)=\infty)$ is *not* recursive.

10.17. PROBLEM. Show that, for any number d, the property $\lambda z(\{z\}(d)=0)$ is recursively undecidable.

10.18. PROBLEM. Let d be any number and let Q be the property such that $Q(z)$ holds iff d belongs to the range of the unary recursive function $\{z\}_1$

[1] The name is justified because Thm. 10.13 and Cor. 10.14 are merely two forms of the same result. (PROBLEM: Deduce Thm. 10.13 from Cor. 10.14.)

[2] That is, a nice unary function. Such a function, being p.r., is necessarily total and hence, by our convention, may be called a *sequence*.

(i.e., $\{z\}(x)=d$ for some x). Show that Q is not recursive. (Get a nice sequence h such that

$$\{h(z)\}(x)=0\cdot\{z\}(x)+x.$$

Show that if $\lambda z Q(h(z))$ were recursive, this would contradict Ex. 10.16.)

We now come to the third and last major result of this section.

10.19. RECURSION THEOREM. *Given any $(n+2)$-ary recursive functional F, we can find a nice sequence φ such that*

$$\{\varphi(z)\}(\xi;\mathfrak{x})=F(\xi;\mathfrak{x},\varphi(z),z).$$

PROOF. By the S_n^m thm. we get a nice sequence σ such that

$$(*) \qquad \{\sigma(y)\}(\xi;\mathfrak{x})=\{y\}(\xi;\mathfrak{x},y).$$

(Apply the theorem to the functional G defined by $G(\xi;\mathfrak{x},y)=\{y\}(\xi;\mathfrak{x},y)$ taking y as parameter.) Next, the S_n^m thm. yields a nice sequence ψ such that

$$(**) \qquad \{\psi(z)\}(\xi;\mathfrak{x},y)=F(\xi;\mathfrak{x},\sigma(y),z).$$

Now let φ be the composition of σ and ψ, i.e., $\lambda y\,[\sigma(\psi(y))]$. Substituting $\psi(z)$ for y in both $(*)$ and $(**)$, we obtain the required identity

$$\{\varphi(z)\}(\xi;\mathfrak{x})=F(\xi;\mathfrak{x},\varphi(z),z).$$

Since φ was obtained by composing two nice sequences, it is clearly nice as well. ∎

As a corollary we have:

10.20. COROLLARY. *Given any $(n+1)$-ary recursive functional F, we can find a number e such that*

$$\{e\}(\xi;\mathfrak{x})=F(\xi;\mathfrak{x},e).$$

PROOF. Obtain F' from F by introducing a redundant variable:

$$F'(\xi;\mathfrak{x},y,z)=F(\xi;\mathfrak{x},y).$$

Now apply Thm. 10.19 to F' and take e as $\varphi(d)$ for arbitrary d. ∎

We shall often refer to Thm. 10.19 and Cor. 10.20 briefly as "the RT". Like the S_n^m thm., these results too can be applied to functions. Thus, if f is a given $(n+2)$-ary recursive function we get a nice sequence φ such that

$$\{\varphi(z)\}(\mathfrak{x})=f(\mathfrak{x},\varphi(z),z).$$

And if f is a given $(n+1)$-ary recursive function we can find a number e such that

$$\{e\}(\mathfrak{x})=f(\mathfrak{x},e).$$

To explain the meaning (and great power) of the RT, let us first suppose that we are asked to write a URIM program $\mathfrak{P}$ whose outputs are to depend in a prescribed way on the sequence ξ (given by an oracle) and n numerical inputs $\mathfrak{x}$, i.e.,

$$\mathfrak{P}(\xi;\mathfrak{x})=F(\xi;\mathfrak{x}),$$

where F is prescribed in advance. We know that this problem has a solution iff F is recursive. Moreover, if we can find a description of F (in the sense of Def. 5.2) then by Thm. 5.5 we can actually construct a program $\mathfrak{P}$ which does the required job. This problem does not therefore pose any theoretical difficulty; in fact, it is precisely the kind of problem which a URIM programmer would be expected to solve.

Now suppose that we are asked to write a program $\mathfrak{P}$ whose outputs are to depend in a prescribed way not only on ξ and $\mathfrak{x}$ as before, but also on $\mathfrak{P}$ itself. (To take a simple example, suppose the outputs represent the costs of running our "computer", including our own fees for writing the program, which depend on the length of the program we write.) This problem is not only much harder, but at first sight we might suspect that it is insoluble. We may object that we cannot reasonably be asked to write a program without knowing in advance what this program is supposed to do; and in the present case what the program is supposed to do depends on the — yet unwritten — program itself. A programmer confronted with such a task might well feel like the wise men of Babylon when they were commanded by Nebuchadnezzar not only to interpret his dream but to guess what the dream was[1]. Like them he might object that this was not the kind of problem he could fairly be expected to solve.

However, let us see whether something can be done about it after all. Since every program has the form $\{e\}$, and all the information about $\{e\}$ is contained in the number e, we may state our problem as follows: Find a value of the unknown y for which the identity

$$\{y\}(\xi;\mathfrak{x})=F(\xi;\mathfrak{x},y)$$

will hold, where F is prescribed in advance. We can see at once that a

[1] See Daniel, Ch. 2.

solution may not exist if F is not a recursive functional. On the other hand, Cor. 10.20 tells us that if F *is* recursive then a solution e *does* exist. (And we can actually find such a solution if we are given a description of F.)

Furthermore, suppose we are asked to supply an *infinite sequence of programs* $\{\varphi(z)\}$ for $z=0,1,2,\ldots$ such that the outputs of the zth program depend not only on the data ξ and $\mathfrak{x}$ but also on the program itself, and the very dependence itself varies with z. Here we must solve an identity of the form

$$\{\eta(z)\}(\xi;\mathfrak{x})=F(\xi;\mathfrak{x},\eta(z),z)$$

for the unknown sequence η, with prescribed F. Thm. 10.19 shows that for recursive F a nice solution φ exists. Moreover, given a description of F we can actually write a "master program" which computes φ.

The RT, in the form of Cor. 10.20, is very useful in situations where we seek an n-ary recursive functional, say G, such that G satisfies a condition of the form

$$G(\xi;\mathfrak{x})=r,$$

where r is an expression involving $\xi,\mathfrak{x}$ and G itself. To find a solution, we replace "G" by "$\{y\}$" on both sides. Our condition then takes the form

$$\{y\}(\xi;\mathfrak{x})=F(\xi;\mathfrak{x},y).$$

If we can show that F here is recursive, then by the RT we can find a value e for y such that the condition holds. Then $\{e\}_n$ can be taken as the required G.

Let us now present a few applications of the RT.

10.21. Example. Let P be the property such that $P(z)$ holds iff z is an index of the nowhere defined unary function, i.e., $\{z\}(x)=\infty$ for all x. We show that P is not recursive.

Put $f(x,y,z)=\{z\}(y)$. Applying the RT to f, we get a nice sequence φ such that $\{\varphi(z)\}(x)=\{z\}(\varphi(z))$. Thus

$$(*)\qquad P(\varphi(z))=(\{z\}(\varphi(z))=\infty).$$

Now let g be defined by cases:

$$g(z)=\begin{cases}0 & \text{if } P(z),\\ \infty & \text{otherwise.}\end{cases}$$

If P were recursive, then g would be recursive as well by Ex. 6.2 and Ex. 10.5. If e is an index of g, the definition of g yields the identity

$$(**) \qquad P(z) = (\{e\}(z) \neq \infty).$$

Putting e for z in $(*)$ and $\varphi(e)$ for z in $(**)$ we get a contradiction.

10.22. PROBLEM. Let Q and R be the properties such that $Q(z)$ holds iff $\{z\}(x) = \infty$ for infinitely many values of x, and $R(z)$ holds iff the function $\lambda x\,[\{z\}(x)]$ has an infinite range (i.e., takes infinitely many different values). Show that Q and R are not recursive.

The results of Ex. 10.21, Prob. 10.22 and many similar results (including some obtained earlier in this section) can also be derived from the following rather general theorem.

10.23. THEOREM. *Let P be a property such that:*

(1) *$P(z)$ holds for some, but not for every, numerical value of z.*

(2) *If a and b are any two indexes of the same unary function (i.e., if the identity $\{a\}(x) = \{b\}(x)$ holds),* then $P(a) = P(b)$.

Then P is not recursive.

PROOF. Take numbers a and b such that $P(a)$ holds but $P(b)$ does not. Define g by

$$g(y) = \begin{cases} b & \text{if } P(y), \\ a & \text{otherwise.} \end{cases}$$

From this definition the identity

$$(3) \quad P(g(y)) = \neg P(y)$$

follows at once. On the other hand, if P were recursive, then g and hence also the binary function $\lambda xy(\{g(y)\}(x))$, would be recursive as well. Applying the RT to this binary function, we get for some number e the identity $\{e\}(x) = \{g(e)\}(x)$. By condition (2) we would therefore have $P(g(e)) = P(e)$, contradicting (3). ∎

We conclude this section with

10.24. SIMULTANEOUS RECURSION THEOREM. *Let F and G be given $(n+2)$-ary recursive functionals. We can find numbers c and d such that*

$$\{c\}(\xi\,;\mathfrak{x}) = F(\xi\,;\mathfrak{x},c,d),$$

$$\{d\}(\xi\,;\mathfrak{x}) = G(\xi,\mathfrak{x},c,d).$$

Proof. By the RT we can find a nice sequence φ such that

$$\{\varphi(z)\}(\xi;\mathfrak{x})=F(\xi;\mathfrak{x},\varphi(z),z).$$

Next, applying the RT to the functional H such that $H(\xi;\mathfrak{x},y)=$ $=G(\xi;\mathfrak{x},\varphi(y),y)$, we find a number d such that

$$\{d\}(\xi;\mathfrak{x})=G(\xi;\mathfrak{x},\varphi(d),d).$$

Putting $c=\varphi(d)$ and substituting d for z, we get the required result. ∎

10.25. **Problem.** Prove the following generalization of the RT. *For any given $(n+m+1)$-ary recursive functional F we can find a nice m-ary function g such that*

$$\{g(z_1,\ldots,z_m)\}(\xi;\mathfrak{x})=F(\xi;\mathfrak{x},g(z_1,\ldots,z_m),z_1,\ldots,z_m).$$

(Let $G(\xi;\mathfrak{x},y,z)=F(\xi;\mathfrak{x},y,(z)_1,\ldots,(z)_m)$. Apply the RT to G and put $z=\exp(p_1,z_1)\ldots\exp(p_m,z_m)$.)

10.26 **Problem.** Given $(n+m+k)$-ary recursive functionals F_i for $i=1,\ldots k$, show how to find k nice functions g_i such that

$$\{g_i(z^*)\}(\xi;\mathfrak{x})=F_i(\xi;\mathfrak{x},g_i(z^*),\ldots,g_m(z^*),z^*),\quad i=1,\ldots,k,$$

where z^* is short for $z_1,\ldots,z_m$. (Use induction on k.)

§ 11. Recursively enumerable sets

From now up to the end of this chapter we confine our attention to *functions* rather than functionals. By *relation* we shall therefore mean a first-order relation. (Much of what we shall do can be generalized to second-order relations as well, but we shall not make use of this.)

By *n-dimensional set* we shall mean any subset of N^n. There is a natural one-to-one correspondence between n-ary relations and n-dimensional sets. To each n-ary relation P there corresponds the n-dimensional set $\{\mathfrak{x}\colon P(\mathfrak{x})=0\}$, which is called the *extension* of P. Conversely, given any $A\subseteq N^n$ we let $\lambda\mathfrak{x}(\mathfrak{x}\in A)$ be the n-ary relation whose extension is A, i.e., we put

$$(\mathfrak{x}\in A)=\begin{cases}0 & \text{if } \mathfrak{x}\in A,\\ 1 & \text{if } \mathfrak{x}\in N^n-A.\end{cases}$$

Through this one-to-one correspondence, any definition or result dealing with n-ary relations can be understood as dealing with n-dimensional sets, and *vice versa*. In what follows we shall often make use of this without special mention. For example, we shall say that an n-dimensional set A is *recursive*, meaning that the corresponding relation is recursive.

Let us note that if P is an n-ary relation having A as its extension, then the extension of $\neg P$ is the *complement* A^c of A relative to N^n (i.e., $A^c = N^n - A$). If Q is another n-ary relation having B as its extension, then clearly the extensions of $P \vee Q$ and $P \wedge Q$ are $A \cup B$ and $A \cap B$ respectively. (See Ex. 6.6.)

We introduce the operation of *(unbounded) existential quantification* which can be applied to any $(n+1)$-ary relation P, to yield an n-ary relation R:

$$R(\mathfrak{x}) = \exists y P(\mathfrak{x}, y).$$

Here $R(\mathfrak{a})$ holds iff there is some number b such that $P(\mathfrak{a}, b)$ holds.

Similarly, *(unbounded) universal quantification* can be applied to P to yield an n-ary relation Q:

$$Q(\mathfrak{x}) = \forall y P(\mathfrak{x}, y).$$

Here $Q(\mathfrak{a})$ holds iff $P(\mathfrak{a}, b)$ holds for every number b.

We shall soon see that these operations are not recursive.

11.1. DEFINITION. A relation is *recursively enumerable* (briefly, *r.e.*) if it is obtained by an existential quantification from a recursive relation. (For a justification of this terminology, see Thm. 11.5 below.)

It is easy to see that a recursive relation is r.e., for

$$P(\mathfrak{x}) = \exists y\, [P(\mathfrak{x}) \wedge (y = y)],$$

and $\lambda \mathfrak{x} y\, [P(\mathfrak{x}) \wedge (y = y)]$ is a recursive relation if P is. But we shall soon show that not every r.e. relation is recursive.

We shall make frequent implicit use of:

11.2. LEMMA. *The following operations applied to r.e. relations yield r.e. relations:*

(1) *composition with total recursive functions*;
(2) *existential quantification*;
(3) *conjunction*;
(4) *disjunction*;
(5) *bounded existential quantification*;
(6) *bounded universal quantification.*

PROOF. (1) If

$$R(z_1, \ldots, z_k) = \exists y P(z_1, \ldots, z_k, y),$$

then

$$R(f_1(\mathfrak{x}), \ldots, f_k(\mathfrak{x})) = \exists y P(f_1(\mathfrak{x}), \ldots, f_k(\mathfrak{x}), y).$$

If P is a recursive relation and $f_1,\ldots,f_k$ are total recursive functions, then

$$\lambda \mathfrak{x} y P(f_1(\mathfrak{x}),\ldots,f_k(\mathfrak{x}),y)$$

is a recursive relation.

(2) Using Ex. 6.13 we have

$$\exists y\, \exists z\, P(\mathfrak{x},y,z) = \exists u\, P(\mathfrak{x},K(u),L(u)),$$

and, if P is recursive, so is $\lambda \mathfrak{x} u P(\mathfrak{x},K(u),L(u))$.

(3) We have

$$\exists y\, P(\mathfrak{x},y) \wedge \exists y\, Q(\mathfrak{x},y) = \exists y\, \exists z\, [P(\mathfrak{x},y) \wedge Q(\mathfrak{x},z)]$$

and the two consecutive existential quantifiers $\exists y\, \exists z$ can be contracted into one as in (2).

(4) Similar to (3).

(5) $\exists w{<}u\, \exists y\, P(\mathfrak{x},w,y) = \exists w\, \exists y\, [(w{<}u) \wedge P(\mathfrak{x},w,y)]$.

(6) $\forall w{<}u\, \exists y\, P(\mathfrak{x},w,y) = \exists z \forall w{<}u\, P(\mathfrak{x},w,(z)_w)$. ∎

By the *graph* of the n-ary function f we mean the $(n+1)$-dimensional set

$$\{\langle \mathfrak{x},f(\mathfrak{x})\rangle :\ \mathfrak{x} \in \mathrm{dom}(f)\}.$$

(Recall that, by the conventions of §1, $\mathrm{dom}(f) = \{\mathfrak{x}:\ f(\mathfrak{x}) \neq \infty\}$.) Note that the graph of f is the extension of the relation $\lambda \mathfrak{x} u(f(\mathfrak{x})=u)$.

The recursion-theoretic interest in r.e. sets[1] stems in part from the following:

11.3. THEOREM. *Let A be the graph of an n-ary function f. Then A is r.e. iff f is recursive.*

PROOF. If f is recursive, we can find an index e of f by the NFT and we have

$$(\langle \mathfrak{x},u\rangle \in A) = \exists y(T(e,\mathfrak{x},y) \wedge (U(y)=u)),$$

which shows that A is an r.e. set.

Conversely, if A is r.e., then for some $(n+2)$-ary recursive relation R we have

$$(\langle \mathfrak{x},u\rangle \in A) = \exists y R(\mathfrak{x},u,y).$$

It is then easy (using Ex. 6.13) to verify the identity

$$f(\mathfrak{x}) = K(\mu z R(\mathfrak{x},K(z),L(z))),$$

from which it follows that f is recursive. ∎

[1] That is, extensions of r.e. relations.

We note that if A is the graph of an n-ary *total* recursive function f then A is *recursive:*

$$(\langle \mathfrak{x},u\rangle \in A) = (f(\mathfrak{x}) \doteq u),$$

(see Ex. 6.7).

11.4. THEOREM. *A relation is recursive iff both it and its negation are recursively enumerable.*

PROOF. If R is a recursive relation, then so is $\neg R$ (see Ex. 6.6). Hence R and $\neg R$ are certainly recursively enumerable.

Conversely, suppose

$$R(\mathfrak{x}) = \exists y P(\mathfrak{x}, y),$$

$$\neg R(\mathfrak{x}) = \exists y Q(\mathfrak{x}, y),$$

where P and Q are recursive relations. Then we can easily verify the identity

$$R(\mathfrak{x}) = P(\mathfrak{x}, \mu y\, [P(\mathfrak{x}, y) \vee Q(\mathfrak{x}, y)]),$$

which shows R to be recursive. ∎

If $f_1, \ldots, f_n$ are unary functions, we say that the n-tuple $\langle f_1, \ldots, f_n\rangle$ *enumerates* the n-dimensional set

$$\{\langle f_1(t), \ldots, f_n(t)\rangle : t \in \mathrm{dom}(f_1) \cap \ldots \cap \mathrm{dom}(f_n)\}.$$

The following theorem provides a justification for the term "r.e. set" and gives useful characterizations of r.e. sets.

11.5. THEOREM. *For any n-dimensional set A, the following four conditions are equivalent:*

(1) *A is an r.e. set;*

(2) *A is the domain of some n-ary recursive function;*

(3) *A is enumerated by an n-tuple of recursive unary functions.*

PROOF. (1)⇒(2). Suppose

$$(\mathfrak{x} \in A) = \exists y R(\mathfrak{x}, y),$$

where R is a recursive relation. Then if f is the n-ary function $\lambda\mathfrak{x}(\mu y R(\mathfrak{x}, y))$, it is clear that f is recursive. Also, it is not difficult to see that $\mathrm{dom}(f) = A$.

(2)⇒(3). Suppose $A = \mathrm{dom}(f)$, where f is an n-ary recursive function. By the NFT, f possesses an index e and we have $f(\mathfrak{x}) = U(\mu y T(e, \mathfrak{x}, y))$, hence it follows that

$$(\mathfrak{x} \in A) = \exists y T(e, \mathfrak{x}, y).$$

We define unary functions f_i for $i = 1,...,n$ as follows:

$$f_i(t) = \begin{cases} (t)_i & \text{if } T(e,(t)_1,...,(t)_n,(t)_0), \\ \infty & \text{otherwise.} \end{cases}$$

Then the f_i are p.r. (see Ex. 6.14). We show that $\langle f_1,...,f_n \rangle$ enumerates A. First, take t such that $T(e,(t)_1,\ldots,(t)_n,(t)_0)$ holds. Then $\exists y T(e,(t)_1,\ldots,(t)_n,y)$ holds as well, so that $\langle (t)_1,\ldots,(t)_n \rangle \in A$. But in this case

$$\langle f_1(t),...,f_n(t) \rangle = \langle (t)_1,...,(t)_n \rangle$$

by the definition of the f_i. On the other hand, if $T(e,(t)_1,...,(t)_n,(t)_0)$ fails to hold then $\langle f_1(t),...,f_n(t) \rangle$ is undefined. Thus, for any value of $t, \langle f_1(t),...,f_n(t) \rangle$ is either in A or undefined. It follows that the n-tuple $\langle f_1,...,f_n \rangle$ enumerates a *subset* of A. Now suppose $\mathfrak{c} \in A$. Then $\exists y T(e,\mathfrak{c},y)$ holds, so that $T(e,\mathfrak{c},b)$ holds for some (unique) number b. For

$$t = 2^b \exp(p_1,c_1)...\exp(p_n,c_n)$$

we see without difficulty that $\langle f_1(t),...,f_n(t) \rangle = \mathfrak{c}$. Thus $\langle f_1,...,f_n \rangle$ enumerates the *whole* of A.

Before we go on, let us note that if we are given some member $\mathfrak{a}$ of A, then we can modify the definition of f_i by putting a_i instead of ∞ in that definition. This makes the f_i p.r. rather than merely recursive.

(3)⇒(1). Let A be enumerated by the n-tuple $\langle f_1,\ldots,f_n \rangle$, where the f_i are recursive unary functions. For each i, let B_i be the graph of f_i. Then

$$(\mathfrak{x} \in A) = \exists t\, [(\langle t,x_1 \rangle \in B_1) \wedge \ldots \wedge (\langle t,x_n \rangle \in B_n)],$$

so that A is r.e. by Thm. 11.3 (and Lemma 11.2). ∎

We shall denote by "$\text{dom}_n\{e\}$" the domain of the n-ary recursive function having e as index. With this notation, Thm. 11.5 yields:

11.6. Corollary. *The n-dimensional r.e. sets are precisely the sets* $\text{dom}_n\{e\}$ *(where* $e = 0,1,\ldots$*). The* $(n+1)$*-ary relation* $\lambda \mathfrak{x} z(\mathfrak{x} \in \text{dom}_n\{z\})$ *is recursively enumerable; in fact* $(\mathfrak{x} \in \text{dom}_n\{z\}) = \exists y\, T(z,\mathfrak{x},y)$.

PROOF. The first assertion is merely a re-statement of the equivalence between (1) and (2) of Thm. 11.5. The second assertion follows from the proof that (2)⇒(3) in the same theorem. ∎

We shall say that e is an *index* of the n-dimensional r.e. set $\text{dom}_n\{e\}$. When we say that an r.e. set is *given* (or *can be found* or *constructed* etc.) we mean that an index of it is given (or can be found or constructed etc.).

Note that if an $(n+1)$-ary recursive relation R is given and the r.e. set A is defined by the identity

$$(\mathfrak{x} \in A) = \exists y R(\mathfrak{x}, y),$$

then (using the proof of Thm. 11.5) we can find an index of A.

11.7. COROLLARY. *Each n-ary r.e. relation is obtained by existential quantification from an* $(n+1)$*-ary p.r. relation.*

PROOF. Obvious from the identity

$$(\mathfrak{x} \in \text{dom}_n\{e\}) = \exists y\, T(e, \mathfrak{x}, y)$$

and the fact that T_n is a p.r. relation. ∎

11.8. COUNTER-EXAMPLE. The property $\lambda z(z \in \text{dom}_1\{z\})$ is r.e. by Cor. 11.6. But its negation cannot be r.e., for in that case we would have for some e

$$(z \notin \text{dom}_1\{z\}) = (z \in \text{dom}_1\{e\})$$

and putting $z = e$ we would get a contradiction. Thus (by Thm. 11.4) the property $\lambda z(z \in \text{dom}_1\{z\})$ cannot be recursive. (This result is merely a reformulation of Ex. 10.7.)

The following result throws additional light on the difference between recursive and r.e. sets.

11.9. THEOREM. *A set* $A \subseteq N$ *is recursive iff* A *is finite or* A *is enumerated by a* monotone increasing *recursive sequence.*

PROOF. Suppose that A is recursive. Then by the RT there is a unary recursive function f such that

$$f(t) = \mu x\, [(x \in A) \wedge \forall z < t\, (f(z) < x)].$$

(To see this, replace "f" on both sides by "$\{e\}$".) It is clear that the successive values of f are the members of A in increasing order, until A is exhausted — at which point the values of f become ∞. Thus f enumerates A; and if A is infinite f is total, i.e., a sequence.

It is clear that a finite set is recursive. (If $A = \{a_1, \ldots, a_k\}$, then

$$(x \in A) = (x \neq x) \vee (x = a_1) \vee \ldots \vee (x = a_k).)$$

Finally, suppose that A is enumerated by the monotone recursive sequence φ. Then we have

$$(x \in A) = \exists t \leqslant x\, (\varphi(t) = x),$$

so that A is recursive. ∎

11.10. PROBLEM. Show that every infinite n-dimensional r.e. set B has an infinite recursive subset. (First deal with the case $n=1$. Then, for $n>1$, consider the one-dimensional set

$$B' = \{\exp(p_1, x_1) \ldots \exp(p_n, x_n) : \mathfrak{x} \in B\}.)$$

Using Thm. 11.5 and Cor. 11.6, we can adapt the S_n^m-thm. and the RT to r.e. sets. We devote the rest of the present section to this task.

11.11. THEOREM. *For any given $(n+m)$-ary r.e. relation R we can find a nice m-ary function g such that*

$$(\mathfrak{x} \in \mathrm{dom}_n\{g(y_1, \ldots, y_m)\}) = R(\mathfrak{x}, y_1, \ldots, y_m).$$

Also,

$$(\mathfrak{x} \in \mathrm{dom}_n\{S_n^m(z, y_1, \ldots, y_m)\}) = (\langle \mathfrak{x}, y_1, \ldots, y_m \rangle \in \mathrm{dom}_{n+m}\{z\}).$$

PROOF. By Thm. 11.5, we can find an $(n+m)$-ary recursive function f such that $\mathrm{dom}(f)$ is the extension of R, i.e.,

$$(\langle \mathfrak{x}, y_1, \ldots, y_m \rangle \in \mathrm{dom}(f)) = R(\mathfrak{x}, y_1, \ldots, y_m).$$

By the S_n^m thm. we find a nice m-ary function g such that

$$\{g(y_1, \ldots, y_m)\}(\mathfrak{x}) = f(\mathfrak{x}, y_1, \ldots, y_m).$$

Thus

$$(\mathfrak{x} \in \mathrm{dom}_n\{g(y_1, \ldots, y_m)\}) = (\langle \mathfrak{x}, y_1, \ldots, y_m \rangle \in \mathrm{dom}(f)).$$

This proves the first part of our theorem. The second part follows at once from Cor. 10.14. ∎

We shall refer to Thm. 11.11 as the *S_n^m theorem for r.e. sets.*

11.12. PROBLEM. Given a unary recursive function f, find a nice sequence φ such that

$$\mathrm{dom}_1\{\varphi(z)\} = \begin{cases} \mathrm{dom}_1\{z\} & \text{if } z \in \mathrm{dom}(f), \\ \emptyset & \text{otherwise.} \end{cases}$$

(Consider the binary relation R such that

$$R(x,z) = (x \in \mathrm{dom}_1\{z\}) \wedge (z \in \mathrm{dom}(f)).)$$

11.13. RECURSION THEOREM FOR R.E. SETS. *Given an $(n+2)$-ary r.e. relation R, we can find a nice sequence φ such that*

$$(\mathfrak{x} \in \mathrm{dom}_n\{\varphi(z)\}) = R(\mathfrak{x},\varphi(z),z).$$

PROOF. Left as an easy exercise for the reader. ∎

§ 12. Diophantine relations

In what follows *polynomial* will mean polynomial with integer (possibly negative) coefficients. But the *variables* of a polynomial are still only allowed to range over N, unless otherwise stated.

We extend to polynomials the notation introduced in Ex. 6.7. Thus, e.g., if p is a polynomial, we regard the expression "$(p(\mathfrak{x})=0)$" as having the value 0 for $\mathfrak{x} \in N^n$ such that $p(\mathfrak{x})$ vanishes, and 1 for all other $\mathfrak{x} \in N^n$. (The use of negative coefficients is merely a convenience, not a necessity, since "$(p(\mathfrak{x})=0)$" can always be re-written as "$(f(\mathfrak{x})=g(\mathfrak{x}))$", where f and g are polynomials with natural number coefficients.)

We shall say that an n-ary relation E is *elementary* if, for some polynomial p in n variables,

$$E(\mathfrak{x}) = (p(\mathfrak{x})=0).$$

Note that if E' is obtained from E by introducing a redundant variable, i.e., $E'(\mathfrak{x},y)=E(\mathfrak{x})$, then E' is elementary as well, because $E'(\mathfrak{x},y) = (p(\mathfrak{x})+0\cdot y=0)$. We shall often make use of this fact without special mention.

12.1. LEMMA. *The disjunction and conjunction of n-ary elementary relations are themselves elementary.*

PROOF. Let p and q be two polynomials in n variables. Then

$$(p(\mathfrak{x})=0) \vee (q(\mathfrak{x})=0) = (p(\mathfrak{x})\cdot q(\mathfrak{x})=0),$$

$$(p(\mathfrak{x})=0) \wedge (q(\mathfrak{x})=0) = ([p(\mathfrak{x})]^2+[q(\mathfrak{x})]^2=0).$$ ∎

An n-ary relation P is *diophantine* if for some $m \geqq 0$ and some $(n+m)$-ary elementary relation E the identity

$$P(\mathfrak{x}) = \exists y_1 \ldots \exists y_m E(\mathfrak{x},y_1,\ldots,y_m)$$

holds. An n-dimensional set A is *diophantine* if it is the extension of a diophantine relation, i.e., the relation $\lambda\mathfrak{x}(\mathfrak{x}\in A)$ is diophantine. Note that an n-ary function f has a diophantine graph (or, as we shall say briefly, *d.g.*) iff $\lambda\mathfrak{x}u(f(\mathfrak{x})=u)$ is a diophantine relation.

It is easy to see that every diophantine relation is recursively enumerable. The rest of this chapter is devoted mainly to proving the converse — every r.e. relation is diophantine. This result yields a (negative) solution to Hilbert's Tenth Problem[1], and is of great importance in itself.

12.2. LEMMA. *Let P be an n-ary relation obtained in one of the following ways:*
(1) *by existential quantification from an* $(n+1)$*-ary diophantine relation;*
(2) *by conjunction from n-ary diophantine relations;*
(3) *by disjunction from n-ary diophantine relations.*
Then P itself is diophantine.
PROOF. Case (1) is obvious. To deal with case (2), let

$$P(\mathfrak{x}) = \exists y_1 \ldots \exists y_k D(\mathfrak{x},y_1,\ldots,y_k) \wedge \exists z_1 \ldots \exists z_m E(\mathfrak{x},z_1,\ldots,z_m),$$

where the relations D and E are elementary. Here we may assume that the variables $y_1\ldots,y_k$ are all different from the variables $z_1,\ldots,z_m$. Then

$$P(\mathfrak{x}) = \exists y_1 \ldots \exists y_k \exists z_1 \ldots \exists z_m (D(\mathfrak{x},y_1,\ldots,y_k) \wedge E(\mathfrak{x},z_1,\ldots,z_m)),$$

so that by Lemma 12.1 P is diophantine. (By introducing redundant variables we can replace D and E by $(n+k+m)$-ary elementary relations D' and E'.)

Case (3) is treated similarly. ∎

12.3. THEOREM. *The p.r. functions described by the symbols* $\mathbf{Z}$,$\mathbf{S}$ *and* $\mathbf{I}_{n,i}$ *(where* $1 \leqslant i \leqslant n$*) have diophantine graphs. A function obtained by composition from functions having d.g.s has a d.g. as well.*
PROOF. The first part of the theorem is left as an easy problem to the reader. Now, suppose g is a k-ary function and $h_1,\ldots,h_k$ are n-ary functions, and let

$$f(\mathfrak{x}) = g(h_1(\mathfrak{x}),\ldots,h_k(\mathfrak{x})).$$

Then

$$(f(\mathfrak{x})=y) = \exists y_1 \ldots \exists y_k [(h_1(\mathfrak{x})=y_1) \wedge \ldots \wedge (h_k(\mathfrak{x})=y_k) \wedge (g(y_1,\ldots,y_k)=y)].$$

Hence, by Lemma 12.2, if $g,h_1,\ldots,h_k$ have d.g.s, so does f. ∎

[1] See §16.

12.4. THEOREM. *If P is a diophantine relation and $g_1,\ldots,g_k$ are total functions with d.g.s, then the relation $\lambda\mathfrak{x}\,[P(g_1(\mathfrak{x}),\ldots,g_k(\mathfrak{x}))]$ is diophantine. A relation with d.g. is diophantine.*

PROOF. The first part of our theorem is proved in the same way as the last part of Thm. 12.3.

Now suppose that P is an n-ary relation with a d.g., i.e., the $(n+1)$-ary relation $\lambda\mathfrak{x}y(P(\mathfrak{x})=y)$ is diophantine. But $P(\mathfrak{x})=(P(\mathfrak{x})=0)$, so that P is diophantine as well. ∎

If we can show that a function obtained by primitive recursion from two total functions with d.g.s has itself got a d.g., then by Thm. 12.3 it will follow that *every* p.r. function has a d.g. (see Def. 5.2 and the text following it). In particular, by Thm. 12.4 it will follow that every p.r. relation is diophantine. Since by Cor. 11.7 every r.e. relation is obtained by existential quantification from a p.r. relation, it will then follow from Lemma 12.2 that every *r.e.* relation is diophantine. Unfortunately, to prove that the class of total functions with d.g.s is closed under the operation of primitive recursion is no easy task, and we shall have to work pretty hard for it. We start with:

12.5. DEFINITION. *Gödel's function* β is the ternary function[1]

$$\lambda xyz\,[\mathrm{rm}(x, y(z+1)+1)].$$

To study the behaviour of the function β, we shall need the following elementary number-theoretic result.

12.6. CHINESE REMAINDER THEOREM. *For $n\geqslant 0$, let $d_0,d_1,\ldots,d_n$ be any positive numbers that are mutually prime (i.e., no two of them have a non-trivial common factor). Let d be their product $d_0\ldots d_n$. Then for any given $n+1$ numbers $r_i<d_i$ (where $i=0,1,\ldots,n$) we can find a number $c<d$ such that $r_i=\mathrm{rm}(c,d_i)$ for all $i\leqslant n$.*

PROOF. Let us say that an $(n+1)$-tuple of numbers $\langle r_0,\ldots,r_n\rangle$ is *good* if $r_i<d_i$ for all $i\leqslant n$. Then there are exactly d different good $(n+1)$-tuples.

Each number c gives us a good $(n+1)$-tuple: $\langle\mathrm{rm}(c,d_0),\ldots,\mathrm{rm}(c,d_n)\rangle$. Allowing c to take all values $<d$, we get d such $(n+1)$-tuples. We shall show that they are all different; hence it follows at once that *every* good $(n+1)$-tuple must be so obtained, as the theorem claims.

Indeed, suppose that $0\leqslant c<c'<d$ and c yields the same $(n+1)$-tuple as c'. Then $\mathrm{rm}(c,d_i)=\mathrm{rm}(c',d_i)$ for all $i\leqslant n$. Therefore $c'-c$ must be divisible

[1] For the definition of rm see Ex. 6.11.

by each d_i, and — since the d_i are mutually prime — it follows that $c'-c$ must be divisible by their product d. But this is impossible, because $0<c'-c<d$. ∎

12.7. THEOREM. *For* $n \geqslant 0$, *let* $r_0, r_1, \ldots, r_n$ *be any given numbers. Then we can find numbers a and c such that* $\beta(c,a,i)=r_i$ *for all* $i \leqslant n$.

PROOF. Let s be any number greater than n and each of the r_i (e.g., take $s=n+r_0+\ldots+r_n+1$). Put $a=s!$ and for each $i \leqslant n$ let $d_i=a(i+1)+1$.

Clearly, $r_i<d_i$. We shall show that the d_i are mutually prime. Suppose that $0 \leqslant i<j \leqslant n$ and d_i had a prime factor p in common with d_j. Then p would divide d_j-d_i, that is, $a(j-i)$. Since p is prime, it would therefore divide a or $j-i$. But p cannot divide a, because we have assumed that p divides $d_i=a(i+1)+1$. Also, p cannot divide $j-i$, because $0<j-i \leqslant n<s$, so that $j-i$ must divide $s!=a$, and we have just seen that p does not divide a.

By the Chinese Remainder Theorem, we find a number c such that $\mathrm{rm}(c,d_i)=r_i$ for all $i \leqslant n$. This is the required result, in view of our definition of the d_i and Def. 12.5. ∎

12.8. LEMMA. *The relations* $\leqslant, <$ *are diophantine. The functions* rm *and* β *have diophantine graphs.*

PROOF. Use the identities

$$(x \leqslant y) = \exists z(x+z=y),$$

$$(x<y) = (x+1 \leqslant y),$$

$$(\mathrm{rm}(x,y)=z) = (y=0) \wedge (z=x) \vee (z<y) \wedge \exists u(x=yu+z),$$

$$\beta(x,y,z) = \mathrm{rm}(x,y(z+1)+1),$$

and apply 12.2, 12.3 and 12.4. ∎

Now let the function f be obtained from total functions g and h by primitive recursion:

$$f(\mathfrak{x},0)=g(\mathfrak{x}),$$

$$f(\mathfrak{x},y+1)=h(\mathfrak{x},y,f(\mathfrak{x},y)).$$

Then, for any $\mathfrak{a} \in N^n$ and numbers b and c, the equality $f(\mathfrak{a},b)=c$ is clearly equivalent to the following condition:

There are numbers $r_0, r_1, \ldots, r_b$ *such that* $r_0=g(\mathfrak{a})$; *and* $h(\mathfrak{a},t,r_t)=r_{t+1}$ *for every* $t<b$; *and* $r_b=c$.

In view of Thm. 12.7 we therefore have the identity

$$(f(\mathfrak{x},y)=z) = \exists u \exists v\, [(\beta(u,v,0)=g(\mathfrak{x})) \wedge \forall t<y(h(\mathfrak{x},t,\beta(u,v,t))=\beta(u,v,t+1)) \wedge (\beta(u,v,y)=z)].$$

If we can prove that the class of diophantine relations is closed under bounded universal quantification, then this identity (together with earlier results of the present section) shows that the class of total functions with d.g.s is closed under primitive recursion. As pointed out after Thm. 12.4, this implies that every r.e. relation is diophantine.[1]

The following three sections (§§13–15) are devoted to proving that the class of diophantine relations is in fact closed under bounded universal quantification. On a first reading of this book, the reader who so wishes may accept this result without proof and go directly to §16.

*§ 13. The Fibonacci sequence

In the present section and in §14, we let φ be the Fibonacci sequence defined in Prob. 6.15:

(1) $\varphi(0)=0, \quad \varphi(1)=1, \quad \varphi(x+2)=\varphi(x+1)+\varphi(x).$

We shall be especially interested in the sequence $\lambda x\,[\varphi(2x)]$. From (1) we see without difficulty

(2) $\varphi(0)=0, \quad \varphi(2)=1, \quad \varphi(2x+4)=3\varphi(2x+2)-\varphi(2x).$

13.1. THEOREM. *$\varphi(2x+2)\geqslant 2^x$ and $\varphi(2x)<3^x$ for all x.*

PROOF. The first inequality follows from (1) by induction on x. (Observe that $\varphi(x+1)\geqslant\varphi(x)$ and hence $\varphi(x+2)\geqslant\varphi(x)$ for all x.)

The second inequality follows by induction on x using (2). ∎

The rest of this section is devoted to proving that the sequence $\lambda x\,[\varphi(2x)]$ has a diophantine graph.[2]

[1] As will be seen in §16, this in turn yields a negative solution to Hilbert's Tenth Problem. In fact, it was precisely the conjecture that Hilbert's problem has a negative solution which led mathematicians to try to show that every r.e. relation is diophantine.

[2] The motivation for proving this result is explained below in §17.

13.2. LEMMA.

$$(u^2-uv-v^2=1)=\exists x\,[(u=\varphi(2x+1))\wedge(v=\varphi(2x))].$$

PROOF. Using (1) it is easy to verify by induction on x that

(3) $$(\varphi(x+1))^2-\varphi(x)\cdot\varphi(x+1)-(\varphi(x))^2=(-1)^x.$$

Next, we show that if u and v are numbers such that

(4) $$(u^2-uv-v^2)^2=1, \quad \text{with } u>0,$$

then

(5) for some x, $u=\varphi(x+1)$ and $v=\varphi(x)$.

We do this by induction on $u+v$. For $v=0$, (4) reduces to $u=1$, so that by (1) we have (5), with $x=0$.

Now assume (4) with $v>0$. Then $u\geqslant v$; otherwise we would get $u^2-uv-v^2\leqslant-2$. We put $u'=v$ and $v'=u-v$; thus $u=u'+v'$ and $v=u'$. Substituting this in (4), we find that

$$(u'^2-u'v'-v'^2)^2=1.$$

Also, $u'=v>0$ by assumption. Since $u'+v'=u<u+v$, we can use the induction hypothesis to conclude that $u'=\varphi(y+1)$ and $v'=\varphi(y)$ for some y. Hence we find

$$u=\varphi(y+1)+\varphi(y)=\varphi(y+2), \qquad v=\varphi(y+1).$$

Thus (5) holds with $x=y+1$.

From (3), and the fact that (4)⇒(5), the assertion of the lemma follows at once. ∎

We shall make use of the identity

(6) $$\varphi(x+y)=\varphi(x+1)\cdot\varphi(y)+\varphi(x)\cdot\varphi(y-1),$$

which holds for all values of x and all positive values of y. This identity is easily verified by induction on y, using (1).

We shall also make use of the fact that *any two consecutive Fibonacci numbers — say $\varphi(x)$ and $\varphi(x+1)$ — are always mutually prime.* This can be seen at once from (1) by induction on x.

We put

$$x|y \;=\; \exists z(xz=y).$$

Thus the relation $|$ is diophantine. (We shall only use the symbol $|$ for divisibility of *natural numbers*, not negative integers.)

13.3. LEMMA. *$\varphi(x)|\varphi(xy)$ holds for all x and y.*
PROOF. For $x=0$ the result is trivial, so we may assume $x>0$. We use induction on y. The cases $y=0$ and $y=1$ are again trivial.

Now let $y=z+1$, where z is positive. Then xz is positive as well and by (6) we have

$$\begin{aligned}\varphi(xy)&=\varphi(x+xz)\\&=\varphi(x+1)\cdot\varphi(xz)+\varphi(x)\cdot\varphi(xz-1).\end{aligned}$$

Using the induction hypothesis we get the required result. ∎

13.4. LEMMA. *If $\varphi(x)|\varphi(y)$ holds and $x\neq 2$, then $x|y$ holds.*
PROOF. We may assume that $y>x>2$, because all other cases are trivial. We can thus put $y=xz+u$, where $u<x$ and $z>0$.

By (6) we have

$$\varphi(y)=\varphi(u+1)\cdot\varphi(xz)+\varphi(u)\cdot\varphi(xz-1).$$

By Lemma 13.3, $\varphi(x)|\varphi(xz)$ holds; also, the consecutive Fibonacci numbers $\varphi(xz-1)$ and $\varphi(xz)$ are mutually prime, hence $\varphi(xz-1)$ and $\varphi(x)$ are mutually prime as well. Thus $\varphi(x)|\varphi(u)$ must hold.

But, since $u<x$ and $x>2$, it is clear that $\varphi(u)<\varphi(x)$. It follows that $\varphi(u)=0$, i.e., $u=0$. ∎

Our next lemma looks a bit ugly, but it contains most of the spade-work needed for the two neat lemmas following it.

13.5. LEMMA. *For $x>0$ and $0<u\leqslant z$, put*

$$f(u,x,z)=(z-u)\cdot\varphi(xu-1)\cdot\varphi(x)+\varphi(xu)\cdot\varphi(x-1).$$

Then $\varphi(x)|z$ holds iff $(\varphi(x))^2|f(u,x,z)$ holds.
PROOF. By induction on u. First, take $u=1$. We have

$$f(1,x,z)=z\varphi(x-1)\cdot\varphi(x),$$

and the result follows from the fact that $\varphi(x)$ and $\varphi(x-1)$ are mutually prime.

Next, let $u=v+1$, where $v>0$. We wish to express $f(u,x,z)$ in terms of

$f(v,x,z)$. To this end, we observe that

$$\varphi(xu-1)=\varphi(xv-1+x)$$
$$=\varphi(xv)\cdot\varphi(x)+\varphi(xv-1)\cdot\varphi(x-1), \qquad \text{by (6);}$$
$$\varphi(xu)=\varphi(xv+x)$$
$$=\varphi(xv+1)\cdot\varphi(x)+\varphi(xv)\cdot\varphi(x-1), \qquad \text{by (6);}$$
$$\varphi(xv+1)=\varphi(xv)+\varphi(xv-1), \qquad \text{by (1).}$$

Using these identities we find without difficulty

$$f(u,x,z)=f(v,x,z)\cdot\varphi(x-1)+r\cdot\varphi(x)\cdot\varphi(xv),$$

where $r=(z-u)\cdot\varphi(x)+\varphi(x-1)$. From this, Lemma 13.3, and the fact that $\varphi(x)$ and $\varphi(x-1)$ are mutually prime, we get the result

$$(\varphi(x))^2|f(u,x,z) \text{ holds iff } (\varphi(x))^2|f(v,x,z) \text{ holds.}$$

This completes the induction. ∎

13.6. LEMMA. *For all x and y, if* $(\varphi(x))^2|\varphi(y)$ *holds then* $\varphi(x)|y$ *holds as well.*

PROOF. We may assume $y>x>2$, because all other cases are trivial.

Suppose $(\varphi(x))^2|\varphi(y)$ holds. By Lemma 13.4 we have $y=xz$, where $z>0$. Using Lemma 13.5 (with $u=z$), we see that $\varphi(x)|z$, and hence also $\varphi(x)|y$, must hold. ∎

13.7. LEMMA. *For all x and y, if* $x\cdot\varphi(x)|y$ *holds then* $(\varphi(x))^2|\varphi(y)$ *holds as well.*

PROOF. The case $y=0$ is trivial, so we may assume $y>0$.

If $x\cdot\varphi(x)|y$ holds, then $x>0$ and $y=xz$ for some $z>0$. Thus $\varphi(x)|z$ holds and from Lemma 13.5 (with $u=z$) we see that $(\varphi(x))^2|\varphi(y)\cdot\varphi(x-1)$ must hold. The required result follows at once, since $\varphi(x)$ and $\varphi(x-1)$ are mutually prime. ∎

13.8. LEMMA. *Put* $d=\varphi(2b)+\varphi(2b+2)$ *and consider the sequence* $\psi=\lambda x[\mathrm{rm}(\varphi(2x),d)]$. *We have:*

(7) $\mathrm{rm}(\varphi(2x),d)=\varphi(2x)$ *if* $b>0$ *and* $x\leqslant b+1$,

(8) $\mathrm{rm}(\varphi(2x),d)=d-\varphi(4b-2x+2)$ *if* $b\leqslant x\leqslant 2b$.

Also, the sequence ψ *is periodic with period* $2b+1$, *i.e.*,

(9) $\mathrm{rm}(\varphi(2(x+2b+1)),d)=\mathrm{rm}(\varphi(2x),d)$.

PROOF. (7) is obvious, since under the conditions stated $\varphi(2x) \leqslant \varphi(2b+2) < d$.

To prove (8), observe that, under the condition stated,

$$0 \leqslant \varphi(b) \leqslant d - \varphi(4b-2x+2) \leqslant d - \varphi(2) < d.$$

Hence it will be enough to prove the congruence

$$\text{(10)} \qquad \varphi(2x) \equiv -\varphi(4b-2x+2) \pmod d$$

for $b \leqslant x \leqslant 2b$. We do this by induction on x. For $x=b$ and $x=b+1$, (10) is immediate from the definition of d. For the induction step, put $x=z+2$, then use (2) in the form

$$\varphi(2z+4) = 3\varphi(2z+2) - \varphi(2z);$$

now apply the induction hypothesis to z and $z+1$, and use (2) again in the form

$$\varphi(4b-2z+2) = 3\varphi(4b-2z) - \varphi(4b-2z-2).$$

Actually, this induction proves (10) for $b \leqslant x \leqslant 2b+1$. Moreover, (10) holds even for $x<b$. For, in this case we put $u=2b-x+1$; then $b+1<u \leqslant 2b+1$, so by what we have already proved (10) holds with x replaced by u. This yields (10) for x as well.

It remains to prove (9). Again, we work with congruences mod d. We can re-state (9) thus:

$$\text{(9')} \qquad \varphi(2x+4b+2) \equiv \varphi(2x).$$

We prove this by induction on x. For $x=0$, we have to show $\varphi(4b+2) \equiv 0$; but this is the same as (10) for $x=0$.

For $x=1$, we must show that $\varphi(4b+4) \equiv 1$. Putting $x=2b$ in (10), we obtain $\varphi(4b) \equiv -1$. But by (2) and by what we have already shown,

$$\varphi(4b+4) = 3\varphi(4b+2) - \varphi(4b) \equiv 0 - (-1) \equiv 1.$$

For the induction step we put $x=z+2$ and use (2) again. ∎

We now define an auxiliary binary function g by the identities

$$\text{(11)} \qquad g(w,0)=0, \qquad g(w,1)=1, \qquad g(w,x+2) = wg(w,x+1) \dot{-} g(w,x).$$

We shall be interested in $g(w,x)$ for $w \geqslant 2$ only. For such w it is easy to see (by induction on x) that $g(w,x+1) > g(w,x)$; hence the "$\dot{-}$" in (11) may be replaced by "$-$".

Using (11), it is easy to see by induction on x that

$$\text{(12)} \qquad g(w,x) \equiv x \pmod{w-2}, \quad \text{for } w \geqslant 2.$$

Also, using (2) and (11) it can easily be seen by induction on x that

(13) $\varphi(2x) \equiv g(w,x) \pmod{w-3}$, for $w \geqslant 2$.

The following lemma is similar to Lemma 13.2.

13.9. LEMMA. *Let $w \geqslant 2$. Then the conditions*

(14) $u^2 - wuv + v^2 = 1$, *with* $u \leqslant v$,

(15) *for some x,* $u = g(w,x)$ *and* $v = g(w,x+1)$,

are equivalent.

PROOF. (15)$\Rightarrow$(14) is easily verified by induction on x, using (11).

To prove (14)$\Rightarrow$(15), we proceed by introduction on $wv-u$. This is positive, since (14) implies $u < v$. For $u = 0$, (14) reduces to $v = 1$; so (15) holds with $x = 0$.

Now let (14) hold with $u > 0$. Re-arranging (14), we have

$$u^2 = 1 + v(wu - v), \quad 0 < u \leqslant v.$$

From this we get $0 \leqslant wu - v < u$. We put $u' = wu - v$ and $v' = u$. Thus $u = v'$ and $v = wv' - u'$. Substituting this in (14) we find that $u'^2 - wu'v' + v'^2 = 1$. Also, as we have seen, $u' = wu - v < u = v'$. Since

$$wv' - u' = v \leqslant (w-1)\, v < wv - u,$$

we may use our induction hypothesis to conclude that $u' = g(w,y)$ and $v' = g(w,y+1)$ for some y. Therefore $u = g(w,y+1)$ and (using (11)) $v = g(w,y+2)$, so (15) holds with $x = y+1$. ∎

We can now prove the main result of this section:

13.10. THEOREM. *The function $\lambda x\,[\varphi(2x)]$ has a diophantine graph.*

PROOF. We show that, for any numbers x and y, $(\varphi(2x) = y)$ holds iff there are numbers r,s,t,u,v,z and w such that

(i) $x \leqslant y < t$,

(ii) $t^2 - tz - z^2 = 1$,

(iii) $r^2 - 2rs - 4s^2 = 1$,

(iv) $t^2 | r$,

(v) $w = 3 + (4s + r)s$,

(vi) $u^2 - wuv + v^2 = 1$,

(vii) $x = \mathrm{rm}(u,t)$,

(viii) $y = \mathrm{rm}(u, 4s + r)$.

By the results of §12, the theorem will then be proved.

Suppose first that, for given x and y, there are numbers r,s,t,u,v,z and w satisfying the eight conditions (i)–(viii).

By Lemmas 13.2 and 13.9 we obtain, from (ii), (iii) and (vi),

$$t=\varphi(a), \qquad r=\varphi(2b+1), \qquad 2s=\varphi(2b), \qquad u=g(w,c),$$

for some a,b and c. Let us put

$$d=\varphi(2b)+\varphi(2b+2);$$

then by (1) we have also

$$d=2\cdot\varphi(2b)+\varphi(2b+1)=4s+r.$$

By (viii) we now have

$$y=\mathrm{rm}(g(w,c),d).$$

But from (v) and (13) we get

$$\varphi(2c)\equiv g(w,c) \pmod{ds},$$

and therefore we can also write

$$y=\mathrm{rm}(\varphi(2c),d).$$

Dividing c by $2b+1$ we have $c=q\cdot(2b+1)+e$, where q is the quotient, and the remainder is $e\leqslant 2b$. By the last part of Lemma 13.8, we now have

$$y=\mathrm{rm}(\varphi(2e),d),$$

because $c\equiv e \pmod{2b+1}$.

We want to know whether $e\leqslant b$ or $b<e\leqslant 2b$. If the latter were the case, then by (8)

$$y=d-\varphi(4b-2e+1)\geqslant d-\varphi(2b)=\varphi(2b+2).$$

But by (i) and (iv)

$$y<t\leqslant r=\varphi(2b+1)\leqslant\varphi(2b+2).$$

Thus we conclude that $e\leqslant b$. If $b=0$, then $d=\varphi(0)+\varphi(2)=1$, hence we have $y=\mathrm{rm}(\varphi(2e),1)=0$ and by (i) also $x=0$; thus $y=\varphi(2x)$ as required. We may therefore assume $b>0$ and use (7), getting

$$y=\varphi(2e).$$

We must show that $x=e$. If $e=0$, then $y=\varphi(0)=0$ and again we see from (i) that $x=0$. If $e>0$, then by Thm. 13.1 and (i)

$$e \leqslant 2^{e-1} \leqslant \varphi(2e)-y<t.$$

From (vii) we have $x=\mathrm{rm}(u,t)=\mathrm{rm}(g(w,c),t)$. But from (iii)–(v) it is easy to verify that $t|w-2$, hence by (12) we get

$$x=\mathrm{rm}(c,t).$$

Also, since $t=\varphi(a)$ and $r=\varphi(2b+1)$, we conclude from (iv) and Lemma 13.6 that $t|2b+1$. Thus, since we know that $e<t$, we have

$$x=\mathrm{rm}(c,t)=\mathrm{rm}(q\cdot(2b+1)+e,t)=e,$$

as required.

Now for the converse. Let x and y be numbers such that $\varphi(2x)=y$. We have to find r,s,t,u,v,z and w to satisfy (i)–(viii).

We choose

$$t=\varphi(24x+1), \qquad z=\varphi(24x).$$

Clearly, (i) is satisfied; and, by Lemma 13.2, (ii) is satisfied as well.

From (1) it is easy to see that $\varphi(a)$ is even iff $a\equiv 0 \pmod 3$. Hence t is odd. Also, it is easy to see that if $a\equiv 1 \pmod 8$ then $\varphi(a)\equiv 1 \pmod 3$. Hence $t\equiv 1 \pmod 3$ and $t(24x+1)-1\equiv 0 \pmod 3$. It follows that $\varphi(t(24x+1)-1)$ is even. Put

$$r=\varphi(t(24x+1)), \qquad s=\tfrac{1}{2}\varphi(t(24x+1)-1).$$

Then (iii) and (iv) are satisfied because of Lemmas 13.2 and 13.7.

Next, choose w as dictated by (v) and take

$$u=g(w,x), \qquad v=g(w,x+1),$$

so that (vi) holds by Lemma 13.9.

As in the first part of this proof, we see from (iii)–(v) that $t|w-2$. Therefore by (12),

$$\mathrm{rm}(u,t)=\mathrm{rm}(g(w,x),t)=\mathrm{rm}(x,t).$$

But by (i) we have $x<t$. Hence $\mathrm{rm}(x,t)=x$ and (vii) is satisfied.

Finally, $4s+r|w-3$ by (v), so (13) yields

$$\mathrm{rm}(u,4s+r)=\mathrm{rm}(g(w,x),4s+r)=\mathrm{rm}(\varphi(2x),4s+r).$$

But $\varphi(2x)=y$, and clearly $y<4s+r$ by (i) and (iv). Thus $\mathrm{rm}(u,4s+r)=y$ and (viii) is satisfied. ∎

*§ 14. The power function

The main result of the present section is that the power function $\lambda yz(y^z)$ has a diophantine graph.

We start by making another excursion into elementary number theory. The equation

(1) $$x^2 - Ay^2 = 1,$$

where A is a given natural number that is not a square, is known (mistakenly) as *Pell's equation*. To *solve* it is to find all natural solutions, i.e., all pairs $\langle x,y\rangle \in N^2$ satisfying (1). We propose to solve (1) in the important special case where

(2) $$A = a^2 - 1, \quad a > 1.$$

The reader should check that, because of (2), $A^{1/2}$ *must be irrational.*

If $\langle x,y\rangle$ and $\langle x',y'\rangle$ are two solutions, then $x < x'$ iff $y < y'$. Thus the solutions can be arranged in increasing order of *both* components. We let $\langle X_z(a), Y_z(a)\rangle$ be the zth solution (in the above ordering), beginning with $z=0$. (We shall soon see that a zth solution exists for all z.) The 0th and 1st solutions are evident: $\langle 1,0\rangle$ and $\langle a,1\rangle$. They are known as the *trivial* and *fundamental* solution respectively. The complete solution is given by:

14.1. THEOREM. *Equation* (1), *subject to condition* (2), *has infinitely many solutions. The* z^{th} *solution is uniquely determined by*

(3) $$X_z(a) + Y_z(a) \cdot A^{1/2} = (a + A^{1/2})^z.$$

PROOF. Let z be any fixed number. By multiplying out $(a+A^{1/2})^z$, and separating even and odd powers of $A^{1/2}$, we obtain

(4) $$x + yA^{1/2} = (a + A^{1/2})^z,$$

where x and y are natural numbers. Moreover, x and y are uniquely determined by (4): if $x + yA^{1/2} = x' + y'A^{1/2}$, then $x = x'$ and $y = y'$ because $A^{1/2}$ is irrational.

If in the right-hand side of (4) we replace $A^{1/2}$ by $-A^{1/2}$, then the terms containing even powers of $A^{1/2}$ are unchanged, while those containing odd powers of $A^{1/2}$ change signs. Thus

(5) $$x - yA^{1/2} = (a - A^{1/2})^z.$$

Multiplying (4) by (5), and using (2), we find that $x^2 - Ay^2 = 1$. Thus the pair $\langle x,y\rangle$ determined by (4) is a solution of (1).

Also, it is clear that in (4) x and y increase monotonically with z. So (4) gives us infinitely many solutions, in the correct order.

It only remains to show that *every* solution satisfies (4) for some z. Let, therefore, $\langle x,y\rangle$ be any solution of (1). Then $x+yA^{1/2}\geqslant x\geqslant 1$. On the other hand, by (2) we have $a+A^{1/2}>1$. It follows that $x+yA^{1/2}$ must lie between two consecutive natural powers of $a+A^{1/2}$, i.e., for some (unique) z we have

$$(6) \qquad (a+A^{1/2})^z \leqslant x+yA^{1/2} < (a+A^{1/2})^{z+1}.$$

By (2) we have

$$(7) \qquad (a+A^{1/2})(a-A^{1/2})=1.$$

Therefore, multiplying (6) by $(a-A^{1/2})^z$, i.e. by $(a+A^{1/2})^{-z}$, we get

$$(8) \qquad 1\leqslant p+qA^{1/2}<a+A^{1/2},$$

where p and q are (uniquely determined) integers such that

$$(9) \qquad (x+yA^{1/2})(a-A^{1/2})^z=p+qA^{1/2}.$$

Here p and q can be obtained by multiplying out the left-hand side, and separating even and odd powers of $A^{1/2}$. If in the left-hand side of (9) we replace $A^{1/2}$ by $-A^{1/2}$, then the terms with even powers of $A^{1/2}$ are unchanged, while those with odd powers change signs:

$$(x-yA^{1/2})(a+A^{1/2})^z=p-qA^{1/2}.$$

Multiplying this equality by (9) and using (7) as well as our assumption that $\langle x,y\rangle$ is a solution of (1), we get

$$1=(p+qA^{1/2})(p-qA^{1/2}).$$

Using this and (7), we can take reciprocals in (8):

$$(10) \qquad a-A^{1/2}<p-qA^{1/2}\leqslant 1.$$

By (7), $a-A^{1/2}=(a+A^{1/2})^{-1}>0$. Hence by (10) we have

$$0<p-qA^{1/2}.$$

Adding this last inequality to the left half of (8) we find that $1<2p$. Since p is an integer we must therefore have $p\geqslant 1$. From this and the right half of (10) it follows that $q\geqslant 0$.

On the other hand, adding the left half of (10) to the right half of (8) and re-arranging, we get $qA^{1/2}<A^{1/2}$; hence $q<1$. Thus q must be 0.

From the right half of (10) it now follows that $p \leqslant 1$. But we have seen above that also $p \geqslant 1$. Hence $p=1$.

Putting $p=1$ and $q=0$ in (9) and using (7), we obtain (4), as claimed. ∎

The following seven lemmas are concerned with various aspects of the behaviour of $X_z(a)$ and $Y_z(a)$ as z and a vary. In the first four of these lemmas a will be kept fixed and only z will vary, so we put, for the sake of brevity,

$$x_z = X_z(a), \qquad y_z = Y_z(a).$$

14.2. LEMMA.

$$x_0 = 1, \qquad y_0 = 0;$$
$$x_1 = a, \qquad y_1 = 1;$$
$$x_{z+2} = 2ax_{z+1} - x_z, \quad y_{z+2} = 2ay_{z+1} - y_z.$$

PROOF. The 0th and 1st solutions have already been obtained above. The remaining pair of equalities are obtained as follows:

$$\begin{aligned} x_{z+2} + y_{z+2}A^{1/2} &= (a+A^{1/2})^{z+2} && \text{by 14.1,} \\ &= (a+A^{1/2})^z(2a(a+A^{1/2}) + A - a^2) \\ &= (a+A^{1/2})^z(2a(a+A^{1/2}) - 1) && \text{by (2),} \\ &= 2a(a+A^{1/2})^{z+1} - (a+A^{1/2})^z \\ &= (2ax_{z+1} - x_z) + (2ay_{z+1} - y_z)A^{1/2} && \text{by 14.1.} \end{aligned}$$

The required result follows from the irrationality of $A^{1/2}$. ∎

14.3. LEMMA. $x_{z+1} = ax_z + Ay_z$.

PROOF.

$$\begin{aligned} x_{z+1} + y_{z+1}A^{1/2} &= (a+A^{1/2})^{z+1} \\ &= (a+A^{1/2})^z(a+A^{1/2}) \\ &= (x_z + y_zA^{1/2})(a+A^{1/2}) \\ &= (ax_z + Ay_z) + (x_z + ay_z)A^{1/2}. \end{aligned}$$ ∎

14.4. LEMMA. $a^z \leqslant x_z \leqslant (2a)^z$.

PROOF. In multiplying out $(a+A^{1/2})^z$, the terms with even power of $A^{1/2}$ include a^z. Since all these terms are non-negative, we have $a^z \leqslant x_z$ by Thm. 14.1.

It follows at once from (2) that $A^{1/2}<a$. Hence, by Thm. 14.1,

$$x_z \leqslant x_z + y_z A^{1/2} = (a + A^{1/2})^z \leqslant (2a)^z.$$ ∎

14.5. LEMMA. $y_z \equiv z \pmod{a-1}$.
PROOF. By induction on z, using Lemma 14.2. ∎

14.6. LEMMA. *Let S be the diophantine relation defined by*

$$S(u,v) = \exists x \exists y \left[(x^2-(u^2-1)(u-1)^2y^2=1) \wedge (x>1) \wedge (u>1) \wedge (v=ux)\right].$$

Then if $S(b,c)$ holds, we have $c \geqslant b^b$. Also, for every $b>1$ we can find some c such that $S(b,c)$ holds.
PROOF. Suppose that $S(b,c)$ holds. Then for some x and y we have

$$x^2-(b^2-1)((b-1)y)^2=1,$$

where

$$x>1, \qquad b>1, \qquad c=bx.$$

It follows that for some z

$$x=X_z(b), \qquad (b-1)y=Y_z(b).$$

Moreover, $z>0$ because $x>1$. From Lemma 14.5 we get

$$0 \equiv (b-1)y \equiv z \pmod{b-1}$$

Therefore z is a positive multiple of $b-1$. Hence, using Lemma 14.4,

$$x=X_z(b) \geqslant b^z \geqslant b^{b-1},$$

so that $c=bx \geqslant b^b$ as claimed.

Now suppose $b>1$. Put

$$x=X_{b-1}(b), \qquad c=bx.$$

Then certainly $x>X_0(b)=1$. Also, by Lemma 14.5 we see that

$$Y_{b-1}(b) \equiv 0 \pmod{b-1},$$

so that $Y_{b-1}(b)$ is a multiple of $b-1$ and for some y we have $Y_{b-1}(b) = (b-1)y$. It now follows that

$$x^2-(b^2-1)(b-1)^2y^2=1$$

and we see that $S(b,c)$ holds. ∎

REMARK. If we define $S(u,v) = (\varphi(2u^2+2) = v)$ instead of as above, then the same result concerning S can be proved much more easily, using Thm. 13.1.

14.7. LEMMA. *Let* $y>1$ *and* $a>y^z$. *Then*

$$X_z(a)\cdot y^z \leqslant X_z(ay) < X_z(a)\cdot(y^z+1).$$

PROOF. We start by using Thm. 14.1. Expanding the right-hand side of (3), ignoring terms with odd powers of $A^{1/2}$ and recalling that $A=a^2-1$, we obtain

$$X_z(a)=\sum\tbinom{z}{2k}a^{z-2k}(a^2-1)^k,$$

where the summation is over all $k\leqslant\frac{1}{2}z$. Hence

$$X_z(a)\cdot y^z=\sum\tbinom{z}{2k}(ay)^{z-2k}(a^2y^2-y^2)^k,$$

$$X_z(ay)=\sum\tbinom{z}{2k}(ay)^{z-2k}(a^2y^2-1)^k,$$

where the summations are again over $k\leqslant\frac{1}{2}z$. The left half of our lemma now follows at once, since our assumptions imply that

$$0<a^2y^2-y^2<a^2y^2-1.$$

By comparing the expressions for $X_z(a)$ and $X_z(ay)$ we see that in order to prove the remaining part of the lemma it is enough to show, for all $k\leqslant\frac{1}{2}z$,

$$y^{z-2k}(a^2y^2-1)^k<(a^2-1)^k(y^z+1).$$

But for $k\leqslant\frac{1}{2}z$ we clearly have $y^{z-2k}(a^2y^2-1)^k\leqslant a^{2k}y^z$; hence it is sufficient to show

$$a^{-2k}(a^2-1)^k>y^z(y^z+1)^{-1}.$$

Now, by the well-known Bernoulli inequality,

$$a^{-2k}(a^2-1)^k=(1-a^{-2})^k\geqslant 1-ka^{-2}.$$

Next, since $k\leqslant z<y^z<a$, we have

$$1-ka^{-2}>1-a^{-1}.$$

Finally, since $y^z+1\leqslant a$, we have

$$1-a^{-1}\geqslant 1-(y^z+1)^{-1}=y^z(y^z+1)^{-1},$$

as required. ∎

14.8. LEMMA. *Let* $y>1$ *and* $a>y^z$. *Then the double inequality*

$$X_z(a)\leqslant X_w(ay)\leqslant X_z(a)\cdot a$$

holds iff $w=z$.

PROOF. If $w=z$ then the double inequality follows directly from Lemma 14.7, using our assumptions $y>1$ and $a\geqslant y^z+1$.

Since $X_w(ay)$ increases with w, it remains to show that

$$X_{z-1}(ay)<X_z(a), \qquad X_{z+1}(ay)>X_z(a)\cdot a$$

(the former only for positive z and the latter for any z). This is easily done, using Lemmas 14.7 and 14.3. We leave the details to the reader. ∎

We now come to the main result of this section:

14.9. THEOREM. *The function $\lambda yz(y^z)$ has a diophantine graph.*

PROOF. We claim: for any numbers x,y and z, the equality $x=y^z$ holds iff:

$z=0$ and $x=1$; or

$z>0$ and $y\leqslant 1$ and $x=y$; or

$z>0$ and $y>1$ and $x>0$ and there are numbers a,b,c,u,v,r and s such that

(i) $y+z<b$,

(ii) $S(b,c)$ holds,

(iii) $x+c+z+3<a$,

(iv) $1<r<\varphi(2a)$,

(v) $r^2-(a^2-1)(z+s(a-1))^2=1$,

(vi) $rx\leqslant u<r(x+1)$,

(vii) $u^2-(a^2y^2-1)v^2=1$.

(In (ii), S is the diophantine relation of Lemma 14.6; and in (iv), φ is the Fibonacci sequence.)

It is clear that once we establish this claim, the theorem will be proved. (Here we are making use of Thm. 13.10.)

To establish our claim, we may assume

$$(*) \qquad z>0, \qquad y>1, \qquad x>0,$$

because the remaining cases are trivial.

Suppose, first, that there are a,b,c,u,v,r and s satisfying conditions (i)–(vii). We have to show that $x=y^z$.

By (v) we have for some n

$$r=X_n(a), \qquad z+s(a-1)=Y_n(a).$$

We would like to show that $n=z$. Using (iv), Thm. 13.1, (iii) and Lemma 14.4 we have

$$1<r<\varphi(2a)<3^a<a^a\leqslant X_a(a).$$

Thus $1<r<X_a(a)$; so by the ordering of the solutions of our Pellian equation it follows that

$$0<n<a.$$

On the other hand, by our assumption ($*$) and by (iii) we also have

$$0<z<a.$$

By Lemma 14.5,

$$z\equiv z+s(a-1)=Y_n(a)\equiv n \pmod{a-1}.$$

Therefore n must in fact be z.

Next, from (vii) we see that $u=X_w(ay)$ for some w. Thus (vi) yields

$$X_z(a)\cdot x\leqslant X_w(ay)<X_z(a)\cdot(x+1);$$

hence using ($*$) and (iii) we have

$$X_z(a)\leqslant X_w(ay)\leqslant X_z(a)\cdot a.$$

At this point we wish to apply Lemma 14.8, so we need to check that $y>1$ and $y^z<a$. Now, $y>1$ by ($*$); also, by (i), (ii), Lemma 14.6 and (iii) we have

$$y^z<b^b\leqslant c<a.$$

Thus we may apply Lemma 14.8 and conclude that $w=z$. We can now re-write (vi) as follows:

$$X_z(a)\cdot x\leqslant X_z(ay)<X_z(a)\cdot(x+1).$$

On the other hand, by Lemma 14.7,

$$X_z(a)\cdot y^z\leqslant X_z(ay)<X_z(a)\cdot(y^z+1).$$

The last two double inequalities mean that $x=\mathrm{q}(X_z(ay), X_z(a))$ and $y^z=\mathrm{q}(X_z(ay), X_z(a))$ respectively (see Ex. 6.11); hence $x=y^z$ as required.

Now for the converse. Retaining our assumption ($*$), we let $x=y^z$. First, we choose b so that (i) is satisfied. In particular, by (i) and ($*$) we have $b>1$, so by Lemma 14.6 we can find c such that (ii) is satisfied as well.

Next, we choose a to satisfy (iii). Moreover, since 2^{a-1} increases with a faster than $(2a)^z$, we can take a so large that $2^{a-1}>(2a)^z$.

Now we put

$$r=X_z(a).$$

Then by Thm. 13.1, by our choice of a, by Lemma 14.4 and by $(*)$ we have

$$\varphi(2a) \geqslant 2^{a-1} > (2a)^z \geqslant X_z(a) > 1,$$

so (iv) is satisfied.

By Lemma 14.5,

$$Y_z(a) = z + s(a-1),$$

where s is an integer. But if s were negative, then by (iii) we would have $Y_z(a) < 0$, which is impossible. Thus s is a natural number. From the definition of $X_z(a)$ and $Y_z(a)$ it follows that (v) is satisfied.

Next, we choose $u = X_z(ay)$. By $(*)$ we have $y > 1$, and by (iii) we have $a > x = y^z$. Thus we may use Lemma 14.7, which yields (vi) by the choice of r and u and our assumption that $x = y^z$.

Finally, we take $v = Y_z(ay)$, so that (vii) is satisfied. ∎

Using Thm. 14.9, we shall now prove that two other very useful functions have diophantine graphs.

14.10. THEOREM. *The binomial coefficient, i.e., the function* $\lambda xy\binom{x}{y}$, *has a diophantine graph.*

PROOF. We use the following definition of the binomial coefficient:

$$\binom{x}{y} = \{\prod_{z<y} (x-z)\}/y!.$$

First let us assume that $0 < y \leqslant x$. We wish to find the quotient

$$\mathrm{q}(2^{xy}(1+2^x)^x, 2^{x^2}).$$

We have

$$2^{xy}(1+2^x)^x = \sum_{z \leqslant x} \binom{x}{z} 2^{x(y+x-z)}.$$

In the sum on the right-hand side, each of the first $y+1$ terms (i.e., those with $z \leqslant y$) is a multiple of 2^{x^2}. All the remaining terms together (those with $z > y$) contribute at most

$$\begin{aligned}\sum_{0<z\leqslant x} \binom{x}{z} 2^{x(x-1)} &= 2^{x(x-1)}((1+1)^x - 1) \\ &= 2^{x(x-1)}(2^x - 1),\end{aligned}$$

which is $< 2^{x^2}$. Thus we have found

$$\mathrm{q}(2^{xy}(1+2^x)^x, 2^{x^2}) = \sum_{z \leqslant y} \binom{x}{z} 2^{x(y-z)}.$$

Similarly we find

$$\mathrm{q}(2^{x(y-1)}(1+2^x)^x, 2^{x^2}) = \sum_{z<y} \binom{x}{z} 2^{x(y-z-1)}.$$

Multiplying this equality by 2^x and substracting from the preceding one we get

$$\mathrm{q}(2^{xy}(1+2^x)^x, 2^{x^2}) - 2^x \cdot \mathrm{q}(2^{x(y-1)}(1+2^x)^x, 2^{x^2}) = \binom{x}{y}.$$

Thus, assuming that $0<y\leqslant x$, we have shown that $z=\binom{x}{y}$ iff there are numbers u and v such that

(i) $u=z+v\cdot 2^x$,
(ii) $u\cdot 2^{x^2}\leqslant 2^{xy}(1+2^x)^x<(u+1)\cdot 2^{x^2}$,
(iii) $v\cdot 2^{x^2}\leqslant 2^{x(y-1)}(1+2^x)^x<(v+1)\cdot 2^{x^2}$.

Bringing in the other cases (i.e., $y=0$ and $x<y$) we now have: $z=\binom{x}{y}$ iff:

$y=0$ and $z=1$; or
$x<y$ and $z=0$; or
$0<y\leqslant x$ and there are u and v satisfying (i), (ii) and (iii).

In view of Thm. 14.9, this yields the required result. ∎

14.11. **Theorem.** *The function $\lambda x(x!)$ has a diophantine graph.*

Proof. We show that

$$x! = \mathrm{q}(r^x, \tbinom{r}{x}), \text{ i.e., } x! \leqslant r^x/\tbinom{r}{x} < x!+1,$$

where r is any number $>(2x)^{x+1}$; e.g., $r=(2x)^{x+1}+1$.

The cases $x=0$ and $x=1$ are trivial, so we may assume $x>1$. We have

$$r^x/\tbinom{r}{x} = x!/(1-r^{-1})(1-2r^{-1})\dots(1-(x-1)r^{-1}).$$

From this identity we get at once

$$x! \leqslant r^x/\tbinom{r}{x}.$$

But from the same identity we also get

$$r^x/\tbinom{r}{x} \leqslant x!/(1-xr^{-1})^x.$$

Now, it is easy to show that for every real number α such that $0<\alpha<\frac{1}{2}$

$$(1-\alpha)^{-1} < 1+2\alpha.$$

Also, if β is real and $0<\beta<1$, then

$$(1+\beta)^x = \sum_{z\leqslant x} \binom{x}{z}\beta^z < 1+\beta\sum_{z\leqslant x}\binom{x}{z} = 1+\beta(1+1)^x = 1+\beta\cdot 2^x.$$

Using these two inequalities with $\alpha = xr^{-1}$ and $\beta = 2xr^{-1}$, we get

$$r^x/\binom{r}{x} < x!(1+2xr^{-1})^x < x!(1+2^{x+1}xr^{-1}).$$

By the choice of r this last quantity is clearly $< x!+1$. Thus we have shown that

$$x! \leqslant r^x/\binom{r}{x} < x!+1,$$

as claimed. Thus $y = x!$ iff

$$y\binom{r}{x} \leqslant r^x < (y+1)\binom{r}{x},$$

where, e.g., $r = (2x)^{x+1}+1$.

In view of Thms. 14.9 and 14.10, this proves the present theorem. ∎

*§ 15. Bounded universal quantification

We shall need one more result stating that a particular function has a d.g.:

15.1. LEMMA. *The binary function f defined by*

$$f(y, t) = \prod_{w<y} (1+(w+1)t)$$

has a d.g.

PROOF. For the time being, let $t \neq 0$. Then it is easy to see that

(1) $$f(y,t) = \binom{\alpha}{y} t^y y!, \quad \text{where} \quad \alpha = y + t^{-1}.$$

Now let a be any positive number. Using Taylor's theorem we develop $(1+a^{-2})^\alpha$ in powers of a^{-2} up to the yth power:

(2) $$(1+a^{-2})^\alpha = \sum_{j \leqslant y} \binom{\alpha}{j} a^{-2j} + \text{remainder}.$$

The remainder *(à la Lagrange)* is

$$\binom{\alpha}{y+1} a^{-2(y+1)} (1+\vartheta a^{-2})^{\alpha-y-1}$$

with $0 < \vartheta < 1$. Since $y < \alpha \leqslant y+1$ by (1), it is easy to see that $0 < \binom{\alpha}{y+1} \leqslant 1$. Also,

$$(1+\vartheta a^{-2})^{\alpha-y-2} \leqslant 1.$$

Therefore the remainder can be re-written in the form $\eta a^{-2(y+1)}$ with $0 < \eta \leqslant 1$. Multiplying (2) by a^{2y+1} we obtain

(3) $$a^{2y+1}(1+a^{-2})^\alpha = \sum_{j \leqslant y} \binom{\alpha}{j} a^{2(y-j)+1} + \eta a^{-1}.$$

Now let us assume $y \neq 0$ (as well as $t \neq 0$). We develop $(1+a^{-2})^\alpha$ up to the $(y-1)^{\text{th}}$ power of a^{-2}. This time the remainder is

$$\binom{\alpha}{y} a^{-2y}(1+\vartheta a^{-2})^{\alpha-y}.$$

Using the fact that $y < \alpha \leqslant y+1$ it is easily seen that $0 < \binom{\alpha}{y} \leqslant y+1$ and $(1+\vartheta a^{-2})^{\alpha-y} < 2$, so this remainder can be re-written as

$$2\zeta a^{-2y}(y+1),$$

with $0 < \zeta < 1$. Multiplying by a^{2y-1} we get

$$\text{(4)} \qquad a^{2y-1}(1+a^{-2})^\alpha = \sum_{j<y} \binom{\alpha}{j} a^{2(y-j)-1} + 2\zeta a^{-1}(y+1).$$

We now put

$$\text{(5)} \qquad a = 2(y+1)!\,t^y.$$

Then all the terms under the summation signs in (3) and (4) are natural numbers (for, when the coefficients $\binom{\alpha}{j}$ are written out as fractions, their denominators divide a). Also the remainder terms in (3) and (4) are <1. Thus, if we put

$$u = \sum_{j \leqslant y} \binom{\alpha}{j} a^{2(y-j)+1} \quad \text{and} \quad v = \sum_{j<y} \binom{\alpha}{j} a^{2(y-j)-1}$$

then u and v are the unique natural numbers such that

$$u \leqslant a^{2y+1}(1+a^{-2})^\alpha < u+1,$$
$$v \leqslant a^{2y-1}(1+a^{-2})^\alpha < v+1.$$

These inequalities can be stated equivalently:

$$\text{(6)} \qquad u^t a^2 \leqslant a^t(a^2+1)^{yt+1} < (u+1)^t a^2,$$

$$\text{(7)} \qquad v^t a^{t+2} \leqslant (a^2+1)^{yt+1} < (v+1)^t a^{t+2}.$$

On the other hand, from the definition of u and v we have at once

$$u - a^2 v = \binom{\alpha}{y} a,$$

so that by (1) and (5) we get

$$\text{(8)} \qquad u = a^2 v + 2(y+1) \cdot z,$$

where $z = f(y,t)$.

Bringing in also the trivial cases $t=0$ and $y=0$ which we have so far left aside, we now have: $f(y,t)=z$ iff:
$yt=0$ and $z=1$; or
$yt>0$ and there are a, u and v satisfying (5)–(8).
By virtue of the results of §14 it follows that f has a d.g. ∎

We can now begin our direct attack on bounded universal quantification. We start with the following special case.

15.2. LEMMA. *Let*

$$P(\mathfrak{x},y)=\forall w<y\ \exists z_1<y\ \dots\ \exists z_m<y\,[f(\mathfrak{x},w,z_1,\dots,z_m)=0],$$

where $m\geqslant 0$ *and* f *is a polynomial. Then* P *is diophantine.*

PROOF. We find a polynomial g with $n+1$ variables and *positive* coefficients such that

(1) $g(\mathfrak{x},y)>y$ for all $\mathfrak{x},y$;

(2) $|f(\mathfrak{x},w,z_1,\dots,z_m)|<g(\mathfrak{x},y)$ whenever $w,z_1,\dots,z_m<y$.

(Such g may be obtained from f as follows: change the sign of all negative coefficients in f, then substitute y for each of the variables $w,z_1,\dots,z_m$ and finally add $y+1$.)

We claim that $P(\mathfrak{x},y)$ holds iff:
$y=0$; or
$y>0$ and there are numbers $s,t,v_1,\dots,v_m$ such that
(i) $t=g(\mathfrak{x},y)!$,
(ii) $1+(s+1)t=\prod_{w<y}(1+(w+1)t)$,
(iii) $1+(s+1)t$ divides $f(\mathfrak{x},s,v_1,\dots,v_m)$,
(iv) $1+(s+1)t\,|\prod_{j<y}(v_i-j)$ for $i=1,\dots,m$.
Once this claim is established, the lemma will follow at once. (Note that (i) and (ii) are of the right form by 14.11 and 15.1. Also, (iii) can be written as

$$\exists u\,[u^2(1+(s+1)t)^2=(f(\mathfrak{x},s,v_1,\dots,v_m))^2].$$

Finally,

$$\prod_{j<y}(v-j)=\binom{v}{y}\cdot y!,$$

so that (iv) is of the right form as well, by the results of §14.)

Since our claim is trivially true for $y=0$, we shall from now on assume $y>0$. Also we fix $\mathfrak{x}\in N^n$.

First, suppose that $s,t,v_1,\ldots,v_m$ are numbers satisfying (i)–(iv). We have to show that $P(\mathfrak{x},y)$ holds. Fix any $w<y$; we must show that $f(\mathfrak{x},w,z_1,\ldots,z_m)=0$ for some $z_1,\ldots,z_m<y$.

By (i) we have $t>0$, hence $1+(w+1)t>1$. Let p be a prime divisor of $1+(w+1)t$, and put

$$(3) \qquad z_i=\mathrm{rm}(v_i,p) \quad \text{for } i=1,\ldots,m.$$

From (ii) and (iv) it follows that $p|\prod_{j<y}(v_i-j)$ for each i. Since p is prime it follows that for each i there is some $j(i)<y$ such that p divides $v_i-j(i)$. Hence, by (3), $z_i=\mathrm{rm}(j(i),p)$, so that $z_i\leqslant j(i)<y$, as required.

Next, since $p|1+(w+1)t$, it follows that p cannot divide t. Hence by (i) we must have $p>g(\mathfrak{x},y)$. Therefore by (2) we get

$$(4) \qquad |f(\mathfrak{x},w,z_1,\ldots,z_m)|<p.$$

On the other hand, since p divides both $1+(w+1)t$ and, by (ii), also $1+(s+1)t$, it must divide their difference $(s-w)t$. We have seen, however, that p cannot divide t, so it must divide $s-w$, i.e., $w\equiv s \pmod p$. Also, by (3), $z_i\equiv v_i \pmod p$ for each i. Thus we have

$$(5) \qquad f(\mathfrak{x},w,z_1,\ldots,z_m)\equiv f(\mathfrak{x},s,v_1,\ldots,v_m) \pmod p.$$

By (ii), (iii) and (5) it now follows that p divides $f(\mathfrak{x},w,z_1,\ldots,z_m)$; but by (4) this means that $f(\mathfrak{x},w,z_1,\ldots,z_m)=0$ as required.

Now for the converse. Suppose that $P(\mathfrak{x},y)$ holds. We are assuming $y>0$, so we must show that there are $s,t,v_1,\ldots,v_m$ satisfying (i)–(iv).

We choose t as dictated by (i). Next, when the product $\prod_{w<y}(1+(w+1)t)$ is multiplied out, we obtain 1 plus terms containing powers of t (there is at least one such term since $y>0$). Hence we can find s such that (ii) is satisfied.

Since $P(\mathfrak{x},y)$ holds, we have for each $w<y$ numbers $z_1(w),\ldots,z_m(w)<y$ such that

$$(6) \qquad f(\mathfrak{x},w,z_1(w),\ldots,z_m(w))=0.$$

Let us fix any i, $1\leqslant i\leqslant m$. We show that there is a number v_i such that

$$(7) \qquad z_i(w)=\mathrm{rm}(v_i,\, 1+(w+1)t) \quad \text{for all } w<y.$$

By the Chinese Remainder Theorem 12.6, it is enough to show that $z_i(w)<1+(w+1)t$ for all $w<y$, and that the divisors $1+(w+1)t$, for different $w<y$, are mutually prime.

Since $z_i(w)<y$ by assumption, we have in fact

$$z_i(w)<g(\mathfrak{x},y)\leq t<1+(w+1)t$$

by (1) and (i).

To show that the $1+(w+1)t$ are mutually prime, suppose that p were a prime divisor of both $1+(u+1)t$ and $1+(w+1)t$, where $u<w<y$. Then p would also divide their difference $(w-u)t$. Thus p divides $w-u$ or t. However,

$$w-u<y<g(\mathfrak{x},y)$$

by (1), so (i) implies that $w-u$ divides t. Hence p must divide t in any case. But this contradicts our assumption that p divides $1+(w+1)t$.

Thus for each $i=1,\ldots,m$ we have a number v_i satisfying (7). It remains to show that our $s,t,v_1,\ldots,v_m$ satisfy (iii) and (iv).

Fix any $w<y$. Then, by (ii), $1+(w+1)t$ must divide $1+(s+1)t$ and hence also the difference $(w-s)t$. But since clearly $1+(w+1)t$ and t are mutually prime, it follows that $1+(w+1)t$ divides $w-s$; i.e.,

$$s\equiv w \pmod{1+(w+1)t}.$$

Also, (7) means that

$$v_i\equiv z_i(w) \pmod{1+(w+1)t}.$$

These facts, together with (6), yield

$$f(\mathfrak{x},s,v_1,\ldots,v_m)=0 \pmod{1+(w+1)t}.$$

In other words, $1+(w+1)t$ divides $f(\mathfrak{x},s,v_1,\ldots,v_m)$ for all $w<y$. But since, as we have seen, these divisors are mutually prime, their product must also divide $f(\mathfrak{x},s,v_1,\ldots,v_m)$. By (ii) this product is $1+(s+1)t$, so that (iii) is satisfied.

Finally, we recall that by assumption $z_i(w)<y$. Hence $v_i-z_i(w)$ is one of the factors in $\prod_{j<y}(v_i-j)$. By (7) it therefore follows that $1+(w+1)t$ divides $\prod_{j<y}(v_i-j)$. Again, since these divisors are mutually prime, their product $1+(s+1)t$ must also divide $\prod_{j<y}(v_i-j)$ and (iv) is satisfied. ∎

Lemma 15.2 can be generalized to the case where the bound on the initial universal quantifier is not necessarily the same as that of the m existential quantifiers:

15.3. LEMMA. *Let*

$$P(\mathfrak{x},u,v)=\forall w<u\ \exists z_1<v\ \ldots\ \exists z_m<v\,[f(\mathfrak{x},w,z_1,\ldots,z_m)=0],$$

where $m\geq 0$ *and* f *is a polynomial. Then* P *is diophantine.*

PROOF. Introducing new variables $w', z'_1, \ldots, z'_m$ and y, we put

$$Q(\mathfrak{x},u,v) = \exists y \, [(y=u+v) \wedge \forall w<y \; \exists w'<y \; \exists z_1<y \; \exists z'_1<y \ldots \exists z_m<y \; \exists z'_m<y \; E(\mathfrak{x},u,v,w,w',z_1,z'_1,\ldots,z_m,z'_m)],$$

where

$$E(\mathfrak{x},u,\ldots) = (u+w'=w) \vee [(f(\mathfrak{x},w,z_1,\ldots,z_m)=0) \wedge (z_1+z'_1+1=v) \wedge \ldots \wedge (z_m+z'_m+1=v)].$$

By Lemma 12.1, E is an elementary relation. Hence by Lemmas 15.2 and 12.2 Q is diophantine. We now proceed to prove the identity

$$P(\mathfrak{x},u,v) = Q(\mathfrak{x},u,v),$$

from which the assertion of our lemma follows at once. Let us fix numbers $\mathfrak{x}$, u and v. First, assume that $P(\mathfrak{x},u,v)$ holds. We choose $y=u+v$, and fix any $w<y$. We must show that there are numbers $w',z_1,z'_1,\ldots,z_m,z'_m<y$ such that $E(\mathfrak{x},u,\ldots)$ holds. There are two cases to consider.

If $u \leqslant w$, we put $w'=w-u$. Then $w' \leqslant w<y$ as required. We take the z_i and the z'_i to be arbitrary numbers $<y$ (e.g., all of them $=w$). By our choice of w' we have $u+w'=w$, so that $E(\mathfrak{x},u,\ldots)$ holds.

If $w<u$, we take w' to be any number $<y$ (e.g., $w'=w$). By assumption $P(\mathfrak{x},u,v)$ holds, so there are $z_1,\ldots,z_m<v$ such that

(1) $\qquad f(\mathfrak{x},w,z_1,\ldots,z_m)=0.$

Note that $z_i<y$, since $v \leqslant u+v=y$. We put

$$z'_i = v-z_i-1 \quad \text{for } i=1,\ldots,m.$$

Then clearly $z'_i<v \leqslant y$, as required. By the choice of the z'_i we have

(2) $\qquad z_i+z'_i+1=v \quad \text{for } i=1,\ldots,m.$

From (1) and (2) it follows that $E(\mathfrak{x},u,\ldots)$ holds, as claimed.

Conversely, suppose that $Q(\mathfrak{x},u,v)$ holds. Given any $w<u$, we must find $z_1,\ldots,z_m<v$ satisfying (1). Now, because $Q(\mathfrak{x},u,v)$ holds, there are numbers $w',z_1,z'_1,\ldots,z_m,z'_m$ such that $E(\mathfrak{x},u,\ldots)$ holds. On the other hand, $u+w' \neq w$ since we were given $w<u$; so the fact that $E(\mathfrak{x},u,\ldots)$ holds means that (1) and (2) must be satisfied. Thus we have (1), where $z_1,\ldots,z_m<v$ because of (2). ∎

We are now ready to prove:

15.4. THEOREM. *The class of diophantine relations is closed under bounded universal quantification.*

PROOF. Let

$$P(\mathfrak{x},u) = \forall w < u \; \exists z_1 \ldots \exists z_m \, [f(\mathfrak{x},w,z_1,\ldots,z_m)=0],$$

where $m \geqslant 0$ and f is a polynomial. We must show that P is diophantine. To this end we put

$$Q(\mathfrak{x},u) = \exists v \; \forall w < u \; \exists z_1 < v \ldots \exists z_m < v \, [f(\mathfrak{x},w,z_1\ldots,z_m)=0].$$

For given numbers $\mathfrak{x}$ and u, if $Q(\mathfrak{x},u)$ holds then clearly $P(\mathfrak{x},u)$ holds as well, since the condition imposed by Q is apparently more stringent than that imposed by P.

Conversely, suppose that $P(\mathfrak{x},u)$ holds, Thus for every $w<u$ we may choose m numbers $z_1(w),\ldots,z_m(w)$ such that

$$f(\mathfrak{x},w,z_1(w),\ldots,z_m(w))=0.$$

Since every finite set of numbers is bounded, we have (for sufficiently big v) $v>z_i(w)$ for all $w<u$ and $i=1,\ldots,m$. The existence of such v means that $Q(\mathfrak{x},u)$ holds.

Thus we have proved that P and Q are the same relation. On the other hand, from the definition of Q and Lemma 15.3 we see that Q is obtained by existential quantification from a diophantine relation. Hence Q (i.e., P) is itself diophantine by Lemma 12.2. ∎

§ 16. The MRDP Theorem and Hilbert's Tenth Problem

The main result of this section is:

16.1. THEOREM. *A relation is r.e. iff it is diophantine.*

PROOF. As pointed out at the end of §12, it follows from Thm. 15.4 that every r.e. relation is diophantine. The converse is obvious. ∎

We shall refer to this result as "the MRDP Theorem". This is an acronym for the names of the four mathematicians — Matijasevič, Robinson, Davis and Putnam — to whose joint efforts it is due. (For further details see §17.)

16.2. REMARK. The proof of the MRDP Theorem is completely constructive. Suppose an n-ary r.e. relation P is given to us by means of an index e. Thus

$$P(\mathfrak{x}) = (\mathfrak{x} \in \mathrm{dom}_n\{e\}) = \exists y T(e,\mathfrak{x},y).$$

(See Cor. 11.6.) Then the proof of the MRDP Theorem (including of course the proofs of all the results leading up to it) provides us with a method for finding a polynomial p and a number m such that

$$P(\mathfrak{x}) = \exists y_1 \ldots \exists y_m [p(\mathfrak{x}, y_1, \ldots, y_m) = 0].$$

(Of course, this method is not really practical.)

A *diophantine equation* is any equation of the form

(1) $f(\mathfrak{x}) = 0,$

where f is a polynomial (with integer coefficients).

Hilbert's Tenth Problem is the problem of devising an algorithm whereby, given any diophantine equation, one could decide whether or not that equation has a solution.

By a *solution* of a diophantine equation (1) one normally means a solution in *integers*, i.e., an n-tuple of integers satisfying (1). However, we shall now show that the nature of the Tenth Problem is not changed if we take *solution* to mean solution in *natural numbers*.

Given a polynomial f in n variables, let us define polynomials g and h, in $4n$ and $2n$ variables, respectively, as follows:

$$g(\mathfrak{x}, \mathfrak{y}, \mathfrak{z}, \mathfrak{u}) = f(x_1^2 + y_1^2 + z_1^2 + u_1^2, \ldots, x_n^2 + y_n^2 + z_n^2 + u_n^2),$$
$$h(\mathfrak{x}, \mathfrak{y}) = f(x_1 - y_1, \ldots, x_n - y_n).$$

By a well-known theorem of Lagrange, every natural number is the sum of four squares. Hence (1) has a solution in natural numbers iff the diophantine equation $g(\mathfrak{x}, \mathfrak{y}, \mathfrak{z}, \mathfrak{u}) = 0$ has a solution in integers. Conversely, every integer is the difference of two natural numbers, so (1) has a solution in integers iff the diophantine equation $h(\mathfrak{x}, \mathfrak{y}) = 0$ has a solution in natural numbers.

We shall therefore take *solution* to mean solution in natural numbers. Under this interpretation, the Tenth Problem is equivalent (as we have just seen) to the problem as originally posed.

It is not difficult to encode polynomials by numbers; to each polynomial f a *code number* $\# f$ is assigned such that coding and decoding are algorithmic (i.e., we have an algorithm for calculating the code number of any given polynomial, as well as an algorithm for finding out whether or not any given number is the code of a polynomial, and if so of which one.) This may be done, e.g., by a method like that used in §8 for encoding URIM programs.

Let P be the property (i.e., unary relation) such that, for any number z,

$P(z)$ holds iff $z = \# f$ for some polynomial f such that equation (1) has a solution.

Then clearly Hilbert's Tenth Problem is equivalent to the decision problem[1] for P. However, we shall now show that P is *recursively undecidable* (i.e., non-recursive). Hence — if we accept Church's Thesis — we must conclude that the decision problem for P is *not* solvable. This constitutes a negative solution to Hilbert's Tenth Problem.

Let R be an r.e. but non-recursive property. For example, using 11.8 we may take

$$R(z) = (z \in \mathrm{dom}_1\{z\}) = \exists y T(z,z,y).$$

By the MRDP Theorem we can find a number n and a polynomial f of $n+1$ variables such that

$$(2) \qquad R(z) = \exists x_1 \ldots \exists x_n [f(\mathfrak{x},z) = 0].$$

For each value of z we get from f a polynomial f_z (of n variables) defined by

$$(3) \qquad f_z(\mathfrak{x}) = f(\mathfrak{x},z).$$

Let the sequence φ be defined by the identity $\varphi(z) = \# f_z$. Clearly, φ is algorithmic. Moreover, for any of the standard methods for encoding polynomials, it is easy to show that φ is recursive.

But from (2), (3) and the definitions of P and φ we get the identity

$$R(z) = P(\varphi(z)).$$

Since R is not recursive, P cannot be recursive.

We conclude this section with two easy consequences of the MRDP Theorem.

16.3. THEOREM. *Given any r.e. set $A \subseteq N$, we can find a polynomial g such that A is the set of all non-negative values assumed by g as its variables range through N.*

PROOF. By the MRDP Theorem we can find n and a polynomial f of $n+1$ variables such that

$$(z \in A) = \exists x_1 \ldots \exists x_n [f(\mathfrak{x},z) = 0].$$

Now put

$$g(\mathfrak{x},z) = z - (z+1)(f(\mathfrak{x},z))^2.$$

It is easy to see that g has the required property. ∎

[1] See discussion following Ex. 10.6.

16.4. THEOREM. *Given any r.e. set $A \subseteq N$, we can find polynomials g and h such that A is the set of all integer values assumed by the rational function g/h as its variables range through N.*

PROOF. With f as in the proof of Thm. 16.2, we put

$$g(\mathfrak{x},z)=z+(f(\mathfrak{x},z))^2,$$

$$h(\mathfrak{x},z)=(z+1)(f(\mathfrak{x},z))^2+1. \qquad \blacksquare$$

Using the last two results we can (at least in principle) obtain a "general algebraic formula" for each r.e. set of numbers. (In particular, we can get such a formula for the prime numbers — a result which had eluded number theorists for several centuries.) However, such formulas would be rather too complicated and difficult to obtain to be of much *practical* use.

§ 17. Historical and bibliographical remarks

An excellent work on recursion theory is ROGERS [1967]. For some historical remarks on the origins of recursion theory see §2.4. of that work.

TURING [1936] was the first to use imaginary computers (which came to be known as *Turing machines*) for characterizing the class of algorithmic functions.

Imaginary computers like URIM (which are known as *register machines*) were invented for the same purpose by several people independently (Shepherdson and Sturgis, Lambek, Minsky) in the late 1950's. The first mention in print is probably LAMBEK [1961] and MINSKY [1961]. These machines gained wider currency following SHEPHERDSON and STURGIS [1963].

Our definition of the class of primitive recursive functions (in §5) is essentially the same as the original definition given by GÖDEL [1931], except that at the time he still called these functions simply "recursive".

The first definition of the class of recursive functions was given by GÖDEL [1934] following a suggestion by J. Herbrand. This definition was proved by KLEENE [1936] to be equivalent to another definition, which is virtually the same as the one used by us (§5).

Church's Thesis was proposed by CHURCH [1936], and an equivalent thesis is implicit in TURING [1936]. For a detailed discussion of the evidence for the thesis see KLEENE [1952].

The Normal Form Theorem (10.1 and 10.3) and the Recursion Theorem (10.19 and 10.20) are due to KLEENE [1936] and [1938]. Thm. 10.23 is due to RICE [1953] and the Simultaneous Recursion Theorem (10.24) is due to SMULLYAN [1961].

The problem of finding an algorithm for deciding the solvability of diophantine equations was included by HILBERT [1900] in his famous list of problems for twentieth century mathematics. (Hilbert explicitly mentioned the possibility that some problems on his list might not have a positive solution. A proof of non-existence of a positive solution would constitute a negative solution.) The first major advance towards a (negative) solution was made by J. ROBINSON [1952]. Using the properties of the solutions of Pell's equation[1], she proved the results reproduced here as Thms. 14.9–14.11. However, her proof depended on the (then still unproved) assumption that there exists a binary diophantine relation which possesses a certain property ("having roughly exponential growth"). This advance was continued by M. DAVIS, H. PUTNAM and J. ROBINSON [1961], who proved the results reproduced here as Lemmas 15.1 and 15.2. But their proof depended on J. Robinson's earlier result, and hence on her unproved assumption. The road from this point to Thm. 16.1 had already been bridged by M. DAVIS [1958], who showed that every r.e. relation P can be represented as

$$P(\mathfrak{x}) = \exists y\, \forall w{<}y\, \exists z_1{<}y \ldots \exists z_m{<}y\, [f(\mathfrak{x}, y, w, z_1, \ldots, z_m) = 0],$$

where f is a polynomial. Another proof of the same result was given also by R. M. ROBINSON [1956]. (These results of Davis and R. M. Robinson are not really essential to the proof of Thm. 16.1, and we have circumvented them *via* Lemma 15.3 and Thm. 15.4.)

The final break-through was made by MATIJASEVIČ [1970]. He proved that the sequence $\lambda x[\varphi(2x)]$ has a diophantine graph (Thm. 13.10). But the well-known growth behaviour of that sequence (Thm. 13.1) is easily seen to imply that the relation $\lambda xy[\varphi(2x) = y]$ has the "roughly exponential growth" required by J. Robinson's assumption. Thus his result bridged the major gap in the proof of Thm. 16.1. Our presentation in §13 is based on MATIJASEVIČ [1971].

Theorems 16.3 and 16.4 were proved (for diophantine sets) by PUTNAM [1960].

[1] The equation was posed by Fermat in 1657 as a challenge to British mathematicians. A partial solution was given by Lord Brouncker. The first complete solution was given by Lagrange in 1767 and later streamlined by Dirichlet. Our treatment (proof of Thm 14.1) follows Dirichlet.

CHAPTER 7

LOGIC—LIMITATIVE RESULTS

The main results proved in this chapter have one feature in common: they display the inherent limitations of formalism and the formal method in mathematics. As such, they are fundamental to any serious discussion of the philosophy of mathematics.

It is significant that these results apply even to a basic, elementary and (it would seem) philosophically unproblematic branch of mathematics — *elementary arithmetic*[1], which is concerned with the natural numbers and the operations of addition and multiplication of natural numbers. Because of this we shall be content to confine ourselves in this chapter to elementary arithmetic, although the results in question can for the most part be readily generalized to other, more elaborate, settings. (Something will be said about this in Ch. 8.)

We shall assume knowledge of Chapters 1,2,3 and 6. Knowledge of Ch. 5 will only be assumed in a few remarks and starred problems.

§ 1. General notation and terminology

Unless otherwise stated, the definitions and conventions made in Chapters 1–3 and 6 remain in force in this chapter.

Throughout this chapter, we take $\mathscr{L}$ to be the first-order language with equality having one individual constant $\mathbf{0}$, one unary function symbol $\mathbf{s}$ and two binary function symbols $\mathbf{+}$ and $\mathbf{\times}$. $\mathscr{L}$ has no other extralogical symbols. We call $\mathscr{L}$ *the first-order language of arithmetic.*

As a concession to common usage, we lay down the following two definitions:

$$(\mathbf{r}+\mathbf{t}) =_{\mathrm{df}} +\mathbf{rt},$$

$$(\mathbf{r}\times\mathbf{t}) =_{\mathrm{df}} \times\mathbf{rt},$$

[1] Often also called *elementary number theory.*

where **r** and **t** are any terms. (These two conventions are additional to, and in the same spirit as, those made in def. 1.5.1.) We use the normal conventions for omitting parentheses.

We let $\mathbf{\Phi}_n$ be the set of all $\mathscr{L}$-formulas whose free variables are among $\mathbf{v}_1,\dots,\mathbf{v}_n$ (i.e., the first n variables of $\mathscr{L}$). In particular, $\mathbf{\Phi}_0$ is the set of all $\mathscr{L}$-sentences.

For any formula $\boldsymbol{\alpha}$ and terms $\mathbf{t}_1,\dots,\mathbf{t}_n$ we write "$\boldsymbol{\alpha}(\mathbf{t}_1,\dots,\mathbf{t}_n)$" instead of "$\boldsymbol{\alpha}(\mathbf{v}_1/\mathbf{t}_1,\dots,\mathbf{v}_n/\mathbf{t}_n)$". Also, if **r** is a term, we write "$\mathbf{r}(\mathbf{t}_1,\dots,\mathbf{t}_n)$" for the result of simultaneously substituting $\mathbf{t}_1,\dots,\mathbf{t}_n$ for $\mathbf{v}_1,\dots,\mathbf{v}_n$, respectively, in **r**.

By recursion on k we define the kth *numeral* $\mathbf{s}_k$:

$$\mathbf{s}_0=\mathbf{0}, \qquad \mathbf{s}_{k+1}=\mathbf{s}\mathbf{s}_k.$$

We shall frequently write, e.g., "$\boldsymbol{\alpha}(\mathbf{s}_a)$" as short for "$\boldsymbol{\alpha}(\mathbf{s}_{a_1},\dots,\mathbf{s}_{a_n})$".

We let $\mathfrak{N}$ be the $\mathscr{L}$-structure with universe N (the set of natural numbers) and with $\mathbf{0}^{\mathfrak{N}}=0$ (the number zero), $\mathbf{s}^{\mathfrak{N}}=\lambda x(x+1)$ (the successor function), $\mathbf{+}^{\mathfrak{N}}=+$ (addition of natural numbers) and $\mathbf{\times}^{\mathfrak{N}}=\times$ (multiplication of natural numbers). We call $\mathfrak{N}$ the *standard* $\mathscr{L}$*-interpretation* or $\mathscr{L}$*-structure*, or *the (first-order) structure of natural numbers.* We say that an $\mathscr{L}$-sentence $\boldsymbol{\alpha}$ is *true* if $\mathfrak{N}\models\boldsymbol{\alpha}$; otherwise we say that $\boldsymbol{\alpha}$ is *false*.

It is easy to verify that

(1.1) $\quad \mathbf{s}_k^{\mathfrak{N}}=k \quad$ for all $k\in N$.

(See remark following Thm. 2.2.1.)

We recall (see remarks following Prob. 2.2.4) that if $\boldsymbol{\alpha}\in\mathbf{\Phi}_n$ then $\mathfrak{N}\models\boldsymbol{\alpha}[a]$ means that $\sigma\models\boldsymbol{\alpha}$ for some (hence for every) valuation σ having $\mathfrak{N}$ as its underlying structure and such that $\mathbf{v}_i^{\sigma}=a_i$ for $i=1,\dots,n$. Using (1.1) and Thm. 2.3.15 it is easy to see that

(1.2) $\quad \mathfrak{N}\models\boldsymbol{\alpha}[a] \;$ iff $\; \mathfrak{N}\models\boldsymbol{\alpha}(\mathbf{s}_a)$

for any $\boldsymbol{\alpha}\in\mathbf{\Phi}_n$. We shall use this frequently without special mention.

In this chapter we mean by *theory* any set $\mathbf{\Sigma}$ of $\mathscr{L}$-sentences closed under first-order deducibility; i.e., $\mathbf{\Sigma}\subseteq\mathbf{\Phi}_0$, and $\boldsymbol{\varphi}\in\mathbf{\Sigma}$ whenever $\boldsymbol{\varphi}\in\mathbf{\Phi}_0$ and $\mathbf{\Sigma}\vdash\boldsymbol{\varphi}$. (Cf. §4 of Ch. 5.)

The set $\mathbf{\Phi}_0$ of *all* sentences constitutes an *inconsistent* theory. Moreover, by Thm. 3.1.18, $\mathbf{\Phi}_0$ is the *only* inconsistent theory; thus a theory $\mathbf{\Sigma}$ is consistent iff $\mathbf{\Sigma}$ is *properly* included in $\mathbf{\Phi}_0$.

Throughout this chapter, we let $\mathbf{\Omega}$ be the set of all true sentences, i.e.,

$$\mathbf{\Omega}=\{\boldsymbol{\varphi}:\ \boldsymbol{\varphi}\in\mathbf{\Phi}_0,\ \mathfrak{N}\models\boldsymbol{\varphi}\}.$$

By the soundness of the predicate calculus (Thm. 3.1.2) $\boldsymbol{\Omega}$ is a theory. Notice that for any $\boldsymbol{\varphi}\in\boldsymbol{\Phi}_0$ we have either $\boldsymbol{\varphi}\in\boldsymbol{\Omega}$ or $\neg\boldsymbol{\varphi}\in\boldsymbol{\Omega}$, but not both. We call $\boldsymbol{\Omega}$ *complete first-order arithmetic*[1].

We shall say that a set of sentences $\boldsymbol{\Sigma}$ is *sound* if $\boldsymbol{\Sigma}\subseteq\boldsymbol{\Omega}$.

Let $\boldsymbol{\Sigma}_0$ be an arbitrary subset of $\boldsymbol{\Phi}_0$. If

$$\boldsymbol{\Sigma}=\{\boldsymbol{\varphi}:\ \boldsymbol{\varphi}\in\boldsymbol{\Phi}_0,\ \boldsymbol{\Sigma}_0\vdash\boldsymbol{\varphi}\},$$

then $\boldsymbol{\Sigma}$ is clearly a theory; in fact, it is the smallest theory which includes $\boldsymbol{\Sigma}_0$. We say that $\boldsymbol{\Sigma}_0$ is *a set of postulates for* $\boldsymbol{\Sigma}$ and that $\boldsymbol{\Sigma}$ is *the theory based on* $\boldsymbol{\Sigma}_0$ *(as a set of postulates)*.

§ 2. Nonstandard models of Ω

Let $^*\mathfrak{N}$ be an arbitrary $\mathscr{L}$-structure with universe $^*N$, designated individual ^{*0}, unary operation $^*s$ and binary operations $^*+$ and $^*\times$ (corresponding to the function symbols $\boldsymbol{+}$ and $\boldsymbol{\times}$ respectively).

By an *embedding* of the standard structure $\mathfrak{N}$ *into* $^*\mathfrak{N}$ we mean a one–one mapping f of N into $^*N$ such that

$$f(0) = {}^*0, \qquad f(m+1) = {}^*s(f(m)),$$
$$f(m+n) = f(m)\ {}^*\!+f(n), \qquad f(mn) = f(m)\ {}^*\!\times f(n),$$

for all m and n.

If f is as above and maps N *onto* $^*N$, then f is called an *isomorphism of* $\mathfrak{N}$ *onto* $^*\mathfrak{N}$. If such an isomorphism exists, $\mathfrak{N}$ and $^*\mathfrak{N}$ are said to be *isomorphic*[2].

2.1. Problem. Let f be an embedding of $\mathfrak{N}$ into $^*\mathfrak{N}$. Let σ be any valuation whose underlying structure is $\mathfrak{N}$, and let $f\sigma$ be the valuation such that the underlying structure of $f\sigma$ is $^*\mathfrak{N}$ and $\mathbf{x}^{f\sigma}=f(\mathbf{x}^\sigma)$ for every variable $\mathbf{x}$. Show that $\mathbf{t}^{f\sigma}=f(\mathbf{t}^\sigma)$ for every term $\mathbf{t}$. In particular, $\mathbf{t}^{*\mathfrak{N}}=f(\mathbf{t}^{\mathfrak{N}})$ for every closed term $\mathbf{t}$.

2.2. Problem[3]. Let f be an isomorphism of $\mathfrak{N}$ onto $^*\mathfrak{N}$, and let σ and $f\sigma$ be as in Prob. 2.1. Show that $\boldsymbol{\alpha}^{f\sigma}=\boldsymbol{\alpha}^\sigma$ for every formula $\boldsymbol{\alpha}$. In particular, $^*\mathfrak{N}\models\boldsymbol{\varphi}$ iff $\mathfrak{N}\models\boldsymbol{\varphi}$, for every sentence $\boldsymbol{\varphi}$.

The last part of Prob. 2.2 means that, if $\mathfrak{N}$ and $^*\mathfrak{N}$ are isomorphic, then $^*\mathfrak{N}$ is also a model of $\boldsymbol{\Omega}$ (i.e., $^*\mathfrak{N}\models\boldsymbol{\varphi}$ for all $\boldsymbol{\varphi}\in\boldsymbol{\Omega}$). Such a $^*\mathfrak{N}$ is called

[1] This agrees with the terminology of Ch. 5; see Lemma 5.4.1.
[2] This terminology is the same as that introduced in §1 (and extended in §7) of Ch. 5.
[3] Cf. Prob. 5.1.1.

a *standard* model of $\mathbf{\Omega}$, while a *nonstandard* model of $\mathbf{\Omega}$ is one which is not isomorphic with $\mathfrak{N}$. We shall now show that nonstandard models of $\mathbf{\Omega}$ actually exist[1].

2.3. THEOREM. *There exists a nonstandard model of* $\mathbf{\Omega}$ *with denumerable universe.*

PROOF. Let

$$\Sigma_n = \{\mathbf{v}_1 \neq \mathbf{s}_i:\ i<n\}, \qquad \Sigma = \bigcup\{\Sigma_n:\ n\in N\}.$$

We want to show that $\mathbf{\Omega}\cup\Sigma$ is satisfiable.

By the Compactness Theorem 3.3.16 it is enough to show that every finite subset of $\mathbf{\Omega}\cup\Sigma$ is satisfiable. But any such finite subset is included in $\mathbf{\Omega}\cup\Sigma_n$ for some n; and $\mathbf{\Omega}\cup\Sigma_n$ is clearly satisfied by any valuation σ whose underlying structure is $\mathfrak{N}$ and such that, say, $\mathbf{v}_1^\sigma = n$.

By Thm. 3.3.15 it follows that $\mathbf{\Omega}\cup\Sigma$ is satisfied by some valuation τ whose underlying structure $^*\mathfrak{N}$ has a finite or denumerable universe. However, $\mathbf{\Omega}$ clearly cannot be satisfied in a finite structure (because $\mathbf{\Omega}$ contains sentences $\mathbf{s}_m \neq \mathbf{s}_n$ for all pairs of different numbers m and n) so that $^*N$ must be denumerable.

It remains to prove that $^*\mathfrak{N}$ is nonstandard. To this end, suppose that f is an embedding of $\mathfrak{N}$ into $^*\mathfrak{N}$; we show that f is not onto. For the valuation τ (mentioned in the preceding paragraph) and for any n we have

$$\tau \models \mathbf{v}_1 \neq \mathbf{s}_n.$$

But

$$\mathbf{s}_n^\tau = \mathbf{s}_n^{*\mathfrak{N}} = f(\mathbf{s}_n^{\mathfrak{N}}) = f(n)$$

(see Prob. 2.1) so we must have $\mathbf{v}_1^\tau \neq f(n)$ for all n. Thus f cannot map N onto $^*N$. ∎

2.4. PROBLEM. Let $^*\mathfrak{N}$ be any model of $\mathbf{\Omega}$. Let f be the mapping of N into $^*N$ defined by: $f(n) = \mathbf{s}_n^{*\mathfrak{N}}$ for all n.

(i) Show that f is one–one. (If $m \neq n$, then the sentence $\mathbf{s}_m \neq \mathbf{s}_n$ belongs to $\mathbf{\Omega}$ and must hold in $^*\mathfrak{N}$.)

(ii) Show that f is an embedding of $\mathfrak{N}$ into $^*\mathfrak{N}$.

(iii) Show that f is the *only* embedding of $\mathfrak{N}$ into $^*\mathfrak{N}$. (Use Prob. 2.1.)

(iv) Prove that $^*\mathfrak{N}$ is a standard model of $\mathbf{\Omega}$ iff $^*N = \{\mathbf{s}_n^{*\mathfrak{N}}:\ n\in N\}$.

[1] By the Upward Löwenheim-Skolem Theorem 5.2.7, $\mathbf{\Omega}$ has models of all nondenumerable infinite cardinalities; such models are clearly nonstandard. *Here* we show that $\mathbf{\Omega}$ has *denumerable* nonstandard models.

Later in this chapter we shall discover some of the properties of non-standard models of $\mathbf{\Omega}$. For the time being we confine ourselves to discussing the implications of the fact that they *exist.* This fact means that $\mathfrak{N}$, the first-order structure of natural numbers, cannot be characterized uniquely (even up to isomorphism) in the corresponding formal language $\mathscr{L}$. For, any sentence (or set of sentences) of $\mathscr{L}$ that holds in $\mathfrak{N}$ also holds in other structures, not isomorphic with $\mathfrak{N}$.

We would like to suggest that this is not accidental and that there is *no* way of characterizing the natural numbers formally. By this we mean, more precisely, that there is *no* set of (finite) expressions in *any* formal language, which are (jointly) true in a structure that can reasonably be identified as *the* structure of natural numbers, but not in any other structure not isomorphic with it.

At first glance this claim may seem a bit too far-reaching. It might be thought that the structure $\mathfrak{N}$ cannot be characterized in the corresponding language $\mathscr{L}$ simply because $\mathfrak{N}$ is too rudimentary and (correspondingly) $\mathscr{L}$ is too poor. It might thus be hoped that by considering additional notions connected with the natural numbers and by introducing corresponding symbols for the notions, one could obtain a more elaborate "structure of natural numbers" and a correspondingly richer language in which this structure could be characterized (up to isomorphism). This objection seems all the more plausible because there is a well-known *informal* way of characterizing the natural numbers; and it might be hoped that this could somehow be formalized. It will therefore be instructive to outline this informal characterization, and to see what happens when one tries to formalize it.

The informal (or, perhaps more precisely, *semi-formal*) characterization we have in mind involves the following notions: *(natural) number, set of (natural) numbers,* the particular number 0 *(zero)*, the unary operation *s (successor)*, the binary operations + and × (*addition* and *multiplication* of numbers) and the binary relation $\in$ (*membership* of a number in a set of numbers).

Seven postulates (the *Peano postulates*) are laid down:

(i) $s(n) \neq 0$ for every *number* n.

(ii) If n and m are *numbers* and $n \neq m$, then $s(n) \neq s(m)$.

(iii) $n+0=n$ for every *number* n.

(iv) $n+s(m)=s(n+m)$ for all *numbers* n and m.

(v) $n \times 0=0$ for every *number* n.

(vi) $n \times s(m)=(n \times m)+n$ for all *numbers* n and m.

(vii) If M is a *set of numbers*, and if $0 \in M$, and if $s(n) \in M$ for every *number* n such that $n \in M$, then $n \in M$ for every *number* n.

Postulate (vii) is of course *the principle of mathematical induction.*

The arithmetic of natural numbers can be developed on the basis of these postulates[1]. Moreover, it can be proved that these postulates uniquely characterize the natural numbers (up to isomorphism). We shall not do so in detail, but merely hint at the proof.

In Prob. 2.4(iv) we saw that a model $^*\mathfrak{N}$ of $\mathbf{\Omega}$ is nonstandard iff in the universe of $^*\mathfrak{N}$ there are objects different from the values of all the numerals $\mathbf{s}_n$. This suggests (correctly) that in order to show that the seven postulates (i)–(vii) do not have "nonstandard models" (i.e., pathological interpretations) we must deduce *from these postulates* that the set $\{0, s(0), s(s(0)),\ldots\}$ contains *all* natural numbers. But this is in fact an immediate consequence of the induction principle (vii).

We observe that the induction principle (vii), which was crucial to the above argument, involves two notions (*set of numbers* and $\in$) which have no counterpart in the structure $\mathfrak{N}$ and do not correspond to any symbol of $\mathscr{L}$. It might therefore be thought that if we augment $\mathfrak{N}$ by adding these notions (and extend $\mathscr{L}$ accordingly) we might be able to characterize the augmented structure uniquely in the extended formal language. Let us see whether this is so.

We start by specifying a structure $\mathfrak{U}$ (in the sense of §2 of Ch. 1) of which postulates (i)–(vii) may reasonably be regarded as an informal (or semiformal) description. The universe U of $\mathfrak{U}$ must contain all natural numbers and sets of natural numbers. Thus $U = N \cup S$, where S is the set of all subsets of N. We take 0 as designated individual of $\mathfrak{U}$. The basic operations of $\mathfrak{U}$ are s, $+$ and $\times$, i.e., successor, addition and multiplication[2]. The basic relations of $\mathfrak{U}$ are the properties (i.e., subsets of U) N and S and the membership relation

$$E = \{\langle a,b \rangle : a \in b \in S\}.$$

Next, we specify the language $\mathscr{L}'$ appropriate for $\mathfrak{U}$. Clearly, $\mathscr{L}'$ must

[1] For a semi-formal development based on a similar set of postulates for the positive integers, see LANDAU [1930].

[2] In order to conform to the format of §2 of Ch. 1, $s(a)$, $a+b$ and $a \times b$ must be defined (and belong to U) even when a or b (or both) are sets of numbers rather than numbers. We therefore define them arbitrarily (e.g. as equal to 0) in these cases. This will not affect matters in any significant way.

be the language obtained from $\mathscr{L}$ by adding two unary predicate symbols **N** and **S** and a binary predicate symbol **E**.

We call $\mathfrak{U}$ the *standard* $\mathscr{L}'$-structure, since it is highly natural to regard $\mathfrak{U}$ as *the* structure of natural numbers corresponding to $\mathscr{L}'$. (Actually, $\mathfrak{U}$ is essentially what is usually called the *second-order* structure of natural numbers and $\mathscr{L}'$ is essentially the corresponding *second-order* language, except that here we have reduced them to first-order entities by the method outlined in §3 of Ch. 1.)

The Peano postulates can now be formalized — i.e., "translated" into $\mathscr{L}'$ — in a straightforward way. For example, (i) is formalized as $\forall \mathbf{x}(\mathbf{Nx} \to \mathbf{sx} \neq \mathbf{0})$, and (vii) as

(1) $$\forall \mathbf{x}\,[\mathbf{Sx} \wedge \mathbf{E0x} \wedge \forall \mathbf{y}\{\mathbf{Eyx} \to \mathbf{Esyx}\} \to \forall \mathbf{y}\{\mathbf{Ny} \to \mathbf{Eyx}\}].$$

It is immediately clear that if there is any hope of characterizing $\mathfrak{U}$ uniquely in $\mathscr{L}'$, the formalized versions of (i)–(vii) must be supplemented by other sentences, formalizing facts that are implicit in the semi-formal version. These include sentences like $\mathbf{N0}$, $\forall \mathbf{x}(\mathbf{Nx} \to \mathbf{Nsx})$, $\forall \mathbf{x} \forall \mathbf{y}\,[\mathbf{Nx} \wedge \mathbf{Ny} \to \mathbf{N(x+y)}]$ and, less trivially, sentences formalizing facts about the property S and relation E, for example:

(2) $$\forall \mathbf{x} \forall \mathbf{y}(\mathbf{Sx} \wedge \mathbf{Eyx} \to \mathbf{Ny}),$$

(3) $$\forall \mathbf{x} \forall \mathbf{y}[\mathbf{Sx} \wedge \mathbf{Sy} \wedge \forall \mathbf{z}(\mathbf{Ezx} \leftrightarrow \mathbf{Ezy}) \to \mathbf{x} = \mathbf{y}].$$

(Sentence (3) is a formalized version of the *extensionality principle* for sets of numbers.)

Instead of chasing after such sentences one by one, let us go the whole hog and consider the set $\mathbf{\Omega}'$ of *all* sentences that hold in $\mathfrak{U}$, i.e.,

$$\mathbf{\Omega}' = \{\boldsymbol{\varphi} \colon \boldsymbol{\varphi} \text{ an } \mathscr{L}'\text{-sentence, } \mathfrak{U} \models \boldsymbol{\varphi}\}.$$

Clearly, $\mathbf{\Omega}'$ contains the formalized versions of the Peano postulates (e.g., the induction postulate (1)) as well as the additional sentences mentioned above such as (2) and (3).

It is easy to define in a natural way the notion of *isomorphism* between two $\mathscr{L}'$-structures (a one–one mapping between their universes, which "respects" the basic operations and relations). We leave the (simple) details to the reader.

Let $^*\mathfrak{U}$ be an arbitrary $\mathscr{L}'$-structure. We use a notation similar to the one used above in connection with $\mathscr{L}$-structures: we let $^*U$ be the universe of $^*\mathfrak{U}$, ^{*0} the designated individual of $^*\mathfrak{U}$, $^*N$ the property (i.e., subset of $^*U$) corresponding to the predicate symbol **N**, etc.

We say that the $\mathscr{L}'$-structure $^*\mathfrak{U}$ is *regular* if every member of $^*S$ is a subset of $^*N$, and $^*E$ is the membership relation between members of $^*N$ and those of $^*S$, i.e.,

$$^*E=\{\langle a,b\rangle : a\in b\in {}^*S\}.$$

The standard structure $\mathfrak{U}$ is clearly regular.

2.5. Problem. Let $^*\mathfrak{U}$ be a model of $\boldsymbol{\Omega}'$. For any a in $^*S$ put

$$f(a)=\{b: \langle b,a\rangle\in {}^*E\}.$$

Show that f is a one–one mapping of $^*S$ into the set of all subsets of $^*N$. (Use the fact that sentences (2) and (3) belong to $\boldsymbol{\Omega}'$.) Hence show that if we replace each a by $f(a)$ we obtain a regular structure isomorphic with $^*\mathfrak{U}$.

By Prob. 2.5, there is no loss of generality in dealing only with those models of $\boldsymbol{\Omega}'$ which are regular.

An $\mathscr{L}'$-structure $^*\mathfrak{U}$ is *full* if it is regular and $^*S$ is the set of *all* subsets of $^*N$.

The informal argument outlined above (which shows that (i)–(vii) characterize the natural numbers uniquely) can now be converted into a proof that every *full* model $^*\mathfrak{U}$ of $\boldsymbol{\Omega}'$ is standard (i.e., isomorphic with $\mathfrak{U}$). For, using the fact that the (formalised) induction postulate (1) belongs to $\boldsymbol{\Omega}'$, we can easily show that

$$^*N=\{\mathbf{s}_n^{*\mathfrak{U}}: n\in N\}.$$

Hence, putting

$$f(n)=\mathbf{s}_n^{*\mathfrak{U}} \quad \text{for all } n\in N,$$

$$f(a)=\{f(n): n\in a\} \quad \text{for all } a\in S,$$

it is easy to verify that f is an isomorphism of $\mathfrak{U}$ onto $^*\mathfrak{U}$. (The details are left as a problem to the reader.)

Does this mean that $\boldsymbol{\Omega}'$ characterizes the standard structure $\mathfrak{U}$ uniquely? Not at all. For it is easy to show that $\boldsymbol{\Omega}'$ has nonstandard models (i.e., models that are not isomorphic with $\mathfrak{U}$). To show that $\boldsymbol{\Omega}'$ has a denumerable nonstandard model we argue as in the proof of Thm. 2.3, except that instead of the formulas $\mathbf{v}_1\neq\mathbf{s}_n$ we use the conjunction formulas $\mathbf{N}\mathbf{v}_1\wedge(\mathbf{v}_1\neq\mathbf{s}_n)$.

Of course, we already know that a nonstandard model of $\boldsymbol{\Omega}'$ cannot be full, so if we simply *decree* that non-full models of $\boldsymbol{\Omega}'$ are to be disregarded, the only remaining model (up to isomorphism) is $\mathfrak{U}$. However, the point is that there is no *formal* way of excluding the non-full models.

The informal characterization of the natural numbers works (or seems to work) only because it tacitly assumes that the notion *set of natural numbers* is interpreted correctly, as referring to *all* subsets of N. Thus it is not an *absolute* characterization but only *relative* to the notion of *power set* (set of all subsets) of a (possibly infinite) set. This latter notion cannot be characterized in a purely formal way; besides, it is considerably more problematic than the notion of natural number. (In this connection cf. Thm. 10.7.9 and Prob. 10.7.10.)

If we believe that part of the task of mathematics is to characterize structures such as $\mathfrak{N}$ (or $\mathfrak{U}$) uniquely up to isomorphism[1], then the above results suggest that *mathematics cannot be completely formalized.*

However, we could take the view that the real task is to study not some structure as such, but rather what can correctly be asserted of it in the appropriate formal language — i.e., in our case, $\mathbf{\Omega}$ rather than $\mathfrak{N}$. From this point of view, the fact that $\mathbf{\Omega}$ has "pathological" (nonstandard) models as well as the intended standard one is not very damaging.

The trouble is that $\mathbf{\Omega}$ was defined (see §1) in terms of $\mathfrak{N}$. Moreover, the definition was purely semantic and thus highly non-constructive. It is therefore natural to enquire whether $\mathbf{\Omega}$ can be characterized in an alternative, more constructive way[2].

Later in the present chapter we shall prove results that throw much light on this problem.

§ 3. Arithmeticity

In this chapter, when we say *function* we mean (unless otherwise indicated) a *function* in the sense of §1 of Ch. 6.

By *(n-ary) relation* we mean (unless otherwise indicated) a *first-order (n-ary) relation* in the sense of §6 of Ch. 6.

3.1. DEFINITION. Let $\mathbf{\Sigma}$ be a theory and let P be an n-ary relation. A formula $\alpha \in \Phi_n$ *weakly represents P in $\mathbf{\Sigma}$* if, for every $\mathfrak{x} \in N^n$,

$$\alpha(\mathbf{s}_{\mathfrak{x}}) \in \mathbf{\Sigma} \quad \text{iff} \quad P(\mathfrak{x}) \text{ holds.}$$

[1] Implicit in this is presumably the Platonistic belief in the objective existence of such structures.

[2] What we have in mind is something like what we achieved for the notion of logical validity; see §5 of Ch. 3.

A formula $\alpha \in \Phi_n$ *strongly represents P in* Σ if, for every $\mathfrak{x} \in N^n$,

$$\alpha(s_{\mathfrak{x}}) \in \Sigma \quad \text{if } P(\mathfrak{x}) \text{ holds,}$$
$$\neg\alpha(s_{\mathfrak{x}}) \in \Sigma \quad \text{otherwise.}$$

P is *weakly/strongly representable* in Σ if it is weakly/strongly represented in Σ by some $\alpha \in \Phi_n$.

The same terminology will also be used in connection with n-dimensional sets (see beginning of §11, Ch. 6).

If α strongly represents P in Σ, then α also weakly represents P in Σ, *provided* Σ *is consistent*. (In the inconsistent theory Φ_0, every relation is strongly representable; but only a trivial relation P, such that $P(\mathfrak{x})$ holds for all $\mathfrak{x} \in N^n$, is weakly representable.)

In the case $\Sigma = \Omega$, α strongly represents P iff α weakly represents P. Therefore in connection with Ω we shall drop the adverbs "weakly" and "strongly".

3.2. Definition. A relation is *arithmetical* if it is representable in Ω.

The rest of this section is devoted to characterizing the class of arithmetical relations.

By *logical operations* (on relations) we mean the propositional operations defined in Ex. 6.6.6 as well as universal and existential quantification defined in the beginning of §11, Ch. 6.

3.3. Lemma. *The class of arithmetical relations is closed under the logical operations.*

Proof. Among the propositional operations it is enough to consider negation and implication, for the others can be reduced to these. But if α and β represent in Ω the n-ary relations P and Q, then clearly $\neg\alpha$ and $\alpha \to \beta$ represent $\neg P$ and $P \to Q$.

If $P(\mathfrak{x}) = \exists y Q(\mathfrak{x}, y)$ and α represents Q in Ω, then clearly $\exists v_{n+1}\alpha$ represents P. Universal quantification can be reduced to this and negation. ∎

3.4. Lemma. *Every n-ary diophantine relation is represented in* Ω *by a formula of the form* $\exists v_{n+1} \ldots \exists v_{n+m}(r = t)$, *where* $m \geq 0$.

Proof. Let P be an n-ary elementary relation. Then (see §12 of Ch. 6)

$$P(\mathfrak{x}) = (f(\mathfrak{x}) = g(\mathfrak{x})),$$

where $f(\mathfrak{x})$ and $g(\mathfrak{x})$ are polynomials in the n variables $\mathfrak{x}$, with coefficients in N. Each such polynomial can be *formalized* (i.e., "translated" into $\mathscr{L}$)

by replacing each x_i by $\mathbf{v}_i$, each coefficient by the corresponding numeral, and $+$ and $\times$ by $\mathbf{+}$ and $\mathbf{\times}$. (For example, $3x_1{}^2x_2+x_3$ can be formalized as $\mathbf{s}_3\mathbf{\times v}_1\mathbf{\times v}_1\mathbf{\times v}_2\mathbf{+v}_3$.) Let $\mathbf{r}$ and $\mathbf{t}$ be terms which formalize $f(\mathfrak{x})$ and $g(\mathfrak{x})$ respectively. Clearly the formula $\mathbf{r=t}$ represents P in $\mathbf{\Omega}$.

Every diophantine relation is obtainable by existential quantifications from an elementary relation. Hence by (the proof of) Lemma 3.3, the present lemma is proved. ▮

3.5. Corollary. *Every r.e. relation, and in particular every recursive relation, is arithmetical.*

Proof. Immediate from Lemma 3.4 and the MRDP Thm. 6.16.1. ▮

In the following problem we sketch an alternative proof of Cor. 3.5, which does not use the MRDP Thm.

3.6. Problem. (i) Verify that the results of §12, Ch. 6 (6.12.2–6.12.4 and 6.12.8) continue to hold if the word "diophantine" is replaced by "arithmetical".

(ii) Show that the class of arithmetical relations is closed under bounded universal quantification. (Observe that $\forall y{<}z\, P(\mathfrak{x},y) = \forall y[(y<z) \rightarrow P(\mathfrak{x},y)]$.)

(iii) Hence show (without the MRDP Thm.) that every r.e. relation is arithmetical. (Argue as in the discussion following Lemma 6.12.8.)

3.7. Theorem. *The class of arithmetical relations is the smallest class containing all recursive relations and closed under the logical operations.*

Proof. By Lemma 3.3. and Cor. 3.5, it is enough to show that if P is an n-ary relation represented in $\mathbf{\Omega}$ by a formula $\boldsymbol{\alpha} \in \mathbf{\Phi}_n$, then P can be obtained from recursive relations by logical operations. We show this by induction on deg $\boldsymbol{\alpha}$.

If $\boldsymbol{\alpha}$ is atomic then P is clearly an elementary (hence p.r. and *a fortiori* recursive) relation.

The cases $\boldsymbol{\alpha} = \neg\boldsymbol{\beta}$ and $\boldsymbol{\alpha} = \boldsymbol{\beta} \rightarrow \boldsymbol{\gamma}$ are left to the reader.

If $\boldsymbol{\alpha} = \forall \mathbf{x}\boldsymbol{\beta}$, we choose a variant $\boldsymbol{\alpha}'$ of $\boldsymbol{\alpha}$ such that $\boldsymbol{\alpha}' = \forall \mathbf{v}_{n+1}\boldsymbol{\beta}'$. Clearly, $\boldsymbol{\alpha}'$ also represents P. Then $\boldsymbol{\beta}'$ belongs to $\mathbf{\Phi}_{n+1}$ and represents in $\mathbf{\Omega}$ an $(n+1)$-ary relation which, by the induction hypothesis, is of the required kind, and from which P is obtained by universal quantification. ▮

3.8. Problem. Verify that Thm. 3.7 continues to hold if the word "recursive" is replaced by "p.r." or "elementary".

3.9. Problem. Show that a relation is arithmetical iff its graph is arithmetical.

3.10. DEFINITION. A function is *arithmetical* if its graph is arithmetical.

3.11. REMARK. A relation is, in particular, a function (whose values are all in the set $\{0,1\}$). By Prob. 3.9, a relation is arithmetical as a function (i.e., in the sense of Def. 3.10) iff it is arithmetical as a relation (i.e., in the sense of Def. 3.2). Thus Defs. 3.2 and 3.10 are compatible.

Notice that, by Thm. 6.12.4 and Prob. 3.6, a relation obtained by composing an arithmetical relation with arithmetical functions is itself arithmetical.

We conclude this section with:

3.12. THEOREM. *Every recursive function is arithmetical.*

PROOF. Immediate from Thm. 6.11.3, Def. 3.10 and Cor. 3.5. ∎

*3.13. PROBLEM. Let D be the collection of all arithmetical total unary functions. Let $\mathscr{F}$ be an ultrafilter over N and put

$$D/\mathscr{F}=\{f/\mathscr{F}:\ f\in D\},\qquad \mathfrak{A}=\mathfrak{N}^N/\mathscr{F}|D/\mathscr{F}.$$

(i) Let $\boldsymbol{\varphi}\in\boldsymbol{\Phi}_{n+1}$ and $f_1,\ldots,f_n\in D$. For any $g\in N^N$ put

$$X(g)=\{k\in N:\ \mathfrak{N}\models\boldsymbol{\varphi}\,[f_1(k),\ldots,f_n(k),g(k)]\}.$$

Show that if $X(g)\in\mathscr{F}$ then there exists an $h\in D$ such that $X(h)\in\mathscr{F}$.

(ii) Show that $\mathfrak{A}\prec\mathfrak{N}^N/\mathscr{F}$. (Use Lemma 5.1.5, Łoś' theorem 5.3.7, and (i).)

(iii) Show that if $\mathscr{F}$ is non-principal then $\mathfrak{A}$ is a denumerable nonstandard model of $\boldsymbol{\Omega}$. (For the nonstandardness, consider the individual $\lambda x(x)/\mathscr{F}$.)

§ 4. Tarski's Theorem

We assign a *code number* $\#\mathbf{t}$ to each term $\mathbf{t}$ as follows:

$$\#\mathbf{0}=1,\qquad \#\mathbf{v}_i=3^i\quad\text{for } i=1,2,3,\ldots,$$

$$\#\mathbf{sr}=2\cdot 3^{\#\mathbf{r}},$$

$$\#(\mathbf{r}+\mathbf{t})=4\cdot 3^{\#\mathbf{r}}\cdot 5^{\#\mathbf{t}},$$

$$\#(\mathbf{r}\times\mathbf{t})=8\cdot 3^{\#\mathbf{r}}\cdot 5^{\#\mathbf{t}}.$$

Also, we assign a *code number* $\#\boldsymbol{\alpha}$ to each formula $\boldsymbol{\alpha}$:

$$\#(\mathbf{r}=\mathbf{t})=16\cdot 3^{\#\mathbf{r}}\cdot 5^{\#\mathbf{t}},$$

$$\#(\neg\boldsymbol{\alpha})=32\cdot 3^{\#\boldsymbol{\alpha}},$$

$$\#(\boldsymbol{\alpha}\rightarrow\boldsymbol{\beta})=64\cdot 3^{\#\boldsymbol{\alpha}}\cdot 5^{\#\boldsymbol{\beta}},$$

and, finally,

$$\#\forall \mathbf{v}_i\alpha = 2^{6+i}\cdot 3^{\#\alpha} \quad \text{for } i=1,2,3,\ldots .$$

For the sake of brevity we write, e.g., "TERM" as short for "code number of (a) term" and similarly in other cases; so that, when a word (or phrase) is printed in small capitals, it should be read with the words "code number of" prefixed to it.

Various functions and relations connected with the syntax of $\mathscr{L}$ are easily seen to be recursive.

4.1. EXAMPLE. Consider the property Tm such that Tm(x) holds iff x is a TERM. This property is uniquely determined by the identity

$$\begin{aligned}\mathrm{Tm}(x) = {} & \exists u<x\,(x=3^u)\\ & \vee \exists u<x\ \exists v<x\ \{\mathrm{Tm}(u)\wedge \mathrm{Tm}(v)\\ & \qquad \wedge[(x=2\cdot 3^u)\vee(x=4\cdot 3^u\cdot 5^v)\\ & \qquad \vee(x=8\cdot 3^u\cdot 5^v)]\}.\end{aligned}$$

The quickest way to verify that Tm is recursive is *via* the Recursion Theorem. If in the above equality we replace "Tm" everywhere by "$\{y\}_1$", we obtain a condition of the form

$$\{y\}(x)=f(x,y),$$

where f is a recursive function[1]. By Cor. 6.10.20 we can find a value e of y for which this condition holds identically in x. For this e we clearly have the identity Tm(x)=$\{e\}(x)$, which shows Tm to be recursive[2]. The same method can be used in other cases considered below.

4.2. PROBLEM. Show that the property Fla, such that Fla(x) holds iff x is a FORMULA, is recursive.

4.3. PROBLEM. Let Vbl be the binary relation such that Vbl(x,y) holds iff $y>0$ and x is a TERM containing the yth variable $\mathbf{v}_y$ or x is a FORMULA containing $\mathbf{v}_y$ free. Show that Vbl is recursive.

4.4. PROBLEM. Let Frm be the binary relation such that Frm(x,y) holds iff x is a FORMULA belonging to Φ_y (i.e., having all its free variables among

[1] Bounded existential quantification was introduced (Ex. 6.6.9) only in connection with relations, but (as observed there) "$\exists u<x$", e.g., can be replaced by "$\Pi_{u<x}$" which is applicable to any function(al).

[2] Actually, Tm as well as other functions and relations introduced below can easily be shown to be primitive recursive, but we shall not make use of this.

$\mathbf{v}_1,\ldots,\mathbf{v}_y$). Verify that

$$\mathrm{Frm}(x,y)=\mathrm{Fla}(x)\wedge\forall z<x\,[\mathrm{Vbl}(x,z)\rightarrow(z\leqslant y)],$$

hence Frm is recursive. (The point here is to verify that the bounded quantifier $\forall z<x$ can be used instead of the unbounded $\forall z$.)

The following example is of very great importance.

4.5. EXAMPLE. Consider the total ternary function sb such that if x is an EXPRESSION (i.e., TERM or FORMULA) and $y>0$ then $\mathrm{sb}(x,y,z)$ is the EXPRESSION obtained from it by substituting the numeral $\mathbf{s}_z$ for the variable $\mathbf{v}_y$; otherwise (i.e. if x is not an EXPRESSION or $y=0$) we let $\mathrm{sb}(x,y,z)=x$.

Let $\mathrm{num}(z)=\#\mathbf{s}_z$. Then

$$\mathrm{num}(0)=1,$$
$$\mathrm{num}(z+1)=2\cdot 3^{\mathrm{num}(z)};$$

so num is recursive. It is now easy to verify the identity

$$\mathrm{sb}(x,y,z)=\begin{cases}\mathrm{num}(z) & \text{if } (x=3^y)\wedge(y>0),\\ f(x,y,z) & \text{if } \mathrm{Vbl}(x,y)\wedge(x\neq 3^y),\\ x & \text{otherwise,}\end{cases}$$

where

$$f(x,y,z)=\exp(2,(x)_0)\cdot\exp(3,\mathrm{sb}((x)_1,y,z))\cdot\exp(5,\mathrm{sb}((x)_2,y,z))$$

and Vbl is the relation defined in Prob. 4.3.

From this identity it follows at once (e.g., by the method of Ex. 4.1) that sb is recursive.

We now define the *diagonal sequence d* by the identity

$$d(x)=\mathrm{sb}(x,1,x).$$

Clearly, d is a recursive sequence. Moreover, for any formula $\boldsymbol{\alpha}$ we have

$$(4.6)\qquad d(\#\boldsymbol{\alpha})=\#[\boldsymbol{\alpha}(\mathbf{s}_{\#\boldsymbol{\alpha}})].$$

The main result of this section is the following theorem, due to Tarski.

4.7. THEOREM. *Let T be the property such that $T(x)$ holds iff $x=\#\boldsymbol{\beta}$ for some $\boldsymbol{\beta}\in\boldsymbol{\Omega}$. Then T is not arithmetical.*

PROOF. The diagonal function d is recursive, so by Thm. 3.12 it is arithmetical. If T were arithmetical, then by Lemma 3.3 its negation $\neg T$ would also be arithmetical. By Remark 3.11 the property $\lambda x[\neg T(d(x))]$

would be arithmetical as well. Let $\boldsymbol{\alpha}$ be a formula $\in \boldsymbol{\Phi}_1$ representing this property in $\boldsymbol{\Omega}$. Thus for every x we have

$$\boldsymbol{\alpha}(\mathbf{s}_x) \in \boldsymbol{\Omega} \quad \text{iff} \quad \neg T(d(x)) \text{ holds.}$$

In particular, giving x the value $\# \boldsymbol{\alpha}$, we have

$$\boldsymbol{\alpha}(\mathbf{s}_{\# \boldsymbol{\alpha}}) \in \boldsymbol{\Omega} \quad \text{iff} \quad \neg T(d(\# \boldsymbol{\alpha})) \text{ holds.}$$

But by (4.6) and the definition of T we have, on the contrary,

$$\neg T(d(\# \boldsymbol{\alpha})) \text{ holds} \quad \text{iff} \quad \boldsymbol{\alpha}(\mathbf{s}_{\# \boldsymbol{\alpha}}) \notin \boldsymbol{\Omega}.$$

This contradiction shows that T cannot be arithmetical. ∎

The T of Thm. 4.7 is the property of being a TRUTH (i.e., a TRUE SENTENCE). Somewhat less precisely, T may be called *the property of truth in arithmetic*. Then Thm. 4.7 may be expressed by saying that the property of truth in arithmetic is not arithmetical.

Let us analyse the proof of Thm. 4.7. If P is an n-ary relation and $\boldsymbol{\alpha}$ represents P in $\boldsymbol{\Omega}$, then for any $\mathfrak{a} \in N^n$ the sentence $\boldsymbol{\alpha}(\mathbf{s}_{\mathfrak{a}})$ may be construed as (formally) asserting that $P(\mathfrak{a})$ holds. For, $\boldsymbol{\alpha}(\mathbf{s}_{\mathfrak{a}})$ is true iff $P(\mathfrak{a})$ holds. In particular, if $\boldsymbol{\alpha}$ were a formula representing in $\boldsymbol{\Omega}$ the property $\lambda x[\neg T(d(x))]$, then $\boldsymbol{\alpha}(\mathbf{s}_{\# \boldsymbol{\alpha}})$ would assert that $\neg T(d(\# \boldsymbol{\alpha}))$ holds. But, by (4.6) and the definition of T, we see that "$\neg T(d(\# \boldsymbol{\alpha}))$ holds" is equivalent to "$\boldsymbol{\alpha}(\mathbf{s}_{\# \boldsymbol{\alpha}})$ is false". Thus $\boldsymbol{\alpha}(\mathbf{s}_{\# \boldsymbol{\alpha}})$ would assert something equivalent to "I am false". Such a sentence cannot exist in $\mathscr{L}$, for it could neither be true nor false; so we conclude that the property in question is not arithmetical.

This has an obvious connection with the well-known *Liar paradox*. (A persons says "I am lying now"; is he lying or telling the truth?...) The difference is that while paradoxical sentences asserting their own falsity are (apparently) possible in the imprecise natural languages with their hazy interpretations, no such sentence can exist in a precisely constructed formal language with a rigorously defined interpretation. Therefore the assumption that T is arithmetical is absurd because it implies the existence of such a sentence in $\mathscr{L}$ under the interpretation $\mathfrak{N}$.

An argument like the one used in the proof of Thm. 4.7 can be applied in a wide variety of cases. Consider any formal language with a particular "standard" interpretation. (In our case these were the language $\mathscr{L}$ with its standard interpretation $\mathfrak{N}$.) It may be possible, using some method of coding, to express in this language certain notions of its own syntax. Suppose that this can be done to the extent that an analogue of the diagonal function can be so expressed. (In our case, the diagonal function was arithmetical, i.e., could be expressed in $\mathscr{L}$ under the standard interpretation $\mathfrak{N}$.) Then the property of being a true sentence of the formal language (under its standard interpretation) cannot be expressed in the language itself. For, if it could, the Liar paradox would be reproduced in that language as we saw in the proof of Thm. 4.7.

This suggests that there cannot exist a formal language which — under some "standard" interpretation — could adequately serve as its own metalanguage; for the syntax and semantics of a formal language cannot both be adequately expressed within the language itself. In particular, the dream of certain philosophers, that some day a precise formal language will be constructed in which *all* scientific notions and theories would be expressible, is most probably unrealizable. For, the syntax and semantics of such a language would surely be part of science, and hence would have to be expressible within that language, leading (as in the proof of Thm 4.7) to the Liar paradox.

In the rest of this section we sketch the proof of a somewhat stronger form of Tarski's Theorem.

4.8. DEFINITION. Let f be an n-ary function, and let $\boldsymbol{\alpha} \in \boldsymbol{\Phi}_{n+1}$. We say that $\boldsymbol{\alpha}$ *numeralwise represents* f in a theory $\boldsymbol{\Sigma}$ if for any $n+1$ numbers $\mathfrak{a}, b$ such that $f(\mathfrak{a}) = b$ we have

$$\forall \mathbf{v}_{n+1} [\boldsymbol{\alpha}(\mathbf{s}_{\mathfrak{a}}) \leftrightarrow \mathbf{v}_{n+1} = \mathbf{s}_b] \in \boldsymbol{\Sigma}.$$

4.9. PROBLEM. Let $\boldsymbol{\alpha}$ numeralwise represent the n-ary function f in the theory $\boldsymbol{\Sigma}$. Let $\boldsymbol{\beta} \in \boldsymbol{\Phi}_1$, and let $\boldsymbol{\beta}'$ be the formula

$$\exists \mathbf{v}_{n+1} [\boldsymbol{\beta}(\mathbf{v}_{n+1}) \wedge \boldsymbol{\alpha}].$$

Verify that if $f(\mathfrak{a}) = b$ then the sentence $\boldsymbol{\beta}(\mathbf{s}_b) \leftrightarrow \boldsymbol{\beta}'(\mathbf{s}_{\mathfrak{a}})$ belongs to $\boldsymbol{\Sigma}$.

4.10. PROBLEM. A formula $\boldsymbol{\gamma} \in \boldsymbol{\Phi}_1$ is called a *truth definition inside* a theory $\boldsymbol{\Sigma}$ if for every sentence $\boldsymbol{\varphi}$ we have

$$\boldsymbol{\gamma}(\mathbf{s}_{\#\boldsymbol{\varphi}}) \leftrightarrow \boldsymbol{\varphi} \in \boldsymbol{\Sigma}.$$

(i) Show that if the diagonal function d is numeralwise representable in $\boldsymbol{\Sigma}$ and $\boldsymbol{\Sigma}$ is consistent, there cannot exist a truth definition inside $\boldsymbol{\Sigma}$. (Use Prob. 4.9 to find a formula $\boldsymbol{\delta}$ such that for all n the sentence $\neg\boldsymbol{\gamma}(\mathbf{s}_{d(n)}) \leftrightarrow \boldsymbol{\delta}(\mathbf{s}_n)$ belongs to $\boldsymbol{\Sigma}$ and take $\boldsymbol{\varphi}$ as $\boldsymbol{\delta}(\mathbf{s}_{\#\boldsymbol{\delta}})$.)

(ii) Use (i) to show that there is no truth definition inside $\boldsymbol{\Omega}$; hence get a new proof for Thm. 4.7.

(iii) Show that if $\boldsymbol{\Sigma}$ is a sound theory, there is no truth definition inside $\boldsymbol{\Sigma}$.

§ 5. Axiomatic theories

In order to continue our discussion of Tarski's Thm. 4.7, we lay down:

5.1. DEFINITION. For any set Σ of sentences we let T_Σ be the property such that $T_\Sigma(x)$ holds iff x is a SENTENCE belonging to Σ.

In this notation, the T of Thm. 4.7 is T_Ω.

We now ask: under what conditons is it reasonable to regard a theory Σ as *axiomatic*? Surely, it is not enough to require that Σ be based on some set of "extralogical axioms", i.e., postulates; because the whole of Σ can always be taken as a set of postulates for itself, so that every theory would be axiomatic. We must lay down some condition on the set Γ of postulates. Ordinarily, one requires (at least) that the sentences of Γ be generated one by one by some mechanical effective procedure. This is tantamount to saying that the set of SENTENCES of Γ is enumerated by some algorithimc function. By Church's thesis and Thm. 6.11.5, this means that T_Γ is recursively enumerable. These considerations lead us to:

5.2. DEFINITION. A theory Σ is *axiomatizable* if there exists a set Γ of postulates for Σ such that T_Γ is recursively enumerable. If such a set Γ of postulates is actually *given* to us so that we can find an r.e. index for T_Γ (in the sense of §11 of Ch. 6), then we say that Σ is *axiomatic*.

5.3. THEOREM. *For any axiomatizable theory Σ there exists a set Δ of postulates for Σ such that T_Δ is recursive.*

PROOF. Let Γ, with r.e. T_Γ, be a set of postulates for Σ. If $\Gamma=\emptyset$, then T_Γ is recursive and there is nothing to prove. If $\Gamma\neq\emptyset$, then by Thm. 6.11.5 there is a total recursive function f that enumerates the SENTENCES of Γ, i.e., we have

$$\Gamma=\{\gamma_n:\ n\in N\}$$

and

$$f(n)=\#\gamma_n \quad \text{for all } n.$$

Let

$$\delta_0=\gamma_0, \qquad \delta_{n+1}=\gamma_{n+1}\wedge\delta_n \quad \text{for all } n.$$

Then $\Delta=\{\delta_n:\ n\in N\}$ is clearly a set of postulates for Σ. Also, if $g(n)=\#\delta_n$, then it is easy to see that g is a monotone increasing recursive function. Therefore, by Thm. 6.11.9, T_Δ is recursive. ∎

5.4. THEOREM. *A theory Σ is axiomatizable iff T_Σ is recursively enumerable.*

PROOF. If T_Σ is r.e., then Σ is axiomatizable because we can take the whole of Σ as a set of postulates.

Conversely, let Γ be a set of postulates for Σ and let T_Γ be recursively enumerable.

Let Ax be the property such that $\mathrm{Ax}(y)$ holds iff y is an AXIOM of the predicate calculus (see §1 of Ch. 3). We leave to the reader the simple (if somewhat tedious) task of verifying that Ax is recursive.

We encode finite sequences of formulas: to the sequence $\alpha_0,\ldots,\alpha_n$ we assign the code number $\exp(p_0, \#\alpha_0)\ldots\exp(p_n, \#\alpha_n)$.

Let Ded_Γ be the property such that $\mathrm{Ded}_\Gamma(y)$ holds iff y is a FIRST-ORDER DEDUCTION from Γ. It is easily verified that

$$\begin{aligned}\mathrm{Ded}_\Gamma(y) = (\mathrm{lh}(y) > 0) & \\ \wedge \forall z < \mathrm{lh}(y)\ \{\mathrm{Ax}((y)_z) \vee T_\Gamma((y)_z) & \\ \vee\ \exists u < z \exists v < z[(y)_u = 64 \cdot \exp(3,(y)_v) \cdot \exp(5,(y)_z)]\}. &\end{aligned}$$

From Lemma 6.11.2 it follows that Ded_Γ is an r.e. property. (We observe, by the way, that if T_Γ is recursive then Ded_Γ is recursive as well.)

Now, we clearly have

$$T_\Sigma(x) = \mathrm{Frm}(x,0) \wedge \exists y\ [\mathrm{Ded}_\Gamma(y) \wedge (x = (y)_{\mathrm{lh}(y) \dot{-} 1})].$$

It follows from Prob. 4.4 and Lemma 6.11.2 that T_Σ is an r.e. property. ∎

5.5. COROLLARY. *Ω is not axiomatizable.*

PROOF. Immediate from Cor. 3.5 and Thms. 4.7 and 5.4. ∎

By Church's Thesis, this means that there can be no effective procedure for generating one by one all the sentences of a set of postulates for Ω. This constitutes a serious difficulty for the formalist view of mathematics.

We conclude this section with a result on representability in axiomatizable theories.

5.6. THEOREM. *Only r.e. relations are weakly representable in an axiomatizable theory.*

PROOF. Let $\alpha \in \Phi_n$, and put

$$f(\mathfrak{x}) = \#[\alpha(\mathbf{s}_\mathfrak{x})].$$

If α weakly represents the n-ary relation P in the theory Σ, then the identity

$$P(\mathfrak{x}) = T_\Sigma(f(\mathfrak{x}))$$

must hold. But if $a = \#\boldsymbol{\alpha}$ then

$$f(\mathfrak{x}) = \mathrm{sb}\,(\ldots\,\mathrm{sb}\,(\mathrm{sb}\,(a, 1, x_1), 2, x_2)\ldots, n, x_n).$$

Since sb is recursive (see Ex. 4.5), f is recursive as well. Therefore, if T_Σ is r.e., P must be r.e. by Lemma 6.11.2. ∎

5.7. Problem. Employ Thm. 5.6 to obtain an alternative proof of Cor. 5.5, not using Tarski's Theorem.

5.8. Problem. Prove that only recursive relations are strongly representable in a consistent axiomatizable theory.

§ 6. Baby arithmetic

In this and the following three sections we shall present four theories, each of which has some special important feature. These theories will all be used later on.

In discussing these theories we shall often leave it to the reader to verify claims of the form $\boldsymbol{\Sigma} \vdash \boldsymbol{\alpha}$. In all these cases it is of course possible to construct a deduction of $\boldsymbol{\alpha}$ from $\boldsymbol{\Sigma}$. However, the reader will find it less tedious to verify (by tableau or directly) that $\boldsymbol{\Sigma} \models \boldsymbol{\alpha}$ and then use the Completeness Theorem 3.3.14.

In the present section we shall study the theory $\boldsymbol{\Pi}_0$, which we shall also call *baby arithmetic*, and which is based on the following postulates:

(6.1) $\quad \mathbf{s}_n \mathbf{+} \mathbf{s}_0 \mathbf{=} \mathbf{s}_n,$

(6.2) $\quad \mathbf{s}_n \mathbf{+} \mathbf{s}_{m+1} \mathbf{=} \mathbf{s}(\mathbf{s}_n \mathbf{+} \mathbf{s}_m),$

(6.3) $\quad \mathbf{s}_n \mathbf{\times} \mathbf{s}_0 \mathbf{=} \mathbf{s}_0,$

(6.4) $\quad \mathbf{s}_n \mathbf{\times} \mathbf{s}_{m+1} \mathbf{=} (\mathbf{s}_n \mathbf{\times} \mathbf{s}_m) \mathbf{+} \mathbf{s}_n,$

for all numbers n and m.

All these postulates are clearly true (i.e., hold in $\mathfrak{N}$), hence $\boldsymbol{\Pi}_0 \subseteq \boldsymbol{\Omega}$, i.e., $\boldsymbol{\Pi}_0$ is sound.

$\boldsymbol{\Pi}_0$ is a very weak theory: it is evidently satisfied in a structure whose universe consists of a single individual. Thus, e.g., even the sentence $\mathbf{s}_0 \mathbf{\neq} \mathbf{s}_1$ is not in $\boldsymbol{\Pi}_0$.

6.5. Problem. Show that, for each n, the only n-ary relations strongly representable in $\boldsymbol{\Pi}_0$ are the trivial ones: N^n and the empty n-ary relation.

Nevertheless, we shall soon see that every r.e. relation is *weakly* representable in $\boldsymbol{\Pi}_0$.

6.6. LEMMA. *Let* $\mathbf{t}$ *be a closed term, and* let $\mathbf{t}^{\mathfrak{N}}=t$. *Then* $(\mathbf{t}=\mathbf{s}_t)\in\mathbf{\Pi}_0$.

PROOF. First, let $\mathbf{t}$ be $\mathbf{s}_n+\mathbf{s}_m$. We must show that $(\mathbf{s}_n+\mathbf{s}_m=\mathbf{s}_{n+m})\in\mathbf{\Pi}_0$ for all n and m. But it is easy to verify by induction on m that $\mathbf{s}_n+\mathbf{s}_m=\mathbf{s}_{n+m}$ is deducible (in the predicate calculus) from postulates (6.1) and (6.2). Next, let $\mathbf{t}$ be $\mathbf{s}_n\times\mathbf{s}_m$. By induction on m we verify that $(\mathbf{s}_n\times\mathbf{s}_m=\mathbf{s}_{nm})\in\mathbf{\Pi}_0$.

Finally, if $\mathbf{t}$ is an arbitrary closed term, it is now easy to verify the assertion of our lemma by induction on deg $\mathbf{t}$. ∎

6.7. LEMMA. *Let* φ *be a sentence of the form* $\exists\mathbf{x}_1\ldots\exists\mathbf{x}_m(\mathbf{r}=\mathbf{t})$ *where* $m\geqslant 0$. *If* $\varphi\in\mathbf{\Omega}$, *then* $\varphi\in\mathbf{\Pi}_0$.

PROOF. By induction on m. For $m=0$, φ is $\mathbf{r}=\mathbf{t}$, where $\mathbf{r}$ and $\mathbf{t}$ are closed terms. If $\varphi\in\mathbf{\Omega}$, then $\mathbf{r}^{\mathfrak{N}}=\mathbf{t}^{\mathfrak{N}}$. Let k be this common value. By Lemma 6.6, $\mathbf{r}=\mathbf{s}_k$ and $\mathbf{t}=\mathbf{s}_k$ are in $\mathbf{\Pi}_0$. But $\mathbf{r}=\mathbf{t}$ is deducible from these two sentences, and hence must be in $\mathbf{\Pi}_0$ as well.

Now let φ be $\exists\mathbf{y}\alpha$, where α is a formula of the form $\exists\mathbf{x}_1\ldots\exists\mathbf{x}_m(\mathbf{r}=\mathbf{t})$ having no free variable other than $\mathbf{y}$. If $\varphi\in\mathbf{\Omega}$, then clearly for some n we must have $\alpha(\mathbf{y}/\mathbf{s}_n)\in\mathbf{\Omega}$. Therefore by the induction hypothesis $\alpha(\mathbf{y}/\mathbf{s}_n)\in\mathbf{\Pi}_0$. But, by Thm. 3.1.11, $\alpha(\mathbf{y}/\mathbf{s}_n)\vdash\exists\mathbf{y}\alpha$. ∎

Lemma 6.7 means that if a diophantine equation is solvable (in natural numbers) then this fact is deducible from the postulates of $\mathbf{\Pi}_0$.

6.8. THEOREM. *For any given* n*-ary r.e. relation* P, *we can find a formula of the form* $\exists\mathbf{v}_{n+1}\ldots\exists\mathbf{v}_{n+m}(\mathbf{r}=\mathbf{t})$ *which weakly represents* P *in every sound theory that includes* $\mathbf{\Pi}_0$.

PROOF. By the MRDP Thm. and Lemma 3.4 we can find a formula α of the required form which represents P in $\mathbf{\Omega}$. Thus, for any $\mathfrak{x}\in N^n$, $P(\mathfrak{x})$ holds iff $\alpha(\mathbf{s}_\mathfrak{x})\in\mathbf{\Omega}$. But $\alpha(\mathbf{s}_\mathfrak{x})$ is a sentence of the form covered by Lemma 6.7. Therefore, if $\mathbf{\Pi}_0\subseteq\mathbf{\Sigma}\subseteq\mathbf{\Omega}$, we have $\alpha(\mathbf{s}_\mathfrak{x})\in\mathbf{\Omega}$ iff $\alpha(\mathbf{s}_\mathfrak{x})\in\mathbf{\Sigma}$. ∎

In §4 we remarked that if α represents in $\mathbf{\Omega}$ the n-ary relation P, then for each $\mathfrak{a}\in N^n$ the sentence $\alpha(\mathbf{s}_\mathfrak{a})$ can be construed as formally asserting that $P(\mathfrak{a})$ holds, since $\alpha(\mathbf{s}_\mathfrak{a})$ is true iff $P(\mathfrak{a})$ in fact holds. Thm. 6.8 tells us that if P is r.e. then (for a properly chosen α) all true formal assertions of this form — but no false ones — can already be deduced (in the predicate calculus) from the postulates of baby arithmetic.

In particular, if f is any n-ary recursive function, then its graph is r.e. by Thm. 6.11.3, hence represented in $\mathbf{\Omega}$ by some formula α of the kind described in Thm. 6.8. The sentence $\alpha(\mathbf{s}_\mathfrak{a},\mathbf{s}_b)$ can then be regarded as formally asserting that $f(\mathfrak{a})$ equals b. All true assertions of this kind — and no false ones —

are deducible in $\mathbf{\Pi}_0$. This result was alluded to in our discussion of Church's thesis (§7 of Ch. 6).

6.9. PROBLEM. Verify that the set of POSTULATES of $\mathbf{\Pi}_0$ (i.e., SENTENCES of the forms (6.1)–(6.4)) is recursive. Thus $\mathbf{\Pi}_0$ is axiomatic.

6.10. THEOREM. *A relation is weakly representable in* $\mathbf{\Pi}_0$ *iff it is recursively enumerable.*

PROOF. Immediate from Prob. 6.9, Thm. 5.6 and Thm. 6.8. ∎

§ 7. Junior arithmetic

In this section we shall study a theory $\mathbf{\Pi}_1$ in which all recursive relations — and only they — are strongly representable.

7.1. DEFINITION. For any terms **r** and **t**, $\mathbf{r} \lessapprox \mathbf{t}$ is the formula $\exists \mathbf{z}(\mathbf{r}+\mathbf{z}=\mathbf{t})$, where **z** is the first[1] variable that occurs neither in **r** nor in **t**.

The theory $\mathbf{\Pi}_1$ — which we shall call *junior arithmetic* — is based on postulates (6.1)–(6.4) plus the following:

$$\mathbf{s}_n \neq \mathbf{s}_m, \tag{7.2}$$

$$\forall \mathbf{v}_1(\mathbf{v}_1 \lessapprox \mathbf{s}_n \leftrightarrow \mathbf{v}_1=\mathbf{s}_0 \lor \ldots \lor \mathbf{v}_1=\mathbf{s}_n), \tag{7.3}$$

$$\forall \mathbf{v}_1(\mathbf{s}_n \lessapprox \mathbf{v}_1 \lor \mathbf{v}_1 \lessapprox \mathbf{s}_n), \tag{7.4}$$

for all n and all $m \neq n$.

Obviously, $\mathbf{\Pi}_1$ is an extension of $\mathbf{\Pi}_0$; it is a *proper* extension because, e.g., $\mathbf{s}_0 \neq \mathbf{s}_1$ belongs to $\mathbf{\Pi}_1$ but not to $\mathbf{\Pi}_0$.

Also, since all the postulates of $\mathbf{\Pi}_1$ are evidently true, $\mathbf{\Pi}_1$ is sound, i.e., included in $\mathbf{\Omega}$.

7.5. PROBLEM. Let $^*\mathfrak{N}$ be a structure with domain $^*N = N \cup \{\infty\}$, where ∞ is some arbitrary object $\notin N$; let $^*0=0$ and let the operations $^*s$, $^*+$ and $^*\times$ of $^*\mathfrak{N}$ be the extensions of the ordinary successor, addition and multiplication such that $^*s(\infty)=0$, $a\,^*\!+b=\infty$ and $a\,^*\!\times b=0$ whenever $a=\infty$ or $b=\infty$ (or both). Show that $^*\mathfrak{N}$ is a model of $\mathbf{\Pi}_1$. Hence show that the sentence $\forall \mathbf{v}_1(\mathbf{s}\mathbf{v}_1 \neq \mathbf{s}_0)$ is not in $\mathbf{\Pi}_1$, so that $\mathbf{\Pi}_1$ is a *proper* sub-theory of $\mathbf{\Omega}$.

7.6. PROBLEM. Exactly the same as Prob. 2.4, but with "$\mathbf{\Omega}$" replaced by "$\mathbf{\Pi}_1$".

[1] If **z** is *any* variable not occurring in **r** or **t**, then $\exists \mathbf{z}(\mathbf{r}+\mathbf{z}=\mathbf{t})$ is a variant of $\mathbf{r} \lessapprox \mathbf{t}$. Below we shall sometimes write "$\mathbf{r} \lessapprox \mathbf{t}$", where, in fact, there should be some variant of $\mathbf{r} \lessapprox \mathbf{t}$. This slight inaccuracy is harmless, since by 3.1.10 mutual variants are provably equivalent.

Below we shall write, e.g., "$\exists \mathbf{x} \leqslant \mathbf{r}\, \alpha$" as short for $\exists \mathbf{x}(\mathbf{x} \leqslant \mathbf{r} \wedge \alpha)$.

7.7. LEMMA. *Let φ be a sentence of the form $\exists \mathbf{x}_1 \leqslant \mathbf{s}_{n_1} \ldots \exists \mathbf{x}_k \leqslant \mathbf{s}_{n_k} (\mathbf{r}=\mathbf{t})$, where $k \geqslant 0$. Then $\varphi \in \Pi_1$ if $\varphi \in \Omega$, and $\neg\varphi \in \Pi_1$ if $\neg\varphi \in \Omega$.*

PROOF. By induction on k. For $k=0$, φ is $\mathbf{r}=\mathbf{t}$, where $\mathbf{r}$ and $\mathbf{t}$ are closed terms. If $\varphi \in \Omega$ then by Lemma 6.7 we have $\varphi \in \Pi_0 \subseteq \Pi_1$. If $\neg\varphi \in \Omega$, then $r \neq t$, where $r = \mathbf{r}^{\mathfrak{N}}$ and $t = \mathbf{t}^{\mathfrak{N}}$. By Lemma 6.6, the sentences $\mathbf{r}=\mathbf{s}_r$ and $\mathbf{t}=\mathbf{s}_t$ are in Π_0, hence in Π_1. By (7.2) the sentence $\mathbf{s}_r \neq \mathbf{s}_t$ is in Π_1. But from these three sentences the sentence $\mathbf{r} \neq \mathbf{t}$ (i.e., $\neg\varphi$) is deducible.

Now let $\varphi = \exists \mathbf{y} \leqslant \mathbf{s}_n \alpha$, where α is a formula of the form

$$\exists \mathbf{x}_1 \leqslant \mathbf{s}_{n_1} \ldots \exists \mathbf{x}_k \leqslant \mathbf{s}_{n_k} (\mathbf{r}=\mathbf{t})$$

having no free variable other than $\mathbf{y}$. First, suppose $\varphi \in \Omega$. Then for some $m \leqslant n$ we must have $\alpha(\mathbf{y}/\mathbf{s}_m) \in \Omega$. Hence by the induction hypothesis,

$$\text{(1)} \qquad \alpha(\mathbf{y}/\mathbf{s}_m) \in \Pi_1.$$

Also, since $m \leqslant n$, the sentence $\mathbf{s}_m \leqslant \mathbf{s}_n$ is deducible from (7.3), and hence belongs to Π_1. Using this and (1) we have

$$\mathbf{s}_m \leqslant \mathbf{s}_n \wedge \alpha(\mathbf{y}/\mathbf{s}_m) \in \Pi_1.$$

From this last sentence we can deduce (see Thm. 3.1.11) $\exists \mathbf{y}(\mathbf{y} \leqslant \mathbf{s}_n \wedge \alpha)$, i.e., φ.

Suppose, on the other hand, that $\neg\varphi \in \Omega$. Then for every $m \leqslant n$ we must have $\neg\alpha(\mathbf{y}/\mathbf{s}_m) \in \Omega$. Hence, by the induction hypothesis, $\neg\alpha(\mathbf{y}/\mathbf{s}_m) \in \Pi_1$. But from the sentences $\neg\alpha(\mathbf{y}/\mathbf{s}_m)$, where $m=0,\ldots,n$, we can deduce

$$\neg\exists \mathbf{y}[(\mathbf{y}=\mathbf{s}_0 \vee \ldots \vee \mathbf{y}=\mathbf{s}_n) \wedge \alpha]$$

and from this and (7.3) we can deduce $\neg\exists \mathbf{y}(\mathbf{y} \leqslant \mathbf{s}_n \wedge \alpha)$, i.e., $\neg\varphi$. ∎

7.8. LEMMA. *Let $\Pi_1 \vdash \alpha(\mathbf{x}_1/\mathbf{s}_{a_1},\ldots,\mathbf{x}_k/\mathbf{s}_{a_k})$. If $m = \max(a_1,\ldots,a_k)$ and $\mathbf{y}$ is a variable different from $\mathbf{x}_1,\ldots,\mathbf{x}_k$, then*

$$\Pi_1, \neg(\mathbf{y} \leqslant \mathbf{s}_m) \vdash \exists \mathbf{x}_1 \leqslant \mathbf{y} \ldots \exists \mathbf{x}_k \leqslant \mathbf{y}\, \alpha.$$

PROOF. Using postulate (7.3) — with m instead of n — we easily verify

$$\Pi_1, \neg(\mathbf{y} \leqslant \mathbf{s}_m) \vdash \mathbf{y} \neq \mathbf{s}_0 \wedge \ldots \wedge \mathbf{y} \neq \mathbf{s}_m.$$

For each $j=1,\ldots,k$ we have $a_j \leqslant m$, so

$$\Pi_1, \neg(\mathbf{y} \leqslant \mathbf{s}_m) \vdash \mathbf{y} \neq \mathbf{s}_0 \wedge \ldots \wedge \mathbf{y} \neq \mathbf{s}_{a_j}.$$

Using postulate (7.3) for $n=a_j$ we now get

$$\Pi_1, \neg(\mathbf{y} \leqslant \mathbf{s}_m) \vdash \neg(\mathbf{y} \leqslant \mathbf{s}_{a_j}).$$

Hence, using postulate (7.4) with $n=a_j$, we have

$$\Pi_1, \neg(\mathbf{y} \leqslant \mathbf{s}_m) \vdash \mathbf{s}_{a_j} \leqslant \mathbf{y} \quad \text{for } j=1,\ldots,k.$$

But from $\boldsymbol{\alpha}(\mathbf{x}_1/\mathbf{s}_{a_1},\ldots,\mathbf{x}_k/\mathbf{s}_{a_k})$ and the formulas $\mathbf{s}_{a_j} \leqslant \mathbf{y}$ (for $j=1,\ldots,k$) we can deduce $\exists \mathbf{x}_1 \leqslant \mathbf{y} \ldots \exists \mathbf{x}_k \leqslant \mathbf{y}\, \boldsymbol{\alpha}$. ∎

The following result is extremely important.

7.9. Lemma. *Given two n-ary r.e. relations P and P′, we can find formulas*

(i) $\boldsymbol{\beta} = \exists \mathbf{v}_{n+1} \leqslant \mathbf{y} \cdots \exists \mathbf{v}_{n+k} \leqslant \mathbf{y}\, (\mathbf{r}=\mathbf{t}),$

(ii) $\boldsymbol{\beta}' = \exists \mathbf{v}_{n+1} \leqslant \mathbf{y} \cdots \exists \mathbf{v}_{n+k} \leqslant \mathbf{y}\, (\mathbf{r}'=\mathbf{t}'),$

(where $\mathbf{y}$ *is* $\mathbf{v}_{n+k+1}$*) such that the formula*

(iii) $\gamma = \exists \mathbf{y}(\boldsymbol{\beta} \wedge \neg \boldsymbol{\beta}')$

belongs to $\boldsymbol{\Phi}_n$*, and for every* $\mathfrak{a} \in N^n$ *we have*

$$\gamma(\mathbf{s}_\mathfrak{a}) \in \Pi_1 \quad \textit{if} \quad P(\mathfrak{a}) \wedge \neg P'(\mathfrak{a}) \textit{ holds},$$
$$\neg\gamma(\mathbf{s}_\mathfrak{a}) \in \Pi_1 \quad \textit{if} \quad \neg P(\mathfrak{a}) \wedge P'(\mathfrak{a}) \textit{ holds}.$$

Proof. By the MRDP thm. and Lemma 3.4 we can find formulas

$$\boldsymbol{\alpha} = \exists \mathbf{v}_{n+1} \cdots \exists \mathbf{v}_{n+k}(\mathbf{r}=\mathbf{t}),$$
$$\boldsymbol{\alpha}' = \exists \mathbf{v}_{n+1} \cdots \exists \mathbf{v}_{n+k}(\mathbf{r}'=\mathbf{t}'),$$

representing in $\boldsymbol{\Omega}$ the relations P and P' respectively. (Without loss of generality we have assumed that the number of bound variables in both cases in the same. This can be arranged by introducing redundant quantifiers as in Prob. 2.2.4.)

Let $\boldsymbol{\beta}$, $\boldsymbol{\beta}'$ and γ be given by (i), (ii) and (iii). Since $\boldsymbol{\alpha}$ and $\boldsymbol{\alpha}'$ are in $\boldsymbol{\Phi}_n$, we have $\gamma \in \boldsymbol{\Phi}_n$. We shall show that γ behaves in the required manner.

First, suppose that $P(\mathfrak{a})$ holds and $P'(\mathfrak{a})$ does not. Since $\boldsymbol{\alpha}$ represents P in $\boldsymbol{\Omega}$, we have $\boldsymbol{\alpha}(\mathbf{s}_\mathfrak{a}) \in \boldsymbol{\Omega}$. Therefore there exist numbers $a_{n+1},\ldots,a_{n+k}$ such that the sentence

$$\mathbf{r}(\mathbf{s}_\mathfrak{a}, \mathbf{s}_{a_{n+1}}, \ldots, \mathbf{s}_{a_{n+k}}) = \mathbf{t}(\mathbf{s}_\mathfrak{a}, \mathbf{s}_{a_{n+1}}, \ldots, \mathbf{s}_{a_{n+k}})$$

also belongs to $\boldsymbol{\Omega}$. From this it follows that if $m=\max(a_{n+1},\ldots,a_{n+k})$ then

$\beta(s_a)(y/s_m) \in \Omega$. But this sentence is of the kind considered in Lemma 7.7, so we have

(1) $\beta(s_a)(y/s_m) \in \Pi_1$.

On the other hand, if p is any number, then $\beta'(s_a)(y/s_p)$ cannot be in Ω; for if it were, the sentence $\alpha'(s_a)$, which is logically entailed by it, would also be in Ω, but α' represents P' in Ω and $P'(a)$ does not hold. Thus for all p we have $\neg\beta'(s_a)(y/s_p) \in \Omega$ and hence, by Lemma 7.7,

(2) $\neg\beta'(s_a)(y/s_p) \in \Pi_1$ for all p.

Using (2) for $p=m$ and (1) we see that the sentence

$$\beta(s_a)(y/s_m) \wedge \neg\beta'(s_a)(y/s_m)$$

belongs to Π_1. But from this sentence we can deduce

$$\exists y\,[\beta(s_a) \wedge \neg\beta'(s_a)],$$

which is $\gamma(s_a)$.

Now suppose that $P(a)$ does not hold but $P'(a)$ does. Since α' represents P' in Ω, we have $\alpha'(s_a) \in \Omega$. It follows that there are numbers $a_{n+1},\ldots,a_{n+k}$ such that the sentence

$$r'(s_a, s_{a_{n+1}}, \ldots, s_{a_{n+k}}) = t'(s_a, s_{a_{n+1}}, \ldots, s_{a_{n+k}})$$

belongs to Ω. By Lemma 6.7, this sentence must belong to Π_0 and hence to Π_1. Taking $m = \max(a_{n+1},\ldots,a_{n+k})$ and using Lemma 7.8 we therefore obtain

(3) $\Pi_1, \neg(y \leqslant s_m) \vdash \beta'(s_a)$.

Also, using the method by which we got (2), we now get

$$\neg\beta(s_a)(y/s_p) \in \Pi_1 \quad \text{for all } p.$$

Using this fact for $p=0,\ldots,m$ together with postulate (7.3) for $n=m$, we have

$$\Pi_1,\ y \leqslant s_m \vdash \neg\beta(s_a).$$

From this and (3) we easily get

$$\Pi_1 \vdash \neg\beta(s_a) \vee \beta'(s_a).$$

Generalizing on $\mathbf{y}$ (see 3.1.5) we obtain

$$\Pi_1 \vdash \forall \mathbf{y}\, [\neg \beta(\mathbf{s}_a) \vee \beta'(\mathbf{s}_a)].$$

But this last sentence is provably equivalent to $\neg \gamma(\mathbf{s}_a)$. ∎

7.10. THEOREM. *Given a recursive relation R, we can find a formula* γ *(of the form described in Lemma 7.9) such that* γ *strongly represents R in any theory that includes* Π_1.

PROOF. In Lemma 7.9, take P and P' to be R and $\neg R$ respectively. Then the lemma shows that γ strongly represents R in Π_1, and hence in any theory that includes Π_1. ∎

7.11. REMARK. Our proof that every recursive relation is strongly representable in Π_1 employed the characterization of r.e. relations (hence also of recursive relations) peovided by the MRDP Theorem. But the same fact can also be proved using, e.g., an older characterization of r.e. relations due to M. Davis and R. Robinson (see §17 of Ch. 6) or even a more direct characterization of recursive relations.

7.12. PROBLEM. Verify that the set of POSTULATES of Π_1 is recursive, so that Π_1 is axiomatic. Hence show that a relation is weakly representable in Π_1 iff it is r.e., and strongly representable in Π_1 iff it is recursive.

7.13. PROBLEM. Let Σ be a theory such that $\Pi_1 \subseteq \Sigma$.

(i) Show that every total recursive function is numeralwise representable in Σ. (If α strongly represents the graph of an n-ary function f, prove that the formula

$$\forall \mathbf{y} \leqslant \mathbf{v}_{n+1}\, [\alpha(\mathbf{v}_{n+1}/\mathbf{y}) \leftrightarrow \mathbf{y} = \mathbf{v}_{n+1}],$$

where $\mathbf{y}$ is, e.g., $\mathbf{v}_{n+2}$, numeralwise represents f.)

(ii) Show that, if Σ is consistent, there cannot exist a truth definition inside Σ. (Use Prob. 4.10.)

§ 8. A finitely axiomatized theory

Whereas Π_0 and Π_1 were based on *infinitely* many postulates, the theory Π_2, which we study in this section, is based on the following nine postulates:

(8.1) $\forall \mathbf{v}_1(\mathbf{s}\mathbf{v}_1 \neq \mathbf{s}_0)$,

(8.2) $\forall \mathbf{v}_1 \forall \mathbf{v}_2(\mathbf{s}\mathbf{v}_1 = \mathbf{s}\mathbf{v}_2 \rightarrow \mathbf{v}_1 = \mathbf{v}_2)$,

(8.3) $\forall \mathbf{v}_1(\mathbf{v}_1 + \mathbf{s}_0 = \mathbf{v}_1)$,

(8.4) $\forall v_1 \forall v_2[v_1+sv_2=s(v_1+v_2)]$,

(8.5) $\forall v_1(v_1 \times s_0=s_0)$,

(8.6) $\forall v_1 \forall v_2(v_1 \times sv_2=v_1 \times v_2+v_1)$,

(8.7) $\forall v_1(v_1 \leqslant s_0 \rightarrow v_1=s_0)$,

(8.8) $\forall v_1 \forall v_2(v_1 \leqslant sv_2 \rightarrow v_1 \leqslant v_2 \vee v_1=sv_2)$,

(8.9) $\forall v_1 \forall v_2(v_1 \leqslant v_2 \vee v_2 \leqslant v_1)$.

8.10. REMARK. The set of POSTULATES of $\mathbf{\Pi}_2$, being finite, is certainly recursive (cf. Thm. 6.11.9) so that $\mathbf{\Pi}_2$ is axiomatic. We say that it is *finitely axiomatized* because the set of its postulates is finite. Note also that $\mathbf{\Pi}_2$ is clearly sound.

8.11. THEOREM. $\mathbf{\Pi}_1 \subseteq \mathbf{\Pi}_2$.

PROOF. We show that the postulates of $\mathbf{\Pi}_1$ belong to $\mathbf{\Pi}_2$.

Postulates (6.1)–(6.4) are clearly deducible from (8.3)–(8.6) respectively.

Postulates (7.2) are $s_n \neq s_m$, where $n \neq m$. Suppose first that $n > m$. We proceed by induction on m. Since $n > m$, the numeral s_n is ss_k, where $k \geqslant m$. For $m=0$, the sentence $ss_k \neq s_m$ is deducible from (8.1). For the induction step, let $m=p+1$. Then $k>p$, so by the induction hypothesis the sentence $s_k \neq s_p$ belongs to $\mathbf{\Pi}_2$. But from this sentence and (8.2) we deduce $ss_k \neq ss_p$, i.e., $s_n \neq s_m$ — as required.

If $n<m$ then by what we have just proved the sentence $s_m \neq s_n$ belongs to $\mathbf{\Pi}_2$. But from this we deduce $s_n \neq s_m$.

Next, consider (7.3). We proceed by induction on n. For $n=0$, we start from (8.3) and deduce $s_0+s_0=s_0$, hence $\exists v_1(s_0+v_1=s_0)$, which by Def. 7.1 is $s_0 \leqslant s_0$, hence

$$\forall v_1(v_1=s_0 \rightarrow v_1 \leqslant s_0).$$

From this last sentence and (8.7) we deduce

$$\forall v_1(v_1 \leqslant s_0 \leftrightarrow v_1=s_0),$$

which is (7.3) for $n=0$.

Now suppose that for some n the sentence (7.3) belongs to $\mathbf{\Pi}_2$. From this sentence we deduce

(1) $\forall v_1(v_1 \leqslant s_n \vee v_1=s_{n+1} \leftrightarrow v_1=s_0 \vee \ldots \vee v_1=s_n \vee v_1=s_{n+1})$.

From (8.8) we deduce

(2) $\forall v_1(v_1 \leqslant s_{n+1} \rightarrow v_1 \leqslant s_n \vee v_1=s_{n+1})$.

Next, from (8.4) we deduce

$$\forall v_1 \forall v_2 (v_1 + v_2 = s_n \rightarrow v_1 + sv_2 = s_{n+1})$$

and hence

$$\forall v_1 [\exists v_2 (v_1 + v_2 = s_n) \rightarrow \exists v_2 (v_1 + v_2 = s_{n+1})].$$

But by Def. 7.1 this sentence is

(3) $$\forall v_1 (v_1 \leqslant s_n \rightarrow v_1 \leqslant s_{n+1}).$$

Also, from (8.3) we deduce $s_{n+1} + s_0 = s_{n+1}$, hence

$$\exists v_1 (s_{n+1} + v_1 = s_{n+1}),$$

which is the same as $s_{n+1} \leqslant s_{n+1}$. But from this we deduce

$$\forall v_1 (v_1 = s_{n+1} \rightarrow v_1 \leqslant s_{n+1}).$$

Using this sentence together with (2) and (3) we deduce

$$\forall v_1 (v_1 \leqslant s_{n+1} \leftrightarrow v_1 \leqslant s_n \vee v_1 = s_{n+1}).$$

From this and (1) we deduce

$$\forall v_1 (v_1 \leqslant s_{n+1} \leftrightarrow v_1 = s_0 \vee \ldots \vee v_1 = s_n \vee v_1 = s_{n+1}),$$

which is (7.3) for $n+1$.

Finally, (7.4) is evidently deducible from (8.9). ∎

8.12. Remark. $\mathbf{\Pi}_1$ is a *proper* sub-theory of $\mathbf{\Pi}_2$ because, e.g., postulate (8.1) does not belong to $\mathbf{\Pi}_1$ (see Prob. 7.5).

8.13. Problem. Let $^*\mathfrak{N}$ be a structure like that of Prob. 7.5, except that $^*s(\infty) = \infty$, $a \,^*\!\times b = \infty$ whenever $a = \infty$ but $b \neq 0$, or $b = \infty$ and a is arbitrary. (As in Prob. 7.5, $\infty \,^*\!\times 0 = 0$, and $a \,^*\!+ b = \infty$ whenever $a = \infty$ or $b = \infty$.) Verify that $^*\mathfrak{N}$ is a model for $\mathbf{\Pi}_2$. Hence show that the sentence $\forall v_1 (sv_1 \neq v_1)$ is not in $\mathbf{\Pi}_2$.

§ 9. First-order Peano arithmetic

The theory $\mathbf{\Pi}$ which we shall now begin to study has as its postulates the six sentences (8.1)–(8.6) and all sentences of the form

(9.1) $$\forall v_2 \ldots \forall v_k [\alpha(s_0) \rightarrow \forall v_1 \{\alpha \rightarrow \alpha(sv_1)\} \rightarrow \forall v_1 \alpha],$$

where $k \geqslant 1$ and $\alpha \in \mathbf{\Phi}_k$. Postulates (9.1) are called *induction* postulates.

The postulates of **Π** were obviously obtained by trying to formalize in $\mathscr{L}$ the seven Peano postulates listed in §2. For this reason **Π** is usually called *first-order Peano arithmetic*. (It must be stressed however that what Peano was aiming at was not **Π** but, essentially, **Ω′** of §2.)

Clearly, **Π** is sound. Also, the induction postulates make it a very powerful theory: for many years no one had been able to find a true sentence that expresses a fact of genuine general mathematical interest and yet does not belong to **Π**. The first such sentences were found by PARIS and HARRINGTON [1977]. (In §11 we shall study older methods for constructing true sentences that do not belong to **Π**, but these methods yield sentences expressing facts that are of purely logical interest.)

9.2. PROBLEM. Verify that the set of POSTULATES of **Π** is recursive; hence **Π** is axiomatic.

We shall now present a few examples of sentences belonging to **Π**.
First, we show that

$$(9.3) \qquad \forall v_1\,[\exists v_2(v_1 = sv_2) \vee v_1 = s_0] \in \mathbf{\Pi}.$$

We let α be the formula inside the square brackets in (9.3). Then it is easy to see that $\vdash \alpha(sv_1)$ and hence

$$\vdash \forall v_1\,[\alpha \to \alpha(sv_1)].$$

Also, we clearly have $\vdash \alpha(s_0)$. Hence, using postulate (9.1), we get (9.3).
From (9.3) and Def. 7.1 we have

$$\forall v_1 \forall v_2\,[v_1 \leqslant v_2 \leftrightarrow \exists v_3(v_1 + sv_3 = v_2) \vee v_1 + s_0 = v_2] \in \mathbf{\Pi}.$$

Hence, using postulates (8.3) and (8.4),

$$(9.4) \qquad \forall v_1 \forall v_2\,[v_1 \leqslant v_2 \leftrightarrow \exists v_3\{s(v_1 + v_3) = v_2\} \vee v_1 = v_2] \in \mathbf{\Pi}.$$

Using postulate (8.1) we now get

$$(9.5) \qquad \forall v_1(v_1 \leqslant s_0 \leftrightarrow v_1 = s_0) \in \mathbf{\Pi}.$$

Next, we shall show that

$$(9.6) \qquad \forall v_1(s_0 \leqslant v_1) \in \mathbf{\Pi}.$$

We observe that $(s_0 \leqslant s_0) \in \mathbf{\Pi}$ by (9.5). Also, using postulate (8.4) we have

$$\forall v_1 \forall v_2(s_0 + v_2 = v_1 \to s_0 + sv_2 = sv_1) \in \mathbf{\Pi},$$

hence

$$\forall v_1(s_0 \leqslant v_1 \rightarrow s_0 \leqslant sv_1) \in \Pi.$$

Using postulate (9.1) with the formula $s_0 \leqslant v_1$ as α, we obtain (9.6).

9.7. PROBLEM. Show that

$$\forall v_1 \forall v_2 [sv_2 + v_1 = s(v_2 + v_1)] \in \Pi.$$

9.8. THEOREM. $\Pi_2 \subseteq \Pi$.

PROOF. We show that the three sentences (8.7)–(8.9) belong to Π.

For (8.7), this follows at once from (9.5). To deal with (8.8), we observe that by (9.4)

$$\forall v_1 \forall v_2 [v_1 \leqslant sv_2 \leftrightarrow \exists v_3 \{s(v_1 + v_3) = sv_2\} \vee v_1 = sv_2] \in \Pi,$$

hence, using postulate (8.2) and Def. 7.1,

$$\forall v_1 \forall v_2 [v_1 \leqslant sv_2 \leftrightarrow v_1 \leqslant v_2 \vee v_1 = sv_2] \in \Pi. \tag{9.9}$$

It follows at once that (8.8) belongs to Π.

To deal with (8.9), we start by observing that

$$\forall v_2(s_0 \leqslant v_2 \vee v_2 \leqslant s_0) \in \Pi, \tag{9.10}$$

because $\forall v_2(s_0 \leqslant v_2) \in \Pi$ by (9.6). Next, from postulates (8.3) and (8.4) we deduce $\forall v_1(v_1 + ss_0 = sv_1)$ and hence

$$\forall v_1(v_1 \leqslant sv_1).$$

Using this and Prob. 9.7, we get, from (9.4),

$$\forall v_1 \forall v_2 [v_1 \leqslant v_2 \rightarrow sv_1 \leqslant v_2 \vee v_2 \leqslant sv_1] \in \Pi. \tag{9.11}$$

Also, from (9.9) we have

$$\forall v_1 \forall v_2(v_2 \leqslant v_1 \rightarrow v_2 \leqslant sv_1) \in \Pi.$$

Combining this with (9.11) we get

$$\forall v_1 \forall v_2(v_1 \leqslant v_2 \vee v_2 \leqslant v_1 \rightarrow sv_1 \leqslant v_2 \vee v_2 \leqslant sv_1) \in \Pi. \tag{9.12}$$

Using postulate (9.1) we get, from (9.10) and (9.12),

$$\forall v_1 \forall v_2(v_1 \leqslant v_2 \vee v_2 \leqslant v_1) \in \Pi,$$

as required. ∎

9.13. PROBLEM. Show that $\forall v_1(sv_1 \neq v_1) \in \Pi$; hence (by Prob. 8.13) Π is a *proper* extension of Π_2.

9.14. PROBLEM. Let $^*\mathfrak{N}$ be any model of $\mathbf{\Pi}$. Since $\mathbf{\Pi}_1 \subseteq \mathbf{\Pi}$, Prob. 7.6 shows that there is an unique embedding f of $\mathfrak{N}$ into $^*\mathfrak{N}$. Thus without loss of generality we may identify $f(n)$ with n for all $n \in N$ and regard $^*\mathfrak{N}$ as an extension of $\mathfrak{N}$ (i.e., we may assume $N \subseteq {}^*N$ and $^*s(n) = n+1$, $n \,^*\!+ m = n+m$, and $n \,^*\!\times m = nm$ for all $n,m \in N$.) For $a,b \in {}^*N$ we define: $a \,^*\!\leqslant b$ if, for some $c \in {}^*N$, $a \,^*\!+ c = b$; also, $a \equiv b$ if, for some $n \in N$, $a \,^*\!+ n = b$ or $b \,^*\!+ n = a$.

(i) Show that $^*\!\leqslant$ is a total ordering on $^*N$.

(ii) Show that $\equiv$ is an equivalence relation on $^*N$.

(iii) Call any of the $\equiv$-classes into which $^*N$ is partitioned by $\equiv$ a *block*. Show that N constitutes a block. Also show that any other block (if there are any) has, under the ordering $^*\!\leqslant$, the same order type as the integers.

(iv) For any blocks A and B, define $A \,^*\!\leqslant B$ if $a \,^*\!\leqslant b$ for all $a \in A$ and $b \in B$. Show that $^*\!\leqslant$ is a total ordering on the set of all blocks.

(v) Show that if $^*\mathfrak{N}$ is nonstandard then the ordering of the blocks is dense (i.e., between any two different blocks there is a third). Show also that N is the first block, but there is no last block.

(Each part of this problem is solved by verifying that $\mathbf{\Pi}$ contains certain sentences. Thus, e.g., in (i) one shows that $^*\!\leqslant$ is transitive by verifying that the associative law of addition

$$\forall \mathbf{v}_1 \forall \mathbf{v}_2 \forall \mathbf{v}_3 \, [(\mathbf{v}_1 + \mathbf{v}_2) + \mathbf{v}_3 = \mathbf{v}_1 + (\mathbf{v}_2 + \mathbf{v}_3)]$$

is in $\mathbf{\Pi}$.)

Note that all these results hold *a fortiori* if $^*\mathfrak{N}$ is a model of $\mathbf{\Omega}$.

9.15. PROBLEM. Let $^*\mathfrak{N}$ be a nonstandard model of $\mathbf{\Pi}$ (i.e., a model of $\mathbf{\Pi}$ that is not isomorphic with $\mathfrak{N}$). As in Prob. 9.14, we regard $^*\mathfrak{N}$ as an extension of $\mathfrak{N}$.

(i) Show that for every $\varphi \in \mathbf{\Phi}_{n+1}$ and all $a_1,\dots,a_n \in {}^*N$ we have

$$^*N - N \neq \{a \in {}^*N \colon {}^*\mathfrak{N} \models \varphi[a_1,\dots,a_n,a]\}.$$

(ii) Prove the so-called "overspill lemma": if $\varphi \in \mathbf{\Phi}_{n+1}$ and $a_1,\dots,a_n \in {}^*N$ and there are infinitely many $a \in N$ such that $^*\mathfrak{N} \models \varphi[a_1,\dots,a_n,a]$, then there is some $b \in {}^*N - N$ such that $^*\mathfrak{N} \models \varphi[a_1,\dots,a_n,b]$.

9.16. PROBLEM. Write $\mathbf{r} < \mathbf{t}$ for $\mathbf{sr} \leqslant \mathbf{t}$. Let γ be the formula

$$\exists \mathbf{v}_2 \exists \mathbf{v}_3 (\mathbf{v}_2 < \mathbf{v}_1 \wedge \mathbf{v}_3 < \mathbf{v}_1 \wedge \mathbf{v}_1 = \mathbf{v}_2 \times \mathbf{v}_3).$$

($\gamma(\mathbf{x})$ stands for "$\mathbf{x}$ is composite"). Let $^*\mathfrak{N}$ be any model of $\mathbf{\Pi}$. Prove that $^*\mathfrak{N}$ is nonstandard iff there is an infinite sequence $a_0, a_1, a_2, \ldots$ of members of $^*N$ such that for all n we have $a_{n+1} = {}^*s(a_n)$ and $^*\mathfrak{N} \models \gamma[a_n]$. (If $^*\mathfrak{N}$ is nonstandard and contains no such sequence, let φ be the formula

$$\forall \mathbf{v}_2 \exists \mathbf{v}_3 [\mathbf{v}_2 < \mathbf{v}_3 \wedge \mathbf{v}_3 < \mathbf{v}_1 + \mathbf{v}_2 \wedge \neg \gamma(\mathbf{v}_3)]$$

and show that

$$^*N - N = \{a \in {}^*N\colon \ {}^*\mathfrak{N} \models \varphi[a]\},$$

contradicting (i) of Prob. 9.15.)

*9.17. Problem. Let $\mathfrak{A}$ be any $\mathscr{L}$-structure. A subset $X \subseteq A$ is said to be *definable* if there is a formula $\varphi \in \Phi_1$ such that

$$X = \{a \in A\colon \ \mathfrak{A} \models \varphi[a]\}.$$

An individual $a \in A$ is *definable* if $\{a\}$ is a definable subset of A. Let $\mathsf{Df}(\mathfrak{A})$ be the set of all definable members of A, and put

$$\mathfrak{Df}(\mathfrak{A}) = \mathfrak{A} | \mathsf{Df}(\mathfrak{A}).$$

If $\mathfrak{A} = \mathfrak{Df}(\mathfrak{A})$ then $\mathfrak{A}$ is said to be *pointwise definable*. (Observe that the standard structure $\mathfrak{N}$ is pointwise definable.)

(i) Prove that if each non-empty definable subset of A has a definable member then $\mathfrak{Df}(\mathfrak{A}) \prec \mathfrak{A}$.

(ii) Show that if $\mathfrak{A}$ is a model of $\mathbf{\Pi}$ then each non-empty definable subset of A has a definable member.

(iii) Show that if $\mathfrak{Df}(\mathfrak{A}) \prec \mathfrak{A}$ then $\mathfrak{Df}(\mathfrak{A})$ is pointwise definable.

(iv) Let $\mathfrak{A}$ and $\mathfrak{A}'$ be elementarily equivalent pointwise definable $\mathscr{L}$-stuctures. Prove that they are isomorphic. (For each $a \in A$ there is $\varphi \in \Phi_1$ such that a is the unique individual satisfying φ in $\mathfrak{A}$. Since $\mathfrak{A} \equiv \mathfrak{A}'$, there is a unique individual a' satisfying φ in $\mathfrak{A}'$. Show that the mapping f defined by $f(a) = a'$ is an isomorphism of $\mathfrak{A}$ onto $\mathfrak{A}'$.)

(v) Use (iv) to show that, up to isomorphism, $\mathfrak{N}$ is the only pointwise definable model of $\mathbf{\Omega}$.

*9.18. Problem. Let $\mathbf{\Sigma}$ be a complete consistent theory such that $\mathbf{\Pi} \subseteq \mathbf{\Sigma}$.

(i) Let $\mathfrak{A}$ be a model of $\mathbf{\Sigma}$. Show that $\mathfrak{Df}(\mathfrak{A})$ is a prime (Prob. 5.6.14) model of $\mathbf{\Sigma}$. (Like Prob. 9.17(iv).) In particular, $\mathfrak{N}$ is a prime model of $\mathbf{\Omega}$.

(ii) Show that $\mathbf{\Sigma}$ has no (finitely) saturated (Prob. 5.6.15) model. (Let $\mathbf{x}|\mathbf{y}$ be a formula expressing "$\mathbf{x}$ divides $\mathbf{y}$". Let P be the set of all prime numbers and for each $X \subseteq P$ let

$$\Delta_X = \{\mathbf{s}_p|\mathbf{v}_1\colon p \in X\} \cup \{\neg \mathbf{s}_p|\mathbf{v}_1\colon p \in P - X\}.$$

In $B_2(\Sigma)$ (Ch 5, §6) show that $\{|\varphi|\colon \varphi \in \Delta_X\}$ can be extended to an ultrafilter U_X such that, if $X \neq X'$, then $U_X \neq U_{X'}$. Now apply Prob. 5.6.15(vii).)

(iii) Show that Σ has no universal (Prob. 5.6.16) model. (Use (ii) and Prob. 5.6.16.)

§ 10. Undecidability

Let $\mathbf{\Sigma}$ be a set of sentences. The *decision problem for* $\mathbf{\Sigma}$ is the problem of finding an algorithm whereby, for each sentence φ, one could determine whether $\varphi \in \mathbf{\Sigma}$ or not. If such an algorithm can be found, the decision problem for $\mathbf{\Sigma}$ is *solvable* and $\mathbf{\Sigma}$ itself is *decidable.*

Since the processes of coding and decoding (i.e., the transition from an expression to its code number and *vice versa*) are obviously effective, the decision problem for $\mathbf{\Sigma}$ is equivalent to the decision problem for the property T_Σ, as described in Ch. 6 (see discussion preceding Ex. 6.10.7). Thus, by Church's Thesis, $\mathbf{\Sigma}$ is decidable iff T_Σ is recursive.

Without committing our technical terminology to Church's Thesis, we shall say that the decision problem for $\mathbf{\Sigma}$ is *recursively solvable* and $\mathbf{\Sigma}$ is *recursively decidable* if T_Σ is recursive. If T_Σ is not recursive, we say that the decision problem for $\mathbf{\Sigma}$ is *recursively unsolvable* and $\mathbf{\Sigma}$ is *recursively undecidable.*

From Tarski's Theorem 4.7 and Cor. 3.5 it follows at once that $\mathbf{\Omega}$ is recursively undecidable. The recursive undecidability of $\mathbf{\Omega}$ as well as the four theories $\mathbf{\Pi}_0$, $\mathbf{\Pi}_1$, $\mathbf{\Pi}_2$ and $\mathbf{\Pi}$ also follows from:

10.1. Theorem. *If* $\mathbf{\Sigma}$ *is a theory in which every recursive property is weakly representable, then* $\mathbf{\Sigma}$ *is recursively undecidable.*

Proof. Similar to that of Thm. 4.7. Suppose T_Σ were recursive. Then the property

$$\lambda x[\neg T_\Sigma(d(x))]$$

would also be recursive, because the diagonal function d is recursive (see Ex. 4.5). Therefore this property would be weakly represented in $\mathbf{\Sigma}$ by some $\boldsymbol{\alpha} \in \mathbf{\Phi}_1$. Thus for each $x \in N$ we must have

$$\boldsymbol{\alpha}(\mathbf{s}_x) \in \mathbf{\Sigma} \quad \text{iff} \quad \neg T_\Sigma(d(x)) \text{ holds.}$$

In particular, giving x the value $\#\alpha$,

$$\alpha(\mathbf{s}_{\#\alpha})\in\Sigma \quad \text{iff} \quad \neg T_\Sigma(d(\#\alpha)) \text{ holds.}$$

But by (4.6) and Def. 5.1 we actually have

$$\neg T_\Sigma(d(\#\alpha)) \text{ holds} \quad \text{iff} \quad \alpha(\mathbf{s}_{\#\alpha})\notin\Sigma.$$

This contradiction shows that T_Σ cannot be recursive. ∎

10.2. Corollary. *If Σ is a sound theory that includes Π_0 (i.e., $\Pi_0\subseteq\Sigma\subseteq\Omega$), then Σ is recursively undecidable.*

Proof. By Thm. 6.8, every r.e. relation — hence certainly every recursive property — is weakly representable in Σ. Thus, by Thm. 10.1, Σ is recursively undecidable. ∎

10.3. Corollary. *If Σ is a consistent theory in which every recursive property is strongly representable, then Σ is recursively undecidable.*

Proof. We have already noted (following Def. 3.1) that if α strongly represents P in a consistent theory Σ, then α also weakly represents P in Σ. ∎

10.4. Corollary. *If Σ is a consistent theory and $\Pi_1\subseteq\Sigma$, then Σ is recursively undecidable.*

Proof. Immediate from Thm. 7.10 and Cor. 10.3. ∎

The recursive undecidability of Π_0, Π_1, Π_2, Π and Ω follow at once from Cor. 10.2. The same results (except for Π_0) follow also from Cor. 10.4. We can use Cor. 10.4 — but not Cor. 10.2 — to prove the recursive undecidability of consistent but unsound theories.

10.5. Theorem. *Let Σ be a theory such that $\Sigma\cup\Pi_2$ is consistent. Then Σ is recursively undecidable.*

Proof. Put

$$\Delta=\{\varphi:\ \varphi\in\Phi_0,\ \Sigma\cup\Pi_2\vdash\varphi\}.$$

Let π be the conjunction of the nine postulates (8.1)–(8.9) of Π_2. It is clear that for any sentence φ we have $\varphi\in\Delta$ iff $(\pi\rightarrow\varphi)\in\Sigma$. Thus

$$T_\Delta(x)=T_\Sigma(64\cdot 3^{\#\pi}\cdot 5^x).$$

From this we see that if T_Σ were recursive then T_Δ would also be recursive.

But Δ is a consistent theory and an extension of Π_2, hence also of Π_1. So, by Cor. 10.4, T_Δ is *not* recursive. ∎

We can now strengthen Cor. 10.2:

10.6. COROLLARY. *Every sound theory is recursively undecidable.*
PROOF. If Σ is sound, then $\Sigma \cup \Pi_2$ is also sound, hence consistent. So, by Thm. 10.5, Σ is recursively undecidable. ∎

As an application of Cor. 10.6 we prove:

10.7. CHURCH'S THEOREM. *Let*

$$\Lambda = \{\varphi : \varphi \in \Phi_0, \vdash \varphi\}.$$

Then Λ *is recursively undecidable.*
PROOF. Immediate from Cor. 10.6, since Λ is clearly a sound theory. ∎

10.8. REMARK. Clearly, Λ is an axiomatic theory (it is based on the empty set of postulates). Thus, by Thm. 5.4, T_Λ is recursively enumerable. Hence there is an effective procedure for generating, one by one, all sentences of Λ, i.e., all logically valid sentences of $\mathscr{L}$. Such a procedure was actually described in §5 of Ch. 3. On the other hand, the fact that we have not been able to devise an algorithm for deciding whether any given sentence is logically valid or not is not accidental. For, by Church's Theorem and Church's Thesis, such an algorithm does not exist (at any rate not for the particular language $\mathscr{L}$ of this chapter).

The result of Thm. 10.7 is often expressed by saying that *the predicate calculus is undecidable.*

In the remaining part of this section we shall consider not only $\mathscr{L}$ (the first-order language of arithmetic) but ofter first-order languages as well.

If $\mathscr{L}'$ is any first-order language, then by *$\mathscr{L}'$-theory* we mean a set of $\mathscr{L}'$-sentences closed under first-order deducibility. The *decision problem* for a set of $\mathscr{L}'$-sentences is formulated in the same way as in the case of $\mathscr{L}$. Thus, it makes sense to say that such a set is *decidable* or *undecidable.*

For $i=1,2$ let $\mathscr{L}_i$ be a first-order language and let Σ_i be a set of $\mathscr{L}_i$-sentences. By a *reduction of* Σ_1 *to* Σ_2 we mean an effective (i.e., algorithmic) mapping f of the set of all $\mathscr{L}_1$-sentences into the set of all $\mathscr{L}_2$-sentences such that for each $\mathscr{L}_1$-sentence φ we have $\varphi \in \Sigma_1$ iff $f(\varphi) \in \Sigma_2$. If such a reduction exists, then Σ_1 is *reducible* to Σ_2.

It is clear that if Σ_1 is undecidable and reducible to Σ_2, then Σ_2 is undecidable as well.

For the rest of this section, we shall assume the correctness of Church's Thesis. Thus the $\mathscr{L}$-theories which we have proved to be *recursively* undecidable, will now be taken to be undecidable. With this as our point of departure, we shall use various reductions to show that certain theories (in other languages) are also undecidable.

10.9. Example. Let $\mathscr{L}_1$ be the language obtained from $\mathscr{L}$ by omitting the unary function symbol **s** and adding a new individual constant **1**. If φ is an $\mathscr{L}$-sentence and a term of the form **sr** occurs in φ, replace that term by **r+1**, and continue this process until all occurrences of **s** have been eliminated. Call the resulting $\mathscr{L}_1$-sentence φ^*. We claim that if $\models \varphi$ then also $\models \varphi^*$. Indeed, let $\mathfrak{U}_1$ be an $\mathscr{L}_1$-structure in which φ^* does not hold. If $+$ and 1 are the binary operation and individual of $\mathfrak{U}_1$ corresponding to $\mathbf{+}$ and **1**, we let $s(a)=a+1$ for every a in the domain of $\mathfrak{U}_1$. By taking s as the operation corresponding to **s**, we obtain an $\mathscr{L}$-structure in which φ fails to hold.

Let Σ_1 be any $\mathscr{L}_1$-theory. Put

$$\Sigma = \{\varphi\colon\ \varphi \in \Phi_0,\ \varphi^* \in \Sigma_1\}.$$

Clearly, * is a reduction of Σ to Σ_1. We claim that Σ is an $\mathscr{L}$-theory. Indeed, if $\varphi \in \Phi_0$ and $\Sigma \vdash \varphi$, then for some $\varphi_1,\ldots,\varphi_k \in \Sigma$ we must have

$$\vdash \varphi_1 \to \ldots \to \varphi_k \to \varphi.$$

Hence, by what we have shown, also

$$\vdash \varphi_1{}^* \to \ldots \to \varphi_k^* \to \varphi^*.$$

But $\varphi_1^*,\ldots,\varphi_k^* \in \Sigma_1$, and Σ_1 is an $\mathscr{L}_1$-theory. Hence $\varphi^* \in \Sigma_1$, so that $\varphi \in \Sigma$.

Next, let $\mathfrak{N}_1$ be the $\mathscr{L}_1$-structure with domain N and with **0**,**1**, $\mathbf{+}$ and $\mathbf{\times}$ interpreted in the obvious way, as 0, 1, addition and multiplication, respectively.

For any $\mathscr{L}$-sentence φ we clearly have $\mathfrak{N} \models \varphi$ iff $\mathfrak{N}_1 \models \varphi^*$.

If the $\mathscr{L}_1$-theory Σ_1 has $\mathfrak{N}_1$ as a model, it now follows that the $\mathscr{L}$-theory Σ defined above is sound (i.e., has $\mathfrak{N}$ as a model). By Cor. 10.6 and Church's thesis, Σ is undecidable. Since * is a reduction of Σ to Σ_1, it follows that Σ_1 is undecidable.

Thus we have shown that every $\mathscr{L}_1$-theory that has $\mathfrak{N}_1$ as a model is undecidable. In particulai, if Λ_1 is the set of all logically valid $\mathscr{L}_1$-sentences, then Λ_1 is undecidable.

10.10. Problem. Let Σ_1 and Σ_2 be $\mathscr{L}'$-theories. We say that Σ_2 is a *finite extension* of Σ_1 if there is a finite set of $\mathscr{L}'$-sentences $\{\varphi_1,\ldots,\varphi_k\}$ such that $\Sigma_1 \cup \{\varphi_1,\ldots,\varphi_k\}$ is a set of postulates for Σ_2.

Show that in this case the mapping *, where $\varphi^* = \varphi_1 \to \ldots \to \varphi_k \to \varphi$, is a reduction of Σ_2 to Σ_1. Hence if Σ_2 is undecidable then so is Σ_1.

10.11. PROBLEM. Let $\mathscr{L}_1$ and $\mathfrak{N}_1$ be as in Ex. 10.9. Let $\mathfrak{Z}$ be *the structure of integers*, i.e., the $\mathscr{L}_1$-structure whose domain is the set of all integers and with the symbols **0**,**1**, + and × interpreted in the obvious way. Let φ be the formula

$$\exists \mathbf{u} \exists \mathbf{v} \exists \mathbf{x} \exists \mathbf{y} (\mathbf{z} = \mathbf{u} \times \mathbf{u} + \mathbf{v} \times \mathbf{v} + \mathbf{x} \times \mathbf{x} + \mathbf{y} \times \mathbf{y}),$$

where **u**,**v**,**x**,**y**,**z** are distinct variables.

For any sentence α, let α^* be the relativization of α to φ, with **z** as chosen variable (cf. §12 of Ch. 2).

(i) Verify that for any $\mathscr{L}_1$-sentence α we have $\mathfrak{N}_1 \models \alpha$ iff $\mathfrak{Z} \models \alpha^*$. (Use Lagrange's theorem that every natural number is the sum of four squares.)

(ii) Let Σ be an $\mathscr{L}_1$-theory such that $\mathfrak{Z}$ is a model for Σ and the sentences $\exists \mathbf{z}\varphi$, $\varphi\{\mathbf{0}\}$, $\varphi\{\mathbf{1}\}$, $\varphi\{+\}$ and $\varphi\{\times\}$ belong to Σ. (For the definition of the latter four sentences, see §12 of Ch. 2.) Let Σ' be the set of all $\mathscr{L}_1$-sentences α such that $\alpha^* \in \Sigma$. Use Cor. 2.12.4 to show that Σ' is an $\mathscr{L}_1$-theory. Hence use (i) and Ex. 10.9 to prove that Σ is undecidable.

(iii) Let Σ be any $\mathscr{L}_1$-theory that has $\mathfrak{Z}$ as a model[1]. Show that Σ is undecidable. (Use (ii) and Prob. 10.10.)

10.12. PROBLEM. Let $\mathscr{L}_1$ and $\mathfrak{Z}$ be as above. Let $\mathscr{L}_2$ be the language obtained from $\mathscr{L}_1$ by omitting the constants **0** and **1**. Let $\mathfrak{Z}_2$ be the $\mathscr{L}_2$-structure which is the same as $\mathfrak{Z}$ except that it has no designated individuals. Let α and β be the formulas

$$\forall \mathbf{v}_2(\mathbf{v}_2 + \mathbf{v}_1 = \mathbf{v}_2), \qquad \forall \mathbf{v}_2(\mathbf{v}_2 \times \mathbf{v}_1 = \mathbf{v}_2),$$

respectively. For any $\mathscr{L}_1$-sentence φ, let φ^* be

$$\forall \mathbf{x} \forall \mathbf{y}[\alpha(\mathbf{x}) \to \beta(\mathbf{y}) \to \varphi'],$$

where **x** and **y** are the first two variables that do not occur in φ, and φ' is obtained from φ by replacing all occurrences of **0** and **1** by occurrences of **x** and **y** respectively.

(i) Verify that for any $\mathscr{L}_1$-sentence we have $\mathfrak{Z} \models \varphi$ iff $\mathfrak{Z}_2 \models \varphi^*$.

(ii) Let Σ_2 be an $\mathscr{L}_2$-theory that has $\mathfrak{Z}_2$ as a model and contains the sentences $\exists! \mathbf{v}_1 \alpha$ and $\exists! \mathbf{v}_1 \beta$. (For the definition of $\exists!$ see Def. 2.10.1.)

[1] In particular, Σ can be taken, e.g., as the set of all $\mathscr{L}_1$-sentences which hold in all rings with identity.

Let $\boldsymbol{\Sigma}_1$ be the $\mathscr{L}_1$-theory based on the set of postulates $\boldsymbol{\Sigma}_2 \cup \{\boldsymbol{\alpha}(\mathbf{0}), \boldsymbol{\beta}(\mathbf{1})\}$. Show that * is a reduction of $\boldsymbol{\Sigma}_1$ to $\boldsymbol{\Sigma}_2$ and $\boldsymbol{\Sigma}_1$ has $\mathfrak{Z}$ as a model. Hence, by part (iii) of Prob. 10.11, $\boldsymbol{\Sigma}_2$ is undecidable.

(iii) Let $\boldsymbol{\Sigma}_2$ be any $\mathscr{L}_2$-theory that has $\mathfrak{Z}_2$ as a model. Show that $\boldsymbol{\Sigma}_2$ is undecidable. (Use (ii) and Prob. 10.10.)

10.13. PROBLEM. Let $\mathscr{L}_2$ and $\mathfrak{Z}_2$ be as in Prob. 10.12. Let $\mathscr{L}_3$ be the first-order language with equality whose only extralogical symbols are two ternary predicate symbols **S** and **P**. Let $\mathfrak{Z}_3$ be the $\mathscr{L}_3$-structure whose domain is the set of all integers and in which **S** and **P** are interpreted as the set of all triples of the forms $\langle i,j,i+j\rangle$ and $\langle i,j,ij\rangle$ respectively.

Let $\boldsymbol{\varphi}$ be any $\mathscr{L}_2$-sentence. By Lemma 2.10.4 we can find an $\mathscr{L}_2$-sentence $\boldsymbol{\varphi}'$ logically equivalent to $\boldsymbol{\varphi}$ such that any atomic subformula of $\boldsymbol{\varphi}'$ which contains $\mathbf{+}$ or $\mathbf{\times}$ has the form $\mathbf{x+y=z}$ or $\mathbf{x\times y=z}$. Replace all such subformulas of $\boldsymbol{\varphi}'$ by **Sxyz** and **Pxyz** respectively, and let $\boldsymbol{\varphi}^*$ be the resulting $\mathscr{L}_3$-sentence.

(i) Verify that for each $\mathscr{L}_2$-sentence $\boldsymbol{\varphi}$ we have $\mathfrak{Z}_2 \models \boldsymbol{\varphi}$ iff $\mathfrak{Z}_3 \models \boldsymbol{\varphi}^*$.

(ii) Let $\boldsymbol{\alpha}$ and $\boldsymbol{\beta}$ be the $\mathscr{L}_3$-sentences $\forall \mathbf{v}_1 \forall \mathbf{v}_2 \exists! \mathbf{v}_3 \mathbf{S}\mathbf{v}_1\mathbf{v}_2\mathbf{v}_3$ and $\forall \mathbf{v}_1 \forall \mathbf{v}_2 \exists! \mathbf{v}_3 \mathbf{P}\mathbf{v}_1\mathbf{v}_2\mathbf{v}_3$. Prove that for any $\mathscr{L}_2$-sentence $\boldsymbol{\varphi}$ we have $\models \boldsymbol{\varphi}$ iff $\models \boldsymbol{\alpha} \rightarrow \boldsymbol{\beta} \rightarrow \boldsymbol{\varphi}^*$. (Argue as in the proof of Thm. 2.10.5.)

(iii) Prove that if $\boldsymbol{\Sigma}$ is an $\mathscr{L}_3$-theory that has $\mathfrak{Z}_3$ as a model, then $\boldsymbol{\Sigma}$ is undecidable. (Consider the finite extension $\boldsymbol{\Sigma}'$ of $\boldsymbol{\Sigma}$ based on the postulates $\boldsymbol{\Sigma} \cup \{\boldsymbol{\alpha},\boldsymbol{\beta}\}$, with $\boldsymbol{\alpha}$ and $\boldsymbol{\beta}$ as in (ii). Show that * is a reduction to $\boldsymbol{\Sigma}'$ of some $\mathscr{L}_2$-theory which has $\mathfrak{Z}_2$ as a model.)

10.14. PROBLEM. Let $\mathscr{L}_3$ and $\mathfrak{Z}_3$ be as in Prob. 10.13. Let $\mathscr{L}_4$ be the first-order language with equality whose only extralogical symbol is a quaternary predicate symbol **R**. Let $\mathfrak{Z}_4$ be the $\mathscr{L}_4$-structure whose domain is the set of integers and in which **R** is interpreted as the set of all quadruples of the forms $\langle i,j,i+j,i+j\rangle$ and $\langle i,j,ij,k\rangle$, where $k \neq ij$.

For any $\mathscr{L}_3$-sentence $\boldsymbol{\varphi}$, let $\boldsymbol{\varphi}^*$ be the $\mathscr{L}_4$-sentence obtained by replacing each atomic subformula **Sxyz** by **Rxyzz** and each atomic subformula **Pxyz** by $\exists \mathbf{u}(\mathbf{u} \neq \mathbf{z} \wedge \mathbf{Rxyzu})$, where **u** is the first variable different from **x**,**y** and **z**.

(i) Verify that for each $\mathscr{L}_3$-sentence $\boldsymbol{\varphi}$ we have $\mathfrak{Z}_3 \models \boldsymbol{\varphi}$ iff $\mathfrak{Z}_4 \models \boldsymbol{\varphi}^*$. Also, if $\models \boldsymbol{\varphi}$ then $\models \boldsymbol{\varphi}^*$.

(ii) Prove that if $\boldsymbol{\Sigma}$ is an $\mathscr{L}_4$-theory that has $\mathfrak{Z}_4$ as a model, then $\boldsymbol{\Sigma}$ is undecidable. (Show that * is a reduction to $\boldsymbol{\Sigma}$ of some $\mathscr{L}_3$-theory that has $\mathfrak{Z}_3$ as a model.)

10.15. PROBLEM. Let $\mathscr{L}_4$ be as in Prob. 10.14, and let $\mathscr{L}_5$ be the first-order language with equality whose only extralogical symbols are an individual constant **c** and a binary predicate symbol **Q**. Put **Sx** and **Px** for **Qxc** and **Qcx**, respectively. Let $\boldsymbol{\alpha}$ be the sentence

$$\exists x Sx \wedge \forall x(x = c \vee Px \vee Sx) \wedge \neg Pc \wedge \neg Sc \wedge \forall x \neg (Px \wedge Sx)$$
$$\wedge\ \forall x[Px \to \exists ! u(Su \wedge Qux) \wedge \exists ! v(Sv \wedge Qxv)]$$
$$\wedge\ \forall u \forall v[Su \wedge Sv \to \exists ! x(Px \wedge Qux \wedge Qxv) \wedge \neg Quv].$$

For each $\mathscr{L}_4$-sentence $\boldsymbol{\varphi}$, let $\boldsymbol{\varphi}^*$ be the $\mathscr{L}_5$-sentence $\boldsymbol{\alpha} \to \boldsymbol{\beta}$ where $\boldsymbol{\beta}$ is obtained from $\boldsymbol{\varphi}$ in two steps, as follows: first, relativize $\boldsymbol{\varphi}$ to **Sz**, with **z** as chosen variable (see §12 of Ch. 2); then replace each atomic subformula **Ruvwz** by

$$\forall x \forall y(Px \wedge Py \wedge Qxy \wedge Qux \wedge Qxv \wedge Qwy \wedge Qyz),$$

where **x** and **y** are the first two variables different from **u**,**v**,**w** and **z**.

(i) Let $\mathfrak{U}$ be an $\mathscr{L}_4$-structure. Let $\mathfrak{B}$ be an $\mathscr{L}_5$-structure obtained as follows. The individuals of $\mathfrak{B}$ are those of $\mathfrak{U}$, plus a new object (u,v) for every ordered pair $\langle u,v\rangle$ of individuals of $\mathfrak{U}$, plus one additional object c. We take c as the designated individual of $\mathfrak{B}$. As $\mathbf{Q}^{\mathfrak{B}}$ we take the set of all pairs $\langle (u,v),(w,z)\rangle$ where $\langle u,v,w,z\rangle \in \mathbf{R}^{\mathfrak{U}}$, as well as all pairs $\langle c,(u,v)\rangle$, $\langle u,c\rangle$, $\langle u,(u,v)\rangle$ and $\langle (u,v),v\rangle$, where u and v are any individuals of $\mathfrak{U}$. Verify that $\mathfrak{U} \models \boldsymbol{\varphi}$ iff $\mathfrak{B} \models \boldsymbol{\varphi}^*$.

(ii) For $i=4,5$ let Λ_i be the set of all logically valid $\mathscr{L}_i$-sentences. Show that * is a reduction of Λ_4 to Λ_5. Hence, by part (ii) of Prob. 10.14, Λ_5 is undecidable.

(iii) Let $\mathscr{L}_6$ be like $\mathscr{L}_5$ but without the constant **c**, and let Λ_6 be the set of all logically valid $\mathscr{L}_6$-sentences. Show that Λ_5 is reducible to Λ_6, hence Λ_6 is undecidable. (Use 3.1.13.)

(iv) Let $\mathscr{L}_7$ be a first-order language without equality, whose only extra-logical symbols are two binary predicate symbols. Let Λ_7 be the set of all logically valid $\mathscr{L}_7$-sentences. Prove that Λ_7 is undecidable. (Use Thm. 2.11.2 to find a reduction of Λ_6 to a finite extension of Λ_7.)

§ 11. Incompleteness

In this section we again confine our attention to the first-order language of arithmetic, $\mathscr{L}$.

A theory $\boldsymbol{\Sigma}$ is *complete* if it is consistent and for every sentence $\boldsymbol{\alpha}$ we have $\boldsymbol{\alpha} \in \boldsymbol{\Sigma}$ or $\neg\boldsymbol{\alpha} \in \boldsymbol{\Sigma}$. (Cf. Lemma 5.4.1.)

Among the consistent theories, complete theories are maximal with respect to inclusion: clearly, no complete theory can be properly included in a consistent theory. In particular, $\mathbf{\Omega}$ is the only sound and complete theory; for $\mathbf{\Omega}$ is evidently complete, and any sound theory is, by definition, included in $\mathbf{\Omega}$.

If $\mathbf{\Sigma}$ is a consistent theory, then by Thm. 3.3.13 some $\mathscr{L}$-structure $\mathfrak{U}$ is a model for $\mathbf{\Sigma}$. If $\mathbf{\Sigma}'$ is the set of *all* sentences holding in $\mathfrak{U}$, then $\mathbf{\Sigma}'$ is a complete theory, and $\mathbf{\Sigma} \subseteq \mathbf{\Sigma}'$. Therefore $\mathbf{\Sigma}$ itself is complete iff $\mathbf{\Sigma} = \mathbf{\Sigma}'$.

Thus, a consistent theory $\mathbf{\Sigma}$ is complete iff it is the set of all sentences holding in some $\mathscr{L}$-structure $\mathfrak{U}$. (Cf. Prob. 5.4.2.)

From now on we shall confine our attention to *axiomatizable* theories.

11.1. EXAMPLE. Let $\mathbf{\Sigma}$ be the set of all sentences that hold in the $\mathscr{L}$-structure whose universe consists of a single individual. (There is only one such structure, up to isomorphism.) By the above, $\mathbf{\Sigma}$ is complete. Also, $\mathbf{\Sigma}$ can be characterized as the set of all sentences deducible from the single postulate $\forall \mathbf{v}_1(\mathbf{v}_1 = \mathbf{0})$, so $\mathbf{\Sigma}$ is axiomatizable.

11.2. PROBLEM. Let $\mathfrak{U}$ be any finite $\mathscr{L}$-structure. Show that the set of all sentences that hold in $\mathfrak{U}$ is an axiomatizable complete theory. (Cf. Prob. 5.2.12 (iii).)

The following theorem and (in a somewhat different version) the one after it are due to Gödel.

11.3. THEOREM. *Every sound axiomatizable theory is incomplete.*

PROOF. By Cor. 5.5, $\mathbf{\Omega}$ is not axiomatizable. Hence every sound axiomatizable theory is *properly* included in $\mathbf{\Omega}$ and must therefore be incomplete. ∎

This result merely highlights the message of Cor. 5.5: no complete axiomatization of $\mathbf{\Omega}$ is possible.

If $\mathbf{\Sigma}$ is a sound incomplete theory, there must exist a true sentence $\boldsymbol{\varphi}$ such that $\boldsymbol{\varphi} \notin \mathbf{\Sigma}$. The proof of the following theorem yields a method for constructing such a sentence $\boldsymbol{\varphi}$, for any sound axiomatic theory.

11.4. THEOREM. *Given any sound axiomatic theory* $\mathbf{\Sigma}$, *we can find a true sentence* $\boldsymbol{\varphi}$ *of the form* $\forall \mathbf{x}_1 \ldots \forall \mathbf{x}_k(\mathbf{r} \neq \mathbf{t})$ *such that* $\boldsymbol{\varphi} \notin \mathbf{\Sigma}$.

PROOF. Since $\mathbf{\Sigma}$ is given to us as an axiomatic theory, we can assume by Def. 5.2 that we are given the r.e. property $T_{\mathbf{\Gamma}}$, where $\mathbf{\Gamma}$ is a set of postulates for $\mathbf{\Sigma}$. From $T_{\mathbf{\Gamma}}$ we can construct the r.e. property $T_{\mathbf{\Sigma}}$ as in the proof of Thm. 5.4. Hence we can also construct the r.e. property $\lambda x T_{\mathbf{\Sigma}}(d(x))$, where d is the diagonal sequence of Ex. 4.5.

By the MRDP Theorem and Lemma 3.4, we can find a formula $\boldsymbol{\alpha} \in \boldsymbol{\Phi}_1$ of the form

$$\exists \mathbf{v}_2 \dots \exists \mathbf{v}_m (\mathbf{r} = \mathbf{t}),$$

where $m \geqslant 1$, such that $\boldsymbol{\alpha}$ represents $\lambda x T_\Sigma(d(x))$ in $\boldsymbol{\Omega}$. It follows that the formula

$$\boldsymbol{\beta} = \forall \mathbf{v}_2 \dots \forall \mathbf{v}_m (\mathbf{r} \neq \mathbf{t}),$$

which is logically equivalent to $\neg \boldsymbol{\alpha}$, represents $\lambda x [\neg T_\Sigma(d(x))]$ in $\boldsymbol{\Omega}$. Thus

$$\boldsymbol{\beta}(\mathbf{s}_{\#\boldsymbol{\beta}}) \in \boldsymbol{\Omega} \quad \text{iff} \quad \neg T_\Sigma(d(\#\boldsymbol{\beta})) \text{ holds.}$$

But by (4.6) and the definition of T_Σ we see that

$$\neg T_\Sigma(d(\#\boldsymbol{\beta})) \text{ holds iff } \boldsymbol{\beta}(\mathbf{s}_{\#\boldsymbol{\beta}}) \notin \boldsymbol{\Sigma}.$$

Let $\boldsymbol{\varphi}$ be the sentence $\boldsymbol{\beta}(\mathbf{s}_{\#\boldsymbol{\beta}})$. Then by what we have just shown,

$$\boldsymbol{\varphi} \in \boldsymbol{\Omega} \quad \text{iff} \quad \boldsymbol{\varphi} \notin \boldsymbol{\Sigma}.$$

We cannot have $\boldsymbol{\varphi} \in \boldsymbol{\Sigma}$, because then $\boldsymbol{\varphi} \notin \boldsymbol{\Omega}$, contrary to the soundness of $\boldsymbol{\Sigma}$. Hence we have $\boldsymbol{\varphi} \notin \boldsymbol{\Sigma}$ and $\boldsymbol{\varphi} \in \boldsymbol{\Omega}$, i.e., $\boldsymbol{\varphi}$ is true but is not in $\boldsymbol{\Sigma}$. ∎

Note that the sentence $\boldsymbol{\varphi}$ constructed in the above proof asserts the unsolvability (in natural numbers) of a certain diophantine equation.

In the proof of Thm. 4.7 we showed that *if* T_Ω were arithmetical then we could find a sentence asserting its own falsity (or, at any rate, something equivalent to its own falsity). Since such a sentence cannot exist, we concluded that T_Ω is not arithmetical. Here, on the other hand, we have used the same method to construct a sentence $\boldsymbol{\varphi}$ which asserts something equivalent to "I do not belong to $\boldsymbol{\Sigma}$". There is nothing contradictory about the existence of $\boldsymbol{\varphi}$ (we actually have shown how to construct it!). Moreover, $\boldsymbol{\varphi}$ cannot be false because this would contradict the soundness of $\boldsymbol{\Sigma}$. Thus $\boldsymbol{\varphi}$ is true.

While Thm. 4.7 was proved by showing that if T_Ω were arithmetical then the Liar paradox could be reproduced in $\mathscr{L}$, the proof of 11.4 merely skirts the paradox.

Having disposed of the problem of completeness for sound axiomatizable theories, we now turn to axiomatizable theories that are consistent but not necessarily sound.

11.5. THEOREM. *Every axiomatizable complete theory is recursively decidable.*

PROOF. Let $\boldsymbol{\Sigma}$ be an axiomatizable complete theory. Since $\boldsymbol{\Sigma}$ is axiomatizable, T_Σ is r.e. by Thm. 5.4. On the other hand, $\neg T_\Sigma(x)$ holds iff x is not a SENTENCE or x is a SENTENCE whose negation belongs to Σ. Thus we have the identity

$$\neg T_\Sigma(x) = \neg \mathrm{Frm}(x,0) \vee T_\Sigma(32 \cdot 3^x).$$

Hence, by Prob. 4.4 and Lemma 6.11.2, $\neg T_\Sigma$ is recursively enumerable. Since both T_Σ and $\neg T_\Sigma$ are r.e., it follows from Thm. 6.11.4 that T_Σ is recursive. ∎

Note that from Thm. 11.5 and Cor. 10.6 we get another proof of Thm. 11.3. Note also that the argument of Thm. 11.5 can be used to prove the recursive decidability of the various axiomatic complete theories in §4 of Ch. 5.

Is every axiomatizable and recursively decidable theory necessarily complete? A negative answer is provided by:

11.6. PROBLEM. Let Σ be the set of all sentences deducible from the single postulate

$$\forall \mathbf{v}_1(\mathbf{v}_1 = \mathbf{s}_0 \vee \mathbf{v}_1 = \mathbf{s}_1).$$

Show that Σ is the intersection of finitely many (actually: 5/3) different theories, each of which is axiomatizable and complete by Prob. 11.2, and hence recursively decidable. Hence show that Σ is recursively decidable and incomplete.

The following two theorems were proved, in a weaker form, by Gödel. A version which is intermediate between Gödel's original one and the one given below is due to Rosser.

11.7. THEOREM. *Every consistent axiomatizable theory that includes Π_1 is incomplete.*

PROOF. Immediate from Cor. 10.4 and Thm. 11.5. ∎

In the proof of the following theorem — a strengthened version of Gödel's celebrated First Incompleteness Theorem — we show how, for any given consistent axiomatic theory Σ such that $\Pi_1 \subseteq \Sigma$, one can actually construct a sentence φ such that $\varphi \notin \Sigma$ and $\neg\varphi \notin \Sigma$.

11.8. FIRST INCOMPLETENESS THEOREM. *Given any consistent axiomatic theory Σ that includes Π_1, we can find a formula $\gamma \in \Phi_1$ (of the form described in Lemma 7.9) such that the sentences $\gamma(\mathbf{s}_{\#\gamma})$ and $\neg\gamma(\mathbf{s}_{\#\gamma})$ do not belong to Σ.*

PROOF. As in the proof of Thm. 11.4, we can construct the r.e. property T_Σ. Put

$$P(x)=T_\Sigma(32\cdot 3^{d(x)}), \qquad P'(x)=T_\Sigma(d(x)),$$

and construct γ as in Lemma 7.9 (for $n=1$).

Suppose we had $\gamma(\mathbf{s}_{\#\gamma})\in\Sigma$. By the consistency of Σ we get $\neg\gamma(\mathbf{s}_{\#\gamma})\notin\Sigma$. But

$$\#[\gamma(\mathbf{s}_{\#\gamma})]=d(\#\gamma), \qquad \#[\neg\gamma(\mathbf{s}_{\#\gamma})]=32\cdot 3^{d(\#\gamma)}.$$

Hence $\neg P(\#\gamma)\wedge P'(\#\gamma)$ holds. By Lemma 7.9 it follows that

$$\neg\gamma(\mathbf{s}_{\#\gamma})\in\mathbf{\Pi}_1\subseteq\Sigma,$$

contrary to the assumed consistency of Σ.

Now suppose $\neg\gamma(\mathbf{s}_{\#\gamma})\in\Sigma$. By the consistency of Σ we get $\gamma(\mathbf{s}_{\#\gamma})\notin\Sigma$. Hence in this case $P(\#\gamma)\wedge\neg P'(\#\gamma)$ holds, so that by Lemma 7.9 $\gamma(\mathbf{s}_{\#\gamma})\in\Sigma$. This again contradicts the consistency of Σ. ∎

At first sight it might seem that Thms. 11.7 and 11.8 add little of interest to Thms. 11.3 and 11.4. After all, we are really interested in theories that are not merely consistent but sound. And, for a sound theory Σ, Thms. 11.7 and 11.8 yield roughly the same results as Thms. 11.3 and 11.4, but only subject to the additional condition that $\mathbf{\Pi}_1\subseteq\Sigma$.

However, notice that the condition of *soundness* (which is imposed in Thms. 11.3 and 11.4) is purely semantic. A theory Σ is sound iff all its sentences are true, i.e., $\Sigma\subseteq\mathbf{\Omega}$; and we know that $\mathbf{\Omega}$ cannot be characterized in a reasonably constructive way. On the other hand, the conditions imposed in Thms. 11.7 and 11.8 are much less problematic. The condition $\mathbf{\Pi}_1\subseteq\Sigma$ means that certain sentences of a rather simple form ((6.1)–(6.4) and (7.2)–(7.4)) belong to Σ. To show that this condition holds, it is enough to describe a method for constructing deductions of these sentences from a given set of postulates for Σ. (Actually, since $\mathbf{\Pi}_1\subseteq\mathbf{\Pi}_2$, it is enough to exhibit deductions of the nine postulates of $\mathbf{\Pi}_2$.) Also, the condition that Σ is consistent simply means that the sentence $\mathbf{0}\neq\mathbf{0}$ does not belong to Σ, i.e., that no deduction from a given set of postulates for Σ terminates with this sentence.

Thus the advantage of Thms. 11.7 and 11.8 is that the notions mentioned in them are far more elementary, constructive and "tangible" than the notion of soundness.

This fact is utilized in the proof of Gödel's Second Incompleteness Theorem, which we now proceed to outline.

We start by noting that the proof of Thm. 11.8 works also if $\boldsymbol{\Sigma}$ is assumed to be axiomatizable rather than axiomatic, except that in this case we can only say that the formula γ *exists*, not that we can actually *find* it.

Now, let $\boldsymbol{\Sigma}$ be any axiomatizable theory that includes first-order Peano arithmetic $\boldsymbol{\Pi}$. Define P and P' as in the proof of Thm. 11.8, and let $\boldsymbol{\alpha}$, $\boldsymbol{\alpha}'$, $\boldsymbol{\beta}$, $\boldsymbol{\beta}'$ and γ be as in Lemma 7.9 (for $n=1$). Since $\boldsymbol{\Pi}_1 \subseteq \boldsymbol{\Pi} \subseteq \boldsymbol{\Sigma}$, the second part of the proof of Thm. 11.8 establishes that

(1) *if* $\boldsymbol{\Sigma}$ *is consistent, then* $\neg\gamma(\mathbf{s}_{\#\gamma}) \notin \boldsymbol{\Sigma}$.

Let us look for an $\mathscr{L}$-sentence that formalizes assertion (1). Let $k = \#(\mathbf{0} \neq \mathbf{0})$. (In fact, $k = 32 \cdot 3^{240}$.) To say that $\boldsymbol{\Sigma}$ is consistent is clearly tantamount to saying that the sentence $\mathbf{0} \neq \mathbf{0}$ is not in $\boldsymbol{\Sigma}$, i.e., that $T_{\Sigma}(k)$ does not hold. Since $\mathbf{0} \neq \mathbf{0}$ is a sentence, it is easy to see that $d(k) = k$. Thus the statement that $\boldsymbol{\Sigma}$ is consistent is equivalent to the statement that $T_{\Sigma}(d(k))$ does not hold, i.e., that $P'(k)$ does not hold. But the formula $\boldsymbol{\alpha}'$ represents P' in $\boldsymbol{\Omega}$. (See proof of Lemma 7.9.) Therefore, if we put $\boldsymbol{\varphi} = \neg\boldsymbol{\alpha}'(\mathbf{s}_k)$ then

(2) $\boldsymbol{\Sigma}$ *is consistent iff* $\boldsymbol{\varphi} \in \boldsymbol{\Omega}$.

It is natural to regard $\boldsymbol{\varphi}$ as the $\mathscr{L}$-sentence asserting the consistency of $\boldsymbol{\Sigma}$.

Next, we observe that the statement that $\neg\gamma(\mathbf{s}_{\#\gamma}) \notin \boldsymbol{\Sigma}$ means that $P(\#\gamma)$ does not hold. Since $\boldsymbol{\alpha}$ represents P in $\boldsymbol{\Omega}$, we have

(3) $\neg\gamma(\mathbf{s}_{\#\gamma}) \notin \boldsymbol{\Sigma}$ *iff* $\neg\boldsymbol{\alpha}(\mathbf{s}_{\#\gamma}) \in \boldsymbol{\Omega}$.

From (2) and (3) it follows that (1) is equivalent to the statement that the sentence

(4) $\boldsymbol{\varphi} \rightarrow \neg\boldsymbol{\alpha}(\mathbf{s}_{\#\gamma})$

is true, i.e., belongs to $\boldsymbol{\Omega}$. Since we have actually established (1), we conclude that the sentence (4) in fact belongs to $\boldsymbol{\Omega}$.

It turns out that (4) belongs not only to $\boldsymbol{\Omega}$ but to $\boldsymbol{\Pi}$. This can be established by analysing in detail the whole chain of steps in the informal proof of (1) and showing that this proof can be formalized as a deduction of (4) from the postulates of $\boldsymbol{\Pi}$. (This is made possible by the great strength of these postulates.) This detailed analysis is beyond the scope of our book, and we ask reader to accept on trust the fact that (4) belongs to $\boldsymbol{\Pi}$.

Next, let m be any number. Inspecting the formulas $\boldsymbol{\alpha}$, $\boldsymbol{\beta}$ and γ (see proof of Lemma 7.9) we easily see that $\neg\boldsymbol{\alpha}(\mathbf{s}_m) \vdash \neg\exists \mathbf{y}\boldsymbol{\beta}(\mathbf{s}_m)$ and hence

$\neg\alpha(s_m) \vdash \neg\gamma(s_m)$. Using this for $m = \#\gamma$, and the fact that (4) belongs to $\mathbf{\Pi}$, we see that

$$(5) \qquad (\varphi \to \neg\gamma(s_{\#\gamma})) \in \mathbf{\Pi}.$$

Now, we have assumed that $\mathbf{\Pi} \subseteq \mathbf{\Sigma}$. If $\mathbf{\Sigma}$ is consistent, then φ cannot belong to $\mathbf{\Sigma}$, because then by (5) we would get $\neg\gamma(s_{\#\gamma}) \in \mathbf{\Sigma}$, contradicting (1). Thus we have:

11.9. SECOND INCOMPLETENESS THEOREM. *Let $\mathbf{\Sigma}$ be an axiomatizable theory that includes first-order Peano arithmetic. If $\mathbf{\Sigma}$ is consistent, then the $\mathscr{L}$-sentence asserting this is not in $\mathbf{\Sigma}$.* ∎

This result can be extended to $\mathscr{L}'$-theories $\mathbf{\Sigma}'$ in languages other than $\mathscr{L}$, provided that appropriate "translations" or counterparts of the postulates of $\mathbf{\Pi}$ are in $\mathbf{\Sigma}'$. Speaking somewhat loosely, we may therefore say that if $\mathbf{\Gamma}$ is an axiomatic theory (in any formal language) which is at least as strong as $\mathbf{\Pi}$ and if $\mathbf{\Gamma}$ is consistent, then this fact cannot be deduced in $\mathbf{\Gamma}$.

This constitutes a grave difficulty for the formalist view of mathematics. According to the formalist doctrine (as embodied, e.g., in the Hilbert Programme), classical mathematics — more precisely, those parts of it that refer to infinite collections — cannot be justified by virtue of its alleged *meaning* but must be taken to be an axiomatic *formal* theory (or perhaps a *collection* of such formal theories).

In order to be sure that a formal theory will not lead us to contradiction, we need to prove its consistency. And one will only be convinced by a proof that *can* be justified by virtue of the meaning of the concepts used in it.

But by the Second Incompleteness Theorem it turns out that if classical mathematics is taken to include, e.g., first-order Peano arithmetic (and there is very little left if it is *not* included), then any consistency proof for such an axiomatic formal theory is — at least in some sense — less elementary that the methods which the formal theory formalizes. It would seem that the consistency proof itself is at least as much in need of justification as the formal theory whose consistency is being proved.

A conceivable way out of this dilemma would be to demarcate the boundary — between methods of proof that are supposed to be convincing by virtue of their meaning, and methods that are regarded merely as formal manipulations unjustified by meaning — in such a way that the methods needed for a consistency proof for classical mathematics would fall just on the "right" side of the boundary. The Second Incompleteness Theorem implies that the task of inventing a satisfactory demarcation of this kind is tremendously difficult. But it is not hopeless.

*11.10. PROBLEM. (i) Show that there are $2^{\aleph_0}$ countable models of $\mathbf{\Pi}$, no two of which are elementarily equivalent. (Using the First Incompleteness Theorem, get for each $f \in 2^N$, i.e., for each infinite sequence of 0's and 1's, a consistent theory $\mathbf{\Sigma}_f$ such that $\mathbf{\Pi} \subseteq \mathbf{\Sigma}_f$ and such that if $f \neq f'$ then there is a sentence $\boldsymbol{\varphi}$ for which $\boldsymbol{\varphi} \in \mathbf{\Sigma}_f$ and $\neg \boldsymbol{\varphi} \in \mathbf{\Sigma}_{f'}$.)

(ii) Show that there are $2^{\aleph_0}$ pointwise definable models of $\mathbf{\Pi}$ no two of which are elementarily equivalent. (Use (i) and Prob. 9.17.) This result is in sharp contrast with the last result of Prob. 9.17.

§ 12. Historical and bibliographical remarks

The existence of non-standard models of $\mathbf{\Omega}$ was proved by SKOLEM [1934]. He constructed such a model using a method very much like that of Prob. 3.13.

A rigorous semi-formal characterization of the natural numbers was first given by DEDEKIND [1888].

PEANO [1889] proposed his famous postulates, formalized in a language very much like the language $\mathscr{L}'$ of §2.

Results which include Thm. 4.7 and Prob. 4.10 were contained in a paper published by Tarski in Polish in 1933, of which TARSKI [1935] is a German translation. (English translation in TARSKI [1956].)

The undecidability of the predicate calculus (Thm. 10.7) is due to CHURCH [1936a]. For other undecidability results see TARSKI, MOSTOWSKI and ROBINSON [1953] and DAVIS [1965]. Additional results on undecidability, as well as positive solutions of various decision problems, abound in the literature and are too numerous to be listed here.

The results of §11 are due to GÖDEL [1931]. For his First Incompleteness Theorem, Gödel required the theory in question to be ω-consistent, i.e., if a sentence $\neg \forall \mathbf{x} \boldsymbol{\alpha}$ belongs to the theory, then for some number n the sentence $\boldsymbol{\alpha}(\mathbf{x}/\mathbf{s}_n)$ does not belong to the theory. This requirement is strictly stronger than consistency. ROSSER [1936] gave a modified proof, in which only consistency is assumed.

A proof of the Second Incompleteness Theorem can be found in FEFERMAN [1960].

CHAPTER 8

RECURSION THEORY (CONTINUED)

In this chapter we shall discuss a selection of topics in recursion theory, most of which arise or acquire their significance from ideas studied in Ch. 7. Thus, while the bulk of this chapter can be understood by a reader who has only read Ch. 6, a proper appreciation of the material requires some knowledge of Ch. 7.

§ 1. The arithmetical hierarchy

In this section *relation* will mean a second-order relation in the sense of §6 of Ch. 6. Recall that first-order relations are also included, as a special case.

We define the class of *arithmetical relations* as the smallest class that contains all recursive relations and is closed under the logical operations, i.e., the propositional operations (defined in Ex. 6.6.6) as well as (unbounded) existential and universal quantification (defined at the beginning of §11 of Ch. 6).

For first-order relations this definition is equivalent to Def. 7.3.2, in view of Thm. 7.3.7.

There is a neat and useful classification of arithmetical relations, based on the following:

1.1. DEFINITION. By induction on m we define classes Σ^0_m, Π^0_m and Δ^0_m as follows:

Σ^0_0 is the class of all recursive relations;

for all m, Π^0_m consists of all relations which are negations of relations belonging to Σ^0_m;

for all m, $\Delta^0_m = \Sigma^0_m \cap \Pi^0_m$;

for all m, Σ^0_{m+1} consists of all relations obtained by an existential quantification from relations belonging to Π^0_m.

The collection of all classes Σ^0_m, Π^0_m and Δ^0_m is the *arithmetical hierarchy*.

The superscript "0" in the above notation is designed to distinguish between the arithmetical hierarchy and other hierarchies (whose classes are denoted similarly, but with different superscripts) which are studied in recursion theory and other branches of mathematics.

Evidently, a relation belonging to any class in the arithmetical hierarchy is arithmetical. We shall soon see that the converse is also true.

From Def. 1.1 it is clear that $\Pi^0_0 = \Sigma^0_0$, since a relation is recursive iff its negation is recursive. Also, the first-order relations belonging to Σ^0_1 are precisely the r.e. relations.

Further, since

$$\neg \exists y Q(\xi;\mathfrak{x},y) = \forall y \neg Q(\xi;\mathfrak{x},y),$$

it follows that, for all m, Π^0_{m+1} consists of all relations obtainable by a universal quantification from relations in Σ^0_m.

Thus, if in each clause of Def. 1.1 we interchange the symbols "Σ" and "Π" and the words "existential" and "universal", the resulting statement is true. The same holds for any theorem deduced from Def. 1.1. This fact is *the principle of duality* for the arithmetical hierarchy. (We do not pursue this matter in rigorous detail because we shall only use duality in an informal way.)

By induction on m it is now easy to show that an n-ary relation P is in Σ^0_m iff the identity

$$P(\xi;\mathfrak{x}) = \exists y_1 \forall y_2 \exists y_3 \ldots R(\xi;\mathfrak{x},y_1,\ldots,y_m)$$

holds for some recursive relation R. Here the quantifiers alternate between existential and universal, beginning (on the left) with the former.

The dual of the above statement yields a general form of Π^0_m relations.

We shall now show that the arithmetical classes enjoy certain closure properties.

1.2. LEMMA. *Let $H_1,\ldots,H_k$ be total recursive n-ary functionals. If*

$$Q(\xi;\mathfrak{x}) = P(\xi;H_1(\xi;\mathfrak{x}),\ldots,H_k(\xi;\mathfrak{x})),$$

and P belongs to some class in the arithmetical hierarchy, then Q belongs to the same class.

PROOF. For the class Σ^0_0 $(=\Pi^0_0)$ the lemma is obvious. For Σ^0_m and Π^0_m with $m>0$, the result follows by induction on m. The result for Δ^0_m is then immediate from the definition of this class. ∎

1.3. **Problem.** Show that if

$$Q(\xi;\mathfrak{x}, z_1,\ldots,z_m)=P(\lambda y H(\xi;y, z_1,\ldots,z_m);\mathfrak{x}),$$

where H is a total recursive functional, and if P belongs to some class in the arithmetical hierarchy, then Q belongs to the same class. (For the classes Σ^0_m and Π^0_m proceed by induction on m. Use Prob. 6.5.4 or Prob. 6.8.7.)

1.4. **Lemma.** *For all m, the class* Σ^0_{m+1} *is closed under existential quantification. Dually,* Π^0_{m+1} *is closed under universal quantification.*
Proof. Similar to the proof of part (2) of Lemma 6.11.2. ∎

1.5. **Lemma.** *Every class in the arithmetical hierarchy is closed under disjunction, conjunction and bounded quantification (both existential and universal). For every m, the class* Δ^0_m *is closed under negation, and hence under all propositional operations.*
Proof. For Σ^0_m and Π^0_m proceed by induction on m. The case $m=0$ is covered by Ex. 6.6.6 and Ex. 6.6.9. In the induction step, for Σ^0_{m+1} argue as in the proof of Lemma 6.11.2, and for Π^0_{m+1} by duality. For Δ^0_m the results then follow by the definition of this class. ∎

1.6. **Lemma.** $\Sigma^0_m \cup \Pi^0_m \subseteq \Delta^0_{m+1}$ *for all m.*
Proof. By induction on m. For $m=0$ we observe that

$$\begin{aligned} P(\xi;\mathfrak{x}) &= \exists y[P(\xi;\mathfrak{x}) \wedge (y=y)] \\ &= \forall y[P(\xi;\mathfrak{x}) \wedge (y=y)], \end{aligned}$$

so that, if P is recursive, it belongs to both Σ^0_1 and Π^0_1, hence to Δ^0_1.

For the induction step, let $P\in\Sigma^0_{m+1}$. Then $P(\xi;\mathfrak{x})=\exists y Q(\xi;\mathfrak{x},y)$, where $Q\in\Pi^0_m$. By the induction hypothesis we have $Q\in\Delta^0_{m+1}\subseteq\Pi^0_{m+1}$, hence $P\in\Sigma^0_{m+2}$. Also, $\lambda y(y=y)$ is recursive, i.e., belongs to Σ^0_0; but by the induction hypothesis

$$\Sigma^0_0\subseteq\Delta^0_1\subseteq\Sigma^0_1\subseteq\Delta^0_2\subseteq\ldots\subseteq\Sigma^0_{m+1}.$$

Since

$$P(\xi;\mathfrak{x})=\forall y[P(\xi;\mathfrak{x})\wedge(y=y)],$$

it follows from Lemma 1.5 that $P\in\Pi^0_{m+2}$. Thus $P\in\Sigma^0_{m+2}\cap\Pi^0_{m+2}=\Delta^0_{m+2}$, and we have shown that $\Sigma^0_{m+1}\subseteq\Delta^0_{m+2}$. Dually, also $\Pi^0_{m+1}\subseteq\Delta^0_{m+2}$. ∎

In the sequel we shall often make implicit use of the results obtained so far in this section.

We can now see that any arithmetical relation belongs to Δ^0_m for sufficiently large m. The recursive relations constitute the class $\Sigma^0_0 = \Sigma^0_0 \cap \Pi^0_0 = \Delta^0_0$. If P is in Δ^0_m, then $\neg P$ too is in Δ^0_m; and if P' is in $\Delta^0_{m'}$ then both P and P' are in $\Delta^0_{\max(m,m')}$ so that the disjunction $P \vee P'$ also belongs to this class. The other propositional operations can be reduced to negation and disjunction. Finally, both universal and existential quantification map Δ^0_m into Δ^0_{m+1}.

Thus the arithmetical hierarchy is a cumulative classification of all arithmetical relations.

We shall now obtain a convenient canonical form for relations in Σ^0_{m+1} and in Π^0_{m+1}.

1.7. ENUMERATION THEOREM. *Let P be an n-ary relation. Then P belongs to Σ^0_{m+1} iff there is a number e such that*

$$P(\xi;\mathfrak{x}) = \exists y_1 \forall y_2 \ldots \forall y_m \exists y_{m+1} T_{n+m}(\xi;\, e, \mathfrak{x}, y_1, \ldots, y_{m+1})$$

or

$$P(\xi;\mathfrak{x}) = \exists y_1 \forall y_2 \ldots \exists y_m \forall y_{m+1} \neg T_{n+m}(\xi;\, e, \mathfrak{x}, y_1, \ldots, y_{m+1}),$$

according as m is even or odd. Similarly for Π^0_{m+1}, interchanging "$\exists$" with "$\forall$" and "T_{n+m}" with "$\neg T_{n+m}$".

PROOF. It is enough to prove the result for Σ^0_1, since the other cases then follow easily by induction on m, using Def. 1.1.

In Cor. 6.11.6 we have already got the required result for *first-order* relations. (For, the r.e. relations are clearly the same as the first-order relations belonging to Σ^0_1.) We shall use the same method in the present proof.

First, observe that if

(1) $$P(\xi;\mathfrak{x}) = \exists y T(\xi; e, \mathfrak{x}, y)$$

then $P \in \Sigma^0_1$ because T_n is a p.r. relation, and hence belongs to Π^0_0.

Conversely, suppose that $P \in \Sigma^0_1$. Then

(2) $$P(\xi;\mathfrak{x}) = \exists y R(\xi; \mathfrak{x}, y),$$

where R is a recursive relation. Put

(3) $$F(\xi;\mathfrak{x}) = \mu y R(\xi; \mathfrak{x}, y).$$

Then F is clearly a recursive functional and hence by the NFT (Thm. 6.10.1)

F has an index e, for which

(4) $$F(\xi;\mathfrak{x})=U(\mu yT(\xi;e,\mathfrak{x},y)).$$

From (2), (3) and (4) we easily get (1). ∎

Put $Q(\xi;\mathfrak{x},z)=\exists yT(\xi;z,\mathfrak{x},y)$. Then Q is clearly in Σ_1^0. By Thm. 1.7, Q *enumerates* all n-ary relations in Σ_1^0, in the sense that as e runs through N, the relation P defined by (1) in the above proof runs through all n-ary relations in Σ_1^0.

Similarly, for each n and each of the classes Σ_{m+1}^0 and Π_{m+1}^0, Thm. 1.7 gives us an $(n+1)$-ary relation belonging to the class in question, which enumerates all n-ary relations in this class. Hence the name of the theorem.

1.8. PROBLEM. Show that, for all n and m, there does *not* exist an $(n+1)$-ary Δ_m^0 relation which enumerates all n-ary Δ_m^0 relations. In particular, since $\Sigma_0^0=\Pi_0^0=\Delta_0^0$, Thm. 1.7 cannot be extended to Σ_0^0 and Π_0^0.

1.9. HIERARCHY THEOREM. *For every m we can find a relation belonging to Σ_{m+1}^0 but not to Π_{m+1}^0 (hence its negation belongs to Π_{m+1}^0 but not to Σ_{m+1}^0).*

PROOF. Thm. 1.7 gives us a binary Σ_{m+1}^0 relation Q which enumerates all unary Σ_{m+1}^0 relations. Put

$$P(\xi;x)=Q(\xi;x,x).$$

Then P is in Σ_{m+1}^0 by Lemma 1.2. On the other hand, if P were in Π_{m+1}^0 then $\neg P$ would be in Σ_{m+1}^0, so for some e we would have the identity

$$\neg Q(\xi;x,x)=Q(\xi;x,e).$$

For $x=e$ this yields a contradiction. ∎

The Hierarchy Theorem is so called because it shows that the cumulative arithmetical hierarchy is *non-degenerate*, in the sense that we get something new each time we go up the hierarchy. For, if P is in Σ_{m+1}^0 but not in Π_{m+1}^0 then $P\notin\Delta_{m+1}^0$ and hence $P\notin\Sigma_m^0\cup\Pi_m^0$.

The rest of this section is devoted to an important characterization of Δ_{m+1}^0 in terms of Σ_m^0 and Π_m^0. We begin with:

1.10. DEFINITION. Let Φ be any class of functionals. The *recursive closure* of Φ (briefly, $\mathscr{R}\Phi$) is the class of all functionals obtainable from the functionals of Φ and the basic functionals Z, S, A and $I_{n,i}$ (for all n and i such that $1\leqslant i\leqslant n$) by a finite number of applications of the basic operations:

composition, primitive recursion and minimization[1]. If $\mathscr{R}\Phi=\Phi$, we say that Φ is *recursively closed.*

1.11. REMARK. The following facts are easily established and will be used without special mention:

(i) $\mathscr{R}\emptyset$ is the class of recursive functionals.
(ii) $\mathscr{R}\Phi$ is the smallest recursively closed class that includes Φ.
(iii) If $\Psi\subseteq\mathscr{R}\Phi$, then $\mathscr{R}\Psi\subseteq\mathscr{R}\Phi=\mathscr{R}(\Psi\cup\Phi)$.
(iv) $F\in\mathscr{R}\Phi$ iff $F\in\mathscr{R}\Phi_0$ for some *finite* $\Phi_0\subseteq\Phi$.

1.12. LEMMA. *$\Delta^0_{m+1}\subseteq\mathscr{R}\Pi^0_m$ for all m.*
PROOF. Let P be an n-ary Δ^0_{m+1} relation. Then P is both Σ^0_{m+1} and Π^0_{m+1}. In other words, both P and $\neg P$ are Σ^0_{m+1}. Thus

$$P(\xi;\mathfrak{x})=\exists yQ(\xi;\mathfrak{x},y),$$
$$\neg P(\xi;\mathfrak{x})=\exists yR(\xi;\mathfrak{x},y),$$

where Q and R are Π^0_m. We easily verify (as in the proof of Thm. 6.11.4) that

$$P(\xi;\mathfrak{x})=Q(\xi;\mathfrak{x},\mu y[Q(\xi;\mathfrak{x},y)\vee R(\xi;\mathfrak{x},y)]).$$

Thus P is obtained by recursive operations from Q and R, and hence belongs to $\mathscr{R}\Pi^0_m$. ∎

1.13. THEOREM. $\Delta^0_1=\Pi^0_0$ $(=\Sigma^0_0=\Delta^0_0)$.
PROOF. Since Π^0_0 $(=\Sigma^0_0)$ is the set of recursive relations, $\Pi^0_0\subseteq\mathscr{R}\emptyset$; therefore $\mathscr{R}\Pi^0_0\subseteq\mathscr{R}\emptyset$, so by Lemma 1.12 all Δ^0_1 relations are recursive. Thus $\Delta^0_1\subseteq\Pi^0_0$. By Lemma 1.6, $\Delta^0_1=\Pi^0_0$. ∎

1.14. LEMMA. *Let H be a total unary functional. Let G be any n-ary functional obtained from H and the basic functionals Z,S and the $I_{m,i}$ (but not A) by a finite number of applications of the basic operations: composition, primitive recursion and minimization. Then for some n-ary recursive functional F,*

$$G(\xi;\mathfrak{x})=F(\lambda yH(\xi;y);\mathfrak{x}).$$

PROOF. By induction on the total number k of applications of the basic operations made in order to obtain G from H, Z, S and the $I_{m,i}$.

For $k=0$, we must show that if G is H, Z, S or $I_{n,i}$ then the conclusion of our lemma is correct. For H, this follows from the identity

$$H(\xi;x)=A(\lambda yH(\xi;y);x).$$

[1] For the definition of these basic functionals and operations, see Def. 6.5.2.

The cases of Z, S and the $I_{n,i}$ are trivial. (E.g., for S we have $S(\xi;x) = S(\lambda y H(\xi;y);x)$.)

In the induction step we must show that if G is obtained by one application of composition, primitive recursion or minimization from functionals satisfying the conclusion of our lemma, then G has this property as well. This is left as a simple exercise to the reader. ∎

1.15. LEMMA. *Let P be an n-ary relation. If $P \in \mathscr{R}\Pi^0_m$, then $P \in \Delta^0_{m+1}$.*

PROOF. Since $P \in \mathscr{R}\Pi^0_m$, it follows that P is in the recursive closure of some finite subset of Π^0_m. Thus $P \in \mathscr{R}\{Q_0,\ldots,Q_k\}$, where the Q_i are Π^0_m relations. Now, let Q' be defined by the identity

$$Q'(\xi;x,y) = (A(\xi;x) = y).$$

Then Q' is clearly a recursive relation, and so certainly belongs to Π^0_m. Therefore we assume without loss of generality that Q' is one of the Q_i.

It follows that P is obtainable *via* the three basic operations from the Q_i and the basic functionals *other than A*. For, A itself can be obtained from Q' by minimization:

$$A(\xi;x) = \mu y Q'(\xi;x,y),$$

and Q' is one of the Q_i.

Next, the Q_i may be replaced by *unary* Π^0_m relations. Suppose, e.g., that Q_i is binary. Put

$$M_i(\xi;x) = Q_i(\xi;(x)_0, (x)_1).$$

Then M_i is a unary Π^0_m relation. Also, Q_i may be recovered from M_i by composing the latter with a p.r. function:

$$Q_i(\xi;x,y) = M_i(\xi;2^x \cdot 3^y).$$

A similar device can be used if Q_i is ternary, etc.

Thus P can be obtained *via* the three basic operations from unary Π^0_m relations $M_0,\ldots,M_k$ and the basic functionals (excluding A). Let us put

$$M(\xi;x) = \exp(p_0, M_0(\xi;x))\ldots\exp(p_k, M_k(\xi;x)).$$

Then each M_i can be obtained by composing a p.r. function with the functional M:

$$M_i(\xi;x) = (M(\xi;x))_i.$$

Therefore P can be obtained *via* the three basic operations from M and the basic functionals (excluding A). By Lemma 1.14 it follows that

$$P(\xi;\mathfrak{x})=F(\lambda y M(\xi;y);\mathfrak{x}),$$

where F is some recursive functional. Let e be an index of F. By the NFT we have

$$P(\xi;\mathfrak{x})=U(\mu y T^*(\prod_{t\leqslant y}\exp(p_t,M(\xi;t)),e,\mathfrak{x},y)).$$

Hence it is easy to verify the identities

(1) $$P(\xi;\mathfrak{x})=\exists y\,\exists z[K(\xi;y,z)\wedge T^*(z,e,\mathfrak{x},y)\wedge(U(y)=0)],$$

(2) $$P(\xi;\mathfrak{x})=\forall y\,\forall z[K(\xi;y,z)\wedge T^*(z,e,\mathfrak{x},y)\rightarrow(U(y)=0)],$$

where $K(\xi;y,z)=(\prod_{t\leqslant y}\exp(p_t,M(\xi,t))=z)$.

If we can prove that the relation K is Δ^0_{m+1}, then (1) shows that P is obtained by existential quantifications from a Σ^0_{m+1} relation, hence P is Σ^0_{m+1}. Similarly, (2) shows that P is Π^0_{m+1}. Hence P is Δ^0_{m+1}.

To prove that K is Δ^0_{m+1} we observe that

(3) $$K(\xi;y,z)=(\mathrm{lh}(z)\leqslant y+1)\wedge\forall t\leqslant y\,[(z)_t=M(\xi;t)].$$

Let us put

$$R(\xi;t,u)=(u=M(\xi;t)).$$

Clearly, if we can prove that the relation R is Δ^0_{m+1} then (3) shows that K is Δ^0_{m+1} as well.

To show that R is Δ^0_{m+1} we observe that by the definition of M

$$R(\xi;t,u)=(\mathrm{lh}(u)\leqslant k+1)\wedge((u)_0=M_0(\xi;t))\wedge\ldots\wedge((u)_k=M_k(\xi;t)).$$

Thus to prove that R is Δ^0_{m+1} it is enough to show that the relations $R_0,\ldots,R_k$ defined by the identities

$$R_i(\xi;t,v)=(v=M_i(\xi;t))\quad\text{for } i=0,\ldots,k$$

are Δ^0_{m+1}. However, from this definition of the R_i it follows that

$$R_i(\xi;t,v)=(v=0)\wedge M_i(\xi;t)\vee(v=1)\wedge\neg M_i(\xi;t).$$

Since the M_i are Π^0_m, hence Δ^0_{m+1}, this shows that the R_i are Δ^0_{m+1} as well. ∎

1.16. THEOREM. *A relation belongs to* Δ^0_{m+1} *iff it belongs to* $\mathscr{R}(\Sigma^0_m\cup\Pi^0_m)$.

PROOF. Observe that $\Sigma^0_m\subseteq\mathscr{R}\Pi^0_m$, since any Σ^0_m relation can be got by negation from a Π^0_m relation. Hence $\mathscr{R}(\Sigma^0_m\cup\Pi^0_m)=\mathscr{R}\Pi^0_m$, and our theorem follows at once from Lemmas 1.12 and 1.15. ∎

*§ 2. A result concerning T_Ω

The ideas of §1 can easily be extended to cover relations with more than one sequence argument (cf. §9 of Ch. 6). Thus, it makes sense to say that an (m,n)-ary relation (a relation having m sequence arguments and n numerical arguments) is arithmetical.

Now, we introduce two new operations: existential and universal quantification over a *sequence variable* (rather than over a numerical variable). These operations are defined in the obvious way.

Using these new operations, one defines classes Σ_k^1, Π_k^1 and Δ_k^1 for all $k>0$, as follows: Σ_1^1 is the class of all (m,n)-ary relations obtained from arithmetical relations by one existential quantification over a sequence variable. For all $k>0$, Π_k^1 is the class of all negations of relations in Σ_k^1; and Σ_{k+1}^1 is the class of relations obtained from Π_k^1 relations by one existential quantification over a sequence variable. Finally, $\Delta_k^1=\Sigma_k^1\cap\Pi_k^1$.

The classes Σ_k^1, Π_k^1 and Δ_k^1 constitute the *analytical hierarchy*.

The class Δ_1^1 is of special importance; the role it plays in the analytical hierarchy is analogous to that played by Δ_1^0 (which, by Thm. 1.13, is the class of recursive relations) in the arithmetical hierarchy. A relation belonging to Δ_1^1 is said to be *hyperarithmetical.*

We shall not study the analytical hierarchy in this book. But we shall now sketch a proof that the property T_Ω defined in Ch. 7 is hyperarithmetical.

We shall use the same terminology and notation as in Ch. 7. In particular, recall that $\mathscr{L}$ is the first-order language of arithmetic, $\mathbf{\Omega}$ is the set of all $\mathscr{L}$-sentences which are true (i.e., hold in the standard structure $\mathfrak{N}$) and T_Ω is the property of being a SENTENCE of $\mathbf{\Omega}$.

Let $\mathbf{\Phi}=\{\varphi_i\colon i\in N\}$ be a sequence of $\mathscr{L}$-sentences. We say that $\mathbf{\Phi}$ is a *truth sequence* if the following six conditions hold:

(i) If $\neg\neg\varphi\in\mathbf{\Phi}$, then $\varphi\in\mathbf{\Phi}$.

(ii) If $(\varphi\rightarrow\psi)\in\mathbf{\Phi}$, then $\neg\varphi\in\mathbf{\Phi}$ or $\psi\in\mathbf{\Phi}$.

(iii) If $\neg(\varphi\rightarrow\psi)\in\mathbf{\Phi}$, then $\varphi\in\mathbf{\Phi}$ and $\neg\psi\in\mathbf{\Phi}$.

(iv) If $\forall\mathbf{x}\alpha\in\mathbf{\Phi}$, then $\alpha(\mathbf{x}/\mathbf{s}_n)\in\mathbf{\Phi}$ for all n.

(v) If $\neg\forall\mathbf{x}\alpha\in\mathbf{\Phi}$, then $\neg\alpha(\mathbf{x}/\mathbf{s}_n)\in\mathbf{\Phi}$ for some n.

(vi) If φ belongs to $\mathbf{\Phi}$ and has the form $\mathbf{r}=\mathbf{t}$ or $\mathbf{r}\neq\mathbf{t}$ (where $\mathbf{r}$ and $\mathbf{t}$ are closed terms), then φ is true (i.e., $\varphi\in\mathbf{\Omega}$).

We now define a (1,0)-ary relation Ts as follows: Ts (ξ) holds iff there is a truth sequence $\{\varphi_i\colon i\in N\}$ such that $\xi(i)=\#\varphi_i$ for all i.

It is then an easy matter to verify that Ts is arithmetical. (In connection with clause (vi) of the definition of *truth sequence,* note that while the property T_Ω of being a TRUE SENTENCE is not arithmetical, the property of being a TRUE SENTENCE of the form $\mathbf{r}{=}\mathbf{t}$ or $\mathbf{r}{\neq}\mathbf{t}$ is recursive.)

Moreover, it is easy to show by induction on deg $\boldsymbol{\varphi}$ that $\boldsymbol{\varphi}\in\boldsymbol{\Omega}$ iff $\boldsymbol{\varphi}$ belongs to some truth sequence. Thus

$$(1) \qquad T_\Omega(x) = \exists\xi\, \exists y\, [\mathrm{Ts}\,(\xi) \wedge (\xi(y)=x)].$$

Also, a sentence $\boldsymbol{\varphi}$ belongs to $\boldsymbol{\Omega}$ iff $\neg\boldsymbol{\varphi}\notin\boldsymbol{\Omega}$. Hence

$$(2) \qquad T_\Omega(x) = \forall\xi\, \forall y\, [\mathrm{Frm}\,(x,0) \wedge \{\neg\mathrm{Ts}\,(\xi) \vee (\xi(y) \neq 32\cdot 3^x)\}].$$

From (1) and (2) it follows at once that T_Ω is hyperarithmetical.

§ 3. Encoded theories

Many of the results proved in Ch. 7 involve both logic and recursion theory. We shall now set up a framework which generalizes and abstracts from the particular logical setting of Ch. 7, and in which we can formulate and prove generalized and rather abstract versions of some of those results.

3.1. DEFINITION. An *encoded theory* is a triple $\langle A,\varphi,f\rangle$ where $A\subseteq N$, φ is a recursive sequence and f is a binary total recursive function.

If $\boldsymbol{\Sigma}$ is a theory in the sense of Ch. 7 (a deductively closed set of sentences in the first-order language of arithmetic), then the corresponding encoded theory is $\langle A,\varphi,f\rangle$, where

$$A = \{x\colon\ T_\Sigma(x) \text{ holds}\} = \#[\Sigma],$$

$$\varphi(x) = 32\cdot 3^x,$$

$$f(x,y) = \mathrm{sb}\,(x,1,y).$$

Thus A is the set of all code numbers of sentences in $\boldsymbol{\Sigma}$ (see Def. 7.5.1); and if x is a code number of a formula $\boldsymbol{\alpha}$ then $\varphi(x)$ is the code number of the negation of $\boldsymbol{\alpha}$ (see §4 of Ch. 7) and $f(x,y)$ is the code number of the formula obtained from $\boldsymbol{\alpha}$ be substituting the yth numeral for the first variable (see Ex. 7.4.5).

More generally, we may consider an arbitrary formal language $\mathscr{L}$ (which may or may not be a first-order language). We suppose that $\mathscr{L}$ has certain objects called "expressions" which are coded by natural numbers. Let certain expressions be designated as "provable" (or "true"); let A be the set of code numbers of all such expressions. Suppose also that we have a recursive sequence φ and a binary total recursive function f such that whenever x is the code number of an expression α then $\varphi(x)$ and $f(x,y)$ are code numbers of expressions which in some appropriate sense can be regarded as the "negation of α" and the "y^{th} instance of α" respectively. Then $\langle A,\varphi,f\rangle$ is the encoded theory corresponding to this situation.

(In many results concerning encoded theories $\langle A,\varphi,f\rangle$ the sequence φ plays no role whatsoever. Such results can be applied to languages which have no negation.)

We shall need the following definitions.

3.2. Definition. An encoded theory $\langle A,\varphi,f\rangle$ is *consistent* if $A\cap\varphi[A]=\emptyset$, *recursively decidable* if A is recursive, *axiomatizable* if A is an r.e. set[1]. An *extension* of an encoded theory $\langle A,\varphi,f\rangle$ is any encoded theory $\langle A',\varphi,f\rangle$ with $A\subseteq A'$.

3.3. Definition. Let $\langle A,\varphi,f\rangle$ be an encoded theory and let $M\subseteq N$. Then M is *weakly representable* in $\langle A,\varphi,f\rangle$ if there is a number n such that the identity

$$(f(n,x)\in A)=(x\in M)$$

holds. M is *strongly representable* in $\langle A,\varphi,f\rangle$ if there is a number n such that for every x

$$f(n,x)\in A \quad \text{if} \quad x\in M,$$

$$\varphi f(n,x)\in A \quad \text{if} \quad x\notin M.$$

(Cf. Def. 7.3.1.)

In what follows, if $M\subseteq N$ then we put $M^{c}=N-M$.

Generalized versions of several results of Ch. 7 can be obtained as corollaries to the following:

3.4. Theorem. *Let $\langle A,\varphi,f\rangle$ be an encoded theory in which every Σ^0_m (or every Π^0_m) set of numbers is weakly representable. Then A is not in Π^0_m (not in Σ^0_m, respectively).*

Proof. Suppose A is in Π^0_m. Then clearly A^{c} is in Σ^0_m and by 1.2 the set

$$M=\{x\colon f(x,x)\in A^{c}\}$$

must also be in Σ^0_m. If every Σ^0_m set is weakly representable in the encoded theory, then for some number n we have

$$(f(n,x)\in A)=(x\in M).$$

[1] Cf. Thm. 7.5.4.

Using the definition of M and taking n as the value of x we have $f(n,n) \in A$ iff $f(n,n) \in A^c$, which is absurd.

The same argument works also when "Σ^0_m" and "Π^0_m" are interchanged. ∎

In particular, if every arithmetical set of numbers is weakly representable in $\langle A,\varphi,f\rangle$, then by Thm. 3.4 we see that $A \notin \Pi^0_m$ for *all m*. Thus by what we have shown in §1, A is not arithmetical. This is a generalized abstract version of Thm. 7.4.7.

For $m=0$ Thm. 3.4 says that if every recursive set of numbers is weakly representable in an encoded theory, then the theory is recursively undecidable. This is a generalized form of Thm. 7.10.1.

3.5. Problem. Let $\langle A,\varphi,f\rangle$ be a consistent encoded theory in which every Σ^0_m (or every Π^0_m) set of numbers is strongly representable. Show that $\varphi^{-1}[A]$ is not in Σ^0_m (not in Π^0_m, respectively).

§ 4. Inseparable pairs of sets

Two sets A and B of natural numbers are *recursively separable* if there are disjoint recursive sets C and D such that $A \subseteq C$ and $B \subseteq D$. If such C and D exist, then without loss of generality we may assume that $D = C^c$. Thus A and B are recursively separable iff there is a recursive set C that includes A and is disjoint from B. If A and B are not recursively separable, we say that they are *recursively inseparable.*

The notion of recursive separability arises in a natural way in connection with encoded theories:

4.1. Theorem. *An encoded theory $\langle A,\varphi,f\rangle$ has a recursively decidable and consistent extension iff A and $\varphi[A]$ are recursively separable.*

Proof. If $\langle A',\varphi,f\rangle$ is a recursively decidable and consistent extension of $\langle A,\varphi,f\rangle$ then from Def. 3.2 we see at once that A' is a recursive set that includes A and is disjoint from $\varphi[A']$, hence also from $\varphi[A]$.

Conversely, if C is a recursive set that includes A and is disjoint from $\varphi[A]$, put

$$A' = \{x\colon\ x \in C,\ \varphi(x) \notin C\}.$$

It is easy to see that $\langle A',\varphi,f\rangle$ is a recursively decidable consistent extension of $\langle A,\varphi,f\rangle$. ∎

It is easy to see that if A and $\varphi[A]$ are recursively separable, then so are A and $\varphi^{-1}[A]$. In view of this and Thm. 4.1, the following theorem can be regarded as a generalized version of Cor. 7.10.3.

4.2. THEOREM. *Let $\langle A,\varphi,f\rangle$ be an encoded theory in which every recursive set of numbers is strongly representable. Then A and $\varphi^{-1}[A]$ are recursively inseparable.*

PROOF. Let C be a set of numbers including A and disjoint from $\varphi^{-1}[A]$. It follows from Def. 3.3 that every set which is strongly representable in $\langle A,\varphi,f\rangle$ is weakly representable in $\langle C,\varphi,f\rangle$. But then by Thm. 3.4 (with $m=0$) we see that C cannot be recursive. ∎

4.3. REMARK. Thm. 4.2 is of interest only if the encoded theory $\langle A,\varphi,f\rangle$ is consistent — otherwise, the sets A and $\varphi^{-1}[A]$ intersect and are trivially recursively inseparable. Note also that if $\langle A,\varphi,f\rangle$ is axiomatizable then A is r.e. and it is easy to see that $\varphi^{-1}[A]$ is r.e. as well.

If $\mathbf{\Sigma}$ is a consistent (and axiomatizable) theory in the sense of Ch. 7, and if $\mathbf{\Sigma}$ includes "junior arithmetic" $\mathbf{\Pi}_1$, then the corresponding encoded theory $\langle A,\varphi,f\rangle$ satisfies the condition of Thm. 4.2 (see Thm. 7.7.10). Hence A and $\varphi^{-1}[A]$ are a pair of disjoint (and r.e.) recursively inseparable sets. We shall soon strengthen this result.

4.4. DEFINITION. Two sets of numbers A and B are *effectively inseparable* if there is a recursive binary function g with the property that whenever a and b are numbers such that

$$A\subseteq \mathrm{dom}_1\{a\}, \qquad B\subseteq \mathrm{dom}_1\{b\},$$

$$\mathrm{dom}_1\{a\}\cap \mathrm{dom}_1\{b\}=\emptyset,$$

then $g(a,b)$ is defined and $g(a,b)\notin \mathrm{dom}_1\{a\}\cup \mathrm{dom}_1\{b\}$.

Clearly, if A and B are effectively inseparable, they are also recursively inseparable.

For our first example of a pair of effectively inseparable and disjoint r.e. sets we shall need two lemmas.

4.5. SEPARATION LEMMA. *Let A and B be given n-dimensional r.e. sets. Then we can find disjoint n-dimensional r.e. sets A' and B' such that $A-B\subseteq A'$, $B-A\subseteq B'$, and $A\cup B=A'\cup B'$.*

PROOF. By 6.11.6 we can find indexes i and j for A and B respectively. Thus

$$(\mathfrak{x}\in A)=\exists y T(i,\mathfrak{x},y),$$

$$(\mathfrak{x}\in B)=\exists y T(j,\mathfrak{x},y).$$

Now define A' and B' by the identities

$$(\mathfrak{x} \in A') = \exists y[T(i,\mathfrak{x},y) \wedge \neg \exists z \leqslant y\ T(j,\mathfrak{x},z)],$$

$$(\mathfrak{x} \in B') = \exists y[T(j,\mathfrak{x},y) \wedge \neg \exists z < y\ T(i,\mathfrak{x},z)].$$

The idea behind this definition is this. An n-tuple $\mathfrak{x}$ belongs to A iff some (unique) computation code y "puts $\mathfrak{x}$ in A", i.e., y is such that $T(i,\mathfrak{x},y)$ holds. Similarly with "A" and "i" replaced by "B" and "j". Now A' is defined as the set of all $\mathfrak{x} \in A$ such that the computation code putting $\mathfrak{x}$ in A is smaller than the computation code (if any) putting $\mathfrak{x}$ in B. And B' is defined as the set of all $\mathfrak{x} \in B$ such that the computation code (if any) putting $\mathfrak{x}$ in A is not smaller than the computation code putting $\mathfrak{x}$ in B.

It is easy to see that A' and B' are as required. ∎

4.6. Symmetric Lemma. *Let*

$$C = \{z\colon\ z \in \mathrm{dom}_1\{K(z)\}\},$$

$$D = \{z\colon\ z \in \mathrm{dom}_1\{L(z)\}\},$$

where K and L are the functions defined in Ex. 6.6.13. *Then C–D and D–C are effectively inseparable.*

Proof. Suppose that

$$(1) \qquad C - D \subseteq \mathrm{dom}_1\{a\}, \qquad D - C \subseteq \mathrm{dom}_1\{b\}.$$

We show that $J(b,a) \in \mathrm{dom}_1\{a\}$ iff $J(b,a) \in \mathrm{dom}_1\{b\}$.

Indeed, suppose that $J(b,a)$ belongs to $\mathrm{dom}_1\{a\}$ but not to $\mathrm{dom}_1\{b\}$. Since $b = K(J(b,a))$ and $a = L(J(b,a))$, it follows from the definition of C and D that $J(b,a) \in D - C$. Hence by (1) we have $J(b,a) \in \mathrm{dom}_1\{b\}$, contrary to our assumption. Similarly, the assumption that $J(b,a)$ belongs to $\mathrm{dom}_1\{b\}$ but not to $\mathrm{dom}_1\{a\}$ leads to contradiction.

If $\mathrm{dom}_1\{a\}$ and $\mathrm{dom}_1\{b\}$ are disjoint, then $J(b,a)$ cannot belong to both, so it belongs to neither. ∎

4.7. Example. Let C and D be as in Lemma 4.6. Clearly, these sets are r.e., so by Lemma 4.5 we find disjoint r.e. sets A and B such that $C - D \subseteq A$ and $D - C \subseteq B$. Since $C - D$ and $D - C$ are effectively inseparable, A and B must also be effectively inseparable.

4.8. Problem. Let

$$A = \{z\colon\ \{z\}(z) = 0\}, \qquad B = \{z\colon\ \{z\}(z) = 1\}.$$

It is easy to see that A and B are r.e. and disjoint. Prove that they are effectively inseparable. (Find a nice function g such that $\{g(x,y)\}(z)=1$ whenever $z \in \mathrm{dom}_1\{x\}-\mathrm{dom}_1\{y\}$, and $\{g(x,y)\}(z)=0$ whenever $z \in \mathrm{dom}_1\{y\}-\mathrm{dom}_1\{x\}$. Show that g has the property required in Def. 4.4.)

From a given pair of effectively inseparable sets, new pairs may be obtained using the following:

4.9. LEMMA. *Let A and B be effectively inseparable, and let ϱ be any recursive sequence. Then $\varrho[A]$ and $\varrho[B]$ are effectively inseparable.*

PROOF. Let g be as in Def. 4.4. Using Thm. 6.11.11 we get a nice sequence φ such that

$$(z \in \mathrm{dom}_1\{\varphi(x)\}) = (\varrho(z) \in \mathrm{dom}_1\{x\}).$$

Now put $h(x,y)=\varrho g(\varphi(x), \varphi(y))$. It is easy to verify that h has the property required in Def. 4.4 (with respect to $\varrho[A]$ and $\varrho[B]$). ∎

Note that if A and B are r.e. then so are $\varrho[A]$ and $\varrho[B]$.

We shall now apply the above results to encoded theories. Let C and D be sets of numbers, and let $\langle A,\varphi,f\rangle$ be an encoded theory. We shall say that the pair $\langle C,D\rangle$ is *representable in* $\langle A,\varphi,f\rangle$ if for some n

$$f(n,x) \in A \quad \text{for all } x \in C,$$

$$\varphi f(n,x) \in A \quad \text{for all } x \in D.$$

We now have:

4.10. THEOREM. *If the pair $\langle C,D\rangle$ is representable in the encoded theory $\langle A,\varphi,f\rangle$, and C, D are effectively inseparable, then A, $\varphi^{-1}[A]$ are effectively inseparable.*

PROOF. Let n be as above, and put $\varrho(x)=f(n,x)$. Then $\varrho[C]\subseteq A$ and $\varrho[D]\subseteq\varphi^{-1}[A]$. By Lemma 4.9, $\varrho[C]$ and $\varrho[D]$ are effectively inseparable, hence so are A and $\varphi^{-1}[A]$. ∎

4.11. EXAMPLES. Let Σ be a theory (in the sense of Ch. 7) that includes "junior arithmetic" $\mathbf{\Pi}_1$ and let $\langle A,\varphi,f\rangle$ be the corresponding encoded theory. From Lemma 7.7.9 it follows at once that if C and D are r.e. sets of numbers then $\langle C-D, D-C\rangle$ is representable in $\langle A,\varphi,f\rangle$. In particular, we can take C and D to be those of Lemma 4.6. Thus by Thm. 4.10 the sets A, $\varphi^{-1}[A]$ are effectively inseparable.[1]

[1] This is in fact the recursion-theoretic content of the First Incompleteness Theorem (Thm. 7.11.8).

The same result holds for a wide variety of encoded theories. For example, in any of the usual formalizations of axiomatic set theory (such as the one studied in Ch. 10 of this book) it is possible to deduce appropriate translations of all the sentences of "junior arithmetic". (For the formalized set theory of Ch. 10 the method is indicated in §2 of that chapter.) Therefore, if $\langle A,\varphi,f\rangle$ is an encoded version of such a theory, A and $\varphi^{-1}[A]$ are effectively inseparable.

§ 5. Productive and creative sets; reducibility

The notion of *r.e. set* can be regarded as a recursion-theoretic analogue of the set-theoretic notion of *countable set*. Thus a set is countable iff it is the range of some mapping whose domain is included in N; and by Thm. 6.11.5 a set of numbers is r.e. iff it is the range of some *recursive* unary function.

Other useful recursion-theoretic notions may be defined by pursuing the same kind of analogy.

What should be the recursion-theoretic analogue of the notion of *uncountable set?* At first sight it may seem that if r.e. sets are analogous to countable sets then non-r.e. sets are similarly analogous to uncountable sets. But this is of course wrong. What we are looking for is the notion of set which not only fails to be r.e., but whose failure to be r.e. is actually "demonstrated in a recursive way".

To find the right analogue, let us analyse how a set A may be proved to be uncountable. Perhaps the most direct way of proving this is to show that *for every countable set B such that $B \subseteq A$ there is a member of A not belonging to B.* The required analogue is obtained from the above statement by replacing "*every countable set*" by "*every given r.e. set*" (where "given" means, as usual, given *via* an index) and by insisting that the required member of $A-B$ be obtained by a *recursive function* from the given index of the subset B. These considerations lead to the following definition of *productive set*, which is the correct recursion-theoretic analogue of uncountable set.

5.1. DEFINITION. A set $A \subseteq N$ is *productive* if there is a unary recursive function f such that $f(e) \in A - \mathrm{dom}_1\{e\}$ for every number e such that $\mathrm{dom}_1\{e\} \subseteq A$. Such a function f is called a *productive function for A.*

Note that by this definition $f(e)$ has to be defined when $\mathrm{dom}_1\{e\} \subseteq A$, but not necessarily otherwise.

A productive set is *a fortiori* not r.e., so the lowest class in the arithmetical hierarchy in which we may expect to find a productive set is the class Π^0_1.

5.2. EXAMPLE. Let

$$A=\{z:\ z\notin \mathrm{dom}_1\{z\}\}.$$

Now, since $(z\in A^c)=(z\in \mathrm{dom}_1\{z\})$, it follows from 6.11.6 that A^c is r.e., hence A is Π_1^0. We show that A is productive by proving that the identity function $\lambda z(z)$ is productive for A. Indeed, we have

$$(z\in A)=(z\notin \mathrm{dom}_1\{z\}).$$

Thus the number e belongs to exactly one of the two sets A and $\mathrm{dom}_1\{e\}$. In particular, if $\mathrm{dom}_1\{e\}\subseteq A$ then e must belong to A but not to $\mathrm{dom}_1\{e\}$. (Cf. Ex. 6.11.8.)

5.3. PROBLEM. Let A, B be sets of numbers such that B is r.e. and $A\cap B$ is productive. Prove that A is productive. (Obtain a nice sequence φ for which the identity $\mathrm{dom}_1\{\varphi(z)\}=\mathrm{dom}_1\{z\}\cap B$ holds.)

5.4. PROBLEM. Let $\mathrm{ran}\{e\}$ be the range of the unary function $\lambda x[\{e\}(x)]$.

(i) Find nice sequences φ and ψ such that for every z

$$\mathrm{dom}_1\{z\}=\mathrm{ran}\{\varphi(z)\},\qquad \mathrm{ran}\{z\}=\mathrm{dom}_1\{\psi(z)\}.$$

(Use the S_n^m thm. to get φ for which the identity $\{\varphi(z)\}(x)=0\cdot\{z\}(x)+x$ holds. To obtain ψ use 6.11.11.)

(ii) Show that a set A of numbers is productive iff there is a unary recursive function g such that $g(e)\in A-\mathrm{ran}\{e\}$ whenever $\mathrm{ran}\{e\}\subseteq A$. (Use φ and ψ of (i). Show that if g satisfies the condition then $g\varphi$ is a productive function for A; and if f is a productive function for A then $f\psi$ satisfies the condition.)

(iii) Show that the set $\{z:\ z\notin \mathrm{ran}\{z\}\}$ is Π_1^0 and productive.

5.5. THEOREM. *Every productive set has a productive sequence (i.e., a total productive function).*

PROOF. Let f be a productive function for the set A. Let i be an index of f. Put

$$R(x,z)=\exists y T(i,z,y)\wedge(x\in \mathrm{dom}_1\{z\}).$$

Then R is clearly r.e., and by applying to it Thm. 6.11.11 we obtain a nice sequence φ such that

$$(x\in \mathrm{dom}_1\{\varphi(z)\})=\exists y T(i,z,y)\wedge(x\in \mathrm{dom}_1\{z\}).$$

This can be re-written as follows:

$$\text{(1)} \qquad \text{dom}_1\{\varphi(z)\} = \begin{cases} \text{dom}_1\{z\} & \text{if } f(z) \neq \infty, \\ \emptyset & \text{if } f(z) = \infty. \end{cases}$$

(This is a solution of Prob. 6.11.12.). Now let j be an index of the function $f\varphi$ (i.e., $\lambda z[f(\varphi(z))]$). Put

$$\text{(2)} \qquad g(z) = U(\mu y[T(i,z,y) \vee T(j,z,y)]).$$

Clearly, g is recursive. Also, by the choice of i and j it is clear that, if at least one of the values $f(z)$ and $f\varphi(z)$ is defined, then $g(z)$ is defined and equals one of these two values.

As a matter of fact, for each z at least one of the values $f(z)$, $f\varphi(z)$ *must* be defined. For, if $f(z) = \infty$ then by (1) we have $\text{dom}_1\{\varphi(z)\} = \emptyset \subseteq A$; and, since f is a productive function for A, $f\varphi(z)$ must be defined (and, incidentally, must belong to A).

Thus g is a *total* recursive function. We shall show that it is productive for A.

Suppose $\text{dom}_1\{z\} \subseteq A$. Then, since f is productive for A, we have

$$f(z) \in A - \text{dom}_1\{z\}.$$

In particular $f(z)$ must be defined, so that by (1) we have $\text{dom}_1\{\varphi(z)\} = \text{dom}_1\{z\} \subseteq A$. Hence, using again the fact that f is productive for A,

$$f\varphi(z) \in A - \text{dom}_1\{\varphi(z)\} = A - \text{dom}_1\{z\}.$$

But we have seen that $g(z)$ must equal $f(z)$ or $f\varphi(z)$, so in either case we have $g(z) \in A - \text{dom}_1\{z\}$, as required. ∎

The above result can be strengthened still further:

5.6. THEOREM. *Every productive set has a monotone increasing productive sequence.*

PROOF. Let A be a productive set. By Thm. 5.5 we may assume that A has a productive sequence, say φ.

By Thm. 6.11.11 we get a nice sequence τ such that

$$(x \in \text{dom}_1\{\tau(z)\}) = (x \in \text{dom}_1\{z\}) \vee (x = \varphi(z)).$$

This can be re-written thus:

$$\text{(1)} \qquad \text{dom}_1\{\tau(z)\} = \text{dom}_1\{z\} \cup \{\varphi(z)\}.$$

Now put

(2) $\tau^0(z)=z, \qquad \tau^{u+1}(z)=\tau\tau^u(z).$

Clearly, the binary function $\lambda zu\,[\tau^u(z)]$ defined by (2) is p.r. and certainly recursive. Putting $\tau^u(z)$ for z in (1) we get

(3) $\mathrm{dom}_1\{\tau^{u+1}(z)\}=\mathrm{dom}_1\{\tau^u(z)\}\cup\{\varphi\tau^u(z)\}.$

Now suppose that z is such that $\mathrm{dom}_1\{z\}\subseteq A$. Then for all u we have

(4) $\mathrm{dom}_1\{\tau^u(z)\}\subseteq A,$

(5) $\varphi\tau^u(z)\in A-\mathrm{dom}_1\{\tau^u(z)\}.$

These two facts are proved simultaneously by induction on u. For $u=0$ we have (4) by our assumption that $\mathrm{dom}_1\{z\}\subseteq A$; and (5) holds because φ is productive for A. Next, if we assume (4) and (5) for a given u, then by (3) we get (4) for $u+1$ as well; and since φ is productive for A we also have (5) for $u+1$.

Let us outline how we are going to obtain a monotone productive sequence ψ for A. The value $\psi(z)$ will be defined by induction on z. If z is such that $\mathrm{dom}_1\{z\}\nsubseteq A$, then all we need to ensure is that $\psi(z)$ is greater than the previous values of ψ. This is easily done. However, if $\mathrm{dom}_1\{z\}\subseteq A$ we must also ensure that $\psi(z)\in A-\mathrm{dom}_1\{z\}$. Now, if $\mathrm{dom}_1\{z\}\subseteq A$ we know that (5) holds. From this together with (3) it follows at once that the values $\varphi\tau^u(z)$ for $u=0,1,2,\ldots$ are all different and all belong to $A-\mathrm{dom}_1\{z\}$. Among these infinitely many different values there must be such that are greater than $\psi(z-1)$ and we shall choose such a value as $\psi(z)$.

Now let us fill in the details. Put

$$f(z,w)=\mu y\,[\exists u<y\,[\varphi\tau^u(z)=\varphi\tau^y(z)]\vee(\varphi\tau^y(z)>w)].$$

Clearly, f is recursive. We show that it is total. Let us fix z and w and consider two cases.

Case 1: The values $\varphi\tau^u(z)$ for $u=0,1,2\ldots$ are all different. Then for every y the first disjunct in the definition of f is false. But for some y the second disjunct must hold; so $f(z,w)$ is the least such y. In particular, in this case we have $\varphi\tau^{f(z,w)}(z)>w$.

Case 2: The values $\varphi\tau^u(z)$ for $u=0,1,2\ldots$ are not all different. Then the first disjunct in the definition of f must hold for some y (while the second disjunct may or may not hold for some y). Thus $f(z,w)$ is defined in this case as well.

We now define

$$\psi(0)=\varphi(0),$$

$$\psi(z+1)=\max(\psi(z)+1,\ \varphi\tau^{f(z+1,\psi(z))}(z+1)).$$

(For the definition of max see Ex. 6.6.5.) From this definition it is immediate that ψ is recursive and monotone. It remains to show that ψ is productive for A. Since $\psi(0)=\varphi(0)$, we only need to consider the behaviour of ψ for *positive* values of its argument.

Suppose therefore that $\mathrm{dom}_1\{z+1\}\subseteq A$. Then, as we have shown above, the values $\varphi\tau^u(z+1)$ for $u=0,1,2,\ldots$ are all different and they all belong to $A-\mathrm{dom}_1\{z+1\}$. Therefore in the definition of $f(z+1,\psi(z))$ we have Case 1, and hence

$$\varphi\tau^{f(z+1,\psi(z))}(z+1)>\psi(z).$$

It now follows from the definition of $\psi(z+1)$ that

$$\psi(z+1)=\varphi\tau^{f(z+1,\psi(z))}(z+1).$$

Thus $\psi(z+1)$ has the form $\varphi\tau^u(z+1)$ and must therefore be in $A-\mathrm{dom}_1\{z+1\}$, as required. ∎

5.7. Problem. Show that every productive set has an infinite recursive subset. (Using 6.10.15, get a monotone recursive sequence φ such that $\mathrm{dom}_1\{\varphi(z)\}=\emptyset$ for all z. Let ψ be a monotone productive sequence for the set A. Show that $\psi\varphi$ enumerates an infinite recursive subset of A.)

5.8. Definition. A set of numbers is *creative* if it is r.e. and its complement is productive.

A productive set is a set whose failure to be r.e. is "proved in a recursive way" — by means of a productive function. Similarly, we may say that a creative set is an r.e. set whose failure to be recursive is "proved in a recursive way" — by means of a productive function for its complement.

5.9. Example. The set $C=\{z:\ z\in\mathrm{dom}_1\{z\}\}$ is creative by Ex. 5.2.

5.10. Problem. Prove that if $\langle A,\varphi,f\rangle$ is an encoded theory of the kind discussed in Thm. 4.10 and Ex. 4.11, and $\langle A,\varphi,f\rangle$ is consistent and axiomatizable, then A is creative. (Show that if A and B are disjoint and effectively inseparable r.e. sets, then each of them is creative.)

5.11. DEFINITION. Let A and B be sets of numbers. *A many–one reduction of A to B* is a recursive sequence ϱ for which the identity

$$(x \in A) = (\varrho(x) \in B)$$

holds, i.e., $A = \varrho^{-1}[B]$. If such a ϱ exists then A is *many–one reducible to B*.

We shall usually omit the words "many–one" and say simply "reduction" and "reducible".

We shall apply the notion of reducibility to the study of productive and creative sets.

5.12. THEOREM. *If A is productive and reducible to B, then B is productive.*
PROOF. Let ϱ be a reduction of A to B. Using the S^m_n thm. for r.e. sets (6.11.11) we find a nice sequence φ such that

$$(x \in \mathrm{dom}_1\{\varphi(z)\}) = (\varrho(x) \in \mathrm{dom}_1\{z\}).$$

Thus for every e we have

$$(1) \qquad \varrho^{-1}[\mathrm{dom}_1\{e\}] = \mathrm{dom}_1\{\varphi(e)\}.$$

Also, since ϱ is a reduction of A to B we have

$$(2) \qquad \varrho^{-1}[B] = A.$$

Let f be a productive function for A. We show that $\varrho f \varphi$ is a productive function for B.

Let $\mathrm{dom}_1\{e\} \subseteq B$. Then from (1) and (2) it follows that $\mathrm{dom}_1\{\varphi(e)\} \subseteq A$. Since f is productive for A, the value $f\varphi(e)$ is defined and belongs to $A - \mathrm{dom}_1\{\varphi(e)\}$. By (1) and (2) it now follows that $\varrho f \varphi(e)$ is in $B - \mathrm{dom}_1\{e\}$, as required. ∎

5.13. COROLLARY. *If A is creative and reducible to an r.e. set B, then B is creative.*
PROOF. A reduction of A to B is clearly also a reduction of A^c to B^c. ∎

5.14. EXAMPLE. Let $B = \{z\colon \mathrm{dom}_1\{z\} = N\}$. Thus B is the set of all indexes of recursive sequences. (Cf. Prob. 6.10.10). We shall show that B is productive by reducing to it the set A of Ex. 5.2. Let f be defined by

$$(1) \qquad f(z,y) = \begin{cases} y & \text{if } \neg T(z,z,y), \\ \infty & \text{otherwise.} \end{cases}$$

By Ex. 6.10.5 f is recursive, so we can use the S_n^m thm. to get a nice sequence ϱ such that

(2) $$\{\varrho(z)\}(y)=f(z,y).$$

We claim that ϱ is a reduction of A to B, i.e., that for all z

(3) $$z\notin \mathrm{dom}_1\{z\} \quad \text{iff} \quad \mathrm{dom}_1\{\varrho(z)\}=N.$$

Recall that $(x\in \mathrm{dom}_1\{z\})=\exists y T(z,x,y)$. Thus for given z we have $z\notin \mathrm{dom}_1\{z\}$ iff $\neg T(z,z,y)$ holds for all y; so that (3) follows at once from (1) and (2).

Note that $(z\in B)=\forall x(x\in \mathrm{dom}_1\{z\})$, so that B belongs to class Π_2^0 in the arithmetical hierarchy. Now, level 2 in this hierarchy is the lowest in which we can hope to find a productive set whose complement is also productive. (Sets of level 1 are r.e. sets and their complements.) We now show that our B is in fact such a set: B^c is productive as well.

By Thm. 6.11.13 (the RT for r.e. sets) we can find a nice sequence φ such that

$$(x\in \mathrm{dom}_1\{\varphi(z)\})=(\varphi(z)\in \mathrm{dom}_1\{z\}).$$

Thus

$$\mathrm{dom}_1\{\varphi(z)\}=\begin{cases} N & \text{if } \varphi(z)\in \mathrm{dom}_1\{z\},\\ \emptyset & \text{otherwise.}\end{cases}$$

We therefore have the identity

$$(\varphi(z)\in B)=(\varphi(z)\in \mathrm{dom}_1\{z\}),$$

from which it follows at once that φ is a productive sequence for B^c.

5.15. **Problem.** Prove that the set $\{z\colon \mathrm{dom}_1\{z\}\neq\emptyset\}$ is creative.

5.16. **Problem.** Prove that the set $\{z\colon \mathrm{ran}\{z\}=N\}$ is a Π_2^0 productive set whose complement is also productive. (Show that the φ of Prob. 5.4 is a reduction to these sets of the sets considered in 5.14.)

The notion of reducibility arises naturally in connection with encoded theories. If $\langle A,\varphi,f\rangle$ and $\langle B,\psi,g\rangle$ are encoded theories, then a reduction of A to B is a uniform recursive method of reducing the decision problem for the first theory to that for the second. (This is obviously closely connected with the notion of reducibility defined in the discussion preceding Ex. 7.10.9. In fact, the results of 7.10.9–7.10.15 can easily be rephrased in terms of reducibility — in our present sense — of the corresponding encoded theories.)

5.17. THEOREM. *Let $\langle A,\varphi,f\rangle$ be an encoded theory in which every r.e. set of numbers is weakly representable. Then A^c is productive. In particular, if $\langle A,\varphi,f\rangle$ is also axiomatizable, then A is creative.*

PROOF. By Ex. 5.9 the set $C=\{z\colon z\in \mathrm{dom}_1\{z\}\}$ is creative. In particular, since C is r.e., it is weakly representable in $\langle A,\varphi,f\rangle$. Therefore for some n the identity

$$(f(n,x)\in A)=(x\in C)$$

holds. Thus $\lambda x[f(n,x)]$ is a reduction of C to A, hence also of C^c to A^c. By Thm. 5.12, A^c must be productive. ∎

5.18. PROBLEM. Let $\langle A,\varphi,f\rangle$ be an encoded theory in which every Π_1^0 set of numbers is weakly representable. Show that A is productive.

The following two problems assume familiarity with Ch. 7.

5.19. PROBLEM. Show that $\#[\mathbf{\Omega}]$ and $(\#[\mathbf{\Omega}])^c$ are productive.

5.20. PROBLEM. Let $\mathbf{\Sigma}$ be a theory in the sense of Ch. 7.

(i) Prove that if $\mathbf{\Sigma}$ is sound and includes "baby arithmetic" $\mathbf{\Pi}_0$, then $(\#[\mathbf{\Sigma}])^c$ is productive. (Use Thm. 7.6.8 and Thm. 5.17.)

(ii) Prove that if $\mathbf{\Sigma}$ is sound then $(\#[\mathbf{\Sigma}])^c$ is productive; hence if $\mathbf{\Sigma}$ is sound and axiomatizable then $\#[\mathbf{\Sigma}]$ is creative. (Let $\mathbf{\Delta}$ be the set of all sentences deducible from $\mathbf{\Sigma}\cup\mathbf{\Pi}_2$. As in the proof of Thm. 7.10.5, obtain a reduction of $\#[\mathbf{\Delta}]$ to $\#[\mathbf{\Sigma}]$.)

5.21. REMARK. Let $\mathscr{L}$ be any first-order language and let $\mathbf{\Sigma}$ be an $\mathscr{L}$-theory (i.e., a deductively closed set of $\mathscr{L}$-sentences). Call $\mathbf{\Sigma}$ *creative* if the set of all code numbers of sentences of $\mathbf{\Sigma}$ (under any of the standard methods of encoding $\mathscr{L}$-expressions by numbers) is creative. The results of this section yield a whole spectrum of creative theories, widely different in their (apparent) mathematical intricacy. At one end of the scale, we see from Prob. 5.20 that the set of all logically true sentences in the first-order language of arithmetic is creative. Using the reductions of 7.10.9–7.10.15 we see that the same holds for the set of logically true sentences in various other first-order languages. At the other end of the scale, we see from Prob. 5.10 that if $\mathbf{\Sigma}$ is any of the usual formalizations of axiomatic set theory then $\mathbf{\Sigma}$ is creative provided it is consistent. In between these two extremes we have, e.g., first-order Peano arithmetic which is creative by Prob. 5.10 as well as by Prob. 5.20.

§ 6. One–one reducibility; recursive isomorphisms

We say that A is *one–one reducible to* B if there is a one–one reduction of A to B, i.e., a recursive sequence ϱ which does not take any value more than once and such that $A=\varrho^{-1}[B]$.

6.1. THEOREM. *Any r.e. set of numbers is one–one reducible to the complement of any productive set, hence to any creative set.*

PROOF. Let A be an r.e. set of numbers and let B^c be productive. By Thm. 5.6, B^c has a monotone productive sequence ψ. Thus for any z we have

$$(1) \qquad \psi(z)\in B^c - \mathrm{dom}_1\{z\} \quad \text{if } \mathrm{dom}_1\{z\}\subseteq B^c.$$

By 6.11.13 (the RT for r.e. sets) we get a nice — and hence recursive and monotone — sequence φ such that

$$(x\in \mathrm{dom}_1\{\varphi(z)\})=(x=\psi\varphi(z))\wedge(z\in A),$$

which can be re-stated in the form

$$(2) \qquad \mathrm{dom}_1\{\varphi(z)\}=\begin{cases}\{\psi\varphi(z)\} & \text{if } z\in A,\\ \emptyset & \text{otherwise.}\end{cases}$$

Now, $\psi\varphi$ is clearly recursive and monotone, and hence one–one. We shall show that $\psi\varphi$ reduces A to B.

First, let $e\in A$. Then, by (2), $\mathrm{dom}_1\{\varphi(e)\}$ is the singleton $\{\psi\varphi(e)\}$. Hence $\psi\varphi(e)\notin B^c-\mathrm{dom}_1\{\varphi(e)\}$, and it follows from (1) that $\mathrm{dom}_1\{\varphi(e)\}\nsubseteq B^c$. This means that $\psi\varphi(e)\in B$.

Conversely, if $e\notin A$, then, by (2), $\mathrm{dom}_1\{\varphi(e)\}=\emptyset\subseteq B^c$ and we see from (1) that $\psi\varphi(e)\in B^c$, i.e. $\psi\varphi(e)\notin B$. ∎

It follows from Thm. 6.1 that any axiomatic formal theory is reducible (in the sense of §10 of Ch. 7) to any creative theory: e.g., to any of the theories mentioned in Remark 5.21.

Two sets of numbers are *many–one equivalent* if each of them is many–one reducible to the other. The notion of *one–one equivalence* is defined in a similar way.

It is easy to verify that both many–one equivalence and one–one equivalence are indeed equivalence relations between sets of numbers. The following theorem shows that the creative sets constitute an equivalence class under both relations.

6.2. THEOREM. *Let A be creative, and let B any set of numbers. Then the following three conditions are equivalent:*

(1) *B is creative;*
(2) *A and B are one–one equivalent;*
(3) *A and B are many–one equivalent.*

PROOF. (1) ⇒ (2) by Thm. 6.1, and (2) ⇒ (3) trivially.

Now assume (3). Then *B* is reducible to the creative — and hence r.e. — set *A*. From this it follows at once that *B* is r.e. as well. Since *A* is reducible to *B*, Cor. 5.13 shows that *B* is creative. ∎

A *recursive permutation* is a recursive sequence which takes every value exactly once[1].

Sets of numbers *A* and *B* are said to be *recursively isomorphic* if there is a recursive permutation φ which maps *A* onto *B*. (It is easy to see that in this case, if ψ is defined by $\psi(x)=\mu y(\varphi(y)=x)$, then ψ is a recursive permutation mapping *B* onto *A*.)

The following remarkable result, due to Myhill, is the recursion-theoretic analogue of the Schröder–Bernstein theorem.

6.3. THEOREM. *Two sets of numbers are recursively isomorphic iff they are one–one equivalent.*

PROOF. If *A* and *B* are recursively isomorphic, then they are obviously one–one equivalent.

To prove the converse, suppose *A* and *B* are one–one equivalent, and let φ and ψ be one–one reductions of *A* to *B* and *B* to *A* respectively.

We seek unary recursive functions *f*, *g* and binary recursive functions *h*, *k* satisfying the following identities:

$$(1)\qquad f(x)=\begin{cases}\mu u[\neg\exists y<x\{f(y)\doteq u\}] & \text{if } x \text{ is even,}\\ h(x,\mu u[\neg\exists y<x\{f(y)\doteq h(x,u)\}]) & \text{if } x \text{ is odd;}\end{cases}$$

$$(2)\qquad g(x)=\begin{cases}\mu u[\neg\exists y<x\{g(y)\doteq u\}] & \text{if } x \text{ is odd,}\\ k(x,\mu u[\neg\exists y<x\{g(y)\doteq k(x,u)\}]) & \text{if } x \text{ is even;}\end{cases}$$

$$(3)\qquad h(x,0)=\psi g(x),\qquad h(x,u+1)=\psi g(\mu y[f(y)\doteq h(x,u)]);$$

$$(4)\qquad k(x,0)=\varphi f(x),\qquad k(x,u+1)=\varphi f(\mu y[g(y)\doteq k(x,u)]).$$

Such functions can be obtained, using Prob. 6.10.26 as follows. In the above identities, replace "*f*", "*g*", "*h*" and "*k*" by "$\{e_1\}$", "$\{e_2\}$", "$\{e_3\}$" and "$\{e_4\}$", respectively. Then Prob. 6.10.26 yields numbers e_1, e_2, e_3 and

[1] It is easy to verify that the recursive permutations constitute a group under the operation of composition. It can be argued that the recursion-theoretically significant notions are precisely those that are invariant under this group.

e_4 for which the resulting identities all hold. We now only have to define

$$f(x)=\{e_1\}(x), \qquad g(x)=\{e_2\}(x),$$

$$h(x,u)=\{e_3\}(x,u), \qquad k(x,u)=\{e_4\}(x,u).$$

We shall now prove that, for all y and y',

(5) $f(y)\neq\infty$, and if $y\neq y'$ then $f(y)\neq f(y')$;

(6) $g(y)\neq\infty$, and if $y\neq y'$ then $g(y)\neq g(y')$;

(7) $f(y)\in A$ iff $g(y)\in B$.

We proceed by induction. Assume that (5), (6) and (7) hold for all y and y' which are smaller than some number x; we shall prove that (5), (6) and (7) hold for all $y,y'\leqslant x$.

Suppose first that x is even. Then by (1), $f(x)$ is defined as the least number not belonging to the set $\{f(y) : y<x\}$. Therefore (5) holds for all $y,y'\leqslant x$.

Next, we prove that for all w the following statement holds:

(8) If for all $v<w$ the values $k(x,v)$ are distinct and belong to the set $\{g(y) : y<x\}$, then $k(x,w)$ is defined and is different from all these $k(x,v)$ with $v<w$.

For $w=0$, (8) merely claims that $k(x,0)$ is defined. But by (4) we have $k(x,0)=\varphi f(x)$ and we have already shown that $f(x)$ is defined.

Now let $w=u+1$. Suppose that all the values $k(x,v)$ for $v\leqslant u$ are distinct and belong to the set $\{g(y) : y<x\}$. Since (6) is assumed to hold for all $y,y'<x$, it follows that for each $v\leqslant u$ there is a *unique* $y_v<x$ such that $g(y_v)=k(x,v)$. Also, these y_v are all distinct, since the $k(x,v)$ are distinct.

By (4) we now have

(9) $k(x,v+1)=\varphi f(y_v)$ for all $v\leqslant u$.

Since (5) holds for all $y, y'<x$ (and even, as we have shown, for all $y, y'\leqslant x$) it follows that $f(y_u)$ is defined. Hence $k(x,u+1)$ is defined by (9). It remains to show that $k(x, u+1)\neq k(x,v)$ for all $v\leqslant u$.

Suppose that $k(x, u+1)=k(x,0)$. Then by (9) and (4) this would mean that $\varphi f(y_u)=\varphi f(x)$. Since φ is one–one, it would follow that $f(y_u)=f(x)$. But, since $y_u<x$, this contradicts the fact that (5) holds for all $y, y'\leqslant x$.

Suppose next that $k(x,u+1)=k(x,v+1)$ for some $v<u$. Then by (9) this would mean $\varphi f(y_u)=\varphi f(y_v)$. Since φ is one–one and y_u, y_v are distinct numbers $<x$, we would again get a contradiction.

Thus we have shown that $k(x,u+1)$ is defined and different from $k(x,v)$ for all $v \leqslant u$, and the proof of (8) is complete.

Let us put

(10) $$u(x)=\mu u[\neg \exists y<x\{g(y) \doteq k(x,u)\}].$$

Then, for our particular x, it follows from (8) that $u(x) \neq \infty$, since otherwise the successive values $k(x,u)$ for $u=0, 1, 2, \ldots$ would all have to be distinct members of the finite set $\{g(y) : y<x\}$, which is clearly impossible.

Turning now to (2) and recalling that x was assumed to be even, we see that $g(x)=k(x,u(x))$. Thus, by what we have just shown, $g(x)$ is defined. Moreover, from (10) it follows that $g(x) \neq g(y)$ for all $y<x$. Thus (6) holds for all $y, y' \leqslant x$.

Next, we want to show that $f(x) \in A$ iff $g(x) \in B$. Since we know that $g(x)=k(x,u(x))$, it will be enough to show that

(11) $$f(x) \in A \quad \text{iff} \quad k(x,u) \in B$$

for all $u \leqslant u(x)$. We proceed by induction on u.

For $u=0$ we have (11) because $k(x,0)=\varphi f(x)$ by (4), and φ is a reduction of A to B.

Now suppose that (11) has been established for some $u<u(x)$. Since $u<u(x)$, it follows from (10) that

(12) $$k(x,u)-g(y_u)$$

for some unique $y_u<x$. But (7) was assumed for all $y<x$. Hence by (11) and (12) we have

(13) $$f(x) \in A \quad \text{iff} \quad f(y_u) \in A.$$

Also, by (4) and (12) we have $k(x, u+1)=\varphi f(y_u)$; hence, since φ is a reduction of A to B,

$$f(y_u) \in A \quad \text{iff} \quad k(x,u+1) \in B.$$

From this and (13) we see that (11) holds for $u+1$ as well.

Thus we have established that $f(x) \in A$ iff $g(x) \in B$ and hence (7) holds for all $y \leqslant x$.

So far we have assumed x to be even. If x is odd, the treatment is exactly the same except that f, k, φ and A exchange roles with g, h, ψ and B respectively.

Thus (5), (6) and (7) have been established for *all* y and y'.

From (5) we see that f is a one–one sequence (it does not take a value more than once). Also, from the first part of (1), which deals with the case that x is even, we see that f takes each value at least once. Thus f is a recursive permutation. Similarly, g is a recursive permutation.

We put

$$\varrho(x)=g(\mu y(f(y)=x)),$$

i.e., $\varrho=gf^{-1}$. Then ϱ is clearly a recursive permutation, and it follows at once from (7) that ϱ maps A onto B. ∎

From Thms. 6.2 and 6.3 it follows that any two creative sets are recursively isomorphic. Thus, if $\mathbf{\Sigma}_1$ and $\mathbf{\Sigma}_2$ are two creative theories, then their respective sets of code numbers can be obtained from each other by a recursive permutation. From a purely recursion-theoretic viewpoint it would therefore seem that the differences in mathematical intricacy between such theories (see Remark 5.21) are illusory. However, it is also correct to say that there is much more to a formal theory than meets the recursion-theorist's eye.

§ 7. Turing degrees

Let φ be any sequence, not necessarily recursive. We recall (see §5 of Ch. 6) that an n-ary function f is *φ-recursive* if the identity $f(x)=F(\varphi;x)$ holds for some recursive functional F.

Using Prob. 6.5.4 (see also Prob. 6.8.7) it is easy to see that if ψ is φ-recursive and χ is ψ-recursive then χ is φ-recursive.

Let us say that sequences φ and ψ are *recursively equivalent* if φ is ψ-recursive and ψ is φ-recursive. It is easily seen that recursive equivalence is indeed an equivalence relation. The classes into which the collection N^N of all sequences is partitioned by this relation are called *degrees of unsolvability*, or *Turing degrees*, or, briefly, *degrees*. We define deg φ to be the degree to which the sequence φ belongs.

A partial ordering $\leqslant$ of degrees is defined as follows: $\deg\psi\leqslant\deg\varphi$ if ψ is φ-recursive. (This definition is legitimate, since, if ψ' and φ' are recursively equivalent to ψ and φ respectively, and ψ is φ-recursive then ψ' is φ'-recursive.)

If neither $\deg\psi\leqslant\deg\varphi$ nor $\deg\varphi\leqslant\deg\psi$, then $\deg\varphi$ and $\deg\psi$ are said to be *incomparable*.

The recursive sequences constitute a single degree, denoted by O. For every φ we clearly have $\mathrm{O}\leqslant\deg\varphi$.

7.1. PROBLEM. Show that a function f is φ-recursive iff $f \in \mathscr{R}\{\varphi\}$. (See Def. 1.10.) Hence show that $\deg \psi \leqslant \deg \varphi$ iff $\mathscr{R}\{\psi\} \subseteq \mathscr{R}\{\varphi\}$.

7.2. PROBLEM. Show that for each φ there are at most denumerably many degrees $\leqslant \deg \varphi$, but the collection of *all* degrees has the cardinality of the continuum.

7.3. PROBLEM. Prove that any two degrees have a supremum.

The study of degrees constitutes an important branch of recursion theory. Unfortunately, it is beyond the scope of this book; the reader interested in this topic should consult the references in §9 below. Here and in the next section we shall barely skim the surface: we shall prove two results in order to illustrate some of the methods used in the study of degrees.

For any finite sequence of numbers $\langle a_0, \ldots, a_{k-1} \rangle$ we define $B(a_0, \ldots, a_{k-1})$ to be the set

$$\{\varphi \in N^N: \ \varphi(i) = a_i \text{ for all } i < k\}.$$

In particular, if $k=0$ (so that the finite sequence is in fact empty) this set is the whole of N^N.

With this notation we prove:

7.4. LEMMA. *For any non-recursive sequence ψ, finite sequence of numbers $\langle a_0, \ldots, a_{k-1} \rangle$ and number e, there exist a number $m \geqslant k$ and numbers $a_k, \ldots, a_m$ such that for every $\varphi \in B(a_0, \ldots, a_{k-1}, a_k, \ldots, a_m)$ we have*

$$\varphi \neq \lambda x[\{e\}(\psi; x)], \qquad \psi \neq \lambda x[\{e\}(\varphi; x)].$$

PROOF. We let $a_k = 0$, unless $\{e\}(\psi; k) = 0$ in which case we put $a_k = 1$. This choice of a_k ensures that — no matter how $a_{k+1}, \ldots, a_m$ will be chosen — we shall have, for every $\varphi \in B(a_0, \ldots, a_m)$, the inequality $\varphi(k) \neq \{e\}(\psi; k)$, and hence $\varphi \neq \lambda x[\{e\}(\psi; x)]$. It therefore remains to choose m and $a_{k+1}, \ldots, a_m$ such that $\psi \neq \lambda x[\{e\}(\varphi; x)]$ for every $\varphi \in B(a_0, \ldots, a_m)$.

We define a set $A \subseteq N^2$ as follows. For any given x and z, $\langle x, z \rangle \in A$ holds iff for some φ

(1) $$\varphi \in B(a_0, \ldots, a_k), \qquad \{e\}(\varphi; x) = z.$$

Now, suppose that x and z are such that $\langle x, z \rangle \in A$, and let φ be a sequence satisfying (1). Recall that by the NFT (6.10.1) we have

$$\{e\}(\varphi; x) = U(\mu y T^*(\tilde{\varphi}(y), e, x, y)),$$

where

$$\tilde{\varphi}(y) = \prod_{t \leqslant y} \exp(p_t, \varphi(t)).$$

By (1) there is therefore a (unique) y such that

$$T^*(\tilde{\varphi}(y), e, x, y) \wedge (U(y)=z).$$

Now let

$$w=\prod \exp(p_t, \varphi(t)),$$

where the product is taken for all $t \leqslant \max(k, y)$. We clearly have

(2) $$((w)_0=a_0) \wedge \ldots \wedge ((w)_k=a_k),$$

(3) $$T^*(\prod_{t \leqslant y} \exp(p_t,(w)_t), e, x, y) \wedge (U(y)=z).$$

Conversely, if y and w are such that (2) and (3) hold, put $\varphi=\lambda t[(w)_t]$ and it is easy to see that φ satisfies (1). Thus we have shown that $\langle x,z\rangle \in A$ iff there are y and w such that (2) and (3) hold. It follows that A is an r.e. set.

Since ψ is non-recursive, the graph of ψ must be different from A by Thm. 6.11.3. Therefore at least one of the following must be the case:

(i) For some x, $\langle x,\psi(x)\rangle \notin A$.

(ii) For some x and z, $\langle x,z\rangle \in A$ but $\psi(x) \neq z$.

Suppose first that (i) holds, and let x be a number such that $\langle x,\psi(x)\rangle \notin A$. Then by the definition of A we have $\{e\}(\varphi; x) \neq \psi(x)$ for every $\varphi \in B(a_0,\ldots,a_k)$. Hence in this case we can put $m=k$, and the proof is complete.

Now assume (ii) and let x and z be numbers such that $\langle x,z\rangle \in A$ but $\psi(x) \neq z$. Then, again by the definition of A, there is some $\varphi \in B(a_0,\ldots,a_k)$ such that for some y we have

$$T^*(\tilde{\varphi}(y), e, x, y) \wedge (U(y)=z).$$

We put $m=\max(k,y)$ and $a_i=\varphi(i)$ for $i=k+1,\ldots,m$. Then if φ' is *any* sequence in $B(a_0,\ldots,a_m)$, we have $\tilde{\varphi}'(y)=\tilde{\varphi}(y)$ and hence $\{e\}(\varphi'; x)= =z \neq \psi(x)$. ∎

7.5. THEOREM. *If ψ is a non-recursive sequence, then there exists an uncountable set Ψ of sequences such that $\psi \in \Psi$ and*

(∗) *if ψ_1, ψ_2 are distinct members of Ψ, then* deg ψ_1 *and* deg ψ_2 *are incomparable.*

PROOF. Consider the family of *all* $\Psi \subseteq N^N$ satisfying condition (∗). This family is partially ordered by inclusion, and it is easy to verify that the hypothesis of Zorn's Lemma is satisfied. Also, the singleton $\{\psi\}$ belongs to the family, since in this case $(*)$ holds trivially. Therefore by Zorn's Lemma there exists a maximal set Ψ satisfying $(*)$ and such that $\{\psi\}\subseteq\Psi$, i.e., $\psi\in\Psi$. We shall show that Ψ is uncountable.

Suppose we had $\Psi=\{\psi_j\colon j\in N\}$. We shall define an infinite sequence $\{a_i\colon i\in N\}$ by stages. Suppose that up to the nth stage $a_0,\ldots,a_{k-1}$ have already been defined. Then at the nth stage we define $a_k,\ldots,a_m$ for some $m\geqslant k$, as follows. Put $K(n)=e$ and $L(n)=j$. Thus $n=J(e,j)$. Now, it is clear that ψ_j is non-recursive. (If $\psi_j=\psi$, then ψ_j is non-recursive by assumption. Otherwise $\deg\psi_j$ and $\deg\psi$ are incomparable by $(*)$ and again ψ_j cannot be recursive.) Using Lemma 7.4 we define $a_k,\ldots,a_m$ such that for every $\varphi\in B(a_0,\ldots,a_m)$ we have

$$\varphi\neq\lambda x[\{e\}(\psi_j;x)],\qquad \psi_j\neq\lambda x[\{e\}(\varphi;x)].$$

In this way a_i is eventually defined for every i. We let $\varphi(i)=a_i$ for all i. Then $\deg\varphi$ is incomparable with $\deg\psi_j$ for all j. For, if for some j $\deg\varphi$ and $\deg\psi_j$ were comparable, then for some e we should have $\varphi=\lambda x[\{e\}(\psi_j;x)]$ or $\psi_j=\lambda x[\{e\}(\varphi;x)]$, but this was prevented at the $J(e,j)^{\text{th}}$ stage in the definition of the a_i.

Since Ψ is a proper subset of $\Psi\cup\{\varphi\}$, this contradicts the maximality of Ψ. ∎

Our proof that there exists an uncountable set of pairwise incomparable degrees used the axiom of choice (in the form of Zorn's Lemma). However, the same result — and even the existence of a set of power $2^{\aleph_0}$ of pairwise incomparable degrees — can be proved without using the axiom of choice. (For references see §9.)

In the following problem we outline a topological version of the proof of Thm. 7.5.

7.6. PROBLEM. Define the *distance* $d(\varphi,\psi)$ between two different sequences φ and ψ to be $(\mu x[\varphi(x)\neq\psi(x)]+1)^{-1}$, and $d(\varphi,\varphi)=0$.

(i) Show that with this distance function, N^N becomes a complete metric space.

(ii) Show that, if ψ is a non-recursive sequence, the set of all sequences φ such that $\deg\psi$ is comparable with $\deg\varphi$ is of the first category, i.e., a countable union of nowhere dense sets. (Use Lemma 7.4.)

(iii) Use the Baire category theorem to prove Thm. 7.5. (Obtain a maximal Ψ by Zorn's Lemma, as in the original proof. Use (ii) to show that, if Ψ were countable, then the whole space N^N would be of the first category, contrary to Baire's theorem.)

*§ 8. Post's problem and its solution

If A is a set of numbers, the corresponding property $\lambda x(x \in A)$ is, by our conventions, a sequence of 0's and 1's. We define the *degree of A* (briefly, deg A) to be the degree of $\lambda x(x \in A)$.

If A is an r.e. set of numbers, then deg A is called an *r.e.* degree. Note that if deg A is r.e. it does not necessarily follow that A is r.e.; for example, the complement of an r.e. set — which may not itself be r.e. — always has an r.e. degree, because deg A=deg A^c for every A.

It is easy to see that, if A is many-one reducible to B, then deg $A \leqslant$ deg B. Therefore by Thm. 6.1 all creative sets have the same degree, which is $\geqslant$ any r.e. degree. The degree of creative sets is denoted by O′.

All the r.e. sets which we have met up to this point were either recursive or creative. Thus from what we have learnt so far we do not yet know *whether there exist r.e. degrees other than* O *(the degree of all recursive sets) and* O′. This problem was posed by Post in 1944 and remained open for twelve years.

Post's problem is of importance in connection with encoded theories. It is reasonable to take deg A as a measure of the difficulty of the decision problem for A. Thus Post's problem is equivalent to the problem of whether there exists an axiomatizable encoded theory whose decision problem, while being recursively unsolvable, is less difficult than that of a creative theory.

Post's problem was solved independently, and almost simultaneously, by the American Friedberg and the Russian Mučnik, both of whom were young students at the time. They showed, in fact, how to construct two r.e. sets whose degrees are incomparable. It is clear that these two degrees must be different from both O and O′. (The same method can be used to construct an infinite set of pairwise incomparable r.e, degrees.)

We shall now present Friedberg's construction.

We wish to construct two r.e. properties (i.e., sequences of 0's and 1's) α and β such that deg α and deg β are incomparable.

For deg α and deg β to be incomparable, we must ensure that for every

e we have

$$\alpha \neq \lambda x[\{e\}(\beta;x)], \qquad \beta \neq \lambda x[\{e\}(\alpha;x)].$$

We shall achieve this by defining, together with α and β, a sequence φ such that for every number e we have

$$\alpha(\varphi(2e))=0 \quad \text{iff} \quad \{e\}(\beta;\varphi(2e))=1,$$

$$\beta(\varphi(2e+1))=0 \quad \text{iff} \quad \{e\}(\alpha;\varphi(2e+1))=1.$$

We shall define α, β and φ "by stages". At the nth stage we define, by induction on n, properties α_n and β_n and a sequence φ_n; these will serve as our "n^{th} approximations" to α, β and φ respectively.

For $n=0$ we put

$$\alpha_0(x)=1, \qquad \beta_0(x)=1, \qquad \varphi_0(z)=2^z.$$

Now suppose that α_n, β_n and φ_n have been defined. To define α_{n+1}, β_{n+1} and φ_{n+1}, we distinguish two cases, each divided into two subcases.

Case 1: $(n)_0$ is even; say $(n)_0=2e$. We let β_{n+1} be the same as β_n. To define α_{n+1} and φ_{n+1}, we search for a number $b<n$ satisfying the condition

$$(*) \qquad T(\beta_n;\, e,\varphi_n(2e),b) \wedge (U(b)=1) \wedge \neg\alpha_n(\varphi_n(2e)).$$

Recall that by the NFT (6.10.1) there is at most one such b (it is the computation code for $\langle\beta_n;\, e,\varphi_n(2e)\rangle$).

Subcase 1′. If there is some (unique) $b<n$ satisfying $(*)$, we put

$$\alpha_{n+1}(x)=\alpha_n(x) \vee (x=\varphi_n(2e)),$$

$$\varphi_{n+1}(z)=\begin{cases} 3^b \cdot \varphi_n(z) & \text{if } z \text{ is odd and } > 2e,\\ \varphi_n(z) & \text{otherwise.}\end{cases}$$

Subcase 1″. If there is no $b<n$ satisfying $(*)$, we let α_{n+1} and φ_{n+1} be the same as α_n and φ_n respectively.

Case 2: $(n)_0$ is odd; say $(n)_0=2e+1$. We let α_{n+1} be the same as α_n. Also, we consider the condition

$$(**) \qquad T(\alpha_n;\, e,\varphi_n(2e+1),b) \wedge (U(b)=1) \wedge \neg\beta_n(\varphi_n(2e+1)).$$

Subcase 2′. If there is some $b<n$ satisfying $(**)$ — in which case this b is unique — we put

$$\beta_{n+1}(x)=\beta_n(x) \vee (x=\varphi_n(2e+1)),$$

$$\varphi_{n+1}(z)=\begin{cases} 3^b \cdot \varphi_n(z) & \text{if } z \text{ is even and } > 2e+1,\\ \varphi_n(z) & \text{otherwise.}\end{cases}$$

Subcase 2″. If there is no such b, we put $\beta_{n+1}=\beta_n$, $\varphi_{n+1}=\varphi_n$.

It is easy to see (e.g., using Prob. 6.10.26) that $\lambda nx\,[\alpha_n(x)]$ and $\lambda nx\,[\beta_n(x)]$ are recursive binary relations and $\lambda nz\,[\varphi_n(z)]$ is a recursive binary function. (In fact, they are p.r.: but we shall not need this.)

We put

$$\alpha(x)=\exists n\alpha_n(x), \qquad \beta(x)=\exists n\beta_n(x).$$

Clearly, α and β are r.e. properties.

From the definition of the α_n it is clear that if a number x has the property α_n (i.e., if $\alpha_n(x)$ vanishes) then x has the property α_{n+1} as well. Among the numbers having the property α_{n+1} there is at most one "new" number, which does not have the property α_n. In fact, such a new number exists iff n falls under subcase 1′, in which case the new number is $\varphi_n(2e)$. Since there is no number with the property α_0, it follows by induction that for each n there are only finitely many numbers with the property α_n. Nevertheless, from what we have just said and from the definition of α it follows at once that if a number has the property α then that number also has the property α_n for all sufficiently large n. Similar facts hold also for the β_n and β.

Let us now examine the definitions of α_n, β_n and φ_n more closely. We shall first explain these definitions informally; later on we shall show formally that they actually work: the degrees of α and β are incomparable.

Suppose that $(n)_0$ is even, say $(n)_0=2e$, so that we are in case 1. What we *try* to do in this case is to ensure that $\alpha\neq\lambda x[\{e\}(\beta;x)]$ by allowing $\alpha(\varphi(2e))$ to hold (i.e., to be 0) iff $\{e\}(\beta;\varphi(2e))=1$. However, at this stage we do not yet have β and $\varphi(2e)$ but only their n^{th} approximations β_n and $\varphi_n(2e)$; so we use these instead. Thus we try to compute $\{e\}(\beta_n;\varphi_n(2e))$, and if we find that its value is 1, we allow $\alpha_{n+1}(\varphi_n(2e))$ to hold. Then $\alpha(\varphi_n(2e))$ will certainly hold. (If $\alpha_n(\varphi_n(2e))$ already holds, then $\alpha_{n+1}(\varphi_n(2e))$ will hold automatically, and we do not have to do anything. We have to add $\varphi_n(2e)$ as a new number only if $\alpha_n(\varphi_n(2e))$ does not hold. This is why the clause $\neg\alpha_n(\varphi_n(2e))$ is included in ($*$).)

Of course, we must not make the definition of α_{n+1} dependent on an *indefinite* search for a computation of the value $\{e\}(\beta_n;\varphi_n(2e))$, because this value may not be defined, in which case α_{n+1} would not be recursive. Therefore we confine our search for a computation code b to numbers $<n$. Since there are infinitely many numbers $m>n$ with $(m)_0=(n)_0=2e$, we hope to resume our search at those later m^{th} stages, so that if the computation code b exists at all, it will eventually be discovered even if it is not smaller than our present n.

However, here we have a snag: if $m > n$ and $(m)_0 = (n)_0 = 2e$, then at the future m^{th} stage we shall be searching for a computation of $\{e\}(\beta_m; \varphi_m(2e))$ rather than of $\{e\}(\beta_n; \varphi_n(2e))$. How can we overcome this snag? Suppose that $\{e\}(\beta_n; \varphi_n(2e))$ is defined and that the corresponding computation code is b. Then the computation of $\{e\}(\beta_n; \varphi_n(2e))$ depends not on the *whole* of β_n but only on $\tilde{\beta}_n(b)$ — i.e., on the first $b+1$ values of β_n. Now, if n is sufficiently large, then any number $\leqslant b$ that has the property β will already have the property β_n, so that $\tilde{\beta}_n(b) = \tilde{\beta}(b)$. And if $m > n$ then $\tilde{\beta}_m(b) = \tilde{\beta}(b) = \tilde{\beta}_n(b)$. Therefore the computation of $\{e\}(\beta_n; \varphi_n(2e))$ will be the same as that of $\{e\}(\beta_m; \varphi_n(2e))$. If we can arrange matters so that for sufficiently large n we have also $\varphi_m(2e) = \varphi_n(2e)$ for all $m > n$, then we will have overcome the snag; because for sufficiently large n the computation of $\{e\}(\beta_n; \varphi_n(2e))$ will be the same as the computation of $\{e\}(\beta_m; \varphi_m(2e))$ for any $m > n$.

For any given n and z, let us say that $\varphi_n(z)$ is *pegged* if $\varphi_m(z) = \varphi_n(z)$ for all $m > n$. The above analysis of what happens in case 1 and a similar analysis of case 2 suggest that the φ_n should be defined in such a way that for every z there is some n for which $\varphi_n(z)$ is pegged. But why not take $\varphi_n(z)$ to be independent of n in the first place? To answer this question let us take another look at the definitions of α_n, β_n and φ_n.

Suppose again that $(n)_0 = 2e$, so that we are in case 1. Suppose also that condition $(*)$ holds for some $b < n$. This means that we are in subcase 1′; we then allow $\varphi_n(2e)$ to have the property α_{n+1} and hence also the property α. Thus, we *would* be sure that $\alpha \neq \lambda x[\{e\}(\beta; x)]$ *if* we can arrange that $\{e\}(\beta; \varphi_n(2e)) = 1$. Now, what we do know from $(*)$ is that $\{e\}(\beta_n; \varphi_n(2e)) = 1$. Also, since b is the corresponding computation code, we know that the computation of $\{e\}(\beta_n; \varphi_n(2e))$ depends only on the values $\beta_n(x)$ for $x \leqslant b$. Therefore all will be well *if* we can ensure that $\beta(x) = \beta_n(x)$ for all $x \leqslant b$.

Now, by looking at case 2 we see that every number with the property β but not β_n must be of the form $\varphi_m(z)$, where z is odd and $m > n$. Thus, if z is odd, we would like to make sure that $\varphi_m(z) > b$ for all $m > n$. Actually, we take $\varphi_m(z)$ to be non-decreasing in m, so we only need to ensure that $\varphi_{n+1}(z) > b$ for odd z. Of course, if for some odd z it happens that $\varphi_n(z) \leqslant b$, then $\varphi_n(z)$ must *not* be pegged if we want to have $\varphi_{n+1}(z) > b$.

Thus we are faced with two conflicting requirements. On the one hand, for each z we want $\varphi_n(z)$ to be pegged for some n. On the other hand, whenever n falls under subcase 1′ we would like to ensure that $\varphi_{n+1}(z)$ is sufficiently large for odd z. (And, symmetrically, whenever n falls under subcase 2′ we want to ensure that $\varphi_{n+1}(z)$ is sufficiently large for even z.)

The conflict between these requirements is resolved as follows. In subcase 1′ we do not insist that $\varphi_{n+1}(z) > b$ for *all* odd z but only for those odd z which are $> 2e$. For the other values of z we let $\varphi_{n+1}(z)$ be the same as $\varphi_n(z)$. We do a similar thing in subcase 2′. A simple argument (which we shall present below) shows that this procedure ensures that for each z there is some n such that $\varphi_n(z)$ is pegged. Also, if n falls under subcase 1′ *and is so large that $\varphi_n(2e)$ is already pegged* — which is the only case that really matters — it turns out that $\beta_n(x) = \beta(x)$ for all $x \leqslant b$, as required. (A similar fact also holds in subcase 2′.)

There is another point concerning the definition of the φ_n which requires an explanation. If $(n)_0 = 2e$ and n is so large that $\varphi_n(2e)$ is pegged, we want $\varphi_n(2e)$ to have the property α *only* if $\{e\}(\beta;\varphi_n(2e)) = 1$. Now, if we had $\varphi_n(2e) = \varphi_m(2d)$ for some m and some $d \neq e$, then $\varphi_n(2e)$ could get the property α "by mistake", i.e., even if $\{e\}(\beta;\varphi_n(2e)) \neq 1$; because it may happen that $\{d\}(\beta;\varphi_m(2d)) = 1$. We therefore want to ensure that $\varphi_n(z) \neq \varphi_m(y)$ whenever $z \neq y$. This is done by defining $\varphi_n(z)$ in such a way that we have $(\varphi_n(z))_0 = z$ for all n and z. (In fact, $\varphi_n(z)$ always has the form $2^z \cdot 3^u$ for some u.)

Having explained the definitions of the α_n and β_n, and especially the intricacies in the definition of the φ_n — on which the whole construction hinges — let us now show formally that the construction actually works.

8.1. LEMMA. *For every z there is an n such that $\varphi_n(z)$ is pegged.*

PROOF. We use induction on z. Suppose that z is odd. We know that any number having the property α must have the property α_n for all sufficiently large n. Now, by the induction hypothesis there are only finitely many different numbers of the form $\varphi_m(y)$ with $y < z$. Therefore, if n is *sufficiently large,* then any among these numbers that have the property α must also have the property α_n.

For all such large n we must have $\varphi_{n+1}(z) = \varphi_n(z)$. Otherwise, n would fall under subcase 1′, and $(n)_0 = 2e < z$, because this is the only case in which $\varphi_{n+1}(z) \neq \varphi_n(z)$ for odd z. But, in subcase 1′, $\varphi_n(2e)$ is allowed to have the property α_{n+1}, hence also α, while it does *not* have the property α_n. Since $2e < z$, this contradicts the "sufficient largeness" of n.

A similar argument applies to even z. ∎

For any given z we put $\varphi(z) = \varphi_n(z)$, where n is taken so large that $\varphi_n(z)$

is pegged. (Of course, the least n for which $\varphi_n(z)$ is pegged depends on z.) By Lemma 8.1, $\varphi(z)$ is uniquely defined for each z.

We now prove:

8.2. THEOREM. *The r.e. properties α and β have incomparable degrees.*

PROOF. To prove that α is not β-recursive, we show that for every e

$$\alpha(\varphi(2e)) \text{ holds iff } \{e\}(\beta;\varphi(2e))=1.$$

First suppose that $\{e\}(\beta;\varphi(2e))=1$. Then for some (unique) b we have

$$(1) \qquad T(\beta;\, e,\varphi(2e),\, b) \wedge (U(b)=1).$$

We take $n=2^{2e}\cdot m$, where m is odd. We choose m so large that $n>b$, and $\varphi_n(2e)$ is already pegged, and $\tilde{\beta}(b)=\tilde{\beta}_n(b)$ (i.e., every number $<b$ having the property β already has the property β_n). Then (1) can be written in the form

$$T(\beta_n;\, e,\varphi_n(2e),\, b) \wedge (U(b)=1).$$

Thus if $\varphi_n(2e)$ does *not* have the property α_n then $(*)$ holds; and, since $b<n$ and $(n)_0=2e$, it follows that n falls under subcase 1'. Then $\varphi_n(2e)$ is made to have the property α_{n+1}, and hence also α.

If $\varphi_n(2e)$ *does* have the property α_n, then again it has the property α. Thus in any event $\alpha(\varphi_n(2e))$ holds. But $\varphi_n(2e)=\varphi(2e)$ by the choice of n.

Conversely, suppose that $\alpha(\varphi(2e))$ holds. Then there is some (unique) n such that $\alpha_n(\varphi(2e))$ does not hold but $\alpha_{n+1}(\varphi(2e))$ does. This n must therefore fall under subcase 1' and

$$(2) \qquad \varphi(2e)=\varphi_n((n)_0).$$

But the φ_n are defined in such a way that $\varphi_m(y)$ and $\varphi_n(z)$ are the same only if $y=z$ (in fact, $(\varphi_n(z))_0=z$). Also, $\varphi(2e)=\varphi_m(2e)$ for sufficiently large m. Therefore by (2) we have $(n)_0=2e$. It now also follows from (2) that $\varphi_n(2e)$ must be pegged.

Since n falls under subcase 1', there exists a unique $b<n$ such that $(*)$ holds. In particular, since $\varphi_n(2e)$ is pegged, we can replace $\varphi_n(2e)$ in $(*)$ by $\varphi(2e)$ and obtain

$$(3) \qquad T(\beta_n;\, e,\varphi(2e),\, b) \wedge (U(b)=1).$$

(We have omitted the third conjunct because we do not need it.) If we can show that $\tilde{\beta}_n(b)=\tilde{\beta}(b)$ then it will follow at once from (3) that $\{e\}(\beta;\varphi(2e))=1$, as required.

Suppose that $\tilde{\beta}_n(b)\neq\tilde{\beta}(b)$. Then there exists some $t\leqslant b$ such that t has the property β but not β_n. Therefore there is some $m\geqslant n$ such that t has the property β_{m+1} but not β_m. Thus m falls under subcase 2′. Hence we have not only $m\geqslant n$ but actually $m>n$, since n falls under subcase 1′. Also, we must have $t=\varphi_m(2d+1)$, where $(m)_0=2d+1$.

Since $m>n$, we have

$$(4) \qquad t=\varphi_m(2d+1)\geqslant\varphi_{n+1}(2d+1).$$

Now, either $2d+1>2e$ or $2e>2d+1$. If $2d+1>2e$, then (since n falls under subcase 1′) we have $\varphi_{n+1}(2d+1)=3^b\cdot\varphi_n(2d+1)>b$ and by (4) we get $t>b$, contradicting the assumption that $t\leqslant b$.

On the other hand, if $2e>2d+1$ then (since m falls under subcase 2′) we have $\varphi_{m+1}(2e)=3^c\cdot\varphi_m(2e)$, where c is some computation code and hence is positive. Therefore $\varphi_{m+1}(2e)>\varphi_m(2e)$, and since $m>n$ we have also $\varphi_{m+1}(2e)>\varphi_n(2e)$. But this contradicts the fact that $\varphi_n(2e)$ is pegged.

This completes the proof that α is not β-recursive. Symmetrically, we can show that β is not α-recursive. ∎

8.3. PROBLEM. Show that the sequence φ defined above is not recursive. (Find a nice sequence ψ such that the identity $\{\psi(z)\}(\beta;x)=\neg\beta(z)$ holds, and show that $\beta(z)=\alpha\varphi(2\cdot\psi(z))$.)

§ 9. Historical and bibliographical remarks

For a wealth of information on all the topics touched upon in this chapter (as well as on other topics in recursion theory) we refer the reader to the treatise ROGERS [1967]. The topics of §§3–6 are studied in detail by SMULLYAN [1961], to whom our own treatment owes much. Two books devoted entirely to degrees of unsolvability (touched upon in §§7 and 8) are SACKS [1963] and SHOENFIELD [1971].

The arithmetical hierarchy was first studied by KLEENE [1943], who proved the Enumeration Theorem (Thm. 1.7) and the Hierarchy Theorem (Thm. 1.9) as well as various other results, some of which are included in §1. Broadly similar results were obtained independently, but published only after the second world war, by MOSTOWSKI [1947]. Thm. 1.16 is due to POST [1948]. The result of §2 is due to HILBERT and BERNAYS [1939].

The study of creative sets and various notions of reducibility was started by Post [1944]. Important results — including the remarkable fact that one-one equivalence implies recursive isomorphism, reproduced here as Thm. 6.3 — were proved by Myhill [1955]. Productive sets were studied by Dekker [1955].

The notion of degree of unsolvability was introduced by Post [1944], following the ideas of Turing [1939]. The result reproduced here as Thm. 7.5 is due to Shoenfield [1960].

Post [1944] posed the question that came to be known as "Post's problem". It was solved by Friedberg [1957] and Mučnik [1956].

CHAPTER 9

INTUITIONISTIC FIRST-ORDER LOGIC

This chapter is a brief — and rather sketchy — introduction to the logic which governs mathematical statements when these are interpreted in a *constructive* (as opposed to *structural*) way. The resulting system of logic will be compared with the classical system developed in the first three chapters of this book. We shall therefore assume familiarity with the material covered there, including the method of first-order tableaux.

§ 1. Preliminary discussion

In §2 of Ch. 1 we remarked that a great many mathematical statements are about structures. With this in mind, we have set up first-order formal languages, in which such statements can be formalized. The structural interpretation of sentences of such a language was then precisely formulated in the BSD (2.1.1).

Underlying this approach is the idea that structures are given to us as — or *as if* they were — entities existing "out there", finished and complete before we ever come to use them in our semantic analysis of formulas. For each $\mathscr{L}$-structure, any $\mathscr{L}$-sentence has a definite truth value — it is either *true* or *false* in that structure, independently of whether we actually know (or shall ever know) which of the two is the case.

Note that a structure $\mathfrak{U}$ is a certain system of classes: the domain U of $\mathfrak{U}$ is an arbitrary non-empty class; each basic n-ary relation is a sub-class of U^n; and each basic n-ary operation is completely determined by its graph, which is a sub-class of U^{n+1}. Except in relatively trivial cases, some or all of these classes are infinite. Thus, a structural interpretation of mathematical statements commits us to regarding systems of (infinite) classes, as — or *as if* they were — finished pre-existing entities.

Now, it is philosophically doubtful as to what extent such a view is justified.[1] Moreover, certain important mathematical statements neither require a structural interpretation nor lend themselves to it without losing some of their content.

Consider the following well-known mathematical statement:

(1.1) *For any given natural number n we can find a prime number greater than n.*

Of course, this statement can be formalized in a suitable first-order language. For example, in the language $\mathscr{L}$ of Ch. 7, and using the notation of Prob. 7.9.16, we can write

(1.2) $\forall v_2 \exists v_1[\neg\gamma \wedge v_2 < v_1]$;

and we may wish to say that statement (1.1) *means* that the sentence (1.2) expresses a truth about the structure $\mathfrak{N}$ of natural numbers, or that (1.2) holds in $\mathfrak{N}$. But in this we would be doing an injustice to (1.1); because what it says (and means) is that *we* are able to *do* something, and this message gets lost in the claim that (1.2) holds in $\mathfrak{N}$.

In fact (1.1) claims that a certain *construction* can be made. We prove this claim by prescribing how to make the constuction and showing that it yields the required result. (E.g., thus: *Calculate* $n!+1$ *and, using a previously established construction which yields for any number* >1 *the least prime factor of that number, find the least prime factor p of* $n!+1$. *This p is the result of the new construction. We already know from the previous construction that p is prime and divides* $n!+1$. *Therefore, by another previous result, p cannot divide n! and hence must be* $>n$.)

Moreover, to understand (1.1) — *and its proof* — we do not need to assume the pre-existence of the structure $\mathfrak{N}$, or of any infinite class, as a finished entity. What we do need to assume is that we can construct natural numbers 0, 1, 2 and so on without ever having to stop,[2] but without ever having constructed "all" natural numbers; and that we can perform certain operations (which are also constructions) on constructions made previously.

Under this *constructive* point of view — which in the case of (1.1) seems most natural — a statement such as (1.1) conveys more information (and also information of a different *kind*) than does the claim that a corresponding formal sentence such as (1.2) holds in some infinite structure such as $\mathfrak{N}$.

[1] The philosophical position which claims that mathematical entities, including infinite classes, actually exist as ideal objects independent of the human mind is known as *Platonism*.

[2] In this there is, of course, an idealization; because life is too short.

Most mathematicians recognize this and attribute special importance to constructive statements and proofs.[1]

Since the constructive interpretation of a statement conveys more information than the structural one, it should come as no surprise that it may require more effort to produce a constructive proof than a classical (non-constructive) proof of the same statement. In other words, certain arguments which occur frequently in classical proofs are not acceptable from a constructive viewpoint, and cannot be used in constructive proofs.

As an example of this, consider the statement:

(1.3) *There are irrational real numbers a and b such that a^b is rational.*

From a classical point of view, we can prove (1.3) as follows. Let $b=2^{1/2}$. It is well known that b is irrational. If b^b is rational, put $a=b$ and we are through. If b^b is irrational, put $a=b^b$; then $a^b=2$, which is rational.

However, if we want to interpret (1.3) constructively, then it must be re-phrased (or re-interpreted) as a claim that we can *construct* two *particular* real numbers a and b, and *prove* that they are irrational and that a^b is rational. The above proof does not actually construct the required a and b. After reading and understanding it, we still do not know any particular pair of numbers with the required property. To convert this proof into a constructive proof of (the constructive version of) statement (1.3), one would need to make a much deeper study of the numbers $b=2^{1/2}$ and b^b. (In fact it can be shown that b^b is irrational.)

In this chapter we propose to pursue some of the consequences of the constructive point of view. We shall not discard first-order languages as a means for *formalizing* mathematical statements, but we shall try to see what happens when a *structural* interpretation of formal sentences is replaced by a purely *constructive* one.

Just as the structural approach suggested the so-called *classical* first-order logic developed earlier in this book, so the constructive approach will naturally suggest a somewhat different logic governing first-order sentences. This is known as *intuitionistic* first-order logic.

[1] This was also the attitude we adopted earlier in this book. Although the interpretation chosen for the *object language* $\mathscr{L}$ was structural, results *about* $\mathscr{L}$ were, in most cases where this was possible, formulated and proved constructively.

§ 2. Philosophical remark

In taking the constructive viewpoint seriously and studying intuitionistic logic, we are not committing ourselves to the *intuitionistic philosophy of mathematics* of Brouwer and his more orthodox disciples. This philosophy claims, among other things: that the constructive interpretation of mathematical statements is the *only* legitimate one; that any mathematical statement or proof which cannot be understood constructively is meaningless and must be rejected; that mathematical activity is essentially subjective and consists in mental constructions in the mind of the *individual* mathematician, rather than a primarily *social* activity concerned with objective facts independent of any individual mind; that it is in principle a pre-linguistic or extra-linguistic activity and therefore ideally no language is required for *doing* mathematics but only for *recording* and for *transmitting* it from one individual mind to another; and that when mathematical ideas are expressed in linguistic form (even in a formal language) they are necessarily distorted and lose some of their rigour and sharpness.

Most mathematicians and philosophers of mathematics do not accept these views. While admitting that a constructive approach is always valuable, and sometimes even more natural than a structural one, they do not reject the latter either, and regard the fruitful and intricate interplay between the two as the very soul of mathematics. For Platonists, the structural approach is justified in itself. Others, who are neither intuitionists nor Platonists, believe (or hope) that, while infinite structures may not really exist as Platonic objects, it is possible to justify using them *as if* they did.

§ 3. Constructive meaning of sentences

Let us sketch some of the features of a constructive interpretation of formal sentences.

The variables occurring in a sentence are taken to range over *constructions* of some prescribed kind. Let us call these the *basic* constructions. For example, in the case considered in §1, if we want (1.2) to be a formal version of (1.1), we take the natural numbers as basic constructions. As we have already pointed out, if there are infinitely many basic constructions, we do not think of them as constituting a *finished* collection, but as being generated one by one, as and when required. The same applies to other constructions discussed below.

Besides the basic constructions we also consider constructions of more complicated types. For example, there are constructions that can be applied to one or several given basic constructions, to yield yet another basic construction. More generally, a construction can be applied to zero or more constructions of specified kinds, to yield other constructions.[1]

Under a constructive interpretation, each formal sentence asserts that a construction satisfying certain conditions can be made. *Proving* a sentence consists in performing such a construction. Thus, a *proof* of a given sentence is nothing but a construction whose performability the given sentence asserts. For example, under a suitable constructive interpretation, the sentence (1.2) asserts the possibility of making — and is proved by actually making — a construction which can be applied to any number (i.e., any basic construction) n, to yield a number p together with a proof (which is also a construction) that p is prime and a proof that $p>n$.

Thus, under a constructive interpretation, instead of saying "the sentence α asserts that such-and-such a construction can be made" we can say equivalently "such-and-such a construction is a proof of α" or "α is proved by such-and-such a construction".

Note the difference: under a structural interpretation (an $\mathscr{L}$-structure) the assertion made by an $\mathscr{L}$-sentence is quite a separate matter from the possibility of proving the assertion; but under a constructive interpretation these two matters are inseparable.

§ 4. Constructive interpretations

In order to see what constructive meanings should be attributed to the connectives and quantifiers, we shall have to proceed more formally.

We let $\mathscr{L}$ be a first-order language. For the sake of simplicity *we shall assume throughout this chapter except in* §13 *that* $\mathscr{L}$ *is a language without equality (and hence without function symbols other than constants).*

The variables of $\mathscr{L}$ are enumerated in a fixed *alphabetic ordering*: $\mathbf{v}_1$, $\mathbf{v}_2$, $\mathbf{v}_3$, etc.

Since we cannot presuppose that the constructive meanings of $\vee$ and $\wedge$ can be reduced to those of $\neg$ and $\rightarrow$ as in classical logic, we take all these

[1] Complex constructions occur frequently in mathematics, and have already cropped up many times in this book. For example, in proving Lemma 1.8.6 we set up a construction that can be applied to two given propositional tableaux of specified kinds, to yield a third propositional tableau. Tableaux themselves are constructions, of course. In proving Thm. 6.5.5. we set up a construction that can be applied to any description of a functional to yield a program for computing that functional. Descriptions and programs are themselves constructions.

four connectives as primitive symbols (cf. §15 of Ch. 1). Similarly, we take both quantifiers $\exists$ and $\forall$ as primitive.

Formulas are formed in the usual way. We assume that the definitions of the various syntactical notions (e.g., *free occurrence* of a variable in a formula, *substitution, alphabetic change*, etc.) have been adapted to the present setting. In this chapter the *degree* of a formula $\boldsymbol{\alpha}$ (briefly, $\deg \boldsymbol{\alpha}$) is taken to be the total number of occurrences of the four connectives and two quantifiers in $\boldsymbol{\alpha}$. A *constructive interpretation* $\mathfrak{C}$ of $\mathscr{L}$ consists of the following ingredients.

(1) A prescription for generating certain constructions, called the *basic* constructions or *individuals*[1] of $\mathfrak{C}$.

(2) A construction that can be applied to any constant $\mathbf{a}$ of $\mathscr{L}$, to yield an individual $\mathbf{a}^{\mathfrak{C}}$ of $\mathfrak{C}$.

(3) A decision procedure that can be applied to any construction f and any k-ary predicate symbol $\mathbf{P}$ of $\mathscr{L}$ and any k-tuple $\langle b_1,\ldots,b_k\rangle$ of individuals of $\mathfrak{C}$, whereby we can decide whether or not f is a *proof* of the *atomic statement* $\mathbf{P}^{\mathfrak{C}}(b_1,\ldots,b_k)$.

Let $\mathfrak{C}$ be a constructive interpretation. Let $\mathbf{x}_1,\ldots,\mathbf{x}_n$ be distinct variables, and let $\boldsymbol{\alpha}$ be any formula all of whose free variables are among $\mathbf{x}_1,\ldots,\mathbf{x}_n$. With any given n-tuple $\langle a_1,\ldots,a_n\rangle$ of individuals of $\mathfrak{C}$ associate a *statement* $\boldsymbol{\alpha}^{\mathfrak{C}}[\mathbf{x}_1/a_1,\ldots,\mathbf{x}_n/a_n]$. Intuitively, this is the statement made by the formula $\boldsymbol{\alpha}$ under the interpretation $\mathfrak{C}$, when $a_1,\ldots,a_n$ are taken as the values of $\mathbf{x}_1,\ldots,\mathbf{x}_n$ respectively.

When we deal with a fixed interpretation $\mathfrak{C}$, we shall omit the superscript and write "$\boldsymbol{\alpha}[\mathbf{x}_1/a_1,\ldots,\mathbf{x}_n/a_n]$". Also, when there is no risk of confusion, we shall abbreviate this further and write simply "$\boldsymbol{\alpha}[a_1,\ldots,a_n]$".

To explain the meaning of a statement, we should specify what constructions the statement asserts to be performable. In view of what we saw in §3, this is equivalent to specifying what constructions constitute proofs of the statement.

Proceeding by induction on $\deg \boldsymbol{\alpha}$, we shall now specify what a proof of $\boldsymbol{\alpha}[a_1,\ldots,a_n]$ is. (What follows cannot be regarded as a rigorous *definition* but as an *explanation* which, we hope, will serve as a heuristic guide to the constructive meaning of formulas.)

First, let $\boldsymbol{\alpha}=\mathbf{P}\mathbf{r}_1\ldots\mathbf{r}_k$, where $\mathbf{P}$ is a k-ary predicate symbol and $\mathbf{r}_1,\ldots,\mathbf{r}_k$

[1] We shall use lower case italic letters from the beginning of the alphabet (esp. "a", "b" and "c") to denote these individuals. We reserve "f", "g" and "h" for *arbitrary* constructions.

are terms, i.e., constants or variables. For each $i=1,\ldots,k$ we let b_i be an individual defined as follows. If $\mathbf{r}_i$ is a constant, we put $b_i=\mathbf{r}_i^{\mathfrak{C}}$. If $\mathbf{r}_i$ is a variable, it must be among $\mathbf{x}_1,\ldots,\mathbf{x}_n$; if $\mathbf{r}_i$ is $\mathbf{x}_j$ we put $b_i=a_j$. We now let $\boldsymbol{\alpha}[a_1,\ldots,a_n]$ be the atomic statement $\mathbf{P}^{\mathfrak{C}}(b_1,\ldots,b_k)$. The proofs of this statement are already specified as part of the definition of $\mathfrak{C}$, in clause (3) above.

Now let $\boldsymbol{\alpha}=\neg\boldsymbol{\beta}$. The statement $\boldsymbol{\alpha}[a_1,\ldots,a_n]$ is thus the negation of $\boldsymbol{\beta}[a_1,\ldots,a_n]$. What should a proof of $\boldsymbol{\alpha}[a_1,\ldots,a_n]$ be like? Surely, it must show that $\boldsymbol{\beta}[a_1,\ldots,a_n]$ is unprovable. Constructively, this means that *a proof of* $\boldsymbol{\alpha}[a_1,\ldots,a_n]$ *is a construction g which, when applied to any construction f, yields a proof that f is not a proof of* $\boldsymbol{\beta}[a_1,\ldots,a_n]$ *(plus a proof that g has this property).*

Since this *constructive meaning of negation* is rather tricky, let us illustrate it by analysing a proof of a negation statement presented in Ch. 1. Consider the proof of the consistency of the propositional calculus immediately following Thm. 1.11.3. It consists in showing that *the empty set of formulas does not have a (propositional) confutation.* Suppose therefore that, for any finite set of formulas $\boldsymbol{\Phi}$, the statement $\boldsymbol{\beta}[\boldsymbol{\Phi}]$ says that $\boldsymbol{\Phi}$ has a confutation. A (constructive) proof of $\boldsymbol{\beta}[\boldsymbol{\Phi}]$ is simply a confutation of $\boldsymbol{\Phi}$, which is of course a special kind of construction. From the definition of the notion *confutation of* $\boldsymbol{\Phi}$ we know how to tell whether or not any construction f presented to us is a confutation of $\boldsymbol{\Phi}$. In particular, two of the necessary conditions are:

(i) f should be a tableau having $\boldsymbol{\Phi}$ as its initial node;

(ii) if $\boldsymbol{\Phi}$ does not contain a prime formula and its negation, f must have more nodes than one, and hence $\boldsymbol{\Phi}$ has to contain some formula to which one of the three tableau rules is applicable.

Thus, to show that f is *not* a confutation of $\boldsymbol{\Phi}$, it is enough to show that f fails to satisfy (i) or (ii). Now, the statement that we want to prove is $\neg\boldsymbol{\beta}[\emptyset]$. It is proved by showing that, for any construction f and for $\boldsymbol{\Phi}=\emptyset$, at least one of the conditions (i) and (ii) must break down: if f is not a tableau having $\emptyset$ as its initial node then (i) fails, and if f has $\emptyset$ as its initial node then (ii) fails. Also, every construction f either is or is not a tableau having $\emptyset$ as its initial node (and we can always tell which of these two is the case!), so that for each f we can actually locate one of the two conditions (i) and (ii) which fails.

Let us now return to our inductive description of what a proof of $\boldsymbol{\alpha}[a_1,\ldots,a_n]$ is.

Let $\alpha=\beta\wedge\gamma$. Then *a proof of* $\alpha[a_1,...,a_n]$ *is a construction consisting of two parts (or, an ordered pair of constructions) the first of which is a proof of* $\beta[a_1,...,a_n]$ *and the second a proof of* $\gamma[a_1,...,a_n]$. This explanation is both clear and highly natural, and calls for no further comment.

Next, let $\alpha=\beta\vee\gamma$. *Then a proof of* $\alpha[a_1,...,a_n]$ *is any construction which is a proof of* $\beta[a_1,...,a_n]$ *or a proof of* $\gamma[a_1,...,a_n]$. This is also very natural. But note that it implies that $(\beta\vee\neg\beta)[a_1,...,a_n]$ may not be provable. For we must admit the possibility that, while we cannot find a proof of $\beta[a_1,...,a_n]$, we may not be able to find a proof of $(\neg\beta)[a_1,...,a_n]$ either. After all, a proof of $(\neg\beta)[a_1,...,a_n]$ gives us not merely the *fact* that $\beta[a_1,...,a_n]$ is unprovable, but a *constructive* proof of this fact. Thus the law of the excluded middle *(tertium non datur)* does not generally hold under a constructive interpretation.

Now let $\alpha=\beta\rightarrow\gamma$. The natural thing to do here is to take *a proof of* $\alpha[a_1,...,a_n]$ to be *a construction g which, whenever it is applied to a proof f* of $\beta[a_1,...,a_n]$ *yields a proof of* $\gamma[a_1,...,a_n]$ *(plus a proof that g has this property)*.

To deal with the quantifiers, we introduce the following notation. If $\mathbf{x}$ is the same as $\mathbf{x}_i$, then

$$(*)\qquad \alpha[\mathbf{x}_1/a_1,...,\mathbf{x}_n/a_n][\mathbf{x}/a]$$

is the same as $\alpha[\mathbf{x}_1/b_1,...,\mathbf{x}_n/b_n]$, where $b_j=a_j$ for all $j\neq i$, and $b_i=a$. If $\mathbf{x}$ is different from all the $\mathbf{x}_i$, then $(*)$ is the same as $\alpha[\mathbf{x}_1/a_1,...,\mathbf{x}_n/a_n,\mathbf{x}/a]$.

Now let $\alpha=\forall\mathbf{x}\beta$. Here it is natural to take as *a proof of* $\alpha[a_1,...,a_n]$ *a construction g which, when applied to any individual a, yields a proof of* $\beta[\mathbf{x}_1/a_1,...,\mathbf{x}_n/a_n][\mathbf{x}/a]$ *(plus a proof that g has this property)*.

Finally, let $\alpha=\exists\mathbf{x}\beta$. Here too there is only one natural way to proceed, provided we remember that the constructive meaning of $\exists$ is not merely "there exists..." (exists *where?*) but "we can construct...". *A proof of* $\alpha[a_1,...,a_n]$ *is a construction consisting of two parts (or, an ordered pair of constructions) the first of which is a basic construction (i.e., an individual) a and the second a proof of* $\beta[\mathbf{x}_1/a_1,...,\mathbf{x}_n/a_n][\mathbf{x}/a]$.

The above explanation of the term *proof of* $\alpha[a_1,...,a_n]$ cannot be regarded as a precise definition, for several reasons.

First, it uses notions — especially the notion of *construction* — which are left too vague. To turn our explanation into a definition, one would need to have as a background theory a general theory about constructions.

Second, the explanation is — to use a technical term — very *impredicative*: in the part dealing with negation, for instance, a proof of $(\neg\beta)[a_1,...,a_n]$

is characterized as a construction that yields such-and-such results when applied to *arbitrary* constructions. Now, the arbitrary constructions referred to must, presumably, include also the very construction (proof) which is being characterized. This is somewhat suspect. To mitigate this, one could e.g. distinguish constructions of various *levels* and characterize a construction only in terms of its effect on constructions of levels lower than itself. A formula of higher degree should have proofs of higher level.

These and other difficulties can be overcome, at least to some extent, and our explanation can be converted into something much more like a real definition; but this is beyond the scope of the present book.[1] Nevertheless, even as they stand our explanations should be helpful as a heuristic guide and a partial justification for the development in the following sections.

4.1. PROBLEM. Let all the free variables of $\boldsymbol{\alpha}$ be among the distinct variables $\mathbf{x}_1,\ldots,\mathbf{x}_n$, and let $\mathbf{x}$ be another variable. Check that, for any individuals $a_1,\ldots,a_n$, a, a construction is a proof of $\boldsymbol{\alpha}[\mathbf{x}_1/a_1,\ldots,\mathbf{x}_n/a_n,\mathbf{x}/a]$ iff it is a proof of $\boldsymbol{\alpha}[\mathbf{x}_1/a_1,\ldots,\mathbf{x}_n/a_n]$.

4.2. PROBLEM. Let $\boldsymbol{\alpha}$ be as in Prob. 4.1, and let $\boldsymbol{\alpha}'$ be a variant of $\boldsymbol{\alpha}$. Check that a construction is a proof of $\boldsymbol{\alpha}[a_1,\ldots,a_n]$ iff it is a proof of $\boldsymbol{\alpha}'[a_1,\ldots,a_n]$.

4.3. PROBLEM. Let $\mathbf{t}$ be any term (i.e., a variable or a constant), and let the distinct variables $\mathbf{x}_1,\ldots,\mathbf{x}_n$ include all the free variables of $\boldsymbol{\alpha}(\mathbf{x}/\mathbf{t})$. Show how to obtain from any proof of $(\boldsymbol{\alpha}(\mathbf{x}/\mathbf{t}))[a_1,\ldots,a_n]$ a proof of $(\exists\mathbf{x}\boldsymbol{\alpha})[a_1,\ldots,a_n]$.

§ 5. Intuitionistic tableaux

We introduce a new kind of constructive statement, called "$\blacktriangleright$statements" These are statements about finite sets of formulas.

Let $\boldsymbol{\Phi}=\{\boldsymbol{\varphi}_1,\ldots,\boldsymbol{\varphi}_k\}$ be a finite set of formulas and let $\boldsymbol{\psi}$ be a formula. Let all the free variables of $\boldsymbol{\Phi}$ and $\boldsymbol{\psi}$ be among the distinct variables $\mathbf{x}_1,\ldots,\mathbf{x}_n$. Then the statement

$$\boldsymbol{\Phi}\blacktriangleright\boldsymbol{\psi}$$

means that we can perform a construction f such that when applied to any given constructive interpretation $\mathfrak{C}$, any n given individuals $a_1,\ldots,a_n$ of $\mathfrak{C}$ and any k given proofs $f_1,\ldots,f_k$ of $\boldsymbol{\varphi}_1^{\mathfrak{C}}[\mathbf{x}_1/a_1,\ldots,\mathbf{x}_n/a_n],\ldots,\boldsymbol{\varphi}_k^{\mathfrak{C}}[\mathbf{x}_1/a_1,\ldots,\mathbf{x}_n/a_n]$, respectively, f yields a proof of $\boldsymbol{\psi}^{\mathfrak{C}}[\mathbf{x}_1/a_1,\ldots,\mathbf{x}_n/a_n]$. (In particular, in the case $k=0$, for any given $\mathfrak{C}$ and $a_1,\ldots,a_n$, the construction f must yield a proof of $\boldsymbol{\psi}^{\mathfrak{C}}[\mathbf{x}_1/a_1,\ldots,\mathbf{x}_n/a_n]$.)

[1] The interested reader should consult the works referred to at the end of this chapter.

Also, the statement

$$\Phi \blacktriangleright$$

means that $\Phi \blacktriangleright \psi$ for every formula ψ.

If $\Phi \blacktriangleright \psi$, we say that ψ *follows intuitionistically* from Φ. In particular, if $\emptyset \blacktriangleright \psi$, we write this as "$\blacktriangleright \psi$" and say that ψ is *intuitionistically valid.* If $\Phi \blacktriangleright$, we say that Φ is *intuitionistically contradictory.*

It is clear that the statement

(5.1) $\Phi, \psi \blacktriangleright \psi$

always holds.

The following thirteen $\blacktriangleright$ rules can be verified easily, using the explanation of §4. In each case, a proof of the $\blacktriangleright$ statement below the horizontal line can be obtained by a rather simple construction from any given proof(s) of the $\blacktriangleright$ statement(s) above the line. In all cases, Φ is any finite set of formulas, and α and β are any formulas. The name of each rule appears on the left. In the $\blacktriangleright\exists$ rule and the $\forall\blacktriangleright$ rule, **t** is any term. In the $\exists\blacktriangleright$ rule and the $\blacktriangleright\forall$ rule, **y** is a variable *which is not free in any of the formulas below the line. Here as well as in the rest of this chapter* we let the letter "ξ" stand for any formula or for the empty string of symbols. (In the latter case, $\Phi \blacktriangleright \xi$ is simply $\Phi \blacktriangleright$.)

$$\vee\blacktriangleright: \quad \frac{\Phi, \alpha\vee\beta, \alpha \blacktriangleright \xi; \quad \Phi, \alpha\vee\beta, \beta \blacktriangleright \xi}{\Phi, \alpha\vee\beta \blacktriangleright \xi} \qquad \blacktriangleright\vee_1: \quad \frac{\Phi \blacktriangleright \alpha}{\Phi \blacktriangleright \alpha\vee\beta}$$

$$\blacktriangleright\vee_2: \quad \frac{\Phi \blacktriangleright \beta}{\Phi \blacktriangleright \alpha\vee\beta}$$

$$\wedge\blacktriangleright: \quad \frac{\Phi, \alpha\wedge\beta, \alpha, \beta \blacktriangleright \xi}{\Phi, \alpha\wedge\beta \blacktriangleright \xi} \qquad \blacktriangleright\wedge: \quad \frac{\Phi \blacktriangleright \alpha; \quad \Phi \blacktriangleright \beta}{\Phi \blacktriangleright \alpha\wedge\beta}$$

$$\rightarrow\blacktriangleright: \quad \frac{\Phi, \alpha\rightarrow\beta \blacktriangleright \alpha; \quad \Phi, \alpha\rightarrow\beta, \beta \blacktriangleright \xi}{\Phi, \alpha\rightarrow\beta \blacktriangleright \xi} \qquad \blacktriangleright\rightarrow: \quad \frac{\Phi, \alpha \blacktriangleright \beta}{\Phi \blacktriangleright \alpha\rightarrow\beta}$$

$$\neg\blacktriangleright: \quad \frac{\Phi, \neg\alpha \blacktriangleright \alpha}{\Phi, \neg\alpha \blacktriangleright \xi} \qquad \blacktriangleright\neg: \quad \frac{\Phi, \alpha \blacktriangleright}{\Phi \blacktriangleright \neg\alpha}$$

$$\exists\blacktriangleright: \quad \frac{\Phi, \exists \mathbf{x}\alpha, \alpha(\mathbf{x}/\mathbf{y}) \blacktriangleright \xi}{\Phi, \exists \mathbf{x}\alpha \blacktriangleright \xi} \qquad \blacktriangleright\exists: \quad \frac{\Phi \blacktriangleright \alpha(\mathbf{x}/\mathbf{t})}{\Phi \blacktriangleright \exists \mathbf{x}\alpha}$$

$$\forall\blacktriangleright: \quad \frac{\Phi, \forall \mathbf{x}\alpha, \alpha(\mathbf{x}/\mathbf{t}) \blacktriangleright \xi}{\Phi, \forall \mathbf{x}\alpha \blacktriangleright \xi} \qquad \blacktriangleright\forall: \quad \frac{\Phi \blacktriangleright \alpha(\mathbf{x}/\mathbf{y})}{\Phi \blacktriangleright \forall \mathbf{x}\alpha}$$

The first nine rules are the *propositional* $\blacktriangleright$ rules; the last four are the *quantifier* $\blacktriangleright$ rules.

We leave the verification of these rules to the reader.

We have justified (5.1) and the ► rules in terms of the heuristic explanations of §4. We now adopt an *axiomatic* approach: we take (5.1) as an *axiom scheme* and the thirteen ► rules as *rules of inference*. From now on we shall only assert a ► statement if we can derive it from instances of (5.1) by a finite number of applications of the ► rules.

While the explanations of §4 are certainly not acceptable as a rigorous definition, scheme (5.1) and the thirteen ► rules are generally accepted as correct principles of intuitionistic first-order logic. (Thus, any correct rigorous version of the ideas of §4 should vindicate (5.1) and the thirteen rules.) Also, all principles of intuitionistic first-order logic at present generally accepted, can be derived from (5.1) by means of the thirteen rules.

We shall now set up a very efficient tableau method for deriving correct ► statements.

We introduce a new symbol – *(minus)*. By the *negative* of the formula α we mean the expression $-\alpha$. (This should not be confused with the *negation* $\neg\alpha$ of α.) The minus sign is allowed to occur at most once in each expression, and only in front of a whole formula. (Therefore no brackets are needed for writing negatives. For example, in $-\alpha\wedge\beta$ the minus must apply to the *whole* formula $\alpha\wedge\beta$ and not just to α.) By "±formula" we mean formula or negative of a formula.

An *intuitionistic propositional tableau* has at its initial node a finite set of formulas and at most one negative. There are nine *propositional rules* for extending a given tableau. They are presented schematically below. In each case the name of the rule appears on the left. The meaning of the stars that accompany seven of the rules will be explained later.

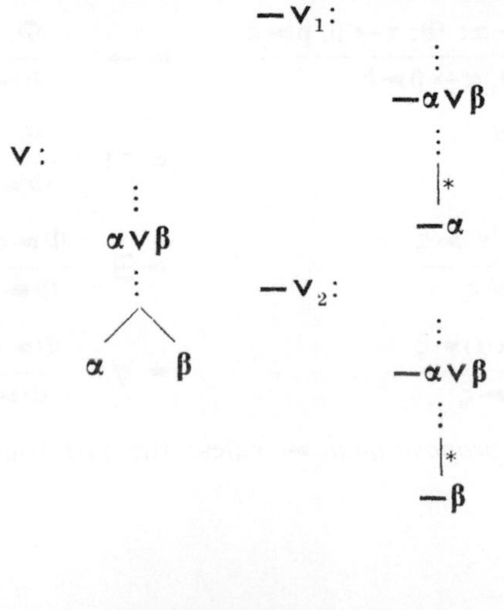

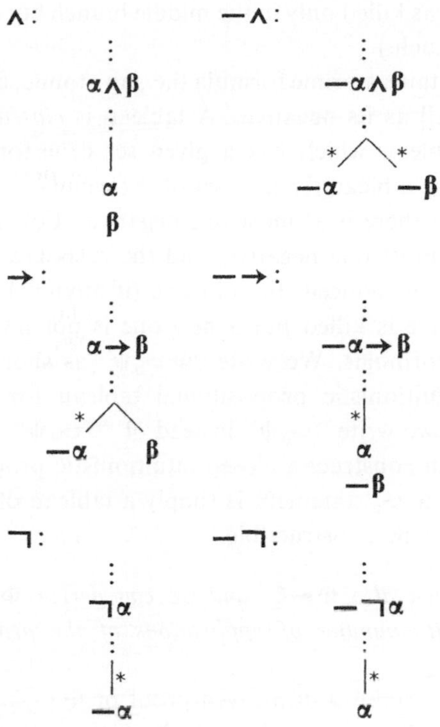

The function of the star is to "kill" any negative occurring earlier in the branch. For example, in the tableau

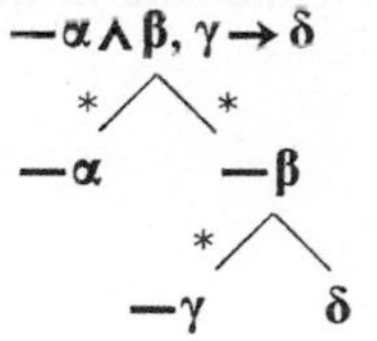

the $\pm$formulas of the leftmost branch are $\gamma \rightarrow \delta$ and $-\alpha$, but *not* $-\alpha \wedge \beta$, which has been "killed" by the star accompanying the edge that leads to $-\alpha$; the $\pm$formulas of the middle branch are $\gamma \rightarrow \delta$ and $-\gamma$, but not $-\alpha \wedge \beta$ nor $-\beta$ because they have been "killed" by the first and second stars, respectively, of this branch; the $\pm$formulas of the rightmost branch are $\gamma \rightarrow \delta$, $-\beta$, δ because $-\alpha \wedge \beta$ has been "killed" by the only star of this branch; but $-\beta$ *does* belong to this branch because there is no star below $-\beta$ in *this* branch. *Notice that a star kills a negative only in the branch in which the star occurs; the same negative may continue to live in other branches.*

(Thus, in our example, $-\beta$ was killed only in the middle branch but continues to live in the rightmost branch.)

A branch is *closed* if it contains a prime formula (i.e., an atomic, existential or universal formula) as well as its negative. A tableau is *closed* if all its branches are closed. A tableau which has a given set of $\pm$formulas as its initial node is said to be a tableau *for* that set of $\pm$formulas.

Notice that in any branch there is at most one negative. For, the initial node is allowed to have at most one negative, and the rules are such that whenever a new negative is introduced, the old one (if any) is killed. (In the $-\neg$ rule the old negative is killed but a new one is not introduced.)

Let Φ be a finite set of formulas. We write "$\Phi \rhd_0 \psi$" as short for "we can construct a closed intuitionistic propositional tableau for $\Phi, -\psi$". In particular, if Φ is empty we write "$\rhd_0 \psi$" instead of "$\emptyset \rhd_0 \psi$". We write "$\Phi \rhd_0$" as short for "we can construct a closed intuitionistic propositional tableau for Φ". A *proof* of a $\rhd_0$ statement is simply a tableau of the kind that the statement asserts to be constructible.

5.2. THEOREM. *If* $\Phi \rhd_0 \xi$ *then also* $\Phi \blacktriangleright \xi$; *and we can derive* $\Phi \blacktriangleright \xi$ *from instances of* (5.1) *by a finite number of applications of the propositional* $\blacktriangleright$ *rules.*

PROOF. By induction on the depth d of a given proof of $\Phi \rhd_0 \xi$.

If $d=0$, then (ξ must be a prime formula and) Φ must contain ξ. Thus $\Phi \blacktriangleright \xi$ is an instance of (5.1).

If $d>0$, we distinguish nine cases according to the tableau rule that gave rise to level 1 in the given proof. For example, if the rule in question is the $\rightarrow$ rule then the given proof begins thus:

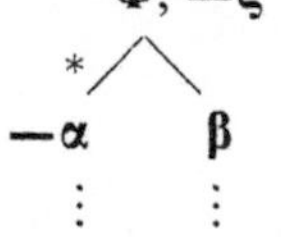

where $\alpha \rightarrow \beta \in \Phi$. (If ξ is the empty string of symbols, then the initial node consists of Φ only.) From this proof we obtain, by amalgamating the initial node with the nodes $\{-\alpha\}$ and $\{\beta\}$ respectively, proofs of $\Phi \rhd_0 \alpha$ and $\Phi, \beta \rhd_0 \xi$ and the depths of these proofs are $<d$. By the induction hypothesis, we can derive $\Phi \blacktriangleright \alpha$ and $\Phi, \beta \blacktriangleright \xi$ from (5.1) by finitely many applications of the propositional $\blacktriangleright$ rules. But since $\alpha \rightarrow \beta \in \Phi$, we get from these two $\blacktriangleright$ statements the statement $\Phi \blacktriangleright \xi$ by one application of the $\rightarrow\blacktriangleright$ rule.

The other eight cases are similar. ∎

5.3. PROBLEM. Show that $\Phi, \psi \rhd_0 \psi$ for any formula ψ. Thus a branch that contains *any* formula (not necessarily prime) as well as its negative is "as good as closed". (Use induction on deg ψ.)
5.4. PROBLEM. Prove the converse of Thm. 5.2: if we can derive $\Phi \blacktriangleright \xi$ from (5.1) using the nine propositional $\blacktriangleright$ rules, then $\Phi \rhd_0 \xi$.

In the following problem — and in the sequel, whenever we make a comparison with classical logic — it should be remembered that we now have four primitive connectives and two primitive quantifiers. In particular, whenever we refer to classical propositional logic, we assume that it is treated as sketched in §15 of Ch. 1.

5.5. PROBLEM. Show that if $\Phi \rhd_0$ then we can construct a propositional confutation (in the sense of Ch. 1) of Φ; and that if $\Phi \rhd_0 \psi$ then we can construct a propositional confutation of $\Phi, \neg\psi$, hence $\Phi \vdash_0 \psi$. (Examine what happens to our present nine tableau rules if the minus sign is replaced by negation.)
5.6. PROBLEM. Show that:
(i) $\rhd_0 \alpha \vee \beta$ iff $\rhd_0 \alpha$ or $\rhd_0 \beta$.
(ii) $\alpha \rhd_0 \beta$ iff $\rhd_0 \alpha \to \beta$.
(iii) $\rhd_0 \alpha \vee \neg\alpha$ and $\rhd_0 \neg\neg\alpha \to \alpha$ are impossible if α is a prime formula.
5.7. PROBLEM. Show that for all α and β
(i) $\rhd_0 \alpha \to \neg\neg\alpha$,
(ii) $\rhd_0 \neg\neg\neg\alpha \to \neg\alpha$,
(iii) $\rhd_0 \neg\neg(\neg\neg\alpha \to \alpha)$,
(iv) $\rhd_0 \neg\neg(\alpha \vee \neg\alpha)$,
(v) $\neg\neg(\alpha \to \beta) \rhd_0 \neg\neg\alpha \to \neg\neg\beta$,
(vi) $\neg\neg\alpha \to \neg\neg\beta \rhd_0 \neg\neg(\alpha \to \beta)$,
(vii) $\alpha \to \beta, \neg\beta \rhd_0 \neg\alpha$,
(viii) $\neg\alpha \vee \beta \rhd_0 \alpha \to \beta$,
(ix) $\rhd_0 \neg\neg[(\neg\alpha \to \beta) \to (\neg\alpha \to \neg\beta) \to \alpha]$.
(Note that even if a formula has been used once to extend a branch, it may have to be used again in some extension of that branch.)

We shall now sketch a method whereby we can effectively decide, for each formula α, whether or not $\rhd_0 \alpha$.

Let a formula α be given. Let Φ be the set of all subformulas of α. Then Φ is finite and we can actually construct it.

We consider statements of the form $\Psi \blacktriangleright \xi$, where $\Psi \subseteq \Phi$ and ξ is the empty string or $\in \Phi$. There are finitely many such statements.

Using Thm. 5.2 and Prob. 5.4 it is easy to see that $\rhd_0 \alpha$ iff there is a sequence $\{\Psi_i \blacktriangleright \xi_i\colon 1 \leqslant i \leqslant n\}$ of statements of the above form, such that

(1) for each i, the statement $\Psi_i \blacktriangleright \xi_i$ is an instance of (5.1) or follows from one or two earlier statements in the sequence by a single application of a propositional $\blacktriangleright$ rule.

(2) $\Psi_n = \emptyset$ and $\xi_n = \alpha$.

(3) All the statements in the sequence are distinct.

There are only finitely many sequences satisfying condition (3), and we can check whether or not there is among them a sequence satisfying the other two conditions as well.

(By the same method we can also decide, for each finite set Φ of formulas and for each ξ, whether or not $\Phi \rhd_0 \xi$.)

First-order intuitionistic tableaux are constructed like propositional ones, except that there are four additional *quantifier* rules for extending branches. Also, the rule for closing a branch is different. The four new rules are represented schematically as follows:

$$\exists\colon \quad \begin{array}{c} \vdots \\ \exists x\alpha \\ \vdots \\ | \\ \alpha(x/y) \end{array} \qquad\qquad -\exists\colon \quad \begin{array}{c} \vdots \\ -\exists x\alpha \\ \vdots \\ |^{*} \\ -\alpha(x/t) \end{array}$$

$$\forall\colon \quad \begin{array}{c} \vdots \\ \forall x\alpha \\ \vdots \\ | \\ \alpha(x/t) \end{array} \qquad\qquad -\forall\colon \quad \begin{array}{c} \vdots \\ -\forall x\alpha \\ \vdots \\ |^{*} \\ -\alpha(x/y) \end{array}$$

Here $\mathbf{t}$ is any term (i.e., variable or constant) and $\mathbf{y}$ is any variable *which is not free in any of the $\pm$formulas of the branch which is being extended.* (The free variables of a negative $-\beta$ are the same as those of β.) The particular $\mathbf{y}$ used in an application of the $\exists$ rule or the $-\forall$ rule is called the *critical* variable of that application.

A branch is *closed* if it contains an atomic formula as well as its negative.

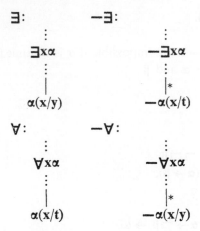

If Φ is a finite set of formulas, we write "$\Phi \rhd \psi$" as short for "we can construct a closed intuitionistic first-order tableau for $\Phi, -\psi$"; and we write "$\Phi \rhd$" as short for "we can construct a closed intuitionistic first-order tableau for Φ". A *proof* of a $\rhd$ statement is a tableau of the kind whose constructibility that statement asserts.

5.8. THEOREM. *If* $\Phi \rhd \xi$ *then* $\Phi \blacktriangleright \xi$; *and we can derive* $\Phi \blacktriangleright \xi$ *from* (5.1) *by a finite number of applications of the thirteen* $\blacktriangleright$ *rules.*

PROOF. Similar to Thm. 5.2. ∎

5.9. PROBLEM. Show that $\Phi, \psi \rhd \psi$ for any formula ψ. Thus a branch containing *any* formula and its negative is "as good as closed". Hence show that if $\Phi \rhd_0 \xi$ then also $\Phi \rhd \xi$.

5.10. PROBLEM. Prove the converse of Thm. 5.8.

5.11. PROBLEM. Show that if $\Phi \rhd$ then we can construct a first-order confutation (in the sense of Ch. 2) of Φ; and if $\Phi \rhd \psi$ then we can construct a first-order confutation for $\Phi, \neg\psi$, hence $\Phi \vdash \psi$.

5.12. PROBLEM. Show that, if α is atomic and contains $\mathbf{x}$, it is impossible that $\rhd \neg\neg\forall \mathbf{x}(\alpha \vee \neg\alpha)$; also it is impossible that $\rhd \neg\neg\forall \mathbf{x}(\neg\neg\alpha \rightarrow \alpha)$.

5.13. PROBLEM. Show that $\rhd \alpha \vee \beta$ iff $\rhd \alpha$ or $\rhd \beta$. Also, $\alpha \rhd \beta$ iff $\rhd \alpha \rightarrow \beta$.

5.14. PROBLEM. Show that $\rhd \exists \mathbf{x}\alpha$ iff we can find a term $\mathbf{t}$ for which $\rhd \alpha(\mathbf{x}/\mathbf{t})$.

5.15. PROBLEM. Show that for all α:

(i) $\rhd \neg\neg\forall \mathbf{x}\alpha \rightarrow \forall \mathbf{x}\neg\neg\alpha$, $\rhd \exists \mathbf{x}\neg\neg\alpha \rightarrow \neg\neg\exists \mathbf{x}\alpha$;

(ii) $\rhd \forall \mathbf{x}\neg\alpha \rightarrow \neg\exists \mathbf{x}\alpha$, $\rhd \neg\exists \mathbf{x}\alpha \rightarrow \forall \mathbf{x}\neg\alpha$;

(iii) $\rhd \exists \mathbf{x}\neg\alpha \rightarrow \neg\forall \mathbf{x}\alpha$.

5.16. PROBLEM. Let α be an atomic formula containing $\mathbf{x}$. Show that it is impossible that $\rhd \neg\forall \mathbf{x}\alpha \rightarrow \exists \mathbf{x}\neg\alpha$. What about $\rhd \forall \mathbf{x}\neg\neg\alpha \rightarrow \neg\neg\forall \mathbf{x}\alpha$ and $\rhd \neg\neg\exists \mathbf{x}\alpha \rightarrow \exists \mathbf{x}\neg\neg\alpha$?

The following three results will be needed in §7.

5.17. PROBLEM. Let $\mathbf{y}_1, \ldots, \mathbf{y}_n$ be any variables. Show that if $\Phi \rhd \xi$ then we can find a proof of $\Phi \rhd \xi$ in which none of these variables is used as a critical variable. (Prove a result similar to Lemma 2.5.2, then proceed as in the proof of Lemma 2.5.3.)

5.18. PROBLEM. Let Ψ be a finite set of formulas, and let ψ be a formula. Show that if $\Phi \rhd \xi$ then we can find a proof T of $\Phi \rhd \xi$ such that Ψ can be added to the initial node of T, yielding a proof of $\Phi \cup \Psi \rhd \xi$. Also show that if $\Phi \rhd$ then we can find a proof T' of $\Phi \rhd$ such that by adding $-\psi$ to the initial node of T' we get a proof of $\Phi \rhd \psi$. (Use 5.17.)

5.19. PROBLEM. Show that, for any variable $\mathbf{z}$ and any term $\mathbf{s}$, we can transform any proof of $\Phi \rhd \xi$ into a proof of $\Phi(\mathbf{z}/\mathbf{s}) \rhd \xi(\mathbf{z}/\mathbf{s})$, where $\xi(\mathbf{z}/\mathbf{s})$ is the empty string if ξ is empty. (Proceed as in 2.5.4.)

§ 6. Kripke's semantics

There are various methods of interpreting $\mathscr{L}$-formulas in such a way that the rules of intuitionistic logic — but not all the rules of classical logic — turn out to be sound. One of these methods, which is rather simple and very appealing intuitively, is due to Kripke. We shall present Kripke's semantics in a purely structural manner, without attempting to bring it into line with the constructive viewpoint[1]. Rather, we shall give Kripke's semantics an independent, but structural, heuristic justification. Also, we shall not deal separately with propositional logic, but start at once with first-order logic.

In what follows, when we say that $\mathbf{t}$ is a *term of* $\boldsymbol{\alpha}$ we mean that $\mathbf{t}$ is a constant occurring in $\boldsymbol{\alpha}$ or a variable free in $\boldsymbol{\alpha}$. We use a similar terminology for sets of formulas.

A *Kripke System* $\mathfrak{K}$ for $\mathscr{L}$ consists of the following ingredients:

(1) A non-empty collection $\Sigma_{\mathfrak{K}}$, whose members are called the *states* of $\mathfrak{K}$.

(2) A partial ordering $\leqslant_{\mathfrak{K}}$ of $\Sigma_{\mathfrak{K}}$.

(3) A mapping $T_{\mathfrak{K}}$ which assigns to every state $A \in \Sigma_{\mathfrak{K}}$ a non-empty set $T_{\mathfrak{K}}(A)$ of $\mathscr{L}$-terms, in such a way that if $A \leqslant_{\mathfrak{K}} B$ then $T_{\mathfrak{K}}(A) \subseteq T_{\mathfrak{K}}(B)$.

(4) A mapping $F_{\mathfrak{K}}$ which assigns to each state $A \in \Sigma_{\mathfrak{K}}$ a (possibly empty) set $F_{\mathfrak{K}}(A)$ of atomic $\mathscr{L}$-formulas such that if $\boldsymbol{\alpha} \in F_{\mathfrak{K}}(A)$ then all the terms of $\boldsymbol{\alpha}$ belong to $T_{\mathfrak{K}}(A)$, and if $A \leqslant_{\mathfrak{K}} B$ then $F_{\mathfrak{K}}(A) \subseteq F_{\mathfrak{K}}(B)$.

When we deal with one particular Kripke system $\mathfrak{K}$, we shall omit the subscript, and write simply "Σ", "$\leqslant$", "T" and "F".

Intuitively, we interpret the states of $\mathfrak{K}$ as "states in the progress of knowledge", and the partial order $\leqslant$ as "possible succession in time": $A \leqslant B$ means that state B may possibly follow state A. At each state A, we "know" certain terms — those belonging to $T(A)$ — and we "know" that certain atomic formulas — those belonging to $F(A)$ — are true. We assume that we never "forget", so that what we know at state A we shall still know at any state that may follow A. Also, we cannot know that an atomic formula is true without knowing at the same time the terms of that formula. In view of this intuitive explanation, conditions (1)–(4) above are quite natural. (The stipulations that $\Sigma \neq \emptyset$ and $T(A) \neq \emptyset$ for all $A \in \Sigma$ are made to exclude trivial cases.)

Let $\mathfrak{K}$ be any Kripke system for $\mathscr{L}$. We shall now define a relation $\Vdash_{\mathfrak{K}}$ between states of $\mathfrak{K}$ and $\pm$formulas of $\mathscr{L}$. (When there is no risk of

[1] For a discussion of this question see references quoted at the end of this chapter.

confusion, we write simply "$\Vdash$", omitting the subscript.) If $A \Vdash \alpha$ holds, we say that α is *forced (to be true)* at A or that A *forces* α. Intuitively, $A \Vdash \alpha$ means that at the state A the formula α is definitely known to be true; and $A \Vdash -\alpha$ means that at the state A the formula α is understood but not known to be true.

6.1. KRIPKE'S SEMANTIC DEFINITION.

(1) If α is atomic, then

$$A \Vdash \alpha \quad \text{iff } \alpha \in F(A).$$

(2) If $\alpha = \beta \vee \gamma$, then

$A \Vdash \alpha$ iff all the terms of α are in $T(A)$, and $A \Vdash \beta$ or $A \Vdash \gamma$.

(3) If $\alpha = \beta \wedge \gamma$, then

$$A \Vdash \alpha \quad \text{iff } A \Vdash \beta \text{ and } A \Vdash \gamma.$$

(4) If $\alpha = \beta \rightarrow \gamma$, then

$A \Vdash \alpha$ iff all the terms of α are in $T(A)$, and $B \Vdash \gamma$ whenever $B \geqslant A$ and $B \Vdash \beta$.

(5) If $\alpha = \neg \beta$, then

$A \Vdash \alpha$ iff all the terms of α are in $T(A)$, and $B \nVdash \beta$ whenever $B \geqslant A$.

(6) If $\alpha = \exists \mathbf{x} \beta$, then

$$A \Vdash \alpha \quad \text{iff } A \Vdash \beta(\mathbf{x}/\mathbf{t}) \text{ for some term } \mathbf{t}.$$

(7) If $\alpha = \forall \mathbf{x} \beta$, then

$$A \Vdash \alpha \quad \text{iff } B \Vdash \beta(\mathbf{x}/\mathbf{t}) \text{ whenever } B \geqslant A \text{ and } \mathbf{t} \in T(B).$$

(8) If α is any formula, then

$A \Vdash -\alpha$ iff all the terms of α are in $T(A)$, but $A \nVdash \alpha$.

We shall refer to this definition briefly as "KSD".

If Φ is a set of $\pm$formulas, we write $A \Vdash \Phi$ provided A forces every $\pm$formula in Φ.

6.2. LEMMA. *Suppose that* $A \Vdash \alpha$. *Then all the terms of* α *are in* $T(A)$. *Also, if* $B \geqslant A$ *then* $B \Vdash \alpha$.

PROOF. Easy, by induction on deg α. ∎

Some clauses in KSD call for comment. Suppose we are at state A. Intuitively, we say that we "understand" $\boldsymbol{\alpha}$ if all the terms of $\boldsymbol{\alpha}$ are in $T(A)$. To be sure that $\boldsymbol{\beta} \rightarrow \boldsymbol{\gamma}$ is true, we must understand this formula, and we must also be certain that if at any future time (i.e., at any state B that may follow A) we come to accept $\boldsymbol{\beta}$ as true, then at that time we also accept $\boldsymbol{\gamma}$ as true; hence clause (4). To be sure that $\neg\boldsymbol{\beta}$ is true, we must understand $\boldsymbol{\beta}$ and be certain that $\boldsymbol{\beta}$ will never be accepted as true; hence clause (5). Finally, to be sure that $\forall \mathbf{x}\boldsymbol{\beta}$ is true, we must be certain that whenever any term $\mathbf{t}$ comes to be known, then $\boldsymbol{\beta}(\mathbf{x}/\mathbf{t})$ shall be accepted as true; hence clause (7). The other clauses in KSD call for no special comment.

Note that if $A \Vdash \neg\boldsymbol{\alpha}$ then $A \Vdash -\boldsymbol{\alpha}$; but the converse is not necessarily true.

From now on, when we say "Kripke system", without any further qualification, we mean Kripke system for $\mathscr{L}$ *or for any language obtained from* $\mathscr{L}$ *by adding new constants.*

We say that a set $\boldsymbol{\Phi}$ of $\pm$formulas is *enforceable* if there exists a Kripke system $\mathfrak{K}$ and a state A of $\mathfrak{K}$ such that $A \Vdash \boldsymbol{\Phi}$.

6.3. THEOREM. *If a set* $\boldsymbol{\Phi}$ *of formulas is satisfiable* (*in the sense of Def.* 2.1.5), *then* $\boldsymbol{\Phi}$ *is enforceable.*

PROOF. Let σ be a valuation such that $\sigma \models \boldsymbol{\Phi}$, and let U be the universe of σ. By adding new constants, if necessary, we get a language $\mathscr{L}'$ and an $\mathscr{L}'$-expansion[1] σ' of σ such that for every $a \in U$ there is a constant $\mathbf{a}$ in $\mathscr{L}'$ for which $\mathbf{a}^{\sigma'} = a$.

We define a Kripke system for $\mathscr{L}'$. The system will have just one state, A. As $T(A)$ we take the set of *all* $\mathscr{L}'$-terms. As $F(A)$ we take the set of *all* atomic $\mathscr{L}'$-formulas satisfied by σ'. If $\boldsymbol{\alpha}$ is any $\mathscr{L}'$-formula, then by induction on deg $\boldsymbol{\alpha}$ it is easy to see that $A \Vdash \boldsymbol{\alpha}$ iff $\sigma' \models \boldsymbol{\alpha}$. In particular, $A \Vdash \boldsymbol{\Phi}$. ∎

6.4. PROBLEM. Show that if $\boldsymbol{\Phi}$ is finite, then in the proof of Thm. 6.3 there is no need to extend $\mathscr{L}$. (Use Thm. 3.3.1 to show that without loss of generality we can assume that for each $a \in U$ there is a variable $\mathbf{x}$ such that $\mathbf{x}^{\sigma} = a$.)

6.5. COUNTER-EXAMPLE. The converse of Thm. 6.3 does not always hold. Let $\mathbf{P}$ be a unary predicate symbol of $\mathscr{L}$, and consider the following Kripke system:

$\Sigma = \{A_1, A_2, A_3, \ldots\}$;

$A_n \leqslant A_m$ iff $n \leqslant m$;

[1] For the meaning of this, see §9 of Ch. 2.

for all n the set $T(A_n)$ consists of all the variables;
$F(A_n) = \{\mathbf{Pv}_i : 1 \leqslant i < n\}$.

Then for all n we have $A_n \Vdash -\mathbf{Pv}_n$; also $A_{n+1} \Vdash \mathbf{Pv}_n$, hence $A_n \Vdash -\neg\mathbf{Pv}_n$. Therefore $A_n \Vdash -\mathbf{Pv}_n \vee \neg\mathbf{Pv}_n$, so

$$A_n \Vdash -\forall\mathbf{x}[\mathbf{Px} \vee \neg\mathbf{Px}].$$

Since this is the case for *every n*, we have also

$$A_n \Vdash \neg\forall\mathbf{x}[\mathbf{Px} \vee \neg\mathbf{Px}].$$

6.6. PROBLEM. Let **P** be a unary predicate symbol of $\mathscr{L}$. Show that the sentence

$$\neg[\forall\mathbf{x}\neg\neg\mathbf{Px} \rightarrow \neg\neg\forall\mathbf{x}\mathbf{Px}]$$

is enforceable.

We write "$\mathbf{\Phi} \Vdash \alpha$" when we want to assert that $\mathbf{\Phi}, -\alpha$ is *not* enforceable. This means that for every Kripke system $\mathfrak{K}$ and state A of $\mathfrak{K}$, if $A \Vdash \mathbf{\Phi}$ and all the terms of α are in $T(A)$, then $A \Vdash \alpha$. In particular, if $\mathbf{\Phi} = \emptyset$ we write "$\Vdash \alpha$" and say that α is K-*valid*. This means that for any Kripke system $\mathfrak{K}$ and any state A of $\mathfrak{K}$ we have $A \Vdash \alpha$ provided all the terms of α are in $T(A)$.

6.7. THEOREM. *Let* $\mathbf{\Phi}$ *be any set of formulas. If* $\mathbf{\Phi} \Vdash \alpha$ *then* $\mathbf{\Phi} \models \alpha$. *In particular, if* $\Vdash \alpha$ *then* $\models \alpha$.

PROOF. If $\mathbf{\Phi} \not\models \alpha$, then $\mathbf{\Phi}, \neg\alpha$ is satisfiable. By Thm. 6.3 there is a state A of some Kripke system such that $A \Vdash \mathbf{\Phi}, \neg\alpha$, and hence $A \Vdash \mathbf{\Phi}, -\alpha$. Thus $\mathbf{\Phi} \not\Vdash \alpha$. ∎

6.8. COUNTER-EXAMPLES. The converse of Thm. 6.7 does not always hold. From 6.5 we see at once that the following logically valid formulas are not K-valid:

$$\mathbf{Px} \vee \neg\mathbf{Px}, \qquad \forall\mathbf{x}[\mathbf{Px} \vee \neg\mathbf{Px}], \qquad \neg\neg\forall\mathbf{x}[\mathbf{Px} \vee \neg\mathbf{Px}].$$

From 6.6 we see that the logically valid sentences

$$\forall\mathbf{x}\neg\neg\mathbf{Px} \rightarrow \neg\neg\forall\mathbf{x}\mathbf{Px}, \qquad \neg\neg[\forall\mathbf{x}\neg\neg\mathbf{Px} \rightarrow \neg\neg\forall\mathbf{x}\mathbf{Px}]$$

are not K-valid.

6.9. PROBLEM. Let **P** and **Q** be unary predicate symbols, and let **x** and **y** be distinct variables. Show that the logically valid formula

$$\forall\mathbf{x}[\mathbf{Py} \vee \mathbf{Qx}] \rightarrow \mathbf{Py} \vee \forall\mathbf{x}\mathbf{Qx}$$

is not K-valid. (Consider two states A, B where $A \leqslant B$. Put $F(A)=\{\mathbf{y}\}$, $T(B)=\{\mathbf{x},\mathbf{y}\}$; $F(A)=\{\mathbf{Qy}\}$, $F(B)=\{\mathbf{Py}, \mathbf{Qy}\}$. Show that the negative of the above formula is forced by A.)

In what follows we shall need to consider two simple operations on Kripke systems.

First, let $\mathfrak{K}$ be any Kripke system, let $\mathbf{t}$ be a term that does not belong to $T(A)$ for any state A of $\mathfrak{K}$, and let $\mathbf{s}$ be an arbitrary term. We let $\mathfrak{K}(\mathbf{s}/\mathbf{t})$ be the Kripke system obtained from $\mathfrak{K}$ when $\mathbf{s}$ is replaced by $\mathbf{t}$. (Thus the states of $\mathfrak{K}(\mathbf{s}/\mathbf{t})$ and their partial ordering are the same as in $\mathfrak{K}$; but whenever $\mathbf{s}$ belongs to $T_{\mathfrak{K}}(A)$, we replace $\mathbf{s}$ by $\mathbf{t}$ in $T_{\mathfrak{K}}(A)$ and in all the atomic formulas of $F_{\mathfrak{K}}(A)$.) Now let $\boldsymbol{\alpha}$ be any formula such that neither $\mathbf{s}$ nor $\mathbf{t}$ are among the terms of $\boldsymbol{\alpha}$. It is easy to see that

$$A \Vdash \boldsymbol{\alpha}(\mathbf{x}/\mathbf{s}) \text{ in } \mathfrak{K} \quad \text{iff } A \Vdash \boldsymbol{\alpha}(\mathbf{x}/\mathbf{t}) \text{ in } \mathfrak{K}(\mathbf{s}/\mathbf{t}).$$

The same applies also to $-\boldsymbol{\alpha}(\mathbf{x}/\mathbf{s})$ and $-\boldsymbol{\alpha}(\mathbf{x}/\mathbf{t})$.

Next, let $\mathfrak{K}$, $\mathbf{s}$ and $\mathbf{t}$ be as above. We let $\mathfrak{K}(\mathbf{s},\mathbf{t})$ be the Kripke system obtained from $\mathfrak{K}$ be "introducing $\mathbf{t}$ as a synonym for $\mathbf{s}$". More explicitly, we leave the states of $\mathfrak{K}$ and their partial ordering unchanged; but whenever $\mathbf{s}$ belongs to $T_{\mathfrak{K}}(A)$ we add $\mathbf{t}$ as well, and we add to $F_{\mathfrak{K}}(A)$ all formulas obtained from a formula of $F_{\mathfrak{K}}(A)$ by replacing one or more occurrences of $\mathbf{s}$ by occurrences of $\mathbf{t}$. Now let $\boldsymbol{\alpha}$ be any formula such that $\mathbf{t}$ is not a term of $\boldsymbol{\alpha}(\mathbf{x}/\mathbf{s})$. It is easy to see that

$$A \Vdash \boldsymbol{\alpha}(\mathbf{x}/\mathbf{s}) \text{ in } \mathfrak{K} \quad \text{iff } A \Vdash \boldsymbol{\alpha}(\mathbf{x}/\mathbf{t}) \text{ in } \mathfrak{K}(\mathbf{s},\mathbf{t}).$$

The same applies also to $-\boldsymbol{\alpha}(\mathbf{x}/\mathbf{s})$ and $-\boldsymbol{\alpha}(\mathbf{x}/\mathbf{t})$.

One more bit of notation: if $\boldsymbol{\Phi}$ is a set of $\pm$formulas, we put $\boldsymbol{\Phi}^+$ for the set of all formulas in $\boldsymbol{\Phi}$.

Finally, let us agree to say that a branch in an intuitionistic tableau is *enforceable* if the set of $\pm$formulas of the branch is enforceable.

We are now ready to prove:

6.10. LEMMA. *Let an enforceable branch in a first-order intuitionistic tableau be extended to one or two new branches by one of the thirteen rules. Then the new branch — or, if there are two, at least one of them — is enforceable.*

PROOF. Let $\boldsymbol{\Phi}$ be the set of $\pm$formulas of the given branch. We consider the cases of the quantifier rules, leaving the propositional rules to the reader.

First, take the $\exists$ rule. Here the $\pm$formulas of the new branch are

$\mathbf{\Phi}, \boldsymbol{\alpha}(\mathbf{x}/\mathbf{y})$, where $\exists \mathbf{x}\boldsymbol{\alpha} \in \mathbf{\Phi}$ and $\mathbf{y}$ is not free in any $\pm$formula of $\mathbf{\Phi}$. Let A be a state in some Kripke system $\mathfrak{K}$ such that $A \Vdash \mathbf{\Phi}$. Since $\exists \mathbf{x}\boldsymbol{\alpha} \in \mathbf{\Phi}$, it follows from KSD that for some $\mathbf{s} \in T(A)$ we have $A \Vdash \boldsymbol{\alpha}(\mathbf{x}/\mathbf{s})$. If $\mathbf{s}$ happens to be $\mathbf{y}$, we are through. If $\mathbf{s} \neq \mathbf{y}$, we may assume without loss of generality that $\mathbf{y} \notin T(B)$ for all states B of $\mathfrak{K}$. (If this is not so, we consider $\mathfrak{K}(\mathbf{y}/\mathbf{a})$ instead of $\mathfrak{K}$, where $\mathbf{a}$ is a new constant that does not belong to $T(B)$ for any state B of $\mathfrak{K}$. By what we have seen above, A will force $\mathbf{\Phi}$ in $\mathfrak{K}(\mathbf{y}/\mathbf{a})$ as well.) Now consider $\mathfrak{K}(\mathbf{s},\mathbf{y})$. Since A forces $\mathbf{\Phi}, \boldsymbol{\alpha}(\mathbf{x}/\mathbf{s})$ in $\mathfrak{K}$, it follows that A must force $\mathbf{\Phi}, \boldsymbol{\alpha}(\mathbf{x}/\mathbf{y})$ in $\mathfrak{K}(\mathbf{s},\mathbf{y})$.

Next, consider the $-\exists$ rule. Here the $\pm$formulas of the new branch are $\mathbf{\Phi}^+, -\boldsymbol{\alpha}(\mathbf{x}/\mathbf{t})$, where $-\exists \mathbf{x}\boldsymbol{\alpha} \in \mathbf{\Phi}$. Again, let A be a state in some Kripke system $\mathfrak{K}$ such that $A \Vdash \mathbf{\Phi}$. Since $-\exists \mathbf{x}\boldsymbol{\alpha} \in \mathbf{\Phi}$, it follows from KSD that $A \Vdash -\boldsymbol{\alpha}(\mathbf{x}/\mathbf{s})$ for every $\mathbf{s} \in T(A)$. If $\mathbf{t} \in T(A)$, we are through. Otherwise, we may assume that $\mathbf{t} \notin T(B)$ for *all* states B of $\mathfrak{K}$. (If this is not so, we can choose a new constant $\mathbf{a}$ and consider $\mathfrak{K}(\mathbf{t}/\mathbf{a})$ instead of $\mathfrak{K}$.) Now take any $\mathbf{s} \in T(A)$ and consider $\mathfrak{K}(\mathbf{s},\mathbf{t})$. It is clear that in $\mathfrak{K}(\mathbf{s},\mathbf{t})$ the state A forces $\mathbf{\Phi}$ (hence also $\mathbf{\Phi}^+$) as well as $-\boldsymbol{\alpha}(\mathbf{x}/\mathbf{t})$.

The case of the $\forall$ rule is similar to that of the $-\exists$ rule.

Finally, take the $-\forall$ rule. Here the $\pm$formulas of the new branch are $\mathbf{\Phi}^+, -\boldsymbol{\alpha}(\mathbf{x}/\mathbf{y})$, where $-\forall \mathbf{x}\boldsymbol{\alpha} \in \mathbf{\Phi}$ and $\mathbf{y}$ is not free in $\mathbf{\Phi}$. Now let B be a state in some Kripke system $\mathfrak{K}$ such that $B \Vdash \mathbf{\Phi}$. Since $-\forall \mathbf{x}\boldsymbol{\alpha} \in \mathbf{\Phi}$, it follows from KSD that for some A, where $B \leqslant A$, and some $\mathbf{s} \in T(A)$, we have $A \Vdash -\boldsymbol{\alpha}(\mathbf{x}/\mathbf{s})$. Also, since $B \leqslant A$, it follows by Lemma 6.2 that $A \Vdash \mathbf{\Phi}^+$. From this point on we can argue as in the case of the $\exists$ rule. ∎

We can now show that the method of first-order intuitionistic tableaux is *sound* relative to Kripke's semantics:

6.11. THEOREM. *Let* $\mathbf{\Phi}$ *be a finite set of formulas. If* $\mathbf{\Phi} \rhd \boldsymbol{\psi}$, *then* $\mathbf{\Phi} \Vdash \boldsymbol{\psi}$. *Also, if* $\mathbf{\Phi} \rhd$ *then* $\mathbf{\Phi}$ *is not enforceable.*

PROOF. Suppose $\mathbf{\Phi} \nVdash \boldsymbol{\psi}$. Then, by definition, $\mathbf{\Phi}, -\boldsymbol{\psi}$ is enforceable. By Lemma 6.10 it follows that in any first-order intuitionistic tableau for $\mathbf{\Phi}, -\boldsymbol{\psi}$ at least one branch must be enforceable. But then this branch cannot contain a formula together with its negative, and hence cannot be closed. It is therefore impossible that $\mathbf{\Phi} \rhd \boldsymbol{\psi}$.

The second part of the theorem is proved similarly. ∎

6.12. PROBLEM. Consider the following method of *Beth tableaux*. We modify the method studied above in three ways. First, we allow the initial node to contain more than one negative. Second, instead of the two rules

$-\vee_1$ and $-\vee_2$ we have one $-\vee$ rule:

$$\begin{array}{l} \vdots \\ -\alpha\vee\beta \\ \vdots \\ \vdots \\ | \\ -\alpha \\ -\beta \end{array}$$

Note that this rule is *not* starred. Third, we remove the stars from the rules $-\wedge$, $\rightarrow$, $\neg$ and $-\exists$ (so that only the three rules $-\rightarrow$, $-\neg$ and $-\forall$ remain starred).

If Φ is a set of $\pm$formulas, we write "$\Phi \rhd_B$" as short for "we can construct a closed Beth tableau for Φ".

Prove that if $\Phi \rhd_B$ then Φ is not enforceable. (Note that the star in the three starred rules can be interpreted as "transition to a possible future state". The proof in detail is like that of 6.11.)

§ 7. The Elimination Theorem for intuitionistic tableaux

In §§8–9 we shall introduce intuitionistic versions of the propositional and predicate calculi, and we shall want to show in a constructive way that they are equivalent to the methods of intuitionistic propositional and first-order tableaux. For this purpose — just as in the corresponding classical case — we shall need an elimination theorem. The main work is done in the following:

7.1. Elimination Lemma. *Let Φ be a finite set of formulas. If $\Phi,\delta \rhd \xi$ and $\Phi \rhd \delta$, then $\Phi \rhd \xi$.*

Proof. Let T_1 and T_2 be given proofs of $\Phi, \delta \rhd \xi$ and $\Phi \rhd \delta$ respectively. Thus T_1 and T_2 are closed first-order intuitionistic tableaux for $\Phi,\delta, -\xi$ and $\Phi, -\delta$. (Here and in what follows, if ξ happens to be the empty string of symbols then $-\xi$ too is considered to be the empty string.)

Let r be the least number such that δ is not used in T_1 below the rth level (neither to extend a branch nor to close one). If $r=0$, then either δ is not used *at all* in T_1 — in which case T_1 proves $\Phi \rhd \xi$, and we are through — or δ is used at level 0 only, to close T_1 at once. But in this latter case δ must be (atomic and) the same as ξ, so that T_2 is a proof of $\Phi \rhd \xi$ — and again we are through. Thus we may assume $r>0$.

Similarly, let s be the least number such that $-\delta$ is not used in T_2 below the sth level. If $-\delta$ is not used at all in T_2, then T_2 proves $\Phi \rhd$, hence we have

$\Phi \rhd \xi$ by Prob. 5.18. If $-\delta$ is only used at the 0th level, to close T_2 at once, then δ must be (atomic and) in Φ, so that T_1 is actually a proof of $\Phi \rhd \xi$. Thus we may assume $s>0$. So $r+s \geqslant 2$.

Our lemma will be proved by induction on deg δ; this is the *primary induction* of our proof. But *within* the induction step of our primary induction we shall use a *secondary induction* on $r+s$.

To help the reader find his way through this double induction, we present here a schematic plan of our proof:

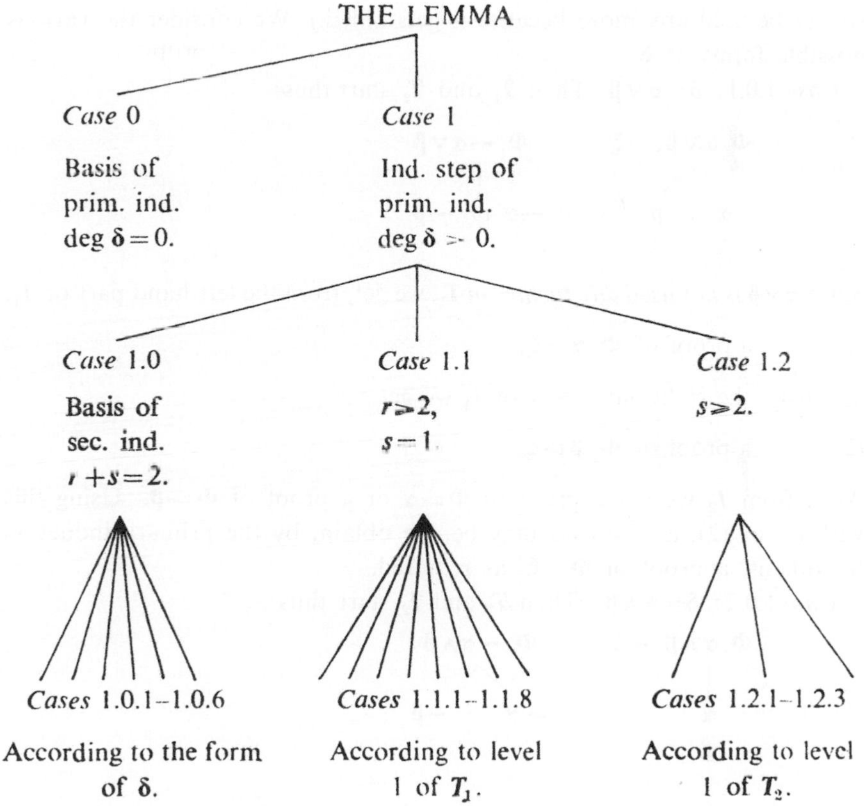

Now for the actual proof.

Case 0: deg $\delta=0$. Then δ is atomic and the only way in which it can be used in T_1 is for closing branches in which $-\delta$ crops up. But since we have Φ at the initial node of T_1, and we are given a proof T_2 of $\Phi \rhd \delta$, we can replace the use of δ in T_1 by an appeal to T_2; for by T_2 any branch of T_1 in which $-\delta$ crops up is as good as closed without using δ. (Here we are employing Prob. 5.18 again. Below we shall often use that result, without special mention.) Thus we can get rid of $\boldsymbol{\delta}$ in T_1 and obtain a proof of $\boldsymbol{\Phi} \rhd \boldsymbol{\xi}$, as required.

Case 1: $\boldsymbol{\delta}$ is not atomic. Then $\boldsymbol{\delta}$ can only be used for one of the six rules for formulas (the rules without minus) and $-\boldsymbol{\delta}$ can only be used for the corresponding minus rule (or, in the case of disjunction, one of the *two* corresponding minus rules). We now start our secondary induction.

Case 1.0: $r+s=2$, so that $r=s=1$. Thus $\boldsymbol{\delta}$ and $-\boldsymbol{\delta}$ are used to yield level 1 of T_1 and T_2 respectively, but are never used after that. (In fact, $-\boldsymbol{\delta}$ *cannot* be used any more because it gets killed.) We consider the various possible forms of $\boldsymbol{\delta}$.

Case 1.0.1: $\boldsymbol{\delta}=\boldsymbol{\alpha}\vee\boldsymbol{\beta}$. Then T_1 and T_2 start thus:

$$\begin{array}{ccc} & \boldsymbol{\Phi}, \boldsymbol{\alpha}\vee\boldsymbol{\beta}, -\boldsymbol{\xi} & \\ & \diagup \quad \diagdown & \\ \boldsymbol{\alpha} & & \boldsymbol{\beta} \\ \vdots & & \vdots \end{array} \qquad\qquad \begin{array}{c} \boldsymbol{\Phi}, -\boldsymbol{\alpha}\vee\boldsymbol{\beta} \\ |* \\ -\boldsymbol{\alpha} \text{ or } -\boldsymbol{\beta} \\ \vdots \end{array}$$

Since $\boldsymbol{\alpha}\vee\boldsymbol{\beta}$ *is not used any further in* T_1, we get, from the left-hand part of T_1,

(1) a proof of $\boldsymbol{\Phi}, \boldsymbol{\alpha} \rhd \boldsymbol{\xi}$,

and from the right-hand part of T_1 we get

(2) a proof of $\boldsymbol{\Phi}, \boldsymbol{\beta} \rhd \boldsymbol{\xi}$.

Also, from T_2 we get a proof of $\boldsymbol{\Phi} \rhd \boldsymbol{\alpha}$ or a proof of $\boldsymbol{\Phi} \rhd \boldsymbol{\beta}$. Using this with (1) or (2), as the case may be, we obtain, by the primary induction hypothesis, a proof of $\boldsymbol{\Phi} \rhd \boldsymbol{\xi}$, as required.

Case 1.0.2: $\boldsymbol{\delta}=\boldsymbol{\alpha}\wedge\boldsymbol{\beta}$. Then T_1 and T_2 start thus:

$$\begin{array}{c} \boldsymbol{\Phi}, \boldsymbol{\alpha}\wedge\boldsymbol{\beta}, -\boldsymbol{\xi} \\ | \\ \boldsymbol{\alpha} \\ \boldsymbol{\beta} \\ \vdots \end{array} \qquad\qquad \begin{array}{ccc} & \boldsymbol{\Phi}, -\boldsymbol{\alpha}\wedge\boldsymbol{\beta} & \\ & {}^*\diagup \quad \diagdown^* & \\ -\boldsymbol{\alpha} & & -\boldsymbol{\beta} \\ \vdots & & \vdots \end{array}$$

From T_1 we get

(3) a proof of $\boldsymbol{\Phi}, \boldsymbol{\alpha}, \boldsymbol{\beta} \rhd \boldsymbol{\xi}$.

From the right-hand part of T_2 we get a proof of $\boldsymbol{\Phi} \rhd \boldsymbol{\beta}$ and hence a proof of $\boldsymbol{\Phi}, \boldsymbol{\alpha} \rhd \boldsymbol{\beta}$. From this and (3) we get, by our primary induction hypothesis,

(4) a proof of $\boldsymbol{\Phi}, \boldsymbol{\alpha} \rhd \boldsymbol{\xi}$.

Also, from the left-hand part of T_2 we get a proof of $\Phi \rhd \alpha$; and using this and (4) we get, again by the primary induction hypothesis, a proof of $\Phi \rhd \xi$.

Case 1.0.3: $\delta = \alpha \to \beta$. T_1 and T_2 start thus:

$$
\begin{array}{ccc}
\multicolumn{3}{c}{\Phi, \alpha \to \beta, -\xi} \\
& {}^{*}\diagup \quad \diagdown & \\
-\alpha & & \beta \\
\vdots & & \vdots
\end{array}
\qquad
\begin{array}{c}
\Phi, -\alpha \to \beta \\
|^{*} \\
\alpha \\
-\beta \\
\vdots
\end{array}
$$

From T_1 we get

(5) a proof of $\Phi \rhd \alpha$,

(6) a proof of $\Phi, \beta \rhd \xi$.

Also, from T_2 we get a proof of $\Phi, \alpha \rhd \beta$. Using this and (5) we get, by the primary induction hypothesis, a proof of $\Phi \rhd \beta$; and from this and (6) we get a proof of $\Phi \rhd \xi$.

Case 1.0.4: $\delta = \neg\alpha$. T_1 and T_2 start thus:

$$
\begin{array}{c}
\Phi, \neg\alpha, -\xi \\
|^{*} \\
-\alpha \\
\vdots
\end{array}
\qquad
\begin{array}{c}
\Phi, -\neg\alpha \\
|^{*} \\
\alpha \\
\vdots
\end{array}
$$

We get proofs of $\Phi \rhd \alpha$ and $\Phi, \alpha \rhd$ and hence, by the primary induction hypothesis, a proof of $\Phi \rhd$ and hence a proof of $\Phi \rhd \xi$.

Case 1.0.5: $\delta = \exists x\alpha$. T_1 and T_2 start thus:

$$
\begin{array}{c}
\Phi, \exists x\alpha, -\xi \\
| \\
\alpha(x/y) \\
\vdots
\end{array}
\qquad
\begin{array}{c}
\Phi, -\exists x\alpha \\
|^{*} \\
-\alpha(x/t) \\
\vdots
\end{array}
$$

Here y is not free in $\Phi, \exists x\alpha, \xi$. By Prob. 5.17 we may assume that y is not bound in α, so that in performing the substitution $\alpha(x/y)$ no alphabetic changes are made. From T_1 we obtain a proof of $\Phi, \alpha(x/y) \rhd \xi$. Hence by Prob. 5.19, substituting t for y, we get a proof of $\Phi, \alpha(x/t) \rhd \xi$. Using this and the proof of $\Phi \rhd \alpha(x/t)$, which we get from T_2, we obtain, by the induction hypothesis, a proof of $\Phi \rhd \xi$.

Case 1.0.6: $\delta = \forall x\alpha$. This is treated by the same method as *case* 1.0.5.

We have now established the basis of our secondary induction and we must start the secondary induction step. Here we assume $r+s>2$. We distinguish two cases: either $s=1$ and $r\geqslant 2$ (case 1.1) or $s\geqslant 2$ and $r\geqslant 1$ (case 1.2).

Case 1.1: $s=1$ and $r\geqslant 2$. We consider the various ways in which level 1 of T_1 could have been obtained. There are two possibilities. *First*, level 1 of T_1 was obtained by using some $\pm$formula other than $\boldsymbol{\delta}$ (i.e., a formula of $\boldsymbol{\Phi}$ or $-\boldsymbol{\xi}$); this possibility is covered by cases 1.1.1–1.1.2 below. *Second*, level 1 of T_1 was obtained by using $\boldsymbol{\delta}$; this possibility is covered by cases 1.1.3–1.1.8.

Case 1.1.1: Level 1 of T_1 was obtained by applying a non-splitting rule (i.e., any rule except $\vee$, $-\wedge$, $\rightarrow$) to a formula of $\boldsymbol{\Phi}$ or to $-\boldsymbol{\xi}$. Here there are ten sub-cases to consider, according to which rule was used, but they are all treated by the same method. Consider, e.g., the sub-case where the rule in question is $-\rightarrow$. Then T_1 starts thus:

$$\begin{array}{c}\boldsymbol{\Phi}, \boldsymbol{\delta}, -\boldsymbol{\xi} \\ |{*} \\ \boldsymbol{\alpha} \\ -\boldsymbol{\beta} \\ \vdots\end{array}$$

where $\boldsymbol{\xi}=\boldsymbol{\alpha}\rightarrow\boldsymbol{\beta}$. From this we get

(7) a proof of $\boldsymbol{\Phi}, \boldsymbol{\alpha}, \boldsymbol{\delta} \rhd \boldsymbol{\beta}$.

Moreover, if r' is the least number such that $\boldsymbol{\delta}$ is not used in the proof (7) below the $(r')^{\text{th}}$ level, then $r'=r-1$. (This is because, when (7) is obtained from T_1, level r of T_1 becomes level $r-1$ of (7).)

Also, the given proof T_2 of $\boldsymbol{\Phi} \rhd \boldsymbol{\delta}$ can be converted, *without changing* s into a proof of $\boldsymbol{\Phi}, \boldsymbol{\alpha} \rhd \boldsymbol{\delta}$. From this and (7) we get, using our *secondary* induction hypothesis,

(8) a proof of $\boldsymbol{\Phi}, \boldsymbol{\alpha} \rhd \boldsymbol{\beta}$.

We now construct a proof of $\boldsymbol{\Phi} \rhd \boldsymbol{\xi}$. We start, as in T_1, by applying the $-\rightarrow$ rule to $-\boldsymbol{\xi}$ (i.e., to $-\boldsymbol{\alpha}\rightarrow\boldsymbol{\beta}$), thus:

$$\begin{array}{c}\boldsymbol{\Phi}, -\boldsymbol{\xi} \\ |{*} \\ \boldsymbol{\alpha} \\ -\boldsymbol{\beta}\end{array}$$

and this is as good as closed because we have (8).

The other nine sub-cases are left to the reader.

Case 1.1.2: Level 1 of T_1 was obtained by applying one of the rules $\vee, -\wedge, \rightarrow$ to a formula of Φ or to $-\xi$. Consider, e.g., the sub-case where the rule in question is $\rightarrow$. Then T_1 starts thus:

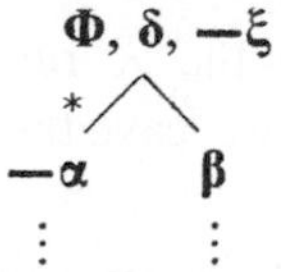

where $\alpha \rightarrow \beta \in \Phi$. From the left-hand part of T_1 we get

(9) a proof of $\Phi, \delta \rhd \alpha$,

and the least number r' such that δ is not used in (9) below the $(r')^{\text{th}}$ level is smaller than r.

From (9) and T_2, by the secondary induction hypothesis we get

(10) a proof of $\Phi \rhd \alpha$.

Similarly, from the right-hand part of T_1 together with T_2 we get, by the secondary induction hypothesis,

(11) a proof of $\Phi, \beta \rhd \xi$.

We construct a proof of $\Phi \rhd \xi$ as follows. We start by using the formula $\alpha \rightarrow \beta \in \Phi$ as in T_1. By (10) and (11) this is as good as closed.

The remaining two sub-cases (those of $\vee$ and $-\wedge$) are similar and are left to the reader.

Case 1.1.3: Level 1 of T_1 was obtained by applying the $\vee$ rule to δ. Then T_1 starts thus:

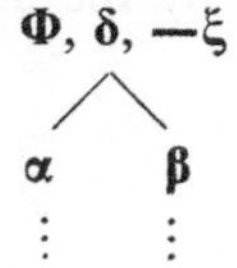

where $\delta = \alpha \vee \beta$. But in this case δ need not be used ever again in T_1. For example, if further down in the left-hand part of T_1 we have again

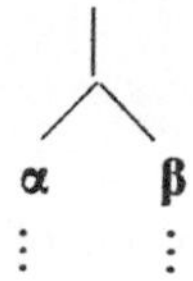

then we can cut out this new node $\{\beta\}$ as well as all the nodes below it, and also the new node $\{\alpha\}$ (but *not* the nodes below it) because we already have α above this part of T_1. Thus by cutting out redundant parts of T_1 we can reduce r to 1. Now we are back to case 1.0, which we have covered before.

Case 1.1.4: Level 1 of T_1 was obtained by applying the $\wedge$ rule to δ. This is similar to (but slightly simpler than) case 1.1.3 and we leave the details to the reader.

Case 1.1.5: Level 1 of T_1 was obtained by applying the $\rightarrow$ rule to δ. Then T_1 begins thus:

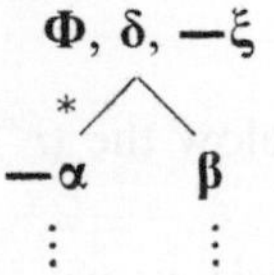

where $\delta = \alpha \rightarrow \beta$. Here we may assume that δ is not used again in the right-hand part of T_1, because any such use would be redundant, as explained in case 1.1.3. But in the left-hand part of T_1 it *may* be necessary to use δ again, because $-\alpha$ may get killed. Thus from the right-hand part of T_1 we get

(12) a proof of $\Phi, \beta \rhd \xi$,

and from the left-hand part of T_1 we only get

(13) a proof of $\Phi, \delta \rhd \alpha$.

But if r' is the least number such that δ is not used in (13) after the $(r')^{\text{th}}$ level, then $r' < r$. Therefore from T_2 and (13) we get, by the secondary induction hypothesis,

(14) a proof of $\Phi \rhd \alpha$.

We now construct a proof of $\Phi, \delta \rhd \xi$ as follows. We start by applying the $\rightarrow$ rule to δ, getting

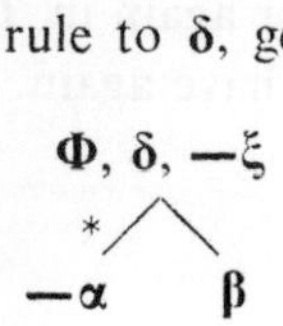

and, since we have (12) and (14), we can continue this proof of $\Phi, \delta \rhd \xi$

without ever using $\boldsymbol{\delta}$ *again.* Thus we have reduced r to 1, and we are back to case 1.0, which we have already covered.

Case 1.1.6: Level 1 of T_1 was obtained by applying the $\neg$ rule to $\boldsymbol{\delta}$. This is similar to (but slightly simpler than) case 1.1.5, and we leave it to the reader.

Case 1.1.7: Level 1 of T_1 was obtained by applying the $\exists$ rule to $\boldsymbol{\delta}$. Then T_1 begins thus:

$$\begin{array}{c} \boldsymbol{\Phi}, \boldsymbol{\delta}, -\boldsymbol{\xi} \\ | \\ \boldsymbol{\alpha}(\mathbf{x}/\mathbf{y}) \\ \vdots \end{array}$$

where $\boldsymbol{\delta} = \exists \mathbf{x} \boldsymbol{\alpha}$ and $\mathbf{y}$ is not free in $\boldsymbol{\Phi}, \boldsymbol{\delta}, \boldsymbol{\xi}$. By Prob. 5.17 we may assume that $\mathbf{y}$ is not bound in $\boldsymbol{\alpha}$, so that when we substitute $\mathbf{y}$ for $\mathbf{x}$ in $\boldsymbol{\alpha}$ no alphabetic changes are made. From T_1 we get

(15) a proof of $\boldsymbol{\Phi}, \boldsymbol{\alpha}(\mathbf{x}/\mathbf{y}), \boldsymbol{\delta} \rhd \boldsymbol{\xi}$,

and if r' is the least number such that $\boldsymbol{\delta}$ is not used in (15) after the $(r')^{\text{th}}$ level, then $r' = r - 1$.

Also, the proof T_2 of $\boldsymbol{\Phi} \rhd \boldsymbol{\delta}$ can be converted, *without changing s*, into a proof of $\boldsymbol{\Phi}, \boldsymbol{\alpha}(\mathbf{x}/\mathbf{y}) \rhd \boldsymbol{\delta}$. From this and (15) we get, by the secondary induction hypothesis,

(16) a proof of $\boldsymbol{\Phi}, \boldsymbol{\alpha}(\mathbf{x}/\mathbf{y}) \rhd \boldsymbol{\xi}$.

Now let us examine T_2. Since we have assumed $s = 1$, T_2 must start thus:

$$\begin{array}{c} \boldsymbol{\Phi}, -\boldsymbol{\delta} \\ |^{*} \\ -\boldsymbol{\alpha}(\mathbf{x}/\mathbf{t}) \\ \vdots \end{array}$$

where $\mathbf{t}$ is some term. This yields

(17) a proof of $\boldsymbol{\Phi} \rhd \boldsymbol{\alpha}(\mathbf{x}/\mathbf{t})$.

Also, by Prob. 5.19 we can substitute $\mathbf{t}$ for $\mathbf{y}$ in (16) and get a proof of $\boldsymbol{\Phi}, \boldsymbol{\alpha}(\mathbf{x}/\mathbf{t}) \rhd \boldsymbol{\xi}$. From this and (17) we get, by the primary induction hypothesis, a proof of $\boldsymbol{\Phi} \rhd \boldsymbol{\xi}$.

Case 1.1.8: Level 1 of T_1 was obtained by applying the $\forall$ rule to $\boldsymbol{\delta}$.

Then T_1 starts thus:

$$\begin{array}{l}\mathbf{\Phi}, \boldsymbol{\delta}, -\boldsymbol{\xi}\\ \quad| \\ \boldsymbol{\alpha}(\mathbf{x}/\mathbf{t})\\ \quad\vdots\end{array}$$

where $\mathbf{t}$ is a term and $\boldsymbol{\delta}=\forall\mathbf{x}\boldsymbol{\alpha}$. From this we get

(18) a proof of $\mathbf{\Phi}, \boldsymbol{\alpha}(\mathbf{x}/\mathbf{t}), \boldsymbol{\delta} \rhd \boldsymbol{\xi}$,

and if r' is the least number such that $\boldsymbol{\delta}$ is not used in (18) below the $(r')^{\text{th}}$ level, then $r'=r-1$.

Also, the proof T_2 of $\mathbf{\Phi} \rhd \boldsymbol{\delta}$ can be converted, without changing s, into a proof of $\mathbf{\Phi}, \boldsymbol{\alpha}(\mathbf{x}/\mathbf{t}) \rhd \boldsymbol{\delta}$. From this and (18) we get, by the secondary induction hypothesis,

(19) a proof of $\mathbf{\Phi}, \boldsymbol{\alpha}(\mathbf{x}/\mathbf{t}) \rhd \boldsymbol{\xi}$.

Now consider T_2. Since $s=1$, T_2 must start thus:

$$\begin{array}{l}\mathbf{\Phi}, -\boldsymbol{\delta}\\ \quad|^{*} \\ -\boldsymbol{\alpha}(\mathbf{x}/\mathbf{y})\\ \quad\vdots\end{array}$$

where $\mathbf{y}$ is not free in $\mathbf{\Phi}$ nor in $\boldsymbol{\delta}$. Again, we may assume that the substitution of $\mathbf{y}$ for $\mathbf{x}$ in $\boldsymbol{\alpha}$ does not require any alphabetic change. Thus T_2 yields a proof of $\mathbf{\Phi} \rhd \boldsymbol{\alpha}(\mathbf{x}/\mathbf{y})$; and by Prob. 5.19 we may substitute $\mathbf{t}$ for $\mathbf{y}$ and get a proof of $\mathbf{\Phi} \rhd \boldsymbol{\alpha}(\mathbf{x}/\mathbf{t})$. From this and (19) we get, by the primary induction hypothesis, a proof of $\mathbf{\Phi} \rhd \boldsymbol{\xi}$.

We have now covered all possible subcases of case 1.1.

Case 1.2: $s \geqslant 2$. We consider level 1 of T_2. The only ways in which this level could have been obtained is by applying one of the rules $\vee$, $\wedge$, $\rightarrow$, $\exists$, $\forall$ to a formula of $\mathbf{\Phi}$, because any other move would kill $-\boldsymbol{\delta}$ at once — contradicting our assumption that $s \geqslant 2$. (The $\rightarrow$ rule kills $-\boldsymbol{\delta}$ only on one branch, so it must be considered as a possibility.) The rules $\wedge$, $\exists$, $\forall$ will be considered together (case 1.2.1) and each of the rules $\vee$, $\rightarrow$ will be considered separately (cases 1.2.2–1.2.3).

Case 1.2.1: Level 1 of T_2 was obtained by applying one of the rules

$\wedge$, $\exists$, $\forall$ to some formula $\alpha \in \Phi$. Then T_2 starts thus:

$$\begin{array}{c} \Phi, -\delta \\ | \\ \Psi \\ \vdots \end{array}$$

where Ψ is a set of one or two formulas (one in the case of $\exists$ and $\forall$, two in the case of $\wedge$). From T_2 we get

(20) a proof of $\Phi \cup \Psi \rhd \delta$;

and if s' is the least number such that $-\delta$ is not used in (20) below the s'^{th} level, then $s' = s-1$.

Also, from T_1 we get, without increasing r, a proof of $\Phi \cup \Psi, \delta \rhd \xi$. From this and (20) we get, by the secondary induction hypothesis,

(21) a proof of $\Phi \cup \Psi \rhd \xi$.

We construct a proof of $\Phi \rhd \xi$ as follows. Starting as in T_2, we use $\alpha \in \Phi$ to obtain

$$\begin{array}{c} \Phi, -\xi \\ | \\ \Psi \end{array}$$

and this is as good as closed by (21).

Case 1.2.2: Level 1 of T_2 was obtained by applying the rule $\vee$ to some formula of Φ. Then T_2 starts thus:

$$\begin{array}{ccc} & \Phi, -\delta & \\ / & & \backslash \\ \alpha & & \beta \\ \vdots & & \vdots \end{array}$$

where $\alpha \vee \beta \in \Phi$. We consider the left-hand and right-hand parts of T_2 in turn, and using the same method as in case 1.2.1 we get

(22) proofs of $\Phi, \alpha \rhd \xi$ and $\Phi, \beta \rhd \xi$.

We construct a proof of $\Phi \rhd \xi$ as follows. Starting as in T_2, we apply the rule $\vee$ to $\alpha \vee \beta$. By (22) this is as good as closed.

Case 1.2.3: Level 1 of T_2 was obtained by applying the rule $\rightarrow$ to some formula of $\mathbf{\Phi}$. Then T_2 starts thus:

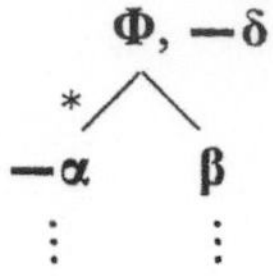

where $\boldsymbol{\alpha} \to \boldsymbol{\beta} \in \mathbf{\Phi}$. From the left-hand part of T_2 we get

(23) a proof of $\mathbf{\Phi} \rhd \boldsymbol{\alpha}$.

And from the right-hand part of T_2 we get, as in case 1.2.1,

(24) a proof of $\mathbf{\Phi}, \boldsymbol{\beta} \rhd \boldsymbol{\xi}$.

Now start a proof of $\mathbf{\Phi} \rhd \boldsymbol{\xi}$ by applying the rule $\to$ to $\boldsymbol{\alpha} \to \boldsymbol{\beta}$:

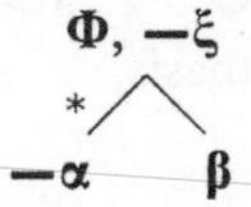

By (23) and (24) this is as good as closed. ∎

7.2. Corollary. *If* $\mathbf{\Phi}, \boldsymbol{\delta} \rhd_0 \boldsymbol{\xi}$ *and* $\mathbf{\Phi} \rhd_0 \boldsymbol{\delta}$, *then* $\mathbf{\Phi} \rhd_0 \boldsymbol{\xi}$.
Proof. Same as proof of 7.1, except that the basis of the primary induction is the case where $\boldsymbol{\delta}$ is a *prime* (rather than atomic) formula, and all consideration of the quantifier rules is left out. ∎

We now introduce (temporarily) an additional fourteenth rule for extending intuitionistic tableaux. This rule, called "the $\pm$ rule" is analogous to the EM-rule of classical tableaux and is represented schematically as:

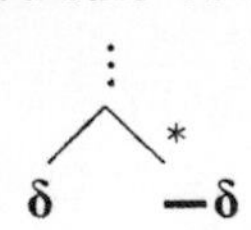

where $\boldsymbol{\delta}$ is any formula. However:

7.3. Elimination Theorem. *Given a proof of* $\mathbf{\Phi} \rhd \boldsymbol{\xi}$ *(or of* $\mathbf{\Phi} \rhd_0 \boldsymbol{\xi}$*) which uses the* $\pm$ *rule, we can construct a proof of the same statement without using that rule.*
Proof. Similar to 1.8.7. ∎

7.4. Problem. Show that if $\mathbf{\Phi} \rhd \boldsymbol{\alpha}$ and $\mathbf{\Phi} \rhd \boldsymbol{\alpha} \to \boldsymbol{\beta}$, then $\mathbf{\Phi} \rhd \boldsymbol{\beta}$. Similarly with "$\rhd_0$" instead of "$\rhd$". (Apply $\pm$ rule with $\boldsymbol{\delta} = \boldsymbol{\alpha} \to \boldsymbol{\beta}$, then use 7.3.)
7.5. Problem. Show that if $\mathbf{\Phi} \rhd \boldsymbol{\alpha}$ and $\mathbf{\Phi} \rhd \neg \boldsymbol{\alpha}$ then $\mathbf{\Phi} \rhd$. Similarly with "$\rhd_0$" instead of "$\rhd$". (Apply $\pm$ rule with $\boldsymbol{\delta} = \neg \boldsymbol{\alpha}$, then use 7.3.)

§ 8. Intuitionistic propositional calculus

As *axioms* of the *intuitionistic propositional calculus* we take all formulas of the following forms:

(8.1) $\alpha \to \beta \to \alpha,$

(8.2) $(\alpha \to \beta \to \gamma) \to (\alpha \to \beta) \to \alpha \to \gamma,$

(8.3) $\alpha \to \beta \to \alpha \wedge \beta,$

(8.4) $\alpha \wedge \beta \to \alpha,$

(8.5) $\alpha \wedge \beta \to \beta,$

(8.6) $\alpha \to \alpha \vee \beta,$

(8.7) $\beta \to \alpha \vee \beta,$

(8.8) $(\alpha \to \gamma) \to (\beta \to \gamma) \to \alpha \vee \beta \to \gamma,$

(8.9) $(\alpha \to \beta) \to (\alpha \to \neg \beta) \to \neg \alpha,$

(8.10) $\neg \alpha \to \alpha \to \beta.$

As rule of inference we take *modus ponens.*

Deduction in this calculus is defined in the usual way. We write "$\Phi \vdash_{I0} \psi$" to indicate that we can construct a deduction of ψ from Φ in this calculus. Also, we write "$\Phi \vdash_{I0}$" as short for "for some formula α, both $\Phi \vdash_{I0} \alpha$ and $\Phi \vdash_{I0} \neg \alpha$". Using axiom (8.10) it is easy to show that if $\Phi \vdash_{I0}$ then $\Phi \vdash_{I0} \alpha$ for *every* formula α. If $\Phi \vdash_{I0}$ we say that Φ is *inconsistent.*

The Deduction Theorem — if $\Phi, \alpha \vdash_{I0} \beta$ *then* $\Phi \vdash_{I0} \alpha \to \beta$ — is proved in the usual way. (Note that Ax I and Ax II of the classical propositional calculus presented in §10 of Ch. 1 — which are the only ones needed to prove the Deduction Theorem — are present here as (8.1) and (8.2).)

8.11. PROBLEM. Show that if φ is an axiom of the intuitionistic propositional calculus then $\rhd_0 \varphi$.

8.12. THEOREM. *For any finite set* Φ *of formulas,* $\Phi \rhd_0 \xi$ *iff* $\Phi \vdash_{I0} \xi$.
PROOF. Suppose $\Phi \rhd_0 \xi$. We proceed by induction on the depth d of a given proof T of $\Phi \rhd_0 \xi$.

If $d=0$, then ξ (is atomic and) belongs to Φ. So $\Phi \vdash_{I0} \xi$ trivially.

If $d>0$, we consider nine cases, according to which propositional rule accounts for level 1 of T. All nine cases are very easy, and the following hints should suffice.

For the $\vee$ rule, use the Deduction Theorem and (8.8). For $-\vee_1$ and $-\vee_2$ use (8.6) and (8.7) respectively. For $\wedge$ use the Deduction Theorem (8.4) and (8.5). For $-\wedge$ use (8.3). The case of the $\rightarrow$ rule is trivial. For $-\rightarrow$ use the Deduction Thm. For $\neg$ use (8.10). For $-\neg$ use the Deduction Theorem (twice) and (8.9).

In all nine cases we find that $\boldsymbol{\Phi} \vdash_{10} \xi$.

Now assume that $\boldsymbol{\Phi} \vdash_{10} \xi$, and first take the case that ξ is a formula. Then we have a deduction $\alpha_1, \ldots, \alpha_n$ of ξ from $\boldsymbol{\Phi}$. Using Probs. 5.3, 7.4 and 8.11 it is easy to show by induction on $k=1,\ldots,n$ that $\boldsymbol{\Phi} \rhd_0 \alpha_k$. For $k=n$ we get $\boldsymbol{\Phi} \rhd_0 \xi$ as required.

Finally, suppose that $\boldsymbol{\Phi} \vdash_{10}$. Then for some (indeed, for all) α we have $\boldsymbol{\Phi} \vdash_{10} \alpha$ and $\boldsymbol{\Phi} \vdash_{10} \neg\alpha$. Thus, by what we have already shown, $\boldsymbol{\Phi} \rhd_0 \alpha$ and $\boldsymbol{\Phi} \rhd_0 \neg\alpha$. Hence $\boldsymbol{\Phi} \rhd_0$ by Prob. 7.5. ∎

In the sequel we shall often use Thm. 8.12 without special mention.

8.13. **Problem.** Prove the consistency of the intuitionistic propositional calculus, i.e., the impossibility of $\emptyset \vdash_{10}$.

§ 9. Intuitionistic predicate calculus

As *axioms* of the *intuitionistic predicate calculus* we take all formulas of the following eight groups:

(9.1) All intuitionistic propositional axioms, i.e., all instances of (8.1)–(8.10);

(9.2) $\forall \mathbf{x}(\alpha \rightarrow \beta) \rightarrow \forall \mathbf{x}\alpha \rightarrow \forall \mathbf{x}\beta$;

(9.3) $\forall \mathbf{x}(\alpha \rightarrow \beta) \rightarrow \exists \mathbf{x}\alpha \rightarrow \exists \mathbf{x}\beta$;

(9.4) $\alpha \rightarrow \forall \mathbf{x}\alpha$, provided $\mathbf{x}$ is not free in α;

(9.5) $\exists \mathbf{x}\alpha \rightarrow \alpha$, provided $\mathbf{x}$ is not free in α;

(9.6) $\forall \mathbf{x}\alpha \rightarrow \alpha(\mathbf{x}/\mathbf{t})$, provided $\mathbf{t}$ is free for $\mathbf{x}$ in α;

(9.7) $\alpha(\mathbf{x}/\mathbf{t}) \rightarrow \exists \mathbf{x}\alpha$ provided $\mathbf{t}$ is free for $\mathbf{x}$ in α;

(9.8) All generalizations of axioms of the preceding seven groups.

As rule of inference we adopt *modus ponens*.

Deduction is defined as usual. We use "$\vdash_{\mathrm{I}}$" for this calculus in the same way as we use "$\vdash_{10}$" for the calculus of §8.

The *Deduction Theorem* is proved in the usual way.

9.9. PROBLEM. Show that $\rhd\varphi$ for every axiom φ of the intuitionistic predicate calculus.

Using Prob. 9.9 it is quite easy to show that if $\Phi \vdash_I \xi$, where Φ is a finite set of formulas, then also $\Phi \rhd \xi$. But in order to prove the converse of this (as well as for other purposes) we shall need a few simple technical results. Since the work here is very similar to the work done in §1 of Ch. 3, we can proceed rather quickly.

9.10. THEOREM. *Let* $\mathbf{x}$ *be a variable which is not free in* Φ *or* ξ*; then*

(i) *if* $\Phi \vdash_I \alpha$, *then* $\Phi \vdash_I \forall\mathbf{x}\alpha$;

(ii) *if* $\Phi, \alpha \vdash_I \xi$, *then* $\Phi, \exists\mathbf{x}\alpha \vdash_I \xi$.

PROOF. A counterpart of Thm. 3.1.4 can be proved — in exactly the same way — for our present calculus. Hence (i).

To prove (ii), let us first take the case where ξ is a formula. If $\Phi, \alpha \vdash_I \xi$, then by the Deduction Theorem and (i) we have $\Phi \vdash_I \forall\mathbf{x}(\alpha \to \xi)$. Hence, using ax. (9.3) we get $\Phi, \exists\mathbf{x}\alpha \vdash_I \exists\mathbf{x}\xi$; and by ax. (9.5) we have $\Phi, \exists\mathbf{x}\alpha \vdash_I \xi$, as required.

Now let ξ be empty. If $\Phi, \alpha \vdash_I$, then in particular $\Phi, \alpha \vdash_I \neg\exists\mathbf{x}\alpha$. Thus, by what we have already proved, $\Phi, \exists\mathbf{x}\alpha \vdash_I \neg\exists\mathbf{x}\alpha$ and therefore $\Phi, \exists\mathbf{x}\alpha \vdash_I$ as required. ∎

9.11. THEOREM. *If* $\beta \sim \beta'$, *then* β *and* β' *are provably equivalent, i.e.,* $\beta \vdash_I \beta'$ *and* $\beta' \vdash_I \beta$.

PROOF. A counterpart of Lemma 3.1.9 can be proved — in a very similar way — for our present calculus. Hence it is enough to show that if $\mathbf{z}$ is not free in α but is free for $\mathbf{x}$ in α, then:

(1) $\forall\mathbf{x}\alpha$ and $\forall\mathbf{z}[\alpha(\mathbf{x}/\mathbf{z})]$ are provably equivalent;

(2) $\exists\mathbf{x}\alpha$ and $\exists\mathbf{z}[\alpha(\mathbf{x}/\mathbf{z})]$ are provably equivalent.

Now, (1) is proved exactly as in Thm. 3.1.10. To prove (2), we notice that, using ax. (9.7) we have $\alpha(\mathbf{x}/\mathbf{z}) \vdash_I \exists\mathbf{x}\alpha$, hence by Thm. 9.10 (ii) we have

$$\exists\mathbf{z}[\alpha(\mathbf{x}/\mathbf{z})] \vdash_I \exists\mathbf{x}\alpha.$$

Since alphabetic changes are reversible, we can show similarly that $\exists\mathbf{x}\alpha \vdash_I \exists\mathbf{z}[\alpha(\mathbf{x}/\mathbf{z})]$. ∎

9.12. PROBLEM. Show that for *every* α, $\mathbf{x}$ and $\mathbf{t}$:

(i) $\vdash_I \forall\mathbf{x}\alpha \to \alpha(\mathbf{x}/\mathbf{t})$;

(ii) $\vdash_I \alpha(\mathbf{x}/\mathbf{t}) \to \exists\mathbf{x}\alpha$.

(For (i) proceed exactly as in 3.1.11. But (ii) *cannot* be proved exactly as in 3.1.11, because now $\exists$ is *not* defined in terms of $\forall$.)

9.13. THEOREM. *Let* **c** *be a constant that does not occur in* Φ, α *or* ξ; *then:*
(i) *if* $\Phi \vdash_I \alpha(\mathbf{x}/\mathbf{c})$, *then* $\Phi \vdash_I \forall \mathbf{x}\alpha$;
(ii) *if* $\Phi, \alpha(\mathbf{x}/\mathbf{c}) \vdash_I \xi$, *then* $\Phi, \exists \mathbf{x}\alpha \vdash_I \xi$.

PROOF. (i) is proved exactly as in Thm. 3.1.12.

To prove (ii), suppose first that ξ is a formula. Let **y** be a variable that does not occur in α or ξ. Then

$$\alpha(\mathbf{x}/\mathbf{c}) = \alpha(\mathbf{x}/\mathbf{y})(\mathbf{y}/\mathbf{c}).$$

Therefore, if $\Phi, \alpha(\mathbf{x}/\mathbf{c}) \vdash_I \xi$, we have by the Deduction Theorem and (i),

$$\Phi \vdash_I \forall \mathbf{y}[\alpha(\mathbf{x}/\mathbf{y}) \to \xi].$$

Using (9.3) and (9.5) we obtain

$$\Phi, \exists \mathbf{y}[\alpha(\mathbf{x}/\mathbf{y})] \vdash_I \xi.$$

But by Thm. 9.11 we know that $\exists \mathbf{y}[\alpha(\mathbf{x}/\mathbf{y})]$ and $\exists \mathbf{x}\alpha$ are provably equivalent, hence $\Phi, \exists \mathbf{x}\alpha \vdash_I \xi$ as required.

If ξ is empty, we proceed as in 9.10. ∎

We can now show that the intuitionistic predicate calculus is equivalent to the method of intuitionistic first-order tableaux:

9.14. THEOREM. *Let* Φ *be finite. Then* $\Phi \rhd \xi$ *iff* $\Phi \vdash_I \xi$.

PROOF. Suppose $\Phi \rhd \xi$. We proceed exactly as in Thm. 8.12, except that we now have to consider four additional cases, corresponding to the quantifier rules:

In the case of rule $\exists$, the induction hypothesis is

$$\Phi, \alpha(\mathbf{x}/\mathbf{y}) \vdash_I \xi,$$

where $\exists \mathbf{x}\alpha \in \Phi$ and **y** is not free in Φ or ξ. By Thm. 9.10 (ii) we get

$$\Phi, \exists \mathbf{y}[\alpha(\mathbf{x}/\mathbf{y})] \vdash_I \xi.$$

But $\exists \mathbf{y}[\alpha(\mathbf{x}/\mathbf{y})]$ is a variant of $\exists \mathbf{x}\alpha$, hence provably equivalent to it by Thm. 9.11. Since we already have $\exists \mathbf{x}\alpha$ in Φ, it follows that $\Phi \vdash_I \xi$, as required.

In the case of the $-\exists$ rule, the induction hypothesis is

$$\Phi \vdash_I \alpha(\mathbf{x}/\mathbf{t}),$$

where $\xi = \exists \mathbf{x}\alpha$. By Prob. 9.12 (ii) we get $\Phi \vdash_I \xi$.

In the case of rule $\forall$, the induction hypothesis is

$$\mathbf{\Phi}, \boldsymbol{\alpha}(\mathbf{x}/\mathbf{t}) \vdash_{\mathrm{I}} \xi,$$

where $\forall \mathbf{x}\boldsymbol{\alpha} \in \mathbf{\Phi}$. But then by Prob. 9.12 (i) we have $\mathbf{\Phi} \vdash_{\mathrm{I}} \boldsymbol{\alpha}(\mathbf{x}/\mathbf{t})$, hence $\mathbf{\Phi} \vdash_{\mathrm{I}} \xi$.

Finally, in the case of the $-\forall$ rule, the induction hypothesis is

$$\mathbf{\Phi} \vdash_{\mathrm{I}} \boldsymbol{\alpha}(\mathbf{x}/\mathbf{y}),$$

where $\xi = \forall \mathbf{x}\boldsymbol{\alpha}$ and $\mathbf{y}$ is not free in $\mathbf{\Phi}$ or ξ. By Thm. 9.10(i) we get $\mathbf{\Phi} \vdash_{\mathrm{I}} \forall \mathbf{y}[\boldsymbol{\alpha}(\mathbf{x}/\mathbf{y})]$ and hence, by Thm. 9.11, $\mathbf{\Phi} \vdash_{\mathrm{I}} \forall \mathbf{x}\boldsymbol{\alpha}$, i.e., $\mathbf{\Phi} \vdash_{\mathrm{I}} \xi$.

Now suppose, conversely, that $\mathbf{\Phi} \vdash_{\mathrm{I}} \xi$. To show that $\mathbf{\Phi} \rhd \xi$, proceed as in Thm. 8.12, but this time using Prob. 9.9 instead of Prob. 8.11. ∎

In the sequel we shall often use Thm. 9.14 without special mention.

As an application, we prove the following result, due to Rasiowa and Sikorski.

9.15. THEOREM. *Suppose that* $\vdash_{\mathrm{I}} \exists \mathbf{x}\boldsymbol{\alpha}$. *If* $\exists \mathbf{x}\boldsymbol{\alpha}$ *has terms (i.e., free variables or constants), then* $\vdash_{\mathrm{I}} \boldsymbol{\alpha}(\mathbf{x}/\mathbf{t})$ *for some such term* $\mathbf{t}$. *If* $\exists \mathbf{x}\boldsymbol{\alpha}$ *has no terms, then* $\vdash_{\mathrm{I}} \forall \mathbf{x}\boldsymbol{\alpha}$.

PROOF. By Prob. 5.14, $\vdash_{\mathrm{I}} \boldsymbol{\alpha}(\mathbf{x}/\mathbf{t})$ for some term $\mathbf{t}$. If $\mathbf{t}$ is a term of $\exists \mathbf{x}\boldsymbol{\alpha}$, we are through.

If $\mathbf{t}$ is not a term of $\exists \mathbf{x}\boldsymbol{\alpha}$ — and, in particular, if $\exists \mathbf{x}\boldsymbol{\alpha}$ has no terms then either $\mathbf{t}$ is a variable $\mathbf{y}$ not free in $\exists \mathbf{x}\boldsymbol{\alpha}$, or $\mathbf{t}$ is a constant $\mathbf{c}$ not occurring in $\exists \mathbf{x}\boldsymbol{\alpha}$. In the first case we have $\vdash_{\mathrm{I}} \forall \mathbf{y}[\boldsymbol{\alpha}(\mathbf{x}/\mathbf{y})]$ by Thm. 9.10(i) and hence $\vdash_{\mathrm{I}} \forall \mathbf{x}\boldsymbol{\alpha}$ by Thm. 9.11. In the second case we have $\vdash_{\mathrm{I}} \forall \mathbf{x}\boldsymbol{\alpha}$ by Thm. 9.13(i) In both cases we have $\vdash_{\mathrm{I}} \boldsymbol{\alpha}(\mathbf{x}/\mathbf{s})$ for *every* term $\mathbf{s}$, by Prob. 9.12(i). ∎

9.16. REMARK. Another consequence of Thm. 9.14 is that deducibility is invariant with respect to language. Suppose that $\mathbf{\Phi}$ is a set of $\mathscr{L}$-formulas and ξ is an $\mathscr{L}$-formula or the empty string. Let $\mathscr{L}'$ be an extension of $\mathscr{L}$, and suppose that $\mathbf{\Phi} \vdash_{\mathrm{I}} \xi$ in $\mathscr{L}'$. Then $\mathbf{\Phi}_0 \vdash_{\mathrm{I}} \xi$ in $\mathscr{L}'$, where $\mathbf{\Phi}_0$ is some finite subset of $\mathbf{\Phi}$. Hence by Thm. 9.14 we have $\mathbf{\Phi}_0 \rhd \xi$ in $\mathscr{L}'$. If $\boldsymbol{T}$ is a proof of $\mathbf{\Phi}_0 \rhd \xi$ in $\mathscr{L}'$, then all the symbols occurring in $\boldsymbol{T}$, except perhaps some constants, must be in $\mathscr{L}$. Now, any constant which occurs in $\boldsymbol{T}$ and does not belong to $\mathscr{L}$ can be replaced by a variable, provided that variable does not already occur in $\boldsymbol{T}$. In this way we obtain a proof of $\mathbf{\Phi}_0 \rhd \xi$ which is entirely in $\mathscr{L}$. Therefore by Thm. 9.14 we have $\mathbf{\Phi}_0 \vdash_{\mathrm{I}} \xi$ in $\mathscr{L}$ and hence $\mathbf{\Phi} \vdash_{\mathrm{I}} \xi$ in $\mathscr{L}$.

9.17. PROBLEM. Prove the consistency of the intuitionistic predicate calculus, i.e., the impossibility of $\emptyset \vdash_{\mathrm{I}}$.

§ 10. Completeness

In this section our aim is to prove the completeness of the intuitionistic predicate calculus relative to Kripke's semantics.

We shall adopt here a non-constructive attitude. Thus *in this section* when we say, e.g., $\mathbf{\Phi} \vdash_{\mathrm{I}} \boldsymbol{\alpha}$, we mean that there *exists* a deduction of $\boldsymbol{\alpha}$ from $\mathbf{\Phi}$, not that we necessarily know how to find such a deduction. However, note that under this non-constructive interpretation the results of §9 continue to hold.

If $\boldsymbol{C}$ is a set of new individual constants not belonging to $\mathscr{L}$, we let $\mathscr{L}(\boldsymbol{C})$ be the language obtained from $\mathscr{L}$ by adding the constants in $\boldsymbol{C}$.

Throughout this section we let λ be the cardinality of $\mathscr{L}$ (see Def. 3.3.10). Then, by Thm. 3.3.11, λ is also the cardinality of the set of all $\mathscr{L}$-formulas.

A set $\mathbf{\Phi}$ of $\mathscr{L}$-formulas will be called *strongly consistent in* $\mathscr{L}$ if the following four conditions hold:

(a) $\mathbf{\Phi}$ is consistent, i.e., $\mathbf{\Phi} \nvdash_{\mathrm{I}}$.

(b) Whenever $\boldsymbol{\alpha}$ is an $\mathscr{L}$-formula such that $\mathbf{\Phi} \vdash_{\mathrm{I}} \boldsymbol{\alpha}$, then $\boldsymbol{\alpha} \in \mathbf{\Phi}$.

(c) Whenever $\boldsymbol{\alpha} \vee \boldsymbol{\beta} \in \mathbf{\Phi}$, then $\boldsymbol{\alpha} \in \mathbf{\Phi}$ or $\boldsymbol{\beta} \in \mathbf{\Phi}$.

(d) Whenever $\exists \mathbf{x} \boldsymbol{\alpha} \in \mathbf{\Phi}$, then also $\boldsymbol{\alpha}(\mathbf{x}/\mathbf{t}) \in \mathbf{\Phi}$ for some term $\mathbf{t}$.

10.1. LEMMA. *Let* $\boldsymbol{C}$ *be a set, of cardinality* λ, *of constants not belonging to* $\mathscr{L}$. *Let* $\mathbf{\Phi}$ *be a set of* $\mathscr{L}$*-formulas and let* $\boldsymbol{\gamma}$ *be an* $\mathscr{L}$*-formula such that* $\mathbf{\Phi} \nvdash_{\mathrm{I}} \boldsymbol{\gamma}$. *Then there exists a set* $\mathbf{\Psi}$ *of* $\mathscr{L}(\boldsymbol{C})$*-formulas such that* $\mathbf{\Psi}$ *is strongly consistent in* $\mathscr{L}(\boldsymbol{C})$, *and* $\mathbf{\Phi} \subseteq \mathbf{\Psi}$, *and* $\mathbf{\Psi} \nvdash_{\mathrm{I}} \boldsymbol{\gamma}$.

PROOF. The cardinality of $\mathscr{L}(\boldsymbol{C})$ is clearly λ. We fix a well-ordering $\{\boldsymbol{\varphi}_\xi \colon \xi < \lambda\}$ of all $\mathscr{L}(\boldsymbol{C})$-formulas. By transfinite recursion we define for each $\zeta \leqslant \lambda$ a set $\mathbf{\Phi}_\zeta$ of $\mathscr{L}(\boldsymbol{C})$-formulas such that:

(1) $\mathbf{\Phi}_\eta \subseteq \mathbf{\Phi}_\zeta$ for all $\eta < \zeta$,

(2) $\mathbf{\Phi}_\zeta \nvdash_{\mathrm{I}} \boldsymbol{\gamma}$,

(3) only finitely many, or at most $|\zeta|$ new constants (i.e., constants of $\boldsymbol{C}$) occur in $\mathbf{\Phi}_\zeta$.

For $\zeta = 0$ we put $\mathbf{\Phi}_0 = \mathbf{\Phi}$. Then (1)—(3) clearly hold.

Now, let $0 < \varrho \leqslant \lambda$, and suppose that for all $\zeta < \varrho$ the $\mathbf{\Phi}_\zeta$ have been defined in accordance with (1)–(3).

If ϱ is a limit ordinal (in particular if $\varrho = \lambda$) we put

$$\mathbf{\Phi}_\varrho = \bigcup \{\mathbf{\Phi}_\zeta \colon \zeta < \varrho\}.$$

It is easy to see that (1)–(3) hold for ϱ.

If ϱ is a successor, say $\varrho = \zeta + 1$, we distinguish four cases:

Case 1. If $\mathbf{\Phi}_\zeta$, $\boldsymbol{\varphi}_\zeta \vdash_{\mathrm{I}} \boldsymbol{\gamma}$, we put $\mathbf{\Phi}_{\zeta+1} = \mathbf{\Phi}_\zeta$. Then (1)–(3) hold automatically for $\zeta+1$.

Case 2. If $\mathbf{\Phi}_\zeta$, $\boldsymbol{\varphi}_\zeta \nvdash_{\mathrm{I}} \boldsymbol{\gamma}$ and $\boldsymbol{\varphi}_\zeta$ is neither a disjunction formula nor an existential formula, we put

$$\mathbf{\Phi}_{\zeta+1} = \mathbf{\Phi}_\zeta \cup \{\boldsymbol{\varphi}_\zeta\}.$$

Here too (1)–(3) hold for $\zeta+1$.

Case 3. If $\mathbf{\Phi}_\zeta$, $\boldsymbol{\varphi}_\zeta \nvdash_{\mathrm{I}} \boldsymbol{\gamma}$ and $\boldsymbol{\varphi}_\zeta = \boldsymbol{\alpha} \vee \boldsymbol{\beta}$, then we must have $\mathbf{\Phi}_\zeta$, $\boldsymbol{\varphi}_\zeta$, $\boldsymbol{\alpha} \nvdash_{\mathrm{I}} \boldsymbol{\gamma}$ or $\mathbf{\Phi}_\zeta, \boldsymbol{\varphi}_\zeta, \boldsymbol{\beta} \nvdash_{\mathrm{I}} \boldsymbol{\gamma}$, otherwise by ax. (8.8) we would have $\mathbf{\Phi}_\zeta$, $\boldsymbol{\varphi}_\zeta \vdash_{\mathrm{I}} \boldsymbol{\gamma}$, contrary to assumption. If $\mathbf{\Phi}_\zeta, \boldsymbol{\varphi}_\zeta, \boldsymbol{\alpha} \nvdash_{\mathrm{I}} \boldsymbol{\gamma}$, we put

$$\mathbf{\Phi}_{\zeta+1} = \mathbf{\Phi}_\zeta \cup \{\boldsymbol{\varphi}_\zeta, \boldsymbol{\alpha}\};$$

otherwise, we put

$$\mathbf{\Phi}_{\zeta+1} = \mathbf{\Phi}_\zeta \cup \{\boldsymbol{\varphi}_\zeta, \boldsymbol{\beta}\}.$$

Again, (1)–(3) must hold for $\zeta+1$.

Case 4. If $\mathbf{\Phi}_\zeta, \boldsymbol{\varphi}_\zeta \nvdash_{\mathrm{I}} \boldsymbol{\gamma}$ and $\boldsymbol{\varphi}_\zeta = \exists \mathbf{x} \boldsymbol{\alpha}$, then by (3) we can find a new constant $\mathbf{c}$ that does not occur in $\mathbf{\Phi}_\zeta$ or $\boldsymbol{\varphi}_\zeta$ and for some such $\mathbf{c}$ we put

$$\mathbf{\Phi}_{\zeta+1} = \mathbf{\Phi}_\zeta \cup \{\boldsymbol{\varphi}_\zeta, \boldsymbol{\alpha}(\mathbf{x}/\mathbf{c})\}.$$

Then (1) and (3) clearly hold for $\zeta+1$; also, by Thm. 9.13(ii) we see that (2) must hold for $\zeta+1$.

We put

$$\mathbf{\Psi} = \mathbf{\Phi}_\lambda,$$

and it is easy to verify that $\mathbf{\Psi}$ has all the required properties. First, it is clear that $\mathbf{\Phi} \subseteq \mathbf{\Psi}$ and $\mathbf{\Psi} \nvdash_{\mathrm{I}} \boldsymbol{\gamma}$, hence also $\mathbf{\Psi} \nvdash_{\mathrm{I}}$. Next, if $\boldsymbol{\alpha}$ is an $\mathscr{L}(C)$-formula such that $\mathbf{\Psi} \vdash_{\mathrm{I}} \boldsymbol{\alpha}$, then $\mathbf{\Psi}, \boldsymbol{\alpha} \nvdash_{\mathrm{I}} \boldsymbol{\gamma}$ and $\boldsymbol{\alpha} = \boldsymbol{\varphi}_\zeta$ for some $\zeta < \lambda$. Thus $\boldsymbol{\alpha}$ was put in $\mathbf{\Phi}_{\zeta+1}$. Similarly, if $\boldsymbol{\alpha} \vee \boldsymbol{\beta} \in \mathbf{\Psi}$, then $\boldsymbol{\alpha} \vee \boldsymbol{\beta} = \boldsymbol{\varphi}_\zeta$ for some ζ and $\boldsymbol{\alpha}$ or $\boldsymbol{\beta}$ was put into $\mathbf{\Phi}_{\zeta+1}$. Finally, if $\exists \mathbf{x} \boldsymbol{\alpha} \in \mathbf{\Psi}$ then $\exists \mathbf{x} \boldsymbol{\alpha} = \boldsymbol{\varphi}_\zeta$ for some ζ and for some new constant $\mathbf{c}$ we have put $\boldsymbol{\alpha}(\mathbf{x}/\mathbf{c})$ in $\mathbf{\Phi}_{\zeta+1}$. ∎

We define a sequence of languages $\{\mathscr{L}_n \colon n \in N\}$ as follows. $\mathscr{L}_0$ is our original language $\mathscr{L}$; and for each n we put

$$\mathscr{L}_{n+1} = \mathscr{L}_n(C_n),$$

where C_n is a set, of cardinality λ, of constants not belonging to $\mathscr{L}_n$. We let $\mathscr{L}'$ be the union of all these languages, i.e.,

$$\mathscr{L}' = \mathscr{L}(C),$$

where $C = \bigcup\{C_n \colon n \in N\}$. The cardinality of C is clearly λ.

Note that, if a set $\boldsymbol{\Psi}$ of $\mathscr{L}_n$-formulas is strongly consistent in $\mathscr{L}_n$, then for every formula $\boldsymbol{\alpha}$ of $\mathscr{L}_n$ we have $\boldsymbol{\alpha} \rightarrow \boldsymbol{\alpha} \in \boldsymbol{\Psi}$, hence all the constants of $\mathscr{L}_n$ must actually occur in $\boldsymbol{\Psi}$, and $\boldsymbol{\Psi}$ cannot at the same time be strongly consistent in $\mathscr{L}_m$ with $m<n$.

We now define a Kripke system $\mathfrak{K}$ as follows. The states of $\mathfrak{K}$ will be sets of formulas: if for some n — which, by what we have just seen, must necessarily be unique — $\boldsymbol{\Psi}$ is a set of $\mathscr{L}_n$-formulas and is strongly consistent in $\mathscr{L}_n$, we take $\boldsymbol{\Psi}$ as a state of $\mathfrak{K}$. For that unique n we let $T(\boldsymbol{\Psi})$ be the set of all terms of $\mathscr{L}_n$; and we let $F(\boldsymbol{\Psi})$ be the set of all atomic formulas belonging to $\boldsymbol{\Psi}$. The states are partially ordered by inclusion: $\boldsymbol{\Psi} \leqslant \boldsymbol{\Omega}$ iff $\boldsymbol{\Psi} \subseteq \boldsymbol{\Omega}$.

$\mathfrak{K}$ is clearly a Kripke system for $\mathscr{L}'$.

10.2. Lemma. *For each state $\boldsymbol{\Psi}$ of $\mathfrak{K}$ and for each $\mathscr{L}'$-formula $\boldsymbol{\alpha}$ we have $\boldsymbol{\Psi} \Vdash_{\mathfrak{K}} \boldsymbol{\alpha}$ iff $\boldsymbol{\alpha} \in \boldsymbol{\Psi}$.*

Proof. Let n be the unique number such that $\boldsymbol{\Psi}$ is strongly consistent in $\mathscr{L}_n$. We proceed by induction on $\deg \boldsymbol{\alpha}$ and distinguish seven cases, corresponding to the seven "positive" clauses of KSD. Cases 1,2,3 and 6 ($\boldsymbol{\alpha}$ atomic, $\boldsymbol{\alpha}=\boldsymbol{\beta} \vee \boldsymbol{\gamma}$, $\boldsymbol{\alpha}=\boldsymbol{\beta} \wedge \boldsymbol{\gamma}$, $\boldsymbol{\alpha}=\exists \mathbf{x} \boldsymbol{\beta}$) are routine and are left to the reader. We deal in detail with the other three cases.

Case 4: $\boldsymbol{\alpha}=\boldsymbol{\beta} \rightarrow \boldsymbol{\gamma}$. First, suppose $\boldsymbol{\alpha} \in \boldsymbol{\Psi}$. Then $\boldsymbol{\alpha}$ is in $\mathscr{L}_n$ and hence all the terms of $\boldsymbol{\alpha}$ are in $T(\boldsymbol{\Psi})$. Now let $\boldsymbol{\Omega} \geqslant \boldsymbol{\Psi}$; then there is some $m \geqslant n$ such that $\boldsymbol{\Omega}$ is strongly consistent in $\mathscr{L}_m$. Also, $\boldsymbol{\alpha} \in \boldsymbol{\Omega}$ because $\boldsymbol{\Psi} \subseteq \boldsymbol{\Omega}$. Let $\boldsymbol{\Omega} \Vdash_{\mathfrak{K}} \boldsymbol{\beta}$; we have to show that $\boldsymbol{\Omega} \Vdash_{\mathfrak{K}} \boldsymbol{\gamma}$. But by the induction hypothesis we have $\boldsymbol{\beta} \in \boldsymbol{\Omega}$, and hence $\boldsymbol{\Omega} \vdash_{\mathrm{I}} \boldsymbol{\gamma}$ by *modus ponens*. Since $\boldsymbol{\Omega}$ is strongly consistent in $\mathscr{L}_m$, we must have $\boldsymbol{\gamma} \in \boldsymbol{\Omega}$; hence $\boldsymbol{\Omega} \Vdash_{\mathfrak{K}} \boldsymbol{\gamma}$ by the induction hypothesis. Thus $\boldsymbol{\Psi} \Vdash_{\mathfrak{K}} \boldsymbol{\alpha}$.

Conversely, let $\boldsymbol{\alpha} \notin \boldsymbol{\Psi}$. If $\boldsymbol{\alpha}$ is not in $\mathscr{L}_n$, then not all the terms of $\boldsymbol{\alpha}$ belong to $T(\boldsymbol{\Psi})$; hence $\boldsymbol{\Psi} \nVdash_{\mathfrak{K}} \boldsymbol{\alpha}$. If $\boldsymbol{\alpha}$ *is* in $\mathscr{L}_n$, then $\boldsymbol{\Psi} \nvdash_{\mathrm{I}} \boldsymbol{\alpha}$ because $\boldsymbol{\Psi}$ is strongly consistent in $\mathscr{L}_n$. Hence, by the Deduction Theorem, $\boldsymbol{\Psi}, \boldsymbol{\beta} \nvdash_{\mathrm{I}} \boldsymbol{\gamma}$. By Lemma 10.1 we extend $\boldsymbol{\Psi} \cup \{\boldsymbol{\beta}\}$ to a set $\boldsymbol{\Omega}$ of $\mathscr{L}_{n+1}$-formulas which is strongly consistent in $\mathscr{L}_{n+1}$ and such that $\boldsymbol{\Omega} \nvdash_{\mathrm{I}} \boldsymbol{\gamma}$, hence $\boldsymbol{\gamma} \notin \boldsymbol{\Omega}$. By the induction hypothesis we see that $\boldsymbol{\Omega} \Vdash_{\mathfrak{K}} \boldsymbol{\beta}$ and $\boldsymbol{\Omega} \nVdash_{\mathfrak{K}} \boldsymbol{\gamma}$. Since $\boldsymbol{\Psi} \leqslant \boldsymbol{\Omega}$, we have $\boldsymbol{\Psi} \nVdash_{\mathfrak{K}} \boldsymbol{\alpha}$.

Case 5: $\boldsymbol{\alpha}=\neg \boldsymbol{\beta}$. Suppose $\boldsymbol{\alpha} \in \boldsymbol{\Psi}$. Then clearly all the terms of $\boldsymbol{\alpha}$ are in $T(\boldsymbol{\Psi})$. Let $\boldsymbol{\Omega} \geqslant \boldsymbol{\Psi}$; we have to show that $\boldsymbol{\Omega} \nVdash_{\mathfrak{K}} \boldsymbol{\beta}$. But, since $\boldsymbol{\alpha} \in \boldsymbol{\Psi} \subseteq \boldsymbol{\Omega}$ and $\boldsymbol{\Omega}$ is consistent, it follows that $\boldsymbol{\beta} \notin \boldsymbol{\Omega}$; hence $\boldsymbol{\Omega} \nVdash_{\mathfrak{K}} \boldsymbol{\beta}$ by the induction hypothesis. Thus $\boldsymbol{\Psi} \Vdash_{\mathfrak{K}} \boldsymbol{\alpha}$.

Conversely, suppose $\boldsymbol{\alpha} \notin \boldsymbol{\Psi}$. If $\boldsymbol{\alpha}$ is not in $\mathscr{L}_n$, then (as in case 4) we see that $\boldsymbol{\Psi} \nVdash_{\mathfrak{K}} \boldsymbol{\alpha}$. If $\boldsymbol{\alpha}$ *is* in $\mathscr{L}_n$, then $\boldsymbol{\Psi} \nvdash_{\mathrm{I}} \boldsymbol{\alpha}$ because of the strong consistency of Ψ in $\mathscr{L}_n$. Now, it is easy to see that $\beta \to \neg\beta \vdash_{I0} \neg\beta$ (see ax. (8.9)), and therefore, recalling that $\alpha = \neg\beta$, we see that $\Psi \nvdash_I \beta \to \neg\beta$; hence $\Psi, \beta \nvdash_{I0} \neg\beta$. By Lemma 10.1 we extend $\Psi \cup \{\beta\}$ to a set Ω strongly consistent in $\mathscr{L}_{n+1}$. Then by the induction hypothesis $\Omega \Vdash_{\mathfrak{K}} \beta$, and since $\Psi \leqslant \Omega$ we have $\Psi \nVdash_{\mathfrak{K}} \alpha$.

Case 7: $\alpha = \forall x\beta$. Let $\alpha \in \Psi$. If $\Omega \geqslant \Psi$, then also $\alpha \in \Omega$ and hence by Prob. 9.12(i) we have $\Omega \vdash_I \beta(x/t)$ for every term $t \in T(\Omega)$. By the strong consistency of Ω (in some $\mathscr{L}_m$ with $m \geqslant n$) we get $\beta(x/t) \in \Omega$; hence $\Omega \Vdash_{\mathfrak{K}} \beta(x/t)$ by the induction hypothesis. Thus $\Psi \Vdash_{\mathfrak{K}} \alpha$.

Conversely, let $\alpha \notin \Psi$. If α is not in $\mathscr{L}_n$, we see (as in the previous cases) that $\Psi \nVdash_{\mathfrak{K}} \alpha$. If α *is* in $\mathscr{L}_n$, then again $\Psi \nvdash_I \alpha$. Take some $c \in C_n$. Then c cannot occur in Ψ or α, because the constants of C_n were taken *not* to be in $\mathscr{L}_n$. Thus by Thm. 9.13(i) we have $\Psi \nvdash_I \beta(x/c)$. By Lemma 10.1, we extend Ψ to a set Ω, strongly consistent in $\mathscr{L}_{n+2}$, such that $\Omega \nvdash_I \beta(x/c)$, hence $\beta(x/c) \notin \Omega$. Thus $\Omega \nVdash_{\mathfrak{K}} \beta(x/c)$ by the induction hypothesis; and since $\Psi \leqslant \Omega$ we have $\Psi \nVdash_{\mathfrak{K}} \alpha$. ∎

10.3. STRONG COMPLETENESS THEOREM. *There exists a Kripke system $\mathfrak{K}$ for the language $\mathscr{L}(C)$, where C is a set, of cardinality λ, of constants not belonging to $\mathscr{L}$, such that for any set Φ of $\mathscr{L}$-formulas and any $\mathscr{L}$-formula γ such that $\Phi \nvdash_I \gamma$, some state of $\mathfrak{K}$ forces $\Phi \cup \{-\gamma\}$. In particular, if $\Phi \nvdash_I$ then some state of $\mathfrak{K}$ forces Φ.*

PROOF. Let $\mathfrak{K}$ be the Kripke system defined above. If Φ and γ are in $\mathscr{L}$ and $\Phi \nvdash_I \gamma$, then by Lemma 10.1 we extend Φ to a set Ψ of $\mathscr{L}_1$-formulas which is strongly consistent in $\mathscr{L}_1$ and such that $\Psi \nvdash_I \gamma$, hence $\gamma \notin \Psi$. Then Ψ is a state of $\mathfrak{K}$, and by Lemma 10.2 we clearly have $\Psi \Vdash_{\mathfrak{K}} \Phi \cup \{-\gamma\}$.

In particular, if $\Phi \nvdash_I$ then $\Phi \nvdash_I \gamma$ for some $\mathscr{L}$-formula γ. ∎

10.4. THEOREM. *For any set Φ of $\mathscr{L}$-formulas and any $\mathscr{L}$-formula γ, we have $\Phi \vdash_I \gamma$ iff $\Phi \Vdash \gamma$. Also, $\Phi \vdash_I$ iff Φ is not enforceable.*

PROOF. If $\Phi \vdash_I \gamma$, then for some finite $\Phi_0 \subseteq \Phi$ we have $\Phi_0 \vdash_I \gamma$, hence $\Phi_0 \Vdash \gamma$ by Thm. 6.11. Therefore $\Phi \Vdash \gamma$. Conversely, if $\Phi \nvdash_I \gamma$ then $\Phi \cup \{-\gamma\}$ is enforceable by Thm. 10.3; thus $\Phi \nVdash \gamma$.

The proof that $\Phi \vdash_I$ iff Φ is not enforceable is similar. ∎

Note that in Thm. 10.4 we have not used the full power of Thm. 10.3. For by Thm. 10.3 we have a certain *universal* Kripke system $\mathfrak{K}$ which works for *every* Φ and γ.

It is doubtful whether Kripke's semantics is — or can be turned into — a satisfactory explication of the constructivist heuristic outlined in §4.

Nevertheless, because of the results of the present section, it is very useful both heuristically and technically. Many mathematicians are so used to (or perhaps brainwashed into) the purely structural mode of thinking that they find it difficult to test directly the constructive validity of a given formula. The heuristic of Kripke's semantics provides them with a convenient structural *Ersatz*. On the technical level, Kripke systems are useful for obtaining independence (=unprovability) results for intuitionistic (and even classical) formal theories. As a very simple example, observe that by 6.8 we have

$$\nvdash_{\mathrm{I}} \neg\neg[\forall x \neg\neg Px \rightarrow \neg\neg \forall x Px]$$

(cf. Prob. 5.16), and by 6.9 we have

$$\nvdash_{\mathrm{I}} \forall x[Py \vee Qx] \rightarrow Py \vee \forall x Qx.$$

§ 11. Translations from classical to intuitionistic logic

It is easy to see (directly or *via* Prob. 5.5) that if $\boldsymbol{\Phi} \vdash_{\mathrm{I}0} \alpha$ then also $\boldsymbol{\Phi} \vdash_0 \alpha$. By 5.6(iii) the converse of this is false. Similarly, we easily see (directly or *via* Prob. 5.11) that if $\boldsymbol{\Phi} \vdash_{\mathrm{I}} \alpha$ then $\boldsymbol{\Phi} \vdash \alpha$. The converse of this is false by 5.16.

It therefore seems that the intuitionistic propositional and predicate calculi are strictly weaker than their classical counterparts. However, this is the case only if we insist on comparing each formula in the classical calculus with the *same* formula in the corresponding intuitionistic calculus. But this comparison is unfair, because the structural interpretation of some of the logical symbols is very different from their constructive interpretation.

In this section we shall show (in a constructive way) that it is possible to *translate* classical formulas in such a way that a formula is provable in the classical propositional or predicate calculus iff its translation is provable in the corresponding intuitionistic calculus. Thus the classical calculi can be "interpreted" within their intuitionistic counterparts. We begin with the propositional case.

We employ the version of the classical propositional calculus formulated in §15 of Ch. 1. Notice that the axioms of this calculus are the same as those of the *intuitionistic* propositional calculus, except that the two schemes (8.9) and (8.10) are replaced by the single scheme

$$(\neg\alpha \rightarrow \beta) \rightarrow (\neg\alpha \rightarrow \neg\beta) \rightarrow \alpha. \tag{11.1}$$

For any set Φ of formulas we put $\neg\Phi = \{\neg\varphi : \varphi \in \Phi\}$. We then have the following result, due to Glivenko:

11.2. THEOREM. *For any set* Φ *of formulas and any formula* φ *we have* $\Phi \vdash_0 \varphi$ *iff* $\neg\neg\Phi \vdash_{I0} \neg\neg\varphi$.

PROOF. Suppose $\Phi \vdash_0 \varphi$. We proceed by induction on the length of a given (classical) deduction of φ from Φ.

If φ is one of the axioms (8.1)–(8.8), then trivially $\vdash_{I0} \varphi$, hence $\vdash_{I0} \neg\neg\varphi$ by Prob. 5.7(i); therefore $\neg\neg\Phi \vdash_{I0} \neg\neg\varphi$.

If φ is an axiom of the form (11.1), then $\vdash_{I0} \neg\neg\varphi$ by Prob. 5.7(ix), and again we have $\neg\neg\Phi \vdash_{I0} \neg\neg\varphi$.

If $\varphi \in \Phi$, then trivially $\neg\neg\Phi \vdash_{I0} \neg\neg\varphi$.

Finally, if φ is obtained from two earlier formulas by *modus ponens*, we get $\neg\neg\Phi \vdash_{I0} \neg\neg\varphi$ by the induction hypothesis and Prob. 5.7(v).

Conversely, if $\neg\neg\Phi \vdash_{I0} \neg\neg\varphi$ then clearly $\neg\neg\Phi \vdash_0 \neg\neg\varphi$; hence $\Phi \vdash_0 \varphi$ by Thm. 1.10.5. ∎

11.3. PROBLEM. Show that $\neg\Phi \vdash_0 \neg\varphi$ iff $\neg\Phi \vdash_{I0} \neg\varphi$. (Use parts (i) and (ii) of Prob. 5.7.) Also, show that if Φ is consistent in the intuitionistic propositional calculus, it is consistent in the classical calculus as well.

By Thm. 11.2, the mapping f defined by $f(\varphi) = \neg\neg\varphi$ is a *faithful translation* from the classical propositional calculus into its intuitionistic counterpart[1].

Another faithful translation is suggested in the following:

11.4. PROBLEM. For any formula φ, let φ' be defined by induction on deg φ as follows:

$\varphi' = \neg\neg\varphi$ for every prime formula φ,

$(\varphi \vee \psi)' = \varphi' \vee \psi'$,

$(\varphi \wedge \psi)' = \varphi' \wedge \psi'$,

$(\varphi \rightarrow \psi)' = \varphi' \rightarrow \psi'$,

$(\neg\varphi)' = \neg(\varphi')$.

Also put $\Phi' = \{\varphi' : \varphi \in \Phi\}$.

(i) Let φ be any formula obtained from its prime components without using disjunction. Show that $\neg\neg\varphi \vdash_{I0} \varphi'$ and $\varphi' \vdash_{I0} \neg\neg\varphi$. (Proceed by induction on deg φ, using Thm. 11.2; for the case of implication use also Prob. 5.7.(vi).)

(ii) Let φ be as in (i) and let Φ be a set of formulas of the same kind

[1] The term *faithful* here refers to the purely formal business of deducibility, *not* to meaning.

(i.e., obtained from their prime components without using disjunction). Show that $\Phi \vdash_0 \varphi$ iff $\Phi' \vdash_{I0} \varphi'$.

(iii) Show that if α is prime then $\nvdash_{I0} \neg\neg\alpha \vee \neg\neg\neg\alpha$. Thus (ii) does not work in the presence of disjunction.

Since every formula can be transformed into a classically equivalent formula not containing disjunction, Prob. 11.4(ii) yields another faithful translation from the classical propositional calculus into its intuitionistic counterpart.

We turn now to the predicate calculus. Together with $\mathscr{L}$, we consider the language $\mathscr{L}^*$ obtained from $\mathscr{L}$ by excluding the universal quantifier.

In view of what was said in §4 of Ch. 3, the classical predicate calculus in $\mathscr{L}^*$ can be based on the following seven groups of axioms:

(11.5) all $\mathscr{L}^*$-formulas of the forms (8.1)–(8.8);

(11.6) all $\mathscr{L}^*$-formulas of the form (11.1);

(11.7) $\neg\exists x \neg(\alpha \to \beta) \to \neg\exists x \neg\alpha \to \neg\exists x \neg\beta$;

(11.8) $\neg\exists x \neg(\alpha \to \beta) \to \exists x \alpha \to \exists x \beta$;

(11.9) $\alpha \to \neg\exists x \neg\alpha$, provided x is not free in α;

(11.10) $\neg\exists x \neg\alpha \to \alpha(x/t)$, provided t is free for x in α;

(11.11) $\neg\exists x_1 \neg\neg\exists x_2 \neg \ldots \neg\exists x_k \neg\alpha$, where $k \geqslant 1$, and $x_1, \ldots, x_k$ are any variables (not necessarily distinct) and α is any axiom of the preceding six groups.

The only rule of inference is *modus ponens*.

11.12. THEOREM. *Let Φ be any set of $\mathscr{L}^*$-formulas, and let φ be an $\mathscr{L}^*$-formula. Then $\Phi \vdash \varphi$ iff $\neg\neg\Phi \vdash_I \neg\neg\varphi$.*

PROOF. Like the proof of 11.2, except that now we also have to verify that $\vdash_I \neg\neg\varphi$ for each $\mathscr{L}^*$-formula φ belonging to the groups (11.7)–(11.11). This can easily be done by tableaux, and is left to the reader. ∎

11.13. PROBLEM. For Φ and φ as in 11.12, show that $\neg\Phi \vdash \neg\varphi$ iff $\neg\Phi \vdash_I \neg\varphi$. Also, show that if such Φ is consistent in the intuitionistic predicate calculus, it is classically consistent as well.

Since every $\mathscr{L}$-formula is classically equivalent to an $\mathscr{L}^*$-formula, Thm. 11.12 provides us with a faithful translation from the classical predicate calculus into its intuitionistic counterpart. Another translation is suggested in the following:

11.14. PROBLEM. For any $\mathscr{L}$-formula φ, let φ' be the formula obtained from φ when each atomic subformula α of φ is replaced by $\neg\neg\alpha$, and let φ^* be the formula obtained from φ when each universal quantifier $\forall x$ is replaced by $\neg\exists x\neg$. For any set Φ of $\mathscr{L}$-formulas let $\Phi' = \{\varphi' : \varphi \in \Phi\}$.

(i) Show that if φ is an $\mathscr{L}$-formula not containing the symbols $\vee$ and $\exists$, then $\neg\neg\varphi^* \vdash_I \varphi'$ and $\varphi' \vdash_I \neg\neg\varphi^*$. (Proceed as in Prob. 11.4(i). In the case where φ is a universal formula, use Prob. 5.15.)

(ii) Let φ be as in (i) and let Φ be a set of formulas of the same kind (not containing $\vee$ and $\exists$). Show that $\Phi \vdash \varphi$ iff $\Phi' \vdash_I \varphi'$. (Use (i) and Thm. 11.12.)

Again, every $\mathscr{L}$-formula can be transformed into a classically equivalent formula without the symbols $\vee$ and $\exists$. Thus we get from Prob. 11.14 a faithful translation from the classical into the intuitionistic predicate calculus. This result is due to Gödel.

The results of this section show that the intuitionistic propositional and predicate calculi are equal in strength to their classical counterparts. However, intuitionistic logic is certainly *richer* than classical logic, because the former makes distinctions that the latter fails to make. Thus, e.g., $\neg\neg\alpha$ is distinguished from α (Prob. 5.6(iii)) and $\neg\forall x\alpha$ is distinguished from $\exists x\neg\alpha$ (Prob. 5.16). There are many such distinctions; but, as the following problem suggests, the most important one is that between $\neg\neg\alpha$ and α — for without it the intuitionistic calculi collapse into their classical counterparts. It will also be seen that the absence of the law of the excluded middle (Probs. 5.6(iii) and 5.12) is equally crucial.

11.15. PROBLEM. (i) Show that if we add the axiom scheme

$$\neg\neg\alpha \to \alpha$$

to the intuitionistic propositional and predicate calculi, we obtain (equivalent versions of) their classical counterparts. (For the propositional case use 11.2; for the predicate case use 11.12 and 5.15.)

(ii) Prove the same result for the scheme $\alpha \vee \neg\alpha$ instead of $\neg\neg\alpha \to \alpha$.

§ 12. The Interpolation Theorem

In this section we prove, for the intuitionistic predicate calculus, an important result known as *Craig's Interpolation Theorem* (or *Lemma*). Originally, Craig proved this result for the *classical* predicate calculus, and it was only later extended by Schütte to the intuitionistic case. (In §13 we shall derive the original classical version from the intuitionistic one.)

In what follows, $\mathbf{\Phi}$ and $\mathbf{\Psi}$ are finite sets of $\mathscr{L}$-formulas and ξ, as usual, is an $\mathscr{L}$-formula or the empty string.

We say that a formula $\boldsymbol{\delta}$ is an *interpolant for* $\langle\mathbf{\Phi}; \mathbf{\Psi}, \xi\rangle$ if the following three conditions hold:

(a) Every free variable of $\boldsymbol{\delta}$ is free in both $\mathbf{\Phi}$ and $\mathbf{\Psi}\cup\{\xi\}$, and every extralogical symbol (constant or predicate symbol) occurring in $\boldsymbol{\delta}$ occurs in both $\mathbf{\Phi}$ and $\mathbf{\Psi}\cup\{\xi\}$;

(b) $\mathbf{\Phi}\rhd\boldsymbol{\delta}$;

(c) $\mathbf{\Psi}, \boldsymbol{\delta}\rhd\xi$.

Condition (a) is expressed more briefly by saying that $\boldsymbol{\delta}$ is in the *common vocabulary* of $\mathbf{\Phi}$ and $\mathbf{\Psi}\cup\{\xi\}$.

Note that, if $\mathbf{\Phi}$ and $\mathbf{\Psi}\cup\{\xi\}$ do not have any predicate symbol in common, then no formula $\boldsymbol{\delta}$ can fulfil condition (a), so in this case we cannot have an interpolant for $\langle\mathbf{\Phi}; \mathbf{\Psi}, \xi\rangle$.

We say that *we can interpolate in* $\langle\mathbf{\Phi};\mathbf{\Psi},\xi\rangle$ if we can do at least one of the following three things:

(1) find (a formula $\boldsymbol{\delta}$ and prove that it is) an interpolant for $\langle\mathbf{\Phi}; \mathbf{\Psi}, \xi\rangle$.

(2) prove $\mathbf{\Phi}\rhd$;

(3) prove $\mathbf{\Psi}\rhd\xi$.

Note that if there is a predicate symbol $\mathbf{P}$ occurring in both $\mathbf{\Phi}$ and $\mathbf{\Psi}\cup\{\xi\}$, then if we can do (2) or (3) we can also do (1). Indeed, let γ be any sentence whose only extralogical symbol is $\mathbf{P}$. If we can do (2), then $\boldsymbol{\delta}=\gamma\wedge\neg\gamma$ will serve for (1) because (as can easily be seen) $\gamma\wedge\neg\gamma\rhd_0$. If we can do (3), then $\boldsymbol{\delta}=\gamma\rightarrow\gamma$ will serve for (1) because $\rhd_0\gamma\rightarrow\gamma$. Thus we had to admit (2) and (3) as separate possibilities only because $\mathbf{\Phi}$ and $\mathbf{\Psi}\cup\{\xi\}$ may not have a predicate symbol in common.

The main work of this section is done in

12.1. Lemma. *Given a proof* $\boldsymbol{T}$ *of* $\mathbf{\Phi}\cup\mathbf{\Psi}\rhd\xi$, *we can interpolate in* $\langle\mathbf{\Phi};\mathbf{\Psi},\xi\rangle$.

Proof. By induction on the depth d of $\boldsymbol{T}$.

If $d=0$, then ξ (is atomic and) belongs to $\mathbf{\Phi}\cup\mathbf{\Psi}$. If $\xi\in\mathbf{\Phi}$, then ξ is easily seen to be an interpolant for $\langle\mathbf{\Phi}; \mathbf{\Psi}, \xi\rangle$. If $\xi\in\mathbf{\Psi}$, then we can trivially prove $\mathbf{\Psi}\rhd\xi$.

Now let $d>0$. We distinguish several cases, according to the way in which level 1 of $\boldsymbol{T}$ was obtained. Strictly speaking, we ought to consider nineteen cases (because level 1 of $\boldsymbol{T}$ could have been obtained by applying any one of the six "positive" rules to a formula of $\mathbf{\Phi}$, or any one of these six rules to a formula of $\mathbf{\Psi}$, or any one of the seven "negative" rules to $-\xi$).

But by grouping together cases whose treatment is identical or very similar, we are left with eleven cases.

Case 1: Level 1 of $\boldsymbol{T}$ was obtained by applying rule $\wedge$ or $\exists$ to some formula of $\boldsymbol{\Phi}$. Then $\boldsymbol{T}$ starts thus:

$$\begin{array}{l} \boldsymbol{\Phi}, \boldsymbol{\Psi}, -\xi \\ \quad | \\ \boldsymbol{\Sigma} \\ \quad \vdots \end{array}$$

where $\boldsymbol{\Sigma}$ is a set of one or two formulas. From $\boldsymbol{T}$ we obtain a proof, with depth $d-1$, of $(\boldsymbol{\Phi} \cup \boldsymbol{\Sigma}) \cup \boldsymbol{\Psi} \rhd \xi$. By the induction hypothesis, we can interpolate in $\langle \boldsymbol{\Phi} \cup \boldsymbol{\Sigma}; \boldsymbol{\Psi}, \xi \rangle$. There are three subcases:

Subcase 1.1: We have got an interpolant $\boldsymbol{\delta}$ for $\langle \boldsymbol{\Phi} \cup \boldsymbol{\Sigma}; \boldsymbol{\Psi}, \xi \rangle$. Notice that the extralogical symbols of $\boldsymbol{\Sigma}$ must occur in $\boldsymbol{\Phi}$. Also, if $\boldsymbol{\Sigma}$ was obtained in $\boldsymbol{T}$ by the $\exists$ rule, the critical variable $\mathbf{y}$ involved does not occur free in the initial node; in particular, $\mathbf{y}$ is not free in $\boldsymbol{\Psi} \cup \{\xi\}$. Hence $\mathbf{y}$ cannot be free in $\boldsymbol{\delta}$, because $\boldsymbol{\delta}$ is an interpolant for $\langle \boldsymbol{\Phi} \cup \boldsymbol{\Sigma}; \boldsymbol{\Psi}, \xi \rangle$. It is now easy to see that $\boldsymbol{\delta}$ is also an interpolant for $\langle \boldsymbol{\Phi}; \boldsymbol{\Psi}, \xi \rangle$. (To get a proof of $\boldsymbol{\Phi} \rhd \boldsymbol{\delta}$, start by obtaining $\boldsymbol{\Sigma}$ from $\boldsymbol{\Phi}$ as in $\boldsymbol{T}$, and then use the proof of $\boldsymbol{\Phi} \cup \boldsymbol{\Sigma} \rhd \boldsymbol{\delta}$ which we possess by the assumption of the present subcase.)

Subcase 1.2: We can prove $\boldsymbol{\Phi} \cup \boldsymbol{\Sigma} \rhd$. Then we can get a proof of $\boldsymbol{\Phi} \rhd$. (Start by obtaining $\boldsymbol{\Sigma}$ as in $\boldsymbol{T}$.)

Subcase 1.3: We can prove $\boldsymbol{\Psi} \rhd \xi$. Then we are through.

Case 2: Level 1 of $\boldsymbol{T}$ was obtained by applying rule $\wedge$ or $\exists$ to a formula of $\boldsymbol{\Psi}$. This is similar to case 1, except that now the induction hypothesis is that we can interpolate in $\langle \boldsymbol{\Phi}; \boldsymbol{\Psi} \cup \boldsymbol{\Sigma}, \xi \rangle$, where $\boldsymbol{\Sigma}$ is the node of level 1 in $\boldsymbol{T}$.

Case 3: Level 1 of $\boldsymbol{T}$ was obtained by applying rule $\neg$ to a formula of $\boldsymbol{\Phi}$. Then $\boldsymbol{T}$ starts thus:

$$\begin{array}{l} \boldsymbol{\Phi}, \boldsymbol{\Psi}, -\xi \\ \quad |* \\ -\alpha \\ \quad \vdots \end{array}$$

where $\neg\alpha \in \boldsymbol{\Phi}$. We obtain a proof, with depth $d-1$, of $\boldsymbol{\Psi} \cup \boldsymbol{\Phi} \rhd \alpha$. Hence, by the induction hypothesis, we can interpolate in $\langle \boldsymbol{\Psi}; \boldsymbol{\Phi}, \alpha \rangle$. We have three subcases:

Subcase 3.1: We have got an interpolant $\boldsymbol{\delta}$ for $\langle \boldsymbol{\Psi}; \boldsymbol{\Phi}, \boldsymbol{\alpha} \rangle$. Then $\neg\boldsymbol{\delta}$ is easily seen to be an interpolant for $\langle \boldsymbol{\Phi}; \boldsymbol{\Psi}, \boldsymbol{\xi} \rangle$. (To get a proof of $\boldsymbol{\Phi} \rhd \neg\boldsymbol{\delta}$, start by using $-\neg\boldsymbol{\delta}$, then use $\neg\boldsymbol{\alpha}$ and employ the proof of $\boldsymbol{\Phi}, \boldsymbol{\delta} \rhd \boldsymbol{\alpha}$ that we possess by assumption. To prove $\boldsymbol{\Psi}, \neg\boldsymbol{\delta} \rhd \boldsymbol{\xi}$, use $\neg\boldsymbol{\delta}$ and employ the proof of $\boldsymbol{\Psi} \rhd \boldsymbol{\delta}$ that we possess by assumption.)

Subcase 3.2: We can prove $\boldsymbol{\Psi} \rhd$. Then we can also prove $\boldsymbol{\Psi} \rhd \boldsymbol{\xi}$.

Subcase 3.3: We can prove $\boldsymbol{\Phi} \rhd \boldsymbol{\alpha}$. Then we can prove $\boldsymbol{\Phi} \rhd$. (Start by using $\neg\boldsymbol{\alpha}$.)

Case 4: Level 1 of $\boldsymbol{T}$ was obtained by applying rule $\neg$ to a formula of $\boldsymbol{\Psi}$, or one of the rules $-\vee_1$, $-\vee_2$, $-\rightarrow$, $-\neg$, $-\forall$ to $-\boldsymbol{\xi}$. These are all treated in a very similar way. As an example, we shall do in detail the case of rule $-\forall$ which is slightly more problematic than the other five. In this case $\boldsymbol{T}$ starts thus:

$$\begin{array}{c} \boldsymbol{\Phi}, \boldsymbol{\Psi}, -\boldsymbol{\xi} \\ |\,* \\ -\boldsymbol{\alpha}(\mathbf{x}/\mathbf{y}) \\ \vdots \end{array}$$

where $\boldsymbol{\xi} = \forall \mathbf{x} \boldsymbol{\alpha}$ and $\mathbf{y}$ is not free in $\boldsymbol{\Phi}$, $\boldsymbol{\Psi}$ or $\boldsymbol{\xi}$. By the induction hypothesis, we can interpolate in $\langle \boldsymbol{\Phi}; \boldsymbol{\Psi}, \boldsymbol{\alpha}(\mathbf{x}/\mathbf{y}) \rangle$. Three subcases arise:

Subcase 4.1: We have got an interpolant $\boldsymbol{\delta}$ for $\langle \boldsymbol{\Phi}; \boldsymbol{\Psi}, \boldsymbol{\alpha}(\mathbf{x}/\mathbf{y}) \rangle$. Then $\mathbf{y}$ cannot be free in $\boldsymbol{\delta}$ (because $\mathbf{y}$ is not free in $\boldsymbol{\Phi}$) and it is easy to see that $\boldsymbol{\delta}$ is also an interpolant for $\langle \boldsymbol{\Phi}; \boldsymbol{\Psi}, \boldsymbol{\xi} \rangle$. (To prove $\boldsymbol{\Psi}, \boldsymbol{\delta} \rhd \boldsymbol{\xi}$, start by using $-\boldsymbol{\xi}$ just as in $\boldsymbol{T}$, then employ the proof of $\boldsymbol{\Psi}, \boldsymbol{\delta} \rhd \boldsymbol{\alpha}(\mathbf{x}/\mathbf{y})$ which we are supposed to possess.)

Subcase 4.2: We can prove $\boldsymbol{\Phi} \rhd$. Done.

Subcase 4.3: We can prove $\boldsymbol{\Psi} \rhd \boldsymbol{\alpha}(\mathbf{x}/\mathbf{y})$. Then we can also prove $\boldsymbol{\Psi} \rhd \boldsymbol{\xi}$. (Start by using $-\boldsymbol{\xi}$, with $\mathbf{y}$ as critical variable.)

Case 5: Level 1 of $\boldsymbol{T}$ was obtained by rule $\vee$ applied to a formula of $\boldsymbol{\Phi}$. Then $\boldsymbol{T}$ starts thus:

$$\begin{array}{c} \boldsymbol{\Phi}, \boldsymbol{\Psi}, -\boldsymbol{\xi} \\ \diagup \quad \diagdown \\ \boldsymbol{\alpha} \qquad \boldsymbol{\beta} \\ \vdots \qquad \vdots \end{array}$$

where $\boldsymbol{\alpha} \vee \boldsymbol{\beta} \in \boldsymbol{\Phi}$. By the induction hypothesis, we can interpolate in both $\langle \boldsymbol{\Phi} \cup \{\boldsymbol{\alpha}\}; \boldsymbol{\Psi}, \boldsymbol{\xi} \rangle$ and $\langle \boldsymbol{\Phi} \cup \{\boldsymbol{\beta}\}; \boldsymbol{\Psi}, \boldsymbol{\xi} \rangle$. Here there are five possible subcases:

Subcase 5.1: We possess interpolants γ and $\boldsymbol{\delta}$ for $\langle \boldsymbol{\Phi} \cup \{\boldsymbol{\alpha}\}; \boldsymbol{\Psi}, \boldsymbol{\xi} \rangle$ and

$\langle\Phi\cup\{\beta\};\Psi,\xi\rangle$ respectively. Then $\gamma\vee\delta$ is an interpolant for $\langle\Phi;\Psi,\xi\rangle$. (To prove $\Phi\rhd\gamma\vee\delta$, start by using $\alpha\vee\beta$, then apply to $-\gamma\vee\delta$ rule $-\vee_1$ in one branch and rule $-\vee_2$ in the other, as follows:

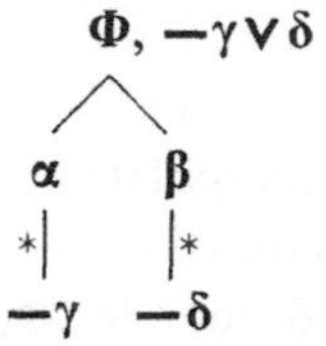

Now employ the proofs of $\Phi,\alpha\rhd\gamma$ and $\Phi,\beta\rhd\delta$ that we have got by assumption. To prove $\Psi,\gamma\vee\delta\rhd\xi$, start by using $\gamma\vee\delta$.)

Subcase 5.2: We have got an interpolant γ for $\langle\Phi\cup\{\alpha\};\Psi,\xi\rangle$ and a proof of $\Phi,\beta\rhd$. Then γ is also an interpolant for $\langle\Phi;\Psi,\xi\rangle$. (To prove $\Phi\rhd\gamma$, start by using $\alpha\vee\beta$, then utilize the proofs of $\Phi,\alpha\rhd\gamma$ and $\Phi,\beta\rhd$ that we are supposed to possess.)

Subcase 5.3: We have got a proof of $\Phi,\alpha\rhd$ and an interpolant for $\langle\Phi\cup\{\beta\};\Psi,\xi\rangle$. This is similar to subcase 5.2.

Subcase 5.4: We can prove $\Phi,\alpha\rhd$ and $\Phi,\beta\rhd$. Then we can prove $\Phi\rhd$. (Start by using $\alpha\vee\beta$.)

Subcase 5.5: We can prove $\Psi\rhd\xi$. Done.

Case 6: Level 1 of T was obtained by applying rule $\vee$ to a formula of Ψ. Then T starts thus:

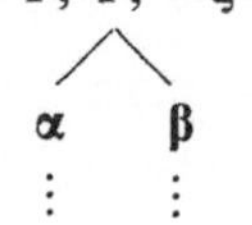

where $\alpha\vee\beta\in\Psi$. By the induction hypothesis we can interpolate in both $\langle\Phi;\Psi\cup\{\alpha\},\xi\rangle$ and $\langle\Phi;\Psi\cup\{\beta\},\xi\rangle$. Here there are five subcases, analogous to those of case 5:

Subcase 6.1: We have got interpolants γ and δ for $\langle\Phi;\Psi\cup\{\alpha\},\xi\rangle$ and $\langle\Phi;\Psi\cup\{\beta\},\xi\rangle$ respectively. Then $\gamma\wedge\delta$ is an interpolant for $\langle\Phi;\Psi,\xi\rangle$. (To prove $\Phi\rhd\gamma\wedge\delta$, start by using $-\gamma\wedge\delta$. To prove $\Psi,\gamma\wedge\delta\rhd\xi$, start by using $\gamma\wedge\delta$, then use $\alpha\vee\beta$.)

The other four subcases of the present case 6 are treated like the analogous subcases of case 5. We leave this to the reader.

Case 7: Level 1 of T was obtained by applying rule $\rightarrow$ to a formula of Φ.

Then T starts thus:

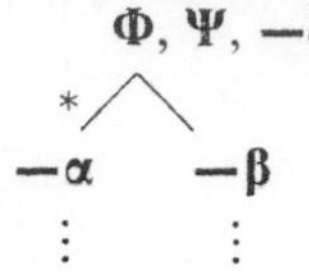

where $\boldsymbol{\alpha \to \beta} \in \boldsymbol{\Phi}$. By the induction hypothesis we can interpolate in $\langle \boldsymbol{\Psi}; \boldsymbol{\Phi}, \boldsymbol{\alpha} \rangle$ and $\langle \boldsymbol{\Phi} \cup \{\boldsymbol{\beta}\}; \boldsymbol{\Psi}, \xi \rangle$. Again there are five possible subcases:

Subcase 7.1: We have got interpolants γ and $\boldsymbol{\delta}$ for $\langle \boldsymbol{\Psi}; \boldsymbol{\Phi}, \boldsymbol{\alpha} \rangle$ and $\langle \boldsymbol{\Phi} \cup \{\boldsymbol{\beta}\}; \boldsymbol{\Psi}, \xi \rangle$ respectively. Then $\gamma \to \boldsymbol{\delta}$ is an interpolant for $\langle \boldsymbol{\Phi}; \boldsymbol{\Psi}, \xi \rangle$. (To prove $\boldsymbol{\Phi} \rhd \gamma \to \boldsymbol{\delta}$, start by using $-\gamma \to \boldsymbol{\delta}$, then use $\boldsymbol{\alpha \to \beta}$. To prove $\boldsymbol{\Psi}, \gamma \to \boldsymbol{\delta} \rhd \xi$, start by using $\gamma \to \boldsymbol{\delta}$.)

Subcase 7.2: We have got an interpolant γ for $\langle \boldsymbol{\Psi}; \boldsymbol{\Phi}, \boldsymbol{\alpha} \rangle$ and a proof of $\boldsymbol{\Phi}, \boldsymbol{\beta} \rhd$. Then $\neg\gamma$ is an interpolant for $\langle \boldsymbol{\Phi}; \boldsymbol{\Psi}, \xi \rangle$. (To prove $\boldsymbol{\Phi} \rhd \neg\gamma$, start by using $-\neg\gamma$, then use $\boldsymbol{\alpha \to \beta}$. To prove $\boldsymbol{\Psi}, \neg\gamma \rhd \xi$, start by using $\neg\gamma$.)

Subcase 7.3: We have got a proof of $\boldsymbol{\Phi} \rhd \boldsymbol{\alpha}$ and an interpolant $\boldsymbol{\delta}$ for $\langle \boldsymbol{\Phi} \cup \{\boldsymbol{\beta}\}; \boldsymbol{\Psi}, \xi \rangle$. Then $\boldsymbol{\delta}$ is also an interpolant for $\langle \boldsymbol{\Phi}; \boldsymbol{\Psi}, \xi \rangle$. (To prove $\boldsymbol{\Phi} \rhd \boldsymbol{\delta}$ start by using $\boldsymbol{\alpha \to \beta}$.)

Subcase 7.4: We have got proofs of $\boldsymbol{\Phi} \rhd \boldsymbol{\alpha}$ and $\boldsymbol{\Phi}, \boldsymbol{\beta} \rhd$. Then we can prove $\boldsymbol{\Phi} \rhd$. (Start by using $\boldsymbol{\alpha \to \beta}$.)

Subcase 7.5: We have got a proof of $\boldsymbol{\Psi} \rhd$ or of $\boldsymbol{\Psi} \rhd \xi$. In either case we can prove $\boldsymbol{\Psi} \rhd \xi$.

Case 8: Level 1 of T was obtained by applying rule $\to$ to a formula of $\boldsymbol{\Psi}$ or rule $-\wedge$ to $-\xi$. These two are treated in almost the same way. As an example, we do in some detail the case of rule $\to$. In this case T starts thus:

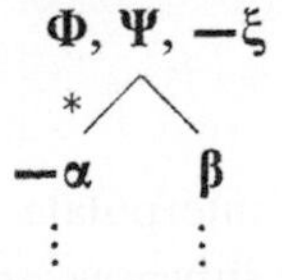

where $\boldsymbol{\alpha \to \beta} \in \boldsymbol{\Psi}$. By the induction hypothesis we can interpolate in $\langle \boldsymbol{\Phi}; \boldsymbol{\Psi}, \boldsymbol{\alpha} \rangle$ and $\langle \boldsymbol{\Phi}; \boldsymbol{\Psi} \cup \{\boldsymbol{\beta}\}, \xi \rangle$. Again five subcases arise:

Subcase 8.1: We have got interpolants γ and $\boldsymbol{\delta}$ for $\langle \boldsymbol{\Phi}; \boldsymbol{\Psi}, \boldsymbol{\alpha} \rangle$ and $\langle \boldsymbol{\Phi}; \boldsymbol{\Psi} \cup \{\boldsymbol{\beta}\}, \xi \rangle$ respectively. Then $\gamma \wedge \boldsymbol{\delta}$ is an interpolant for $\langle \boldsymbol{\Phi}; \boldsymbol{\Psi}, \xi \rangle$. (To prove $\boldsymbol{\Phi} \rhd \gamma \wedge \boldsymbol{\delta}$, start by using $-\gamma \wedge \boldsymbol{\delta}$. To prove $\boldsymbol{\Psi}, \gamma \wedge \boldsymbol{\delta} \rhd \xi$, start by using $\gamma \wedge \boldsymbol{\delta}$, then use $\boldsymbol{\alpha \to \beta}$.)

Subcase 8.2: We possess an interpolant γ for $\langle \boldsymbol{\Phi}; \boldsymbol{\Psi}, \boldsymbol{\alpha} \rangle$ and a proof of $\boldsymbol{\Psi}, \boldsymbol{\beta} \rhd \xi$. Then γ is easily seen to be an interpolant for $\langle \boldsymbol{\Phi}; \boldsymbol{\Psi}, \xi \rangle$.

Subcase 8.3: We have got a proof of $\Psi \rhd \alpha$ and an interpolant δ for $\langle \Phi; \Psi \cup \{\beta\}, \xi \rangle$. Then δ is easily seen to be an interpolant for $\langle \Phi; \Psi, \xi \rangle$.

Subcase 8.4: We can prove $\Psi \rhd \alpha$ and $\Psi, \beta \rhd \xi$. Then we can prove $\Psi \rhd \xi$. (Start by using $\alpha \rightarrow \beta$.)

Subcase 8.5: We can prove $\Phi \dot{\rhd}$. Done.

The final three cases require a little more finesse.

Case 9: Level 1 of T was obtained by applying rule $\forall$ to a formula of Φ. Then T starts thus:

$$\begin{array}{l} \Phi, \Psi, -\xi \\ \quad | \\ \alpha(x/t) \\ \quad \vdots \end{array}$$

where $\mathbf{t}$ is a term and $\forall x \alpha \in \Phi$. By the induction hypothesis we can interpolate in $\langle \Phi \cup \{\alpha(x/t)\}; \Psi, \xi \rangle$. Three subcases are possible:

Subcase 9.1: We have got an interpolant δ for $\langle \Phi \cup \{\alpha(x/t)\}; \Psi, \xi \rangle$. So we can prove both $\Psi, \delta \rhd \xi$ and $\Phi, \alpha(x/t) \rhd \delta$ and hence (since $\forall x \alpha \in \Phi$) also $\Phi \rhd \delta$. Thus, if δ is in the common vocabulary of Φ and $\Psi \cup \{\xi\}$, then δ is an interpolant for $\langle \Phi; \Psi, \xi \rangle$.

But δ may *fail* to be in the common vocabulary of Φ and $\Psi \cup \{\xi\}$. This happens if $\mathbf{t}$ is a term (i.e., a constant or a free variable) of δ — and hence necessarily of $\Psi \cup \{\xi\}$ — but is *not* a term of Φ. In this case we proceed as follows. We choose a variable $\mathbf{y}$ that does not occur in δ and put $\gamma = \delta(t/y)$ (if $\mathbf{t}$ is a constant, this is defined in Def. 2.3.13). Then we have $\delta = \gamma(y/t)$. Now, since we can prove $\Phi \rhd \delta$, i.e., $\Phi \rhd \gamma(y/t)$ and since $\mathbf{t}$ is assumed *not* to be a term of Φ, we obtain — by Thms. 9.10(i) and 9.11 (if $\mathbf{t}$ is a variable) or by Thm. 9.13(i) (in case $\mathbf{t}$ is a constant) — a proof of $\Phi \rhd \forall y \gamma$. Also, since we can prove $\Psi, \delta \rhd \xi$, i.e., $\Psi, \gamma(y/t) \rhd \xi$, we can obviously prove $\Psi, \forall y \gamma \rhd \xi$. Thus $\forall y \gamma$, which does not have the offending term $\mathbf{t}$, is an interpolant for $\langle \Phi; \Psi, \xi \rangle$.

Subcase 9.2: We can prove $\Phi, \alpha(x/t) \rhd$. Then, since $\forall x \alpha \in \Phi$, we can also prove $\Phi \rhd$.

Subcase 9.3: We can prove $\Psi \rhd \xi$. Done.

Case 10: Level 1 of T was obtained by applying rule $\forall$ to a formula of Ψ. Then T starts thus:

$$\begin{array}{l} \Phi, \Psi, -\xi \\ \quad | \\ \alpha(x/t) \\ \quad \vdots \end{array}$$

where $\forall x\alpha \in \Psi$. By the induction hypothesis we can interpolate in $\langle\Phi; \Psi \cup \{\alpha(x/t)\}, \xi\rangle$. Here again there are three subcases:

Subcase 10.1: We have an interpolant δ for $\langle\Phi; \Psi \cup \{\alpha(x/t)\}, \xi\rangle$. So we can prove $\Phi \rhd \delta$ and $\Psi, \alpha(x/t), \delta \rhd \xi$, and (since $\forall x\alpha \in \Psi$) also $\Psi, \delta \rhd \xi$. Thus, if δ is in the common vocabulary of Φ and $\Psi \cup \{\xi\}$, it is an interpolant for $\langle\Phi; \Psi, \xi\rangle$.

But δ fails to be in the common vocabulary of Φ and $\Psi \cup \{\xi\}$ if t is a term of δ — and hence necessarily of Φ — but not of Ψ. In this case we define y and γ as in subcase 9.1. and using the results of §9 we see without difficulty that $\exists y\gamma$ is an interpolant for $\langle\Phi; \Psi, \xi\rangle$.

The other two subcases of the present case are similar to those of case 9.

Case 11: Level 1 of T is obtained by applying rule $-\exists$ to $-\xi$. This is very similar to case 10. In the problematic situation arising in subcase 11.1, we define y and γ as before and then see without difficulty that $\exists y\gamma$ is an interpolant for $\langle\Phi; \Psi, \xi\rangle$. The other details are left to the reader. ∎

We shall say that *we can interpolate in* $\langle\Phi; \alpha\rangle$ if we can interpolate in $\langle\Phi; \emptyset, \alpha\rangle$. We then have:

12.2. INTERPOLATION THEOREM. *Given a proof of* $\Phi \rhd \alpha$, *we can interpolate in* $\langle\Phi; \alpha\rangle$. ∎

Stated more fully, and replacing "$\rhd$" by "$\vdash_I$", the Interpolation Theorem asserts that if we have $\Phi \vdash_I \alpha$ then we can do at least one of the following three things:

(1) Find a formula δ in the common vocabulary of Φ and α and prove that $\Phi \vdash_I \delta$ and $\delta \vdash_I \alpha$ (in which case δ is an *interpolant* for $\langle\Phi; \alpha\rangle$).

(2) Prove that $\Phi \vdash_I$.

(3) Prove that $\vdash_I \alpha$.

12.3. COROLLARY. *If* $\Phi \vdash_I \alpha$ *but* Φ *and* α *have no predicate symbol in common, then* $\Phi \vdash_I$ *or* $\vdash_I \alpha$. ∎

§ 13. Some results in classical logic

In this section we shall use Thm. 12.2 to derive the classical version of the Interpolation Theorem and from the latter we shall derive certain other important results for classical first-order logic.

From now on we let $\mathscr{L}$ be an *arbitrary* first-order language, *possibly with equality and function symbols*. Since we are concerned here with classical logic, we shall revert to regarding only $\neg, \rightarrow$ and $\forall$ as primitive. The other connectives and $\exists$ are introduced as abbreviations (see Def. 1.5.1).

We also revert to the terminology of Ch. 3. Thus, e.g., by *deduction* we mean deduction in the classical first-order predicate calculus.

Let $\mathbf{\Phi}$ be a set of formulas and let $\boldsymbol{\beta}$ be a formula. We say that a formula $\boldsymbol{\delta}$ is in the *common vocabulary of* $\mathbf{\Phi}$ *and* $\boldsymbol{\beta}$ if the following three conditions hold:

(1) Every variable free in $\boldsymbol{\delta}$ is free in both $\mathbf{\Phi}$ and $\boldsymbol{\beta}$.

(2) Every one of the extralogical symbols (i.e., function symbols, including constants, and predicate symbols other than the equality symbol $=$) occurring in $\boldsymbol{\delta}$ occurs also in both $\mathbf{\Phi}$ and $\boldsymbol{\beta}$.

(3) If $=$ occurs in $\boldsymbol{\delta}$ then $\mathbf{\Phi} \cup \{\boldsymbol{\beta}\}$ contains $=$ or some function symbol other than a constant.

We say that $\boldsymbol{\delta}$ is an *interpolant for* $\langle \mathbf{\Phi}; \boldsymbol{\beta} \rangle$ if $\boldsymbol{\delta}$ is in the common vocabulary of $\mathbf{\Phi}$ and $\boldsymbol{\beta}$, and both $\mathbf{\Phi} \vdash \boldsymbol{\delta}$ and $\boldsymbol{\delta} \vdash \boldsymbol{\beta}$.

We say that $\langle \mathbf{\Phi}; \boldsymbol{\beta} \rangle$ *has the interpolation property* if there is an interpolant for $\langle \mathbf{\Phi}; \boldsymbol{\beta} \rangle$, or $\mathbf{\Phi}$ is inconsistent, or $\vdash \boldsymbol{\beta}$.

13.1. INTERPOLATION THEOREM. *If* $\mathbf{\Phi} \vdash \boldsymbol{\beta}$, *then* $\langle \mathbf{\Phi}; \boldsymbol{\beta} \rangle$ *has the interpolation property.*

PROOF. Since any deduction of $\boldsymbol{\beta}$ from $\mathbf{\Phi}$ uses only finitely many formulas of $\mathbf{\Phi}$, we can assume without loss of generality that $\mathbf{\Phi}$ is finite. Then we can replace $\mathbf{\Phi}$ by a conjunction of all its formulas. So from now on we shall assume that $\mathbf{\Phi}$ consists of a single formula $\boldsymbol{\alpha}$.

We distinguish three cases.

Case 1: Both $\boldsymbol{\alpha}$ and $\boldsymbol{\beta}$ contain neither $=$ nor function symbols other than constants. Then our theorem follows easily from Thm. 12.2, using Thm. 11.12 or Prob. 11.14(ii).

Case 2: $\boldsymbol{\alpha}$ or $\boldsymbol{\beta}$ may contain $=$ but neither of them contains function symbols other than constants.

If $\mathbf{P}$ is an n-ary predicate symbol, let $\gamma_{\mathbf{P}}$ be the sentence

$$\forall \mathbf{v}_1 \ldots \forall \mathbf{v}_{2n}(\mathbf{v}_1 = \mathbf{v}_{n+1} \to \ldots \to \mathbf{v}_n = \mathbf{v}_{2n} \to \mathbf{P}\mathbf{v}_1 \ldots \mathbf{v}_n \to \mathbf{P}\mathbf{v}_{n+1} \ldots \mathbf{v}_{2n}).$$

We let $\boldsymbol{\alpha}'$ be the formula

$$\forall \mathbf{v}_1(\mathbf{v}_1 = \mathbf{v}_1) \wedge \gamma_{\mathbf{P}_1} \wedge \ldots \wedge \gamma_{\mathbf{P}_k} \wedge \boldsymbol{\alpha},$$

where $\mathbf{P}_1, \ldots, \mathbf{P}_k$ are all the distinct predicate symbols (including $=$) occurring in $\boldsymbol{\alpha}$. We let $\boldsymbol{\beta}'$ be the formula

$$\forall \mathbf{v}_1(\mathbf{v}_1 = \mathbf{v}_1) \wedge \gamma_{\mathbf{Q}_1} \wedge \ldots \wedge \gamma_{\mathbf{Q}_m} \to \boldsymbol{\beta},$$

where $\mathbf{Q}_1, \ldots, \mathbf{Q}_m$ are all the distinct predicate symbols occurring in $\boldsymbol{\beta}$.

Since $\alpha \vdash \beta$, it is easy to see that $\alpha' \vdash' \beta'$, where $\vdash'$ denotes deducibility without using the axioms of equality (i.e., treating $=$ as if it were an extralogical predicate symbol). By case 1, $\langle\{\alpha'\}; \beta'\rangle$ has the interpolation property (with $=$ treated as extralogical). Three subcases arise:

Subcase 2.1: There is an interpolant δ for $\langle\{\alpha'\}; \beta'\rangle$, for which $\alpha' \vdash' \delta$ and $\delta \vdash' \beta'$. Then it is easy to see that $\alpha \vdash \delta$, $\delta \vdash \beta$, and δ is an interpolant for $\langle\{\alpha\}; \beta\rangle$.

Subcase 2.2: $\{\alpha'\}$ is inconsistent when $=$ is treated as extralogical. Then $\{\alpha\}$ is clearly inconsistent in the ordinary sense. (Actually in this case too we have an interpolant: e.g., $\forall x(x \neq x)$.)

Subcase 2.3: $\vdash' \beta'$. Then clearly $\vdash \beta$. (In this case as well we have an interpolant: e.g., $\forall x(x=x)$.)

Case 3: α or β may contain function symbols other than constants. We proceed by induction on the number of such function symbols occurring in α or β. Let $\mathbf{f}$ be an n-ary function symbol occurring in α or β, where $n>0$.

By Lemma 2.10.4, we can transform α into a logically equivalent formula α^* in which $\mathbf{f}$ only occurs in the form

$$\mathbf{f}x_1 \ldots x_n = y,$$

where $x_1, \ldots, x_n, y$ are distinct variables. The procedure described in the proof of Lemma 2.10.4 is such that α^* has the same free variables and extralogical symbols as α. Also, it is not difficult to see — even without the Completeness Theorem — that α and α^* are *provably* equivalent. Similarly, we transform β into β^*.

Next, we introduce an $(n+1)$-ary predicate symbol $\mathbf{P}$ that occurs neither in α nor in β. We define α' as follows. If $\mathbf{f}$ does not occur in α, then α' is α. If $\mathbf{f}$ occurs in α, then α' is

$$\forall v_1 \ldots \forall v_n \exists! v_{n+1} \mathbf{P} v_1 \ldots v_n v_{n+1} \wedge \alpha^{**},$$

where α^{**} is obtained from α^* when each atomic part of the form $\mathbf{f}x_1 \ldots x_n = y$ is replaced by $\mathbf{P}x_1 \ldots x_n y$. Also, we define β' as follows. If β does not contain $\mathbf{f}$, then $\beta' = \beta$. If β contains $\mathbf{f}$, then β' is

$$\forall v_1 \ldots \forall v_n \exists! v_{n+1} \mathbf{P} v_1 \ldots v_n v_{n+1} \rightarrow \beta^{**},$$

where β^{**} is obtained from β^* as α^{**} was obtained from α^*.

Now, from the fact that $\alpha \vdash \beta$ it follows that $\alpha' \vdash \beta'$. To see this, we note that $\{\alpha, \neg\beta\}$ is unsatisfiable; hence — as in the proof of Thm. 2.10.5 — we can see that $\{\alpha', \neg\beta'\}$ is unsatisfiable, so $\alpha' \vdash \beta'$ by the Completeness Theorem. (We can in fact avoid using the Completeness Theorem and argue in a constructive way: it is not difficult to show that any confutation of $\{\alpha, \neg\beta\}$ can be transformed into a confutation of $\{\alpha', \neg\beta'\}$.)

Since **f** does not occur in α' or β', the induction hypothesis implies that $\langle\{\alpha'\}; \beta'\rangle$ has the interpolation property. Again there are three subcases:

Subcase 3.1: There is an interpolant γ for $\langle\{\alpha'\}; \beta'\rangle$. Let δ be the formula obtained from γ when each subformula $\mathbf{Pt}_1\ldots\mathbf{t}_n\mathbf{s}$ is replaced by $\mathbf{ft}_1\ldots\mathbf{t}_n=\mathbf{s}$. Then δ is an interpolant for $\langle\{\alpha\}; \beta\rangle$. This can be seen by showing, as in the proof of Thm. 2.10.5, that $\{\alpha, \neg\delta\}$ and $\{\delta, \neg\beta\}$ are unsatisfiable; or constructively, by showing that confutations of $\{\alpha', \neg\gamma\}$ and $\{\gamma, \neg\beta'\}$ can be transformed into confutations of $\{\alpha, \neg\delta\}$ and $\{\delta, \neg\beta\}$.

Subcase 3.2: $\{\alpha'\}$ is inconsistent. Then it is not difficult to show (either as in Thm. 2.10.5 and using the Completeness Thm., or constructively) that $\{\alpha\}$ is inconsistent.

Subcase 3.3: $\vdash\beta'$. Then — by the same method as before — we can show that $\vdash\beta$. ∎

In what follows, Σ is any set of $\mathscr{L}$-sentences and **P** is an extralogical n-ary predicate symbol of $\mathscr{L}$, but not the *only* predicate symbol of $\mathscr{L}$.

We say that **P** is *explicitly Σ-definable,* if we have an $\mathscr{L}$-formula β whose free variables are among $\mathbf{v}_1,\ldots,\mathbf{v}_n$ such that **P** does not occur in β and

(13.2) $\quad \Sigma\vdash \mathbf{Pv}_1\ldots\mathbf{v}_n \leftrightarrow \beta.$

Now let **P′** be an n-ary predicate symbol *not* belonging to $\mathscr{L}$. For each $\mathscr{L}$-formula α we let α' be the formula obtained from α when every occurrence of **P** is replaced by **P′**. For a set Φ of $\mathscr{L}$-formulas we put $\Phi'=\{\varphi' : \varphi\in\Phi\}$. We say that **P** is *implicitly Σ-definable* if

(13.3) $\quad \Sigma\cup\Sigma'\vdash \mathbf{Pv}_1\ldots\mathbf{v}_n \leftrightarrow \mathbf{P'v}_1\ldots\mathbf{v}_n.$

Semantically speaking, this means that if $\mathfrak{U}$ and $\mathfrak{U}'$ are any $\mathscr{L}$-structures which are models of Σ, and which have the same universe and agree on the interpretation of all extralogical symbols other than **P**, then $\mathfrak{U}$ and $\mathfrak{U}'$ must also agree on the interpretation of **P**.

The following result is known as Beth's Definability Theorem (for predicate symbols)

13.4. THEOREM. **P** *is explicitly Σ-definable iff it is implicitly Σ-definable.*
PROOF. First, assume that **P** is explicitly Σ-definable, so that for an appropriate $\mathscr{L}$-formula β we have (13.2). Then clearly $\Sigma'\vdash\mathbf{P'v}_1\ldots\mathbf{v}_n\leftrightarrow\beta'$ and (13.3) follows instantly.

Conversely, let **P** be implicitly Σ-definable, so that (13.3) holds. Without loss of generality we may assume Σ to be finite. Then we can replace Σ by a conjunction of all its sentences. So we may assume that Σ consists of a single sentence σ. Thus we have

$$\sigma, \sigma' \vdash \mathbf{P}\mathbf{v}_1 \ldots \mathbf{v}_n \leftrightarrow \mathbf{P}'\mathbf{v}_1 \ldots \mathbf{v}_n .$$

Hence, in particular,

(1) $$\sigma, \mathbf{P}\mathbf{v}_1 \ldots \mathbf{v}_n \vdash \sigma' \rightarrow \mathbf{P}'\mathbf{v}_1 \ldots \mathbf{v}_n .$$

By the Interpolation Theorem 13.1, at least one of the following three cases must arise:

Case 1: There is an interpolant β for $\langle\{\sigma, \mathbf{P}\mathbf{v}_1 \ldots \mathbf{v}_n\}, \sigma' \rightarrow \mathbf{P}'\mathbf{v}_1 \ldots \mathbf{v}_n\rangle$. Then β contains neither **P** nor **P**′, and has no free variables other than $\mathbf{v}_1, \ldots, \mathbf{v}_n$. Also

$$\sigma, \mathbf{P}\mathbf{v}_1 \ldots \mathbf{v}_n \vdash \beta, \qquad \beta \vdash \sigma' \rightarrow \mathbf{P}'\mathbf{v}_1 \ldots \mathbf{v}_n .$$

It follows that $\sigma \vdash \mathbf{P}\mathbf{v}_1 \ldots \mathbf{v}_n \rightarrow \beta$ and $\sigma' \vdash \beta \rightarrow \mathbf{P}'\mathbf{v}_1 \ldots \mathbf{v}_n$, hence clearly also $\sigma \vdash \beta \rightarrow \mathbf{P}\mathbf{v}_1 \ldots \mathbf{v}_n$. Thus $\sigma \vdash \mathbf{P}\mathbf{v}_1 \ldots \mathbf{v}_n \leftrightarrow \beta$, so we have (13.2).

Case 2: $\{\sigma, \mathbf{P}\mathbf{v}_1 \ldots \mathbf{v}_n\}$ is inconsistent. Since **P** is not the only predicate symbol of $\mathscr{L}$, we can find an $\mathscr{L}$-sentence γ which does not contain **P**. Let $\beta = \gamma \wedge \neg\gamma$. Then it is easy to see that $\sigma \vdash \mathbf{P}\mathbf{v}_1 \ldots \mathbf{v}_n \leftrightarrow \beta$, so again we have (13.2).

Case 3: $\vdash \sigma' \rightarrow \mathbf{P}'\mathbf{v}_1 \ldots \mathbf{v}_n$. Then clearly also $\vdash \sigma \rightarrow \mathbf{P}\mathbf{v}_1 \ldots \mathbf{v}_n$. Let γ be any $\mathscr{L}$-sentence not containing **P**, and let $\beta = \gamma \rightarrow \gamma$. It is easy to see that $\sigma \vdash \mathbf{P}\mathbf{v}_1 \ldots \mathbf{v}_n \leftrightarrow \beta$, as required. ∎

13.5. PROBLEM (Beth's Definability Theorem for function symbols). Let **f** be an n-ary function symbol of $\mathscr{L}$. We say that **f** is *explicitly Σ-definable* if there is an $\mathscr{L}$-formula β which does not contain **f** and with no free variables other than $\mathbf{v}_1, \ldots, \mathbf{v}_{n+1}$ such that $\Sigma \vdash \mathbf{f}\mathbf{v}_1 \ldots \mathbf{v}_n = \mathbf{v}_{n+1} \leftrightarrow \beta$.

Let **f**′ be an n-ary function symbol *not* belonging to $\mathscr{L}$. For any $\mathscr{L}$-formula α, let α' be the formula obtained when **f** is replaced by **f**′. Let $\Sigma' = \{\sigma' : \sigma \in \Sigma\}$. Then **f** is *implicitly Σ-definable* if

$$\Sigma \cup \Sigma' \vdash \mathbf{f}\mathbf{v}_1 \ldots \mathbf{v}_n = \mathbf{f}'\mathbf{v}_1 \ldots \mathbf{v}_n .$$

Prove that **f** is explicitly Σ-definable iff it is implicitly Σ-definable.

13.6. PROBLEM (A. Robinson's Consistency Theorem). Let Σ_1 and Σ_2 be consistent sets of $\mathscr{L}$-sentences. Suppose that, for every $\mathscr{L}$-sentence φ such that all the extralogical symbols occurring in φ also occur in both Σ_1 and Σ_2, we have: if $\Sigma_1 \vdash \varphi$ then $\Sigma_2 \nvdash \neg\varphi$. Prove that $\Sigma_1 \cup \Sigma_2$ is consistent.

§ 14. Historical and bibliographical remarks

For a brief outline of intuitionism and its history see KNEEBONE [1963]. A more detailed outline is in FRAENKEL, BAR-HILLEL and LEVY [1973]. A very readable explanation and defence of intuitionistic views is in HEYTING [1972]. An introduction to the technical aspects of intuitionistic mathematics is in TROELSTRA [1969].

BISHOP [1967] is an impressively successful attempt to develop various branches of analysis by constructive — though not specifically intuitionistic — means.

An outline of a general theory of constructions, of the kind needed to make the explanations of §4 more precise, is proposed by KREISEL [1965]. A detailed rigorous development is GOODMAN [1970].

The tableau method of §5 is taken from FITTING [1969]; it is Smullyan's adaptation of the Gentzen-type system $G3$ of KLEENE [1952].

KRIPKE [1965] proposed and investigated the semantics presented here in §6. Broadly similar ideas were proposed by BETH [1956] and GRZEGORCZYK [1964]. The possibility of giving a more constructive character to arguments which employ Kripke's semantics is outlined by Smorynski in TROELSTRA [1973]. However, Smorynski rejects the claim that Kripke's semantics is a heuristically plausible explication of intuitionistic reasoning. In papers to be published, H. de Swart and W. Veldman give a constructive version of Kripke systems and provide a constructive proof of the Completeness Theorem.

Our proof of the Elimination Theorem in §7 is an adaptation of the proof of the analogous result for $G3$ in KLEENE [1952].

The first version of the intuitionistic propositional calculus is in HEYTING [1930] and the first version of the intuitionistic predicate calculus is in HEYTING [1930a]. However, Heyting was anticipated to some extent by KOLMOGOROV [1925]. The calculus presented here in §8 is taken from KLEENE [1952]. The version of the intuitionistic predicate calculus in §9 was suggested to us by D. H. J. de Jongh. It is equivalent to Heyting's.

The first proof of the completeness of the intuitionistic predicate calculus relative to Kripke's semantics is due to KRIPKE [1965]. The proof presented in §10 is adapted from FITTING [1969]. Similar proofs were invented independently by ACZEL [1967] and THOMASON [1968].

The earliest version of a translation of classical into intuitionistic logic is due to KOLMOGOROV [1925].

A wealth of information on intuitionistic formal systems of logic, arithmetic and analysis can be found in TROELSTRA [1973].

FITTING [1969] applies Kripke's semantics to obtain independence proofs for formalized set theory.

The Interpolation Theorem was proved for the classical case by CRAIG [1957] and extended to the intuitionistic case by SCHÜTTE [1962].

The Definability Theorems 13.4 and 13.5 go back to PADOA [1900], who stated them without proof, as obvious. The first proof for first-order logic was given by BETH [1953]. The Consistency Theorem (Prob. 13.6) was proved by A. ROBINSON [1956], who used it to give another proof of the Definability Theorems.

CHAPTER 10

AXIOMATIC SET THEORY

In this chapter we set up and develop a system of axiomatic set theory and eventually prove Gödel's celebrated result that, if the axioms of set theory are consistent, then the Axiom of Choice and the Generalized Continuum Hypothesis may be adjoined without destroying that consistency.

§ 1. Basic developments

The *language of set theory* is a first-order language $\mathscr{L}$ with equality. For the purposes of this chapter it will be convenient to modify somewhat the conventions of Ch. 1, §5 governing the metalinguistic symbols we have heretofore used to denote $\mathscr{L}$-symbols and their combinations. Thus: we assume that the individual *variables* of $\mathscr{L}$ are enumerated in a fixed alphabetic sequence $v_0, v_1, \ldots$ (starting now with v_0 rather than v_1!) We shall use lightface lower case italic letters (possibly with subscripts) as metalinguistic symbols ranging over these variables. *Unless otherwise stated, in any given context, distinct letters of this type are understood to refer to distinct variables.* The *equality symbol* will be denoted by a lightface $=$, the *connectives* by lightface symbols $\wedge$, $\neg$, etc., and the *quantifiers* by lightface symbols $\forall$, $\exists$. With the exception of certain *defined* formulas and terms, we shall as before employ bold lower-case Greek letters $\boldsymbol{\varphi}$, $\boldsymbol{\psi}$, $\boldsymbol{\chi}$, etc., to denote *formulas* and bold lower-case Roman letters $\mathbf{s}$, $\mathbf{t}$, etc., to denote *terms*.

We agree to take negation ($\neg$), conjunction ($\wedge$) and the existential quantifier ($\exists$) as the primitive symbols of $\mathscr{L}$, the others, i.e. $\vee$, $\rightarrow$, $\leftrightarrow$, $\forall$ being defined in terms of these in the usual way (Ch. 1, §14 and Ch. 3, §4). The only *extralogical* symbol of $\mathscr{L}$ is the binary predicate $\in$. If $\mathbf{t}$ and $\mathbf{t}'$ are $\mathscr{L}$-terms, we write $\mathbf{t}\in\mathbf{t}'$ for $\in\mathbf{t}\mathbf{t}'$. $\mathbf{t}\in\mathbf{t}'$ is to be read "$\mathbf{t}$ *belongs* to $\mathbf{t}'$", or "$\mathbf{t}'$ *contains* $\mathbf{t}$" or simply "$\mathbf{t}$ is *in* $\mathbf{t}'$".

We adopt the convention that if a formula φ is first introduced as[1] $\varphi(x_1,\ldots,x_k)$ and $\mathbf{t}_1,\ldots,\mathbf{t}_k$ are terms — which may be either variables or virtual terms as defined in §13 of Ch. 2 (and further explained below) — then $\varphi(\mathbf{t}_1,\ldots,\mathbf{t}_k)$ stands for $\varphi(x_1/\mathbf{t}_1,\ldots,x_k/\mathbf{t}_k)$ (cf. 2.3.14). Similarly, if a term $\mathbf{s}$ is first introduced as $\mathbf{s}(x_1,\ldots,x_k)$, then $\mathbf{s}(\mathbf{t}_1,\ldots,\mathbf{t}_k)$ stands for the result of simultaneously substituting $\mathbf{t}_1,\ldots,\mathbf{t}_k$ for $x_1,\ldots,x_k$ in $\mathbf{s}$.

Defined formulas are abbreviations introduced as in the following scheme:

verbal expression: defined formula $\leftrightarrow_{\mathrm{df}}$ *defining formula,*

where the *defining formula* is the $\mathscr{L}$-formula for which the *defined formula* is to serve as an abbreviation, and the verbal expression (which is sometimes omitted) indicates how the *defined formula* is to be read. The scheme is called the *definition* of its defined formula. We make the following definitions without further delay:

x is not equal to y: $x \neq y \leftrightarrow_{\mathrm{df}} \neg(x=y)$;

x is not in y: $x \notin y \leftrightarrow_{\mathrm{df}} \neg(x \in y)$;

x is a subset of y (or *y includes x*): $x \subseteq y \leftrightarrow_{\mathrm{df}} \forall z(z \in x \rightarrow z \in y)$;

For some x in y, φ: $\exists x \in y \varphi \leftrightarrow_{\mathrm{df}} \exists x(x \in y \wedge \varphi)$;

For all x in y, φ: $\forall x \in y \varphi \leftrightarrow_{\mathrm{df}} \forall x(x \in y \rightarrow \varphi)$;

For some subset u of x, φ: $\exists u \subseteq x \varphi \leftrightarrow_{\mathrm{df}} \exists u(u \subseteq x \wedge \varphi)$;

For all subsets u of x, φ: $\forall u \subseteq x \varphi \leftrightarrow_{\mathrm{df}} \forall u(u \subseteq x \rightarrow \varphi)$;

There is at most one x such that $\varphi(x)$:

$\exists^* x \varphi \leftrightarrow_{\mathrm{df}} \forall x \, \forall y \, [\varphi(x) \wedge \varphi(y) \rightarrow x=y]$, where y is not free in φ.

Extralogical axioms — also called *postulates* — will be introduced as we go along. *Unless otherwise specified*, we write "$\vdash \varphi$" for "the formula φ is a logical consequence of the postulates introduced so far" or, equivalently, for "φ is deducible in the first-order predicate calculus from the postulates introduced so far". Since all the postulates are going to be *sentences* of $\mathscr{L}$, we have

$$\vdash \varphi \Leftrightarrow \vdash \forall x \varphi.$$

Formulas φ for which $\vdash \varphi$ are called *(formal) theorems* of set theory.

[1] It is to be understood that this is merely a device for drawing attention to the variables $x_1, \ldots, x_k$; it does *not* signify that all or indeed any of the x_i occur free in φ.

Now, if set theory is to be more than a mere game with symbols, one must posit[1] a (non-empty) *domain* or *collection* V of objects called *sets*, and a binary relation, *membership*, defined on V. The domain and the membership relation together constitute the *universe of sets*. The postulates — hence also the theorems — of set theory are supposed to be *true* in the universe of sets. In other words, the postulates represent true statements when $\in$ is interpreted as the membership relation and the variables of $\mathscr{L}$ are taken as ranging over V. We may think of the universe of sets as a rough approximation to "Cantor's paradise" of naive set theory.

In accordance with the position we have just advocated, we shall assume that all the formulas of $\mathscr{L}$ actually *refer* to the universe of sets. This enables us to adopt a rather informal method of presenting deductions in $\mathscr{L}$, namely, we deal with the variables of $\mathscr{L}$ as denoting, or ranging over, *sets*, i.e. individuals of V. Thus, e.g., suppose we wish to prove $\vdash \varphi(x)$. We take $\varphi(x)$ as asserting something about "the set x" (whereas strictly speaking we should regard $\varphi(x)$ as expressing a condition on the *value* of x in V). We then proceed to show, using the postulates more or less informally, that "every set x" satisfies the assertion $\varphi(x)$, and hence we conclude $\vdash \varphi(x)$. Of course, all these informal deductions can quite easily (but tediously!) be translated into formal deductions of the first-order predicate calculus in $\mathscr{L}$.

We must distinguish very carefully between *sets* (i.e. objects in V) and *collections* (i.e. pluralities of objects in the intuitive sense). Collections of sets will be called *classes*. To each set x there corresponds a class, called the *extension* of x, which is the collection of all members of x (i.e. objects bearing the membership relation to x). A set and its extension are quite different, but a fundamental part of the intuitive notion of *set*, which we insist on incorporating into our formal theory, is that a set is *uniquely determined* by its extension, that is, no two different sets have the same extension. This is expressed by our first postulate, namely the *Axiom of Extensionality*:

Ext: $\forall x\, \forall y\, [\forall z(z \in x \leftrightarrow z \in y) \rightarrow x = y]$.

The proof of our first theorem is easy and is left to the reader.

[1] Evidently we are adopting what amounts to a *Platonist* view on the question of the existence of sets. This view is in some ways a dangerous oversimplification and would certainly need — at least — major revisions if it were to become a component of a serious philosophy of mathematics. Nonetheless we regard the Platonist position — despite its shortcomings — as being closest in spirit to the intuitive picture behind set theory.

1.1. THEOREM. *If x is not free in* φ, *then*

$$\vdash \exists x \forall y\, [y \in x \leftrightarrow \varphi] \rightarrow \exists! x \forall y\, [y \in x \leftrightarrow \varphi]. \qquad \blacksquare$$

The argument used in the informal proof of Cantor's classic result that any set is of strictly smaller cardinality than its power set shows that there cannot exist a one–one correspondence between all *classes* and all *sets*. Thus, not all classes can be extensions of sets.

It might be thought that the reason for this disparity is that the notion of class used here (i.e., a completely arbitrary collection of sets) is too vague and general, and that, perhaps, if we could single out those classes that can be "precisely described", then these classes would be exactly all set extensions. This prompts the following definition.

Let φ be a formula whose free variables are among $y, y_1, \ldots, y_k$ (where $k \geqslant 0$; the variables $y_1, \ldots, y_k$ are to be regarded as parameters). Assign sets $a_1, \ldots, a_k$ as values to the parameters $y_1, \ldots, y_k$, respectively. Then the given formula φ and the parameter values $a_1, \ldots, a_k$ are said to *define* a class, namely, the collection of all sets a such that φ holds when y takes the value a (and the parameters take their assigned values). A class obtained in this way is said to be *definable*.

Using informal set-theoretic reasoning it is easy to show that if one assumes the universe of sets to be infinite (as we shall want to do in any case!), then the collection of all classes which are definable in the above sense is not more numerous than the universe of sets and hence not more numerous than the collection of extensions of all sets. Moreover, the extension of any given set is definable by means of a formula with one parameter: namely, take the formula $y \in y_1$ and assign the given set as value to y_1. It is therefore tempting to postulate that *every* definable class is the extension of a set. This assertion may be formalized as the following scheme:

(1.2) $\qquad \forall y_1 \ldots \forall y_k \exists x \forall y\, [y \in x \leftrightarrow \varphi]$,

where φ is any formula with free variables among $y, y_1, \ldots, y_k$. This scheme is the (unrestricted) *Comprehension Axiom*.

Unfortunately, however, (1.2) is untenable even when $k=0$, because it leads to the well-known *Russell paradox* as follows. Take φ to be $y \notin y$. Then by (1.2) we get $\exists x \forall y\, (y \in x \leftrightarrow y \notin y)$. From this we immediately obtain $\exists x (x \in x \leftrightarrow x \notin x)$, which is logically false.

Accordingly, we cannot adopt (1.2) in general. But because of its naturalness we shall adopt as postulates certain special cases of (1.2) for particular formulas φ.

The *Axiom of Replacement* (briefly, **Rep**) which we shall now adopt is not a single axiom but an *axiom scheme.* For each formula φ with free variables among $x, y, y_1, \ldots, y_k$ we take as a postulate the corresponding instance of **Rep**, namely

Rep: $\quad \forall y_1 \ldots \forall y_k \, \forall u [\forall x \in u \, \exists^* y \varphi \rightarrow \exists z \, \forall y [y \in z \leftrightarrow \exists x \in u \varphi]]$,

where[1] z does not occur free in φ.

It is easy to see in what sense **Rep** is a particular case of the Comprehension Axiom. If we ignore the initial $k+1$ quantifiers, **Rep** is an implication. The antecendent says that φ defines a partial single-valued mapping on the extension of u, while the consequent says that the class of all images of members of u is the extension of some set z. Thus **Rep** asserts that any class which is the class of all images of the members of some given set under a definable partial single-valued mapping is itself the extension of a set.

1.3. THEOREM. *If z is not free in ψ, then*:

(i) $\vdash \forall u \, \exists z \, \forall y [y \in z \leftrightarrow y \in u \wedge \psi]$.

(ii) $\vdash \exists z \, \forall y [\psi \rightarrow y \in z] \rightarrow \exists z \, \forall y [y \in z \leftrightarrow \psi]$.

PROOF. For (i), put φ for $x=y \wedge \psi$. It is then clear that we have $\vdash \forall x \in u \, \exists^* y \varphi$. Hence, by **Rep**, we conclude that

$$\vdash \exists z \, \forall y [y \in z \leftrightarrow \exists x \in u \varphi].$$

But obviously we have

$$\vdash \exists x \in u \varphi \leftrightarrow y \in u \wedge \psi,$$

so we obtain $\vdash \exists z \, \forall y [y \in z \leftrightarrow y \in u \wedge \psi]$ as required.

(ii) follows easily from (i). ∎

The scheme of sentences of the form

Sep: $\quad \forall y_1 \ldots \forall y_k \, \forall u \, \exists z \, \forall y [y \in z \leftrightarrow y \in u \wedge \psi]$,

where z is not free in ψ is known as the *Axiom of Separation.* In the present axiomatic system it does not have to be introduced as a separate postulate because, as we have shown in Thm. 1.3, it is deducible from **Rep**.

[1] Strictly speaking, we should specify z uniquely, e.g. as the *first* variable not free in φ. However, here and in similar contexts below, we omit such niceties.

1.4. PROBLEM. Show that V is not the extension of any set.

We now establish the existence of a unique set with empty extension.

1.5. THEOREM. $\vdash \exists! z\, \forall y(y \in z \leftrightarrow y \neq y)$.

PROOF. In 1.3(i) take ψ to be $y \neq y$. Then we have

$$\vdash \exists z\, \forall y(y \in z \leftrightarrow y \neq y \wedge y \in u).$$

But

$$\vdash (y \neq y \wedge y \in u) \leftrightarrow y \neq y,$$

since both sides are logically false. Hence

$$\vdash \exists z\, \forall y(y \in z \leftrightarrow y \neq y),$$

and the theorem follows by 1.1. ∎

From now on we shall often use the method explained in §13 of Ch. 2 to introduce *virtual terms*. In each case, we have to select (in the notation of §13 of Ch. 2) a formula $\boldsymbol{\alpha}$ with one free variable, and a second formula $\boldsymbol{\varphi}$. We shall *always* take $\boldsymbol{\alpha}$ to be the formula $\forall y(y \in z \leftrightarrow y \neq y)$. Note that by 1.5 we have $\boldsymbol{\Phi} \models \exists! z\boldsymbol{\alpha}$, where $\boldsymbol{\Phi}$ is the collection of all the postulates introduced so far.[1] As $\boldsymbol{\varphi}$ we shall select various formulas from time to time — beginning, in fact, with $\boldsymbol{\alpha}$ itself. Recall that while in practice we can manipulate virtual terms as ordinary terms in some extension of $\mathscr{L}$, we nevertheless regard a formula containing such terms as an abbreviation for a suitable $\mathscr{L}$-formula. Thus, in particular, a formula containing virtual terms introduced previously can perfectly well be used as the $\boldsymbol{\varphi}$ for introducing new virtual terms.

We define the *free variables* of a term $\mathbf{t}$ inductively as follows: if $\mathbf{t}$ is a variable x, the only free variable of $\mathbf{t}$ is x, and if $\mathbf{t}$ is $\iota y\boldsymbol{\varphi}$, then the free variables of $\mathbf{t}$ consist of all free variables of $\boldsymbol{\varphi}$ with the exception of y.

Abbreviated notations for terms are introduced by means of a method similar to that used for introducing defined formulas, viz., according to the scheme:

verbal expression: *defined term* $=_{\mathrm{df}}$ *defining term*,

where the *defining term* is the (virtual) $\mathscr{L}$-term for which the *defined term* is to serve as an abbreviation, and the verbal expression (which is sometimes

[1] As a matter of fact it is easy to see from the proofs of 1.3 and 1.5 that it is enough take $\boldsymbol{\Phi}$ to be **Ext** plus a single instance of **Rep**, namely the instance where the formula $\boldsymbol{\varphi}$ is $x = y \wedge y \neq y$.

omitted) indicates how the *defined term* is to be read. The scheme is called a *definition* of its defined term.

We now define:

$$\textit{the empty set or zero}: \mathbf{0} =_{\text{df}} \iota z[\forall y(y\in z \leftrightarrow y\neq y)].$$

The term $\mathbf{0}$ is called the *zeroth numeral.*

1.6. LEMMA.

(i) $\vdash y\in \mathbf{0} \leftrightarrow y\neq y.$

(ii) $\vdash z=\mathbf{0} \leftrightarrow \forall y(y\notin z).$

(iii) $\vdash y=\iota x\varphi(x) \leftrightarrow [\exists! x\varphi(x) \wedge \varphi(y)] \vee [\neg\exists! x\varphi(x) \wedge y=\mathbf{0}].$

(iv) $\vdash \exists! x\varphi(x) \rightarrow \varphi(\iota x\varphi).$

PROOF. This lemma follows immediately from the above definitions and Prob. 2.13.3. ∎

We see from (iii) of this lemma that $\iota x\varphi(x)$ is the unique x such that $\varphi(x)$ if such an x exists, or $\mathbf{0}$ if not.

We now define

$$\textit{the set of } x \textit{ such that } \varphi: \{x : \varphi\} =_{\text{df}} \iota z[\forall x(x\in z \leftrightarrow \varphi)],$$

where z is not free in φ. (Note that x is *not free* in the term $\{x : \varphi\}$.)

Terms of the form $\{x : \varphi\}$ are called *abstraction terms*. An abstraction term $\{x : \varphi(x)\}$ is said to be *legitimate* if we have,

$$\vdash y\in \{x : \varphi(x)\} \leftrightarrow \varphi(y). \tag{1.7}$$

Thus the term $\{x : \varphi(x)\}$ is legitimate iff the extension of the set it denotes is precisely the class defined by φ.

We observe that $\{x: x\in y\}$ is legitimate and $\vdash \{x: x\in y\}=y$.

1.8. LEMMA. *The abstraction term* $\{x: \varphi(x)\}$ *is legitimate iff*

$$\vdash \exists z\, \forall x[\varphi(x) \rightarrow x\in z].$$

PROOF. Necessity follows immediately from (1.7). Conversely, suppose that

$$\vdash \exists z\, \forall x[\varphi(x) \rightarrow x\in z].$$

Then, by 1.3(ii) and 1.1 we have

$$\vdash \exists! z\, \forall x[x\in z \leftrightarrow \varphi(x)],$$

so $\{x: \varphi(x)\}$ is legitimate by 1.6(iv). ∎

Lemma 1.8 tells us that the abstraction term $\{x: \varphi(x)\}$ is legitimate iff the class defined by $\varphi(x)$ is included in the extension of some set.

We shall be introducing many abstraction terms in the future, and they will all be legitimate. In some cases when a new abstraction term is introduced, we shall leave the reader to verify its legitimacy. In practice this will amount to nothing more than an application of Lemma 1.8 and the postulates that will have been introduced.

Our next postulate is the *Axiom of Union*

Union: $\forall z \, \exists x \, \forall y [\exists u \in z (y \in u) \leftrightarrow y \in x]$.

This postulate asserts that for each set z there is a (unique) set whose extension is precisely the collection of all members of members of z.

We define

$$\textit{union of } z\text{: } \bigcup z =_{\text{df}} \{y: \exists u \in z (y \in u)\}.$$

Thus $\bigcup z$ is the unique set whose extension is the collection of all members of members of z.

We now introduce our next postulate, the *Power Set Axiom*

Pow: $\forall z \, \exists x \, \forall y [y \subseteq z \leftrightarrow y \in x]$.

This postulate asserts that for each set z there is a (unique) set whose extension is the collection of all subsets of z.

PROBLEM. Show that it would have been enough to postulate **Union** and **Pow** with "$\rightarrow$" in place of "$\leftrightarrow$".

We define

$$\textit{power set of } z\text{: } \mathsf{P}z =_{\text{df}} \{y: y \subseteq z\}.$$

Thus $\mathsf{P}z$ is the unique set whose extension is the collection of all subsets of z.

We also put

$$\textit{one}\text{: } \mathbf{1} =_{\text{df}} \mathsf{P}\mathbf{0},$$

$$\textit{two}\text{: } \mathbf{2} =_{\text{df}} \mathsf{P}\mathbf{1}.$$

The terms **1** and **2** are called the *first* and *second numerals*, respectively.

1.9. THEOREM.

(i) $\vdash \mathbf{0} \neq \mathbf{1}$.

(ii) $\vdash x \in \mathbf{1} \leftrightarrow x = \mathbf{0}$.

(iii) $\vdash x \in \mathbf{2} \leftrightarrow x = \mathbf{0} \vee x = \mathbf{1}$.

PROOF. Left to the reader. ∎

1.10. THEOREM. $\vdash \exists x\, \forall y[y \in x \leftrightarrow y = u \vee y = v]$.
PROOF. Let $\varphi(z,y)$ be the formula

$$[(z=\mathbf{0} \wedge y=u) \vee (z=\mathbf{1} \wedge y=v].$$

By 1.9 we have $\vdash \mathbf{0} \neq \mathbf{1}$, so it follows that $\vdash \forall z \in \mathbf{2}\, \exists^* y \varphi(z,y)$. Hence, by **Rep**,

$$\vdash \exists x\, \forall y[y \in x \leftrightarrow \exists z \in \mathbf{2} \varphi(z,y)],$$

which immediately gives the required result. ∎

We now define:

unordered pair of u and v: $\{u,v\} =_{\mathrm{df}} \{y : y=u \vee y=v\}$;
singleton of u: $\{u\} =_{\mathrm{df}} \{u,u\}$;
union of u and v: $u \cup v =_{\mathrm{df}} \bigcup \{u,v\}$;
intersection of u and v: $u \cap v =_{\mathrm{df}} \{x : x \in u \wedge x \in v\}$;
complement of v in u: $u - v =_{\mathrm{df}} \{x : x \in u \wedge x \notin v\}$;
ordered singleton of u: $\langle u \rangle =_{\mathrm{df}} u$;
ordered pair of u and v: $\langle u,v \rangle =_{\mathrm{df}} \{\{u\}, \{u,v\}\}$;
Cartesian product of u and v: $u \times v =_{\mathrm{df}} \{x : \exists y \in u \exists z \in v(x=\langle y,z \rangle)\}$.

To verify the legitimacy of the defining term in the definition of the Cartesian product, notice that

$$\vdash y \in u \wedge z \in v \rightarrow \langle y,z \rangle \in \mathsf{PP}(u \cup v)$$

and apply 1.8.

Proceeding inductively, we define, for $n \geqslant 3$,

$$\{u_1,\ldots,u_n\} =_{\mathrm{df}} \{u_1,\ldots,u_{n-1}\} \cup \{u_n\},$$
$$u_1 \cup \ldots \cup u_n =_{\mathrm{df}} (u_1 \cup \ldots \cup u_{n-1}) \cup u_n,$$
$$u_1 \cap \ldots \cap u_n =_{\mathrm{df}} (u_1 \cap \ldots \cap u_{n-1}) \cap u_n,$$
$$\langle u_1,\ldots,u_n \rangle =_{\mathrm{df}} \langle \langle u_1,\ldots,u_{n-1} \rangle, u_n \rangle,$$
$$u_1 \times \ldots \times u_n =_{\mathrm{df}} (u_1 \times \ldots \times u_{n-1}) \times u_n.$$

Notice that by our previous definitions we already have

$$\{u_1,u_2\} = \{u_1\} \cup \{u_2\},$$
$$\langle u_1,u_2 \rangle = \langle \langle u_1 \rangle, u_2 \rangle.$$

We also put

$$u^n =_{\mathrm{df}} u \times \ldots \times u \quad \text{(with } n \text{ factors).}$$

1.11. THEOREM.

(i) $\vdash \langle u_1,\ldots,u_n\rangle = \langle v_1,\ldots,v_n\rangle \leftrightarrow u_1 = v_1 \wedge \ldots \wedge u_n = v_n$.

(ii) $\vdash u_1 \times \ldots \times u_n = \{y : \exists x_1 \in u_1 \ldots \exists x_n \in u_n (y = \langle x_1,\ldots,x_n\rangle)\}$.

PROOF. Left to the reader. ∎

We also put

(1.12) *the set of x in u such that* $\boldsymbol{\varphi}$: $\{x \in u : \boldsymbol{\varphi}\} =_{\mathrm{df}} \{x : x \in u \wedge \boldsymbol{\varphi}\}$.

(The defining term here is legitimate by 1.3(i).)

Let $\mathbf{t}$ be a term which is not a variable, and let $\boldsymbol{\varphi}$ be any formula. We put

(1.13) $\{\mathbf{t} : x_1 \in u_1 \wedge \ldots \wedge x_n \in u_n \wedge \boldsymbol{\varphi}\} =_{\mathrm{df}} \{z : \exists x_1 \in u_1 \ldots \exists x_n \in u_n [z = \mathbf{t} \wedge \boldsymbol{\varphi}]\}$,

where z is not free in $\mathbf{t}$ nor in $\boldsymbol{\varphi}$.

To verify the legitimacy of the defining term in this definition, put

$$\boldsymbol{\psi} \leftrightarrow_{\mathrm{df}} \exists x_1 \in u_1 \ldots \exists x_n \in u_n [z = \mathbf{t} \wedge \boldsymbol{\varphi} \wedge y = \langle x_1,\ldots,x_n\rangle].$$

Then clearly $\vdash \forall y \in u_1 \times \ldots \times u_n\, \exists^* z \boldsymbol{\psi}$. Hence, by **Rep**,

(1) $\vdash \exists w\, \forall z [\exists y \in u_1 \times \ldots \times u_n \boldsymbol{\psi} \rightarrow z \in w]$.

Also, from the definition of $\boldsymbol{\psi}$,

(2) $\vdash \exists x_1 \in u_1 \ldots \exists x_n \in u_n [z = \mathbf{t} \wedge \boldsymbol{\varphi}] \rightarrow \exists y \in u_1 \times \ldots \times u_n \boldsymbol{\psi}$.

From (1), (2) and 1.8 the desired result follows easily.

§ 2. Ordinals

We define

epsilon well-orders x:

$$\mathrm{Ew}(x) \leftrightarrow_{\mathrm{df}} \forall y \in x\, \forall z \in x [y = z \vee y \in z \vee z \in y]$$
$$\wedge\, \forall u \subseteq x [u \neq \mathbf{0} \rightarrow \exists y \in u\, \forall z \in u [z \notin y]].$$

The second clause in this definition requires that every non-empty subset of x has a member which is "minimal" with respect to the membership relation.

We also put

x is transitive: $\mathrm{Trans}(x) \leftrightarrow_{\mathrm{df}} \forall y \in x [y \subseteq x]$,

x is an ordinal: $\mathrm{Ord}(x) \leftrightarrow_{\mathrm{df}} \mathrm{Ew}(x) \wedge \mathrm{Trans}(x)$.

2.1. PROBLEM. Show that $\vdash \mathrm{Ord}(\mathbf{0}) \wedge \mathrm{Ord}(\mathbf{1}) \wedge \mathrm{Ord}(\mathbf{2})$.

We shall use lower case Greek letters — chiefly $\alpha, \beta, \gamma, \lambda, \xi, \eta$ — as a new kind of variable ranging over the ordinals. This means that $\forall\alpha\varphi(\alpha)$ and $\exists\alpha\varphi(\alpha)$ stand for $\forall x[\mathrm{Ord}(x) \to \varphi(x)]$ and $\exists x[\mathrm{Ord}(x) \wedge \varphi(x)]$, respectively, where x is a variable not occurring in $\varphi(\alpha)$. Also, if $\alpha, \beta, \gamma,\ldots$ are free in $\varphi(\alpha,\beta,\gamma,\ldots)$, then the assertion $\vdash \varphi(\alpha,\beta,\gamma,\ldots)$ stands for

$$\vdash \mathrm{Ord}(x) \wedge \mathrm{Ord}(y) \wedge \mathrm{Ord}(z) \wedge \ldots \to \varphi(x,y,z,\ldots),$$

where $x,y,z,\ldots$ are variables not occurring in $\varphi(\alpha,\beta,\gamma,\ldots)$. Expressions like $\exists!\alpha\varphi(\alpha)$, $\iota\alpha\varphi(\alpha)$ and $\{\alpha\colon \varphi(\alpha)\}$ are defined similarly.

We define

$$x \textit{ is smaller than } y\colon x<y \leftrightarrow_{\mathrm{df}} x\in y,$$

$$\textit{successor of } x\colon x+1 =_{\mathrm{df}} x\cup\{x\},$$

but we shall use these notations for ordinals only. We write $\exists\alpha<\beta\varphi$ for $\exists\alpha[\alpha<\beta \wedge \varphi]$, etc.

Our next theorem tabulates some of the basic facts about ordinals.

2.2. THEOREM.

(i) $\vdash \alpha\notin\alpha$.
(ii) $\vdash \mathrm{Ord}(\alpha+1) \wedge \alpha<\alpha+1$.
(iii) $\vdash y\in\alpha \to \mathrm{Ord}(y)$.
(iv)[1] $\vdash \gamma<\beta<\alpha \to \gamma<\alpha$.
(v) $\vdash y\subseteq\alpha \wedge \mathrm{Trans}(y) \to y=\alpha \vee y\in\alpha$.
(vi)[2] $\vdash \alpha\leqslant\beta \vee \beta\leqslant\alpha$.
(vii) $\vdash \neg(\alpha<\beta<\alpha) \wedge \neg(\alpha<\alpha)$.
(viii) $\vdash \alpha<\beta \leftrightarrow \alpha+1\leqslant\beta$.
(ix) $\vdash \forall y\in x\, \mathrm{Ord}(y) \to \mathrm{Ew}(x) \wedge \mathrm{Ord}(\bigcup x)$.
(x) $\vdash \forall y\in x\, \mathrm{Ord}(y) \to [\bigcup x\leqslant\beta \leftrightarrow \forall\alpha\in x(\alpha\leqslant\beta)]$.

PROOF. Before we begin, let us define

$$y \textit{ is a minimal element of } x\colon \mathrm{Min}(y,x) \leftrightarrow_{\mathrm{df}} y\in x \wedge \forall z\in x[z\notin y].$$

(i) If $y\in\alpha$, then $y\notin y$. For if not, then $\{y\}$ would be a subset of α with no minimal element, contradicting the definition of an ordinal. Thus $\vdash \alpha\in\alpha \to \alpha\notin\alpha$, whence $\vdash \alpha\notin\alpha$.

(ii) is straightforward, and we leave its proof as an exercise to the reader.

[1] We write $x\in y\in z$ for $x\in y \wedge y\in z$, and $\gamma<\beta<\alpha$ for $\gamma<\beta \wedge \beta<\alpha$.
[2] We write $\alpha\leqslant\beta$ for $\alpha<\beta \vee \alpha=\beta$, and $\exists\alpha\leqslant\beta\varphi$ for $\exists\alpha[\alpha\leqslant\beta \wedge \varphi]$, etc.

(iii) Let $y \in \alpha$. Since Trans(α), we have $y \subseteq \alpha$, and since also Ew(α), it follows immediately that Ew(y). It remains to show that Trans(y), i.e., given that $u \in z \in y$ we must show that $u \in y$. Since Trans(α), we have $z \in \alpha$ and $u \in \alpha$, and since Ew(α), we have either $y \in u$ or $y = u$ or $u \in y$. Either of the first two alternatives implies that $\{u,z,y\}$ has no minimal member, which is impossible. Hence $u \in y$ and (iii) follows.

(iv) follows at once from Trans(α).

(v) Suppose $y \subseteq \alpha$ and Trans(y). Put $u = \alpha - y$. If $u = \mathbf{0}$, then $y = \alpha$. If $u \neq \mathbf{0}$, then u has a minimal member v. Since $v \in u \subseteq \alpha$, we have $v \in \alpha$. To show that $y \in \alpha$ it will suffice to show that $y = v$. If $z \in v$, then $z \in \alpha$, since we have Trans(α); but by the minimality of v in u we cannot have $z \in u$. Thus by the definition of u we must have $z \in y$. Conversely, let $z \in y$. Then $z \in \alpha$, and, since Ew(α), we have $z = v$ or $v \in z$ or $z \in v$. Now $z = v$ would imply $v \in y$, which is impossible, while $v \in z$ would, in the presence of the assumption Trans(y), also imply $v \in y$. Thus the only remaining possibility is $z \in v$.

(vi) Consider the set $\alpha \cap \beta$. It is clearly transitive and so, by (iii) and (v), an ordinal, γ say. Applying (v) again, we see that $\gamma \leqslant \alpha$ and $\gamma \leqslant \beta$. Then $\gamma = \alpha$ or $\gamma = \beta$, for $\gamma \neq \alpha$ and $\gamma \neq \beta$ implies, by (v), $\gamma \in \alpha \cap \beta = \gamma$ which, according to (i), is impossible.

(vii) follows immediately from (i) and (iv).

(viii) is a straightforward consequence of (ii), (vi) and (vii), which we leave as an exercise to the reader.

(ix) Suppose $\forall y \in x$ Ord(y), and let $u \subseteq x$, $u \neq \mathbf{0}$. We claim that u has a minimal element. Since $u \neq \mathbf{0}$, we can choose $\alpha \in u$. If $\alpha \cap u = \mathbf{0}$, then clearly α is a minimal element of u, while if $\alpha \cap u \neq \mathbf{0}$, then a minimal element of $\alpha \cap u$ is easily seen to be a minimal element of u. This proves the claim, which, together with (iv) and (vi), gives Ew(x).

Now let $z = \bigcup x$. Clearly we have Trans(z). Moreover, it is equally clear that $\forall y \in z$ Ord(y), so that Ew(z) by the first part of the proof. Hence Ord(z).

(x) Suppose $\forall y \in x$ Ord(y), and let $z = \bigcup x$. Then Ord(z) by (ix). If $\alpha \in x$, then $\alpha \subseteq z$, so that $\alpha \leqslant z$ by (v). Thus if $\bigcup x \leqslant \beta$ then $\forall \alpha \in x(\alpha \leqslant \beta)$. Conversely if $\forall \alpha \in x(\alpha \leqslant \beta)$, then $\forall \alpha \in x(\alpha \subseteq \beta)$ so that $z \subseteq \beta$, whence by (v) $z \leqslant \beta$. ∎

It follows immediately from (ix) and (vii) of this theorem that each non-empty set x of ordinals has a unique minimal member. We call this the *least* member of x.

We now show that there is no set whose extension includes the class of all ordinals (cf. Prob. 1.4.).

2.3. THEOREM. $\vdash \neg \exists x \forall y[\mathrm{Ord}(y) \to y \in x]$.

PROOF. Suppose $\exists x \forall y[\mathrm{Ord}(y) \to y \in x]$. Then, by 1.3(ii), there is a set x such that $\forall y[\mathrm{Ord}(y) \leftrightarrow y \in x]$. We have $\mathrm{Ord}(x)$ by 2.2(ix) and (iii), and it follows that $x \in x$. But this contradicts 2.2(i). ∎

2.4. THEOREM. *If β does not occur free in $\varphi(\alpha)$, then:*

(i) $\vdash \exists \alpha \varphi(\alpha) \to \exists! \alpha[\varphi(\alpha) \wedge \forall \beta < \alpha \neg \varphi(\beta)]$.

(ii) $\vdash \forall \alpha[\forall \beta < \alpha \varphi(\beta) \to \varphi(\alpha)] \to \forall \alpha \varphi(\alpha)$.

PROOF. (ii) is an easy consequence of (i), so we merely prove (i).

Assume $\exists \alpha \varphi(\alpha)$, and choose γ so that $\varphi(\gamma)$. Let

$$u = \{\beta\colon\ \beta < \gamma \wedge \varphi(\beta)\}.$$

Then $u \subseteq \gamma$. If $u = \mathbf{0}$, then γ is easily seen to satisfy $\varphi(\gamma) \wedge \forall \beta < \gamma \neg \varphi(\beta)$. On the other hand, if $u \neq \mathbf{0}$, then the least member of u satisfies the above condition. Uniqueness follows from 2.2.(vi). ∎

2.4(ii) is called the *principle of transfinite induction (on the ordinals)*. It is a generalization of the familiar principle of induction on the natural numbers.

We now define

least ordinal α such that $\varphi(\alpha)$: $\mu\alpha\varphi(\alpha) =_{\mathrm{df}} \iota\alpha[\varphi(\alpha) \wedge \forall \beta < \alpha \neg \varphi(\beta)]$,

where β is not free in $\varphi(\alpha)$, and

x is a limit ordinal: $\mathrm{Lim}(x) \leftrightarrow_{\mathrm{df}} \mathrm{Ord}(x) \wedge x \neq \mathbf{0} \wedge \forall \alpha(x \neq \alpha + 1)$.

2.5. PROBLEM. Show that:

(i) $\vdash \mathrm{Lim}(\beta) \wedge \alpha < \beta \to \alpha + 1 < \beta$.

(ii) $\vdash \forall y \in x\, \mathrm{Ord}(y) \to \bigcup x = \mu\alpha[\forall \beta \in x(\beta \leqslant \alpha)]$.

(iii) If β, λ do not occur free in $\varphi(\alpha)$, then

$$\vdash \varphi(0) \wedge \forall \alpha\, [\varphi(\alpha) \to \varphi(\alpha+1)] \wedge$$
$$\wedge \forall \lambda\, [\mathrm{Lim}(\lambda) \wedge \forall \beta < \lambda \varphi(\beta) \to \varphi(\lambda)] \to \forall \alpha \varphi(\alpha).$$

We now introduce the *natural numbers*. We define

x is a natural number: $\mathrm{N}(x) \leftrightarrow_{\mathrm{df}} \mathrm{Ord}(x) \wedge \forall \alpha \leqslant x \neg \mathrm{Lim}(\alpha)$.

We shall use lower case italic letters from the middle of the alphabet — chiefly i, j, k, m, n — as variables ranging over the natural numbers in exactly the same way as we have been using Greek letters as variables ranging over the ordinals. We shall also continue to use these letters metamathematically (e.g. as indices) but in each case it should be clear from the context what usage is intended.

We have already defined the numerals **0**, **1** and **2**. For $n \geqslant 3$ we define the nth *numeral* recursively by putting $\mathbf{n} =_{\text{df}} \mathbf{k} \cup \{\mathbf{k}\}$, where $k = n-1$.

2.6. THEOREM.

(i) $\vdash \mathrm{N}(\mathbf{n})$ *for each natural number n.*

(ii) $\vdash \mathrm{N}(n+1) \wedge [\alpha < n \rightarrow \mathrm{N}(\alpha)]$.

(iii) $\vdash \exists n \varphi(n) \rightarrow \exists! n\, [\varphi(n) \wedge \forall m < n \neg \varphi(m)]$, *where m is not free in* $\varphi(n)$.

(iv) $\vdash \varphi(\mathbf{0}) \wedge \forall n\, [\varphi(n) \rightarrow \varphi(n+1)] \rightarrow \forall n \varphi(n)$.

(v) $\vdash \mathbf{0} \in x \wedge \forall y \in x\, [y \cup \{y\} \in x] \rightarrow \forall n\, [n \in x]$.

PROOF. (i) and (ii) are easily proved; we leave this task to the reader.

To prove (iii), take a Greek variable, say α, which is not free in $\varphi(n)$, and put $\mathrm{N}(\alpha) \wedge \varphi(\alpha)$ for $\varphi(\alpha)$ in 2.4(i).

To prove (iv), assume $\neg \forall n \varphi(n)$. Then by (iii) there must be a natural number n such that

$$\neg \varphi(n) \wedge \forall m < n \varphi(m).$$

If $n = \mathbf{0}$, then $\neg \varphi(\mathbf{0})$. On the other hand, if $n \neq \mathbf{0}$ then, since $\neg \mathrm{Lim}(n)$, we must have $n = \alpha + 1$ for some α. Then $\alpha < n$ and so $\mathrm{N}(\alpha)$ by (ii). Thus $\mathrm{N}(\alpha) \wedge \varphi(\alpha) \wedge \neg \varphi(\alpha+1)$, which implies

$$\neg \forall n [\varphi(n) \rightarrow \varphi(n+1)].$$

Finally, (v) is obtained by taking $\varphi(n)$ to be the formula $n \in x$ in (iv). ∎

2.6(v) is the (second-order) *induction axiom* for natural numbers. The other Peano axioms, viz. $\forall n [n+1 \neq \mathbf{0}]$, $\forall n \forall m [n+1 = m+1 \rightarrow n = m]$ are easily derivable as theorems of set theory. Also, the axioms for addition and multiplication of natural numbers become provable when these operations are suitably defined.

2.7. THEOREM. $\vdash \exists x [\mathbf{0} \in x \wedge \forall y \in x [y \cup \{y\} \in x]] \leftrightarrow \exists x\, \forall y [y \in x \leftrightarrow \mathrm{N}(y)]$.

PROOF. Assuming the left-hand side we deduce $\exists x\, \forall n [n \in x]$ from 2.6(v). 1.3(ii) now gives the right-hand side. The converse follows immediately from 2.6(ii). ∎

The *Axiom of Infinity* (briefly: **Inf**) is the left-hand side of the formula proved in 2.7, namely

Inf: $\exists x [\mathbf{0} \in x \wedge \forall y \in x [y \cup \{y\} \in x]]$.

By 2.7 this is tantamount to postulating the existence of a set whose extension is the class of all natural numbers, so we can define

$$\textit{the set of natural numbers}: \omega =_{\text{df}} \{y: \mathrm{N}(y)\}.$$

2.8. THEOREM. $\vdash \mathrm{Lim}(\omega)$.

PROOF. Since $\vdash \forall x \in \omega\, \mathrm{Ord}(x)$, we have $\vdash \mathrm{Ew}(\omega)$ by 2.2(ix). By 2.6(ii) we have $\vdash \mathrm{Trans}(\omega)$; therefore $\vdash \mathrm{Ord}(\omega)$.

If $\alpha < \omega$, i.e. $\alpha \in \omega$, then $\mathrm{N}(\alpha)$ and so $\neg \mathrm{Lim}(\alpha)$. If we had $\neg \mathrm{Lim}(\omega)$ as well, then $\neg \exists \alpha \leqslant \omega\, \mathrm{Lim}(\alpha)$, so that we would have $\mathrm{N}(\omega)$ and hence $\omega \in \omega$, which contradicts 2.2(i). ∎

We now make the following definitions:

$$\textit{f of x}: f`x =_{\text{df}} \iota y[\langle x,y\rangle \in f];$$

$$\textit{f is a function}: \mathrm{Fun}(f) \leftrightarrow_{\text{df}} \forall x \in f\, \exists u\, \exists v[\langle u,v\rangle = x] \wedge \forall u\, \forall v\, \forall w[\langle u,v\rangle \in f \wedge \langle u,w\rangle \in f \rightarrow v = w];$$

$$\textit{domain of f}: \mathrm{dom}(f) =_{\text{df}} \{u: \exists v[\langle u,v\rangle \in f]\};$$

$$\textit{range of f}: \mathrm{ran}(f) =_{\text{df}} \{v: \exists u[\langle u,v\rangle \in f]\};$$

$$\textit{restriction of f to x}: f{\restriction}x =_{\text{df}} f \cap [x \times \mathrm{ran}(f)].$$

To legitimize the defining terms in the definitions of dom(f) and ran(f), we observe that

$$\vdash \langle u,v\rangle \in f \rightarrow u \in \bigcup\bigcup f \wedge v \in \bigcup\bigcup f.$$

For each term $\mathbf{t}(y)$ we also define[1]

$$\mathbf{t}|x =_{\text{df}} \{\langle y, \mathbf{t}(y)\rangle : y \in x\}.$$

Clearly we have, for any term $\mathbf{t}(y)$,

$$\vdash \mathrm{Fun}(\mathbf{t}|x) \wedge \mathrm{dom}(\mathbf{t}|x) = x.$$

The reader will doubtless be familiar with the process of constructing functions by (course of values) recursion on the natural numbers. This process yields a function defined on the natural numbers whose value at a natural number n is determined by n and the behaviour of the function at all $m < n$. We are now going to show that a similar procedure may be

[1] y need not be the only variable free in $\mathbf{t}$, but it will always be clear from the context which variable is intended to play the role of y.

employed in set theory to construct *terms* which behave in a prescribed way on the class of all ordinals.

2.9. THEOREM. *For each term* $\mathbf{s}(y,z)$ *we can construct a term* $\mathbf{t}(x)$ *such that*[1]

$$\vdash \mathbf{t}(\alpha) = \mathbf{s}(\mathbf{t}|\alpha, \alpha).$$

PROOF. Let us define

f meets the recursive conditions up to x (with respect to **s**):

$$\mathrm{Rec}(f,x) \leftrightarrow_{\mathrm{df}} \mathrm{Fun}(f) \wedge x \cup \{x\} \subseteq \mathrm{dom}(f) \wedge \forall \alpha \in \mathrm{dom}(f)[f`\alpha = \mathbf{s}(f \restriction \alpha, \alpha)].$$

Then clearly we have

(1) $\vdash \mathrm{Rec}(f,\gamma) \wedge \alpha \leqslant \gamma \rightarrow f`\alpha = \mathbf{s}(f \restriction \alpha, \alpha) \wedge \mathrm{Rec}(f,\alpha).$

Moreover, we have

(2) $\vdash \mathrm{Rec}(f,\gamma) \wedge \mathrm{Rec}(g,\gamma) \rightarrow f`\gamma = g`\gamma.$

For suppose (2) fails for some γ. Then by 2.4(i) there is a least ordinal γ_0 for which it fails. Then we have

$$\mathrm{Rec}(f,\gamma_0) \wedge \mathrm{Rec}(g,\gamma_0) \wedge f`\gamma_0 \neq g`\gamma_0,$$

and therefore

$$\mathbf{s}(f \restriction \gamma_0, \gamma_0) = f`\gamma_0 \neq g`\gamma_0 = \mathbf{s}(g \restriction \gamma_0, \gamma_0).$$

Accordingly $f \restriction \gamma_0 \neq g \restriction \gamma_0$ and so there must be some $\beta < \gamma_0$ for which $f`\beta \neq g`\beta$. But by (1) we have $\mathrm{Rec}(f,\beta)$ and $\mathrm{Rec}(g,\beta)$. This contradicts the assumption that γ_0 is the least ordinal satisfying these conditions, and (2) follows.

Now put $\mathbf{t}(x)$ for the term

$$\iota y\,[\forall f[\mathrm{Rec}(f,x) \rightarrow f`x = y]].$$

Then we have

(3) $\vdash \exists f\, \mathrm{Rec}(f,\alpha) \rightarrow \mathbf{t}(\alpha) = \mathbf{s}(\mathbf{t}|\alpha, \alpha).$

For, if $\mathrm{Rec}(f,\alpha)$, then from (2) and the definition of **t** we see that $\mathbf{t}(\alpha) = f`\alpha$.

[1] Of course, y and z are not assumed to be the only free variables of **s**, nor is x assumed to be the only free variable of **t**. In fact, the **t** constructed in the following proof has a s free variables x and all the free variables of **s** other than y and z.

Also, if $\beta<\alpha$, then $\mathrm{Rec}(f,\beta)$ so that $\mathbf{t}(\beta)=f`\beta$. But then we get

$$\mathbf{t}(\alpha)=f`\alpha=\mathbf{s}(f\restriction\alpha,\alpha)=\mathbf{s}(\mathbf{t}|\alpha,\alpha)$$

as claimed.

Now, by (3), we will obtain $\vdash\mathbf{t}(\alpha)=\mathbf{s}(\mathbf{t}|\alpha,\alpha)$ if we can show that

$$\vdash\exists f\,\mathrm{Rec}(f,\alpha).$$

To prove this we use the principle of transfinite induction. Assume that $\forall\beta<\alpha\,\exists f\,\mathrm{Rec}(f,\beta)$. Now put g for

$$\mathbf{t}|\alpha\cup\{\langle\alpha,\mathbf{s}(\mathbf{t}|\alpha,\alpha)\rangle\}.$$

Clearly we have

$$\mathrm{Fun}(g)\wedge\mathrm{dom}(g)=\alpha\cup\{\alpha\}.$$

If $\beta<\alpha$, then $g`\beta=\mathbf{t}(\beta)$ and $g`\gamma=\mathbf{t}(\gamma)$ for all $\gamma<\beta$. Since $\exists f\,\mathrm{Rec}(f,\beta)$ for all $\beta<\alpha$, by (3) we have $\mathbf{t}(\beta)=\mathbf{s}(\mathbf{t}|\beta,\beta)$ for all $\beta<\alpha$, which immediately gives $g`\beta=\mathbf{s}(g\restriction\beta,\beta)$ for all $\beta<\alpha$. Finally, we have

$$g`\alpha=\mathbf{s}(\mathbf{t}|\alpha,\alpha)=\mathbf{s}(g\restriction\alpha,\alpha).$$

Hence $\mathrm{Rec}(g,\alpha)$, which completes the proof. ∎

Thm. 2.9 is called the principle of *ordinal* (or *transfinite*) *recursion*, and any term $\mathbf{t}$ satisfying the condition specified in the theorem is said to be *constructed (from* $\mathbf{s}$*) by recursion on* α.

2.10. PROBLEM. (i) Show that, if $\mathbf{t}_1$ and $\mathbf{t}_2$ are both constructed from $\mathbf{s}$ by recursion on α, then

$$\vdash\forall\alpha[\mathbf{t}_1(\alpha)=\mathbf{t}_2(\alpha)].$$

(ii) Let $\mathbf{s}(y,z)$ be any term. Show that

$$\vdash\forall\alpha\,\exists f[\mathrm{Fun}(f)\wedge\mathrm{dom}(f)=\alpha\wedge\forall\beta<\alpha[f`\beta=\mathbf{s}(f\restriction\beta,\beta)]].$$

Many different forms of ordinal recursion may be reduced to the form given in 2.9. For example,

2.11. THEOREM. *Given three terms* $\mathbf{s}_0, \mathbf{s}_1(y,z)$ *and* $\mathbf{s}_2(y,z)$, *we can construct a term* $\mathbf{t}(x)$ *such that*

$$\vdash\mathbf{t}(0)=\mathbf{s}_0\wedge[\alpha=\beta+1\rightarrow\mathbf{t}(\alpha)=\mathbf{s}_1(\mathbf{t}(\beta),\beta)]$$
$$\wedge[\mathrm{Lim}(\alpha)\rightarrow\mathbf{t}(\alpha)=\mathbf{s}_2(\mathbf{t}|\alpha,\alpha)].$$

PROOF. Let $\mathbf{s}(y)$ be the term

$$\iota v[[y=\mathbf{0} \wedge v=\mathbf{s}_0]$$
$$\vee \exists\beta \in \mathrm{dom}(y)\,[\mathrm{dom}(y)=\beta+1 \wedge v=\mathbf{s}_1(y\text{‘}\beta,\beta)]$$
$$\vee [\mathrm{Lim}(\mathrm{dom}(y)) \wedge (v=\mathbf{s}_2(y,\mathrm{dom}(y))]$$
$$\vee [\neg \mathrm{Ord}(\mathrm{dom}(y)) \wedge v=\mathbf{0}]].$$

Then by Thm. 2.9 we can construct a term $\mathbf{t}$ such that $\vdash \mathbf{t}(\alpha)=\mathbf{s}(\mathbf{t}|\alpha)$, and it is an easy task to verify that $\mathbf{t}$ satisfies the required conditions. ∎

Observe that if, in applying Thm. 2.11, one is interested only in $\mathbf{t}(n)$, i.e. when the argument is a natural number, then $\mathbf{s}_2$ is irrelevant and may be taken to be any term whatsoever.

The form of ordinal recursion given in Thm. 2.11 is frequently used to define addition and multiplication of ordinals. For example, to define $\gamma+\alpha$, we take γ as a parameter and use recursion on α, taking $\mathbf{s}_0$ to be γ, $\mathbf{s}_1(y,z)$ to be $y+1$ and $\mathbf{s}_2(y,z)$ to be $\bigcup \mathrm{ran}(y)$. We leave it to the reader to define $\gamma \cdot \alpha$.

We now consider a more general form of ordinal recursion. Let us call a term $\mathbf{r}(y)$ a *bounded ordinal term* if

$$\vdash \forall y\, \mathrm{Ord}(\mathbf{r}(y)) \wedge \forall\alpha\, \exists z\, \forall y[\mathbf{r}(y) \leqslant \alpha \rightarrow y \in z].$$

This condition means that $\mathbf{r}(y)$ is always an ordinal and for each α the class of sets y with $\mathbf{r}(y) \leqslant \alpha$ is the extension of a set.

2.12. THEOREM. *Given a bounded ordinal term* $\mathbf{r}(y)$, *and a term* $\mathbf{s}(z_1,z_2)$, *we can construct a term* $\mathbf{t}(x)$ *such that*

$$\vdash \mathbf{t}(x)=\mathbf{s}(\mathbf{t}|\{y:\mathbf{r}(y)<\mathbf{r}(x)\},x).$$

PROOF. By ordinal recursion we can obtain a term $\mathbf{u}(x)$ such that

(1) $$\vdash \mathbf{u}(\alpha)=\{\langle x,\mathbf{s}(\bigcup\{\mathbf{u}(\beta):\ \beta<\alpha\},x)\rangle : \mathbf{r}(x)=\alpha\}.$$

(We leave it to the reader to verify that this recursion can be reduced to that given in Thm. 2.9.) It is then clear that, for each α, $\mathbf{u}(\alpha)$ is a function with domain $\{x:\ \mathbf{r}(x)=\alpha\}$. If we put

$$\mathbf{v}(\alpha)=\bigcup\{\mathbf{u}(\beta):\beta<\alpha\},$$

then $\mathbf{v}(\alpha)$ is a function with domain $\{x:\ \mathbf{r}(x)<\alpha\}$. Now define $\mathbf{t}(x)$ by

(2) $$\mathbf{t}(x)=\mathbf{u}(\mathbf{r}(x))\text{‘}x.$$

If $\mathbf{r}(y)<\alpha$, then we have, by the definition of $\mathbf{t}$ and $\mathbf{v}$,

$$\mathbf{t}(y)=\mathbf{u}(\mathbf{r}(y))\text{`}y=\mathbf{v}(\alpha)\text{`}y.$$

This implies, again using the definition of $\mathbf{v}$,

(3) $\qquad \vdash \mathbf{t}|\{y:\ \mathbf{r}(y)<\alpha\}=\bigcup\{\mathbf{u}(\beta):\ \beta<\alpha\}.$

(1), (2) and (3) now give

$$\begin{aligned}\vdash \mathbf{t}(x)&=\mathbf{u}(\mathbf{r}(x)),x\\&=\mathbf{s}(\bigcup\{\mathbf{u}(\beta):\ \beta<\mathbf{r}(x)\},x)\\&=\mathbf{s}(\mathbf{t}|\{y:\ \mathbf{r}(y)<\mathbf{r}(x)\},x)\end{aligned}$$

as required. ∎

The term $\mathbf{t}(x)$ obtained in Thm. 2.12 is said to be *constructed by recursion on* $\mathbf{r}(x)$.

§ 3. The Axiom of Regularity

If we apply Thm. 2.9 to the term

$$\mathbf{s}(y)=\bigcup\{\mathsf{P}x:\ x\in \mathrm{ran}(y)\},$$

we obtain a term $\mathbf{t}$ such that

$$\vdash \mathbf{t}(\alpha)=\bigcup\{\mathsf{P}\mathbf{t}(\xi)\ :\ \xi<\alpha\}.$$

Writing R_α for $\mathbf{t}(\alpha)$, we have

(3.1) $\qquad \vdash R_\alpha=\bigcup\{\mathsf{P}R_\xi\ :\ \xi<\alpha\}.$

It is then clear that

(3.2) $\qquad \vdash R_\alpha=\{x:\ \exists\xi<\alpha(x\subseteq R_\xi)\}.$

The intuitive idea behind the construction of the sets R_α is as follows. We think of the ordinal α as the αth "stage" in the process of "collecting" sets. R_α is then the set of all sets "collected" at "stage" α. We see from (3.2) that R_α is in fact the set of all sets x all of whose *members* have been "collected" at some "stage" *before* α. The family of all R_α is called the *cumulative hierarchy*.

We now define

$$x \textit{ is regular}:\ \mathrm{Reg}(x) \leftrightarrow_{\mathrm{df}} \exists\alpha(x\in R_\alpha).$$

It is easy to verify that $\vdash \mathrm{Reg}(x) \leftrightarrow \exists\alpha(x \subseteq R_\alpha)$. We also define

$$\textit{rank of } x\text{:}\ \varrho(x) =_{\mathrm{df}} \mu\alpha[x \subseteq R_\alpha].$$

3.3. THEOREM.

(i) $\vdash [\alpha \leqslant \beta \to R_\alpha \subseteq R_\beta] \wedge [\alpha < \beta \to R_\alpha \in R_\beta]$.

(ii) $\vdash \mathrm{Trans}\,(R_\alpha)$.

(iii) $\vdash R_0 = \mathbf{0} \wedge R_{\alpha+1} = \mathsf{P} R_\alpha$.

(iv) $\vdash \mathrm{Lim}(\lambda) \to R_\lambda = \bigcup\{R_\xi\colon\ \xi < \lambda\}$.

(v) $\vdash \mathrm{Reg}(\alpha) \wedge \varrho(\alpha) = \alpha$.

(vi) $\vdash \mathrm{Reg}(x) \wedge y \in x \to \mathrm{Reg}(y) \wedge \varrho(y) < \varrho(x)$.

(vii) $\vdash \forall y \in x\, \mathrm{Reg}(y) \to \mathrm{Reg}(x)$.

PROOF. (i) follows immediately from (3.1).

(ii) If $x \in R_\alpha$, then by (3.2), $x \subseteq R_\xi$ for some $\xi < \alpha$, whence $x \subseteq R_\alpha$ by (i).

(iii) The first conjunct follows immediately from (3.1). By (3.1) we have $\vdash R_{\alpha+1} = \mathsf{P} R_\alpha \cup R_\alpha$ and, using (ii), we see that $\vdash R_\alpha \subseteq \mathsf{P} R_\alpha$. (iii) follows.

(iv) Assume that $\mathrm{Lim}(\lambda)$. By (i), we have $\bigcup\{R_\xi\colon\ \xi < \lambda\} \subseteq R_\lambda$. But

$$R_\lambda = \bigcup\{\mathsf{P} R_\xi\colon\ \xi < \lambda\} = \bigcup\{R_{\xi+1}\colon\ \xi < \lambda\}$$

by (iii), and since $\mathrm{Lim}(\lambda)$, we have $\xi < \lambda \to \xi + 1 < \lambda$ by 2.5(i). Hence

$$\bigcup\{R_{\xi+1}\colon\ \xi < \lambda\} \subseteq \bigcup\{R_\xi\colon\ \xi < \lambda\}$$

and (iv) follows.

(v) By (3.1), it suffices to show that

$$\vdash \alpha \subseteq R_\alpha \wedge \alpha \notin R_\alpha.$$

We establish the first conjunct by transfinite induction. Assume

$$\forall \xi < \alpha (\xi \subseteq R_\xi);$$

then $\forall \xi < \alpha (\xi \in R_{\xi+1})$, so that

$$\alpha \subseteq \bigcup\{R_{\xi+1}\colon\ \xi < \alpha\} = \bigcup\{\mathsf{P} R_\xi\colon\ \xi < \alpha\} = R_\alpha,$$

as required. To show that $\alpha \notin R_\alpha$, assume the contrary, and let β the least ordinal such that $\beta \in R_\beta$. Then by (3.2), $\beta \subseteq R_\xi$ for some $\xi < \beta$, and it follows that $\xi \in R_\xi$, contradicting the choice of β.

(vi) Assume

$$\mathrm{Reg}(x) \wedge y \in x.$$

Reg(y) follows immediately from (ii). We have $x \subseteq R_{\varrho(x)}$ by definition of $\varrho(x)$ and therefore $y \subseteq R_\xi$ for some $\xi < \varrho(x)$. Accordingly, $\varrho(y) \leqslant \xi < \varrho(x)$.

(vii) Assume

$$\forall y \in x \, \mathrm{Reg}(y);$$

let

$$\alpha = \bigcup \{\varrho(y) : y \in x\}.$$

By 2.5(ii), we have $\forall y \in x[\varrho(y) \leqslant \alpha]$, so that $\forall y \in x[y \subseteq R_\alpha]$. Hence $\forall y \in x[y \in R_{\alpha+1}]$ and it follows that $x \subseteq R_{\alpha+1}$. Thus Reg(x). ∎

Let $\varphi(x)$ be a formula in which the variable z is not free. We write

$$\mathbf{Trans}_{\varphi(x)} \leftrightarrow_{\mathrm{df}} \forall x \, \forall z[\varphi(x) \wedge z \in x \rightarrow \varphi(z)].$$

$\varphi(x)$ is said to be *transitive* in x if $\vdash \mathbf{Trans}_{\varphi(x)}$. (When the identity of the variable x in question is clear from the context, we shall simply write $\mathbf{Trans}_\varphi$ and say that φ is *transitive*.) If φ is the formula $x \in y$ then it is clear that

$$\vdash \mathbf{Trans}_{\varphi(x)} \leftrightarrow \mathrm{Trans}(y).$$

It follows from 3.3(vi) that Reg(x) is a *transitive formula:* this fact will be of great use to us later on.

In Thm. 2.11, put x for $\mathbf{s}_0$ and $\bigcup y$ for $\mathbf{s}_1$. We then obtain a term $\mathbf{t}$ for which

$$\vdash \mathbf{t}(0) = x \wedge \mathbf{t}(n+1) = \bigcup \mathbf{t}(n).$$

We define

$$\textit{Transitive closure of } x\text{:} \ \mathrm{TC}(x) =_{\mathrm{df}} \bigcup \{\mathbf{t}(n) : n \in \omega\}.$$

We now show that TC(x) is the *least transitive set including x.*

3.4. THEOREM.

$$\vdash x \subseteq \mathrm{TC}(x) \wedge \mathrm{Trans}(\mathrm{TC}(x)) \wedge \forall y[\mathrm{Trans}(y) \wedge x \subseteq y \rightarrow \mathrm{TC}(x) \subseteq y].$$

PROOF. Clearly $x \subseteq \mathrm{TC}(x)$. If $z \in \mathrm{TC}(x)$, then $z \in \mathbf{t}(n)$ for some $n \in \omega$. Hence

$$z \subseteq \bigcup \mathbf{t}(n) = \mathbf{t}(n+1) \subseteq \mathrm{TC}(x).$$

Trans(TC(x)) follows.

Now suppose that $\mathrm{Trans}(y) \wedge x \subseteq y$. We show by induction on n that

$$\forall n \in \omega[\mathbf{t}(n) \subseteq y],$$

from which it follows that $\mathrm{TC}(x) \subseteq y$. We have $\mathbf{t}(\mathbf{0}) = x \subseteq y$. If $\mathbf{t}(n) \subseteq y$ then, since $\mathrm{Trans}(y)$, we have $\mathbf{t}(n+1) = \bigcup \mathbf{t}(n) \subseteq y$, which completes the proof. ∎

We are now in a position to prove the important

3.5. THEOREM. $\vdash \forall x\, \mathrm{Reg}(x) \leftrightarrow \forall x[x \neq \mathbf{0} \rightarrow \exists y \in x(x \cap y = \mathbf{0})]$.
PROOF. Assume $\forall x\, \mathrm{Reg}(x)$, and let $x \neq \mathbf{0}$. Let α be the least of the ranks of members of x and let y be a member of x satisfying $\varrho(y) = \alpha$. Then $x \cap y = \mathbf{0}$, for, if $z \in x \cap y$, then by 3.3(vi), $\varrho(z) < \varrho(y) = \alpha$, contradicting the definition of α.

Conversely, suppose that $\exists x \neg \mathrm{Reg}(x)$; choose x to satisfy $\neg \mathrm{Reg}(x)$. Put

$$z = \{y \in \mathrm{TC}(x) : \ \neg \mathrm{Reg}(y)\}.$$

Then $z \neq \mathbf{0}$, for since $\neg \mathrm{Reg}(x)$ there must, by 3.3(vii), exist $y \in x$ such that $\neg \mathrm{Reg}(y)$, and it is then clear that $y \in z$. Moreover, for each $y \in z$ we have $z \cap y \neq \mathbf{0}$. For, if $y \in z$, then $\neg \mathrm{Reg}(y)$, so by 3.3(vii) there is $u \in y$ such that $\neg \mathrm{Reg}(u)$. But then $u \in y \in z$, so that $u \in \mathrm{TC}(x)$ and hence $u \in z$. It follows that $z \cap y \neq \mathbf{0}$, completing the proof. ∎

The sentence

Reg: $\quad \forall x[x \neq \mathbf{0} \rightarrow \exists y \in x(x \cap y = \mathbf{0})]$

is called the *Axiom of Regularity*. Thm. 3.5 asserts that **Reg** is equivalent to the assertion that all sets are regular, hence the name.

The Axiom of Regularity has considerable simplifying power, as we shall see, but it is by no means intuitively obvious. Indeed, it differs from the preceding postulates — with the exception of the Axiom of Extensionality — in not being an instance of the Axiom of Comprehension. Accordingly, *we shall not use it in our proofs until we have established its consistency with the other postulates,* a task which we now turn to.

Let $\varphi(x)$ be any formula. Recall that in §12 of Ch. 2 we defined the *relativization* $\boldsymbol{\alpha}^*$ of a formula $\boldsymbol{\alpha}$ to $\varphi(x)$ (with x as chosen variable). *We agree to write* $\boldsymbol{\alpha}^{(\varphi)}$ *for* $\boldsymbol{\alpha}^*$ *from now on.*

It is easy to see that if $\boldsymbol{\alpha}$ is a formula with free variables among $x_1,\dots,x_k$, and $a_1,\dots,a_k$ are sets such that $\varphi(a_i)$ holds for all i, $1 \leqslant i \leqslant k$, then $\boldsymbol{\alpha}^{(\varphi)}(a_1,\dots,a_k)$ says that $\boldsymbol{\alpha}$ holds in the class defined by $\varphi(x)$ when $\in$ is interpreted as the membeiship relation and $x_1,\dots,x_k$ are assigned the values $a_1,\dots,a_k$.

Now suppose that we are given a set of sentences $\boldsymbol{\Sigma}$, and a sentence $\boldsymbol{\tau}$. $\boldsymbol{\tau}$ is said to be *consistent relative to* $\boldsymbol{\Sigma}$ if the consistency of $\boldsymbol{\Sigma}$ implies that of $\Sigma \cup \{\tau\}$. Our next theorem gives a sufficient condition for this to be the case.

3.6. THEOREM. *Suppose that there is a formula $\varphi(x)$ with exactly one free variable x such that*:

(i) $\Sigma \vdash \exists x \varphi(x)$;

(ii) $\Sigma \vdash \sigma^{(\varphi)}$ for all σ in Σ;

(iii) $\Sigma \vdash \tau^{(\varphi)}$.

Then τ is consistent relative to Σ.

PROOF. By Cor. 2.12.4 we have

$$(1) \qquad \sigma \text{ is logically valid } \Rightarrow [\exists x \varphi(x) \rightarrow \sigma^{(\varphi)}] \text{ is logically valid,}$$

for any sentence σ. Now suppose that $\Sigma \cup \{\tau\}$ is inconsistent. Then there is a finite subset $\{\sigma_1,\ldots,\sigma_n\}$ of Σ such that the sentence $\sigma_1 \wedge \ldots \wedge \sigma_n \rightarrow \neg\tau$ is logically valid. Hence, by (1) and the properties of relativization, the sentence

$$(2) \qquad \exists x \varphi(x) \rightarrow [\sigma_1^{(\varphi)} \wedge \ldots \wedge \sigma_n^{(\varphi)} \rightarrow \neg\tau^{(\varphi)}]$$

is logically valid. But (ii) implies that $\Sigma \vdash \sigma_1^{(\varphi)} \wedge \ldots \wedge \sigma_n^{(\varphi)}$, and this, together with (i) and the logical validity of (2) gives $\Sigma \vdash \neg\tau^{(\varphi)}$. Hence, in view of (iii), Σ is inconsistent. ∎

Observe that the proof of Thm. 3.6 provides an explicit method of converting a proof of an inconsistency from $\Sigma \cup \{\tau\}$ into a proof of an inconsistency from Σ.

We can now prove:

3.7. THEOREM. **Reg** *is consistent relative to the previous postulates.*

PROOF. We apply Thm. 3.6, with Reg(x) as the formula $\varphi(x)$. Naturally, **Reg** will not be used as a postulate in any deduction we make in the present proof.

First, since $\vdash \text{Reg}(\mathbf{0})$ by 3.3(v), we have $\vdash \exists x \text{Reg}(x)$.

Next, we deal with each postulate in order:

(i) *Extensionality.* We have to show that

$$\vdash \text{Reg}(x) \wedge \text{Reg}(y) \wedge \forall z[\text{Reg}(z) \rightarrow [z \in x \leftrightarrow z \in y]] \rightarrow x = y.$$

Assume the antecendent. If $z \in x$, then Reg(z) by the transitivity of Reg; so $z \in y$. Similarly, $z \in y$ implies $z \in x$; hence $x = y$.

(ii) *Replacement.* Let φ be a formula with free variables among

$x, y, y_1, \ldots, y_k$. We have to show that

(1) $$\vdash \mathrm{Reg}(y_1) \wedge \ldots \wedge \mathrm{Reg}(y_k) \wedge \mathrm{Reg}(u) \wedge \forall x \in u[\mathrm{Reg}(x) \rightarrow \exists^* y[\mathrm{Reg}(y) \wedge \psi]] \rightarrow \exists z[\mathrm{Reg}(z) \wedge \forall y[\mathrm{Reg}(y) \rightarrow [y \in z \leftrightarrow \exists x \in u[\mathrm{Reg}(x) \wedge \psi]]]],$$

where ψ is $\varphi^{(\mathrm{Reg})}$.

Assume the antecendent of (1) (i.e. the part of the formula preceding the second $\rightarrow$). Since we are assuming $\mathrm{Reg}(u)$, 3.3(vi) implies that $\forall x \in u\, \mathrm{Reg}(x)$, so we have

$$\forall x \in u\, \exists^* y[\mathrm{Reg}(y) \wedge \psi].$$

Hence, by **Rep** we deduce that there is a z such that

$$\forall y[y \in z \leftrightarrow \exists x \in u[\mathrm{Reg}(y) \wedge \psi]].$$

Using 3.3(vii) we infer from this that

(2) $\mathrm{Reg}(z)$.

Also, we see that

(3) $\forall y[\mathrm{Reg}(y) \rightarrow [y \in z \leftrightarrow \exists x \in u \psi]]$.

But since we have $\mathrm{Reg}(u)$, the transitivity of Reg gives

$$\exists x \in u \psi \leftrightarrow \exists x \in u[\mathrm{Reg}(x) \wedge \psi].$$

Thus (3) is equivalent to

(4) $\forall y[\mathrm{Reg}(y) \rightarrow [y \in z \leftrightarrow \exists x \in u[\mathrm{Reg}(x) \wedge \psi]]]$.

(2) and (4) imply the consequent of (1).

(iii) *Union*. We have to show that

(5) $\vdash \mathrm{Reg}(z) \rightarrow \exists x[\mathrm{Reg}(x) \wedge \forall y[\mathrm{Reg}(y) \rightarrow [y \in x \leftrightarrow \exists u \in z[\mathrm{Reg}(u) \wedge y \in u]]]]$.

Assume $\mathrm{Reg}(z)$, and put $x = \bigcup z$. Then we have $\mathrm{Reg}(x)$ by the transitivity of Reg and 3.3(vii). Moreover, we have, by definition of x,

$$y \in x \leftrightarrow \exists u \in z[y \in u],$$

and hence, since Reg is transitive,

$$y \in x \leftrightarrow \exists u \in z[\mathrm{Reg}(u) \wedge y \in u].$$

(5) follows.

(iv) *Power set.* We have to show that

(6) $\vdash \mathrm{Reg}(z) \to \exists x[\mathrm{Reg}(x) \wedge \forall y[\mathrm{Reg}(y) \to [y \in x \leftrightarrow \forall u[\mathrm{Reg}(u) \wedge u \in y \to u \in z]]]]$.

Assume Reg(z), and put $x = \mathsf{P}z$. Then, applying 3.3(vii) twice, we see that Reg(x).

Also, by definition of x, we have

$$y \in x \leftrightarrow \forall u[u \in y \to u \in z],$$

and hence, since Reg is transitive,

$$\mathrm{Reg}(y) \to [y \in x \leftrightarrow \forall u[\mathrm{Reg}(u) \wedge u \in y \to u \in z]].$$

(6) follows.

(v) *Infinity.* Put $\varphi(x)$ for the formula

$$\exists y[\forall z[z \notin y] \wedge y \in x]$$
$$\wedge \forall y[y \in x \to \exists u[\forall v[v \in u \leftrightarrow v \in y \vee v = y] \wedge u \in x]].$$

Then the Axiom of Infinity is equivalent to the sentence $\exists x \varphi(x)$. Thus we have to prove the existence of a *regular* set x with the following two properties:

(7) $$\exists y[\mathrm{Reg}(y) \wedge \forall z[\mathrm{Reg}(z) \to z \notin y] \wedge y \in x],$$

(8) $$\forall y[\mathrm{Reg}(y) \to y \in x$$
$$\to \exists u[\mathrm{Reg}(u) \wedge \forall v[\mathrm{Reg}(v) \to [v \in u \leftrightarrow v \in y \vee v = y]] \wedge u \in x]].$$

We claim that ω is such an x. First observe that $\vdash \mathrm{Reg}(\omega)$ by 3.3(v). Moreover, taking y to be $\mathbf{0}$ we see that this y is regular, has no members — let alone regular members — and belongs to ω. Thus ω has property (7).

Also, by the definition of ω we have

$$\forall y \in \omega\, \exists u \in \omega\, \forall v[v \in u \leftrightarrow v \in y \vee v = y];$$

using the transitivity of Reg, it follows that ω satisfies (8).

To complete the proof of the theorem we have to show that $\vdash \mathbf{Reg}^{(\mathrm{Reg})}$, i.e.

(9) $$\vdash \mathrm{Reg}(x) \wedge x \neq \mathbf{0} \to \exists y \in x[\mathrm{Reg}(y) \wedge \forall z[\mathrm{Reg}(z) \wedge z \in y \to z \notin x]].$$

Suppose that Reg(x) and $x \neq \mathbf{0}$; let α be the least of the ranks of all members of x, and let y be a member of x with rank α. Then Reg(y) by the transitivity of Reg. Moreover, if $z \in y$ then $\varrho(z) < \alpha$ by 3.3(vi), so that $z \notin x$ by the definition of α. This gives (9), and the proof is complete. ∎

Now that we have established the relative consistency of **Reg**, *we hereby adopt it as a postulate*[1] *and use it in our proofs.* The next three problems should be approached with this fact in mind.

3.8. PROBLEM. Show that:

(i) $\vdash x \notin x$.

(ii) $\vdash \neg \exists x_1 \ldots \exists x_n [x_1 \in x_2 \wedge x_2 \in x_3 \wedge \ldots \wedge x_{n-1} \in x_n \wedge x_n \in x_1]$.

(iii) $\vdash \neg \exists f [\mathrm{Fun}(f) \wedge \mathrm{dom}(f) = \omega \wedge \forall n [f`(n+1) \in f`n]]$.

(iv) $\vdash \mathrm{Ord}(x) \leftrightarrow \forall y \in x\, \forall z \in x [y = z \vee y \in z \vee z \in y] \wedge \mathrm{Trans}(x)$.

(v) $\vdash \varrho(x) = \alpha \leftrightarrow x \subseteq R_\alpha \wedge x \notin R_\alpha$.

3.9. PROBLEM. Let $\psi(x)$ be any formula. Show that:

(i) $\vdash \mathbf{Reg}^{(\psi)}$.

(ii) $\vdash \mathbf{Trans}_\psi \rightarrow \forall x [\psi(x) \rightarrow [\mathrm{Ord}(x) \leftrightarrow \mathrm{Ord}^{(\psi)}(x)]]$.

(iii) $\vdash \mathbf{Trans}_\psi \rightarrow [[\exists \alpha \varphi(\alpha)]^{(\psi)} \leftrightarrow \exists \alpha [\psi(\alpha) \wedge \varphi^{(\psi)}(\alpha)]]$
$\wedge [[\forall \alpha \varphi(\alpha)]^{(\psi)} \leftrightarrow \forall \alpha [\psi(\alpha) \rightarrow \varphi^{(\psi)}(\alpha)]]$,

where $\varphi(y)$ is any formula.

(For (ii), show that the equivalence stated in Prob. 3.8(iv) follows from **Reg** alone; then use (i).)

3.10. PROBLEM. Prove the *principle of induction on rank*: if y does not occur free in $\varphi(x)$, then

$$\vdash \forall x [\forall y [\varrho(y) < \varrho(x) \rightarrow \varphi(y)] \rightarrow \varphi(x)] \rightarrow \forall x \varphi(x).$$

From Thm. 3.5 and the Axiom of Regularity it follows immediately that $\vdash \forall x\, \mathrm{Reg}(x)$, i.e. every set is regular. It is useful to envisage the universe of (regular) sets as a striated cone (see Fig. 6). The root of the cone represents the empty set, the subcones — OAB for example — bounded by the horizontal lines represent the R_α's, and the vertical "spine" represents the class of ordinals.

The rest of this section is devoted to proving some technical results which will be of importance later on.

For any formula φ and any variable z, we define $\varphi^{(z)}$ to be the relativization of φ to the formula $x \in z$, with x as chosen variable (Ch. 2, §12). Note that if the free variables of φ are among $x_1, \ldots, x_n$, then the free variables of $\varphi^{(z)}$ are among $x_1, \ldots, x_n, z$. If $\mathbf{t}$ is a term, we write $\varphi^{(\mathbf{t})}$ for the result of substituting $\mathbf{t}$ for z in $\varphi^{(z)}$.

[1] Our previous remarks imply that in adopting **Reg** we are merely confining our attention to *regular* sets. Experience shows that — so far at least — nothing of mathematical interest is lost by this restriction.

We define:

f is an injection:

$$\mathrm{Inj}(f) \leftrightarrow_{\mathrm{df}} \mathrm{Fun}(f) \wedge \forall x \in \mathrm{dom}(f)\ \forall y \in \mathrm{dom}(f)[f`x = f`y \rightarrow x = y].$$

f is an $\in$*-isomorphism of u onto v*:

$$\mathrm{Isom}(f,u,v) \leftrightarrow_{\mathrm{df}} \mathrm{Inj}(f) \wedge \mathrm{dom}(f) = u \wedge \mathrm{ran}(f) = v$$
$$\wedge \forall x \in u\ \forall y \in u[x \in y \leftrightarrow f`x \in f`y].$$

u is extensional:

$$\mathrm{Ex}(u) \leftrightarrow_{\mathrm{df}} \forall x \in u\ \forall y \in u[x \cap u = y \cap u \rightarrow x = y].$$

It is easy to verify that

$$\vdash \mathrm{Ex}(u) \leftrightarrow \mathbf{Ext}^{(u)},$$

$$\vdash \mathrm{Trans}(u) \rightarrow \mathrm{Ex}(u).$$

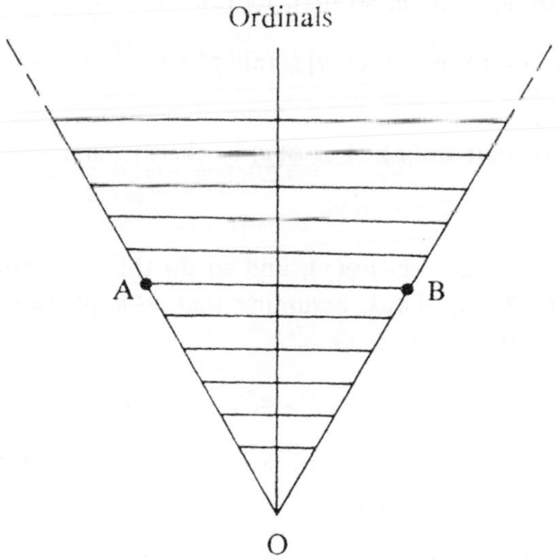

Fig. 6

Our next result asserts that there is an $\in$-isomorphism of any given *extensional* set onto a *transitive* set.

3.11. THEOREM (MOSTOWSKI'S COLLAPSING LEMMA).

$$\vdash \mathrm{Ex}(u) \rightarrow \exists v\ \exists f[\mathrm{Trans}(v) \wedge \mathrm{Isom}(f,u,v)].$$

PROOF. By 3.8(v), $\varrho(y)$ is a bounded ordinal term and so we may apply Thm. 2.12, taking $\mathbf{s}$ to be the term $\operatorname{ran}(z_1 \upharpoonright (u \cap z_2))$. We then obtain a term $\mathbf{t}(x, u)$ such that

(1) $\qquad \vdash \mathbf{t}(x,u) = \{\mathbf{t}(y,u) : y \in x \cap u\}.$

We put

$$f = \{\langle x, \mathbf{t}(x,u)\rangle : x \in u\}, \qquad v = \{\mathbf{t}(x,u) : x \in u\}.$$

Clearly we have

$$\vdash \operatorname{Fun}(f) \wedge \operatorname{dom}(f) = u \wedge \operatorname{ran}(f) = v.$$

Moreover, it follows immediately from (1) that

(2) $\qquad \vdash \forall x \in u[f`x = \{f`y : y \in x \cap u\}].$

Now suppose that Ex(u). We claim that under these conditions we have Trans(v) and Isom(f,u,v). To prove the first assertion, suppose that $z \in v$. Then $z = f`x$ for some $x \in u$, so that, by (2),

$$z = f`x = \{f`y : y \in x \cap u\} \subseteq \operatorname{ran}(f) = v.$$

Hence Trans(v).

We next show that Inj(f). Let $\varphi(y)$ be the formula

(3) $\qquad \forall x \in u[f`x = f`y \rightarrow x = y].$

We have to show that $\forall y \in u \varphi(y)$, and to do this we argue by induction on rank (Prob. 3.10). Thus, assuming that $y \in u$, $\varrho(y) = \alpha$ and $\varphi(z)$ holds for all $z \in u$ such that $\varrho(z) < \alpha$, i.e.

(4) $\qquad \forall z \in u[\varrho(z) < \alpha \rightarrow \forall x \in u[f`x = f`z \rightarrow x = z]],$

we have to show that $\varphi(y)$, i.e. (3). Since we are assuming Ex(u), in order to prove (3) it suffices to prove

(5) $\qquad \forall x \in u[f`x = f`y \rightarrow x \cap u = y \cap u].$

Let $x \in u$ and suppose that $f`x = f`y$. If $z \in x \cap u$, then since $f`x = f`y$ it follows from (2) that $f`z = f`w$ for some $w \in y \cap u$. But then $\varrho(w) < \alpha$ by 3.3(vi) and so $z = w$ by (4), whence $z \in y \cap u$. On the other hand, if $w \in y \cap u$, then, as before, (2) implies that $f`w = f`z$ for some $z \in x \cap u$. Again we have $\varrho(w) < \alpha$ by 3.3(vi) and $w = z$ by (4), so that $w \in x \cap u$. It follows that $x \cap u = y \cap u$, which proves (5).

It remains to show that

$$\forall x \in u\, \forall y \in u[x \in y \leftrightarrow f`x \in f`y].$$

Suppose that $x \in u$ and $y \in u$. If $x \in y$, then $f`x \in f`y$ by (2). Conversely, if $f`x \in f`y$, then, by (2), $f`x = f`z$ for some $z \in y$. Since Inj(f), we have $x=z$, so that $x \in y$. ∎

Notice that in the proof of Thm. 3.11 we have actually shown that the $\in$-isomorphism f can be uniformly defined from u. More precisely, we have shown that we can construct a term $\mathbf{t}(x,u)$ such that

$$\vdash \mathrm{Ex}(u) \to \mathrm{Trans}(\mathbf{t}[u]) \wedge \mathrm{Isom}(\mathbf{t}|u,\, u,\, \mathbf{t}[u]),$$

where

$$\mathbf{t}[u] = \{\mathbf{t}(x,u) : x \in u\}, \qquad \mathbf{t}|u = \{\langle x, \mathbf{t}(x,u)\rangle : x \in u\}.$$

REMARK. Let u be an extensional set which is not transitive. For each $x \in u$, $x \cap u$ is in general a *proper* subset of x, because if $x = x \cap u$ for *all* $x \in u$, then obviously u would be transitive. The set $x - x \cap u$ is just so much "empty space" as far as u is concerned. Now, if f is an $\in$-isomorphism of u onto a transitive set v, then $f`x = f`x \cap v$ for each $x \in u$, and thus we may say that $f`x$ is "densely packed" with respect to $v$. Thus the effect of $f$ is to "collapse" each $x \in u$ onto the "densely packed" set $f`x$. For this reason f is often called a *collapsing isomorphism*, and v the *transitive collapse* of u.

3.12. PROBLEM. Show that

$$\vdash \mathrm{Isom}(f,u,v) \wedge \mathrm{Trans}(v) \to \forall x \in u[f`x = \{f`y : y \in x \cap u\}].$$

3.13. PROBLEM. Show that the $\in$-isomorphism f and the transitive set v whose existence is established in Thm. 3.11 are *unique*. (Argue by induction on rank, using 3.12.)

3.14. PROBLEM. Show that

$$\vdash \mathrm{Ex}(u) \wedge \mathrm{Trans}(x) \wedge x \subseteq u \wedge \mathrm{Isom}(f,u,v) \wedge \mathrm{Trans}(v) \to \forall y \in x[f`y = y].$$

§ 4. Cardinality and the Axiom of Choice

In this section we presuppose a slender acquaintance with cardinal arithmetic (see, e.g. HALMOS [1960] or ROTMAN and KNEEBONE [1961]).

We define

$$x \textit{ is equipollent with } y : x \approx y \leftrightarrow_{\mathrm{df}} \exists f[\mathrm{Inj}(f) \wedge \mathrm{dom}(f) = x \wedge \mathrm{ran}(f) = y].$$

4.1. THEOREM. $\vdash x \approx x \wedge [x \approx y \to y \approx x] \wedge [x \approx y \wedge y \approx z \to x \approx z]$. ∎

4.2. THEOREM.

(i) $\vdash \exists y \subseteq \mathsf{P}x(x \approx y) \wedge \neg \exists z \subseteq x(z \approx \mathsf{P}x)$.

(ii) $\vdash x \approx y' \wedge y' \subseteq y \wedge y \approx x' \wedge x' \subseteq x \rightarrow x \approx y$. ∎

4.2(i) and (ii) are the well-known theorems of *Cantor* and *Schröder–Bernstein* respectively.

We now introduce our last postulate – the *Axiom of Choice*

AC: $\forall x \exists f[\mathrm{Fun}(f) \wedge \mathrm{dom}(f) = x \wedge \forall y \in x[y \neq \mathbf{0} \rightarrow f`y \in y]]$.

A function f whose domain is x and such that, whenever y is a non-empty member of x, f "chooses" a member of y (i.e. $f`y \in y$) is called a *choice function for x*. The axiom of choice postulates that each set has a choice function. Most mathematicians accept the axiom of choice as intuitively true — which is just as well, since it plays a well-nigh indispensible role in modern mathematics. Some, however, distrust it because of its highly non-constructive character.[1] It asserts the existence of an object — a choice function — without indicating how this object is to be constructed or, indeed, characterizing it in any other way. It is therefore a pure existence statement. For this reason it is customary not to use it without giving notice; in any case we want to discuss its consistency relative to the other postulates. Accordingly, we shall depart from our usual convention and write "**AC**" below the deducibility symbol $\vdash$ whenever our proof of the formula in question depends on the Axiom of Choice. Similarly, when introducing a defined term or formula whose meaning depends on the Axiom of Choice, we shall write (**AC**) in the right-hand margin.

One of the most important applications of the Axiom of Choice is in the proof of the following theorem, which asserts that each set is equipollent with an ordinal (i.e., can be well-ordered).

4.3. THEOREM. $\vdash_{\mathbf{AC}} \forall x \exists \alpha[\alpha \approx x]$.

PROOF. We apply Thm. 2.9 with $f`(x - \mathrm{ran}(y))$ as $\mathbf{s}(y,z)$. We then get a term, $\mathbf{t}$ say, such that, if f is a choice function for $\mathsf{P}x$, then

(1) $$\forall \alpha[\mathbf{t}(\alpha) = f`(x - \{\mathbf{t}(\beta) : \beta < \alpha\})].$$

Assuming **AC**, such a choice function f exists, and for it we have (1).

Put

$$u_\alpha = x - \{\mathbf{t}(\beta) : \beta < \alpha\};$$

[1] It is obviously not an instance of the Axiom of Comprehension!

then $\mathbf{t}(\alpha)=f`u_\alpha$. Clearly, we have $u_\alpha \subseteq x$. If $u_\alpha \neq \mathbf{0}$, then $\mathbf{t}(\alpha)=f`u_\alpha \in u_\alpha$, so $\mathbf{t}(\alpha) \in x$; also, it follows from the definition of u_α that if $\beta<\alpha$ then $\mathbf{t}(\beta) \notin u_\alpha$, so that $\mathbf{t}(\alpha) \neq \mathbf{t}(\beta)$.

Suppose now that $u_\alpha \neq \mathbf{0}$ for *all* ordinals α. Then by what we have just seen, if α is any ordinal and $y=\mathbf{t}(\alpha)$, then $y \in x$ and

$$\alpha = \iota\beta[y=\mathbf{t}(\beta)].$$

Thus the set $\{\iota\beta[y=\mathbf{t}(\beta)] : y \in x\}$ includes all the ordinals, in contradiction with Thm. 2.3. We conclude that $u_\alpha=\mathbf{0}$ for some ordinal α. Then, if α is the *least* such ordinal and $\beta<\alpha$, we must have $u_\beta \neq \mathbf{0}$, so that $\mathbf{t}(\beta) \neq \mathbf{t}(\gamma)$ for all $\gamma<\beta$. Also, since $u_\alpha=\mathbf{0}$, we have

$$x=\{\mathbf{t}(\beta) : \beta<\alpha\},$$

so $\mathbf{t}|\alpha$ is a one–one map of α onto x. ∎

We now define

$$\textit{cardinality of } x\text{: } |x| =_{\mathrm{df}} \mu\alpha[\alpha \approx x].$$

Thus $|x|$ is the least ordinal which is equipollent with x, provided such an ordinal exists. Notice that (in the absence of **AC**), a set which is not equipollent with an ordinal has cardinality $\mathbf{0}$.

The following results are immediate consequences of 4.1, 4.2 and 4.3:

(4.4) $\quad \vdash_{\mathbf{AC}} x \approx y \leftrightarrow |x|=|y|$,

(4.5) $\quad \vdash |\alpha| \leqslant \alpha$,

(4.6) $\quad \vdash_{\mathbf{AC}} |x| < |\mathsf{P}x|$.

We define

$$x \textit{ is a cardinal}\text{: } \mathrm{Card}(x) \leftrightarrow_{\mathrm{df}} \mathrm{Ord}(x) \wedge x=|x|.$$

It follows from (4.5) that a cardinal is an ordinal which is not equipollent with any smaller ordinal. It is also clear that we have

(4.7) $\quad \vdash \mathrm{Card}(n) \wedge \mathrm{Card}(\omega)$.

Next, we prove:

4.8. THEOREM. $\vdash_{\mathbf{AC}} \forall y \in x\, \mathrm{Card}(y) \rightarrow \exists\alpha[\mathrm{Card}(\alpha) \wedge \forall\beta \in x[\beta<\alpha]]$.

PROOF. Assume $\forall y \in x\, \mathrm{Card}(y)$. Then in particular $\forall y \in x\, \mathrm{Ord}(y)$ and so, by 2.2.(x),

$$\forall\beta \in x[\beta \leqslant \bigcup x].$$

Putting

$$\alpha = |\mathsf{P}\bigcup x|,$$

we have Card(α) and it follows from 4.6 that $\forall\beta\in x[\beta<\alpha]$. ∎

Thm. 4.8 asserts that, assuming **AC**, for any set x of cardinals there is a cardinal α which exceeds all the members of x. Moreover, we may assume that $\omega\leqslant\alpha$, for if $\alpha<\omega$, then we may replace α by ω. We now apply 2.9, taking for $\mathbf{s}(y,z)$ the term

$$\boldsymbol{\mu}\beta[\mathrm{Card}(\beta)\wedge\omega\leqslant\beta\wedge\beta\notin\mathrm{ran}(y)].$$

We obtain a term $\mathbf{t}(x)$ such that

$$\vdash_{\mathbf{AC}}\mathbf{t}(\alpha)=\boldsymbol{\mu}\beta[\mathrm{Card}(\beta)\wedge\omega\leqslant\beta\wedge\beta\notin\{\mathbf{t}(\gamma):\ \gamma<\alpha\}].$$

We define

Aleph α : $\aleph_\alpha =_{\mathrm{df}} \mathbf{t}(\alpha)$. **(AC)**

4.9. THEOREM.

(i) $\vdash_{\mathbf{AC}}\mathrm{Card}(\aleph_\alpha)\wedge\omega\leqslant\aleph_\alpha\wedge[\alpha<\beta\rightarrow\aleph_\alpha<\aleph_\beta]\wedge\aleph_0=\omega$.

(ii) $\vdash_{\mathbf{AC}}\mathrm{Card}(\beta)\wedge\omega\leqslant\beta\rightarrow\exists\alpha[\beta=\aleph_\alpha]$.

PROOF. (i) is a simple consequence of the definition of $\aleph_\alpha$, and we leave its proof to the reader.

(ii) It follows from (i) that $\aleph_\alpha\neq\aleph_{\alpha'}$ whenever $\alpha\neq\alpha'$, and a straightforward application of **Rep** and Thm. 2.3 shows that there is no set which contains all the $\aleph_\alpha$. In particular, if β is a cardinal such that $\omega\leqslant\beta$, then there must be an ordinal α such that $\aleph_\alpha\notin\beta$. Thus $\beta\leqslant\aleph_\alpha$. If $\beta=\aleph_\alpha$, we are through. If on the other hand $\beta<\aleph_\alpha$, then since $\aleph_\alpha$ is the least cardinal $\geqslant\omega$ not in $\{\aleph_\gamma:\ \gamma<\alpha\}$, we must have $\beta\in\{\aleph_\gamma:\ \gamma<\alpha\}$, which gives the required result. ∎

Calling a cardinal β *infinite* if $\omega\leqslant\beta$, Thm. 4.9 implies that $\aleph_0, \aleph_1,\dots,\aleph_\alpha,\dots$ enumerates the class of all infinite cardinals. Thus, given an infinite cardinal it makes sense to ask precisely *where* it appears in the sequence of $\aleph_\alpha$'s. Consider, for example, the cardinal $|\mathsf{P}\aleph_0|$. We have $\aleph_0<|\mathsf{P}\aleph_0|$, so that $\aleph_1\leqslant|\mathsf{P}\aleph_0|$ by the definition of $\aleph_1$. Cantor was firmly convinced that $|\mathsf{P}\aleph_0|$ is actually *equal* to $\aleph_1$, but never succeeded in proving it. The statement

$$|\mathsf{P}\aleph_0|=\aleph_1$$

is called the *Continuum Hypothesis* (briefly, **CH**) because, as is well-known, $\mathsf{P}\aleph_0$ is equipollent with the continuum, i.e. the set of all real numbers. Its truth or falsity is still an open problem. However, in 1938 Gödel showed that both it and the axiom of choice are consistent relative to the other postulates of set theory. In fact, he demonstrated the relative consistency of a much stronger assertion than the Continuum Hypothesis, namely the so-called *Generalized Continuum Hypothesis* (briefly, **GCH**). This is the statement

$$\forall\alpha[|P\aleph_\alpha|=\aleph_{\alpha+1}].$$

We are eventually going to prove Gödel's result.

Now Gödel's pioneering work still left open the possibility that **CH** or even **GCH** is actually a consequence of the other postulates. However, in 1963 P. J. Cohen showed that **CH** (and hence **GCH**) as well as **AC** are not provable from the other postulates provided these are themselves consistent. Thus **CH** and **AC** are *completely independent* of the other postulates. The proof of Cohen's result is, unfortunately, beyond the scope of this book. Readers interested in finding out more about Cohen's work are advised to consult COHEN [1966], BELL [1977] or JECH [1971].

§ 5. Reflection Principles

Let $\varphi_1,\ldots,\varphi_n$ be a sequence of formulas, and let $x_1,\ldots,x_m$ be a list – in alphabetical order – of all variables that occur free in any φ_i, $1\leqslant i\leqslant n$. We define

u reflects $\varphi_1,\ldots,\varphi_n$:

$$\mathbf{Refl}_{\varphi_1,\ldots,\varphi_n}(u) \leftrightarrow_{\mathrm{df}} \forall x_1\in u\ldots\forall x_m\in u[[\varphi_1\leftrightarrow\varphi_1^{(u)}]\wedge\ldots\wedge[\varphi_n\leftrightarrow\varphi_n^{(u)}]].$$

In this section we prove several results which assert that there are arbitrarily large sets which reflect all the members of a given finite sequence of formulas. An assertion of this kind is called a *reflection principle.* We now formulate two such principles.

The *First Reflection Principle* is the scheme

RP$_1$: $\quad\forall y_1\ldots\forall y_k\,\exists u[y_1\in u\wedge\ldots\wedge y_k\in u\wedge \mathrm{Trans}(u)\wedge\mathbf{Refl}_{\varphi_1,\ldots,\varphi_n}(u)]$,

where $\varphi_1,\ldots,\varphi_n$ are any formulas.

The *Second Reflection Principle* is the scheme

RP$_2$: $\quad\forall\alpha\,\exists\beta[\alpha<\beta\wedge\mathbf{Refl}_{\varphi_1,\ldots,\varphi_n}(R_\beta)]$,

where $\varphi_1,\ldots,\varphi_n$ are any formulas.

REMARK. Taking a single sentence σ in **RP**$_2$, we get

$$\forall\alpha\,\exists\beta[\alpha<\beta\wedge\sigma\leftrightarrow\sigma^{(R_\beta)}].$$

We may think of $\boldsymbol{\sigma}$ as expressing a (first-order) property of the universe of sets and $\boldsymbol{\sigma}^{(R_\beta)}$ as expressing the same property of the extension of $\mathbf{R}_\beta$. Accordingly, $\mathbf{RP}_2$ implies that *each first-order property possessed by the universe is also possessed by the extensions of arbitrarily large sets.* In particular, *there is no property expressible in the language of set theory which distinguishes the universe from the extensions of all of its members.*

It is easy to see that, for each choice of $\varphi_1,\dots,\varphi_n$,

(5.1) $\vdash \mathbf{RP}_2 \to \mathbf{RP}_1$.

We shall prove a theorem which yields $\mathbf{RP}_2$ as a particular case. We shall need:

5.2. Lemma. *Let* $\mathbf{t}(x)$ *be a term such that*:
(i) $\vdash \alpha \leqslant \beta \to \mathbf{t}(\alpha) \subseteq \mathbf{t}(\beta)$;
(ii) $\vdash \mathrm{Lim}(\lambda) \to \mathbf{t}(\lambda) = \bigcup\{\mathbf{t}(\xi)\colon \xi < \lambda\}$.
Then

$$\vdash \forall y \in x\, \mathrm{Ord}(y) \wedge x \neq \mathbf{0} \to \mathbf{t}(\bigcup x) = \bigcup\{\mathbf{t}(\alpha)\colon \alpha \in x\}.$$

Proof. Assume (i), (ii) and $\forall y \in x\, \mathrm{Ord}(y) \wedge x \neq \mathbf{0}$. That

$$\bigcup\{\mathbf{t}(\alpha)\colon \alpha \in x\} \subseteq \mathbf{t}(\bigcup x)$$

follows immediately from (i). If x has a greatest element, β say, then $\bigcup x = \beta$, and we have

$$\mathbf{t}(\bigcup x) = \mathbf{t}(\beta) = \bigcup\{\mathbf{t}(\alpha)\colon \alpha \in x\}.$$

On the other hand, if x has no greatest element, then it is not difficult to see that $\bigcup x$ is a limit ordinal. Therefore, by (ii),

$$\mathbf{t}(\bigcup x) = \bigcup\{\mathbf{t}(\alpha)\colon \alpha \in \bigcup x\}$$
$$\subseteq \bigcup\{\mathbf{t}(\alpha)\colon \alpha \in x\} \quad \text{by (i).} \qquad \blacksquare$$

5.3. Theorem. *Let* $\mathbf{t}(x)$ *be a term such that* $x_1,\dots,x_m$ *are not free in* $\mathbf{t}(x)$ *and*
(i) $\vdash \alpha \leqslant \beta \to \mathbf{t}(\alpha) \subseteq \mathbf{t}(\beta)$;
(ii) $\vdash \mathrm{Lim}(\lambda) \to \mathbf{t}(\lambda) = \bigcup\{\mathbf{t}(\xi)\colon \xi < \lambda\}$.
Put $T(x)$ *for the formula* $\exists \alpha[x \in \mathbf{t}(\alpha)]$. *Then, for any formulas* $\varphi_1,\dots,\varphi_n$ *with free variables among* $x_1,\dots,x_m$ *we have*

$$\vdash \forall \alpha\, \exists \beta[\alpha < \beta \wedge \forall x_1 \in \mathbf{t}(\beta) \dots \forall x_m \in \mathbf{t}(\beta)[[\varphi_1^{(T)} \leftrightarrow \varphi_1^{(\mathbf{t}(\beta))}] \wedge \dots \wedge [\varphi_n^{(T)} \leftrightarrow \varphi_n^{(\mathbf{t}(\beta))}]]].$$

Proof. We may clearly assume without loss of generality that each subformula of each formula in the list $\varphi_1,\dots,\varphi_n$ also occurs in the list. We also assume that the list is enumerated in such a way that the existential formulas occupy the first p places.

Let $1 \leq j \leq p$. Then φ_j is of the form

$$\exists y \psi_j(z_1,\dots,z_k,y),$$

where $z_1,\dots,z_k$ are exactly all the free variables of φ_j, and are therefore among $x_1,\dots,x_m$. We put $\mathbf{s}_j(z_1,\dots,z_k)$ for the term

$$\boldsymbol{\mu}\gamma[\exists y \in \mathbf{t}(\gamma)\psi_j^{(T)}(z_1,\dots,z_k,y)].$$

Thus if for given $z_1,\dots,z_k$ there is a y such that

$$T(y) \wedge \psi_j^{(T)}(z_1,\dots,z_k,y),$$

then $\mathbf{s}_j(z_1,\dots,z_k)$ is the least γ such that a y of this kind can be found in $\mathbf{t}(\gamma)$. We now put $\mathbf{s}_j^*(u)$ for

$$\bigcup\{\mathbf{s}_j(z_1,\dots,z_k)\colon\ z_1 \in u \wedge \dots \wedge z_k \in u\}.$$

It follows immediately from (i) that

$$\vdash \bigcup\{\mathbf{t}(\mathbf{s}_j(z_1,\dots,z_k))\colon\ z_1 \in u \wedge \dots \wedge z_k \in u\} \subseteq \mathbf{t}(\mathbf{s}_j^*(u)).$$

Thus, if $z_1,\dots,z_k$ are in u and there is some y such that $T(y) \wedge \psi_j^{(T)}(z_1,\dots,z_k,y)$, then such a y can already be found in $\mathbf{t}(\mathbf{s}_j^*(u))$. Hence

$$\vdash \forall z_1 \in u \dots \forall z_k \in u[\exists y[T(y) \wedge \psi_j^{(T)}] \leftrightarrow \exists y \in \mathbf{t}(\mathbf{s}_j^*(u))\psi_j^{(T)}],$$

and therefore, since $z_1,\dots,z_k$ are among $x_1,\dots,x_m$, we have, *a fortiori*,

$$\vdash \forall x_1 \in u \dots \forall x_m \in u[\exists y[T(y) \wedge \psi_j^{(T)}] \leftrightarrow \exists y \in \mathbf{t}(\mathbf{s}_j^*(u))\psi_j^{(T)}].$$

Next, putting $\mathbf{s}^*(u)$ for

$$\mathbf{s}_1^*(u) \cup \dots \cup \mathbf{s}_p^*(u),$$

we have, for the same reasons as before,

$$(1) \qquad \vdash \forall x_1 \in u \dots \forall x_m \in u[\exists y[T(y) \wedge \psi_j^{(T)}] \leftrightarrow \exists y \in \mathbf{t}(\mathbf{s}^*(u))\psi_j^{(T)}]$$

for $j=1,2,\dots p$.

We apply 2.11 with $\alpha+1$ as $\mathbf{s}_0$ and $\mathbf{s}^*(\mathbf{t}(y))$ as $\mathbf{s}_1(y,z)$ to obtain a term $\mathbf{r}(x)$ for which

$$\vdash \mathbf{r}(\mathbf{0}) = \alpha+1 \wedge \mathbf{r}(n+1) = \mathbf{s}^*(\mathbf{t}(\mathbf{r}(n))).$$

Put β for

$$\bigcup\{\mathbf{r}(n)\colon\ n \in \omega\}.$$

Then clearly $\alpha < \beta$; we claim that

$$(2) \qquad \forall x_1 \in \mathbf{t}(\beta) \ldots \forall x_m \in \mathbf{t}(\beta)[\varphi_j^{(T)} \leftrightarrow \varphi_j^{(\mathbf{t}(\beta))}]$$

for all $j=1,\ldots,n$. Once this claim has been proved the theorem follows immediately.

We prove (2) by induction on the complexity of φ_j.

(a) If φ_j is atomic, then $\varphi_j^{(T)}$ and $\varphi_j^{(\mathbf{t}(\beta))}$ are both identical with φ_j, and (2) is clear.

(b) If φ_j is a conjunction, then the two conjuncts are included in the list $\varphi_1,\ldots,\varphi_n$. So φ_j is $\varphi_q \wedge \varphi_r$. By inductive hypothesis we already have (2) with q and r in place of j, and from this we easily obtain (2) for φ_j.

(c) If φ_j is a negation, the proof of (2) is like that in case (b).

(d) If φ_j is existential, then $j \leqslant p$, and φ_j has the form $\exists y \psi_j$. Since ψ_j is a subformula of φ_j, it must occur in the list $\varphi_1,\ldots,\varphi_n$, so by inductive hypothesis we have

$$\forall x_1 \in \mathbf{t}(\beta) \ldots \forall x_m \in \mathbf{t}(\beta)[\psi_j^{(T)} \leftrightarrow \psi_j^{(\mathbf{t}(\beta))}],$$

and hence

$$\forall x_1 \in \mathbf{t}(\beta) \ldots \forall x_m \in \mathbf{t}(\beta)[\exists y \in \mathbf{t}(\beta) \psi_j^{(T)} \leftrightarrow \exists y \in \mathbf{t}(\beta) \psi_j^{(\mathbf{t}(\beta))}].$$

Now $\exists y \in \mathbf{t}(\beta) \psi_j^{(\mathbf{t}(\beta))}$ is $\varphi_j^{(\mathbf{t}(\beta))}$, so to get (2) for φ_j we have to show that

$$\forall x_1 \in \mathbf{t}(\beta) \ldots \forall x_m \in \mathbf{t}(\beta)[\varphi_j^{(T)} \leftrightarrow \exists y \in \mathbf{t}(\beta) \psi_j^{(T)}],$$

i.e. that

$$(3) \qquad \forall x_1 \in \mathbf{t}(\beta) \ldots \forall x_m \in \mathbf{t}(\beta)[\exists y[T(y) \wedge \psi_j^{(T)}] \leftrightarrow \exists y \in \mathbf{t}(\beta) \psi_j^{(T)}].$$

Suppose, therefore, that $x_1,\ldots,x_m$ are all in $\mathbf{t}(\beta)$. By the definition of β and Lemma 5.2, we have

$$\mathbf{t}(\beta) = \bigcup \{\mathbf{t}(\mathbf{r}(n)) \colon n \in \omega\}.$$

Hence for each $i=1,\ldots,m$ there must be $n_i \in \omega$ such that $x_i \in \mathbf{t}(\mathbf{r}(n_i))$. Let k be an n_i for which the ordinal $\mathbf{r}(n_i)$ is greatest $(i=1,\ldots,m)$. Then $x_i \in \mathbf{t}(\mathbf{r}(k))$ for $i=1,\ldots,m$.

By (1), we have

$$\exists y[T(y) \wedge \psi_j^{(T)}] \leftrightarrow \exists y \in \mathbf{t}(\mathbf{s}^*(\mathbf{t}(\mathbf{r}(k)))) \psi_j^{(T)},$$

and, since $\vdash \mathbf{s}^*(\mathbf{t}(\mathbf{r}(k))) = \mathbf{r}(k+1)$, we get

$$(4) \qquad \exists y[T(y) \wedge \psi_j^{(T)}] \leftrightarrow \exists y \in \mathbf{t}(\mathbf{r}(k+1)) \psi_j^{(T)}.$$

But $\mathbf{r}(k+1) \leqslant \beta$ so that $\mathbf{t}(\mathbf{r}(k+1)) \subseteq \mathbf{t}(\beta)$. Hence (4) gives, *a fortiori*,

$$\exists y[T(y) \wedge \psi_j^{(T)}] \leftrightarrow \exists y \in \mathbf{t}(\beta)\psi_j^{(T)}.$$

Since $x_1,\dots,x_m$ were arbitrary members of $\mathbf{t}(\beta)$, we immediately obtain (3). This completes the proof. ∎

Taking R_α for $\mathbf{t}(\alpha)$ in 5.3 immediately gives, using **Reg**:

5.4. COROLLARY. *If* $\boldsymbol{\sigma}$ *is any instance of* $\mathbf{RP}_2$, *then* $\vdash \boldsymbol{\sigma}$. ∎

Hence, by (5.1) we get

5.5. COROLLARY. *If* $\boldsymbol{\sigma}$ *is any instance of* $\mathbf{RP}_1$, *then* $\vdash \boldsymbol{\sigma}$. ∎

We are now going to show that the First Reflection Principle can be employed as an alternative postulate for set theory. More precisely, we show that if from our original system of postulates we drop the Axioms of Infinity, Union, and Replacement, and substitute instead the Axiom of Separation (1.4) and the First Reflection Principle, we obtain a system of postulates equivalent to the original one.

Our original theory, which is based on the postulates **Ext**, **Rep**, **Union**, **Pow**, **Inf** and **Reg**, is called *Zermelo–Fraenkel set theory* and is denoted by **ZF**. The theory obtained by adding **AC** to **ZF** is denoted by **ZFC**.

The theory whose postulates consist of **Ext**, **Pow**, **Reg**, **Sep**, and $\mathbf{RP}_1$ we call *Levy–Montague set theory* and is denoted by **LM**.

Let us write $\vdash_{\mathbf{LM}} \boldsymbol{\varphi}$ for "$\boldsymbol{\varphi}$ is deducible from the postulates of **LM**". Then we have:

5.6. THEOREM. *For any formula* $\boldsymbol{\varphi}$,

$$\vdash \boldsymbol{\varphi} \Leftrightarrow \vdash_{\mathbf{LM}} \boldsymbol{\varphi}.$$

In other words, **ZF** *and* **LM** *are equivalent theories.*

PROOF. It clearly suffices to show

(1) $\vdash \boldsymbol{\sigma}$ for every postulate $\boldsymbol{\sigma}$ of **LM**;

(2) $\vdash_{\mathbf{LM}} \boldsymbol{\tau}$ for every postulate $\boldsymbol{\tau}$ of **ZF**.

To prove (1), we run through the postulates of **LM**. If $\boldsymbol{\sigma}$ is **Ext**, **Pow**, or **Reg**, then $\boldsymbol{\sigma}$ is a postulate of **ZF**, and so, *a fortiori*, $\vdash \boldsymbol{\sigma}$. If $\boldsymbol{\sigma}$ is an instance of **Sep**, then $\vdash \boldsymbol{\sigma}$ by 1.3(i). Finally, if $\boldsymbol{\sigma}$ is an instance of $\mathbf{RP}_1$ then $\vdash \boldsymbol{\sigma}$ by 5.5. This proves (1).

To prove (2) we consider the postulates of **ZF** in turn.

(a) **Ext**, **Pow**, and **Reg**. Tiivially we have $\vdash_{\mathbf{LM}} \mathbf{Ext} \wedge \mathbf{Pow} \wedge \mathbf{Reg}$ since all three are postulates of **LM**.

(b) **Rep**. We must show that, if $\varphi(x,y)$ is a formula with free variables among $x,y,y_1,\ldots,y_k$, then, if z does not occur free in φ,

(3) $\vdash_{\mathbf{LM}} \forall x \in u\, \exists^* y \varphi(x,y) \to \exists z\, \forall y [y \in z \leftrightarrow \exists x \in u \varphi(x,y)]$.

Suppose that, for a particular choice of $y_1,\ldots,y_k$, which we assume to be fixed throughout the argument, we have

(4) $\forall x \in u\, \exists^* y \varphi(x,y)$.

Then, by $\mathbf{RP}_1$, which is a postulate of **LM**, there is a set v such that $\mathrm{Trans}(v)$; $u,y_1,\ldots,y_k \in v$ and

(5) $\forall x \in v\, \forall y \in v [\varphi(x,y) \leftrightarrow \varphi^{(v)}(x,y)]$,

(6) $\forall x \in v [\exists y \varphi(x,y) \leftrightarrow \exists y \in v \varphi^{(v)}(x,y)]$.

From (5) and (6) it follows that

(7) $\forall x \in v [\exists y \varphi(x,y) \leftrightarrow \exists y \in v \varphi(x,y)]$.

Using **Sep** — which is a postulate of **LM** — we see that there exists a set z such that

$$\forall y [y \in z \leftrightarrow y \in v \wedge \exists x \in u \varphi(x,y)].$$

We claim that

(8) $y \in z \leftrightarrow \exists x \in u \varphi(x,y)$.

Clearly we have

$$y \in z \to \exists x \in u \varphi(x,y).$$

Conversely, if $\exists x \in u \varphi(x,y)$, then

(9) $\varphi(x',y)$

for some $x' \in u$, whence $\exists w \varphi(x',w)$. Since $\mathrm{Trans}(v)$ and $u \in v$, it follows that $x' \in v$ and (7) then implies that $\exists w \in v \varphi(x',w)$. Thus we have $\varphi(x',y')$ for some $y' \in v$. It follows from (4) and (9) that $y=y'$, so that $y \in v$, whence $y \in z$. This proves (8), and (3) follows.

(c) **Union**. We have to show that

(10) $\vdash_{\mathbf{LM}} \exists x\, \forall y [y \in x \leftrightarrow \exists u \in z (y \in u)]$.

By $\mathbf{RP}_1$, for each z there is v such that $\text{Trans}(v) \wedge z \in v$. Using **Sep** we see that there is a set x such that

$$\forall y[y \in x \leftrightarrow y \in v \wedge \exists u \in z(y \in u)].$$

It is now easy to show, using $\text{Trans}(v)$, that

$$\forall y[y \in x \leftrightarrow \exists u \in z(y \in u)],$$

and (10) follows.

(d) **Inf**. Since **Ext**, **Rep** and **Pow** are theorems of **LM**, we can prove within **LM** the existence of the unordered pair $\{u,v\}$ of any sets u,v (see Thm. 1.10 and the definition immediately following it). Hence, since **Union** is a theorem of **LM**, we can prove within **LM** the existence of the set $u \cup \{u\}$ for any set u. In other words, we have

(11) $\qquad \vdash_{\mathbf{LM}} \forall u\, \exists v\, \forall x[x \in v \leftrightarrow x \in u \vee x = u].$

Using **Sep** and **Ext** we can prove within **LM** the existence of the unique empty set $\mathbf{0}$ (see 1.5). Accordingly, by $\mathbf{RP}_1$ and (11) we can prove within **LM** the existence of a set w satisfying

$$\text{Trans}(w) \wedge \mathbf{0} \in w \wedge \forall u \in w\, \exists v \in w\, \forall x \in w[x \in v \leftrightarrow x \in u \vee x = u].$$

But this implies (using $\text{Trans}(w)$),

$$\mathbf{0} \in w \wedge \forall u \in w\, \exists v \in w\, \forall x[x \in v \leftrightarrow x \in u \vee x = u]$$

i.e.

$$\mathbf{0} \in w \wedge \forall u \in w[u \cup \{u\} \in w].$$

Hence

$$\vdash_{\mathbf{LM}} \exists w[\mathbf{0} \in w \wedge \forall u \in w[u \cup \{u\} \in w]],$$

i.e. $\vdash_{\mathbf{LM}}$ **Inf**. ∎

§ 6. The formalization of satisfaction

Let $\boldsymbol{\varphi}$ be a formula of $\mathscr{L}$, u a set and $\mathfrak{x}$ a sequence of elements of u. In Ch. 5 we defined the (metamathematical) notion of $\mathfrak{x}$ *sarisfying* $\boldsymbol{\varphi}$ in any $\mathscr{L}$-structure, in particular in the structure $\langle u, \in|u\rangle$, where $\in|u$ is the $\in$-relation restricted to u, i.e.

$$\in|u = \{\langle y,z\rangle \in u \times u:\ y \in z\}.$$

Since $\boldsymbol{\varphi}$ has only finitely many fiee variables, we may take $\mathfrak{x}$ to be an *eventually constant* sequence on u; i.e. a sequence which assumes a constant value after some point.

We are going to show that the notion of satisfaction of a formula of $\mathscr{L}$ in a structure of the form $\langle u, \in|u\rangle$ can be expressed in $\mathscr{L}$. Now any assertion expressible in $\mathscr{L}$ is a statement about *sets*, but formulas are not sets. So if we are to succeed in our attempt we must first find some way of replacing *formulas* by *sets*. We do this by associating each formula of $\mathscr{L}$ with a uniquely defined (legitimate) *closed term* of $\mathscr{L}$.

We recall that the individual vaiiables of $\mathscr{L}$ are assumed to be enumerated in a sequence $v_0, v_1, v_2, \ldots$. For each formula φ of $\mathscr{L}$ we define the term $\ulcorner\varphi\urcorner$ by induction on the complexity of φ as follows. We put

$$\ulcorner v_i = v_j \urcorner =_{\mathrm{df}} \langle \mathbf{0}, \mathbf{i}, \mathbf{j} \rangle$$

$$\ulcorner v_i \in v_j \urcorner =_{\mathrm{df}} \langle \mathbf{1}, \mathbf{i}, \mathbf{j} \rangle$$

$$\ulcorner \varphi \wedge \psi \urcorner =_{\mathrm{df}} \langle \mathbf{2}, \ulcorner\varphi\urcorner, \ulcorner\psi\urcorner \rangle$$

$$\ulcorner \neg\varphi \urcorner =_{\mathrm{df}} \langle \mathbf{3}, \ulcorner\varphi\urcorner \rangle$$

$$\ulcorner \exists v_i \varphi \urcorner =_{\mathrm{df}} \langle \mathbf{4}, \mathbf{i}, \ulcorner\varphi\urcorner \rangle.$$

It is clear that this prescription assigns a unique closed term $\ulcorner\varphi\urcorner$ to each formula φ.

We now apply 2.11 to yield a term $\mathbf{t}(n)$ for which

$$\vdash \mathbf{t}(\mathbf{0}) = \mathbf{2} \times \omega \times \omega$$
$$\wedge\, \mathbf{t}(n+1) = \mathbf{t}(n) \cup [\{\mathbf{2}\} \times \mathbf{t}(n) \times \mathbf{t}(n)] \cup [\{\mathbf{3}\} \times \mathbf{t}(n)] \cup [\{\mathbf{4}\} \times \omega \times \mathbf{t}(n)].$$

We put $\mathbf{F}_n$ for $\mathbf{t}(n)$, and define

the set of $\ulcorner$*formulas*$\urcorner$: $\mathbf{F} =_{\mathrm{df}} \bigcup\{\mathbf{F}_n : n \in \omega\}$.

6.1. THEOREM. *For each formula* φ *of* $\mathscr{L}$,

$$\vdash \ulcorner\varphi\urcorner \in \mathbf{F}.$$

PROOF. This theorem is proved by a straightforward induction on the complexity of φ; details are left to the reader. ∎

We now turn to the problem of formalizing the satisfaction definition in $\mathscr{L}$. We define

the set of eventually constant sequences on u:

$$\mathrm{ec}(u) =_{\mathrm{df}} \{x : \mathrm{Fun}(x) \wedge \mathrm{dom}(x) = \omega \wedge \mathrm{ran}(x) \subseteq u$$
$$\wedge\, \exists m \exists y \in u \forall n \geqslant m [x\text{‘}n = y]\}.$$

$$x(y/z) =_{\mathrm{df}} [x - \{\langle y, x\text{‘}y \rangle\}] \cup \{\langle y, z \rangle\}.$$

Thus, for each sequence x and each natural number i, $x(i/z)$ is the sequence obtained from x by replacing $x`i$ by z.

6.2. LEMMA.

$$
\begin{aligned}
\vdash \exists! f[\mathrm{Fun}(f) \wedge \mathrm{dom}(f) = \mathbf{F} \\
\wedge \forall i \, \forall j [f`\langle \mathbf{0},i,j\rangle = \{x \in \mathrm{ec}(u) \colon \; x`i = x`j\} \\
\wedge f`\langle \mathbf{1},i,j\rangle = \{x \in \mathrm{ec}(u) \colon \; x`i \in x`j\} \\
\wedge \forall v \in \mathbf{F} \, \forall w \in \mathbf{F}[f`\langle \mathbf{2},v,w\rangle = f`v \cap f`w \\
\wedge f`\langle \mathbf{3},v\rangle = \mathrm{ec}(u) - f`v \\
\wedge \forall i [f`\langle \mathbf{4},i,v\rangle = \{x \in \mathrm{ec}(u) : \exists z \in u[x(i/z) \in f`v]\}]]].
\end{aligned}
$$

PROOF. The formula we are trying to prove is of the form $\exists! f\varphi(u,f)$. Let $\psi(n,u,f)$ be the formula obtained from φ by replacing the first occurrence of $\mathbf{F}$ by $\mathbf{F}_n$ and the second and third occurrences by $\mathbf{F}_{n-1}$ (where we define $\mathbf{F}_{-1}$ to be $\mathbf{0}$). We prove by induction on n that, given u, for all $n \in \omega$ there is a unique function f satisfying $\psi(n,u,f)$.

If $n = \mathbf{0}$, we simply define f on $\mathbf{F}_0$ such that

$$f`\langle \mathbf{0},i,j\rangle = \{x \in \mathrm{ec}(u) \colon \; x`i = x`j\},$$

$$f`\langle \mathbf{1},i,j\rangle = \{x \in \mathrm{ec}(u) \colon \; x`i \in x`j\}.$$

Clearly f defined in this way is the unique function satisfying $\psi(\mathbf{0},u,f)$.

Now suppose that the condition holds for n; we show that it holds for $n+1$. Let f be the unique function satisfying $\psi(n,u,f)$. Then f is defined on $\mathbf{F}_n$, and we can extend f to $\mathbf{F}_{n+1}$ by putting

$$f`\langle \mathbf{2},v,w\rangle = f`v \cap f`w,$$

$$f`\langle \mathbf{3},v\rangle = \mathrm{ec}(u) - f`v,$$

$$f`\langle \mathbf{4},i,v\rangle = \{x \in \mathrm{ec}(u) \colon \; \exists z \in u[x(i/z) \in f`v]\}.$$

Clearly the function f extended to $\mathbf{F}_{n+1}$ in this way satisfies $\psi(n+1,u,f)$. Since any such function is uniquely determined by its restriction to $\mathbf{F}_n$, it follows that the function f is unique. This completes the induction step.

We now define g to be the union of all the functions f defined above. Clearly g is the unique function satisfying $\varphi(u,g)$. ∎

Putting

$$\text{(6.3)} \qquad \mathrm{S}(u,v) =_{\mathrm{df}} \iota y[\forall f[\varphi(u,f) \to y = f`v]],$$

where $\boldsymbol{\varphi}$ is the formula defined in the proof of Lemma 6.2, we see that, informally speaking, if $v \in \mathbf{F}$, then $\mathrm{S}(u,v)$ is the set of all $x \in \mathrm{ec}(u)$ which satisfy v in $\langle u, \in |u\rangle$. We now define:

x satisfies the formula v in u:

$$\mathbf{Sat}(x,v,u) \leftrightarrow_{\mathrm{df}} v \in \mathbf{F} \wedge x \in \mathrm{S}(u,v).$$

Our next theorem asserts that $\mathbf{Sat}(x,\ulcorner\varphi\urcorner,u)$ is a formalization of the statement: "x satisfies $\boldsymbol{\varphi}$ in the structure $\langle u, \in |u\rangle$".

6.4. THEOREM. *For any indexes i,j and any formulas* $\boldsymbol{\varphi},\boldsymbol{\psi}$ *we have*

$$\begin{aligned}\vdash x \in \mathrm{ec}(u) \to [&(\mathbf{Sat}(x,\ulcorner v_i = v_j\urcorner,u) \leftrightarrow x`\mathbf{i} = x`\mathbf{j})\\ &\wedge (\mathbf{Sat}(x,\ulcorner v_i \in v_j\urcorner,u) \leftrightarrow x`\mathbf{i} \in x`\mathbf{j})\\ &\wedge (\mathbf{Sat}(x,\ulcorner \varphi \wedge \psi\urcorner,u) \leftrightarrow \mathbf{Sat}(x,\ulcorner\varphi\urcorner,u) \wedge \mathbf{Sat}(x,\ulcorner\psi\urcorner,u))\\ &\wedge (\mathbf{Sat}(x,\ulcorner \neg\varphi\urcorner,u) \leftrightarrow \neg\mathbf{Sat}(x,\ulcorner\varphi\urcorner,u))\\ &\wedge (\mathbf{Sat}(x,\ulcorner \exists v_i\varphi\urcorner,u) \leftrightarrow \exists y \in u\,\mathbf{Sat}(x(\mathbf{i}/y),\ulcorner\varphi\urcorner,u))].\end{aligned}$$

PROOF. This is a straightforward consequence of Lemma 6.2 and the definition of the formula **Sat**. We leave the details to the reader. ∎

6.5. THEOREM. *Let* $\boldsymbol{\varphi}$ *be a formula with free variables among* $v_0,\ldots,v_n$. *Then*

$$\vdash \forall x \in \mathrm{ec}(u)\,[\mathbf{Sat}(x,\ulcorner\varphi\urcorner,u) \leftrightarrow \varphi^{(u)}(x`\mathbf{0},\ldots,x`\mathbf{n})].$$

PROOF. A simple induction on the complexity of $\boldsymbol{\varphi}$, using 6.4. We treat the existential case, leaving the other cases to the reader. In fact, if $x \in \mathrm{ec}(u)$, we have (assuming without loss of generality that $i \leqslant n$),

$$\begin{aligned}\mathbf{Sat}(x,\ulcorner\exists v_i\varphi\urcorner,u) &\leftrightarrow \exists y \in u\,\mathbf{Sat}(x(\mathbf{i}/y),\ulcorner\varphi\urcorner,u)\\ &\leftrightarrow \exists y \in u\,\varphi^{(u)}(x`\mathbf{0},\ldots,x`(\mathbf{i}-\mathbf{1}),y,x`(\mathbf{i}+\mathbf{1}),\ldots,x`\mathbf{n})\\ &\leftrightarrow [\exists y\varphi(x`\mathbf{0},\ldots,x`(\mathbf{i}-\mathbf{1}),y,x`(\mathbf{i}+\mathbf{1}),\ldots,x`\mathbf{n})^{(u)}]\\ &\leftrightarrow (\exists v_i\varphi)^{(u)}(x`\mathbf{0},\ldots,x`\mathbf{n}).\end{aligned}$$ ∎

REMARK. It follows immediately from Thm. 6.5 that if $\boldsymbol{\sigma}$ is a *sentence*, then

$$\vdash u \neq \mathbf{0} \to [\mathbf{Val}(\ulcorner\sigma\urcorner,u) \leftrightarrow \sigma^{(u)}], \quad \text{where} \quad \mathbf{Val}(v,u) \leftrightarrow_{\mathrm{df}} \forall x \in \mathrm{ec}(u)\mathbf{Sat}(x,v,u).$$

That is, for $u \neq \mathbf{0}$, the formula $\mathbf{Val}(\ulcorner\sigma\urcorner,u)$ holds iff $\boldsymbol{\sigma}$ is true in the structure

$\langle u, \in|u\rangle$. Accordingly we may say that the formula $\mathbf{Val}(\ulcorner\boldsymbol{\sigma}\urcorner, u)$ expresses the *truth* of $\boldsymbol{\sigma}$ in $\langle u, \in|u\rangle$. It is natural to ask whether there is an $\mathscr{L}$-formula which expresses the truth of sentences in the *whole universe of sets V*. More precisely, is there an $\mathscr{L}$-formula $\mathbf{T}(x)$ such that, for each sentence $\boldsymbol{\sigma}$, we have $\vdash \boldsymbol{\sigma} \leftrightarrow \mathbf{T}(\ulcorner\boldsymbol{\sigma}\urcorner)$? In fact, assuming that **ZF** is consistent, it is easy to modify the argument of Prob. 7.7.13(ii) to show that such a formula **T** cannot exist. Less precisely (but more suggestively!), *truth in V is not definable in* $\mathscr{L}$.

We define

z is an elementary substructure of u:

$$\mathbf{ES}(z,u) \leftrightarrow_{\mathrm{df}} z \subseteq u \wedge \forall v \in \mathbf{F}\, \forall x \in \mathrm{ec}(z)\, [\mathbf{Sat}(x,v,z) \leftrightarrow \mathbf{Sat}(x,v,u)].$$

Note that by 6.5 we have, for any formula $\boldsymbol{\varphi}$ whose free variables are an among $v_0,\ldots,v_n$,

$$\vdash \mathbf{ES}(z,u) \rightarrow \forall v_0 \in z \ldots \forall v_n \in z\, [\boldsymbol{\varphi}^{(z)} \leftrightarrow \boldsymbol{\varphi}^{(u)}].$$

We now prove what amounts to a formalized version of Thm. 5.2.1.

6.6. THEOREM. $\vdash_{\mathbf{AC}} \forall y \subseteq u[\aleph_0 \leq |u| \wedge \aleph_0 \leq |y| \rightarrow \exists z[y \subseteq z \wedge |z| = |y| \wedge \mathbf{ES}(z,u)]$.

PROOF. The proof is nothing more than a formalized version of the proof of Thm. 5.2.1, so we give the merest sketch, leaving the reader to fill in the details.

Let h be a choice function for $\mathrm{P}u$. For each $v \in \mathbf{F}$, $x \in \mathrm{ec}(u)$ and $i \in \omega$ we put $\mathrm{s}(v,x,i)$ for the set

$$\{w \in u\colon \mathbf{Sat}(x(i/w),v,u)\}.$$

Define the sequence $y_0, y_1, \ldots$ of subsets of u inductively as follows: y_0 is y and y_{n+1} is

$$\{h\text{`}\mathrm{s}(v,x,i)\colon\ v \in \mathbf{F} \wedge x \in \mathrm{ec}(y_n) \wedge i \in \omega\}.$$

Then, just as in the proof of Thm. 5.2.1, one verifies that the set $\bigcup\{y_n\colon n \in \omega\}$ satisfies the conditions imposed on z in the theorem. ∎

The last result in this section is proved in just the same way as Prob. 5.1.1; again the details are left to the reader.

6.7. THEOREM. *For any formula* $\boldsymbol{\varphi}$ *whose free variables are all among* $v_0,\ldots,v_n$ *we have*

$$\vdash \mathrm{Isom}(f,x,y) \rightarrow \forall v_0 \in x \ldots \forall v_n \in x[\boldsymbol{\varphi}^{(x)}(v_0,\ldots,v_n) \leftrightarrow \boldsymbol{\varphi}^{(y)}(f\text{`}v_0,\ldots,f\text{`}v_n)].$$

In particular, for any sentence $\boldsymbol{\sigma}$,

$$\vdash \mathrm{Isom}(f,x,y) \rightarrow [\boldsymbol{\sigma}^{(x)} \leftrightarrow \boldsymbol{\sigma}^{(y)}].$$ ∎

§ 7. Absoluteness

A formula φ whose free variables are all among $x_1,\ldots,x_m$ is said to be *absolute* if there is a finite sequence $\sigma_1,\ldots,\sigma_n$ of postulates of **ZF** — called an *absoluteness sequence* for φ — such that, for each formula $\psi(x)$ which does not have free variables among $x_1,\ldots,x_m$,

$$(7.1)\qquad \vdash \mathbf{Trans}_{\psi(x)} \wedge \exists x\psi(x) \wedge \sigma_1^{(\psi)} \wedge \ldots \wedge \sigma_n^{(\psi)} \rightarrow \forall x_1 \ldots \forall x_m[\psi(x_1) \wedge \ldots \wedge \psi(x_m) \rightarrow [\varphi^{(\psi)} \leftrightarrow \varphi]],$$

in which the relativizations of formulas to ψ are understood to be taken with x as chosen variable, the other free variables of ψ being regarded as parameters. (For the definition of $\mathbf{Trans}_{\psi(x)}$, see the material following Thm. 3.3.)

A term $\mathbf{t}$ is said to be *absolute* if for any variable x not free in $\mathbf{t}$ the formula $x=\mathbf{t}$ is absolute.

Informally speaking, the formula $\varphi(x_1,\ldots,x_m)$ (or the term $\mathbf{t}(x_1,\ldots,x_m)$) is absolute if there is a finite sequence $\sigma_1,\ldots,\sigma_n$ of postulates of **ZF** such that whenever X is a non-empty transitive definable class in which $\sigma_1,\ldots,\sigma_n$ hold (when $\in$ is interpreted as the membership relation on X), then, for any members $u_1,\ldots,u_m$ of X, $\varphi(u_1,\ldots,u_m)$ holds in X iff it holds in the universe of all sets (or the value of $\mathbf{t}(u_1,\ldots,u_m)$ in X is the same as its value in the universe of all sets).

7.2. LEMMA. *Suppose φ is an absolute formula whose free variables are all among $x_1,\ldots,x_m$, and let $\sigma_1,\ldots,\sigma_n$ be an absoluteness sequence for φ. Then for any variable y, we have*

$$\vdash \mathrm{Trans}(y) \wedge y \neq \mathbf{0} \wedge \sigma_1^{(y)} \wedge \ldots \wedge \sigma_n^{(y)} \rightarrow \forall x_1 \ldots \forall x_m[x_1 \in y \wedge \ldots \wedge x_m \in y \rightarrow [\varphi^{(y)} \leftrightarrow \varphi]].$$

PROOF. Put $x \in y$ for $\psi(x)$ in (7.1). ∎

We are going to show that a large number of the defined formulas and terms we have introduced in the course of our discussion are absolute.

Let us call a formula of $\mathscr{L}$ *restricted* if all its quantifiers are of the form $\forall x \in y$ or $\exists x \in y$. (N.B.: This refers to formulas in the primitive notation of the language $\mathscr{L}$, i.e., to formulas which *do not contain virtual terms.*)

7.3. LEMMA.

(i) *Any atomic formula is absolute. Also, if φ,φ' are absolute, so are $\neg\varphi$, $\varphi \wedge \varphi'$, $\varphi \vee \varphi'$, $\varphi \rightarrow \varphi'$, $\varphi \leftrightarrow \varphi'$, $\forall x \in y\varphi$ and $\exists x \in y\varphi$. Hence all restricted formulas are absolute.*

(ii) *If* φ *is absolute and* φ' *is a formula with the same free variables as* φ *such that* $\vdash \varphi \leftrightarrow \varphi'$, *then* φ' *is absolute.*

(iii) *Let* φ_1 *and* φ_2 *be absolute formulas such that* $\forall y \varphi_1$ *and* $\exists z \varphi_2$ *have the same free variables. If* φ *is a formula with the same free variables as* $\forall y \varphi_1$ *and* $\exists z \varphi_2$ *such that* $\vdash \varphi \leftrightarrow \forall y \varphi_1$ *and* $\vdash \varphi \leftrightarrow \exists z \varphi_2$, *then* φ *is absolute.*

(iv) *If* $\varphi(x)$ *is absolute and* $\vdash \exists ! x \varphi(x)$, *then the term* $\iota x \varphi(x)$ *is absolute.*

(v) *If* $\varphi(y)$ *and* $\mathbf{t}$ *are absolute, then so is* $\varphi(\mathbf{t})$.

(vi) *If* x *is not free in* $\mathbf{t}$, *and* φ *and* $\mathbf{t}$ *are absolute, then so are* $\forall x \in \mathbf{t} \varphi$ *and* $\exists x \in \mathbf{t} \varphi$.

(vii) *If* $\mathbf{s}(y)$ *and* $\mathbf{t}(y)$ *are absolute, then so is* $\mathbf{s}(\mathbf{t}(y))$.

(viii) *If* $\mathbf{t}$ *is absolute, then so are* $x \in \mathbf{t}$ *and* $\mathbf{t} \in x$.

(ix) *If* $\varphi(y)$ *is an absolute formula, and* $\mathbf{s}$ *and* $\mathbf{t}(x)$ *are absolute terms, where* $\mathbf{t}(x)$ *is not a variable, then the terms* $\{y \in \mathbf{s}: \varphi(y)\}$ *and* $\{\mathbf{t}(x): x \in \mathbf{s}\}$ *are absolute.*

PROOF. (i) is straightforward and is entrusted to the reader.

(ii) Suppose that φ is absolute and that $\vdash \varphi \leftrightarrow \varphi'$. Let $\sigma_1, \ldots, \sigma_n$ be an absoluteness sequence for φ, and let $\tau_1, \ldots, \tau_m$ be the finite sequence of postulates of **ZF** used in some proof of $\varphi \leftrightarrow \varphi'$. It is now a simple exercise to show that φ' is absolute with absoluteness sequence $\sigma_1, \ldots, \sigma_n, \tau_1, \ldots, \tau_m$.

(iii) Assume the hypotheses of (iii). Let $\sigma_1, \ldots, \sigma_n$ be an absoluteness sequence for φ_1 and let $\tau_1, \ldots, \tau_m$ be the finite sequence of postulates of **ZF** used in some proof of $\varphi \leftrightarrow \forall y \varphi_1$. Suppose that the free variables of φ are among $x_1, \ldots, x_k$, and that $\psi(x)$ is a formula which does not have free variables among $x_1, \ldots, x_k$. We may assume that y does not occur free in ψ either, for otherwise we can replace $\forall y \varphi_1$ by a suitable variant $\forall w \varphi_1(w)$, where w does not occur in φ_1 nor among $x_1, \ldots, x_k$. Now suppose that

$$\mathrm{Trans}_{\psi(x)} \wedge \exists x \psi(x) \wedge \sigma_1^{(\psi)} \wedge \ldots \wedge \sigma_n^{(\psi)} \wedge \tau_1^{(\psi)} \wedge \ldots \wedge \tau_m^{(\psi)} \wedge \psi(x_1) \wedge \ldots \wedge \psi(x_k).$$

Then, under these assumptions, we have

$$\begin{aligned} \varphi & \rightarrow \forall y \varphi_1 \\ & \rightarrow \forall y [\psi(y) \rightarrow \varphi_1] \\ & \rightarrow \forall y [\psi(y) \rightarrow \varphi_1^{(\psi)}] \\ & \rightarrow (\forall y \varphi_1)^{(\psi)} \\ & \rightarrow \varphi^{(\psi)}. \end{aligned}$$

A similar argument, using $\vdash \varphi \leftrightarrow \exists z \varphi_2$, shows that, with the requisite assumptions, $\varphi^{(\psi)} \rightarrow \varphi$. Hence φ is absolute as claimed.

(iv) Suppose that $\varphi(x)$ is absolute and $\vdash \exists! x\varphi(x)$. Then we have

$$\vdash x = \iota x\varphi(x) \leftrightarrow \varphi(x),$$

so the required conclusion follows from (ii).

(v) Suppose that $\varphi(y)$ and $\mathbf{t}$ are absolute. We have

$$\vdash \varphi(\mathbf{t}) \leftrightarrow \forall y[y=\mathbf{t} \rightarrow \varphi(y)],$$

$$\vdash \varphi(\mathbf{t}) \leftrightarrow \exists y[y=\mathbf{t} \wedge \varphi(y)],$$

so the absoluteness of $\varphi(\mathbf{t})$ follows from (iii) and (i).

(vi), (vii) and (viii) are immediate consequences of (v) and (i).

(ix) Assume the hypotheses of (ix). We have, by (1.7) and (1.12),

$$\vdash z = \{y \in \mathbf{s}\colon \varphi(y)\} \leftrightarrow \forall y \in z[y \in \mathbf{s} \wedge \varphi(y)] \wedge \forall y \in \mathbf{s}[\varphi(y) \rightarrow y \in z],$$

so that the term $\{y \in \mathbf{s}\colon \varphi(y)\}$ is absolute by (i), (ii), and (vi). Also, by (1.13),

$$\vdash z = \{\mathbf{t}(x)\colon x \in \mathbf{s}\} \leftrightarrow \forall y \in z\, \exists x \in \mathbf{s}[y=\mathbf{t}(x)] \wedge \forall x \in \mathbf{s}\, \exists y \in z[y=\mathbf{t}(x)],$$

so that the term $\{\mathbf{t}(x)\colon x \in \mathbf{s}\}$ is absolute for the same reasons as before. ∎

Using this lemma, we can now verify that each member of the following sequence of defined formulas is absolute. In each case we first write down the formula whose absoluteness is to be verified and then another formula which is either the defining formula of the first or which can be proved equivalent to it. The defining formula will be seen to be absolute by applying Lemma 7.3 and using the fact that earlier formulas in the sequence have been proved absolute. The absoluteness of the given formula then follows from (ii) of 7.3.

(1) $y \subseteq x$: $\forall z \in y[z \in x]$.

(2) $z = \{x,y\}$: $x \in z \wedge y \in z \wedge \forall u \in z[u=x \vee u=y]$.

(3) $z = \{x\}$: $z = \{x,x\}$.

(4) $z = \langle x,y \rangle$: $z = \{\{x\}, \{x,y\}\}$.

(5) $z = \bigcup x$: $\forall y \in x\, \forall u \in y[u \in z] \wedge \forall y \in z\, \exists u \in x[y \in u]$.

(6) $z = x \cup y$: $z = \bigcup\{x,y\}$.

(7) $z = x \cap y$: $z \subseteq x \wedge z \subseteq y \wedge \forall u \in x[u \in y \rightarrow u \in z]$.

(8) $z = x - y$: $z \subseteq x \wedge \forall u \in x[u \notin y \leftrightarrow u \in z]$.

(9) $z = x \times y$: $\forall w \in z\, \exists u \in x\, \exists v \in y[w = \langle u,v \rangle] \wedge \forall u \in x\, \forall v \in y\, \exists w \in z[w = \langle u,v \rangle]$.

(10) $\mathrm{Fun}(f)$: $\forall z \in f\, \exists x \in \bigcup\bigcup f\, \exists y \in \bigcup\bigcup f[z = \langle x,y \rangle]$
$\wedge \forall x \in \bigcup\bigcup f\, \forall y \in \bigcup\bigcup f\, \forall z \in \bigcup\bigcup f[\langle x,y \rangle \in f \wedge \langle x,z \rangle \in f \rightarrow y=z]$

(11) $y = \mathrm{dom}(f)$: $\forall x \in y\, \exists z \in \bigcup\bigcup f[\langle x,z \rangle \in f]$
$\wedge \forall x \in \bigcup\bigcup f\, \forall z \in \bigcup\bigcup f[\langle x,z \rangle \in f \rightarrow x \in y]$.

(12) $y=\mathrm{ran}(f)$: $\forall x\in y\, \exists z\in \bigcup\bigcup f[\langle z,x\rangle \in f]$
$\wedge\, \forall x\in \bigcup\bigcup f\, \forall z\in \bigcup\bigcup f[\langle z,x\rangle\in f \rightarrow x\in y]$.
(13) $y=\mathbf{0}$: $\forall y\in x[y\notin x]$.
(14) $y=x+1$: $y=x\cup\{x\}$.
(15) $y=\mathbf{1},\mathbf{2},\mathbf{3},\mathbf{4}$: $y=\mathbf{0}+1$, $y=\mathbf{1}+1$, $y=\mathbf{2}+1$, $y=\mathbf{3}+1$.
(16) $y=f`x$: $\forall w\in\bigcup\bigcup f[\langle x,w\rangle\in f\leftrightarrow w=y]\wedge\langle x,y\rangle\in f$
$\vee\, \neg\exists z\in\bigcup\bigcup f\,\forall w\in\bigcup\bigcup f[\langle x,w\rangle\in f\leftrightarrow w=z]\wedge y=\mathbf{0}$.
(17) $u=x(z/y)$: $u=[x-\{\langle z,x`z\rangle\}]\cup\{\langle z,y\rangle\}$.
(18) $y=f\restriction x$: $y=f\cap[x\times\mathrm{ran}(f)]$.
(19)[1] $\mathrm{Ord}(x)$: $\forall y\in x\,\forall z\in x[y=z\vee y\in z\vee z\in y]\wedge\forall y\in x[y\subseteq x]$.
(20) $\mathrm{Lim}(x)$: $\mathrm{Ord}(x)\wedge x\neq\mathbf{0}\wedge\forall y\in x[x\neq y+1]$.
(21) $x=\omega$: $\mathrm{Lim}(x)\wedge\forall y\in x\neg\,\mathrm{Lim}(y)$.

We must now extend the notion of absoluteness to formulas and terms with ordinal or number variables. Given $\varphi(x)$, we say that the formula $\varphi(\alpha)$ (or $\varphi(n)$) is *absolute* if $\varphi(x)\wedge\mathrm{Ord}(x)$ (or $\varphi(x)\wedge x\in\omega$) is absolute, and similarly for terms. It follows immediately from (19) and (21) above that if $\varphi(x)$ is absolute, so are $\varphi(\alpha)$, $\varphi(n)$, $\exists n\varphi(n)$ and $\forall n\varphi(n)$.

We next establish the absoluteness of terms defined by ordinal recursion from absolute terms.

7.4. LEMMA. *Let* $\mathbf{s}(y,z)$ *be an absolute term and let* $\mathbf{t}$ *be a term such that* $\vdash\mathbf{t}(\alpha)=\mathbf{s}(\mathbf{t}|\alpha,\alpha)$. *Then the formula* $y=\mathbf{t}(\alpha)$ *is absolute.*

PROOF. Using the absoluteness of the formulas (1)–(21), the absoluteness of $\mathbf{s}(y,z)$ and 7.3, it is not difficult to see that the formula $\mathrm{Rec}(f,x)$ introduced in the proof of Thm. 2.9 is absolute. But by the same proof we have

$$\vdash y=\mathbf{t}(\alpha)\leftrightarrow\forall f[\mathrm{Rec}(f,\alpha)\rightarrow f`\alpha=y],$$

$$\vdash y=\mathbf{t}(\alpha)\leftrightarrow\exists f[\mathrm{Rec}(f,\alpha)\wedge f`\alpha=y].$$

The absoluteness of $y=\mathbf{t}(\alpha)$ now follows from 7.3(iii). ∎

7.5. LEMMA. *Let* $\mathbf{s}_0$ *and* $\mathbf{s}_1(y)$ *be absolute terms, and let* $\mathbf{t}(x)$ *be a term such that*

$$\vdash\mathbf{t}(\mathbf{0})=\mathbf{s}_0\wedge\mathbf{t}(n+1)=\mathbf{s}_1(\mathbf{t}(n)).$$

(See Thm. 2.11.) *Then the formula* $y=\mathbf{t}(n)$ *is absolute.*

PROOF. Let $\varphi(f,x)$ be the formula

$$\mathrm{Fun}(f)\wedge x\cup\{x\}\cup\{\mathbf{0}\}\subseteq\mathrm{dom}(f)\wedge f`\mathbf{0}=\mathbf{s}_0$$

$$\wedge\,\forall n\in\mathrm{dom}(f)[f`(n+1)=\mathbf{s}_1(f`n)].$$

[1] See Prob. 3.8 (iv).

Then, as in the proof of 7.4, it is not difficult to see that

$$\vdash y=\mathbf{t}(n)\leftrightarrow\forall f[\varphi(f,n)\rightarrow f`n=y],$$

$$\vdash y=\mathbf{t}(n)\leftrightarrow\exists f[\varphi(f,n)\wedge f`n=y].$$

Since $\varphi(f,x)$ is absolute by previous results of this section, it follows from 7.3(iii) that $y=\mathbf{t}(n)$ is absolute. ∎

7.6. Corollary. *The term* ec(u) *is absolute.*

Proof. By recursion we define a term $\mathbf{t}(y,u)$ such that

$$\vdash \mathbf{t}(\mathbf{0},u)=\{\omega\times\{x\}:\ x\in u\}\wedge\mathbf{t}(n+1,u)=\{y(n/x):\ x\in u\wedge y\in\mathbf{t}(n,u)\}.$$

This is a recursion of the type considered in 7.5, and, using the appropriate instances of (1)–(21), we see that the terms playing the roles of $\mathbf{s}_0$ and $\mathbf{s}_1$ are absolute. Thus $y=\mathbf{t}(n,u)$ is absolute by 7.5. Also, it is easy to verify that $\mathbf{t}(n,u)$ is the set of all sequences from u which assume a constant value from the nth place on, so that

$$\vdash \mathrm{ec}(u)=\bigcup\{\mathbf{t}(n,u):\ n\in\omega\}.$$

The absoluteness of ec(u) follows. ∎

7.7. Corollary. *The term* **F** *is absolute.*

Proof. In §6 we introduced a term $\mathbf{t}(x)$ for which

$$\vdash \mathbf{t}(\mathbf{0})=2\times\omega\times\omega$$

$$\wedge\ \mathbf{t}(n+1)=\mathbf{t}(n)\cup[\{\mathbf{2}\}\times\mathbf{t}(n)\times\mathbf{t}(n)]\cup[\{\mathbf{3}\}\times\mathbf{t}(n)]\cup[\{\mathbf{4}\}\times\omega\times\mathbf{t}(n)].$$

This is a recursive definition of the type considered in 7.5 and, using the appropriate instances of (1)–(21) we see that the terms playing the roles of $\mathbf{s}_0$ and $\mathbf{s}_1$ are absolute. Therefore $\mathbf{t}(n)$ itself is absolute by 7.5. But by definition we have $\mathbf{F}=\bigcup\{\mathbf{t}(n):\ n\in\omega\}$ and the result follows. ∎

7.8. Lemma. *The formula* **Sat**(x,v,u) *is absolute.*

Proof. Recall that in Lemma 6.2 we proved that

(1) $\vdash \exists! f\varphi(u,f)$

for a certain formula $\varphi(u,f)$. Using the results already proved in this section, it is easy to see that $\varphi(u,f)$ is absolute.

We also recall that in (6.3) we put S(u,v) for

$$\iota y[\forall f[\varphi(u,f)\rightarrow y=f`v]].$$

It follows easily from (1) and 7.3(iv) that $S(u,v)$ is absolute. Since $\mathbf{Sat}(x,v,u)$ was defined to be the formula $v \in \mathbf{F} \wedge x \in S(u,v)$, the required result follows from 7.7. ∎

It is important to observe that there are formulas and terms of $\mathscr{L}$ which are *not* absolute. For example, let us define

$$z \textit{ is countable}: \; C(z) \leftrightarrow_{\mathrm{df}} \exists \alpha \leqslant \omega (z \approx \alpha).$$

Then we have:

7.9. THEOREM. *If* **ZFC** *is consistent*[1], *then the formula* C(*z*) *is* not *absolute.*
PROOF. Suppose that $C(z)$ is absolute; let $\sigma_1,\ldots,\sigma_m$ be an absoluteness sequence for $C(z)$, and let $\tau_1,\ldots,\tau_n$ be a list consisting of all postulates of **ZF** used in a proof of Cantor's theorem that $\exists z \neg C(z)$. Put σ for $\sigma_1 \wedge \ldots \wedge \sigma_m$, τ for $\tau_1 \wedge \ldots \wedge \tau_n$, and $\varphi(y)$ for the formula

$$\mathrm{Trans}(y) \wedge |y| = \aleph_0 \wedge \sigma^{(y)} \wedge \tau^{(y)}.$$

Now $\tau \rightarrow \exists z \neg C(z)$ is a logically valid sentence, so that, by Cor. 2.12.4, the formula

$$\exists x (x \in y) \wedge \tau^{(y)} \rightarrow [\exists z \neg C(z)]^{(y)}$$

is logically valid. Hence

$$(1) \qquad \vdash \varphi(y) \rightarrow [\exists z \neg C(z)]^{(y)}.$$

A straightforward application of $\mathbf{RP}_1$ (see §5) and Thm. 6.6 shows that, assuming **AC**, there is a set u for which

$$(2) \qquad \mathrm{Ex}(u) \wedge |u| = \aleph_0 \wedge \sigma^{(u)} \wedge \tau^{(u)}.$$

By the Mostowski collapsing lemma (3.11), there is a collapsing isomorphism f of u onto a transitive set y. It follows from this, (2) and 6.7 that

$$\mathrm{Trans}(y) \wedge |y| = \aleph_0 \wedge \sigma^{(y)} \wedge \tau^{(y)}.$$

We have thus shown that

$$\vdash_{\mathbf{AC}} \exists y \varphi(y).$$

This, together with (1), gives

$$\vdash_{\mathbf{AC}} \exists y [\varphi(y) \wedge [\exists z \neg C(z)]^{(y)}],$$

[1] Later we shall show that if **ZF** is consistent, so is **ZFC**. Hence we only really need the consistency of **ZF** for Thm. 7.9 to hold.

i.e.,

(3) $\vdash_{AC} \exists y[\varphi(y) \wedge \exists z \in y \neg C(z)^{(y)}]$.

However, by the definition of $\varphi(y)$, we have

(4) $\vdash \varphi(y) \rightarrow \text{Trans}(y) \wedge y \neq 0 \wedge \sigma^{(y)}$,

so, since $\sigma_1, \ldots, \sigma_m$ is an absoluteness sequence for $C(z)$, it follows from (3), (4) and 7.2 that

$$\vdash_{AC} \exists y[\varphi(y) \wedge \exists z \in y \neg C(z)].$$

Hence, using

$$\vdash \varphi(y) \rightarrow \text{Trans}(y) \wedge |y| = \aleph_0,$$

we have

$$\vdash_{AC} \exists y[|y| = \aleph_0 \wedge \exists z \subseteq y \neg C(z)].$$

From this it would follow immediately that **ZFC** is inconsistent. ∎

IMPORTANT REMARK. The argument we have just given is essentially a formal version of an informal argument — due to Skolem — which runs as follows. Suppose **ZF** is consistent. Then, by the Löwenheim–Skolem Theorem it has a countable model $\mathfrak{A} = \langle A, E \rangle$. Now Cantor's Theorem must hold in $\mathfrak{A}$, so there is a member a of A such that $\mathfrak{A} \models \neg C[a]$. For each $x \in A$, let

$$\hat{x} = \{y \in A : yEx\},$$

let

$$A' = \{\hat{x} : x \in A\},$$

let E' be defined on A' by $\hat{x}E'\hat{y} \leftrightarrow xEy$, and let $\mathfrak{A}' = \langle A', E' \rangle$. Using the fact that the Axiom of Extensionality holds in $\mathfrak{A}$, we see that the map $x \mapsto \hat{x}$ establishes an isomorphism of $\mathfrak{A}$ onto $\mathfrak{A}'$. It follows that $\mathfrak{A}' \models \neg C[\hat{a}]$. But clearly we have $\hat{a} \subseteq A$, so $C(\hat{a})$ holds in the universe of sets. Thus $\hat{a}$ is *uncountable* from the point of view of $\mathfrak{A}'$ but is *countable* from the point of view of the universe of sets. This is *Skolem's paradox*, in which a set may be uncountable inside a model of **ZF** but countable "from the outside". (Of course, this simply means that the given model does not contain a function counting the set in question, although such a function can be found somewhere in V.)

7.10. PROBLEM. Show that the term $\mathsf{P}\omega$ is not absolute, so that the term $\mathsf{P}x$ is not absolute.

§ 8. Constructible sets

The sets R_α are determined by the condition:

$$R_0=\mathbf{0},$$

$$R_{\beta+1}=\mathsf{P}R_\beta,$$

$$R_\lambda=\bigcup\{R_\beta:\ \beta<\lambda\} \quad \text{for limit } \lambda.$$

As α runs through the ordinals, the R_α's exhaust the universe of (regular) sets. This universe is thus built up by:

(a) starting from scratch (i.e. from $\mathbf{0}$);

(b) at each successor stage, collecting together all sets whose members have already been collected;

(c) at each limit stage, collecting together all sets which have already been collected;

(d) iterating this process indefinitely (i.e., throughout the ordinals).

We now enquire whether this process can be modified so that instead of obtaining the whole universe we obtain a *smaller* collection in which, however, the axioms of **ZF** still hold.

Now (a) and (c) are extremely natural, and in any case they do not seem to be responsible for the largeness of the resulting collection. We do not, accordingly, propose to modify (a) and (c).

As for (d), it is certainly *partly* responsible for the large size of the resulting collection, because it tells us to continue the process of collecting new sets as long as possible. The obvious way to modify this is to stop the process at some (large) ordinal α. We shall not, however, pursue this possibility here.

We focus our attention on (b): $R_{\beta+1}=\mathsf{P}R_\beta$. This is certainly responsible for a very rapid increase in the size of the R_β's, because, by Cantor's theorem, we have $|\mathsf{P}R_\beta|>|R_\beta|$. It is therefore reasonable to suppose that the largeness of the collection obtained by applying the process (a)–(d) is to a considerable extent due to the great "strength" of the operation P in (b), and we shall therefore try to obtain a smaller collection by replacing P by a weaker operation.

Now, the size of the power set $\mathsf{P}u$ of a given set u is proportional not only to the size of u but also to the "richness" of the entire universe (see Prob. 7.10). We propose to modify the process (a)–(d) by replacing P by another operation D. For a given set u, $\mathsf{D}u$ is to contain only those subsets of u whose existence can be ascertained by, so to speak, examining only u itself, without scanning the whole universe. (These are the so-called *predicatively defined* subsets of u.) Our hope is that the resulting collection — *the universe L of constructible sets* — will be much smaller than the entire universe. On the other hand, since we are going to include in Du every particular subset of u which is definable in **ZF**, it will turn out that L is a model of **ZF**. Moreover, because of the orderly way in which the members of L are constructed, **AC** will hold in L. Finally, since at each stage in the construction of L we are adding as *few* new sets as possible, it will follow that the power set operation in L is as weak as it possibly can be, and so **GCH** will hold in L.

These considerations lead to the construction described in the remainder of the present section.

We first define

the set of all definable subsets of u:

$$\mathsf{D}u =_{\mathrm{df}} \{y\colon\ y \subseteq u \wedge [y = \mathbf{0} \vee \exists v \exists x[v \in \mathbf{F} \wedge x \in \mathrm{ec}(u) \wedge y = \{z \in u\colon\ \mathbf{Sat}(x(0/z), v, u)\}]]\}.$$

Thus, if $u \neq \mathbf{0}$, Du consists of all those subsets of u which are definable in $\langle u, \in|u\rangle$ by a ⌜formula⌝ of $\mathscr{L}$ involving parameters from u, while $\mathsf{D}\mathbf{0} = \{\mathbf{0}\}$.

8.1. LEMMA. *For any formula* φ *of* $\mathscr{L}$ *whose free variables are all among* $v_0, \ldots, v_n$,

$$\vdash u \neq \mathbf{0} \wedge \exists v_1 \in u \ldots \exists v_n \in u[y = \{z \in u\colon\ \varphi^{(u)}(z, v_1, \ldots, v_n)\}] \rightarrow y \in \mathsf{D}u.$$

PROOF. Assume the antecendent and fix $v_1, \ldots, v_n \in u$ so that

$$y = \{z \in u\colon\ \varphi^{(u)}(z, v_1, \ldots, v_n)\}.$$

Define $x \in \mathrm{ec}(u)$ by letting x‘$\mathbf{0}$ be any element of u, x‘$\mathbf{k} = v_k$ for $0 < k \leqslant n$ and x‘$k = x$‘$\mathbf{0}$ for $k \in \omega - \{\mathbf{0}, \ldots, \mathbf{n}\}$. Then, by 6.5,

$$z \in u \rightarrow [\mathbf{Sat}(x(0/z), \ulcorner\varphi\urcorner, u) \leftrightarrow \varphi^{(u)}(z, v_1, \ldots, v_n)].$$

Hence $y = \{z \in u\colon\ \mathbf{Sat}(x(0/z), \ulcorner\varphi\urcorner, u)\}$; since $\ulcorner\varphi\urcorner \in \mathbf{F}$ by 6.1, it follows that $y \in \mathsf{D}u$. ∎

We now apply Thm. 2.9, taking for $\mathbf{s}(y)$ the term $\bigcup\{\mathsf{D}z\colon\ z \in \mathrm{ran}(y)\}$. We obtain a term L_x such that

(8.2) $\vdash L_\alpha = \bigcup\{\mathsf{D}L_\beta\colon\ \beta < \alpha\}$.

8.3. THEOREM.

(i) $\vdash \beta < \alpha \rightarrow [L_\beta \subseteq L_\alpha \wedge L_\beta \in L_\alpha]$.

(ii) $\vdash \mathrm{Trans}(L_\alpha)$.

(iii) $\vdash L_0 = \mathbf{0} \wedge L_{\alpha+1} = \mathrm{D}L_\alpha \wedge [\mathrm{Lim}(\lambda) \rightarrow L_\lambda = \bigcup\{L_\beta \colon \beta < \lambda\}]$.
(iv) $\vdash L_\alpha \subseteq R_\alpha$.

PROOF. (i) If $\beta < \alpha$, then $L_\beta \subseteq L_\alpha$ follows immediately from (8.2). To show that $L_\beta \in L_\alpha$, it is clearly enough to prove that $L_\beta \in \mathrm{D}L_\beta$. If $L_\beta \neq \mathbf{0}$, this follows from 8.1 and the obvious fact that

$$L_\beta = \{x \in L_\beta \colon (x = x)^{(L_\beta)}\}.$$

If $L_\beta = \mathbf{0}$, then $L_\beta \in \{\mathbf{0}\} = \mathrm{D}L_\beta$.

(ii) If $u \in v \in L_\alpha$, then $u \in v \in \mathrm{D}L_\beta$ for some $\beta < \alpha$ by (8.2). Hence $u \in v \subseteq L_\beta$, so that $u \in L_\beta$, whence $u \in L_\alpha$ by (i).

(iii) is proved like 3.3(iii), and (iv) is easily proved by transfinite induction, using the obvious fact that $\vdash \mathrm{D}u \subseteq \mathrm{P}u$. ∎

We now define

$$x \textit{ is constructible} \colon \mathrm{L}(x) \leftrightarrow_{\mathrm{df}} \exists\alpha(x \in L_\alpha),$$

$$\textit{the order of } x \colon \lambda(x) =_{\mathrm{df}} \boldsymbol{\mu}\alpha[x \in \mathrm{D}L_\alpha].$$

Notice that it follows immediately from 8.3(ii) that L(x) *is a transitive formula.*

The class of all sets x satisfying L(x) (together with the membership relation restricted to this class) is called the *constructible universe.*

8.4. THEOREM. $\vdash \mathrm{L}(\alpha) \wedge \lambda(\alpha) = \alpha$.

PROOF. If $\alpha \in L_\alpha$ then, by 8.3(iv), $\alpha \in R_\alpha$, which contradicts 3.3(v). Thus we need only show that $\vdash \alpha \in \mathrm{D}L_\alpha$; we argue by transfinite induction. First, we have $\mathbf{0} \in \{\mathbf{0}\} = \mathrm{D}L_0$. Suppose that $\alpha \neq \mathbf{0}$ and $\beta < \alpha$. By inductive hypothesis, $\beta \in \mathrm{D}L_\beta$, and so $\beta \in L_\alpha$. On the other hand, if $\beta \geq \alpha$ we cannot have $\beta \in L_\alpha$, for this would, in conjunction with Trans(L_α), imply that $\alpha \in L_\alpha$, which we have already shown to be false. Hence

(1) $\quad \alpha = \{x \in L_\alpha \colon \mathrm{Ord}(x)\}.$

But since $\alpha > \mathbf{0}$, it follows from 8.3(i) that $L_\alpha \neq \mathbf{0}$; and $\vdash \mathrm{Trans}(L_\alpha)$ by 8.3(iii). Hence, by 3.9(ii),

(2) $\quad \forall x \in L_\alpha[\mathrm{Ord}(x) \leftrightarrow \mathrm{Ord}^{(L_\alpha)}(x)].$

(1) and (2) now imply that

$$\alpha = \{x \in L_\alpha \colon \mathrm{Ord}^{(L_\alpha)}(x)\},$$

so that $\alpha \in \mathrm{D}L_\alpha$ by 8.1. ∎

8.5. LEMMA. The term Du and *the formulas* $y=L_\alpha$ *and* $y\in L_\alpha$ *are absolute.*
PROOF. The results of §7 and the equivalence

$$\begin{aligned}\vdash w=\mathrm{D}u\leftrightarrow \mathbf{0}\in w\\ &\wedge\forall y\in w\,[y\neq\mathbf{0}\rightarrow y\subseteq u\\ &\wedge\exists v\in\mathbf{F}\,\exists x\in\mathrm{ec}(u)[y=\{z\in u:\ \mathbf{Sat}(x(\mathbf{0}/z),v,u)\}]]\\ &\wedge\forall v\in\mathbf{F}\,\forall x\in\mathrm{ec}(u)\,[\{z\in u:\ \mathbf{Sat}(x(\mathbf{0}/z),v,u)\}\in w]\end{aligned}$$

immediately imply that the term Du is absolute. Thus the term $\bigcup\{\mathrm{D}z:\ z\in\mathrm{ran}(y)\}$ is absolute; since L_α was constructed from this term by ordinal recursion, the formula $y=L_\alpha$ is absolute by 7.4. This and 7.3(viii) imply that the formula $y\in L_\alpha$ is absolute too. ∎

Our next result is the important *reflection principle for the constructible universe.*

8.6. THEOREM. *For any formulas* $\varphi_1,\ldots,\varphi_n$ *whose free variables are among* $x_1,\ldots,x_m$,

$$\vdash\exists\beta[\alpha<\beta\wedge\forall x_1\in L_\beta\ldots\forall x_m\in L_\beta[[\varphi_1^{(\mathrm{L})}\leftrightarrow\varphi_1^{(L_\beta)}]\wedge\ldots\wedge[\varphi_n^{(\mathrm{L})}\leftrightarrow\varphi_n^{(L_\beta)}]].$$

PROOF. By Thm. 8.3 we have

$$\vdash\alpha\leqslant\beta\rightarrow L_\alpha\subseteq L_\beta,$$

$$\vdash\mathrm{Lim}(\lambda)\rightarrow L_\lambda=\bigcup\{L_\xi:\ \xi<\lambda\}.$$

The required result now follows immediately from Thm. 5.3. ∎

We now show that the constructible universe is a model of **ZF**; more precisely, we have:

8.7. THEOREM. *For each postulate* $\boldsymbol{\sigma}$ *of* **ZF**,

$$\vdash\boldsymbol{\sigma}^{(\mathrm{L})}.$$

PROOF. Recall that in §5 we introduced the set of postulates for **LM** and we proved in Thm. 5.6 that **ZF** and **LM** are equivalent. Accordingly, in order to prove the present theorem it suffices to show that

(1) $\vdash\boldsymbol{\tau}^{(\mathrm{L})}$ for every postulate $\boldsymbol{\tau}$ of **LM**.

For suppose that (1) holds, and let $\boldsymbol{\sigma}$ be any postulate of **ZF**. Then by 5.6 we have[1] $\vdash_{\mathbf{LM}}\boldsymbol{\sigma}$, so that, for some finite sequence $\boldsymbol{\tau}_1,\ldots,\boldsymbol{\tau}_n$ of postulates

[1] We recall from §5 that $\vdash_{\mathbf{LM}}\varphi$ stands for "φ is deducible from the postulates of **LM**".

of **LM**, the formula $\tau_1 \wedge \ldots \wedge \tau_n \rightarrow \sigma$ is logically true. Therefore, by 2.12.4, the formula

$$\exists x \mathrm{L}(x) \rightarrow \tau_1^{(\mathrm{L})} \wedge \ldots \wedge \tau_n^{(\mathrm{L})} \rightarrow \sigma^{(\mathrm{L})}$$

is also logically true. But it follows instantly from 8.4 that $\vdash \exists x \mathrm{L}(x)$. Hence,

$$\vdash \tau_1^{(\mathrm{L})} \wedge \ldots \wedge \tau_n^{(\mathrm{L})} \rightarrow \sigma^{(\mathrm{L})},$$

and it now follows from (1) that $\vdash \sigma^{(\mathrm{L})}$.

We prove (1) by considering all the postulates of **LM** in turn. We recall that these are **Ext, Pow, Reg, Sep** and $\mathbf{RP}_1$.

(i) **Ext, Reg**. That (1) holds for these postulates is a straightforward consequence of the transitivity of $\mathrm{L}(x)$; we leave the verification of this to the reader.

(ii) **Pow**. We have to show that

$$\vdash \mathrm{L}(z) \rightarrow \exists x[\mathrm{L}(x) \wedge \forall y[\mathrm{L}(y) \rightarrow [y \in x \leftrightarrow \forall u[\mathrm{L}(u) \wedge u \in y \rightarrow u \in z]]]].$$

The fact that $\mathrm{L}(x)$ is transitive allows us to drop $\mathrm{L}(u)$ is the above formula, and we therefore have to prove

$$\vdash \mathrm{L}(z) \rightarrow \exists x[\mathrm{L}(x) \wedge \forall y[\mathrm{L}(y) \rightarrow [y \in x \leftrightarrow y \subseteq z]]]. \tag{2}$$

Assume $\mathrm{L}(z)$, and put α for

$$\bigcup \{\lambda(y): \mathrm{L}(y) \wedge y \subseteq z\} + 1.$$

It follows instantly that

$$\mathrm{L}(y) \wedge y \subseteq z \rightarrow y \in L_\alpha, \tag{3}$$

so that

$$\{y: \mathrm{L}(y) \wedge y \subseteq z\} = \{y: y \in L_\alpha \wedge y \subseteq z\}.$$

It follows from (3) that $z \in L_\alpha$, and, since L_α is transitive, it is easy to see that

$$y \in L_\alpha \rightarrow [(y \subseteq z)^{(L_\alpha)} \leftrightarrow y \subseteq z].$$

Hence

$$\{y: \mathrm{L}(y) \wedge y \subseteq z\} = \{y: y \in L_\alpha \wedge (y \subseteq z)^{(L_\alpha)}\}.$$

Thus, if we put x for

$$\{y: \mathrm{L}(y) \wedge y \subseteq z\},$$

we deduce from 8.1 that $x \in \mathrm{D}L_\alpha$, so that $\mathrm{L}(x)$. Also, it is clear from the definition of x that

$$\forall y[L(y) \rightarrow [y \in x \leftrightarrow y \subseteq z]],$$

which gives (2).

(iii) **Sep**. Let φ be any formula with free variables among $x, y_1, \ldots, y_k$, and suppose that z is not free in φ. We have to show that

$$(4) \qquad \vdash \mathrm{L}(y_1) \wedge \ldots \wedge \mathrm{L}(y_k) \wedge \mathrm{L}(u) \rightarrow \exists z[\mathrm{L}(z) \wedge \forall x[\mathrm{L}(x) \rightarrow [x \in z \leftrightarrow \varphi^{(\mathrm{L})} \wedge x \in u]]].$$

Assume that $\mathrm{L}(y_1) \wedge \ldots \wedge \mathrm{L}(y_k) \wedge \mathrm{L}(u)$, and put z for

$$\{x \in u \colon \varphi^{(\mathrm{L})}\}.$$

There are ordinals $\alpha, \alpha_1, \ldots, \alpha_k$ such that $u \in L_\alpha \wedge y_1 \in L_{\alpha_1} \wedge \ldots \wedge y_k \in L_{\alpha_k}$. Let γ be the largest of the ordinals $\alpha, \alpha_1, \ldots, \alpha_k$; then clearly $\{u, y_1, \ldots, y_k\} \subseteq L_\gamma$. By 8.6, there exists an ordinal $\beta > \gamma$ such that

$$x \in L_\beta \wedge y_1 \in L_\beta \wedge \ldots \wedge y_k \in L_\beta \rightarrow [\varphi^{(\mathrm{L})} \leftrightarrow \varphi^{(L_\beta)}].$$

Since $\gamma < \beta$, we have $L_\gamma \subseteq L_\beta$, so that $\{u, y_1, \ldots, y_k\} \subseteq L_\beta$. Hence

$$\forall x \in L_\beta[\varphi^{(\mathrm{L})} \leftrightarrow \varphi^{(L_\beta)}].$$

Using the transitivity of L_β and the fact that $u \in L_\beta$, we see that this implies that

$$z = \{x \in L_\beta \colon [x \in u \wedge \varphi]^{(L_\beta)}\}.$$

Therefore, by 8.1, $z \in \mathrm{D}L_\beta = L_{\beta+1}$, so that $\mathrm{L}(z)$. Also, it follows immediately from the definition of z that

$$\mathrm{L}(x) \rightarrow [x \in z \leftrightarrow \varphi^{(\mathrm{L})} \wedge x \in u].$$

(4) follows.

(iv) $\mathbf{RP}_1$. We have to show that

$$(5) \qquad \vdash \mathrm{L}(y_1) \wedge \ldots \wedge \mathrm{L}(y_k) \rightarrow \exists u[\mathrm{L}(u) \wedge y_1 \in u \wedge \ldots \wedge y_k \in u \wedge \mathrm{Trans}^{(\mathrm{L})}(u) \wedge \forall x_1 \in u \ldots \forall x_m \in u[\mathrm{L}(x_1) \wedge \ldots \wedge \mathrm{L}(x_m) \rightarrow [(\varphi_1 \leftrightarrow \varphi_1^{(u)})^{(\mathrm{L})} \wedge \ldots \wedge (\varphi_n \leftrightarrow \varphi_n^{(u)})^{(\mathrm{L})}]]],$$

where $\varphi_1, \ldots, \varphi_n$ are any formulas all of whose free variables are among $x_1, \ldots, x_m$. But, in view of the presence of $\mathrm{L}(u)$ and the transitivity of the formula L(x), it is clear that, in (5), $(\varphi_i \leftrightarrow \varphi_i^{(u)})^{(L)}$ may be replaced by $\varphi_i^{(L)} \leftrightarrow \varphi_i^{(u)}$, $\text{Trans}^{(L)}(u)$ may be replaced by Trans (u), and $L(x_1) \wedge \ldots \wedge L(x_n)$ may be suppressed. Thus we have to prove

$$(6) \qquad \vdash L(y_1) \wedge \ldots \wedge L(y_k) \to \exists u[L(u) \wedge y_1 \in u \wedge \ldots \wedge y_k \in u \wedge \text{Trans}(u) \wedge \forall x_1 \in u \ldots \forall x_m \in u[[\varphi_1^{(L)} \leftrightarrow \varphi_1^{(u)}] \wedge \ldots \wedge [\varphi_n^{(L)} \leftrightarrow \varphi_n^{(u)}]]].$$

So let us assume that $L(y_1) \wedge \ldots \wedge L(y_k)$. Then, as in the proof of (iii), there is an ordinal γ such that $\{y_1, \ldots, y_k\} \subseteq L_\gamma$. By 8.6, there is an ordinal $\beta > \gamma$ such that

$$(7) \qquad \forall x_1 \in L_\beta \ldots \forall x_m \in L_\beta[[\varphi_1^{(L)} \leftrightarrow \varphi_1^{(L_\beta)}] \wedge \ldots \wedge [\varphi_n^{(L)} \leftrightarrow \varphi_n^{(L_\beta)}]].$$

Taking $u = L_\beta$, we see that $y_1 \in u \wedge \ldots \wedge y_k \in u \wedge \text{Trans(u)}$, and this, together with (7), gives (6). ∎

The reader may find it instructive to prove Thm. 8.7 *directly* without going through **LM**.

Let $\psi(x)$ be a formula with one free variable x. We shall say that $\psi(x)$ *defines a transitive model of* **ZF** if

$$\vdash \exists x\psi \wedge \mathbf{Trans}_{\psi(x)} \quad \text{and} \quad \vdash \sigma^{(\psi)}$$

for every postulate σ of **ZF**. Speaking somewhat imprecisely, $\psi(x)$ defines a transitive model of **ZF** iff it is provable in **ZF** that the class defined by ψ is a transitive model of **ZF** (with $\in$ interpreted as the membership relation).

Thm. 8.7 implies that L(x) defines a transitive model of **ZF**. Moreover, we have:

8.8. THEOREM. *Let* $\psi(x)$ *be a formula which defines a transitive model of* **ZF**. *Then*:

(i) $\vdash \forall\alpha\psi(\alpha) \to \forall x[\psi(x) \to [L^{(\psi)}(x) \leftrightarrow L(x)]]$;

(ii) $\vdash \psi(\alpha) \to \psi(L_\alpha)$;

(iii) $\vdash \forall\alpha\psi(\alpha) \to \forall x[L(x) \to \psi(x)]$.

PROOF. (i). We have to show, assuming $\forall\alpha\psi(\alpha)$,

$$(1) \qquad \psi(x) \to [L^{(\psi)}(x) \leftrightarrow L(x)].$$

Now $L^{(\psi)}(x)$ is $(\exists\alpha[x \in L_\alpha])^{(\psi)}$, and by Prob. 3.9(iii), since by assumption $\vdash \exists x\psi \wedge \mathbf{Trans}_{\psi(x)}$, we have

$$\vdash (\exists\alpha[x \in L_\alpha])^{(\psi)} \leftrightarrow \exists\alpha[\psi(\alpha) \wedge [x \in L_\alpha]^{(\psi)}].$$

Hence

$$\vdash \mathrm{L}^{(\psi)}(x) \leftrightarrow \exists\alpha[\psi(\alpha) \wedge [x \in L_\alpha]^{(\psi)}].$$

But we know that the formula $x \in L_\alpha$ is absolute, so, since $\psi(x)$ defines a transitive model of **ZF**, we obtain

$$\vdash \psi(x) \rightarrow [\mathrm{L}^{(\psi)}(x) \leftrightarrow \exists\alpha[\psi(\alpha) \wedge x \in L_\alpha]].$$

But, since we are assuming $\forall\alpha\psi(\alpha)$, we may suppress the $\psi(\alpha)$ and so get

$$\psi(x) \rightarrow [\mathrm{L}^{(\psi)}(x) \leftrightarrow \exists\alpha[x \in \mathrm{L}_\alpha]],$$

i.e. (1).

(ii) Since ψ defines a transitive model of **ZF**, and $\vdash \exists y[y = L_\alpha]$, it follows that $\vdash [\forall\alpha\, \exists y[y = \mathrm{L}_\alpha]]^{(\psi)}$, hence, using Prob. 3.9(iii),

$$\vdash \psi(\alpha) \rightarrow \exists y[\psi(y) \wedge [y = L_\alpha]^{(\psi)}].$$

But the formula $y = L_\alpha$ is absolute, so that we obtain

$$\vdash \psi(\alpha) \rightarrow \exists y[\psi(y) \wedge y = L_\alpha],$$

which implies $\vdash \psi(\alpha) \rightarrow \psi(L_\alpha)$, as required.

(iii) Assume $\forall\alpha\psi(\alpha)$ and $\mathrm{L}(x)$. Then $x \in L_\alpha$ for some ordinal α, and $\psi(L_\alpha)$ by (ii). Hence $\psi(x)$ follows from the transitivity of ψ. ∎

Thm. 8.8.(iii) says that any transitive definable class model of **ZF** that contains all the ordinals includes the constructible universe. On the other hand Thms. 8.4 and 8.7 imply that the constructible universe is a model of this kind. Therefore *the constructible universe is the smallest definable transitive model of* **ZF** *containing all the ordinals*. This is an *invariant* characterization of the constructible universe, independent of the method by which it is defined.

§ 9. The consistency of AC and GCH

The *Axiom of Constructibility* is the sentence

Constr: $\forall x \mathrm{L}(x)$.

We do *not* adopt **Constr** as a postulate. We shall see, however, that it plays an important role in establishing the consistency of **AC** and **GCH**.

9.1. THEOREM. $\vdash \mathbf{Constr}^{(\mathrm{L})}$.

PROOF. By Thm. 8.7, $\mathrm{L}(x)$ defines a transitive model of **ZF** and by Thm. 8.4

we have $\vdash \forall\alpha L(\alpha)$. Hence, by Thm. 8.8(i),

$$\vdash \forall x[L(x) \to [L^{(L)}(x) \leftrightarrow L(x)]],$$

so that

$$\vdash \forall x[L(x) \to L^{(L)}(x)],$$

i.e.

$$\vdash \mathbf{Constr}^{(L)}. \qquad \blacksquare$$

9.2. COROLLARY. **Constr** *is consistent relative to* **ZF**.
PROOF. Since $\vdash L(\mathbf{0})$, we have $\vdash \exists x L(x)$. By 8.7, we have $\vdash \boldsymbol{\sigma}^{(L)}$ for every postulate $\boldsymbol{\sigma}$ of **ZF**, and by 9.1 we have $\vdash \mathbf{Constr}^{(L)}$. The conclusion now follows from 3.6. ∎

Despite the fact that **Constr** is consistent relative to **ZF**, most set theorists do not accept it as an axiom for set theory. For to accept **Constr** would amount to identifying the universe of sets with the constructible universe, and the latter is far too neat and tidy — like a police state. Each constructible set is generated by means of a formula and a finite number of previously constructed sets and can therefore be identified by these "birthmarks". Since there is no reason to suppose that *all* sets can be obtained and identified in this way, **Constr** is to be rejected as a postulate for set theory.[1]

Nonetheless, **Constr** is important because, as we shall see, it implies both **AC** and **GCH**, from which it follows that these latter are both consistent relative to **ZF**.

So, just for the present, we adjoin **Constr** to the postulates of **ZF**. *The resulting theory will be called* **ZFL**. By 9.2, **ZFL** is consistent if **ZF** is, so clearly any theorem of **ZFL** will be consistent relative to **ZF**. We now address ourselves to the task of proving **AC** and **GCH** in **ZFL**. We assume that the reader is familiar with the elementary theory of well-ordered sets; in particular, with the concepts of *initial segment* of a well-ordered set and *lexicographic well-orderings*. (A convenient reference here is KURATOWSKI–MOSTOWSKI [1968].)

9.3. THEOREM. $\mathbf{ZFL} \vdash \mathbf{AC}$.
PROOF. The idea of the proof is to show that the constructible universe can be well-ordered. The proof will be split into 3 parts.

We first claim that, given a well-ordering $\prec$ of a set u, we can construct a well-ordering $\prec^*$ of $\mathrm{ec}(u)$. To see this, for each $x \in \mathrm{ec}(u)$ let $n(x)$ be the

[1] However, note that, by Thm. 9.1, **Constr** holds in the *constructible universe*.

least $n \in \omega$ such that x assumes a constant value at all $m \geqslant n$. Define the relation $\prec^*$ on ec(u) by

$$x \prec^* y \leftrightarrow_{\text{df}} n(x) < n(y) \vee [n(x) = n(y)$$
$$\wedge\, x\text{`}m \prec y\text{`}m \text{ at the least } m \in \omega \text{ for which } x\text{`}m \neq y\text{`}m],$$

for $x, y \in \text{ec}(u)$ and $x \neq y$. It is easy to verify that $\prec^*$ is a well-ordering of ec(u).

Our second claim is that the set $\mathbf{F}$ of ⌜formulas⌝ can be well-ordered. First the set $\mathbf{F}_0 = 2 \times \omega \times \omega$ can be lexicographically well-ordered, using the natural well-ordering of ω. We now argue inductively. If $\mathbf{F}_n$ has been well-ordered by a relation $\prec_n$, one obtains a well-ordering $\prec_{n+1}$ of

$$\mathbf{F}_{n+1} = \mathbf{F}_n \cup [\{2\} \times \mathbf{F}_n \times \mathbf{F}_n] \cup [\{3\} \times \mathbf{F}_n] \cup [\{4\} \times \omega \times \mathbf{F}_n]$$

by well-ordering the last three summands lexicographically (using the given well-ordering $\prec_n$ of $\mathbf{F}_n$), and then putting all the elements of $\mathbf{F}_n$ before those of $[\{2\} \times \mathbf{F}_n \times \mathbf{F}_n] - \mathbf{F}_n$, the elements of this latter set before those of $[\{3\} \times \mathbf{F}_n] - \mathbf{F}_n$, and the elements of this latter before those of $[\{4\} \times \omega \times \times \mathbf{F}_n] - \mathbf{F}_n$. In this way one obtains for each $n \in \omega$ a well-ordering $\prec_n$ of $\mathbf{F}_n$, such that, for $m \leqslant n$, $\prec_m$ is the restriction of $\prec_n$ to $\mathbf{F}_m$, and $\mathbf{F}_m$ is an initial segment of $\mathbf{F}_n$ with respect to $\prec_n$. One can then define a well-ordering $\prec$ on $\mathbf{F}$ by setting, for $u, v \in \mathbf{F}$, $u \neq v$,

$$u \prec v \leftrightarrow_{\text{df}} [\nu(u) < \nu(v)] \vee [\nu(u) = \nu(v) \wedge u \prec_{\nu(u)} v],$$

where, for each $x \in \mathbf{F}$, $\nu(x) = \boldsymbol{\mu} n[x \in \mathbf{F}_n]$.

Finally, we prove the theorem. We construct by ordinal recursion a term $\mathbf{t}(x)$ with the following properties:

(i) $\vdash \mathbf{t}(\alpha)$ *is a well-ordering of* L_α;

(ii) $\vdash$ *if* $\beta \leqslant \alpha$, *then* L_β *is an initial segment of* L_α *with respect to* $\mathbf{t}(\alpha)$;

(iii) $\vdash$ *if* $\beta \leqslant \alpha$, $\mathbf{t}(\beta)$ *is the restriction of* $\mathbf{t}(\alpha)$ *to* L_β.

We proceed somewhat informally, but the construction of $\mathbf{t}$ can easily (but tediously) be recast in the form specified in 2.11.

First, we put

$$\mathbf{t}(\mathbf{0}) = \mathbf{t}(\mathbf{1}) = \mathbf{0}.$$

If $\mathbf{t}(\beta)$ has been defined to meet conditions (i)–(iii) for all $\beta < \lambda$, with λ a limit ordinal, then

$$L_\lambda = \bigcup \{L_\beta \colon \beta < \lambda\}$$

and we put

$$\mathbf{t}(\lambda) = \bigcup\{\mathbf{t}(\beta):\ \beta<\lambda\}.$$

Suppose now that $\alpha=\beta+1$ with $\beta>0$ and $\mathbf{t}(\beta)$ has been defined. Then $L_\alpha = L_\beta \cup \mathrm{D}L_\beta$. Now $\mathbf{t}(\beta)$ is a well-ordering of L_β and so by the first part of the proof we can well-order $\mathrm{ec}(L_\beta)$. Also, by the second part of the proof, we can well-order $\mathbf{F}$. Let $\prec$ be the lexicographic well-ordering of $\mathrm{ec}(L_\beta)\times\mathbf{F}$ obtained from these two well-orderings.

For each $y\in \mathrm{D}L_\beta$ there is $v\in\mathbf{F}$ and $x\in\mathrm{ec}(L_\beta)$ such that

$$(1)\qquad y=\{z\in L_\beta:\ \mathbf{Sat}(x(\mathbf{0}/z),v,L_\beta)\};$$

let $g(y)$ be the $\prec$-least member $\langle x,v\rangle$ of $\mathrm{ec}(L_\beta)\times\mathbf{F}$ such that x and v satisfy (1).

We now define $\mathbf{t}(\alpha)$ by

$$\begin{aligned}\langle y,z\rangle\in\mathbf{t}(\alpha)\leftrightarrow&[[y\in L_\beta\wedge z\in L_\beta\wedge\langle y,z\rangle\in\mathbf{t}(\beta)]\\ &\vee[y\in L_\beta\wedge z\in L_\alpha-L_\beta]\\ &\vee[y\in L_\alpha-L_\beta\wedge z\in L_\alpha-L_\beta\wedge g(y)\prec g(z)]].\end{aligned}$$

It is easy to verify that $\mathbf{t}(\alpha)$ satisfies the required conditions.

We now put $\boldsymbol{\varphi}(x,y)$ for

$$\exists\alpha[\langle x,y\rangle\in\mathbf{t}(\alpha)].$$

Assuming **Constr**, for each set u there is an ordinal α such that $u\subseteq L_\alpha$, and therefore

$$\{\langle x,y\rangle:\ x\in u\wedge y\in u\wedge\boldsymbol{\varphi}(x,y)\}$$

is a well-ordering of u. If for each $u\neq\mathbf{0}$ we put $\mathbf{s}(u)$ for the least member of u under this well-ordering, then

$$\{\langle x,\mathbf{s}(x)\rangle:\ x\neq\mathbf{0}\wedge x\in u\}$$

is a choice function for u. **AC** follows. ∎

We now turn to the problem of deriving **GCH** in **ZFL**. We assume that the reader is familiar with the basic facts about cardinal arithmetic — in particular, the properties of cardinal sums and products — and how to translate these facts into our present formal framework.

9.4. LEMMA.

(i) $\vdash\mathbf{F}\approx\omega$.

(ii) $\vdash_{\mathbf{AC}}\aleph_0\leqslant|u|\rightarrow|\mathrm{ec}(u)|=|u|$.

PROOF. (i) is easy and its proof is left as an exercise to the reader (see the second part of the proof of 9.3).

(ii) Assume $\aleph_0 \leqslant |u|$. For each $x \in u$ let S_x be the set of members of ec(u) which eventually assume the constant value x. Then each member of S_x is uniquely determined by the finite sequence of values it assumes before it assumes the constant value x. It follows that, for each $x \in u$,

$$|S_x| \leqslant |\bigcup\{u^n: n \in \omega\}| \leqslant \Sigma\{|u|^n: n \in \omega\} = |u| \cdot \aleph_0 = |u|.$$

Therefore

$$|\mathrm{ec}(u)| = |\bigcup\{S_x: x \in u\}| \leqslant |u|^2 = |u|.$$

Since we clearly have $|u| \leqslant |\mathrm{ec}(u)|$, (ii) follows. ∎

9.5. LEMMA.

(i) $\vdash \mathrm{Fin}(u) \to \mathrm{Fin}(\mathrm{D}u)$.

(ii) $\vdash_{\mathbf{AC}} \aleph_0 \leqslant |u| \to |\mathrm{D}u| = |u|$.

(iii) $\vdash \alpha < \omega \to \mathrm{Fin}(L_\alpha)$.

(iv) $\vdash_{\mathbf{AC}} \omega \leqslant \alpha \to |L_\alpha| = |\alpha|$,

where Fin(u) *is* $\exists n(u \approx n)$.

PROOF. (i) is an immediate consequence of $\vdash \mathrm{D}u \subseteq \mathrm{P}u$ and (iii) follows easily from (i).

To prove (ii), we first observe that each member of Du is determined by a member of $\mathbf{F}$ and a member of ec(u). Hence, using 9.4, if $\aleph_0 \leqslant |u|$ then

$$|\mathrm{D}u| \leqslant |\mathbf{F} \times \mathrm{ec}(u)| = \aleph_0 \cdot |u| = |u|.$$

Also, it is easy to verify that

$$\vdash x \in u \to \{x\} \in \mathrm{D}u,$$

so that $|u| \leqslant |\mathrm{D}u|$. (ii) follows.

Finally, (iv) is proved by transfinite induction, using (i)–(iii). It follows from 8.4 that $\alpha \subseteq L_\alpha$, so that $|\alpha| \leqslant |L_\alpha|$. Moreover, if $\omega \leqslant \alpha$, then, using the inductive hypothesis and (i)–(iii), we have

$$|L_\alpha| = |\bigcup\{\mathrm{D}L_\beta: \beta < \alpha\}| \leqslant \Sigma\{|\mathrm{D}L_\beta|: \beta < \alpha\} \leqslant |\alpha|^2 = |\alpha|.$$

This proves (iv). ∎

We are now in a position to prove

9.6. THEOREM. $\mathbf{ZFL} \vdash \mathbf{GCH}$.

PROOF. Since we know that $\mathbf{ZFL} \vdash \mathbf{AC}$ we shall use $\mathbf{AC}$ freely. Recall that $\mathbf{GCH}$ is the assertion $\forall\alpha[|\mathrm{P}\aleph_\alpha| = \aleph_{\alpha+1}]$. Since $\vdash_{\mathbf{AC}} \aleph_{\alpha+1} \leqslant |\mathrm{P}\aleph_\alpha|$, in order to prove the theorem it suffices to derive $|\mathsf{P}\aleph_\alpha| \leqslant \aleph_{\alpha+1}$ in **ZFL**; and in view of 9.5(iv), to achieve this it is enough to show that

(*) $\quad \mathbf{ZFL} \vdash \mathsf{P}L_{\aleph_\alpha} \subseteq L_{\aleph_{\alpha+1}}$.

Assume **Constr**, and let $z \subseteq L_{\aleph_\alpha}$. We know from 8.5 that the formula $x \in L_\beta$ is absolute; let $\boldsymbol{\sigma}_1,\ldots,\boldsymbol{\sigma}_n$ be an absoluteness sequence for it. Then, by $\mathbf{RP}_1$, there is a transitive set u such that $L_{\aleph_\alpha} \cup \{z\} \subseteq u$ and

(1) $\quad \mathbf{Constr}^{(u)} \wedge \boldsymbol{\sigma}_1^{(u)} \wedge \ldots \wedge \boldsymbol{\sigma}_n^{(u)}$.

Now, by 9.5(iv), we have

$$|L_{\aleph_\alpha} \cup \{z\}| = |L_{\aleph_\alpha}| = \aleph_\alpha,$$

and so by 6.6 there is an elementary substructure v of u such that $|v| = \aleph_\alpha$ and $L_{\aleph_\alpha} \cup \{z\} \subseteq v$. Because of (1) and the fact that ES(v,u), we have

(2) $\quad \mathbf{Constr}^{(v)} \wedge \boldsymbol{\sigma}_1^{(v)} \wedge \ldots \wedge \boldsymbol{\sigma}_n^{(v)}$.

Also, since u is transitive, it is extensional, so that $\mathbf{Ext}^{(u)}$, whence $\mathbf{Ext}^{(v)}$, and therefore v is extensional. Accordingly, by 3.11, there is an $\in$-isomorphism f of v onto a transitive set y. Since $|v| = \aleph_\alpha$, we have $|y| = \aleph_\alpha$. Moreover, 6.7 and (2) give

(3) $\quad \mathbf{Constr}^{(y)} \wedge \boldsymbol{\sigma}_1^{(y)} \wedge \ldots \wedge \boldsymbol{\sigma}_n^{(y)}$.

Also, since $L_{\aleph_\alpha}$ is a transitive subset of v, 3.14 implies that $f\text{`}x = x$ for every $x \in L_{\aleph_\alpha}$. Hence, by 3.12 we have for our $z \subseteq L_{\aleph_\alpha}$,

$$f\text{`}z = \{f\text{`}x:\ x \in z\} = \{x:\ x \in z\} = z,$$

so that $z \in y$.

By (3) we have $\mathbf{Constr}^{(y)}$, i.e.

$$\forall x \in y [\exists \beta [x \in L_\beta]]^{(y)}.$$

Since y is transitive, this gives, using 3.9(iii),

$$\forall x \in y\ \exists \beta \in y [x \in L_\beta]^{(y)}.$$

Now $x \in L_\beta$ is absolute, and from (3) we see that the members of its absoluteness sequence hold when relativized to the transitive set y, so we get

$$\forall x \in y\ \exists \beta \in y [x \in L_\beta].$$

Since $z\in y$, we obtain in particular $\exists\beta\in y[z\in L_\beta]$. Let β be an ordinal in y such that $z\in L_\beta$. Then, since y is transitive, we have $\beta\subseteq y$, whence $|\beta|\leqslant|y|=\aleph_\alpha$, and so $\beta<\aleph_{\alpha+1}$. Hence $z\in L_\beta\subseteq L_{\aleph_{\alpha+1}}$, and $(*)$ follows. ∎

Notice the similarity between this proof and that of Thm. 7.9.

We conclude from 9.2, 9.3 and 9.6 that

if **ZF** *is consistent, so is* **ZF+AC+GCH**.

§ 10. Problems

(Throughout this section we revert to our original convention and write $\Sigma\vdash\varphi$ for "the formula φ is deducible from the set of sentences Σ in the first-order predicate calculus.")

10.1. Let $\mathbf{ZF}^-$ be **ZF** with **Inf** omitted. Show that

$$\mathbf{ZF}\vdash\sigma^{(R_\omega)}$$

for every postulate σ of $\mathbf{ZF}^-$ and $\mathbf{ZF}\vdash\neg\mathbf{Inf}^{(R_\omega)}\wedge\mathbf{AC}^{(R_\omega)}$. Deduce that, if $\mathbf{ZF}^-$ is consistent, then **Inf** is not a theorem of $\mathbf{ZF}^-$.

*10.2. Let **Z** (resp. **ZC**) be obtained from **ZF** (resp. **ZFC**) by omitting **Rep** and substituting instead **Sep** and the postulate ("pairing axiom")

$$\forall u\,\forall v\,\exists x\,\forall y[y\in x\leftrightarrow y=u\vee y=v].$$

(**Z** is called *Zermelo set theory*.)

(i) Show that

$$\mathbf{ZF}\vdash\mathrm{Lim}(\alpha)\wedge\omega<\alpha\rightarrow\sigma^{(R_\alpha)}\quad(\text{resp. }\mathbf{ZFC}\vdash\mathrm{Lim}(\alpha)\wedge\omega<\alpha\rightarrow\sigma^{(R_\alpha)})$$

for every postulate σ of **Z** (resp. **ZC**).

(ii) Let τ be the sentence $\forall x\,\exists\alpha[x\approx\alpha]$. Show that

$$\mathbf{ZFC}\vdash\mathbf{AC}^{(R_{\omega+\omega})}\wedge\neg\tau^{(R_{\omega+\omega})}.$$

Using (i), deduce that, if **ZF** is consistent, τ is not a theorem of **ZC**.

(iii) Show that, if **ZF** is consistent, then for each theorem σ of **ZF** (resp. **ZFC**), there is a theorem τ of **ZF** (resp. **ZFC**) such that $\mathbf{Z}+\sigma\nvdash\tau$ (resp. $\mathbf{ZC}+\sigma\nvdash\tau$). (If $\mathbf{ZF}\vdash\sigma$, put τ for

$$\exists\alpha[\mathrm{Lim}(\alpha)\wedge\omega<\alpha\wedge\sigma^{(R_\alpha)}].$$

Use $\mathbf{RP}_2$ to show that $\mathbf{ZF}\vdash\boldsymbol{\tau}$. If

$$\beta = \boldsymbol{\mu}\alpha[\mathrm{Lim}(\alpha)\wedge\omega<\alpha\wedge\boldsymbol{\sigma}^{(R_\alpha)}],$$

show that $\mathbf{ZF}\vdash\neg\boldsymbol{\tau}^{(R_\beta)}$, and use (i).)

(iv) Deduce from (iii) that, if **ZF** is consistent, neither **ZF** nor **ZFC** is finitely axiomatizable.

*10.3. A sentence of the form $\boldsymbol{\varphi}^{(R_\omega)}$ is called an *arithmetical sentence*. Show that, if $\boldsymbol{\sigma}$ is an arithmetical sentence and $\mathbf{ZFL}\vdash\boldsymbol{\sigma}$, then $\mathbf{ZF}\vdash\boldsymbol{\sigma}$, and *a fortiori*, if $\mathbf{ZF}+\mathbf{AC}+\mathbf{GCH}\vdash\boldsymbol{\sigma}$, then $\mathbf{ZF}\vdash\boldsymbol{\sigma}$. (Show that $\mathbf{ZF}\vdash R_\omega=L_\omega$, and hence that $\boldsymbol{\sigma}$ is absolute. Let $\boldsymbol{\varphi}_1,\dots,\boldsymbol{\varphi}_n$ be a proof of $\boldsymbol{\sigma}$ from **ZFL**. Show that $\mathbf{ZF}\vdash\boldsymbol{\varphi}_k^{(\mathrm{L})}$ for $k=1,\dots,n$ and then use the absoluteness of $\boldsymbol{\sigma}$ to obtain the result.)

10.4. Let $\mathscr{L}^+$ be a language obtained by adding new individual constants to $\mathscr{L}$, and let $\mathscr{L}^+(\mathbf{a})$ be the language obtained by adding a new individual constant $\mathbf{a}$ to $\mathscr{L}^+$.

(i) Let $\mathbf{t}_1,\dots,\mathbf{t}_n$ be terms of $\mathscr{L}^+$ without free variables, let $\mathbf{ZF}^+$ be a theory in $\mathscr{L}^+$ whose postulates include those of **ZF**, and let $\boldsymbol{\Sigma}$ be the theory in $\mathscr{L}^+(\mathbf{a})$ whose postulates are:

(1) all postulates of $\mathbf{ZF}^+$;

(2) $\mathrm{Trans}(\mathbf{a})\wedge\mathbf{t}_1\in\mathbf{a}\wedge\dots\wedge\mathbf{t}_n\in\mathbf{a}$;

(3) $\mathbf{Refl}_{\boldsymbol{\varphi}}(\mathbf{a})$, for each formula $\boldsymbol{\varphi}$ of $\mathscr{L}$.

Show that $\boldsymbol{\Sigma}$ is a conservative extension of $\mathbf{ZF}^+$, i.e., for each formula $\boldsymbol{\varphi}$ of $\mathscr{L}^+$, $\boldsymbol{\Sigma}\vdash\boldsymbol{\varphi}$ iff $\mathbf{ZF}^+\vdash\boldsymbol{\varphi}$. (Use $\mathbf{RP}_1$.)

(ii) Let $\mathbf{ZFC}^+$ be a theory in $\mathscr{L}^+$ whose postulates include those of **ZFC**, let $\mathbf{t}$ be a term of $\mathscr{L}^+$ without free variables such that

$$\mathbf{ZFC}^+\vdash\mathrm{Trans}(\mathbf{t})\wedge\aleph_0\leq|\mathbf{t}|,$$

and let $\boldsymbol{\Sigma}$ be the theory in $\mathscr{L}^+(\mathbf{a})$ whose postulates are:

(1) all postulates of $\mathbf{ZFC}^+$;

(2) $\mathrm{Trans}(\mathbf{a})\wedge\mathbf{t}\subseteq\mathbf{a}\wedge|\mathbf{t}|=|\mathbf{a}|$;

(3) $\mathbf{Refl}_{\boldsymbol{\varphi}}(\mathbf{a})$, for each formula $\boldsymbol{\varphi}$ of $\mathscr{L}$.

Show that $\boldsymbol{\Sigma}$ is a conservative extension of $\mathbf{ZFC}^+$. (Use $\mathbf{RP}_1$, Thm. 6.6 and the Mostowski collapsing lemma.)

*10.5. (i) Show how to construct an absolute $\mathscr{L}$-formula Post(v) which expresses the statement "v is a ⌜postulate of **ZF**⌝".

(ii) Let Mod(u) be the formula,

$$\forall v[\mathrm{Post}(v)\rightarrow\forall x\in\mathrm{ec}(u)\,\mathbf{Sat}(x,v,u)]\wedge u\neq\mathbf{0}.$$

(Mod(u) then expresses "$\langle u, \in|u\rangle \models \mathbf{ZF}$".) Show that Mod($u$) is absolute.

(iii) Show that, if **ZF** is consistent, then[1]

$$\mathbf{ZF} \nvdash \exists u[\text{Trans}(u) \wedge \text{Mod}(u)]$$

(Suppose $\mathbf{ZF} \vdash \exists u[\text{Trans}(u) \wedge \text{Mod}(u)]$; let u be a set of least rank such that $\text{Trans}(u) \wedge \text{Mod}(u)$. Now observe that $\exists x[\text{Trans}(x) \wedge \text{Mod}(x)]$ holds in $\langle u, \in|u\rangle$, and use (ii) to conclude that **ZF** would be inconsistent.)

*10.6. Let $\boldsymbol{\Sigma}$ be a set of $\mathscr{L}$-sentences. A set u is called a *model* of $\boldsymbol{\Sigma}$ if $\langle u, \in|u\rangle \models \boldsymbol{\Sigma}$. The *index* of u, ind(u), is the least ordinal not in u.

(i) Show that, if u is a transitive model of **ZF**, then ind(u) is a limit ordinal and $\text{ind}(u) = \{\alpha : \alpha \in u\}$.

(ii) Let u be a transitive model of **ZF** and let $\alpha = \text{ind}(u)$. Show that

$$L_\alpha = \{x \in u : L^{(u)}(x)\}.$$

(Use absoluteness of the formula $x \in L_\xi$.) Deduce that L_α is a model of **ZFL**.

(iii) Show that each transitive model of **ZFL** is of the form L_α. (Use (ii).)

*10.7. Assume that there exists a transitive model of **ZF**.

(i) Show that there is an ordinal ξ for which L_ξ is a model of **ZF**.

(ii) Let ξ_0 be the least ordinal for which L_{ξ_0} is a model of **ZF**. L_{ξ_0} is called the *minimal model* of **ZF**. Show that, if u is any transitive model of **ZF**, then $L_{\xi_0} \subseteq u$. (Show that $\xi_0 \leqslant \text{ind}(u)$ and use 10.6(ii).)

(iii) Show that ξ_0 is countable. (Use the Downward Löwenheim–Skolem theorem and the Mostowski Collapsing Lemma.)

(iv) Let $\boldsymbol{\varphi}(x)$ be a transitive $\mathscr{L}$-formula. Show that, if $\mathbf{ZF} \vdash \boldsymbol{\sigma}^{(\varphi)}$ for all postulates $\boldsymbol{\sigma}$ of **ZF**, then it is never the case that[2] $\mathbf{ZF} \vdash \neg\,\mathbf{Constr}^{(\varphi)}$. (Assuming the hypothesis, consider the transitive collapse of $\{x \in L_{\xi_0} : \boldsymbol{\varphi}^{(L_{\xi_0})}(x)\}$.)

*10.8 (i). Show how to construct an absolute term FV(x) such that, for each ⌜formula⌝ v, FV(v) is the set of subscripts of variables free in v. In particular, if $\boldsymbol{\varphi}$ is a formula whose free variables are exactly $v_{n_1}, \ldots, v_{n_k}$, show that

$$\mathbf{ZF} \vdash \text{FV}(\ulcorner \boldsymbol{\varphi} \urcorner) = \{\mathbf{n}_1, \ldots, \mathbf{n}_k\}.$$

[1] Notice that Gödel's Second Incompleteness Theorem (7.11.9) gives the stronger result that, if **ZF** is consistent, then $\mathbf{ZF} \nvdash \exists u\, \text{Mod}(u)$.

[2] This result shows that the method we employed (that of constructing an "inner model") to prove the relative consistency of **Constr** (and hence of **AC** and **GCH**) cannot be used to prove the relative consistency of ¬**Constr** (nor, *a fortiori*, of ¬**AC** or ¬**GCH**).

(ii) Put

$$\mathsf{Df}(u) =_{\mathrm{df}} \{z\in u\colon\ \exists v\in \mathbf{F}\ \exists x\in \mathrm{ec}(u)[\mathrm{FV}(v)=\{\mathbf{0}\}$$
$$\wedge\ \forall y\in u[\mathbf{Sat}(x(\mathbf{0}/y),v,u)\leftrightarrow y=z]]\}.$$

$\mathsf{Df}(u)$ is the set of all definable elements of u (i.e. definable in the structure $\langle u, \in|u\rangle$, cf. Prob. 7.9.17). Let φ be a formula whose free variables are among $v_0,\dots,v_n$. Show that

$$\mathbf{ZF}\vdash x_1\in \mathsf{Df}(u)\wedge \dots \wedge x_n\in \mathsf{Df}(u)\wedge x\in u$$
$$\wedge\ \forall y\in u[\varphi^{(u)}(y, x_1,\dots,x_n)\leftrightarrow y=x]\rightarrow x\in \mathsf{Df}(u).$$

(iii) Prove the *extended reflection principle* of Myhill and Scott: for any formulas $\varphi_1,\dots,\varphi_n$,

$$\mathbf{ZF}\vdash \forall\alpha_1\dots\forall\alpha_m\ \exists\beta[\alpha_1\in \mathsf{Df}(R_\beta)\wedge \dots \wedge \alpha_m\in \mathsf{Df}(R_\beta)\wedge \mathbf{Refl}_{\varphi_1,\dots,\varphi_n}(R_\beta)].$$

(Put $\varphi(\alpha_1,\dots,\alpha_m)$ for the formula

$$\exists\beta[\alpha_1\in \mathsf{Df}(R_\beta)\wedge \dots \wedge \alpha_m\in \mathsf{Df}(R_\beta)\wedge \mathbf{Refl}_{\varphi_1,\dots,\varphi_n}(R_\beta)],$$

and suppose if possible that $\exists\alpha_1\dots\exists\alpha_m\neg\varphi$. Define the ordinals $\gamma_1,\dots,\gamma_m$ inductively by

$$\gamma_i=\mu\alpha_i[\exists\alpha_{i+1}\dots\exists\alpha_m\ \neg\varphi(\gamma_1,\dots,\gamma_{i-1}, \alpha_i,\dots,\alpha_m)]$$

for $i=1,\dots,m$. Then $\neg\varphi(\gamma_1,\dots,\gamma_m)$. Derive a contradiction by using $\mathbf{RP}_2$ and (ii) to prove the existence of an ordinal β such that

$$\gamma_1\in \mathsf{Df}(R_\beta)\wedge \dots \wedge \gamma_m\in \mathsf{Df}(R_\beta)\wedge \mathbf{Refl}_{\varphi_1,\dots,\varphi_n}(R_\beta).)$$

(iv) Let L_{ξ_0} be the minimal model of $\mathbf{ZF}$ (10.7). Show that $\mathsf{Df}(L_{\xi_0})=L_{\xi_0}$, i.e. L_{ξ_0} is pointwise definable (Prob. 7.9.17). (Since L_{ξ_0} is a model of $\mathbf{ZFL}$, it has a definable well-ordering. Using this fact, show that $\mathsf{Df}(L_{\xi_0})$ is an elementary substructure of L_{ξ_0}. If f is the collapsing isomorphism of $\mathsf{Df}(L_{\xi_0})$ onto a transitive set M, show that $M=L_{\xi_0}$ and f is the identity.)

*10.9. Define the term $L(\alpha,u)$ recursively on α by putting

$$L(\mathbf{0},u)=u,$$

$$L(\alpha+1,u)=\mathsf{D}L(\alpha,u),$$

$$L(\lambda,u)=\bigcup\{L(\beta,u)\colon\ \beta<\lambda\}\quad \text{for limit } \lambda.$$

Write $L_\alpha(u)$ for $L(\alpha,u)$, and let Constr(x,u) be the formula

$$\exists\alpha(x\in L_\alpha(u)).$$

A set x satisfying Constr(x,u) is said to be *constructible from u.* Notice that $\mathbf{ZF}\vdash L_\alpha(0)=L_\alpha$.

(i) Show that

$$\mathbf{ZF}\vdash \mathrm{Trans}(u)\rightarrow \mathrm{Trans}(L_\alpha(u))\wedge\alpha\in L_{\alpha+1}(u).$$

(ii) Show that the formula $y=L_\alpha(u)$ is absolute. (Like the proof of the same assertion for $y=L_\alpha$.)

(iii) Show that

$$\mathbf{ZF}\vdash \mathrm{Trans}(u)\rightarrow \boldsymbol{\sigma}^{(\mathrm{Constr}(x,u))}$$

for each postulate $\boldsymbol{\sigma}$ of **ZF**, where the relativization is taken with x as chosen variable. (Like the proof of $\mathbf{ZF}\vdash\boldsymbol{\sigma}^{(\mathrm{L})}$).

(iv) Show that

$$\mathbf{ZF}\vdash \mathrm{Trans}(u)\wedge\exists\alpha[u\approx\alpha]^{(\mathrm{Constr}(x,u))}\rightarrow \mathbf{AC}^{(\mathrm{Constr}(x,u))}.$$

(Like the proof of $\mathbf{ZFL}\vdash\mathbf{AC}$).

(v) Let $\mathbf{Co}(u)$ be the formula $\forall x$ Constr(x,u). Show that

$$\mathbf{ZF}\vdash \mathrm{Trans}(u)\rightarrow \mathbf{Co}(u)^{(\mathrm{Constr}(x,u))}.$$

(Like the proof of $\mathbf{ZF}\vdash\mathbf{Constr}^{(\mathrm{L})}$).

(vi) Show that

$$\mathbf{ZFC}\vdash \mathrm{Trans}(u)\wedge\mathbf{Co}(u)\rightarrow\forall\alpha[|u|\leqslant\aleph_\alpha\rightarrow|\mathrm{P}\aleph_\alpha|=\aleph_{\alpha+1}].$$

(Like the proof of $\mathbf{ZFL}\vdash\mathbf{GCH}$.)

(vii) Show that, if

$$\mathbf{ZF}+\exists x[\neg L(x)\wedge x\subseteq\omega]$$

is consistent, so is

$$\mathbf{ZF}+\exists x[\neg \mathrm{L}(x)\wedge x\subseteq\omega]\wedge\mathbf{AC}\wedge\mathbf{GCH}.$$

(Take $x\subseteq\omega$ such that $\neg\mathrm{L}(x)$; let $u=\omega\cup\{x\}$ and show that

$$\mathbf{ZF}+\exists x[\neg \mathrm{L}(x)\wedge x\subseteq\omega]\wedge\mathbf{AC}\wedge\mathbf{GCH}$$

hold relativized to Constr(x,u).)

(viii) Show that, if u is a transitive set, then the class of all sets x satisfying $\mathrm{Constr}(x,u)$ is the least transitive definable class which includes the class of all ordinals, contains u and is a model of **ZF**.

10.10. Put $\mathrm{OD}(x)$ for the formula $\exists\alpha[x\in \mathrm{Df}(R_\alpha)]$. A set x satisfying $\mathrm{OD}(x)$ is called an *ordinal-definable* set.

(i) Let φ be a formula all of whose free variables are among $v_0,\dots,v_n$. Show that

$$\mathbf{ZF}\vdash \exists\alpha_1\dots\exists\alpha_n\,\forall y[\varphi(y,\alpha_1,\dots,\alpha_n)\leftrightarrow x=y]\rightarrow \mathrm{OD}(x).$$

(Use 10.8(iii) and (ii).)

(ii) Show how to construct a term $\mathbf{t}(x)$ for which

$$\vdash \mathrm{Inj}(\mathbf{t}|\omega)\wedge \mathrm{ran}(\mathbf{t}|\omega)=\mathbf{F}.$$

(iii) Put $\mathrm{Def}(v,z,u)$ for the formula

$$z\in u\wedge v\in\mathbf{F}\wedge \mathrm{FV}(v)=\{\mathbf{0}\}\wedge\exists x\in \mathrm{ec}(u)\,\forall y\in u\big[y=z\leftrightarrow \mathbf{Sat}(x(\mathbf{0}/y),v,u)\big].$$

($\mathrm{Def}(v,z,u)$ expresses "v defines z in the set u".) Also put

$$\mathbf{s}_1(x)=_{\mathrm{df}}\mu\alpha[x\in \mathrm{Df}(R_\alpha)],\qquad \mathbf{s}_2(x)=_{\mathrm{df}}\mu n\,\mathrm{Def}\big(\mathbf{t}(n),x,R_{\mathbf{s}_1(x)}\big),$$

where $\mathbf{t}$ is the term introduced in (ii). Finally, put $x\prec y$ for the formula

$$\mathrm{OD}(x)\wedge\mathrm{OD}(y)\wedge\big[\mathbf{s}_1(x)<\mathbf{s}_1(y)\vee[\mathbf{s}_1(x)=\mathbf{s}_1(y)\wedge\mathbf{s}_2(x)<\mathbf{s}_2(y)]\big].$$

Show that $\prec$ is a well-ordering of the class of all ordinal-definable sets.

10.11. Let Σ be a set of $\mathscr{L}$-sentences which includes all the postulates of **ZF**. (Under these conditions we say that Σ is an *extension* of **ZF**.) Σ is said to have the *selection property* if for each formula φ containing exactly the variable x free there is a formula ψ containing exactly x free such that

$$\Sigma\vdash\exists!x\psi\wedge[\exists x\varphi\rightarrow\exists x(\varphi\wedge\psi)].$$

In other words, Σ has the selection property if each non-empty definable (without parameters) class can be proved in Σ to have a definable element.

(i) Show that $\mathbf{ZF}+\forall x\mathrm{OD}(x)$ is the weakest extension of **ZF** with the selection property. (Use 10.10 to show that $\mathbf{ZF}+\forall x\mathrm{OD}(x)$ has the selection property. If Σ is any extension of **ZF** with the selection property, consider the formula $\neg\mathrm{OD}(x)$.)

(ii) Let $\mathfrak{A}$ be a model of **ZF** and let $\mathfrak{B}=\mathfrak{A}|\mathrm{Df}(A)$. Show that the following conditions are equivalent:

(a) $\mathfrak{A} \models \forall x \mathrm{OD}(x)$;
(b) $\mathbf{Th}(\mathfrak{A})$ has the selection property;
(c) $\mathfrak{B} \prec \mathfrak{A}$.
(Use (i).)

*10.12. Put HOD(x) for the formula $\forall y \in \mathrm{TC}(\{x\})\, \mathrm{OD}(y)$. HOD($x$) is read "$x$ is hereditarily ordinal definable".
(i) Show that

$$\mathbf{ZF} \vdash \mathrm{HOD}(x) \leftrightarrow \mathrm{OD}(x) \wedge \forall y \in x\, \mathrm{HOD}(y),$$

$$\mathbf{ZF} \vdash \forall \alpha \mathrm{HOD}(\alpha).$$

(ii) Show that

$$\mathbf{ZF} \vdash \boldsymbol{\sigma}^{(\mathrm{HOD})}$$

for any postulate $\boldsymbol{\sigma}$ of **ZFC**. Deduce that, if **ZF** is consistent, so is **ZFC**. (Show that, for each postulate of **ZF**, the set whose existence is asserted by the postulate has the property HOD, using (i) and 10.10(i).)
(iii) Show that

$$\mathbf{ZF} \vdash \forall x [\mathrm{L}(x) \rightarrow \mathrm{HOD}(x)].$$

(Use (i), (ii) and Thm. 8.8(iii).)

*10.13. A cardinal $\varkappa$ is said to be (strongly) *inaccessible* (written In($\varkappa$)) if
(a) $\varkappa > \omega$;
(b) for any cardinal $\alpha < \varkappa$, $|\mathrm{P}\alpha| < \varkappa$;
(c) if $x \subseteq \varkappa$ and $|x| < \varkappa$, then $|\bigcup x| < \varkappa$.
A set x is said to be *accessible* (written Acc(x)) if $\varrho(x)$ is less than every inaccessible cardinal. Let $\boldsymbol{\tau}$ be the sentence $\exists \varkappa \mathrm{In}(\varkappa)$.
(i) Show that $\mathbf{ZFC} \vdash \boldsymbol{\sigma}^{(\mathrm{Acc})}$ for every postulate $\boldsymbol{\sigma}$ of **ZFC**, and that $\mathbf{ZFC} \vdash \neg \boldsymbol{\tau}^{(\mathrm{Acc})}$. Deduce that, if **ZFC** is consistent, then the existence of an inaccessible cardinal cannot be proved in **ZFC**.
(ii) Show that $\mathbf{ZFC} \vdash \mathrm{In}(\varkappa) \rightarrow \mathrm{In}^{(\mathrm{L})}(\varkappa)$, and hence that $\mathbf{ZFC} \vdash \boldsymbol{\tau} \rightarrow \boldsymbol{\tau}^{(\mathrm{L})}$. Deduce that, if $\mathbf{ZFC} + \boldsymbol{\tau}$ is consistent, so is $\mathbf{ZFC} + \boldsymbol{\tau} + \mathbf{GCH}$.
(iii) Show that, if $\varkappa$ is inaccessible, then $R_\varkappa$ is a model of **ZFC**.

*10.14. (i) Let $\varkappa$ be an inaccessible cardinal, and let $\alpha < \varkappa$. Show that there is an ordinal β such that $\alpha < \beta < \varkappa$ and $\langle R_\beta, \in | R_\beta \rangle \prec \langle R_\varkappa, \in | R_\varkappa \rangle$. (Define a sequence $\alpha_0, \alpha_1, \ldots < \varkappa$ by

$$\alpha_0 = \alpha + 1,$$

$$\alpha_{n+1} = \mu\gamma[\text{for all formulas } \varphi \text{ whose free variables are all among } v_0, \ldots, v_m \text{ and all } \langle a_0, \ldots, a_{m-1}\rangle \in (R_{\alpha_n})^m, \text{ if } \langle R_\varkappa, \in | R_\varkappa\rangle \models \exists v_m \varphi[a_0, \ldots, a_{m-1}], \text{ then for some } x \in R_\gamma, \langle R_\varkappa, \in | R_\varkappa\rangle \models \varphi[a_0, \ldots, a_{m-1}, x]].$$

Put $\beta = \bigcup\{\alpha_n : n \in \omega\}$ and argue as in the proof of Thm. 5.2.1.)

(ii) Deduce from (i) that, if $\varkappa$ is an inaccessible cardinal, there is $x \subseteq \varkappa$ such that $|x| = \varkappa$, $\bigcup\{R_\beta : \alpha \in x\} = R_\varkappa$ and $\langle R_\alpha, \in | R_\alpha\rangle \prec \langle R_\beta, \in | R_\beta\rangle$ whenever $\alpha, \beta \in x$ and $\alpha < \beta$.

*10.15. Suppose that $\alpha < \beta$ and $\langle R_\alpha, \in | R_\alpha\rangle \prec \langle R_\beta, \in | R_\beta\rangle$. Show that R_α and R_β are both models of **ZF**. (First show that $\mathrm{Lim}(\alpha) \wedge \omega < \alpha$, and use 10.2(i) to show that R_α is a model of **Z**. To see that the **Rep** holds in R_α, suppose that $\varphi(x, y, a_1, \ldots, a_n)$ defines a function, say f, in R_α, using parameters $a_1, \ldots, a_n \in R_\alpha$. Then f must define a function, say g, in R_β which extends f. Show that, for any $a \in \mathbf{R}_\alpha$, $\mathrm{ran}(g \restriction a) \in R_\beta$ and deduce that $\mathrm{ran}(f \restriction a) \in R_\alpha$.)

§ 11. Historical and bibliographical remarks

The notion of (infinite) set was first systematically developed by Cantor in the 1880's, although it had been discussed both by Bolzano in the 1840's and Dedekind in the 1870's. The postulates of **Z** — except for the Axiom of Regularity — were formulated by Zermelo in 1908. The Axiom of Regularity was introduced by Mirimanoff in 1917. The Axiom Scheme of Replacement was roughly indicated by Fraenkel in 1922, but the first-order form of both it and the Axiom Scheme of Separation is due to SKOLEM [1922]. For this reason calling our axiomatic system Zermelo–Fraenkel set theory does an injustice to Skolem, and indeed some authors write "**ZFS**" for our "**ZF**".

The Axiom of Choice was first identified by Zermelo in 1904 (although not christened by him until 1908), and used to give a rigorous proof that each set can be well-ordered.

Our definition of an ordinal in §2 is due to von Neumann (1923), although he was apparently anticipated by Zermelo in an unpublished paper of 1915.

The cumulative hierarchy, the notion of rank, and their use in proving the relative consistency of the Axiom of Regularity was established by von Neumann in 1929. (SKOLEM [1922] had already sketched a different proof of the relative consistency of the axiom.)

The Reflection Principles were first introduced by LEVY [1960] and MONTAGUE [1961].

The formalization of the satisfaction relation given in §6 owes much to unpublished lectures of Dana Scott delivered at the 1965 Leicester Logic Colloquium.

The concept of absoluteness is due to GÖDEL [1940]. Skolem's paradox was formulated and analysed in SKOLEM [1922].

The notion of constructible set and the results of §§8 and 9 are due to GÖDEL [1938] and [1939]. The memoir GÖDEL [1940] contains a different — but essentially equivalent — definition of constructible set.

For further developments in set theory, the reader is advised to consult COHEN [1966], JECH [1971], DRAKE [1974], and BELL [1977]. The book FRAENKEL, BAR-HILLEL and LEVY [1973] contains an excellent account of the foundations of the subject, and a comprehensive bibliography.

CHAPTER 11

NONSTANDARD ANALYSIS

Any structure studied in mathematics may be regarded as an $\mathscr{L}$-structure for a suitably chosen first-order language $\mathscr{L}$. If $\mathfrak{A}$ is an $\mathscr{L}$-structure, then the *theory of* $\mathfrak{A}$ is defined as the set $\mathbf{Th}(\mathfrak{A})$ of all $\mathscr{L}$-sentences $\boldsymbol{\sigma}$ such that $\mathfrak{A} \models \boldsymbol{\sigma}$. In studying $\mathfrak{A}$, mathematicians explore the theory $\mathbf{Th}(\mathfrak{A})$, try to discover new sentences belonging to it, find interconnections between such sentences, etc.

If $\mathfrak{A}$ is infinite, then $\mathbf{Th}(\mathfrak{A})$ has *nonstandard* models, which are not isomorphic to $\mathfrak{A}$. In particular, $\mathfrak{A}$ has elementary extensions which are nonstandard models of $\mathbf{Th}(\mathfrak{A})$. (See, e.g., Thm. 5.2.6.) Since we are interested primarily in $\mathfrak{A}$, we may be tempted to regard these nonstandard models as pathological monsters. However, in 1960 A. Robinson invented a general method for exploiting nonstandard models of $\mathbf{Th}(\mathfrak{A})$ to facilitate the discovery and proof of facts about $\mathfrak{A}$ itself. This method he called *nonstandard analysis*.

Generalizing the proof of Thm. 7.2.3, Robinson obtained, for any $\mathscr{L}$-structure $\mathfrak{A}$, a special kind of elementary extension, called *enlargement* of $\mathfrak{A}$. An enlargement of $\mathfrak{A}$ has certain saturation properties which are not expressible in $\mathscr{L}$ but which can be used to prove assertions of the form $\boldsymbol{\sigma} \in \mathbf{Th}(\mathfrak{A})$. This is analogous to methods used in many branches of mathematics: when we want to study a given structure (e.g., the real numbers, or the Euclidean plane) we may find it convenient to embed it in some larger structure (the complex numbers, or the projective plane) having certain pleasant properties (being algebraically closed, or obeying the principle of duality). The latter structure may aid us in gaining knowledge about the former. The main difference between nonstandard analysis and those other well-known methods is that the relationship between the given structure and its enlargements is characterized in purely logical terms[1].

[1] For a more detailed discussion of this, see §7 below.

In particular, using nonstandard analysis Robinson was able to provide a solid rigorous foundation for the method of "infinitely small" and "infinitely large" quantities in classical analysis. That method, which is intuitively very suggestive, had been widely used during the early stages in the development of the calculus. But, after the failure of repeated attempts to give it consistent justification, it was abandoned in favour of the less intuitive "$\varepsilon - \delta$" method[1].

In this chapter we shall set up the general machinery of nonstandard analysis and then apply it to several mathematical situations.

We shall assume a slender acquaintance with Chapters 4 and 5 and the first four sections of Ch. 10.

§ 1. Enlargements

In this section $\mathscr{L}$ is taken to be an arbitrary first-order language with equality. The connectives $\neg$, $\wedge$ and the existential quantifier $\exists$ are taken to be primitive. The other connectives and $\forall$ are introduced by definition in the usual way.

We denote $\mathscr{L}$-structures by upper-case German letters; the domain (universe) of an $\mathscr{L}$-structure will be assumed to be a set, and will be denoted by the italic counterpart of the German letter denoting the structure.

If $\mathfrak{A}$ is an $\mathscr{L}$-structure and $\mathbf{R}$ is a predicate symbol of $\mathscr{L}$, we take $\mathbf{R}^{\mathfrak{A}}$ to be the corresponding basic relation of $\mathfrak{A}$. Similarly, $\mathbf{f}^{\mathfrak{A}}$ is the basic operation of $\mathfrak{A}$ corresponding to the function symbol $\mathbf{f}$ of $\mathscr{L}$.

We revert to the convention of using **bold type** for *all* symbols of $\mathscr{L}$. We let bold-face Roman letters from the end of the alphabet (usually lower-case and often with subscripts) range over the variables of $\mathscr{L}$. Throughout this chapter we adopt the convention that distinct letters of this kind occurring in the same context refer to distinct variables of $\mathscr{L}$, unless otherwise stated.

If $\boldsymbol{\varphi}$ is an $\mathscr{L}$-formula whose free variables are among $\mathbf{x}_1,\ldots,\mathbf{x}_n$, and if $a_1,\ldots,a_n$ are any individuals of the $\mathscr{L}$-structure $\mathfrak{A}$, we write

(1) $$\mathfrak{A} \models \boldsymbol{\varphi}[\mathbf{x}_1/a_1,\ldots,\mathbf{x}_n/a_n]$$

to assert that $\boldsymbol{\varphi}$ is satisfied in $\mathfrak{A}$ when $\mathbf{x}_1,\ldots,\mathbf{x}_n$ are assigned the values $a_1,\ldots,a_n$ respectively. Note that if for each $i=1,\ldots,n$ we have in $\mathscr{L}$ a constant

[1] However, the older and officially discredited method survived in "lowbrow" mathematics and — as a private heuristic aid — even in "highbrow" mathematics.

$\mathbf{a}_i$ such that $\mathbf{a}_i^{\mathfrak{A}}=a_i$, then (1) is equivalent to

$$\mathfrak{A}\models\varphi(\mathbf{x}_1/\mathbf{a}_1,\ldots,\mathbf{x}_n/\mathbf{a}_n).$$

We recall that $\mathfrak{A}$ is a *substructure* of $\mathfrak{B}$ if $A\subseteq B$, and, for all $\mathbf{R}$ and $\mathbf{f}$ of $\mathscr{L}$, $\mathbf{R}^{\mathfrak{A}}$ and $\mathbf{f}^{\mathfrak{A}}$ are the restrictions to A of $\mathbf{R}^{\mathfrak{B}}$ and $\mathbf{f}^{\mathfrak{B}}$ respectively. In particular, A must be closed under all the operations $\mathbf{f}^{\mathfrak{B}}$.

If C is an arbitrary set, we let $\mathscr{L}_C$ be the language obtained from $\mathscr{L}$ by adding a distinct new constant $\mathbf{c}$ for each $c\in C$. If $\mathfrak{A}$ is an $\mathscr{L}$-structure and $C\subseteq A$, we let $\mathfrak{A}_C$ be the $\mathscr{L}_C$-expansion of $\mathfrak{A}$ obtained by taking each $c\in C$ as the interpretation of the corresponding $\mathbf{c}$. Of particular importance is the case where $C=A$.

1.1. LEMMA. *Let $\mathfrak{A}$ and $\mathfrak{B}$ be $\mathscr{L}$-structures such that $\mathfrak{A}$ is a substructure of $\mathfrak{B}$. Then in order for $\mathfrak{B}_A$ to be an $\mathscr{L}_A$-elementary extension of $\mathfrak{A}_A$ it is enough that for each $\mathscr{L}_A$-sentence σ such that $\mathfrak{A}_A\models\sigma$ we have also $\mathfrak{B}_A\models\sigma$.*
PROOF.[1] Let φ be an $\mathscr{L}_A$-formula, and let the free variables of φ be among $\mathbf{x}_1,\ldots,\mathbf{x}_n$. Let $a_1,\ldots,a_n\in A$. If $\mathfrak{A}_A\models\varphi[\mathbf{x}_1/a_1,\ldots,\mathbf{x}_n/a_n]$, then $\mathfrak{A}_A\models\varphi(\mathbf{x}_1/\mathbf{a}_1,\ldots,\mathbf{x}_n/\mathbf{a}_n)$. Since $\varphi(\mathbf{x}_1/\mathbf{a}_1,\ldots,\mathbf{x}_n/\mathbf{a}_n)$ is an $\mathscr{L}_A$-sentence, it follows that $\mathfrak{B}_A\models\varphi(\mathbf{x}_1/\mathbf{a}_1,\ldots,\mathbf{x}_n/\mathbf{a}_n)$. hence $\mathfrak{B}_A\models\varphi[\mathbf{x}_1/a_1,\ldots,\mathbf{x}_n/a_n]$. On the other hand, if $\mathfrak{A}_A\not\models\varphi[\mathbf{x}_1/a_1,\ldots,\mathbf{x}_n/a_n]$, we consider $\neg\varphi$ and show that $\mathfrak{B}_A\not\models\varphi[\mathbf{x}_1/a_1,\ldots,\mathbf{x}_n/a_n]$. ∎

1.2. DEFINITION. Let $\mathfrak{A}$ be an $\mathscr{L}$-structure. An *$\mathfrak{A}$-concurrent collection* is any collection Φ of $\mathscr{L}_A$-formulas all of which have exactly one free variable, say $\mathbf{x}$, such that Φ is finitely satisfiable in $\mathfrak{A}_A$. In other words, for each finite $\Phi_0\subseteq\Phi$ there exists an $a\in A$ such that $\mathfrak{A}_A\models\varphi[\mathbf{x}/a]$ (or, equivalently, $\mathfrak{A}_A\models\varphi(\mathbf{x}/\mathbf{a})$) for all $\varphi\in\Phi_0$.

1.3. DEFINITION. Let $\mathfrak{A}$ and $\mathfrak{B}$ be $\mathscr{L}$-structures. We say that $\mathfrak{B}$ is an *enlargement* of $\mathfrak{A}$ (in symbols $\mathfrak{A}\ll\mathfrak{B}$) iff $\mathfrak{A}$ is a substructure of $\mathfrak{B}$, and whenever Φ is an $\mathfrak{A}$-concurrent collection then Φ is satisfiable in $\mathfrak{B}_A$. (Thus, if $\mathbf{x}$ is the free variable of Φ, then there is some $b\in B$ such that $\mathfrak{B}_A\models\varphi[\mathbf{x}/b]$ for all $\varphi\in\Phi$.)

1.4. THEOREM. *If $\mathfrak{A}\ll\mathfrak{B}$ then $\mathfrak{A}_A\prec\mathfrak{B}_A$; i.e., $\mathfrak{B}_A$ is an $\mathscr{L}_A$-elementary extension of $\mathfrak{A}_A$.*
PROOF. Let σ be any $\mathscr{L}_A$-sentence. If $\mathfrak{A}_A\models\sigma$, then the singleton $\{\sigma\wedge\mathbf{x}=\mathbf{x}\}$ is trivially $\mathfrak{A}$-concurrent. Thus by Def. 1.3 we must have $\mathfrak{B}_A\models\sigma$. Therefore $\mathfrak{A}_A\prec\mathfrak{B}_A$ by Lemma 1.1. ∎

[1] Cf. Lemma 5.1.6.

1.5. EXISTENCE THEOREM. *For any $\mathscr{L}$-structure $\mathfrak{A}$ there exists $\mathfrak{B}$ such that $\mathfrak{A} \preccurlyeq \mathfrak{B}$.*

PROOF[1]. We let $\mathscr{L}'$ be the language obtained from $\mathscr{L}_A$ by adding, for each $\mathfrak{A}$-concurrent $\boldsymbol{\Phi}$, a distinct new constant $\mathbf{c}_{\boldsymbol{\Phi}}$. If $\boldsymbol{\Phi}$ is $\mathfrak{A}$-concurrent and $\mathbf{x}$ is the free variable of $\boldsymbol{\Phi}$, we put

$$\boldsymbol{\Phi}' = \{\boldsymbol{\varphi}(\mathbf{x}/\mathbf{c}_{\boldsymbol{\Phi}}) : \ \boldsymbol{\varphi} \in \boldsymbol{\Phi}\}.$$

Let

$$\boldsymbol{\Sigma} = \bigcup\{\boldsymbol{\Phi}' : \ \boldsymbol{\Phi} \text{ is } \mathfrak{A}\text{-concurrent}\}.$$

Then $\boldsymbol{\Sigma}$ is a set of $\mathscr{L}'$-sentences. From the fact that each $\boldsymbol{\Phi}$ here is $\mathfrak{A}$-concurrent it follows at once that every finite subset of $\boldsymbol{\Sigma}$ is satisfiable. Thus by the Compactness Theorem (3.3.16 or 5.3.12) $\boldsymbol{\Sigma}$ has some $\mathscr{L}'$-structure $\mathfrak{B}'$ as a model.

Let $\mathfrak{B}$ be the $\mathscr{L}$-reduction of $\mathfrak{B}'$. We define a mapping $a \mapsto a'$ of A into B as follows:

$$a' = \mathbf{a}^{\mathfrak{B}'} \quad \text{for all } a \in A.$$

We claim that this mapping is an embedding of $\mathfrak{A}$ into $\mathfrak{B}$. Indeed, if $a_1 \neq a_2$ then the singleton $\{\mathbf{a}_1 \neq \mathbf{a}_2 \wedge \mathbf{x} = \mathbf{x}\}$ is trivially $\mathfrak{A}$-concurrent and hence $\boldsymbol{\Sigma} \models \mathbf{a}_1 \neq \mathbf{a}_2$. Since $\mathfrak{B}'$ is a model of $\boldsymbol{\Sigma}$, it follows that $\mathbf{a}_1^{\mathfrak{B}'} \neq \mathbf{a}_2^{\mathfrak{B}'}$, as required. In a similar way one shows that, for all $a_1, \ldots, a_n \in A$, for every n-ary function symbol $\mathbf{f}$ of $\mathscr{L}$ and every n-ary predicate symbol $\mathbf{R}$ of $\mathscr{L}$,

$$(\mathbf{f}^{\mathfrak{A}}(a_1, \ldots, a_n))' = \mathbf{f}^{\mathfrak{B}}(a_1', \ldots, a_n'),$$
$$\langle a_1, \ldots, a_n \rangle \in \mathbf{R}^{\mathfrak{A}} \quad \text{iff} \quad \langle a_1', \ldots, a_n' \rangle \in \mathbf{R}^{\mathfrak{B}}.$$

Since we have now shown that $\mathfrak{A}$ can be embedded in $\mathfrak{B}$, we may assume without loss of generality that $\mathfrak{A}$ is actually a substructure of $\mathfrak{B}$. Then from the fact that $\mathfrak{B}'$ is a model of $\boldsymbol{\Sigma}$ it follows immediately that $\mathfrak{A} \preccurlyeq \mathfrak{B}$. ∎

For most of the work in this chapter it will be enough to operate with an arbitrary enlargement of a given $\mathscr{L}$-structure. But occasionally it is useful to have enlargements of a special kind.

1.6. DEFINITION. Let $\mathfrak{A}$ be an $\mathscr{L}$-structure, and let α be an ordinal >0. For each $\beta \leqslant \alpha$ let $\mathfrak{A}^{(\beta)}$ be an $\mathscr{L}$-structure such that

(1) $\mathfrak{A}^{(0)} = \mathfrak{A}$,

(2) $\mathfrak{A}^{(\beta)} \preccurlyeq \mathfrak{A}^{(\beta+1)}$ for all $\beta < \alpha$,

(3) $\mathfrak{A}^{(\beta)} = \bigcup\{\mathfrak{A}^{(\gamma)} : \ \gamma < \beta\}$ for all limit $\beta \leqslant \alpha$.

[1] Cf. Prob. 5.2.12(i).

(More fully, (3) means that, for limit $\beta \leqslant \alpha$,

$$A^{(\beta)} = \bigcup\{A^{(\gamma)}: \gamma < \beta\};$$

and for every n-ary predicate symbol $\mathbf{R}$ of $\mathscr{L}$ we have

$$\mathbf{R}^{\mathfrak{A}^{(\beta)}} = \bigcup\{\mathbf{R}^{\mathfrak{A}^{(\gamma)}}: \gamma < \beta\};$$

and for every n-ary function symbol $\mathbf{f}$ of $\mathscr{L}$ and every $a_1,\ldots,a_n \in A^{(\beta)}$ we have

$$\mathbf{f}^{\mathfrak{A}^{(\beta)}}(a_1,\ldots,a_n) = \mathbf{f}^{\mathfrak{A}^{(\gamma)}}(a_1,\ldots,a_n),$$

where γ is the least ordinal $< \beta$ such that $a_1,\ldots,a_n \in A^{(\gamma)}$.) Then we say that $\mathfrak{A}^{(\alpha)}$ is an *α-enlargement* of $\mathfrak{A}$. If α is a limit ordinal we say that $\mathfrak{A}^{(\alpha)}$ is a *limit enlargement* of $\mathfrak{A}$.

By the Existence Theorem 1.5 it follows that for each positive α there exists an α-enlargement of $\mathfrak{A}$. Note also that a 1-enlargement of $\mathfrak{A}$ is simply an enlargement of $\mathfrak{A}$.

1.7. THEOREM. *Let α be an ordinal >0. If $\mathfrak{B}$ is an α-enlargement of $\mathfrak{A}$ then $\mathfrak{A} \lessdot \mathfrak{B}$.*

PROOF.[1] We proceed by induction on α.

First, let $\alpha = \beta + 1$. The case $\beta = 0$ is trivial. If $\beta > 0$, then, by Def. 1.6, $\mathfrak{C} \lessdot \mathfrak{B}$, where $\mathfrak{C}$ is a β-enlargement of $\mathfrak{A}$. By the induction hypothesis, $\mathfrak{A} \lessdot \mathfrak{C}$. Let $\mathbf{\Phi}$ be an $\mathfrak{A}$-concurrent collection of formulas. Then $\mathbf{\Phi}$ is satisfied in $\mathfrak{C}_A$. But since $\mathfrak{C} \lessdot \mathfrak{B}$, we have, by Thm. 1.4, $\mathfrak{C}_C \prec \mathfrak{B}_C$, and hence certainly $\mathfrak{C}_A \prec \mathfrak{B}_A$. Therefore an individual that satisfies $\mathbf{\Phi}$ in $\mathfrak{C}_A$ must also satisfy it in $\mathfrak{B}_A$. Thus $\mathfrak{A} \lessdot \mathfrak{B}$.

Now let α be a limit ordinal. Then for all $\beta \leqslant \alpha$ we have $\mathscr{L}$-structures $\mathfrak{A}^{(\beta)}$ as in Def. 1.6, with $\mathfrak{B} = \mathfrak{A}^{(\alpha)}$. We claim: *if $\mathfrak{C} = \mathfrak{A}^{(\beta)}$, where $\beta < \alpha$, and if $\boldsymbol{\sigma}$ is an $\mathscr{L}_C$-sentence, then $\mathfrak{C}_C \models \boldsymbol{\sigma}$ iff $\mathfrak{B}_C \models \boldsymbol{\sigma}$.*

This claim is proved by induction on $\deg \boldsymbol{\sigma}$. The cases where $\boldsymbol{\sigma}$ is atomic or a negation sentence or a conjunction sentence are simple and we leave them to the reader.

Let $\boldsymbol{\sigma} = \exists \mathbf{x} \boldsymbol{\varphi}$, where $\boldsymbol{\varphi}$ is an $\mathscr{L}_C$-formula with no free variable other than $\mathbf{x}$. If $\mathfrak{C}_C \models \boldsymbol{\sigma}$, then for some $c \in C$ we have $\mathfrak{C}_C \models \boldsymbol{\varphi}(\mathbf{x}/\mathbf{c})$; hence, by the induction hypothesis on $\deg \boldsymbol{\sigma}$, also $\mathfrak{B}_C \models \boldsymbol{\varphi}(\mathbf{x}/\mathbf{c})$ and therefore $\mathfrak{B}_C \models \boldsymbol{\sigma}$. Conversely, suppose that $\mathfrak{B}_C \models \boldsymbol{\sigma}$. Then for some $b \in B$ we have $\mathfrak{B}_C \models \boldsymbol{\varphi}[\mathbf{x}/b]$. For some $\gamma < \alpha$ we must have $b \in A^{(\gamma)}$ and without loss of generality we may assume $\beta \leqslant \gamma$ (otherwise take β itself instead of γ). Put $\mathfrak{D} = \mathfrak{A}^{(\gamma)}$. Then $\mathfrak{B}_D \models \boldsymbol{\varphi}(\mathbf{x}/\mathbf{b})$ and by the induction hypothesis on $\deg \boldsymbol{\sigma}$ we have

[1] Cf. Prob. 5.2.10.

$\mathfrak{D}_D \models \varphi(\mathbf{x}/\mathbf{b})$, hence $\mathfrak{D}_C \models \boldsymbol{\sigma}$. Now, $\mathfrak{C} = \mathfrak{A}^{(\beta)}$ and $\mathfrak{D} = \mathfrak{A}^{(\gamma)}$ with $\beta \leqslant \gamma < \alpha$. Hence $\mathfrak{C} = \mathfrak{D}$, or $\mathfrak{D}$ is a δ-enlargement of $\mathfrak{C}$ for some $\delta \leqslant \gamma < \alpha$. In the former case we have at once $\mathfrak{C}_C \models \boldsymbol{\sigma}$. In the latter case the induction hypothesis on α implies that $\mathfrak{C} \lessdot \mathfrak{D}$; hence $\mathfrak{C}_C \prec \mathfrak{D}_C$ by Thm. 1.4. Since $\mathfrak{D}_C \models \boldsymbol{\sigma}$, we must have $\mathfrak{C}_C \models \boldsymbol{\sigma}$ as well. This completes the induction on deg $\boldsymbol{\sigma}$ and proves our claim.

By Lemma 1.1. it follows that $\mathfrak{C}_C \prec \mathfrak{B}_C$ for every $\mathfrak{C}$ of the above form, i.e., $\mathfrak{C} = \mathfrak{A}^{(\beta)}$ with $\beta < \alpha$.

Finally, let $\boldsymbol{\Phi}$ be an $\mathfrak{A}$-concurrent collection. Then if $\mathfrak{C} = \mathfrak{A}^{(\beta)}$ with $0 < \beta < \alpha$, our induction hypothesis implies that $\mathfrak{A} \lessdot \mathfrak{C}$ and hence $\boldsymbol{\Phi}$ is satisfiable in $\mathfrak{C}_A$. But we have shown that $\mathfrak{C}_C \prec \mathfrak{B}_C$; therefore certainly $\mathfrak{C}_A \prec \mathfrak{B}_A$, and $\boldsymbol{\Phi}$ is satisfiable in $\mathfrak{B}_A$. Thus $\mathfrak{A} \lessdot \mathfrak{B}$. ∎

1.8. COROLLARY. *If, for all $\beta \leqslant \alpha$, the $\mathscr{L}$-structures $\mathfrak{A}^{(\beta)}$ are as in Def. 1.6 and if $\gamma < \beta \leqslant \alpha$, then $\mathfrak{A}^{(\gamma)} \lessdot \mathfrak{A}^{(\beta)}$.* ∎

§ 2. Zermelo structures and their enlargements

From now on we take $\mathscr{L}$ to be the same as in Ch. 10; i.e., the first-order language with equality and one binary extralogical predicate symbol $\boldsymbol{\in}$. Note, however, that in the present chapter we have reverted to the convention of using **bold type** for all symbols of $\mathscr{L}$.

By a *Zermelo structure* we mean an $\mathscr{L}$-structure $\mathfrak{U}$ such that the basic relation $\boldsymbol{\in}^{\mathfrak{U}}$ of $\mathfrak{U}$ is the membership relation (restricted to the domain U of $\mathfrak{U}$); and U is a non-empty set having the following four properties:

(2.1) transitivity: $a \in b \in U \Rightarrow a \in U$;

(2.2) closure under unordered pairs: $a, b \in U \Rightarrow \{a, b\} \in U$;

(2.3) closure under union: $a \in U \Rightarrow \bigcup a \in U$;

(2.4) closure under power-set: $a \in U \Rightarrow \mathrm{P}a \in U$.

The name "Zermelo structure" is given to $\mathfrak{U}$ because it is a model of Zermelo's set theory, possibly without the axiom of infinity. (See Prob. 10.10.2.) In particular, the axioms of regularity and choice hold in $\mathfrak{U}$ provided we assume them to hold in the universe of sets; and we *do* assume this. (The axiom of regularity is assumed purely for convenience: it simplifies matters but one can do without it.)

For any set A there exists a Zermelo structure $\mathfrak{U}$ such that $A \in U$. For example, we can take $U = R_\alpha$ (see 10.3.1), where α is any limit ordinal greater than the rank of A. It is then easy to verify (cf. Prob. 10.10.2) that U has the required properties.

2.5. PROBLEM. Verify that the domain U of a Zermelo structure has the following properties:

(i) $a \subseteq b \in U \Rightarrow a \in U$;

(ii) $\emptyset \in U$;

(iii) $a,b \in U \Rightarrow a \cup b \in U$;

(iv) $a_1,\ldots,a_n \in U \Rightarrow \{a_1,\ldots,a_n\} \in U$;

(v) $a_1,\ldots,a_n \in U \Rightarrow \langle a_1,\ldots,a_n \rangle \in U$;

(vi) $A,B \in U \Rightarrow A \times B \in U$;

(vii) $A,B \in U \Rightarrow \mathrm{Fun}(A,B) \in U$, where $\mathrm{Fun}(A,B) = B^A =$ the collection of all mappings of A into B.

From now on we let $\mathfrak{U}$ be a fixed but arbitrary Zermelo structure. In applying nonstandard analysis to a given mathematical situation, we shall usually assume — as we are entitled to do — that some particular set belongs to U.

The members of U are called *standard* objects. It will be convenient from now on to reserve the term *set* for standard objects only. If A is a set in the usual sense, but we do not wish to imply that A necessarily belongs to U, we refer to A as a *collection*. Similarly, when $A \subseteq B$, we say that A is a *subset* of B only if we wish to imply that A and B are in U; otherwise, we say that A is a *subcollection* of B, or simply that A is *included in* B.

We also fix a particular enlargement ${}^*\mathfrak{U} = \langle {}^*U, {}^*\in \rangle$ of $\mathfrak{U}$. It is convenient to read the prefix "*" as "*pseudo-*". Unless otherwise stated, ${}^*\mathfrak{U}$ can be an arbitrary enlargement of $\mathfrak{U}$. Occasionally we shall want to assume that ${}^*\mathfrak{U}$ is an α-enlargement for some suitable ordinal α.

It must be stressed that in general ${}^*\in$ is *not* the membership relation, though by Thm. 1.4 its *formal* properties that can be expressed in $\mathscr{L}_U$ are the same as those of membership.

We adopt the convention that if a symbol or an italic letter denotes a member of ${}^*U$ then the same symbol in bold type (or, in the case of a letter, in bold Roman type) denotes the corresponding constant of $\mathscr{L}_{*U}$.

If $\boldsymbol{\sigma}$ is a sentence of $\mathscr{L}_U$, we write "$\models \boldsymbol{\sigma}$" as short for "$\mathfrak{U}_U \models \boldsymbol{\sigma}$". Similarly, if $\boldsymbol{\sigma}$ is a sentence of $\mathscr{L}_{*U}$, we write "${}^*\!\models \boldsymbol{\sigma}$" as short for "${}^*\mathfrak{U}_{*U} \models \boldsymbol{\sigma}$". Note that if $\boldsymbol{\sigma}$ is a sentence of $\mathscr{L}_U$ then ${}^*\!\models \boldsymbol{\sigma}$ iff ${}^*\mathfrak{U}_U \models \boldsymbol{\sigma}$.

For any $\mathscr{L}_U$-sentence $\boldsymbol{\sigma}$ we have, by Thm. 1.4, $\models \boldsymbol{\sigma}$ iff ${}^*\!\models \boldsymbol{\sigma}$. We shall often need to make inferences of the form

(1) $\models \boldsymbol{\sigma}$, hence also ${}^*\!\models \boldsymbol{\sigma}$,

or

(2) ${}^*\!\models \boldsymbol{\sigma}$, hence also $\models \boldsymbol{\sigma}$,

where $\boldsymbol{\sigma}$ is an $\mathscr{L}_U$-sentence. To save space we write

(1′) $(^*) \models \boldsymbol{\sigma}$

and

(2′) $\dagger^* \models \boldsymbol{\sigma}$

as short for (1) and (2) respectively. We call (1) (and its condensed form (1′)) *transference from* $\mathfrak{U}$ *to* $^*\mathfrak{U}$. Similarly, (2) (and its condensed form (2′)) will be called *transference from* $^*\mathfrak{U}$ *to* $\mathfrak{U}$.

The members of $^*U$ are called **sets* (read: *pseudo-sets*). Every set (i.e., member of U) is also a *set because $U \subseteq {}^*U$. But the converse is not true. To see this, observe that, by (2.4), U must be infinite. It follows that $\{\mathbf{x} \neq \mathbf{a} : a \in U\}$ is a $\mathfrak{U}$-concurrent collection of $\mathscr{L}_U$-formulas, and since $\mathfrak{U} \ll {}^*\mathfrak{U}$ there must exist some *set b different from all sets. A *set which is not a set (i.e., any member of $^*U - U$) is said to be *nonstandard*.

For any relation R among sets, defined by means of an $\mathscr{L}_U$-formula $\boldsymbol{\varphi}$, we have automatically a corresponding relation $^*R$ among *sets, which is defined by means of the same $\boldsymbol{\varphi}$. For example, consider the relation $\subseteq$. This relation is defined by means of the formula $\boldsymbol{\varphi}$, where

$$\boldsymbol{\varphi} = \mathbf{x} \subseteq \mathbf{y} = \forall \mathbf{z}(\mathbf{z} \in \mathbf{x} \rightarrow \mathbf{z} \in \mathbf{y});$$

in other words, for any sets a, b we have $a \subseteq b$ iff $\models \mathbf{a} \subseteq \mathbf{b}$ (i.e., $\models \boldsymbol{\varphi}(\mathbf{x}/\mathbf{a}, \mathbf{y}/\mathbf{b})$). Then the relation $^*\subseteq$ is defined as follows: for any *sets a,b we have $a \,^*\!\subseteq b$ iff $^*\!\models \mathbf{a} \subseteq \mathbf{b}$ (i.e., $^*\!\models \boldsymbol{\varphi}(\mathbf{x}/\mathbf{a}, \mathbf{y}/\mathbf{b})$). Thus $a \,^*\!\subseteq b$ iff every *member of a is a *member of b.

Note that if a and b are sets (i.e., standard objects) then, by transference, $a \subseteq b$ iff $a \,^*\!\subseteq b$. More generally, if R and $^*R$ are relations defined on U and $^*U$ respectively by the same $\mathscr{L}_U$-formula, then R must be the restriction of $^*R$ to U.

Note also that $\boldsymbol{\subseteq}$ is a (defined) predicate symbol of $\mathscr{L}$, while $\subseteq$ and $^*\subseteq$ are the corresponding relations on U and $^*U$ respectively. We use the same typographical convention for other defined symbols as well.

Now let $\mathbf{t}$ be a virtual term introduced into $\mathscr{L}_U$ by the method explained in §13 of Ch. 2. (See also the discussion following Thm. 10.1.5. Note that we can use here the virtual terms introduced in Ch. 10, because by (2.1) and Prob. 2.5(ii) we have

$$\models \exists! \mathbf{z}\, \forall \mathbf{y}(\mathbf{y} \in \mathbf{z} \leftrightarrow \mathbf{y} \neq \mathbf{y}),$$

i.e., $\models \exists ! \mathbf{z}\alpha$, where α is the formula used for introducing the virtual terms of Ch. 10.) Suppose $\mathbf{x}_1,\ldots,\mathbf{x}_n$ are the free variables of $\mathbf{t}$. Then we have

$$\models \forall \mathbf{x}_1 \ldots \forall \mathbf{x}_n \exists ! \mathbf{y}(\mathbf{t}=\mathbf{y});$$

and we can define an operation t on U as follows: for any $a_1,\ldots,a_n \in U$ take $t(a_1,\ldots,a_n)$ to be the unique $b \in U$ such that

$$\models \mathbf{t}(\mathbf{x}_1/\mathbf{a}_1,\ldots,\mathbf{x}_n/\mathbf{a}_n)=\mathbf{b}.$$

But by transference we have also

$$^*\!\models \forall \mathbf{x}_1 \ldots \forall \mathbf{x}_n \exists ! \mathbf{y}(\mathbf{t}=\mathbf{y}).$$

Hence we can define an operation $^*t$ on $^*U$ as follows: for any $a_1,\ldots,a_n \in {}^*U$ we let $^*t(a_1,\ldots,a_n)$ be the unique $b \in {}^*U$ such that

$$^*\!\models \mathbf{t}(\mathbf{x}_1/\mathbf{a}_1,\ldots,\mathbf{x}_n/\mathbf{a}_n)=\mathbf{b}.$$

It is easy to see that t is the restriction of $^*t$ to U.

For example, let $\mathbf{t}$ be the term $\mathbf{x} \cup \mathbf{y}$. Then the corresponding operation on U is $\cup$ and we have

$$(^*) \models \forall \mathbf{x}\, \forall \mathbf{y}\, \forall \mathbf{z}(\mathbf{z} \in \mathbf{x} \cup \mathbf{y} \leftrightarrow \mathbf{z} \in \mathbf{x} \vee \mathbf{z} \in \mathbf{y}).$$

Hence the operation $^*\cup$ on $^*U$ is such that for $a,b \in {}^*U$ the *members of $a\,^*\!\cup b$ are the *members of a plus the *members of b. And if both a and b happen to be standard then $a\,^*\!\cup b = a \cup b$.

In the sequel we shall often use, without special comment, relations of the form $^*R$, where R is a familiar relation among sets, and operations of the form $^*t$, where t is a familiar operation on sets. The definitions of such $^*R$ and $^*t$ are obtained automatically from those of R and t by the method just explained.

For any *set a we define $\hat{a}$ (read: "*a hat*" and called *the scope of a*) as the collection $\{b \in {}^*U : b \,^*\!\in a\}$. Thus $\hat{a}$ is a subcollection of $^*U$; in general it is neither a set nor even a *set. The members of $\hat{a}$ are precisely the *members of a. When we want to apply the "hat" (the *scope* symbol) to an expression consisting of more than one letter, we put the hat to the right of the expression. For example: $(a\,^*\!\cup b)\hat{}$.

2.6. **Problem.** Let a,b be *sets. Prove:

(i) $(a\,^*\!\cup b)\hat{} = \hat{a} \cup \hat{b}$;

(ii) $(a\,^*\!\cap b)\hat{} = \hat{a} \cap \hat{b}$;

(iii) $\hat{a} = \hat{b} \Rightarrow a = b$. (Apply transference to the axiom of extensionality.)

(iv) $\hat{a} \subseteq \hat{b}$ iff $a {}^*\!\subseteq b$.

If a is standard, then $a \subseteq \hat{a}$ because every member of a is also a *member of a. When do we have $\hat{a}=a$?

2.7. THEOREM. *Let a be standard. Then $\hat{a}=a$ iff a is finite.*
PROOF. First let $a=\emptyset$. We have $(^*)\models \forall \mathbf{x}(\mathbf{x} \notin \mathbf{0})$; hence $\emptyset$ has no *members, so $\hat{\emptyset}=\emptyset$.

Next, suppose $a=\{b_1,\ldots,b_n\}$, where $n \geqslant 1$. From (2.1) it follows that the b_i are standard; hence we can make the following transference

$$(^*)\models \forall \mathbf{x}(\mathbf{x}\in\mathbf{a} \leftrightarrow \mathbf{x}=\mathbf{b}_1 \vee \ldots \vee \mathbf{x}=\mathbf{b}_n),$$

so the *members of a are precisely $b_1,\ldots,b_n$ and hence $\hat{a}=a$.

Finally, let a be infinite. Then

$$\{\mathbf{x}\in\mathbf{a} \wedge \mathbf{x}\neq\mathbf{b} : b\in a\}$$

is clearly a $\mathfrak{U}$-concurrent collection of formulas. Therefore there exists some $c\in {}^*U$ such that ${}^*\models \mathbf{c}\in\mathbf{a}$ and ${}^*\models \mathbf{c}\neq\mathbf{b}$ for every $b\in a$. Thus c *belongs to a but differs from all the members of a. In other words, $c\in\hat{a}-a$, so that a is a *proper* subcollection of $\hat{a}$. ∎

Let V be a subcollection of ${}^*U$. If $V=\hat{a}$ for some *set a (which, by Prob. 2.6(iii), must be unique) we say that V is *internal*. Otherwise, V is *external*.

2.8. EXAMPLE. Assume that the collection N of natural numbers is a set (i.e., belongs to U). Then $N\subseteq\hat{N}\subseteq{}^*U$. We shall show that N is external.

Let

$$M=\{\langle n, n+1\rangle : n\in N\}.$$

By parts (i) and (vi) of Prob. 2.5 we have $M\in U$. Let **xSy** be the $\mathscr{L}_U$-formula $\langle\mathbf{x},\mathbf{y}\rangle\in\mathbf{M}$, and let S be the corresponding relation on U. Then we clearly have

(1) $\quad (^*)\models \forall\mathbf{z}[\mathbf{0}\in\mathbf{z}\wedge\forall\mathbf{x}\,\exists\mathbf{y}(\mathbf{x}\in\mathbf{z}\rightarrow\mathbf{y}\in\mathbf{z}\wedge\mathbf{xSy})\rightarrow\mathbf{N}\subseteq\mathbf{z}]$.

Now suppose that N is internal, i.e., $N=\hat{A}$ for some *set A. Since $0\in N$, we have

(2) $\quad 0\,{}^*\!\in A$.

Also, for each $n\in N$ there is $m\in N$ such that nSm and hence also $n\,{}^*S\,m$. Thus

(3) $\quad$ for each $n\,{}^*\!\in A$ there is $m\,{}^*\!\in A$ such that $n\,{}^*S\,m$.

From (1), (2) and (3) we get $N \mathrel{{}^*{\subseteq}} A$, hence $\hat{N} \subseteq \hat{A} = N$. Thus $\hat{N} = N$, contrary to Thm. 2.7. Therefore N must be external.

2.9. PROBLEM. Let s be standard and infinite. Show that s is external. (Let b be any denumerable subset of s. Show that b is external, then use the fact that $b = \hat{b} \cap s$.)

2.10. CRITERION. *Let $V \subseteq {}^*U$. Then V is internal iff there is an $\mathscr{L}_{*U}$-formula $\boldsymbol{\varphi}$ with one free variable* **x**, *and a *set s, such that*

(1) $V = \{a \in {}^*U:\ {}^*\models \mathbf{a} \in \mathbf{s} \wedge \boldsymbol{\varphi}(\mathbf{x}/\mathbf{a})\}$.

PROOF. If V is internal, then $V = \hat{s}$ for some *set s. Take $\boldsymbol{\varphi}$ to be the formula $\mathbf{x} = \mathbf{x}$. Then clearly (1) holds.

Conversely, suppose that (1) holds for some $\mathscr{L}_{*U}$-formula $\boldsymbol{\varphi}$ and some *set s. Let $\mathbf{b}_1, \ldots, \mathbf{b}_n$ be all the constants occurring in $\boldsymbol{\varphi}$. Then $\boldsymbol{\varphi} = \boldsymbol{\psi}(\mathbf{y}_1/\mathbf{b}_1, \ldots, \mathbf{y}_n/\mathbf{b}_n)$ for some $\mathscr{L}$-formula $\boldsymbol{\psi}$ whose free variables are $\mathbf{y}_1, \ldots, \mathbf{y}_n, \mathbf{x}$. Then by the axiom of separation (cf. 10.1.3) we have

$$(*) \models \forall \mathbf{y}_1 \cdots \forall \mathbf{y}_n \forall \mathbf{z} \exists \mathbf{u} \forall \mathbf{x} (\mathbf{x} \in \mathbf{u} \leftrightarrow \mathbf{x} \in \mathbf{z} \wedge \boldsymbol{\psi}).$$

In particular, giving $\mathbf{y}_1, \ldots, \mathbf{y}_n, \mathbf{z}$ the values $b_1, \ldots, b_n, s$ respectively, we see that for some *set c

$${}^*\models \forall \mathbf{x} (\mathbf{x} \in \mathbf{c} \leftrightarrow \mathbf{x} \in \mathbf{s} \wedge \boldsymbol{\varphi}).$$

Thus the *members of c are precisely those *sets a such that ${}^*\models \mathbf{a} \in \mathbf{s} \wedge \boldsymbol{\varphi}(\mathbf{x}/\mathbf{a})$. It follows from (1) that $V = \hat{c}$, so that V is internal. ∎

The following easy result is extremely useful:

2.11. ROBINSON'S OVERSPILL LEMMA. *Let s be an infinite set, and let τ be an internal subcollection of $\hat{s}$. If $s \subseteq \tau$ then $(\hat{s} - s) \cap \tau \neq \emptyset$. If $\hat{s} - s \subseteq \tau$ then $s \cap \tau \neq \emptyset$.*

PROOF. The first part follows from the fact that s is external (see Prob. 2.9).

To prove the second part, we show that $\hat{s} - s$ is external. Suppose not; then $\hat{s} - s = \hat{t}$ for some *set t. Hence

$$s = \{a \in {}^*U : {}^*\models \mathbf{a} \in \mathbf{s} \wedge \mathbf{a} \notin \mathbf{t}\}.$$

By Crit. 2.10 it follows that s is internal, contrary to Prob. 2.9. ∎

Let $s \in U$, and let F be the set of all finite subsets of s. ($F \in U$ by (2.4) and Prob. 2.5(i)). If $b \mathrel{{}^*{\in}} F$ then b is a *finite *subset of s. Clearly, such b enjoy the same formal properties as finite sets. More precisely, if $\boldsymbol{\varphi}$ is an $\mathscr{L}_U$-formula with one free variable $\mathbf{x}$ and $\models \boldsymbol{\varphi}(\mathbf{x}/\mathbf{a})$ for every finite set a, then also $^*\!\models \boldsymbol{\varphi}(\mathbf{x}/\mathbf{b})$. Nevertheless, b can have infinitely many *members; i.e., the collection $\hat{b}$ can be infinite.

2.12. THEOREM. *Let $s \in U$, and let F be the set of all finite subsets of s. Then there is some $b \,{}^*\!\!\in F$ such that $s \subseteq \hat{b}$. In particular, if s is infinite, so is $\hat{b}$.*

PROOF. Clearly $\{\mathbf{x} \in \mathbf{F} \wedge \mathbf{c} \in \mathbf{x} : c \in s\}$ is a $\mathfrak{U}$-concurrent collection. The existence of b with the required properties follows at once, since $\mathfrak{U} \preccurlyeq {}^*\mathfrak{U}$. ▌

We now turn to a discussion of mappings. Henceforth we shall reserve the term *function* for mappings which are sets (i.e., belong to U).

Let X and Y be sets. Then Fun(X,Y) is the set of all functions from X into Y (see Prob. 2.5(vii)). If $f \in \text{Fun}(X,Y)$ and $a \in X$, then $f`a$ (read *f of a*, or *f applied to a*) is the value of $f$ at $a$ and is a uniquely determined member of $Y$. (The corresponding virtual term of $\mathscr{L}_U$ is $\mathbf{f`a}$.) We shall usually omit the application sign ` and write simply $fa$ or $f(a)$ instead of $f`a$. (Also, in formal expressions we often write $\mathbf{fa}$ or $\mathbf{f(a)}$ for $\mathbf{f`a}$.)

To the operation Fun on sets and to the operation ` of function application there correspond (as explained earlier in this section) the operation *Fun on *sets and the operation *` of *function *application.

Let X and Y be *sets, and let $f \,{}^*\!\!\in {}^*\text{Fun}(X,Y)$. If $a \,{}^*\!\!\in X$ then $f\,{}^*`a$ (read: *f *applied to a*) is a uniquely determined *member of $Y$. Thus we have a mapping $a \mapsto f\,{}^*`a$ of $\hat{X}$ into $\hat{Y}$. We denote this mapping by f^*. (So, if $a \,{}^*\!\!\in X$, then $f^*`a$ is both f *applied to a and f^* applied to a.) Notice that f and f^* are in general entities of different kinds. The former, f, is a *set of *ordered pairs; in general f is not a mapping. On the other hand, f^* is a mapping and is thus a collection of ordered pairs; in fact

$$f^* = \{\langle a, f^*`a\rangle : a \in \hat{X}\}.$$

(Both f and f^* should be distinguished from $\hat{f}$, which is a collection of *ordered pairs.)

We call f^* the mapping *induced* by f. When there is no risk of confusion we shall write fa or $f(a)$ instead of $f^*`a$.

If f happens to be a function, then the mapping f^* is an extension of f. In this case we call f^* the *natural* extension of f.

2.13. PROBLEM. Let α be an infinite regular cardinal and let ${}^*\mathfrak{U}$ be an α-enlargement of $\mathfrak{U}$. Let A and B be *sets. Let C be a subcollection of $\hat{A}$ such that the cardinality of C is less than α, and let φ be an arbitrary mapping of C into $\hat{B}$. Prove that there exists some $f \,{}^*\!\!\in {}^*\text{Fun}(A,B)$ such that $fc=\varphi(c)$ for all $c\in C$. (Let $\{\mathfrak{U}^{(\beta)} : \beta\leqslant\alpha\}$ be an enlargement chain as described in Def. 1.6, with $\mathfrak{U}^{(0)}=\mathfrak{U}$ and $\mathfrak{U}^{(\alpha)}={}^*\mathfrak{U}$. Choose some $\beta<\alpha$ such that $U^{(\beta)}$ contains A and B and includes C and $\varphi[C]$. Observe that

$$\{\mathbf{x{\in}Fun(A,B){\wedge}x`c{=}b} : c\in C \quad \text{and } b=\varphi(c)\}$$

is a $\mathfrak{U}^{(\beta)}$-concurrent collection.)

We have now finished setting up the general machinery of nonstandard analysis. In the following sections we apply this machinery to various mathematical situations.

§ 3. Filters and monads

Throughout this section we let $X\in U$ be a fixed but arbitrary non-empty set. Since X is standard, it follows that the members of X are standard as well; we call them *points*. (Accordingly, the *members of X — i.e., the members of $\hat{X}$ — will be called **points*.)

By (2.4) and (2.1), PX, PPX as well as all their members are standard.

The reader may visualize X as a conglomeration of corpuscules or little dots — the points of X. When X is extended to $\hat{X}$, new corpuscules are added; they may be visualized as differing in colour from the old ones. The new corpuscules — the nonstandard *members of X — are interspersed in gaps between the old ones as well as outside the old boundary of X. Similarly, every subset A of X is extended to $\hat{A}$ by adding some of the new corpuscules.

We say that $\mathscr{F}$ is a *filter over* X if $\mathscr{F}$ is a non-empty set of subsets of X (i.e., $\emptyset\neq\mathscr{F}\in \mathsf{PP}X$) and has the following two properties:

(1) $A,B\in\mathscr{F}\Rightarrow A\cap B\in\mathscr{F}$,

(2) $X\supseteq B\supseteq A\in\mathscr{F}\Rightarrow B\in\mathscr{F}$.

Note that this definition differs from the one used in Ch. 4 (see beginning of §3 of Ch. 4 and Prob. 4.3.16) in one respect: we now allow $\emptyset\in\mathscr{F}$. But, in the presence of (2), we have $\emptyset\in\mathscr{F}$ iff $\mathscr{F}=\mathsf{P}X$. Thus the only difference between the present definition and that of Ch. 4 is that we now admit PX itself as a filter over X.

Throughout this section, when we say "filter" without any further qualification, we mean filter over X. A filter $\mathscr{F}$ is *proper* if $\mathscr{F}\neq\mathsf{P}X$.

If a filter $\mathscr{F}_1$ is properly included in another filter $\mathscr{F}_2$, we say that $\mathscr{F}_1$

is *coarser* than $\mathscr{F}_2$, and $\mathscr{F}_2$ is *finer* than $\mathscr{F}_1$. Among all filters, $\{X\}$ is the coarsest and PX the finest.

Let $\mathscr{G}$ be *any* set of subsets of X (i.e., $\mathscr{G} \in PPX$). We define $\overline{\mathscr{G}}$ to be the filter *generated* by $\mathscr{G}$. Thus $\overline{\mathscr{G}}$ is the coarsest filter that includes $\mathscr{G}$. If $\mathscr{G}=\emptyset$, then $\overline{\mathscr{G}}=\{X\}$. If $\mathscr{G}\neq\emptyset$, then $\overline{\mathscr{G}}$ is the set of all B such that $A_1\cap\ldots\cap A_n\subseteq$ $\subseteq B\subseteq X$ for some $n\geqslant 1$ and some $A_1,\ldots,A_n\in\mathscr{G}$.

3.1. Definition. For any $\mathscr{G}\in PPX$ we put

$$\mu\mathscr{G}=\{p\in\hat{X}:\ p\in\hat{A} \text{ for all } A\in\mathscr{G}\}.$$

We call $\mu\mathscr{G}$ the *monad* of $\mathscr{G}$.

If $\mathscr{G}=\emptyset$, then $\mu\mathscr{G}=\hat{X}$. If $\mathscr{G}\neq\emptyset$, we can write

$$\mu\mathscr{G}=\bigcap\{\hat{A} : A\in\mathscr{G}\}.$$

Clearly, $\mu\mathscr{G}$ is always a subcollection of $\hat{X}$, so μ is a mapping of PPX into $P\hat{X}$. A subcollection of $\hat{X}$ is said to be *monadic* (or a *monad*) if it is $\mu\mathscr{G}$ for some $\mathscr{G}\in PPX$.

It is easy to see that if $\mathscr{G}_1\subseteq\mathscr{G}_2\in PPX$ then $\mu\mathscr{G}_1\supseteq\mu\mathscr{G}_2$.

3.2. Theorem. *$\mu\mathscr{G}=\mu\overline{\mathscr{G}}$ for every $\mathscr{G}\in PPX$.*

Proof. If $\mathscr{G}=\emptyset$, then $\mu\mathscr{G}=\hat{X}=\mu\{X\}=\mu\overline{\mathscr{G}}$, by Def. 3.1. So now we may assume $\mathscr{G}\neq\emptyset$.

$\mu\mathscr{G}\supseteq\mu\overline{\mathscr{G}}$ because $\mathscr{G}\subseteq\overline{\mathscr{G}}$. Conversely, if $B\in\overline{\mathscr{G}}$ then $A_1\cap\ldots\cap A_n\subseteq B$ for some $n\geqslant 1$ and some $A_1,\ldots,A_n\in\mathscr{G}$. Since $A_1,\ldots,A_n$ and B are all standard, we have also $A_1{}^*\cap\ldots{}^*\cap A_n{}^*\subseteq B$. Hence, by Prob. 2.6, $\hat{A}_1\cap\ldots\cap\hat{A}_n\subseteq\hat{B}$. If $p\in\mu\mathscr{G}$, then $p\in\hat{A}$ for all $A\in\mathscr{G}$, and hence, by what we have just shown, also $p\in\hat{B}$ for all $B\in\overline{\mathscr{G}}$; hence $p\in\mu\overline{\mathscr{G}}$. Thus $\mu\mathscr{G}\subseteq\mu\overline{\mathscr{G}}$. ∎

If $D\,{}^*\!\in\mathscr{G}$ and $\hat{D}\subseteq\mu\mathscr{G}$, we say that D is a *tiny* *member of $\mathscr{G}$. Note that the condition $\hat{D}\subseteq\mu\mathscr{G}$ is equivalent to each of the following two conditions:

$$\hat{D}\subseteq\hat{A}\qquad\text{for all } A\in\mathscr{G},$$

$$D\,{}^*\!\subseteq A\qquad\text{for all } A\in\mathscr{G}.$$

This can be seen at once from Def. 3.1 and Prob. 2.6(iv).

Recall that $\mathscr{G}$ is a *base* for a filter $\mathscr{F}$ if $\overline{\mathscr{G}}=\mathscr{F}$ and for every $A\in\mathscr{F}$ there exists some $B\in\mathscr{G}$ such that $B\subseteq A$. If $\mathscr{G}$ is a base for $\overline{\mathscr{G}}$, we say that $\mathscr{G}$ is a *filter base*. It is easy to see (Prob. 4.3.10) that a set $\mathscr{G}\in PPX$ is a filter base iff $\mathscr{G}\neq\emptyset$ and for every $A,B\in\mathscr{G}$ there exists some $C\in\mathscr{G}$ such that $C\subseteq A\cap B$.

3.3. THEOREM. *Let $\mathscr{G} \in \mathsf{PP}X$. Then $\mathscr{G}$ is a filter base iff it has a tiny *member.*
PROOF. Suppose $\mathscr{G}$ is a filter base. Then

$$\{\mathbf{x} \in \mathfrak{G} \wedge \mathbf{x} \subseteq \mathbf{A}: A \in \mathscr{G}\}$$

is easily seen to be a $\mathfrak{U}$-concurrent collection of formulas. Hence there exists some $D \,{}^*\!\in \mathscr{G}$ such that $D \,{}^*\!\subseteq A$ for all $A \in \mathscr{G}$. D is a tiny *member of $\mathscr{G}$.

Conversely, let $\mathscr{G}$ have a tiny *member D. Then $\mathscr{G} \neq \emptyset$ because $\emptyset$ has no *members. Also, if $A, B \in \mathscr{G}$ then $D \,{}^*\!\subseteq A$ and $D \,{}^*\!\subseteq B$; hence

$$\dagger\, {}^*\!\models \exists \mathbf{x}(\mathbf{x} \in \mathfrak{G} \wedge \mathbf{x} \subseteq \mathbf{A} \cap \mathbf{B}).$$

Thus there exists some $C \in \mathscr{G}$ such that $C \subseteq A \cap B$. It follows that $\mathscr{G}$ is a filter base. ∎

Thm. 3.3. is our second example of a characterization of a standard notion (in this case: being a filter base) by nonstandard means; i.e., by means of notions pertaining to $^*\mathfrak{U}$ rather than $\mathfrak{U}$. (The first example was Thm. 2.7.)

3.4. DEFINITION. Let τ be an arbitrary subcollection of $\hat{X}$. We put

$$\mathsf{F}\tau = \{A \subseteq X : \tau \subseteq \hat{A}\}.$$

Thus F is a mapping of $\mathsf{P}\hat{X}$ into $\mathsf{PP}X$. Moreover, it is easy to verify that $\mathsf{F}\tau$ is a filter; we call it *the filter of* τ. Also, if $\tau \subseteq \sigma \subseteq \hat{X}$ then clearly $\mathsf{F}\tau \supseteq \mathsf{F}\sigma$.

3.5. THEOREM. $\mathsf{F}\mu\mathscr{G} = \overline{\mathscr{G}}$ *for every* $\mathscr{G} \in \mathsf{PP}X$.
PROOF. By Thm. 3.2 and Def. 3.1 we have

$$\mu\mathscr{G} = \mu\overline{\mathscr{G}} \subseteq \hat{A} \quad \text{for all } A \in \overline{\mathscr{G}}.$$

Therefore $\overline{\mathscr{G}} \subseteq \mathsf{F}\mu\mathscr{G}$ by Def. 3.4.

It remains to prove that $\mathsf{F}\mu\mathscr{G} \subseteq \overline{\mathscr{G}}$. Let $A \in \mathsf{F}\mu\mathscr{G}$. Then $\mu\mathscr{G} \subseteq \hat{A}$ by Def. 3.4. Now, $\overline{\mathscr{G}}$ is a filter, and *a fortiori* a filter base; therefore, by Thm. 3.3, $\overline{\mathscr{G}}$ has a tiny *member, say D. Thus

$$D \,{}^*\!\in \overline{\mathscr{G}}, \qquad \hat{D} \subseteq \mu\overline{\mathscr{G}} = \mu\mathscr{G} \subseteq \hat{A}.$$

But then $D \,{}^*\!\subseteq A$, and we have

$$\dagger\, {}^*\!\models \exists \mathbf{x}(\mathbf{x} \in \overline{\mathfrak{G}} \wedge \mathbf{x} \subseteq \mathbf{A}).$$

Thus A has a subset belonging to $\overline{\mathscr{G}}$. But $\overline{\mathscr{G}}$ is a filter, hence $A \in \overline{\mathscr{G}}$ as required. ∎

3.6. COROLLARY. $\mathsf{F}\mu\mathscr{F} = \mathscr{F}$ *for every filter* $\mathscr{F}$. *Distinct filters have distinct monads.* ∎

Cor. 3.6 means that every filter can be recovered from its monad. Thus all the information concerning a filter is encapsulated in its monad. Therefore any statement about filters can, in principle, be replaced by an equivalent statement about their monads. This is often a considerable heuristic simplification, since a filter is a *set of sets* of points, while its monad is merely a collection of *points.

3.7. COROLLARY. *For any filters* $\mathscr{F}_1, \mathscr{F}_2$ *we have* $\mathscr{F}_1 \subseteq \mathscr{F}_2$ *iff* $\mu\mathscr{F}_1 \supseteq \mu\mathscr{F}_2$.
PROOF. If $\mu\mathscr{F}_1 \supseteq \mu\mathscr{F}_2$, then, applying F to both sides and using Cor. 3.6, we have $\mathscr{F}_1 \subseteq \mathscr{F}_2$. The converse is obvious. ∎

3.8. DEFINITION. For any $\tau \subseteq \hat{X}$ we put

$$\tau^- = \mu\mathsf{F}\tau.$$

We call τ^- the monad *generated by* τ or the *monadic closure of* τ.

3.9. THEOREM. *If* $\tau \subseteq \hat{X}$, *then* τ^- *is the smallest monad that includes* τ.
PROOF. By Def. 3.8, τ^- is a monad. Also, by Def. 3.4, $\tau \subseteq \hat{A}$ for all $A \in \mathsf{F}\tau$; therefore, by Def. 3.1, $\tau \subseteq \mu\mathsf{F}\tau = \tau^-$.

Now suppose $\tau \subseteq \mu\mathscr{G}$ for some $\mathscr{G}$. We shall show that $\tau^- \subseteq \mu\mathscr{G}$. Indeed, applying F to both sides of the inclusion $\tau \subseteq \mu\mathscr{G}$, and using 3.5, we get

$$\mathsf{F}\tau \supseteq \mathsf{F}\mu\mathscr{G} = \overline{\mathscr{G}}.$$

It now follows that $\mu\mathsf{F}\tau \subseteq \mu\overline{\mathscr{G}}$. This means that $\tau^- \subseteq \mu\mathscr{G}$. ∎

3.10. COROLLARY. *If* τ *is a monad, then* $\mu\mathsf{F}\tau = \tau^- = \tau$. ∎

3.11. THEOREM. (a) $\mathsf{F}\tau = \mathsf{F}\tau^-$ *for any* $\tau \subseteq \hat{X}$.
(b) *For any monads* σ, τ *we have* $\sigma \subseteq \tau$ *iff* $\mathsf{F}\sigma \supseteq \mathsf{F}\tau$.
PROOF. (a) $\mathsf{F}\tau^- = \mathsf{F}\mu\mathsf{F}\tau = \mathsf{F}\tau$ by Def. 3.8 and Cor. 3.6.
(b) If σ and τ are monads such that $\mathsf{F}\sigma \supseteq \mathsf{F}\tau$, then, applying μ to both sides and using Cor. 3.10, we have $\sigma \subseteq \tau$. The converse is obvious. ∎

To sum up: both μ and F are inclusion-reversing mappings. We have $\overline{\mathscr{G}} = \mathsf{F}\mu\mathscr{G}$ and $\tau^- = \mu\mathsf{F}\tau$ for every $\mathscr{G} \in \mathsf{PP}X$ and $\tau \in \mathsf{P}\hat{X}$. Finally, $\mathsf{F}\mu\mathscr{F} = \mathscr{F}$ for every filter $\mathscr{F}$; and $\mu\mathsf{F}\tau = \tau$ for every monad τ.

We shall now consider filters of certain particular kinds and find their monads.

3.12. EXAMPLE. A *principal* filter is a filter of the form $\{A\}^-$, where $A \subseteq X$. (It is easy to see that $\{A\}^- = \{B\colon A \subseteq B \subseteq X\}$.)

We clearly have $\mu(\{A\}^-)=\mu\{A\}=\hat{A}$. Thus, the monads of principal filters are scopes ("hats") of subsets of X.

In particular, for the improper filter $\mathsf{P}X$ we have $\mathsf{P}X=\{\emptyset\}^-$. Hence $\mu\mathsf{P}X=\hat{\emptyset}=\emptyset$ by Thm. 2.7. At the other extreme we have the filter $\{X\}$, whose monad is $\hat{X}$.

3.13. EXAMPLE. An *ultrafilter* is a proper filter which is not coarser than any other proper filter. The monad of an ultrafilter is called an *ultramonad*. Evidently, ultramonads are characterized by being minimal non-empty monads: a monad τ is an ultramonad iff the only monad properly included in τ is $\emptyset$.

Let $\mathscr{F}$ be an ultrafilter. Take any $p\in\mu\mathscr{F}$. Then $\{p\}\subseteq\mu\mathscr{F}$ and hence $\{p\}^-\subseteq\mu\mathscr{F}$ by Thm. 3.9. Since $\mu\mathscr{F}$ is an ultramonad, we must have $\{p\}^-=\mu\mathscr{F}$. Thus each ultramonad is generated by the singleton of a *point.

Conversely, if p is *any* *point, we shall show that $\{p\}^-$ is an ultramonad. Since $\{p\}^-=\mu\mathsf{F}\{p\}$ by Def. 3.8, it is enough to show that $\mathsf{F}\{p\}$ is an ultrafilter. Let $A\subseteq X$. Then

$$(*)\models \forall \mathbf{x}[\mathbf{x}\in\mathbf{X}\rightarrow(\mathbf{x}\in\mathbf{A}\leftrightarrow\mathbf{x}\notin\mathbf{X}-\mathbf{A})].$$

In particular $p\ {}^*\!\!\in A$ iff $p\ {}^*\!\!\notin X-A$. This means that $\{p\}\subseteq\hat{A}$ iff $\{p\}\nsubseteq(X-A)\hat{}$. Thus, for any $A\subseteq X$, exactly one of the sets A, $X-A$ belongs to $\mathsf{F}\{p\}$. It follows that $\mathsf{F}\{p\}$ is an ultrafilter. (Cf. Thm. 4.3.5.)

We have thus proved that τ is an ultramonad iff $\tau=\{p\}^-$ for some $p\in\hat{X}$. Equivalently, $\mathscr{F}$ is an ultrafilter iff $\mathscr{F}=\mathsf{F}(\{p\}^-)$ for some $p\in\hat{X}$. (By Thm. 3.11 we can write $\mathsf{F}\{p\}$ instead of $\mathsf{F}(\{p\}^-)$.)

By the way, for every proper filter $\mathscr{F}$ we can find an ultrafilter extending $\mathscr{F}$, as follows. Since $\mathscr{F}$ is proper, $\mu\mathscr{F}\neq\emptyset$. Take any $p\in\mu\mathscr{F}$. Then $\{p\}^-\subseteq\mu\mathscr{F}$; and $\mathsf{F}(\{p\}^-)$ is clearly an ultrafilter extending $\mathscr{F}$.

3.14. EXAMPLE. Let $p\in X$. Then $\{p\}\hat{}=\{p\}$ by Thm. 2.7. Therefore by Ex. 3.12 we have $\{p\}=\mu\mathscr{F}$, where $\mathscr{F}$ is the principal filter generated by $\{\{p\}\}$. It follows that $\{p\}^-=\{p\}$, because $\{p\}$ itself is already a monad. Thus we see that, for standard p, $\{p\}^-=\{p\}=$the monad of the principal ultrafilter $\{\{p\}\}^-$.

Conversely, let q be any *point such that the ultrafilter $\mathsf{F}(\{q\}^-)$ is principal. We shall show that q is standard, i.e., $q\in X$. Indeed, by Ex. 3.12 we must have $\{q\}^-=\hat{A}$ for some non-empty $A\subseteq X$. Take any $p\in A$. Then $\{p\}\subseteq A\subseteq\hat{A}=\{q\}^-$. But, since p is standard, it follows from what we have seen above that $\{p\}$ is a monad. Thus $\{p\}=\{q\}^-$, hence $q\in\{p\}$; so $q=p\in X$, as claimed.

We see that, if $q \in \hat{X} - X$, then $\{q\}^-$ must be the monad of a non-principal ultrafilter.

3.15. **Problem.** Prove that if q is nonstandard then $\{q\}^-$ is an infinite collection.

3.16. **Problem.** Let $\mathscr{F}$ be the filter of cofinite subsets of X, i.e.,

$$\mathscr{F} = \{A \subseteq X : X - A \text{ is finite}\}.$$

Prove that $\mu\mathscr{F} = \hat{X} - X$. Hence deduce that an ultrafilter is non-principal iff it is an extension of $\mathscr{F}$.

In Ex. 3.13 we saw that the ultramonads are precisely all collections of the form $\{p\}^-$, with $p \in \hat{X}$. Suppose that the ultramonads $\{p\}^-$ and $\{q\}^-$ have a common member r. Then the ultramonad $\{r\}^-$ is included in both $\{p\}^-$ and $\{q\}^-$. Hence $\{r\}^- = \{p\}^- = \{q\}^-$. Thus distinct ultramonads are disjoint. It follows that the ultramonads partition $\hat{X}$. By Ex. 3.14 and Prob. 3.15, an ultramonad consists of a single point or of infinitely many nonstandard *points.

3.17. **Theorem.** *Let p and q be *points. Then $\{p\}^- = \{q\}^-$ iff $q \mathrel{{}^*{\in}} A$ whenever $p \mathrel{{}^*{\in}} A \subseteq X$.*

Proof. Since the ultramonads partition $\hat{X}$, we have $\{p\}^- = \{q\}^-$ iff $q \in \{p\}^-$. But $\{p\}^- = \mu\mathsf{F}\{p\}$ by Def. 3.8. Hence $q \in \{p\}^-$ iff $q \in \hat{A}$ for every $A \in \mathsf{F}\{p\}$, i.e., $q \mathrel{{}^*{\in}} A$ whenever $p \mathrel{{}^*{\in}} A \subseteq X$. ∎

The following theorem means that $\{p\}^- = \{q\}^-$ iff p and q are "indistinguishable" in $\mathscr{L}_U$.

3.18. **Theorem.** *Let p and q be *points. Then $\{p\}^- = \{q\}^-$ iff whenever $\boldsymbol{\varphi}$ is an $\mathscr{L}_U$-formula with one free variable $\mathbf{x}$ such that ${}^*{\models}\boldsymbol{\varphi}(\mathbf{x}/\mathbf{p})$, we also have ${}^*{\models}\boldsymbol{\varphi}(\mathbf{x}/\mathbf{q})$.*

Proof. Suppose $\{p\}^- = \{q\}^-$; then by Thm. 3.17 we have $q \mathrel{{}^*{\in}} A$ whenever $p \mathrel{{}^*{\in}} A \subseteq X$. Let $\boldsymbol{\varphi}$ be an $\mathscr{L}_U$-formula with one free variable $\mathbf{x}$, such that ${}^*{\models}\boldsymbol{\varphi}(\mathbf{x}/\mathbf{p})$. Put

$$A = \{a : \models \mathbf{a} \in \mathbf{X} \wedge \boldsymbol{\varphi}(\mathbf{x}/\mathbf{a})\}.$$

Clearly, $p \mathrel{{}^*{\in}} A$ and $A \subseteq X$. Hence also $q \mathrel{{}^*{\in}} A$, so we must have ${}^*{\models} \boldsymbol{\varphi}(\mathbf{x}/\mathbf{q})$.

Conversely, suppose $\{p\}^- \neq \{q\}^-$. Then by Thm. 3.17 there is some $A \subseteq X$ such that $p \mathrel{{}^*{\in}} A$ but $q \mathrel{{}^*{\notin}} A$. Since A is standard, $\mathbf{x} \in \mathbf{A}$ is an $\mathscr{L}_U$-formula; and we have ${}^*{\models} \mathbf{p} \in \mathbf{A}$ but not ${}^*{\models} \mathbf{q} \in \mathbf{A}$. ∎

3.19. PROBLEM. Let I be a non-empty collection, and for each $i \in I$ let $\mathscr{G}_i \in \mathsf{PP}X$. Prove:

(i) $\mu\bigcup\{\mathscr{G}_i:\ i\in I\}=\bigcap\{\mu\mathscr{G}_i:\ i\in I\}$.

(ii) $\mu\bigcap\{\mathscr{G}_i:\ i\in I\}=(\bigcup\{\mu\mathscr{G}_i:\ i\in I\})^{-}$.

3.20. PROBLEM. Let $\mathscr{F}_1$ and $\mathscr{F}_2$ be filters. Prove that $\mu(\mathscr{F}_1\cap\mathscr{F}_2)=\mu\mathscr{F}_1\cup\mu\mathscr{F}_2$.

3.21. PROBLEM. (i) Show that under the operation of monadic closure (Def. 3.8) $\hat{X}$ is a topological space. (Use 3.19(i) and 3.20.)

In the remaining parts of the present problem, $\hat{X}$ is taken with this *monadic topology*.

(ii) Show that $\hat{X}$ is compact. (Use 3.19(i) to prove that if a collection of monads has the finite intersection property, it has a non-empty intersection.)

(iii) Show that X is dense in $\hat{X}$.

(iv) Show that $\tau\subseteq\hat{X}$ is clopen iff $\tau=\hat{A}$ for some $A\subseteq X$. (If τ is clopen and $A\in\mathsf{F}\tau$, then $\hat{A}=\tau$ or A intersects every member of $\mathsf{F}(\hat{X}-\tau)$; hence, if $\mathsf{F}\tau$ is non-principal, every member of $\mathsf{F}\tau$ intersects every member of $\mathsf{F}(\hat{X}-\tau)$. Consider tiny *members of $\mathsf{F}\tau$ and $\mathsf{F}(\hat{X}-\tau)$ respectively. Alternatively, use Lemma 4.4.2.)

(v) Let $\mathsf{M}X$ be the collection of all ultramonads: $\mathsf{M}X=\{\{p\}^{-}:\ p\in\hat{X}\}$. Give $\mathsf{M}X$ the quotient topology, so that the *closed* subcollections of $\mathsf{M}X$ are precisely those of the form $\{\{p\}^{-}:\ p\in\tau^{-}\}$ with $\tau\subseteq\hat{X}$. Show that $\mathsf{M}X$ is naturally homeomorphic to the Stone space $\mathsf{SP}X$ of the power-set Boolean algebra $\mathsf{P}X$.

The following result, the deepest in this section, is due to Luxemburg. It was also discovered — independently, but somewhat later — by Hirschfeld. We shall use Hirschfeld's proof.

3.22. THEOREM. *If $\mathscr{F}$ is a non-principal filter, then $\mu\mathscr{F}$ is external.*

PROOF. Suppose that $\mu\mathscr{F}$ is internal; then $\mu\mathscr{F}=\hat{A}$ for some *subset A of X.

Let $\mathscr{B}$ be any strong base for $\mathscr{F}$; thus $\overline{\mathscr{B}}=\mathscr{F}$, and $B_1\cap\ldots\cap B_n\in\mathscr{B}$ whenever $n\geq 1$ and $B_1,\ldots,B_n\in\mathscr{B}$. Since $\mu\mathscr{B}=\mu\overline{\mathscr{B}}=\mu\mathscr{F}=\hat{A}$, we have

(1) $$\hat{A}=\bigcap\{\hat{B}:\ B\in\mathscr{B}\}.$$

We claim that $A\ {}^*\!\!\in\mathscr{B}$. To prove this, we start by choosing a *finite *subset $\mathscr{C}$ of $\mathscr{B}$ such that

(2) $$\mathscr{B}\subseteq\mathscr{C}^{\wedge}.$$

The existence of such $\mathscr{C}$ follows from Thm. 2.12. Next, by Crit. 2.10 the collection $\{C: {}^*\models \mathbf{C} \in \mathscr{C} \wedge \mathbf{A} \subseteq \mathbf{C}\}$ is internal and hence equals $\mathscr{D}\hat{}$ for some *set $\mathscr{D}$. Thus

(3) $\mathscr{D}\hat{} = \{C: C \in \mathscr{C}\hat{} \text{ and } \hat{A} \subseteq \hat{C}\}.$

We put $D = {}^*\bigcap \mathscr{D}$. So a *point p *belongs to D iff p *belongs to every *member of $\mathscr{D}$. By (3) this can be written as

(4) $\hat{D} = \bigcap\{\hat{C}: C \in \mathscr{C}\hat{} \text{ and } \hat{A} \subseteq \hat{C}\}.$

We see at once that $\hat{A} \subseteq \hat{D}$. On the other hand from (1), (2) and (4) it is clear that we have $\hat{D} \subseteq \bigcap\{\hat{B}: B \in \mathscr{B}\}$ and therefore $\hat{D} \subseteq \hat{A}$ by (1). Thus we have $\hat{A} = \hat{D}$, so that $A = D$.

Now, from (3) it is clear that $\mathscr{D}$ is a *subset of $\mathscr{C}$. Since $\mathscr{C}$ was chosen as a *finite *subset of $\mathscr{B}$, it follows that $\mathscr{D}$ too is a *finite *subset of $\mathscr{B}$. (Note: (*)$\models$ **every subset of a finite set is finite!**) But $\mathscr{B}$ is closed under finite intersections hence, by transference, *closed under *finite *intersections. It follows that $A = D \,{}^*\!\in \mathscr{B}$, as claimed.

We now use the fact that $\mathscr{F}$ is non-principal. By Lemma 4.3.11, we have disjoint strong bases $\mathscr{B}_1$ and $\mathscr{B}_2$ for $\mathscr{F}$. But, by what we have just shown,

$$A \in \mathscr{B}_1\hat{} \cap \mathscr{B}_2\hat{} = (\mathscr{B}_1 \cap \mathscr{B}_2)\hat{} = \hat{\emptyset} = \emptyset,$$

which is absurd. So $\mu\mathscr{F}$ must be external. ∎

3.23. Problem. Using Thm. 3.22 and Prob. 3.16, find a new solution for Prob. 2.9.

3.24. Problem. Consider $\hat{X}$ with the monadic topology (Prob. 3.21(i)). Show that for any $A \,{}^*\!\subseteq X$ the following three conditions are equivalent:

(i) $\hat{A}$ is closed,

(ii) $\hat{A}$ is open,

(iii) A is standard.

3.25. Problem. Let $\mathscr{B}$ be a strong filter base, and let $p \in \mu\mathscr{B}$. Show that $\mathscr{B}$ has a tiny *member B such that $p \in \hat{B}$. (Find a *finite *subset $\mathscr{C}$ of $\mathscr{B}$ such that $B \subseteq \hat{C}$ and $p \in \hat{C}$ for every $C \in \mathscr{C}\hat{}$.)

For the rest of this section we assume, in addition to $X \in U$, also $Y \in U$ and $f \in \mathrm{Fun}(X, Y)$.

If $\mathscr{H} \in \mathsf{PP}Y$ and $\tau \subseteq \hat{Y}$ then $\mu\mathscr{H}$ and $\mathsf{F}\tau$ are defined in the obvious way, analogous to Defs. 3.1 and 3.4.

If $\mathscr{G} \in \mathsf{PP}X$ we put $f\mathscr{G} = \{f[A]: A \in \mathscr{G}\}$. Similarly, if $\mathscr{H} \in \mathsf{PP}Y$ we put

$$f^{-1}\mathscr{H} = \{f^{-1}[B]: B \in \mathscr{H}\}.$$

As explained at the end of §2, the function f has a natural extension f^*, which is a mapping of $\hat{X}$ into $\hat{Y}$. When applying f^* to an argument, we omit the superscript "*". In particular, if $\sigma \subseteq \hat{X}$ then $f[\sigma]$ is the collection of all images of members of σ under f^*. Similarly, if $\tau \subseteq \hat{Y}$ then $f^{-1}[\tau]$ is the collection of all members of $\hat{X}$ that are mapped by f^* into τ.

On the other hand, if $A \,{}^*\!\!\subseteq X$ then $f[A]$ is that *subset of Y which is characterized by

$$q \,{}^*\!\!\in f[A] \quad \text{iff} \quad q = fp \text{ for some } p \,{}^*\!\!\in A.$$

Thus $(f[A])\hat{} = f[\hat{A}]$. Similarly, if $B \,{}^*\!\!\subseteq Y$ then $f^{-1}[B]$ is that *subset of X which is characterized by

$$p \,{}^*\!\!\in f^{-1}[B] \quad \text{iff} \quad fp \,{}^*\!\!\in B.$$

Thus $(f^{-1}[B])\hat{} = f^{-1}[\hat{B}]$.

If σ is a monad $\subseteq \hat{X}$, what is $f[\sigma]$? If τ is a monad $\subseteq \hat{Y}$, what is $f^{-1}[\tau]$? The second question is easier, and we deal with it first.

3.26. THEOREM. *$f^{-1}[\mu\mathscr{H}] = \mu f^{-1}\mathscr{H}$ for every $\mathscr{H} \in \mathsf{PP}Y$.*

PROOF.

$$\begin{aligned} p \in f^{-1}[\mu\mathscr{H}] &\Leftrightarrow fp \in \mu\mathscr{H} \\ &\Leftrightarrow fp \in \hat{B} \text{ for all } B \in \mathscr{H} \\ &\Leftrightarrow p \in f^{-1}[\hat{B}] = (f^{-1}[B])\hat{} \text{ for all } B \in \mathscr{H} \\ &\Leftrightarrow p \in \hat{A} \text{ for all } A \in f^{-1}\mathscr{H} \\ &\Leftrightarrow p \in \mu f^{-1}\mathscr{H}. \end{aligned}$$

∎

3.27. THEOREM. *$(f[\mu\mathscr{G}])^- = \mu f\mathscr{G}$ for every filter base $\mathscr{G} \in \mathsf{PP}X$.*

PROOF. We have $\mu\mathscr{G} \subseteq \hat{A}$ for all $A \in \mathscr{G}$. Hence

$$f[\mu\mathscr{G}] \subseteq f[\hat{A}] = (f[A])\hat{}$$

for all $A \in \mathscr{G}$; i.e., $f[\mu\mathscr{G}] \subseteq \hat{B}$ for all $B \in f\mathscr{G}$. Therefore $f[\mu\mathscr{G}] \subseteq \mu f\mathscr{G}$ and by Thm. 3.9 it follows that $(f[\mu\mathscr{G}])^- \subseteq \mu f\mathscr{G}$.

It remains to prove the reverse inclusion. Let $B \in \mathsf{F}f[\mu\mathscr{G}]$; then by Def. 3.4 we have $f[\mu\mathscr{G}] \subseteq \hat{B}$. By Thm. 3.3, $\mathscr{G}$ has a tiny *member D; thus $\hat{D} \subseteq \mu\mathscr{G}$ and hence

$$(f[D])\hat{} = f[\hat{D}] \subseteq f[\mu\mathscr{G}] \subseteq \hat{B}.$$

Therefore $f[D] \,{}^*\!\!\subseteq B$; and since also $D \,{}^*\!\!\in \mathscr{G}$, we have

$$\dagger^* \models \exists \mathbf{x}(\mathbf{x} \in \mathbf{G} \wedge \mathbf{f}[\mathbf{x}] \subseteq \mathbf{B}).$$

So B has a subset of the form $f[A]$, with $A \in \mathscr{G}$. Therefore $B \in (f\mathscr{G})^-$. We have thus shown

$$\mathsf{F}f[\mu\mathscr{G}] \subseteq (f\mathscr{G})^- = \mathsf{F}\mu f\mathscr{G}.$$

Applying μ to both sides we get $(f[\mu\mathscr{G}])^- \supseteq \mu f\mathscr{G}$, as required. ∎

If $^*\mathfrak{U}$ is a limit enlargement, we can prove a stronger result:

3.28. THEOREM. *If* $^*\mathfrak{U}$ *is a limit enlargement of* $\mathfrak{U}$, *then* $f[\mu\mathscr{G}] = \mu f\mathscr{G}$ *for every filter base* $\mathscr{G} \in \mathsf{PP}X$.

PROOF. In view of Thm. 3.27, we only have to show that $\mu f\mathscr{G} \subseteq f[\mu\mathscr{G}]$.

Let $\{\mathfrak{U}^{(\beta)}: \beta \leqslant \alpha\}$ be an enlargement chain as described in Def. 1.6, with $\mathfrak{U}^{(0)} = \mathfrak{U}$ and $\mathfrak{U}^{(\alpha)} = {}^*\mathfrak{U}$, where α is a limit ordinal. Recall that by Cor. 1.8 we have $\mathfrak{U}^{(\gamma)} \lll \mathfrak{U}^{(\beta)}$ whenever $\gamma < \beta \leqslant \alpha$.

Now let $q \in \mu f\mathscr{G}$. Thus, for every $A \in \mathscr{G}$ we have $q \in (f[A])\hat{}$; i.e,

$$^*\models \mathbf{q \in f[A]}.$$

Since α is a limit ordinal, there exists some $\beta < \alpha$ such that $q \in U^{(\beta)}$. Because $\mathfrak{U}^{(\beta)} \prec \mathfrak{U}^{(\alpha)} = {}^*\mathfrak{U}$, we must have

$$\mathfrak{U}^{(\beta)} \models \mathbf{q \in f[A]}.$$

From this and the fact that $\mathscr{G}$ is a filter base it follows that the collection

$$\{\mathbf{fx = q \wedge x \in A}:\ A \in \mathscr{G}\}$$

is $\mathfrak{U}^{(\beta)}$-concurrent. Since $\mathfrak{U}^{(\beta)} \lll \mathfrak{U}^{(\beta+1)}$, there exists some $p \in U^{(\beta+1)}$ such that

$$\mathfrak{U}^{(\beta+1)} \models \mathbf{fp = q \wedge p \in A}$$

for all $A \in \mathscr{G}$. Since $\mathfrak{U}^{(\beta+1)} \prec \mathfrak{U}^{(\alpha)} = {}^*\mathfrak{U}$, we also have $^*\models \mathbf{fp = q \wedge p \in A}$ for all $A \in \mathscr{G}$. Thus $fp = q$, where $p \in \hat{A}$ for all $A \in \mathscr{G}$; hence $p \in \mu\mathscr{G}$ and $q \in f[\mu\mathscr{G}]$, as required. ∎

3.29. PROBLEM. Show that Thm. 3.28 continues to hold if the assumption that $^*\mathfrak{U}$ is a limit enlargement is replaced by the assumption that $\mathscr{G}$ is countable. (Let $\mathscr{G} = \{A_n:\ n \in N\}$. Without loss of generality we may assume $A_{n+1} \subseteq A_n$ for all $n \in N$. If $q \in \mu f\mathscr{G}$, consider the collection $\{A \in \mathscr{G}\hat{}\ :\ q\ {}^*\!\in f[A]\}$ and use the Overspill Lemma 2.11.)

3.30. PROBLEM. Consider $\hat{X}$ and $\hat{Y}$ with the monadic topology (Prob. 3.21). Show that the natural extension f^* of f is clopen (maps clopen sets onto clopen sets) and continuous. Assuming that $^*U$ is a limit enlargement show that f^* is closed (maps closed sets onto closed sets).

3.31. PROBLEM. Let $\mathscr{L}'$ be a first-order language with equality. For the sake of simplicity we assume that the only extralogical symbol of $\mathscr{L}'$ is a binary predicate symbol **R**. (Do not confuse $\mathscr{L}'$ with $\mathscr{L}$. In particular, our $\mathfrak{U}$ and $^*\mathfrak{U}$ are $\mathscr{L}$-structures but in general they are not $\mathscr{L}'$-structures.) Fix a non-empty set $A \in U$ and a binary relation $R \subseteq A^2$. Let $\mathfrak{A}$ be the $\mathscr{L}'$-structure $\langle A, R\rangle$ and let $\mathfrak{A}^*$ be the $\mathscr{L}'$-structure whose domain is $\hat{A}$ and whose binary relation is $\{\langle a,b\rangle : {}^*\models \langle \mathbf{a},\mathbf{b}\rangle \in \mathbf{R}\}$.

(i) Prove that $\mathfrak{A}^*$ is an enlargement and in particular an $\mathscr{L}'$-elementary extension of $\mathfrak{A}$.

(ii) Let $I \in U$ be a non-empty set and let $f,g \in \mathrm{Fun}(I,A)$. Show that if $\mathscr{F}$ is a filter over I then $\{p \in I: fp=gp\} \in \mathscr{F}$ iff $fp=gp$ for all $p \in \mu\mathscr{F}$.

(iii) Let $I \in U$, and let $\mathscr{F}$ be an ultrafilter over I. Consider the ultrapower $\mathfrak{A}^I/\mathscr{F}$. For any $p \in \mu\mathscr{F}$ we define the *Hirschfeld mapping* ψ_p by putting $\psi_p(f/\mathscr{F})=fp$ for each $f \in \mathrm{Fun}(I,A)$. Note that by (ii) this definition is legitimate. Prove that ψ_p is an elementary embedding of $\mathfrak{A}^I/\mathscr{F}$ into $\mathfrak{A}^*$. Thus $\mathfrak{A}^*$ is a "universal envelope" for *all* ultrapowers of the form $\mathfrak{A}^I/\mathscr{F}$, where $I \in U$.

§ 4. Topology

In this section, as in §3, we fix a non-empty $X \in U$. We consider a topology $\mathscr{T}$ on X; so $\mathscr{T} \in \mathsf{PP}X$ and the members of $\mathscr{T}$ are the *open* subsets of X. Unless otherwise indicated, $\mathscr{T}$ is held fixed; and we shall often commit the peccadillo of saying e.g. "the space X" when we actually mean the topological space $\langle X,\mathscr{T}\rangle$.

To prevent confusion, we shall only use the bar $^-$ to denote the closure operation in the space X, *not* for monadic closure in $\hat{X}$, *nor* for the generated filter.

For each point $p \in X$ we let $\mathscr{F}_p$ be the filter of neighbourhoods of p. Thus

$$\mathscr{F}_p=\{A:\ p \in T \subseteq A \text{ for some } T \in \mathscr{T}\}.$$

We put

$$\mu(p)=\mu\mathscr{F}_p,$$

and call $\mu(p)$ *the monad of* p. The mapping $\mu(\cdot)$ thus defined, which maps X into $\mathsf{P}\hat{X}$, is called the *monadology* of the space X. (When we want to stress the dependence of the monadology on $\mathscr{T}$ we write "$\mu_{\mathscr{T}}$"; but most of the time this is not needed, since $\mathscr{T}$ is held fixed or determined by the context.) If $q \in \mu(p)$ we write

$$q \simeq p$$

and say that q is *near* (or *infinitely close* to) p. Note that in general $\simeq$ is *not* an equivalence relation on $\hat{X}$. In fact, since $\mu(p)$ has been defined only for a *point* p, the statement "$q \simeq p$" is meaningful only if p is a *point* (and hence standard); but q can be any **point*, standard or not.

If $q \simeq p$ for some $p \in X$, we say that the *point q is *near-standard*. For any $p \in X$ we have $p \in \mu(p)$. (PROBLEM: Prove this.) Thus every point of X is near-standard. But, as we shall see, a near-standard *point is not necessarily standard.

A *point which is not near-standard is said to be *remote*.

It is helpful to visualize $\mu(p)$ as a very small cluster of dots or corpuscules (*points) around the point p. Since in general $\mu(p)$ is external (Thm. 3.22) it should be visualized as having no sharp boundary. The remote *points do not belong to any $\mu(p)$.

We shall give nonstandard characterizations of various standard topological notions. Many of these characterizations accord so well with intuition that they may give the reader a feeling of *déjà connu*.

4.1. THEOREM. *Let $B \subseteq X$. Then B is open iff $p \in B \Rightarrow \mu(p) \subseteq \hat{B}$ (i.e., $q \simeq p \in B \Rightarrow q \, {}^*\!\in B$). B is closed iff $\mu(p) \cap \hat{B} \neq \emptyset \Rightarrow p \in B$ (i.e., $p \in X$, $q \simeq p$, $q \, {}^*\!\in B \Rightarrow p \in B$).*

PROOF. B is open iff $B \in \mathscr{F}_p$ for every $p \in B$. This means that $\mu\mathscr{F}_p \subseteq \hat{B}$ for every $p \in B$. But $\mu\mathscr{F}_p = \mu(p)$ by definition.

The second half of the theorem follows at once, since B is closed iff $X - B$ is open. ∎

4.2. PROBLEM. Let $A \subseteq X$ and $p \in X$. Show that $p \in \bar{A}$ iff $\mu(p) \cap \hat{A} \neq \emptyset$.

We started by defining the monadology $\mu(\cdot)$ from the topology $\mathscr{T}$. Now we see from Thm. 4.1 that the topology can be recovered from the monadology. Therefore every statement about the topology can, in principle, be replaced by an equivalent statement about the monadology.

Let λ be a mapping of X into $\mathsf{P}\hat{X}$. By what we have just seen, there can be at most one topology on X such that λ is the corresponding monadology. Under what conditions does such a topology exist?

4.3. THEOREM. *The following three conditions are necessary and jointly sufficient in order that λ be the monadology of some topology on X:*

(i) *$\lambda(p)$ is monadic for every $p \in X$.*

(ii) *$p \in \lambda(p)$ for every $p \in X$.*

(iii) *For every $r \in X$, if $A \in \mathsf{F}\lambda(r)$ then there is some $B \in \mathsf{F}\lambda(r)$ such that $A \in \mathsf{F}\lambda(q)$ for each $q \in B$.*

PROOF. The necessity of (i) and (ii) is obvious. The necessity of (iii) follows easily from the fact that if A is a neighbourhood of r then some *open* neighbourhood B of r is included in A.

To prove the sufficiency of (i), (ii) and (iii) we put

$$\mathscr{T}=\{G\subseteq X:\ G\in \mathsf{F}\lambda(r) \text{ for all } r\in G\}.$$

We leave to the reader the simple task of verifying (without using (i), (ii) and (iii)) that $\mathscr{T}$ is a topology on X and $\lambda(p)\subseteq\mu_{\mathscr{T}}(p)$ for each $p\in X$. We shall show that $\mu_{\mathscr{T}}(p)\subseteq\lambda(p)$.

Let $A\in \mathsf{F}\lambda(p)$. We put

$$G=\{q\in X:\ A\in \mathsf{F}\lambda(q)\}.$$

We claim that $p\in G\subseteq A$ and that G is open (i.e., belongs to $\mathscr{T}$).

Evidently, $p\in G$ since we have assumed $A\in \mathsf{F}\lambda(p)$. Also, if $q\in G$ then $A\in \mathsf{F}\lambda(q)$, hence $\lambda(q)\subseteq\hat{A}$. By (ii) we have $q\in\hat{A}$, and since both q and A are standard this means $q\in A$. Thus $G\subseteq A$. To show that G is open, suppose that $r\in G$. Then $A\in \mathsf{F}\lambda(r)$ and by (iii) there is some $B\in \mathsf{F}\lambda(r)$ such that $A\in \mathsf{F}\lambda(q)$ for all $q\in B$. It follows at once that $B\subseteq G$; and since $B\in \mathsf{F}\lambda(r)$, we must also have $G\in \mathsf{F}\lambda(r)$. This shows that $G\in\mathscr{T}$.

Since $p\in G\subseteq A$ and G is open, it follows from Thm. 4.1 that $\mu_{\mathscr{T}}(p)\subseteq\hat{G}\subseteq\hat{A}$. Thus we have established that $\mu_{\mathscr{T}}(p)\subseteq\hat{A}$ for all $A\in \mathsf{F}\lambda(p)$. Hence $\mu_{\mathscr{T}}(p)\subseteq$ $\subseteq\mu \mathsf{F}\lambda(p)$; but from (i) we see that $\mu \mathsf{F}\lambda(p)=\lambda(p)$. ∎

If $q\in\hat{X}$ and there is a *unique* $p\in X$ such that $q\simeq p$, we put $^{\circ}q=p$ and call it the *standard approximation of* q. If no such unique p exists, then $^{\circ}q$ is undefined.

More generally, if $A\ {}^{*}\!\subseteq X$, we put

$$^{\circ}A=\{p\in X:\mu(p)\cap\hat{A}\neq\emptyset\}.$$

$^{\circ}A$ is called *the standard approximation of* A.

4.4. PROBLEM. Let $A\ {}^{*}\!\subseteq X$. Show that each of the following two conditions is sufficient for $^{\circ}A$ to be closed:

(i) $^{*}\mathfrak{U}$ is a limit enlargement.

(ii) For each $p\in X$ the filter $\mathscr{F}_p$ has a countable base.

(Use the *standard* characterization of closed sets. For (i), let $\{\mathfrak{U}^{(\beta)}:\ \beta\leqslant\alpha\}$ be an enlargement chain as described in Def. 1.6, with α a limit ordinal, $\mathfrak{U}^{(0)}=\mathfrak{U}$, $\mathfrak{U}^{(\alpha)}={}^{*}\mathfrak{U}$. Take β such that $A\in U^{(\beta)}$. For (ii), let $p\in({}^{\circ}A)^{-}$ and let $\mathscr{B}=\{B_n:\ n\in N\}$ be a base for $\mathscr{F}_p$. Without loss of generality assume

$B_{n+1} \subseteq B_n$ for all n. Consider the collection $\{B \in \mathscr{B}\hat{} : A \,{}^*\!\cap B \neq \emptyset\}$ and use the Overspill Lemma 2.11.)

We return to the more pleasant pastime of finding nonstandard characterizations of topological concepts.

4.5. PROBLEM. Prove:

(i) A point p is isolated — i.e., $\{p\}$ is open — iff $\mu(p)=\{p\}$. (Use 4.1.)

(ii) X is a T_0 space — i.e., if p,q are distinct points then $p \notin \{q\}^-$ or $q \notin \{p\}^-$ — iff for any distinct points p,q we have $q \notin \mu(p)$ or $p \notin \mu(q)$. (Use 4.2.)

(iii) X is a T_1 space — i.e., $\{p\}$ is closed for every $p \in X$ — iff for any distinct points p,q we have $p \notin \mu(q)$.

4.6. THEOREM. *X is a T_2 (Hausdorff) space iff the monads of distinct points are disjoint.*

PROOF. If X is T_2 and p,q are distinct points, then there are $A \in \mathscr{F}_p$, $B \in \mathscr{F}_q$ such that $A \cap B = \emptyset$. Hence $\hat{A} \cap \hat{B} = (A \cap B)\hat{} = \emptyset$; and since $\mu(p) \subseteq \hat{A}$ and $\mu(q) \subseteq \hat{B}$, also $\mu(p) \cap \mu(q) = \emptyset$.

Conversely, suppose $\mu(p) \cap \mu(q) = \emptyset$. By Thm. 3.3, $\mathscr{F}_p$ and $\mathscr{F}_q$ have tiny *members D and E respectively. Since $\hat{D} \subseteq \mu(p)$ and $\hat{E} \subseteq \mu(q)$, we have $\emptyset = \hat{D} \cap \hat{E} = (D \,{}^*\!\cap E)\hat{}$. Therefore

$$\dagger^* \models \exists x \, \exists y (x \in \mathfrak{F}_p \wedge y \in \mathfrak{F}_q \wedge x \cap y = 0)$$

Thus p and q have disjoint neighbourhoods. ∎

Note that if q is a near-standard *point of a Hausdorff space then from Thm. 4.6 it follows that $^\circ q$ is defined, because the condition $q \in \mu(p)$ determines p uniquely.

4.7. PROBLEM. The space X is *regular* if for each $p \in X$ the filter $\mathscr{F}_p$ is generated by its closed members. Prove that X is regular iff for each point p and each *point $q \notin \mu(p)$ there are disjoint open sets A, B such that $p \in A$ and $q \notin \hat{B}$.

4.8. PROBLEM. X is *normal* if whenever A, B are disjoint closed sets there are disjoint open sets C, D such that $A \subseteq C$ and $B \subseteq D$. Prove that X is normal iff for any *points p, q such that $p \in \hat{A}$, $q \in \hat{B}$ for some disjoint closed sets A, B, there are disjoint open sets C, D such that $p \in \hat{C}$, $q \in \hat{D}$.

4.9. PROBLEM. Let $\mathscr{F}$ be a proper filter over X, and let $p \in X$. Then p is said to be *adherent* to $\mathscr{F}$ if every member of $\mathscr{F}_p$ intersects every member of $\mathscr{F}$. $\mathscr{F}$ is said to *converge* to p if $\mathscr{F}_p \subseteq \mathscr{F}$. Prove:

(i) p is adherent to $\mathscr{F}$ iff $\mu(p) \cap \mu\mathscr{F} \neq \emptyset$ (Use 3.19(i).)

(ii) $\mathscr{F}$ converges to p iff $\mu\mathscr{F} \subseteq \mu(p)$.

Hence prove:

(iii) p is adherent to $\mathscr{F}$ iff $\mathscr{F}$ can be extended to an ultrafilter converging to p.

(iv) If X is a Hausdorff space, $\mathscr{F}$ cannot converge to more than one point.

We now come to the nonstandard characterization of compact sets. This is probably the single most useful tool of nonstandard analysis.

4.10. THEOREM. *Let $C \subseteq X$. Then C is compact iff $\hat{C} \subseteq \bigcup\{\mu(p) : p \in C\}$; i.e., iff for each $q \in \hat{C}$ we have $q \simeq p$ for some $p \in C$.*

PROOF. Suppose first that C is compact and let $q \notin \bigcup\{\mu(p) : p \in C\}$. We must show that $q \notin \hat{C}$. Put

$$\mathscr{B} = \{B : B \in \mathscr{T}, q \notin \hat{B}\}.$$

If $p \in C$ then, by assumption, $q \notin \mu(p)$. Since $\mathscr{F}_p$ is generated by its open members, there must exist some open neighbourhood B of p such that $q \notin \hat{B}$; hence $p \in B \in \mathscr{B}$. Thus $\mathscr{B}$ is an open cover of C. Since C is compact, there are $B_1, \ldots, B_k \in \mathscr{B}$ such that

$$(^*) \models \mathbf{C} \subseteq \mathbf{B}_1 \cup \ldots \cup \mathbf{B}_k.$$

But $q \notin \hat{B}_1 \cup \ldots \cup \hat{B}_k = (B_1 \cup \ldots \cup B_k)\hat{}$. Hence $q \notin \hat{C}$, as required.

Conversely, suppose that C is *not* compact. Then there is an open cover $\mathscr{B}$ of C such that no finite subset of $\mathscr{B}$ covers C. Consequently,

$$\{\mathbf{x} \in \mathbf{C} \wedge \mathbf{x} \notin \mathbf{B} : B \in \mathscr{B}\}$$

is a $\mathfrak{U}$-concurrent collection of formulas. Thus for some $q \in \hat{C}$ we must have $q \in \hat{B}$ for all $B \in \mathscr{B}$. If $p \in C$, then $p \in B$ for some $B \in \mathscr{B}$. Then $\mu(p) \subseteq \hat{B}$ by Thm. 4.1; therefore $q \notin \mu(p)$. Thus $\hat{C}$ is not included in $\bigcup\{\mu(p) : p \in C\}$. ∎

The standard definition of compactness involves quantification over *sets of sets* of points: *For every $\mathscr{B} \in \mathsf{PP}X$, if $\mathscr{B}$ is an open cover of C then some finite part of $\mathscr{B}$ covers C.* The nonstandard characterization, on the other hand, refers only to *points: *Every *point of C is near some point of C.* Thus Thm. 4.10 constitutes a far-reaching simplification (technically as well as heuristically) of the notion of compactness; hence the great usefulness of this result. For this reason nonstandard analysis is particularly helpful in problems that involve compactness in an essential way.

4.11. Example. Let X be a Hausdorff space, and let $C \subseteq X$ be compact. We show that C is closed. Let $q \simeq p$ and $q \in \hat{C}$. By Thm. 4.10 we have $q \simeq r$ for some $r \in C$; hence, by Thm. 4.6, $p = r \in C$. By Thm. 4.1, C is closed.

4.12. Problem. Use Thms. 4.1 and 4.10 to show that each closed subset of a compact set is compact.

4.13. Problem. Use Prob. 4.9 and Thm. 4.10 to show that for any $C \subseteq X$ the following three conditions are equivalent:

(i) C is compact.

(ii) Every proper filter $\mathscr{F}$ such that $C \in \mathscr{F}$ has an adherent point $p \in C$.

(iii) Every ultrafilter $\mathscr{F}$ such that $C \in \mathscr{F}$ converges to some $p \in C$.

4.14. Problem. For any $C \subseteq X$ put

$$\mathscr{F}_C = \{B\colon\ C \subseteq T \subseteq B \text{ for some } T \in \mathscr{T}\}.$$

Prove that $\bigcup\{\mu(p)\colon\ p \in C\} = \mu\mathscr{F}_C$ iff C is compact. (Observe that for *any* C, $\bigcup\{\mu(p)\colon\ p \in C\} \subseteq \mu\mathscr{F}_C$. If C is compact, argue as in the proof of 4.10 to show that the reverse inclusion holds. For the converse, observe that $\hat{C} \subseteq \mu\mathscr{F}_C$.)

4.15. Problem. Let $X \subseteq X' \in U$, and let X' be a Hausdorff compactification of X. Thus X' has a Hausdorff topology $\mathscr{T}'$ under which X' is compact, X is dense in X' and the topology induced on X by $\mathscr{T}'$ is the original topology $\mathscr{T}$ of X. For each $p \in \hat{X}$, let $\varphi(p)$ be the point of X' such that $p \simeq \varphi(p)$ in the monadology of X'. (Since X' is a compact Hausdorff space, $\varphi(p)$ is uniquely defined.)

(i) Show that φ maps $\hat{X}$ *onto* X'.

(ii) Consider $\hat{X}$ in the topology

$$\varphi^{-1}\mathscr{T}' = \{\varphi^{-1}[B]\colon\ B \in \mathscr{T}'\},$$

so that X' is a quotient space of $\hat{X}$, with φ as the quotient mapping. Show that $\varphi^{-1}\mathscr{T}'$ is equal to or coarser than the monadic topology (Prob. 3.21) on $\hat{X}$. (Use 4.12 and 4.14 to show that any subcollection of $\hat{X}$ closed in the topology $\varphi^{-1}\mathscr{T}'$ is the monad of a filter.)

(iii) Show that the topology induced by $\varphi^{-1}\mathscr{T}'$ on X (as a subcollection of $\hat{X}$) is the original topology $\mathscr{T}$. (Use the fact that φ restricted to X is the identity.)

(iv) Show that (in the topology $\varphi^{-1}\mathscr{T}'$) $\hat{X}$ is a compactification of X. (Use (ii), (iii) and parts (ii) and (iii) of Prob. 3.21.)

4.16. Theorem. *Let $A \subseteq X$. In order that $\bar{A}$ be compact it is necessary that each $q \in \hat{A}$ be near-standard. This condition is also sufficient if X is a regular space.*

PROOF. If $\bar{A}$ is compact, then every *point of $\bar{A}$ — and in particular every *point of A — is near a point of $\bar{A}$, and hence near-standard.

Now suppose that X is regular and every $q \in \hat{A}$ is near-standard. Take an arbitrary $r \in \bar{A}\hat{}$; to show that $\bar{A}$ is compact, we must prove that $r \simeq p$ for some $p \in \bar{A}$. Put

$$\mathscr{B} = \{B: r \,{}^*\!\in B \in \mathscr{T}\}.$$

It is easy to see that $\mathscr{B}$ is closed under finite intersections. Also, if $B \in \mathscr{B}$ then $r \in (\bar{A} \cap B)\hat{}$, hence $\bar{A} \cap B \neq \emptyset$; and since B is open, also $A \cap B \neq \emptyset$. It follows that the collection

$$\{\mathbf{x \in A \wedge x \in B}: \ B \in \mathscr{B}\}$$

is $\mathfrak{U}$-concurrent. Hence for some $q \in \hat{A}$ we have $q \in \hat{B}$ for all $B \in \mathscr{B}$.

By assumption, $q \simeq p$ for some $p \in X$. In fact, $p \in \bar{A}$ by Prob. 4.2. We claim that also $r \simeq p$. Indeed, if this were *not* so, then by the regularity of X there would be some *closed* $D \in \mathscr{F}_p$ such that $r \notin \hat{D}$, hence $X - D \in \mathscr{B}$ and therefore $q \notin \hat{D}$. But we *must* have $q \in \hat{D}$, since $q \simeq p$ and $D \in \mathscr{F}_p$. ∎

4.17. COUNTER-EXAMPLE. The condition of Thm. 4.16 may not be sufficient if X is not regular. For instance, let X be the unit square

$$\{\langle a,b \rangle: \ 0 \leqslant a \leqslant 1, \ 0 \leqslant b \leqslant 1\}.$$

As *open* subsets of X we take all sets of the form

$$B - \{\langle 0,b \rangle: \ 0 < b \leqslant r\},$$

where B is open in the usual (metric) topology of X and $0 \leqslant r$. (It is easy to check that this is indeed a topology on X.) The remote *points of X are precisely those $p \,{}^*\!\in \{\langle 0,b \rangle: \ 0 < b \leqslant 1\}$ such that $p \simeq \langle 0,0 \rangle$ in the monadology of the *metric* topology of X. Thus X is not compact. But if

$$A = X - \{\langle 0,b \rangle: \ 0 \leqslant b \leqslant 1\},$$

then every *point of A is near-standard while $\bar{A} = X$.

4.18. PROBLEM. Show that the following four conditions are equivalent:

(i) X is locally compact.

(ii) Every convergent ultrafilter has a compact member.

(iii) Every near-standard *point of X *belongs to some compact subset of X.

(iv) The collection of all remote *points of X is the monad of a filter $\mathscr{F}$, which is generated by the set of all complements of compact sets.

Show also that if X is assumed to be regular then (iv) implies the other three conditions even if the assumption on the way $\mathscr{F}$ is generated is omitted.

We now consider, in addition to X, a non-empty set $Y \in U$ and some fixed topology on Y. If $q \in Y$, we denote the filter of neighbourhoods of q by "$\mathscr{F}_q$" and the monad of q by "$\mu(q)$". No confusion will arise, because it will always be clear from the context which space is being referred to.

4.19. THEOREM. *Let* $f \in \mathrm{Fun}(X,Y)$, *and let* $p \in X$. *Then* f *is continuous at* p *iff* $f[\mu(p)] \subseteq \mu(fp)$.

PROOF. By definition, f is continuous at p iff for every $B \in \mathscr{F}_{fp}$ we have $f^{-1}[B] \in \mathscr{F}_p$. This is equivalent to $f^{-1}\mathscr{F}_{fp} \subseteq \mathscr{F}_p$. This in turn is equivalent to $\mu f^{-1}\mathscr{F}_{fp} \supseteq \mu\mathscr{F}_p = \mu(p)$. By Thm. 3.26, this is the same thing as

$$f^{-1}[\mu(fp)] \supseteq \mu(p), \text{ i.e., } f[\mu(p)] \subseteq \mu(fp). \quad ∎$$

Thm. 4.19 can be rephrased equivalently as follows: *The function* $f \in \mathrm{Fun}(X,Y)$ *is continuous at* $p \in X$ *iff* $q \simeq p \Rightarrow fq \simeq fp$, *i.e., whenever* q *is infinitely close to* p *then* fq *is infinitely close to* fp. This is what mathematicians always wanted to say about continuity but didn't quite know how, without sacrificing rigour. In nonstandard analysis, this highly intuitive characterization of continuity is at the same time completely rigorous.

4.20. EXAMPLE. Let $f \in \mathrm{Fun}(X,Y)$ be continuous, and let $C \subseteq X$ be compact. To show that $f[C]$ is compact, take an arbitrary *point of $f[C]$. It must be of the form fq where $q \,{}^*\!\in C$. Since C is compact, $q \simeq p$ for some $p \in C$. But f is continuous, so $fq \simeq fp \in f[C]$.

4.21. EXAMPLE. Let f,g be continuous $\in \mathrm{Fun}(X,Y)$. Assuming that Y is a Hausdorff space, we show that the set $A = \{p \in X\colon fp \neq gp\}$ is open. Indeed, let $q \simeq p \in A$. Then by the continuity of f and g we have $fq \simeq fp$ and $gq \simeq gp$. But $fp \neq gp$ because $p \in A$. Since Y is a Hausdorff space, it follows that $fq \neq gq$, i.e., $q \in \hat{A}$.

4.22. PROBLEM. Let $f \in \mathrm{Fun}(X,Y)$ and $p \in X$. Then f is said to be *open* at p if $f\mathscr{F}_p \subseteq \mathscr{F}_{fp}$.

(i) Show that if $f[\mu(p)] \supseteq \mu(fp)$ then f is open at p.

(ii) Assuming that ${}^*\mathfrak{U}$ is a limit enlargement of $\mathfrak{U}$ or that $\mathscr{F}_p$ has a countable base, prove the converse of (i).

4.23. PROBLEM. Let $N \in U$, and let $f \in \mathrm{Fun}(N,X)$. (Thus f is a sequence of points.)

(i) Show that f converges to $p \in X$ iff $fn \simeq p$ for *all* $n \in \hat{N} - N$.

(ii) Show that $p \in X$ is a cluster point of f iff $fn \simeq p$ for *some* $n \in \hat{N} - N$.

Now let X be the topological product of a family $\{X_i: i \in I\}$ of topological spaces. Thus each point of X is a function f with I as its domain and $fi \in X_i$ for all $i \in I$.

For each $i \in I$ we have $i \in \{i\} \in \langle i, fi \rangle \in f \in X$, hence $I \in \mathsf{P}\bigcup\bigcup\bigcup X$. Similarly, $X_i \in \mathsf{P}\bigcup\bigcup\bigcup X$ for each $i \in I$. Since we are assuming thet $X \in U$, it follows from (2.3) and (2.4) that I and each of the X_i belong to U.

If $g \in \hat{X}$ and $i \in \hat{I}$, then $gi (= g^{*}`i)$ is uniquely defined and belongs to $(X_i)\hat{}$.

Let $f \in X$ and for each $i \in I$ let π_i be the canonical projection of X onto X_i (so that $\pi_i f = fi$). Then the filter $\mathscr{F}_f$ of all neighbourhoods of f in X is generated by

$$\bigcup\{\pi_i^{-1}\mathscr{F}_{fi}: i \in I\},$$

where $\mathscr{F}_{fi}$ is the filter of neighbourhoods of fi in X_i. Using Prob. 3.19(i) and Thm. 3.26 we have

$$\mu(f) = \bigcap\{\pi_i^{-1}[\mu(fi)]: i \in I\},$$

where $\mu(fi)$ is the monad of fi in the monadology of X_i. We have thus proved:

4.24. THEOREM. *Let $f \in X$ and $g \in \hat{X}$. Then $g \in \mu(f)$ iff $gi \in \mu(fi)$ for all $i \in I$.* ∎

Note that Thm. 4.24 says nothing about gi for $i \in \hat{I} - I$. Indeed, for such i we have not defined $\mu(fi)$, because X_i is not a topological space but only a *topological *space.

4.25. EXAMPLE. Assume that, for each $i \in I$, X_i is a Hausdorff space. We show that X too is a Hausdorff space. Indeed, if $f, g \in X$ and $h \in \mu(f) \cap \mu(g)$, then, by Thm. 4.24, $hi \in \mu(fi) \cap \mu(gi)$ for all $i \in I$. Hence $fi = gi$ for all $i \in I$, so $f = g$.

4.26. EXAMPLE. Assume that, for each $i \in I$, X_i is compact. We prove (Tychonoff's Theorem) that X is compact. Indeed, let $g \in \hat{X}$. For each $i \in I$, $gi \in (X_i)\hat{}$; and since X_i is compact, we can choose $a_i \in X_i$ such that $gi \simeq a_i$. (This requires the axiom of choice. But if the X_i are Hausdorff spaces then the a_i are unique and the axiom of choice is not needed.) Put $fi = a_i$ for all $i \in I$; then $f \in X$ and by Thm. 4.24 we have $g \simeq f$, as required.

§ 5. Topological groups

Where there is an interplay between algebraic and topological notions, nonstandard analysis often provides interesting insights. As an illustration of this, we shall deal with the subject of topological groups. No previous knowledge of this subject is needed for *understanding* our treatment; but some such knowledge is obviously required for *comparing* our treatment with the conventional (standard) approach.

5.1. DEFINITION. A *topological group* is a triple $\langle G, \cdot, \mathscr{T}\rangle$ such that:
(i) $\langle G, \cdot\rangle$ is a group, i.e., $\cdot$ is a group operation on the set G;
(ii) $\langle G, \mathscr{T}\rangle$ is a topological space; i.e., $\mathscr{T}$ is a topology on G;
(iii) The mapping $\langle g,h\rangle \mapsto g\cdot h$ of the product space $G\times G$ into G, and the mapping $g \mapsto g^{-1}$ of G into itself are continuous.

Throughout this section we assume (unless indicated otherwise) that $\langle G, \cdot, \mathscr{T}\rangle$ is a fixed but arbitrary topological group such that $G\in U$. With the customary (and harmless) abuse of terminology, we speak of G itself as a "topological group", when we really want to refer to $\langle G, \cdot, \mathscr{T}\rangle$. We let e be the identity element of G.

The operation $\cdot$ on G can be extended in a natural way to an operation on $\hat{G}$: if $g_1,g_2\in\hat{G}$, we let $g_1\cdot g_2$ be the unique $h\in\hat{G}$ such that $^*\models \mathbf{g_1}\cdot\mathbf{g_2}=\mathbf{h}$. If σ and τ are subcollections of $\hat{G}$, we put

$$\sigma\cdot\tau=\{g\cdot h:\ g\in\sigma,\ h\in\tau\},$$

$$\sigma^{-1}=\{g^{-1}:\ g\in\sigma\}.$$

If A and B are *subsets of G then $A\cdot B$ is that *subset of G whose *members are precisely all products $a\cdot b$ with $a\ {}^*\!\in A$ and $b\ {}^*\!\in B$. Thus $(A\cdot B)\hat{}=\hat{A}\cdot\hat{B}$. Similarly, $(A^{-1})\hat{}=\hat{A}^{-1}$.

From the fact that (under the operation $\cdot$) G is a group, with e as identity element, it follows by transference that (under the operation $\cdot$ as extended above) $\hat{G}$ is also a group, with e as identity element. Thus G is a subgroup of $\hat{G}$. This result evidently holds also for ordinary groups, without any particular topology.

Using Thms. 4.19 and 4.24, we can re-state condition (iii) of Def. 5.1 as follows:

(1) $\mu(g)\cdot\mu(h)\subseteq\mu(g\cdot h)$ for all $g,h\in G$;

(2) $\mu(g)^{-1}\subseteq\mu(g^{-1})$ for all $g\in G$.

Hence we easily get

(3) $\mu(g)\cdot\mu(h)^{-1}\subseteq\mu(g\cdot h^{-1})$ for all $g,h\in G$.

Let J be the collection of all near-standard members of $\hat{G}$. Thus

(4) $J=\bigcup\{\mu(g):\ g\in G\}$.

From (3) and (4) it follows at once that $J \cdot J^{-1} \subseteq J$, hence J is a subgroup of $\hat{G}$.

Also, by (3), $\mu(e) \cdot \mu(e)^{-1} \subseteq \mu(e)$. Hence $\mu(e)$ is a subgroup of J. The members of $\mu(e)$ are called the *infinitesimals* of the topological group G, and $\mu(e)$ is called the *infinitesimal group* of G.

Using (1), (2) and the fact that $g \in \mu(g)$, we get for each $g \in G$,

$$g \cdot \mu(e) \subseteq \mu(g) \cdot \mu(e) \subseteq \mu(g) \subseteq g \cdot \mu(g^{-1}) \cdot \mu(g) \subseteq g \cdot \mu(e).$$

Thus $g \cdot \mu(e) = \mu(g)$. Similarly, $\mu(e) \cdot g = \mu(g)$. It follows that, for each $g \in G$, the right coset and the left coset of g modulo $\mu(e)$ coincide with each other and with $\mu(g)$. In view of (4) *all* cosets modulo $\mu(e)$ in J are of the form $\mu(g)$, and hence $\mu(e)$ is a *normal* subgroup of J. To sum up:

5.2. THEOREM. *The collection J of near-standard members of $\hat{G}$ is a subgroup of $\hat{G}$. The collection of infinitesimals $\mu(e)$ is a normal subgroup of J. The cosets in J modulo $\mu(e)$ are precisely the monads of points $g \in G$.* ∎

5.3. PROBLEM. Prove that G is a Hausdorff space iff $\mu(e) \cap G = \{e\}$. Hence prove that if G is a T_0 space (Prob. 4.5(iii)) then it is a Hausdorff space and in this case the quotient group $J/\mu(e)$ is isomorphic to the group G.

5.4. THEOREM. *Suppose that G is given as a group (without any particular topology). Let τ be a subgroup of $\hat{G}$. Then the two conditions*

(a) *τ is monadic, i.e., $\tau = \mu\mathscr{F}$ for some filter $\mathscr{F}$ over G,*

(b) *$\tau \cdot g = g \cdot \tau$ for every $g \in G$;*

are necessary and jointly sufficient for the existence of a topology on G under which G becomes a topological group with τ as its infinitesimal group.

PROOF. The conditions are clearly necessary. To prove that they are sufficient we put $\lambda(g) = \tau \cdot g$ for every $g \in G$ and show that the three conditions of Thm. 4.3 are fulfilled.

(i) For any $g \in G$, let

$$\mathscr{F} \cdot g = \{A \cdot g\colon\ A \in \mathscr{F}\},$$

where $\mathscr{F}$ is the filter mentioned in (a), for which $\tau = \mu\mathscr{F}$. It is easy to verify that $\mu(\mathscr{F} \cdot g) = \tau \cdot g = \lambda(g)$; so $\lambda(g)$ is monadic.

(ii) Since τ is a subgroup of $\hat{G}$, it contains e. Hence for every $g \in G$ we have $g \in \tau \cdot g = \lambda(g)$.

(iii) For every $g \in G$ and $A \in \mathsf{F}\lambda(g)$ we have to find some $B \in \mathsf{F}\lambda(g)$ such that $A \in \mathsf{F}\lambda(h)$ for each $h \in B$. It is enough to consider the case $g = e$, since the general case can be obtained by translation (i.e., multiplication by g).

We assume therefore that $A \in \mathrm{F}\tau = \mathscr{F}$. If D is a tiny *member of $\mathscr{F}$, then $\hat{D} \subseteq \tau$, and since τ is a group,

$$(D \cdot D)\hat{} = \hat{D} \cdot \hat{D} \subseteq \tau \subseteq \hat{A}.$$

Thus

$$\dagger^* \models \exists x (x \in \mathcal{F} \wedge x \cdot x \subseteq A).$$

It follows that $B \cdot B \subseteq A$ for some $B \in \mathscr{F}$. Therefore $\hat{B} \cdot \hat{B} \subseteq \hat{A}$ and $\tau \subseteq \hat{B}$. Hence, if $h \in B$ we have

$$\lambda(h) = \tau \cdot h \subseteq \hat{B} \cdot \hat{B} \subseteq \hat{A};$$

so that $A \in \mathrm{F}\lambda(h)$, as required.

It now follows from Thm. 4.3 that there is a topology $\mathscr{T}$ on G such that $\mu_{\mathscr{T}}(g) = \tau \cdot g$ for every $g \in G$.

Using (b) and the fact that τ is a group, it is easy to verify that

$$\mu_{\mathscr{T}}(g) \cdot \mu_{\mathscr{T}}(h) = \mu_{\mathscr{T}}(g \cdot h), \qquad \mu_{\mathscr{T}}(g)^{-1} = \mu_{\mathscr{T}}(g^{-1}).$$

Hence G is a topological group under the topology $\mathscr{T}$.

Finally,

$$\mu_{\mathscr{T}}(e) = \tau \cdot e = \tau,$$

so that τ is the infinitesimal group of G. ∎

For arbitrary near-standard h and g, we now re-define "$h \simeq g$" to mean that h and g are congruent modulo $\mu(e)$. This is consistent with the old definition whenever the latter applies — i.e., when g is standard. According to the present definition $\simeq$ is clearly an equivalence relation on the collection of near-standard points.

Conventionally, the structure of the topological group G is determined by specifying:

(i) the purely algebraic group structure (the "multiplication table"); and

(ii) the topology $\mathscr{T}$.

In the nonstandard approach, the topology is completely determined by the monadology; and the latter is completely determined by the infinitesimal group $\mu(e)$, because $\mu(g) = g \cdot \mu(e)$ for every $g \in G$. Thus the structure of the topological group G is now determined by specifying the purely algebraic structure together with the infinitesimal group, which is a suitable subgroup of $\hat{G}$.

Consequently, the nonstandard treatment of topological groups is rather more algebraic in flavour that the conventional treatment. Topological considerations are often replaced by considerations involving the infinitesimal group $\mu(e)$: e.g., calculations with congruences modulo $\mu(e)$. This seems to be a natural explication of the intuition underlying the notion of topological group.

5.5. EXAMPLE. Let $A, C \subseteq G$, where A is closed and C is compact. We show that $A \cdot C$ is closed. By Thm. 4.1 we have to show that if $g \in G$ and $g \simeq p \cdot q$ where $p \in \hat{A}$ and $q \in \hat{C}$, then $g \in A \cdot C$. But, since $q \in \hat{C}$ and C is compact, we have $q \simeq c$ for some $c \in C$. Hence $g \simeq p \cdot c$, so that $g \cdot c^{-1} \simeq p$. Since $g \cdot c^{-1} \in G$, A is closed and $p \in \hat{A}$, it follows that $g \cdot c^{-1} \in A$. Hence $g \in A \cdot C$ as required.

5.6. EXAMPLE. We show that every topological group is a regular space. Given any closed $A \subseteq G$ and $g \in G - A$, we have to show that there are disjoint open sets B and C such that $g \in B$ and $A \subseteq C$. We use the easily established fact that if $D \in \mathscr{F}_e \cap \mathscr{T}$ (i.e., D is an open neighbourhood of e) and E is *any* subset of G, then $E \cdot D$ is open and $E \subseteq E \cdot D$. Now, $\mathscr{F}_e \cap \mathscr{T}$ is a base for the filter $\mathscr{F}_e$; so we can take a *tiny* *member D of $\mathscr{F}_e \cap \mathscr{T}$. Then $g \cdot D$ and $A \cdot D$ are *open, $g \,{}^*\!\!\in D$ and $A \,{}^*\!\!\subseteq A \cdot D$. Also, $g \cdot D$ and $A \cdot D$ are *disjoint. (Otherwise, using the fact that $\hat{D} \subseteq \mu(e)$, we would conclude that $g \simeq q$ for some $q \in \hat{A}$; and since A is closed it would follow that $g \in A$.) Hence

$$\dagger^* \models \exists \mathbf{x}\, \exists \mathbf{y} (\mathbf{x} \in \mathcal{T} \wedge \mathbf{y} \in \mathcal{T} \wedge \mathbf{g} \in \mathbf{x} \wedge \mathbf{A} \subseteq \mathbf{y} \wedge \mathbf{x} \cap \mathbf{y} = \mathbf{0}).$$

Thus there are open disjoint sets B and C such that $g \in B$ and $A \subseteq C$, as required.

5.7. PROBLEM. For each $A \subseteq G$, define subsets A_{d} and A_{s} of $G \times G$ by

$$A_{\mathrm{d}} = \{\langle g,h \rangle : \ h \cdot g^{-1} \in A\},$$

$$A_{\mathrm{s}} = \{\langle g,h \rangle : \ g^{-1} \cdot h \in A\}.$$

The *right uniformity* and *left uniformity* of G are the filters over $G \times G$ which are generated by $\{A_{\mathrm{d}} : A \in \mathscr{F}_e\}$ and $\{A_{\mathrm{s}} : A \in \mathscr{F}_e\}$ respectively. Prove that the two uniformities coincide iff $\mu(e)$ is a normal subgroup of $\hat{G}$.

5.8. PROBLEM. Let H be a subgroup of G.

(i) Show that $\bar{H}$ is a subgroup of G. (Use Prob. 4.2.)

(ii) Show that H is closed iff $\mu(e) \cap \bar{H}\hat{\ } = \mu(e) \cap \hat{H}$.

(iii) Show that H is open iff $\mu(e) \subseteq \hat{H}$, hence show that every open subgroup of G is closed.

5.9. PROBLEM. Assume that $^*\mathfrak{U}$ is a limit enlargement. Let H be a normal subgroup of G, and let f be the quotient map of G onto G/H (so that $f(g)=g\cdot H$ for each $g\in G$).

(i) Show that, if G/H is given the quotient topology, it becomes a topological group, with $f[\mu(e)]$ as infinitesimal group.

(ii) Show that (under the quotient topology) G/H is a Hausdorff space iff H is closed in G.

5.10. PROBLEM. In addition to G, consider another topological group $G'\in U$ with e' as identity element. G is said to be *locally isomorphic* to G' if there is some $V\in\mathscr{F}_e$ and some homeomorphism f of V onto a neighbourhood of e' such that $f(g\cdot h)=f(g)\cdot f(h)$ whenever $g,h,g\cdot h\in V$.

Prove that G is locally isomorphic to G' iff there is some function $f\in U$ such that the natural extension of f maps $\mu(e)$ isomorphically onto the infinitesimal group of G'.

5.11. PROBLEM. Assume that $N\in U$. Let S be the set of all $f\in\text{Fun}(N,G)$ such that $f(n)=e$ for almost all (i.e., all but finitely many) n. Let $\Pi\in\text{Fun}(S,G)$ be defined by $\Pi f=f(0)\cdot f(1)\cdot\ldots\cdot f(n)$, where $f(m)=e$ for all $m>n$.

(i) Let $V\in\mathscr{F}_e$ and let

$$H=\{\Pi f\colon f\in S, f(n)\in V\cup V^{-1} \text{ for all } n\}.$$

Prove that H is an open — and hence closed — subgroup of G. (Use 5.8.)

(ii) Prove that if G is connected then each $g\in\hat{G}$ is a *finite product of infinitesimals: $g=\Pi f$, for some $f\in\hat{S}$ such that $f(n)\in\mu(e)$ for all $n\in\hat{N}$.

§ 6. The real numbers

From now on we assume that $N\in U$. Hence also $R\in U$, where R is the set of real numbers, since each real number is (identified with) a subset of N.

In the usual topology, R is a commutative topological group with respect to addition. The operation of addition extends in a natural way to $\hat{R}$, as explained in §5. Since $\hat{R}$ is clearly a commutative group, the infinitesimal group $\mu(0)$ is a normal subgroup of $\hat{R}$. The members of $\mu(0)$ are the *infinitesimal *reals*. Clearly, 0 is infinitesimal; but by 4.5(i) there are infinitesimals $\neq 0$, and by 4.5(iii) these are all non-standard.

For arbitrary $a,b\in\hat{R}$ we now re-define "$a\simeq b$" to mean that a and b are congruent modulo $\mu(0)$. This is consistent with the definition made in §5 whenever the latter applies, i.e., when a and b are near-standard.

The operation of multiplication, the absolute-value function $r\mapsto|r|$, and the ordering relation $<$ extend in a natural way from R to $\hat{R}$. We shall denote these extensions by the same symbols that are used for R. Since R is an ordered field, it follows easily by transference that $\hat{R}$ is an ordered field. ($\hat{R}$ is clearly non-archimedean, but every $a \in \hat{R}$ has the *archimedean property: $a<n$ for some $n \in \hat{N}$.)

We employ the usual notation for intervals. Thus for $r, s \in R$ we put

$$(r,s)=\{x \in R:\ r<x<s\}, \qquad [r,s]=\{x \in R:\ r \leqslant x \leqslant s\}.$$

Also, we let P be the set of all positive reals.

The set of symmetric intervals $\{(-r,r):\ r \in P\}$ is a base for the filter of neighbourhoods of 0 in R. Thus $\delta \simeq 0$ iff $\delta \in (-r,r)\hat{}$ for all $r \in P$. This means that $|\delta|<r$ for all $r \in P$.

If $a \in \hat{R}$ and $a^{-1} \simeq 0$, then a is called an *infinite *real*. It is easy to see that a is infinite iff $|a|>r$ for all $r \in P$.

A *real a is called *finite* if it is not infinite. This means that $|a| \leqslant r$ for some $r \in P$. We let Φ be the collection of all finite *reals.

6.1. THEOREM. *A *real is finite iff it is near-standard.*

PROOF. Let a be near-standard: $a \simeq r$ for some $r \in R$. Then $a=r+\delta$, where $\delta \simeq 0$, so

$$|a|=|r+\delta| \leqslant |r|+|\delta|<|r|+1;$$

hence a is finite. (The triangle inequality holds in $\hat{R}$ by transference.)

Conversely, if a is finite then $|a| \leqslant r$ for some $r \in P$. Hence $a \in [-r,r]\hat{}$; but $[-r,r]$ is a compact set, hence, by Thm. 4.10, a is near-standard. ∎

6.2. PROBLEM. Let $X \in U$ be a metric space, with distance function (metric) ϱ. We extend ϱ to a mapping of $\hat{X} \times \hat{X}$ into $\hat{R}$ in the usual way. A *point $q \in \hat{X}$ is called *finite* if $\varrho(p,q)$ is finite for some (hence for *every*) point $p \in X$.

(i) Show that, for each $p \in X$, $\mu(p)=\{q \in \hat{X}:\ \varrho(p,q) \simeq 0\}$.

(ii) Show that every near-standard $q \in \hat{X}$ is finite.

(iii) Show that if the converse of (ii) holds then X is locally compact.

6.3. PROBLEM. Let X be as in Prob. 6.2 and let $f \ {}^*\!\in \mathrm{Fun}(N,X)$.

(i) Show that if fn is finite for every infinite $n \in \hat{N}$ then there is some $m \in N$ such that fn is finite for all $n>m$. (Take any $p \in X$; consider the collection

$$\{m:\ {}^*\!\models \mathbf{m \in N \wedge \forall x[x \in N \wedge x > m \rightarrow \varrho(fx,p) < m]}\}$$

and use Robinson's Overspill Lemma 2.11.)

(ii) Let $fn \simeq p$ for some $p \in \hat{X}$ and all $n \in N$. Show that there is some *infinite* $m \in \hat{N}$ such that $fn \simeq p$ for all $n<m$.

Since R is a Hausdorff space, it follows from Thms. 4.6 and 6.1 that ${}^\circ a$ is defined for each $a \in \Phi$. For reasons that will soon be apparent, we call ${}^\circ a$ the *(standard) place* of a. The mapping $a \mapsto {}^\circ a$ is called the *(standard) place mapping*.

6.4. THEOREM. *Φ is a subring (with identity) of $\hat{R}$. The place mapping is a homomorphism of Φ onto the field R of reals. The kernel of this homomorphism is $\mu(0)$, which is therefore a maximal ideal in Φ.*

PROOF. Let $a,b \in \Phi$. Then $|a| \leqslant r$ and $|b| \leqslant s$ for some $r,s \in P$. It follows that $|a-b| \leqslant r+s$ and $|ab| \leqslant rs$, hence $a-b$ and ab are in Φ. Clearly, $1 \in \Phi$; so Φ is a subring of R, with identity.

For each $a,b \in \Phi$ we have $a \simeq {}^\circ a$ and $b \simeq {}^\circ b$. Since addition is a continuous function from $R \times R$ into R, it follows that $a+b \simeq {}^\circ a + {}^\circ b$, hence ${}^\circ(a+b) = {}^\circ a + {}^\circ b$. Similarly ${}^\circ(ab) = {}^\circ a \cdot {}^\circ b$. Also, ${}^\circ r = r$ for every $r \in R$. Hence the place mapping is a homomorphism of Φ onto R.

Finally, it is clear that ${}^\circ a = 0$ iff $a \in \mu(0)$. ∎

By definition, the inverse of every infinite *real is infinitesimal, and hence finite. Also, the infinitesimals are precisely the non-units of Φ (i.e., the members of Φ which have no inverse in Φ). Thus in the terminology of valuation theory (see e.g. JACOBSON [1964]) Thm. 6.4 means that Φ is a *valuation ring* in $\hat{R}$, the place mapping is the *canonical place* of Φ, and R is the *residue field* of Φ.

The *canonical valuation* of the valuation ring Φ provides an interesting explication of the vague classical notion of *order of magnitude* of a given (infinitesimal, finite or infinite) quantity. For any $a,b \in \hat{R}$, let us say that $a \approx b$ if $a = bc$ for some $c \in \Phi - \mu(0)$. It is easy to see that $\approx$ is an equivalence relation. We define $O(a)$ (read: the *order of magnitude* of a) to be the $\approx$-class of a. Clearly, $O(0) = \{0\}$.

6.5. PROBLEM. For any $a,b \in \hat{R}$ define $O(a) \cdot O(b) = O(ab)$; and $O(a) \leqslant O(b)$ if $a = bc$ for some $c \in \Phi$.

(i) Show that these definitions are legitimate (i.e., independent of the choice of "representatives" a and b).

(ii) Show that (under the multiplication defined above) the collection $G = \{O(a)\colon a \in \hat{R} - \{0\}\}$ is a commutative group, with $O(1)$ as identity element.

(iii) Show that the relation $\leqslant$ is a total ordering of the collection of *all* orders of magnitude $G \cup \{O(0)\}$.

(iv) Let $H = \{O(a)\colon a \in \mu(0) - \{0\}\}$. Show that H is closed under multiplication and $G = H \cup \{O(1)\} \cup H^{-1}$.

(v) Show that for any $a \in \hat{R}$, $O(a) \in H$ iff $O(0) < O(a) < O(1)$.
(vi) Show that $O(a+b) \leqslant \max(O(a), O(b))$ for all $a, b \in \hat{R}$.

In the terminology of valuation theory, Prob. 6.5 means that O is a *valuation* on $\hat{R}$. In fact, it is the *canonical* valuation of the valuation ring Φ.

6.6. PROBLEM. Show that the ordering of $\{O(a)\colon a \in \hat{R}\}$ is dense, with first but without last member. (Use Prob. 7.9.14 (v).)

Starting from ideas explained so far in this chapter, various parts of mathematics can be developed in an elegant and intuitively appealing way. For such developments the reader is referred to the literature quoted in §8 below.

We shall conclude our introductory treatment with a few examples.

6.7. EXAMPLE. We prove the intermediate-value theorem for continuous functions. Let f be a real function, defined and continuous in $[0,1]$. Let $f(0) < 0 < f(1)$. Then it is easy to see that if n is any positive natural number, there is a natural number $m < n$ such that

$$(1) \qquad f\left(\frac{m}{n}\right) \leqslant 0 \leqslant f\left(\frac{m+1}{n}\right).$$

Now take an *infinite* $n \in \hat{N}$. By transference, there is $m < n$ such that (1) holds. Since $[0,1]$ is compact, there exists $r \in [0,1]$ such that

$$\frac{m}{n} \simeq r \simeq \frac{m+1}{n}.$$

Because f is continuous at r, we clearly must have $f(r) = 0$.

6.8. PROBLEM. Let f be a real function defined in an interval (r,s) and let $a \in (r,s)$. Show that $f'(a)$ is defined and equal to b iff

$$\frac{f(a+\delta) - f(a)}{\delta} \simeq b$$

for every non-zero infinitesimal δ.

6.9. PROBLEM. Let $G \in U$ be a Hausdorff, locally compact topological group. Let $\mathscr{C}$ be the set of all compact subsets of G. Define a function $\lambda\colon \mathscr{F}_e \times \mathscr{C} \to N$ as follows. Let $B \in \mathscr{F}_e$ and $C \in \mathscr{C}$. Since C is compact and B includes an open set, there exists an $n \in N$ such that for some $g_1, \ldots, g_n \in G$ we have $C \subseteq \bigcup\{B \cdot g_i\colon i = 1, \ldots, n\}$. Let $\lambda(B,C)$ be the *least* such n. Now fix some $K \in \mathscr{F}_e \cap \mathscr{C}$ (such K exists because G is locally compact) and fix also a tiny *member D of $\mathscr{F}_e$.

(i) Show that for each $C \in \mathscr{C}$ we have

$$\lambda(D,C) \leqslant \lambda(D,K) \cdot \lambda(K,C),$$

hence we can define a mapping η: $\mathscr{C} \to R$ by putting, for each $C \in \mathscr{C}$,

$$\eta(C) = {}^{\circ}\left(\frac{\lambda(D,C)}{\lambda(D,K)}\right).$$

(ii) Show that η is a (right) *Haar measure* on G; i.e., $\eta(C \cdot g) = \eta(C)$ for all $C \in \mathscr{C}$ and $g \in G$, and $\eta(C_1 \cup C_2) = \eta(C_1) + \eta(C_2)$ for all disjoint $C_1, C_2 \in \mathscr{C}$ (For the second assertion, show first that if $g \in \hat{G}$ then $\hat{C}_i \cap D \cdot g$ cannot be non-empty for both $i=1$ and $i=2$.)

(iii) Assuming that $g \cdot D \cdot g^{-1} = D$ for all $g \in G$, show that η is also *left* invariant; i.e., $\eta(g \cdot C) = \eta(C)$ for all $C \in \mathscr{C}$ and $g \in G$. (Show first that $\lambda(D,C) = \lambda(D, g \cdot C \cdot g^{-1})$.)

(iv) Assuming that $\mu(e)$ is a normal subgroup of $\hat{G}$, prove that there exists some tiny *member D of $\mathscr{F}_e$ such that $g \cdot D \cdot g^{-1} = D$ for all $g \in G$. (Take *any* tiny *member B of $\mathscr{F}_e$ and show that the collection $\{g \cdot b \cdot g^{-1}$: $b \in \hat{B}, g \in \hat{G}\}$ is internal and equal to $\hat{D}$ where D is as required.)

6.10. PROBLEM. Let G be a commutative, Hausdorff, locally compact topological group "without small subgroups" (i.e., there is a neighbourhood of e which does not include any non-trivial subgroup of G).

A *one-parameter subgroup* of G is a continuous homomorphism of the additive group of reals R into G. Let M be the set of all one-parameter subgroups of G. For all $f,g \in M$ and all $r,s \in R$ we put $(f+g)(r) = f(r) \cdot g(r)$ and $(sf)(r) = f(sr)$. We endow M with the "compact-open" topology, by taking as sub-base all sets of the form $\{f \in M$: $f[C] \subseteq V\}$, where C is compact $\subseteq R$ and V is open $\subseteq G$.

(i) Verify that under the above definitions M is a topological vector space over the reals. Prove also that

$$\mu_M(0) = \{f \in \hat{M}\colon\ f(r) \simeq e \text{ for all } r \in [-1,1]^{\wedge}\},$$

where 0 is the trivial one-parameter subgroup and $\mu_M(0)$ is its monad in the monadology of M.

(ii) Since G is locally compact and has no "small subgroups", we can fix a compact $V \in \mathscr{F}_e$ such that V does not include any non-trivial subgroup of G. We may also assume $V^{-1} = V$ (otherwise, we replace V by $V \cap V^{-1}$). Having fixed such a V, we let X be the set of all $f \in M$ that map $[-1,1]$ into V. Prove that X is a compact neighbourhood of 0 in M, hence M is finite dimensional. (Use (i) to show that $\mu_M(0) \subseteq \hat{X}$ and that every member of $\hat{X}$ is near-standard. Then observe that a Hausdorff topological vector space over the reals is finite dimensional iff it is locally compact.)

(iii) Let ψ be the map of M into G defined by $\psi(f)=f(1)$. Prove that ψ is a continuous homomorphism and show that the natural extension of ψ maps $\mu_M(0)$ one-one into $\mu(e)$. (Note that $\mu_M(0)$ is the infinitesimal group of M, regarded as an additive topological group. Using the assumption that G has no "small subgroups", show that if γ and δ are infinitesimals of G and $\gamma^2=\delta^2$ then $\gamma=\delta$. Hence show that if $f\in\mu_M(0)$ then $f(1)$ uniquely determines the values $f(\pm k/2^n)$ for all $k,n\in\hat{N}$; so f is uniquely determined by $f(1)$.)

(iv) Let V be as in (ii), and let $\gamma\neq e$ be an infinitesimal of G. Prove that there exists a smallest $n\in\hat{N}$ such that $\gamma^{n+1}\notin\hat{V}$.

(v) For $\gamma\simeq e$, $\gamma\neq e$, let $\mathrm{n}(\gamma)$ be the least n of (iv). Prove that if k and m are *integers such that $|k|, |m|\leq\mathrm{n}(\gamma)$ and $k/\mathrm{n}(\gamma)\simeq m/\mathrm{n}(\gamma)$ then γ^k and γ^m are near the same point of V. (Use the fact that V is compact and does not include a non-trivial subgroup of G.)

(vi) Let γ and $\mathrm{n}(\gamma)$ be as above. For each $r\in R$ put

$$f_\gamma(r)={}^\circ(\gamma^k),$$

where k is any *integer such that $k/\mathrm{n}(\gamma)\simeq r$. Prove that $f_\gamma(r)$ is uniquely defined for each $r\in R$ and that f_γ is a one-parameter subgroup of G. (To show that $f_\gamma(r)$ is uniquely defined, use (v). To show that f_γ is continuous at 0 and hence everywhere, let $W\in\mathscr{F}_e$ be closed and symmetric; consider the collection

$$\{n\colon n\in\hat{N} \text{ and } f_\gamma(x)\in\hat{W} \text{ whenever } |x|<2^{-n}\}$$

and use the Overspill Lemma 2.11.)

(vii) Prove that if $f_\gamma(1/\mathrm{n}(\gamma))=\delta$ then $\mathrm{n}(\gamma\cdot\delta^{-1})>\mathrm{n}(\gamma)$.

(viii) Prove that if K is compact $\subseteq G$ and $e\notin K$ then there exists some $n_0\in N$ such that for every $a\in K$ there is a natural $n<n_0$ such that $a^n\notin V$, where V is as above.

(ix) Let X be as in (ii), and let ψ be as in (iii). Put $S=\psi[X]$. Show that S is compact and $\hat{S}\cap\mu(e)=\psi[\mu_M(0)]$, hence in particular $\hat{S}\cap\mu(e)$ is a group.

(x) Let S be as above. Show that $\mu(e)\subseteq\hat{S}$, hence $\mu(e)=\psi[\mu_M(0)]$. (Suppose $\alpha\simeq e$ but $\alpha\notin\hat{S}$. Show that $\alpha\cdot S$ *satisfies the conditions of (viii); hence choose $\gamma\in(\alpha\cdot S)\hat{}$ such that $\mathrm{n}(\gamma)$ is maximal. Using (vii) and (ix) determine δ and show that $\gamma\cdot\delta^{-1}\in(\alpha\cdot S)\hat{}$ contrary to the choice of γ.)

(xi) Show that G is locally isomorphic to a finite-dimensional real vector space. (Use (ii), (iii), (x) and apply Prob. 5.10.)

§ 7. A methodological discussion

Because of limitations of space, we have confined ourselves to an exposition of the rudiments of nonstandard analysis, without exhibiting many deep applications to various branches of mathematics. (Examples of such applications may be found in the literature quoted in the next section.) However, we hope that even the material presented here has convinced the reader that nonstandard analysis provides elegant, intuitively appealing and formally simple characterizations of many (standard) mathematical concepts, and can thereby facilitate the discovery and proof of (standard) mathematical results. It is true that any standard result (i.e., a result which refers to standard concepts only) provable by means of nonstandard analysis can, in principle, also be proved without it[1]. However, the nonstandard proof is often considerably simpler and requires fewer *ad hoc* tricks.

In the preface to the second (1974) edition of A. Robinson [1966], K. Gödel is quoted as making the following enthusiastic appraisal:

"I would like to point out...that nonstandard analysis frequently simplifies substantially the proofs, not only of elementary theorems, but also of deep results. This is true, e.g., also for the proof of the existence of invariant subspaces for compact operators[2], disregarding the improvement of the result; and it is true in an even higher degree in other cases. This state of affairs should prevent a rather common misinterpretation of nonstandard analysis, namely the idea that it is some kind of extravagance or fad of mathematical logicians. Nothing could be farther from the truth."

With this we are in complete agreement. However, he goes on to say:

"Rather there are good reasons to believe that nonstandard analysis, in some version or other, will be the analysis of the future.

"One reason is the just mentioned simplification of proofs, since simplification facilitates discovery. Another, even more convincing reason, is the following: Arithmetic starts with the integers and proceeds by successively enlarging the number system by rational and negative numbers, irrational numbers etc. But the next quite natural step after the reals, namely the introduction of infinitesimals, has simply been omitted. I think, in coming centuries it will be considered a great oddity in the history of mathematics that the first exact theory of infinitesimals was developed 300 years after the invention of the differential calculus."

With this we beg to differ. In our view there is a fundamental difference

[1] This is so because nonstandard analysis does not use any new principle which is irreducible to principles accepted in standard mathematical practice.

[2] Cf. references in the next section.

between the enlarged system of *reals and the classical number systems (the integers, the rationals, the reals, etc.) namely the fact that the latter are *canonical* but the former is *not*. The classical number systems can be characterized (informally or within set theory) uniquely up to isomorphism by virtue of their *mathematical* properties (e.g., the field of rationals is the smallest field containing the integers, and the field of reals is the completion of the field of rationals). On the other hand, there is no known way of singling out a particular enlargement which can plausibly be regarded as canonical; and there is no good reason to believe that a method for obtaining a canonical enlargement will necessarily be invented.

It is therefore not so surprising that ordinary (informal) mathematical practice has discovered the classical number systems, which actually almost force themselves on it, but searched in vain for *the* infinitesimals. There is no such thing as *the* enlarged system of *reals; it depends on the choice of enlargement. Where there is no canonical structure of a given kind, the way mathematicians proceed is by specifying *all* structures of this kind. And the only way to specify all enlargements is by their *formal logical* properties.

For this reason also it seems to us wrong to try totally to replace standard by nonstandard analysis. For example suppose one tries to *define* continuity of a real function f at $r \in R$ by the condition: $f(x) \simeq f(r)$ *for all* $x \simeq r$. This would be a *bad* definition so long as we are not sure that the condition is *invariant* (i.e., independent of the choice of enlargement). But in general the only simple way of proving the invariance of a nonstandard condition is to show that it is equivalent to a standard one. Thus we still need the standard definition of continuity.

We therefore regard nonstandard analysis as an important tool of clarification, exposition and research — often beautiful, sometimes very powerful, but never exclusive.

§ 8. Historical and bibliographical remarks

Nonstandard models of arithmetic were discovered by SKOLEM [1934], but for a long time thereafter they seem to have been locked up as skeletons in the logical cupboard. No serious attempt to study such models — let alone use them as more than pathological counter-examples — was made before HENKIN [1949]. During the 1950s, the study of nonstandard models of arithmetic gradually gathered momentum, and by the end of the decade became quite fashionable. Thus, in the symposium on the foundations of mathematics held at Warsaw in September 1959 (INFINITISTIC METHODS [1961]) at least three papers on the subject were read (one of them by A. Robinson) and the collection BAR-HILLEL et al. [1961] contains two essays on the same subject.

However, these investigations were confined to nonstandard models of (natural number) arithmetic. Also, they were concerned with studying these models as such, or with using them to prove metamathematical results about *formal systems*, rather than mathematical results about *numbers*. Thus they cannot be regarded as belonging to nonstandard analysis, although they undoubtedly prepared the ground for it.

Nonstandard analysis was invented by A. Robinson in the autumn of 1960, when it occurred to him that "the concepts and methods of contemporary Mathematical Logic are capable of providing a suitable framework for the development of the Differential and Integral Calculus by means of infinitely small and infinitely large numbers".[1]

The first published account of the subject is in A. ROBINSON [1961]. Here he still works with the first-order structure of real numbers (whose individuals are the real numbers only) and its proper elementary extensions.

In A. ROBINSON [1962] the treatment is extended to structures having individuals of arbitrary finite types (e.g., sets of reals, sets of sets of reals, etc.). This paper also contains nonstandard proofs of new standard results in complex analysis (strengthened versions of classical results on the distribution of zeros of polynomials and on Julia directions).

A major break-through came in 1964, when Bernstein and Robinson used nonstandard analysis to solve an important open problem in the theory of linear spaces. (See BERNSTEIN and ROBINSON [1966].) A well-known theorem, due to von Neumann and Aronszajn, states that if T is a compact operator in a Hilbert space then T has a (non-trivial) invariant subspace. It was conjectured that the result holds also in the case where T is continuous and non-compact, provided T^2 is compact. Bernstein and Robinson proved that this is in fact the case even if instead of the compactness of T^2 one assumes that of $p(T)$ for some polynomial p.

The characterization of compactness (Thm. 4.10) was discovered by A. Robinson in 1963 (see A. ROBINSON [1965]).

A detailed systematic account of nonstandard analysis, including the above mentioned applications and many others in a wide range of branches of mathematics, is in A. ROBINSON [1966].

[1] See A. ROBINSON [1966].

The spread of Robinson's ideas among analysts was facilitated by LUXEMBURG [1962], which presents an elementary treatment (using an ultrapower of the first-order structure of real or complex numbers) without explicit heavy use of logic. (Instead, each particular instance of transference is carried out separately, as and when required.)

In dealing with entities of higher types, most writers on nonstandard analysis have followed A. ROBINSON [1962, 1966] in using type-theoretic structures, which are technically rather cumbersome. The first treatment using a considerably simpler set-theoretical framework is in MACHOVER [1967]. This is followed also in MACHOVER and HIRSCHFELD [1969] and in the present book. (See also the paper by Robinson and Zakon in LUXEMBURG [1969].)

Important collections of papers on nonstandard analysis are LUXEMBURG [1969], LUXEMBURG and ROBINSON [1972], and HURD and LOEB [1974].

The bulk of the results presented in this chapter are due to A. ROBINSON [1966]. The results of §3 are due to Luxemburg (see his comprehensive paper *A general theory of monads* in LUXEMBURG [1969]; this paper contains also a wealth of results not presented here). Most of the results of §3 were also discovered independently by Hirschfeld and presented in MACHOVER and HIRSCHFELD [1969]. The nonstandard construction of the Haar measure (Prob. 6.9) is due to R. Parikh (in a paper included in LUXEMBURG [1969]). The nonstandard version of the solution of Hilbert's Fifth Problem in the commutative case (Prob. 6.10) is due to Hirschfeld.

REMARK. As noted above, most writers on nonstandard analysis use a type-theoretic framework rather than the set-theoretic framework employed by us. This enables them to assume that the basic relation $^*\in$ of the enlargement is the true membership relation. This is not a very big gain, because a similar assumption *cannot* be made in general for *defined* relations. Moreover, if one insists that $^*\in$ is true membership, then the structure $\mathfrak{U}$ cannot be a substructure of $^*\mathfrak{U}$, although there is a canonical elementary embedding * of $\mathfrak{U}$ into $^*\mathfrak{U}$. Therefore those authors must always distinguish an object $u \in U$ from the corresponding object $^*u \in {}^*U$ (except when u is of lowest type in U). This distinction, although necessary in order to avoid contradictions, is rather cumbersome and is sometimes overlooked.

These points must be borne in mind by the reader who wishes to consult the literature quoted above.

BIBLIOGRAPHY

ACZEL, P.H.G.

[1967] Some results on intuitionistic predicate logic (abstract). *J. Symb. Logic* **32**, 556.

BAR-HILLEL, Y., et al. (eds.)

[1961] *Essays on the Foundations of Mathematics, dedicated to A. A. Fraenkel on his seventieth anniversary.* Magnes Press, Hebrew Univ., Jerusalem.

BARWISE, J. (ed.)

[1977] *Handbook of Mathematical Logic.* North-Holland, Amsterdam.

BELL, J. L.

[1977] *Boolean-Valued Models and Independence Proofs in Set Theory.* Clarendon Press, Oxford, (2nd ed., 1985.)

BELL, J. L., and A. B. SLOMSON

[1969] *Models and Ultraproducts: an Introduction.* North-Holland, Amsterdam.

BENACERRAF, P., and H. PUTNAM (eds.)

[1964] *Philosophy of Mathematics: Selected Readings.* Prentice-Hall, Englewood Cliffs, N. J.

BERNSTEIN, A. R., and A. ROBINSON

[1966] Solution of an invariant subspace problem of K. T. Smith and P. R. Halmos. *Pacific J. Math.* **16**, 421–431. (Abstract in *Notices Am. Math. Soc.* **11**, (1964) 586.)

BETH, E. W.

[1953] On Padoa's method in the theory of definition. *Indag. Math.* **15**, 330–339.

[1955] Semantic entailment and formal derivability. *Mededel. Kon. Ned. Akad. Wetensch. Afd. Letterkunde* N. S. **19**, 309–342.

[1956] Semantic construction of intuitionistic logic. *Mededel. Kon. Ned. Akad. Wetensch. Afd. Letterkunde* N. S. **19**, (13). (Originally in French in: *Colloq. Logique Math. CNRS* (1955).)

BIRKHOFF, G.

[1967] *Lattice Theory* (3rd ed.), Am. Math. Soc. Coll. Pub., Vol. **25**. Am. Math. Soc., Providence, R. I.

BISHOP, E.

[1967] *Foundations of Constructive Analysis.* McGraw-Hill, New York.

BOURBAKI, N.

[1961] *Topologie Générale* (3rd ed.). Hermann, Paris.

CHANG, C. C., and H. J. KEISLER

[1973] *Model Theory.* North-Holland, Amsterdam.

CHURCH, A.

[1936] An unsolvable problem of elementary number theory. *Am. J. Math.*, **58**, 345–363.

[1936a] A note on the Entscheidungsproblem, *J. Symb. Logic* **1**, 40–41. (Reprinted with corrections in DAVIS [1965], pp. 110–115.)

[1956] *Introduction to Mathematical Logic*. Princeton Univ. Press, Princeton, N. J.

COHEN, P. J.

[1966] *Set Theory and the Continuum Hypothesis*. Benjamin, New York.

CRAIG, W.

[1957] Linear reasoning. A new form of the Herbrand–Gentzen theorem. *J. Symb. Logic* **22**, 250–268. See also: *ibid.*, 269–285.

DAVIS, M.

[1958] *Computability and Unsolvability*. McGraw-Hill, New York.

DAVIS, M. (ed.)

[1965] *The Undecidable. Basic papers on undecidable propositions, unsolvable problems, and computable functions*. Raven Press, New York.

DAVIS, M., H. PUTNAM ard J. ROBINSON

[1961] The decision problem for exponential diophantine equations. *Ann. Math.* **74**, 425–436.

DEDEKIND, R.

[1888] *Was sind und was sollen die Zahlen?* Brunswick. (English transl. by W. W. Beman in *Essays on the Theory of Numbers*. Open Court, La Salle, Ill. (1901).)

DEKKER, J. C. E.

[1955] Productive sets. *Trans. Am. Math. Soc.* **78**, 129–149.

DRAKE, F. R.

[1974] *Set Theory: an Introduction to Large Cardinals*. North-Holland, Amsterdam.

DWINGER, P.

[1961] *Introduction to Boolean Algebras*. Physica-Verlag, Würzburg.

EHRENFEUCHT, A., and A. MOSTOWSKI

[1956] Models of axiomatic theories admitting automorphisms. *Fund. Math.* **43**, 50–68.

FEFERMAN, S.

[1952] Review of RASIOWA and SIKORSKI [1951]. *J. Symb. Logic* **17**, 72.

[1960] Arithmetization of metamathematics in a general setting. *Fund. Math.* **49**, 35–92.

FITTING, M. C.

[1969] *Intuitionistic Logic, Model Theory and Forcing*. North-Holland, Amsterdam.

FRAENKEL, A. A.

[1961] *Abstract Set Theory*. North-Holland, Amsterdam.

FRAENKEL, A. A., Y. BAR-HILLEL and A. LEVY

[1973] *Foundations of set theory* (2nd ed.). North-Holland, Amsterdam.

FRAYNE, T., A. MOREL and D. S. SCOTT

[1962] Reduced direct products. *Fund. Math.* **51**, 195–228.

FREGE, G.

[1879] *Begriffsschrift, eine der arithmetischen nachgebildete Formelsprache des reinen Denkens*. (Complete English transl. in VAN HEIJENOORT [1967], pp. 1–82.)

FRIEDBERG, R. M.

[1957] Two recursively enumerable sets of incomparable degrees of unsolvability. *Proc. Nat. Acad. Sci.* **43**, 236–238.

GENTZEN, G.

[1934] Untersuchungen über das Logische Schliessen. *Math. Z.* **39**, 176–210, 405–431. (English transl. in M. E. Szabó (ed.), *The collected Papers of Gerhard Gentzen*, North-Holland, Amsterdam (1969).)

GILLMAN, L., and M. JERISON

[1960] *Rings of Continuous Functions.* Van Nostrand, New York.

GÖDEL, K.

[1930] Die Vollständigkeit der Axiome des logischen Funktionenkalküls. *Monatsh. Math. Phys.* **37**, 349–360. (English transl. in VAN HEIJENOORT [1967], pp. 582–591.)

[1931] Über formal unentscheidbare Sätze der Principia Mathematica und verwandter Systeme I. *Monatsh. Math. Phys.* **38**, 173–198. (English transl. in VAN HEIJENOORT [1967], pp. 596–616.)

[1934] On undecidable propositions of formal mathematical systems. Lecture notes by S. C. Kleene and J. B. Rosser, Inst. for Advanced Study, Princeton, N. J. (Reprinted with corrections in DAVIS [1965].)

[1938] The consistency of the axiom of choice and of the generalized continuum hypothesis. *Proc. Nat. Acad. Sci. U.S.A.* **24**, 556–557.

[1939] Consistency-proof for the generalized continuum hypothesis. *Proc. Nat. Acad. Sci. U.S.A.* **25**, 220–224.

[1940] *The Consistency of the Continuum Hypothesis,* Ann. Math. Studies 3. Princeton Univ. Press, Princeton, N. J.

GOODMAN, N. D.

[1970] A theory of constructions equivalent to arithmetic. In: A. Kino, J. Myhill and R. E. Vesley (eds.), *Intuitionism and Proof Theory,* 101–120. North-Holland, Amsterdam.

GLIVENKO, V.

[1929] Sur quelques points de la logique de M. Brouwer. *Bull. Acad. Roy. Belg. Sci.* (5) **15**, 183–188.

GRZEGORCZYK, A.

[1964] A philosophically plausible formal interpretation of intuitionistic logic. *Indag. Math.* **26**, 596–601.

HALMOS, P.

[1960] *Naive Set Theory.* Van Nostrand, New York.

[1963] *Lectures on Boolean Algebras.* Van Nostrand, New York.

HENKIN, L.

[1949] The completeness of the first-order functional calculus. *J. Symb. Logic* **14**, 159–166.

HEWITT, E.

[1948] Rings of real-valued continuous functions. *Trans. Am. Math. Soc.* **64**, 45–99.

HEYTING, A.

[1930] Die formalen Reglen der intuitionistischen Logik. *Sitzungsber. Preuss. Akad. Wiss. Phys.-Math. Kl.* 42–56.

[1930a] Die formalen Reglen der intuitionistischen Mathematik. *Ibid.*, 57–71, 158–169.

[1972] *Intuitionism: an Introduction* (3rd rev. ed.). North-Holland, Amsterdam.

HILBERT, D.

[1900] Mathematische Probleme. Vortrag, gehalten auf dem internationalen Mathematiker Kongress zu Paris 1900. *Nachr. K. Ges. Wiss. Göttingen Math.-Phys. Kl.*, 253–297. (English transl.: Bull. Am. Math. Soc. **8** (1901–1902) 437–479.)

HINTIKKA, J.

[1955] Form and content in quantification theory. *Acta Phil. Fen* **8**, 7–55.

HURD, A., and P. LOEB

[1974] *Victoria symposium on nonstandard analysis,* Lecture Notes in Mathematics **369**. Springer, Berlin.

INFINITISTIC METHODS

[1961] Proc. Symp. on the Foundations of Mathematics, Warsaw, 2–9 Sept. 1959. Pergamon, New York, and Panstw. Wyd. Nauk.

JACOBSON, N.

[1964] *Lectures in Abstract Algebra,* Vol. 3. Van Nostrand, New York.

JECH, T. J.

[1971] *Lectures in Set Theory,* Lecture Notes in Math. **217**. Springer, Berlin.

KELLEY, J. L.

[1955] *General Topology*. Van Nostrand, New York.

KLEENE, S. C.

[1936] General recursive functions of natural numbers. *Math. Ann.* **112**, 727–742.

[1938] On notation for ordinal numbers. *J. Symb. Logic* **3**, 150–155.

[1943] Recursive predicates and quantifiers. *Trans. Am. Math. Soc.* **53**, 41–73.

[1952] *Introduction to Metamathematics*. North-Holland, Amsterdam, and Van Nostrand, New York.

KNEEBONE, G. T.

[1963] *Mathematical Logic and the Foundations of Mathematics*. Van Nostrand, New York.

KOLMOGOROV, A. N.

[1925] On the principle of the excluded middle. (English transl. in VAN HEIJENOORT [1967], pp. 414–437.)

KREISEL, G.

[1965] Mathematical logic. In: T. L. Saaty (ed.), *Lectures on Modern Mathematics*, Vol. **3**, 95–195. Wiley, New York.

KREISEL, G., and J. L. KRIVINE

[1967] *Elements of Mathematical Logic*. North-Holland, Amsterdam.

KRIPKE, S. A.

[1965] Semantical analysis of intuitionistic logic I, In: J. N. Crossley and M. A. E. Dummett (eds.), *Formal Systems and Recursive Functions*, 92–130. North-Holland, Amsterdam.

KURATOWSKI, K., and A. MOSTOWSKI

[1968] *Set Theory*. North-Holland, Amsterdam.

LAKATOS, I. (ed.)

[1967] *Problems in the Philosophy of Mathematics,* Proc. Int. Colloq. in the Philosophy of Science, London, 1965. North-Holland, Amsterdam.

LAMBEK, J.

[1961] How to program an infinite abacus. *Can. Math. Bull.* **4**, 295–302.

LANDAU, E.

[1930] *Grundlagen der Analysis*. Akad. Verlagsgesellschaft (English transl.: *Foundations of Analysis*. Chelsea, New York (1951).)

LEVY, A.

[1960] Axiom schemata of strong infinity in axiomatic set theory. *Pacific J. Math.* **10**, 223–238.

ŁOS, J.

[1954] On the categoricity in power of elementary deductive systems and some related problems. *Colloq. Math.* **3**, 58–62.

[1955] Quelques remarques, théorèmes et problèmes sur les classes définissables d'algèbres. In: Th. Skolem et al. (eds.), *Mathematical Interpretations of Formal Systems*. North-Holland, Amsterdam.

LÖWENHEIM, L.

[1915] Über Möglichkeiten im Relativkalkül. *Math. Ann.* **76**, 447–470. (English transl. in VAN HEIJENOORT [1967], pp. 228–251.)

LUXEMBURG, W. A. J.

[1962] Non-standard Analysis. Lecture Notes, (Duplicated), Calif. Inst. of Technology.

[1969] (Ed.) *Applications of Model Theory to Algebra, Analysis and Probability*. Holt, Rinehart and Winston, New York.

LUXEMBURG, W. A. J. and A. ROBINSON (eds.)

[1972] *Contributions to Non-standard Analysis*. North-Holland, Amsterdam.

MACHOVER, M.

[1967] *Non-standard analysis without tears*. (Duplicated) Technical Report No. 27, Hebrew Univ., Jerusalem.

MACHOVER, M., and J. HIRSCHFELD

[1969] *Lectures on Non-standard Analysis,* Lecture Notes in Math. **94**. Springer, Berlin.

MAKINSON, D

[1969] On the number of ultrafilters of an infinite Boolean algebra. *Z. Math. Logik.* **15**, 121–122.

MAL'CÉV, A. I.

[1936] Untersuchungen aus dem Gebeite der mathematischen Logik. *Mat. Sb.* **1**, 323–336.

MANSFIELD, R.

[1971] The theory of Boolean ultrapowers. *Ann. Math. Logic* **2**, 279–325.

MATIJASÉVIČ, JU. V.

[1970] Diofantovost pereçislimyh množestv. *Dokl. Akad. Nauk SSSR* **191** (2), 279–282 (English transl.: *Soviet Math. Dokl.* **11** (2), 354–357.)

[1971] Diophantine representation of r.e. predicates. In: J. E. Fenstad (ed.), *Proc. Second Scand. Logic Symp.* North-Holland, Amsterdam.

MINSKY, M. L.

[1961] Recursive unsolvability of Post's problem of "tag" and other topics in the theory of Turing machines. *Ann. Math.* **74**, 437–455.

MONTAGUE, R. M.

[1961] Fraenkel's addition to the axioms of Zermelo. In: Y. BAR-HILLEL et al. [1961].

MOSTOWSKI, A.

[1947] On definable sets of positive integers. *Fund. Math.* **34**, 81–112.

[1958] Quelques observations sur l'usage des méthodes non finitistes dans la métamathématique. In: *Colloq. Intern. du C. R. N. S.*, 1955, Vol. **70**. Paris.

[1966] *Thirty years of foundational studies.* Blackwell, Oxford.

MUČNIK, A. A.

[1956] On the unsolvability of the problem of reducibility in the theory of algorithms. *Dokl. Akad. Nauk SSSR N. S.* **108**, 194–197 (in Russian).

MYHILL, J.

[1955] Creative sets. *Z. Math. Logik Grundl. Math.* **1**, 97–108.

PADOA, A.

[1900] *Introduction logique á une théorie déductive quelconque* (English transl. in VAN HEIJENOORT [1967], pp. 118–123.)

PARIS, J. and L. HARRINGTON

[1977] A mathematical incompleteness in Peano arithmetic. In BARWISE [1977].

PEANO, G.

[1889] *Arithmetices Principia, Nova Methodo Exposita,* Turin. (English transl. in VAN HEIJENOORT [1967], pp. 83–97.)

POST, E. L.

[1944] Recursively enumerable sets of positive integers and their decision problem. *Bull. Am. Math. Soc.* **50**, 284–316.

[1948] Degrees of recursive unsolvability. Prelim. Rept. *Bull. Am. Math. Soc.* **54**, 641–642.

PUTNAM, H.

[1960] An unsolvable problem in number theory. *J. Symb. Logic* **25**, 220–232.

RAMSEY, F. P.

[1930] On a problem in formal logic. *Proc. London Math. Soc.* (2) **30**, 264–286.

RASIOWA, H., and R. SIKORSKI

[1951] A proof of the completeness theorem of Gödel. *Fund. Math.* **37**, 193–200.

RICE, H. G.

[1953] Classes of recursively enumerable sets and their decision problems. *Trans. Am. Math. Soc.* **74**, 358–366.

ROBINSON, A.

[1951] *On the Metamathematics of Algebra.* North-Holland, Amsterdam.

[1956] A result on consistency and its application to the theory of definition. *Indag. Math.* **18**, 47–58.

[1961] Non-standard analysis. *Nederl. Akad. Wetensch. Proc.* (A) **64**, 432–440.

[1962] *Complex function theory over non-archimedean fields.* Tech. Sci. Note No. 30, USAF Contract 61 (052)–187.

[1963] *Introduction to Model Theory and to the metamathematics of Algebra.* North-Holland, Amsterdam.

[1965] Topics in non-archimedean mathematics. *Symp. on Model Theory,* Berkeley, Calif., 1963. North-Holland, Amsterdam.

[1966] *Non-standard Analysis.* North-Holland, Amsterdam.

ROBINSON, J.

[1952] Existential definability in arithmetic. *Trans. Am. Math. Soc.* **72**, 437–449.

ROBINSON, R. M.

[1956] Arithmetical representation of recursively enumerable sets. *J. Symb. Logic* **21**, 162–186.

ROGERS, H., JR.

[1967] *Theory of Recursive Functions and Effective Computability.* McGraw-Hill, New York.

ROSSER, J. B.

[1936] Extensions of some theorems of Gödel and Church. *J. Symb. Logic* **1**, 87-91. (Reprinted in DAVIS [1965], pp. 231-235.)

ROTMAN, B., and G. T. KNEEBONE

[1966] *The Theory of Sets and Transfinite Numbers.* Van Nostrand, New York.

RYLL-NARDZEWSKI, C.

[1959] On theories categorical in power, *Bull. Acad. Polon. Sci. Sér. Sci. Math. Astron. Phys.* **7**, 545-548.

SACKS, G. E.

[1963] *Degrees of unsolvability,* Ann. Math. Studies **55**. Princeton Univ. Press, Princeton, N. J.

SCHÜTTE, K.

[1956] Ein System des verknüpfenden Schliessens *Arch. Math. Logik Grundlagenforsch.* **2**, 56-67.

[1962] Der Interpolationsatz der intuitionistischen Prädikatenlogik. *Math. Ann.* **148**, 192-200.

SHEPHERDSON, J. C., and H. E. STURGIS

[1963] Computability of recursive functions. *J. Assoc. Comput. Mach.* **10**, 217-255.

SHOENFIELD, J. R.

[1960] An uncountable set of incomparable degrees. *Proc. Am. Math. Soc.* **11**, 61-62.

[1967] *Mathematical Logic.* Addison-Wesley, Reading, Mass.

[1971] *Degrees of Unsolvability.* North-Holland, Amsterdam.

SIKORSKI, R.

[1964] *Boolean Algebras* (2nd ed.). Springer, Berlin.

SKOLEM, T.

[1920] Logisch-kombinatorische Untersuchungen über die Erfüllbarkeit oder Beweisbarkeit mathematischer Sätze nebst einem Theoreme über dichte Mengen I. *Skr. Norske Vid.-Akad. Kristiana Mat.-Naturv. Kl.* (4) (English transl. of §1 in VAN HEIJENOORT [1967] 252-263.)

[1922] Einige Bemerkungen zur axiomatischen Begründung der Mengenlehre. *Mat. Kongr. Helsingfors, 4-7 Juli 1922, Den femte Skand. Mat. Kongr., Redogorelse,* 217-232. (English transl. in VAN HEIJENOORT [1967] 290-301.)

[1934] Über die Nicht-charakterisierbarkeit der Zahlenreihe mittels endlich oder abzählbar unendlich vieler Aussagen mit ausschliesslich Zahlenvariablen. *Fund. Math.* **23**, 150-161.

SMULLYAN, R. M.

[1961] *Theory of formal systems,* Ann. Math. Studies **47**. Princeton. Univ. Press, Princeton, N. J.

[1968] *First-order Logic.* Springer, Berlin.

STONE, M. H.

[1936] The representation theorem for Boolean Algebra. *Trans. Am. Math. Soc.* **40**, 37-111.

[1937] Applications of the theory of Boolean rings to general topology. *Trans. Am. Math. Soc.* **41**, 375-481.

TARSKI, A.

[1930] Über einige fundamentale Begriffe der Metamathematik. *C. R. Soc. Sci. Lettres Varsovie* (III) **23**, 22–29. (English transl. in TARSKI [1956], pp. 30–37.)

[1930] Une contribution à la théorie de la mesure. *Fund. Math.* **15**, 42–50.

[1935] Die Wahrheitsbegriff in den formalisierten Sprachen. *Stud. Phil.* (Warsaw) **1**, 261–405. (English transl. in TARSKI [1956], pp. 152–278.)

[1939] Ideale in vollständigen Mengenkörpern. *Fund. Math.* **32**, 45–63.

[1956] *Logic, Semantics, Metamathematics, papers from 1923 to 1938*. Clarendon Press, Oxford.

TARSKI, A., A. MOSTOWSKI and R. M. ROBINSON

[1953] *Undecidable Theories*. North-Holland, Amsterdam.

TARSKI, A., and R. L. VAUGHT

[1957] Arithmetical extensions of relational systems, *Comp. Math.* **13**, 81–102.

THOMASON, R. H.

[1968] On the strong semantical completeness of the intuitionistic predicate calculus. *J. Symb. Logic* **33**, 1–7.

TROELSTRA, A. S.

[1969] *Principles of Intuitionism*, Lecture Notes in Math. **95**. Springer, Berlin.

[1973] (ed.) *Metamathematical Investigations of Intuitionistic Arithmetic and Analysis*, Lecture Notes in Math. **344**. Springer, Berlin.

TURING, A. M.

[1936] On computable numbers, with an application to the Entscheidungsproblem. *Proc. London Math. Soc.* (2) **42**, 230–265; **43**, 544–546.

[1939] Systems of logic based on ordinals. *Proc. London Math. Soc.* (2) **45**, 161–228.

VAN HEIJENOORT, J.

[1967] (ed.) *From Frege to Gödel, a Source Book in Mathematical Logic 1879–1931*. Harvard Univ. Press, Cambridge, Mass.

VAUGHT, R. L.

[1954] Applications of the Löwenheim–Skolem–Tarski theorem to problems of completeness and decidability. *Indag. Math.* **16**, 467–472.

[1961] Denumerable models of complete theories. *Infinitistic Methods*. Pergamon Press, Elmsford, N.Y.

[1974] Model theory before 1945. *Proc. Tarski symposium, AMS Proc. symp. Pure Math.* **25**, 153–172.

WHITEHEAD, A. N., and B. RUSSELL

[1910] *Principia Mathematica*, Vol. 1. Cambridge Univ. Press, London (Vol. 2 appeared in 1912 and Vol. 3 in 1913. Second edition of Vol. 1 in 1925 and Vols. 2 and 3 in 1927.)

GENERAL INDEX*

* Entries for terms and abbreviations refer to the place where they are first explained, defined or re-defined.

INDEX OF SYMBOLS*

* The symbols are listed in order of first occurrence. A symbol used in more than one sense may be listed more than once.

Printed and bound by CPI Group (UK) Ltd, Croydon, CR0 4YY
03/10/2024
01040430-0017